Craig Stanton

Fall '80

CRYSTALLOGRAPHY
and Crystal Chemistry

CRYSTALLOGRAPHY and Crystal Chemistry

an introduction

F. Donald Bloss
Virginia Polytechnic Institute
and State University

Holt, Rinehart and Winston, Inc.
New York, Chicago, San Francisco, Atlanta, Dallas, Montreal, Sydney

Library of Congress Catalog Card Number: 77-136774
ISBN 03-085155-6
Printed in the United States of America
7890 038 09876

To my daughters,
Terry, Janet, and Jill,
in order of appearance

Preface

I have tried to produce a text that is easily understood by undergraduates on their first exposure to crystallography and crystal chemistry. When words alone appeared unable to rescue a topic from obscurity, a drawing was created to fill the gap. The captions accompanying these illustrations are usually extensive so that, even in instances where an astute editor cannot prevent a figure from appearing overleaf from its textual discussion, the figure will not lose its impact.

For the most part the material covered is of common interest to mineralogists, ceramists, chemists, solid-state physicists, and metallurgists. Thus to the extent that the text is self-explanatory, and hopefully most of it is, an instructor can concentrate his lectures on special topics from his particular field. The debatable decision to treat external symmetry prior to the internal symmetry of crystals was made because the operations of external symmetry are more simple and thus more easily grasped by the beginning student. The references that follow some chapters include more detailed treatments or classic papers on topics within that chapter.

Students of mineralogy and geology are not always familiar with the use of tensors to describe physical properties or with the causes and nature of magnetism in crystals. Yet the expanding studies of geophysics and paleomagnetism would seem to demand such a knowledge from them. It is hoped that I have arrived at a simplified presentation of these topics to serve these needs.

The enthusiasm and knowledgability of my colleague and comrade-in-arms, Gerald V. Gibbs, inspired many improvements in this text. Another good friend and colleague, Paul H. Ribbe, also served unflinchingly as critic and sounding board. Professor Otto Kopp of the University of Tennessee carefully read the entire text and eliminated many of the errors that inevitably creep in. My debt to many crystallographers, but especially to Professor Martin Buerger, whose published works have helped shape the course of crystallography during the last decades, will be obvious from the text. Daniel Irwin, Eskander (Alex) Nahouray, and Tso-Hwa Lee of Southern Illinois University plus Sharon Rhudy and R. E. Whittemore of the Virginia Polytechnic Institute and State University drafted the illustrations under my supervision. I am grateful to my colleagues Drs. John Costain, James R. Craig, and Charles Gilbert for reading sections of the galleys and page proofs.

Blacksburg, Virginia F. Donald Bloss

Contents

CRYSTALLOGRAPHY
and Crystal Chemistry

1 External Symmetry

Simple Symmetry Elements

An object that consists of a systematic repetition of identical features is said to have "symmetry." Three main types of external symmetry are recognized: (1) symmetry with respect to a plane, (2) symmetry with respect to a line, or (3) symmetry with respect to a point. As will be seen in the examples below, it is possible for an object to possess more than one type of symmetry.

Symmetry with respect to a plane

Symmetry with respect to a plane is present in an object if an imaginary plane (m in Fig. 1-1A) can be passed through the object's center such that, at *equal* but *opposite distances along any line perpendicular to this plane, identical points occur in or on the object.* Thus the weight lifter in Figure 1-1A is (externally) symmetrical with respect to plane m because he is composed of innumerable pairs of such identical points, a' and a, b' and b, c' and c being but a few. Any object that is symmetrical with respect to a plane is divisible by it into two halves, each one the mirror image of the other. As a consequence, if one of these halves is removed and the symmetry plane is replaced by a mirror, the image in the mirror is a perfect reproduction of the removed half (Fig. 1-1B). Therefore a symmetry plane is often called a *mirror plane* and, using the internationally accepted symbols devised by Hermann and Mauguin, is denoted by m (an abbreviation for mirror).

Many objects possess more than one plane of symmetry. For example, the football field of Figure 1-2A possesses two planes of symmetry, m_1 and m_2. The geometric prism in Figure 1-2B possesses five such planes (unshaded). On the

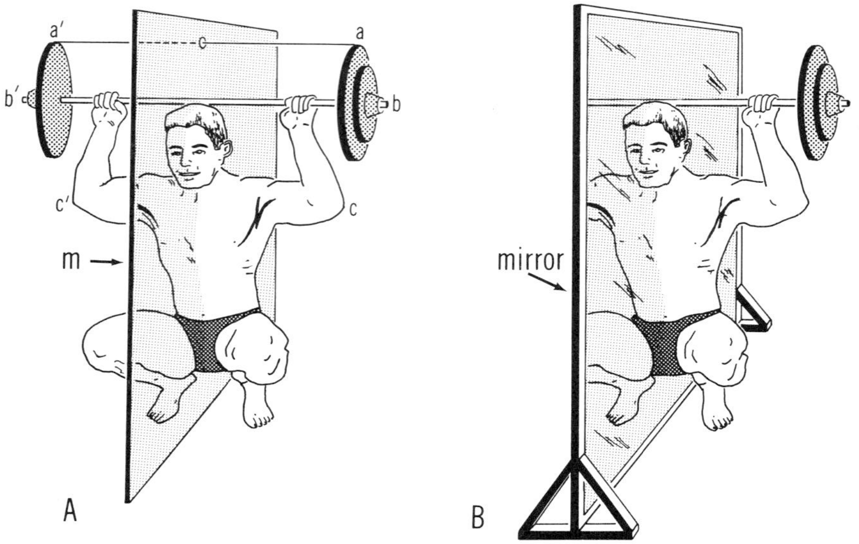

FIGURE 1–1

(A) Gray-shaded plane *m* is a plane of symmetry with respect to the external features of the weight lifter because he is composed of pairs of identical points, *a*′ and *a*, *b*′ and *b*, *c*′ and *c*, and so on, which are at equal but opposite distances along perpendiculars (*a*′*a*, *b*′*b*, *c*′*c*) to plane *m*.
(B) If the weight lifter's right half is removed and a mirror is substituted for symmetry plane *m*, the image of his left half as seen in the mirror is a reproduction of the missing right half.

other hand, a perfect sphere possesses an infinite number of planes of symmetry because any plane through its center, regardless of orientation, will divide the sphere into mirror images.

Symmetry with respect to an axis

Symmetry with respect to an axis is less common in everyday experience; thus hypothetical examples will necessarily be used. Let us assume that an astronaut on a fixed station in space is observing the earth's rotation. Only after a complete (360°) rotation of the earth would the landscape in front of him look the same again; the polar axis of the earth is therefore called a "1-fold axis of symmetry" since the same aspect repeated itself only one time during a 360° rotation. This, of course, amounts to no symmetry at all, because the landscape seen after the full rotation would be identical to the one viewed originally. The Hermann-Mauguin symbol 1 signifies this total lack of symmetry, that is, the possession of only 1-fold axes.

If the earth's surface were a featureless ocean except for two North American continents located where shown in Figure 1-3A, then a 180° turn about its polar axis would cause a complete repetition of the landscape. Crystallographers

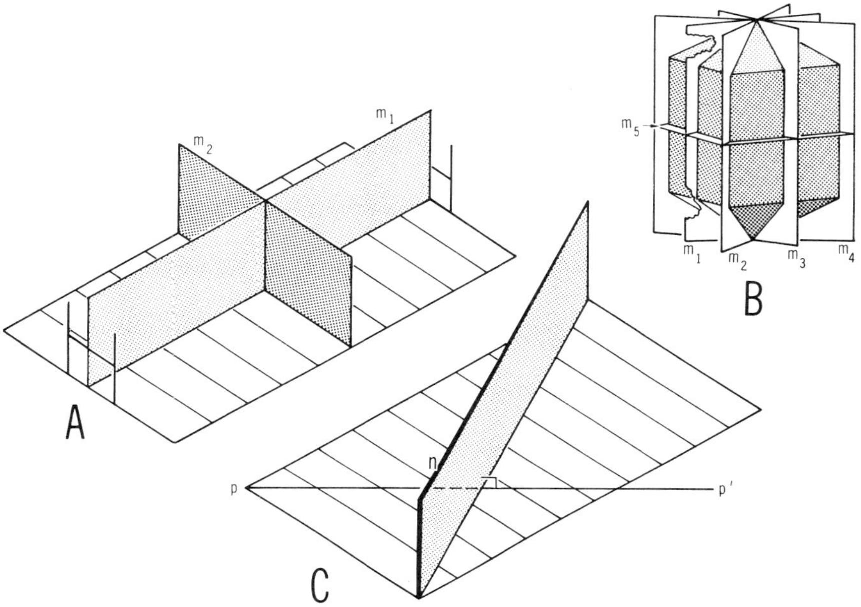

FIGURE 1–2

Objects that possess more than one symmetry plane.
(A) A football field with symmetry planes m_1 and m_2.
(B) A crystal that possesses five symmetry planes (unshaded).
(C) Plane n is not a symmetry plane because it possesses at least one perpendicular, for example, pp', along which the football field contains unlike points (p and p') at equal but opposite distances from this plane.

would refer to the rotational (polar) axis of such an earth as a *diad* or a 2-fold axis of symmetry, assigning it the written symbol 2 and the graphical symbol . We can quickly perceive the duplication of items around a 2-fold axis by sighting directly along it (Fig. 1-3B). Similarly, Figure 1-4 illustrates views directly along the earth's polar axis assuming that this axis is successively a 3-fold, 4-fold, or 6-fold axis of symmetry. The internationally accepted written symbols for these axes are 3, 4, and 6, respectively; the graphical symbols are , , and 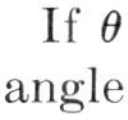.

If θ symbolizes the *repeat angle* for an X-fold axis of symmetry—that is, the angle of rotation about this axis necessary to "repeat the scene"—then, as a brief glance at Figures 1-3 and 1-4 will confirm,

$$\theta = \frac{360°}{X} \qquad \text{(Eq. 1-1)}$$

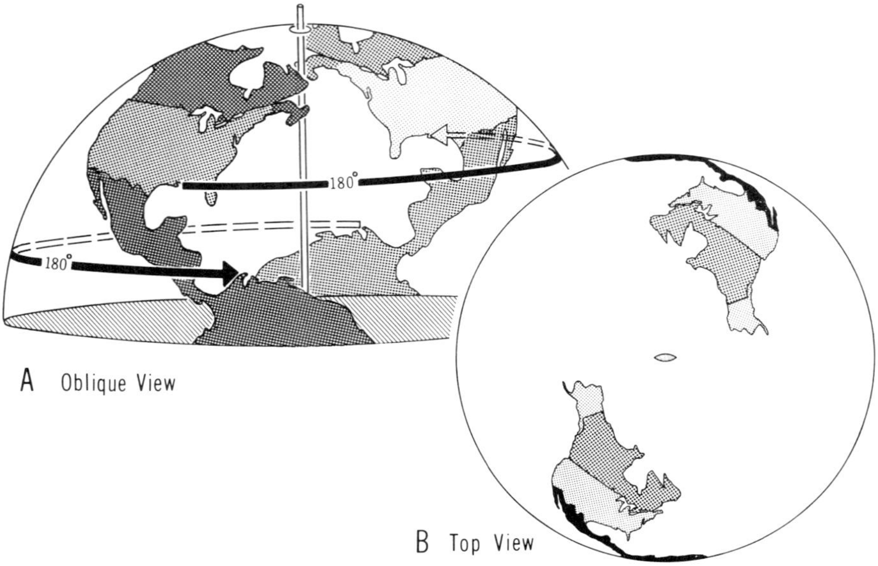

FIGURE 1–3

Views of the Northern Hemisphere if the polar axis were an axis of 2-fold symmetry such that the Americas occur twice.

(A) Oblique view, assuming the earth's interior to be transparent. Note that a 180° rotation on the polar axis would cause the New Orleans at the front to exchange places with the New Orleans at the back. If the observer did not observe the 180° rotation as it was being made, he would never be able to detect that it had occurred.

(B) View down the polar axis to illustrate better the 2-fold symmetry of this axis.

where X represents the axis' written symbol. Applying this equation we observe that for a 1-, 2-, 3-, 4-, or 6-fold axis the repeat angle θ equals 360°, 180°, 120°, 90°, or 60°, respectively. As will be demonstrated later (p. 145), these are the only repeat angles possible for symmetry axes in crystals. As a result a crystal can never possess a 5-fold axis or an axis in which X exceeds 6, for example, a 7- or an 8-fold axis.

Symmetry with respect to a point

Symmetry with respect to a point signifies that equidistant from this point, in opposite directions, identical features will always be encountered. To illustrate, the room in Figure 1-5A is symmetrical with respect to i, its central point. Consequently from *every* point in this room—for example, from a—we can extend a straight line through i to encounter, at an equal distance beyond i, the equivalent point, a'. Similarly, b and b' represent another pair of equivalent points at equal but opposite distances from i. The point with respect to which an object is

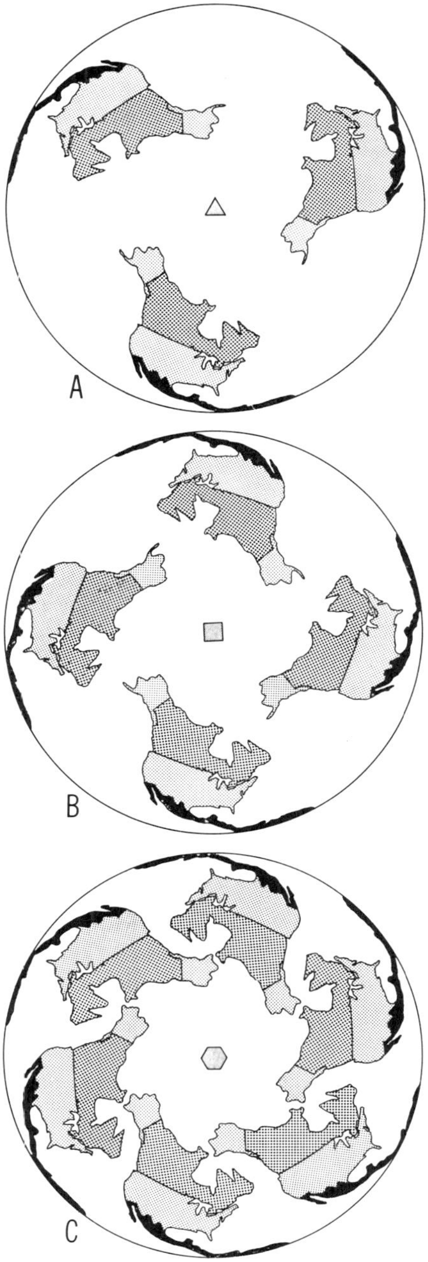

FIGURE 1–4

Views down the polar axis of the earth if this axis were (A) a 3-fold axis of symmetry, (B) a 4-fold axis of symmetry, and (C) a 6-fold axis of symmetry. The polar axis is indicated by the standard symbol in each case.

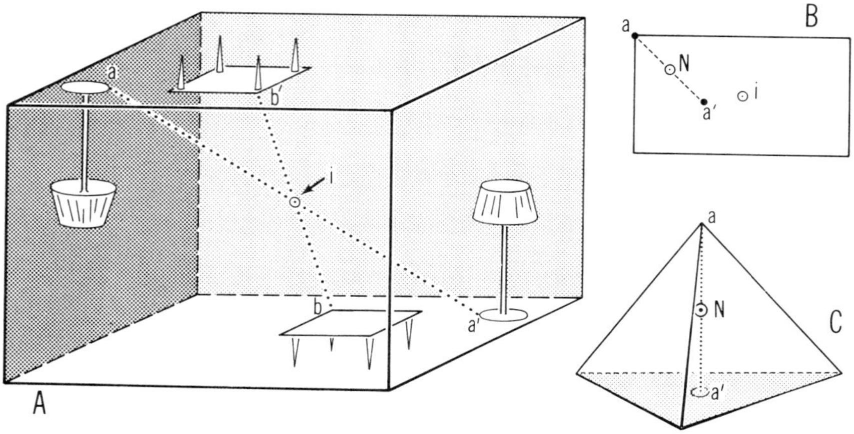

FIGURE 1–5

(A) Appearance of a room if its geometric center i is a center of symmetry. The room consists *entirely* of identical pairs of points, a and a', b and b', and so on, disposed at equal but opposite distances from i.
(B) The necessity of a center of symmetry being located at an object's geometric center i is illustrated for a rectangle. Thus N, a noncentral point, cannot be a center of symmetry because there exists at least one pair of nonidentical points, a and a', at equal but opposite distances from N.
(C) N, the geometric center of the crystal, is in this case not a center of symmetry because there exists at least one pair of nonidentical points, a and a', at equal but opposite distances from it.

symmetrical necessarily coincides with the object's geometric center and is thus called a *center of symmetry*. To elaborate, a rectangle will be symmetrical with respect to its central point (i in Fig. 1-5B) but not for a noncentral point such as N. This becomes obvious if a line is drawn from corner a to (and an equal distance beyond) N to locate point a'; obviously N is not a center of symmetry because point a' is not equivalent to a, being in the rectangle's interior rather than at its corner. We can generalize by saying that *a point is a center of symmetry only if all points equidistant from it, but in opposite directions, are equivalent.* For example, in Figure 1-5A if one lamp were scratched and the other not, point i would cease to be a center of symmetry. Similarly, central point N in the crystal in Figure 1-5C is not a center of symmetry because points a and a' are not equivalent.

Kinds of Symmetry Operations

The geometric movement necessary to bring a given point a_1 into coincidence with a symmetrically equivalent point a_2 is called a *symmetry operation.* In Figure 1-6A, for example, the curved arrows illustrate a single operation for a

2-, 3-, 4-, or 6-fold axis, namely, a 180°, 120°, 90°, or 60° rotation about each respective axis. These operations are called *operations of the first kind* because, applied to all points of left hand a_1 in Figure 1-6A, for example, they "generate" additional left hands only. Similarly, if operative on a right hand, they generate more right hands only. In other words, an operation of the first kind cannot produce a change of hand. In contrast, *operations of the second kind*, namely, the plane of symmetry m and a center of symmetry (symbol: i or $\bar{1}$),[1] produce a

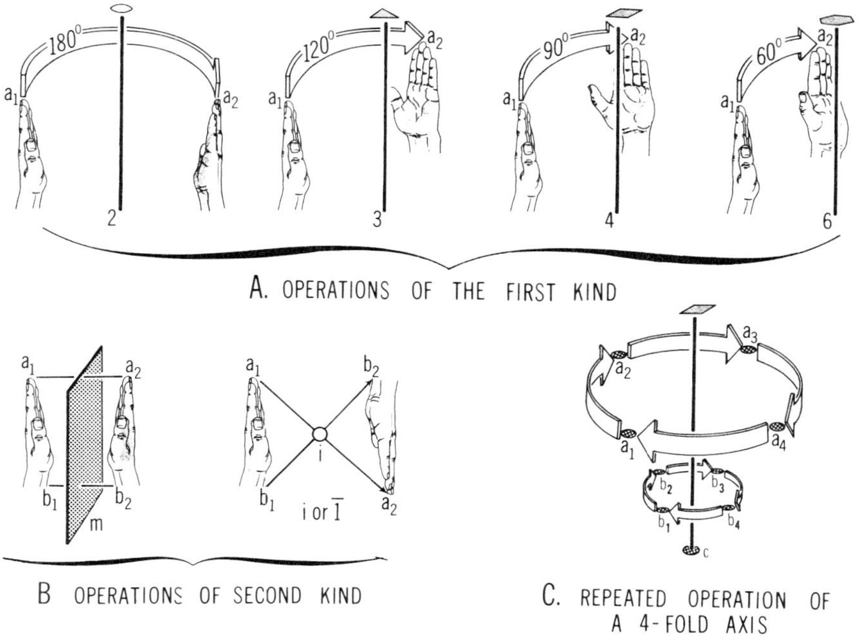

FIGURE 1–6

(A) Illustration of a single operation, that is, the minimum rotation necessary to bring a point a_1 into coincidence with an identical point a_2, for a 2-, 3-, 4-, or 6-fold axis. These operations, applied to all points on the left hand at a_1, could only generate a second left hand at a_2. They are thus called "operations of the first kind."
(B) The operations associable with a plane of symmetry m or a center of symmetry i or $\bar{1}$ are shown to relate point a_1 to a_2 and point b_1 to b_2. These operations, applied to all points on the left hand at a_1, would generate a right hand at a_2. Because they produce a "change of hand," operations m and $\bar{1}$ are called "operations of the second kind."
(C) The four points, a_1 to a_4, are identical points related to each other by rotation about the 4-fold axis. Identical points b_1 to b_4 are similarly related. Note that these b points are closer to each other because the original point b was closer to the 4-fold axis. Thus a point c, which is precisely on this axis, is not related to any other points by the axis's operation.

[1] Read $\bar{1}$ as "bar one."

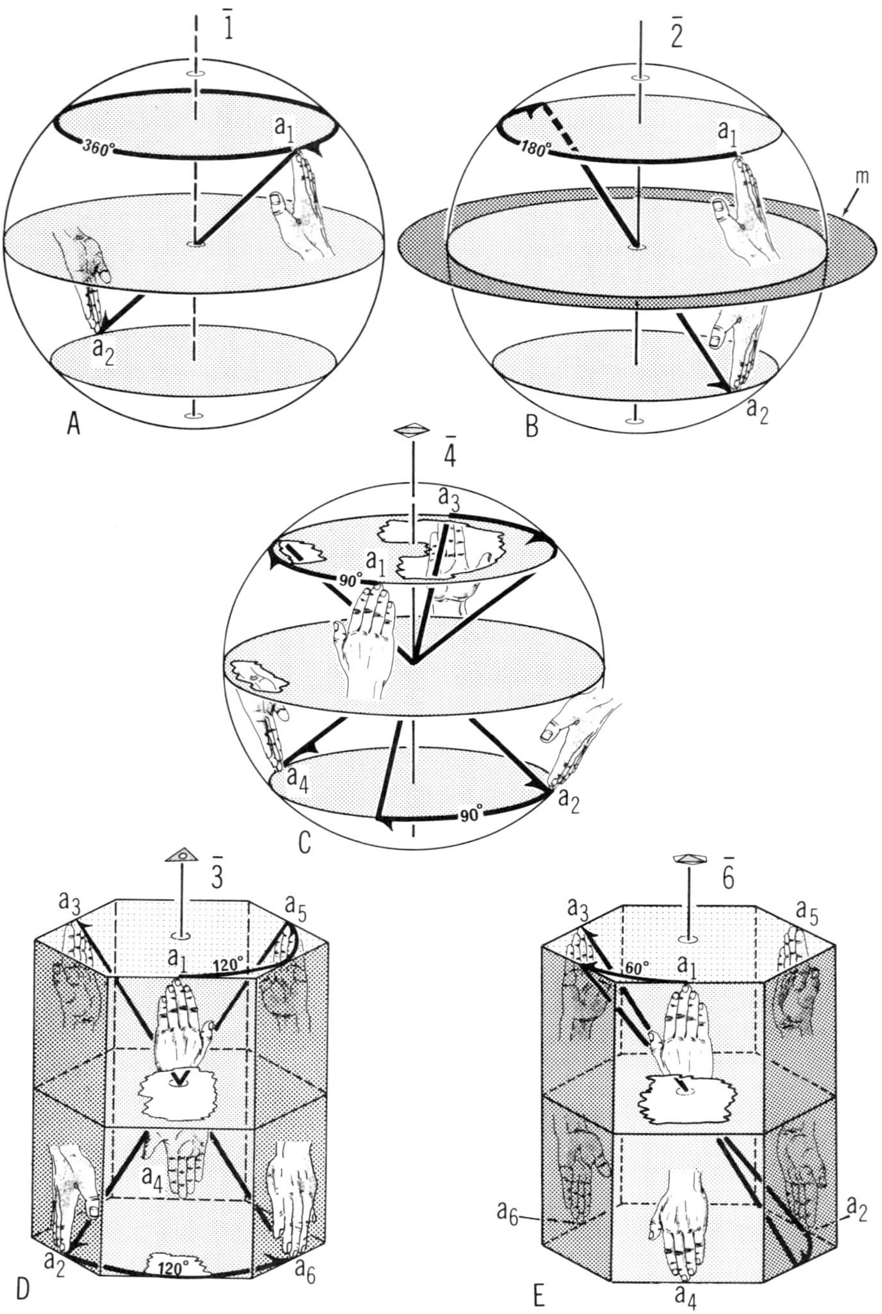

FIGURE 1–7

Illustration of the symmetry operations associable with the five *improper axes* or *inversion axes*. To go from point a_1 to identical point a_2 requires, in each case, not change of hand as Figure 1-6B illustrates. The operation associated with a *plane of symmetry* is called *reflection* and involves dropping a perpendicular to the plane from some given point a_1 (or b_1) and proceeding an equal distance beyond the plane to locate equivalent point a_2 (or b_2). The operation of a *center of symmetry* is called *inversion* and is indicated by the arrows (a_1a_2 and b_1b_2) that pass through the hollow circle denoting the center of symmetry i in Figure 1-6B. Thus arrow a_1a_2 indicates that if we draw a straight line from a_1 to i (and an equal distance beyond), then an equivalent point a_2 is located. Either operation, reflection across a plane m or inversion through a center i, will generate a right hand from a left hand, or vice versa.

The repeated operation of an X-fold axis on a point (or object) usually produces $X - 1$ additional points (or objects) equivalent to the first. For example, the 4-fold axis in Figure 1-6C operates on point a_1 to repeat it as points a_2, a_3, and a_4. The four points, a_1, a_2, a_3, a_4, constitute a set of symmetrically equivalent points generally called *equivalent points* or *equipoints*. The closer the original point is to the axis, the closer together the equipoints will be. Compare, for example, set b_1, b_2, b_3, and b_4 with set a_1, a_2, a_3, and a_4. For a point c located *on* the axis, no additional equipoints are generated; hence c constitutes the entire set.

The Five Improper Axes

The 1-, 2-, 3-, 4-, and 6-fold axes already discussed represent operations of the first kind and hence are sometimes called *proper axes*. Besides these five proper axes, crystallographers generally recognize five compound axes wherein the operation of a 1-, 2-, 3-, 4-, or 6-fold axis is combined with the operation of

---

only the rotation associated with the corresponding proper axes but also an inversion through the center of the object.

(A) For illustrative purposes only, the $\bar{1}$ axis is drawn here as if it coincided with the dashed line. In fact, it lacks a directional sense and any direction outward from the sphere's center could have been chosen as an axis for the 360° rotation that preceded the inversion operation whereby a_1 was related to a_2. Thus a $\bar{1}$ axis yields the same result and is essentially identical to i, a center of symmetry located at the sphere's center.

(B) A $\bar{2}$ axis, unlike a $\bar{1}$ axis, occurs along a particular direction. The plane normal to this $\bar{2}$ axis, the shaded equatorial plane, is always a plane of symmetry m. Either the $\bar{2}$ operation or the mirror plane perpendicular to it serves to relate hand a_1 to a_2. Thus a $\bar{2}$ axis is equal to a mirror plane normal to it.

(C) The $\bar{4}$ operation involves a rotation of 360°/4, then inversion through the center to relate a_1 to a_2, then a_2 to a_3, and so on.

(D) The $\bar{3}$ operation similarly involves a rotation of 360°/3 followed by inversion through the center to relate a_1 to a_2. The rotational direction for this $\bar{3}$ axis has been reversed from that illustrated for the $\bar{4}$ axis to illustrate that direction of rotation is irrelevant so long as it is maintained in the same direction.

(E) A $\bar{6}$ axis yields the same results as a 3-fold axis plus a plane of symmetry normal to this 3-fold axis.

inversion through a central point on the axis as illustrated in Figure 1-7. These axes, respectively symbolized as $\bar{1}$, $\bar{2}$, $\bar{3}$, $\bar{4}$, or $\bar{6}$ (read "bar one, bar two", and so on), are called *inversion axes* or *improper axes* because they represent operations of the second kind that produce right hands from left hands.

By definition a $\bar{1}$ axis involves the 360° rotation of a 1-fold axis followed by inversion through the object's center (Fig. 1-7A). Thus a $\bar{1}$ axis produces the same result as does possession of a center of symmetry i. Consequently the symbol for a center of symmetry is either i (for inversion) or $\bar{1}$, the symbol $\bar{1}$ being preferred.

The operation of a $\bar{2}$ axis includes a 180° rotation about the axis plus an inversion through a central point on this axis. The heavy arrows in Figure 1-7B indicate that point a_1 is related to a_2 by a 180° rotation followed by inversion. This operation, applied individually to every point on the hand at a_1, requires the second hand at a_2 to be present. However, hands a_1 and a_2 in Figure 1-7B are also related by a symmetry plane (heavily shaded) perpendicular to the $\bar{2}$ axis. In short, a $\bar{2}$ axis produces the same results as would a plane of symmetry perpendicular to it. Consequently, the $\bar{2}$ symbol (and operation) is rarely used; crystallographers prefer the concept of a plane of symmetry but realize that if an object possesses a plane of symmetry m, it must also possess a $\bar{2}$ axis perpendicular to this plane.

We can generalize by saying that the operation associated with an inversion or $\bar{X}$ axis (where X stands for 1, 2, 3, 4, or 6) includes a rotation of $360°/X$ followed by an inversion through the object's center. Thus a $\bar{4}$ axis (Fig. 1-7C) involves a 90° rotation plus an inversion so as to relate points a_1 to a_2, a_2 to a_3, a_3 to a_4, and a_4 to a_1; if this operation is applied to *all* the points of the hand at a_1, the hands at a_2, a_3, and a_4 are thereby "generated." Unlike the $\bar{2}$ axis, whose results could be duplicated by a perpendicular plane of symmetry, a $\bar{4}$ axis (graphic symbol, ) represents a unique symmetry operation unlike any of the previously discussed symmetry elements.

A $\bar{3}$ axis (graphic symbol, ) involves a 120° rotation plus an inversion. In Figure 1-7D, for example, such an operation relates points a_1 to a_2, a_2 to a_3, a_3 to a_4, and, ultimately, a_6 to a_1; if this operation is applied to *all* the points of the hand at a_1, the five additional hands at a_2, a_3, a_4, a_5, and a_6 are necessarily present. Actually, a 3-fold axis plus a center of symmetry yield results identical to a $\bar{3}$ axis. For example, a 3-fold axis in place of the $\bar{3}$ axis in Figure 1-7D would operate on the hand at a_1 to "generate" those at a_3 and a_5; then if the center of the geometric figure is a center of symmetry, the operation of inversion "generates" hand a_4 from a_1, a_6 from a_3, and a_2 from a_5. The following conclusions can now be drawn: (1) an object that possesses a 3-fold axis will not possess a center of symmetry; (2) conversely, if an object possesses a center of symmetry, then all 3-fold axes present on it necessarily become $\bar{3}$ axes; (3) an object that possesses a $\bar{3}$ axis invariably possesses a center of symmetry. Because the symmetry operation of a $\bar{3}$ axis is relatively complex, its existence is most easily demonstrated by showing that the object possesses (a) a 3-fold axis and (b) a center of symmetry.

The operation of a $\bar{6}$ axis (graphic symbol,) involves a 60° rotation followed by an inversion. This operation, as illustrated in Figure 1-7E, would thus relate the hand at a_1 to that at a_2, a_2 to a_3, a_3 to a_4, and ultimately a_6 to a_1. We can see from Figure 1-7E that a $\bar{6}$ axis produces the same results as would a 3-fold axis with a plane of symmetry perpendicular to it.

Combinations of Symmetry Elements

The ten basic symmetry elements, 6, 4, 3, 2, 1, $\bar{6}$, $\bar{4}$, $\bar{3}$, $\bar{2}$ (= m), and $\bar{1}$ (= i), or combinations in which three or more of these pass through the center of a crystal, are sufficient to describe the external symmetry of any crystal in the universe. The reasons for the restriction of the external symmetry operations in crystals to these ten basic elements are discussed later (p. 145). Surprisingly, in addition to the ten basic elements there can only be 22 different combinations of them that can pass through a central point in a crystal and be compatible to the crystal's symmetry. The limitation of compatible combinations to 22 arises because many combinations are duplicates of others; the equivalence of a $\bar{6}$ axis (Fig. 1-7E) to a 3-fold axis with a symmetry plane perpendicular to it is one such example. Furthermore, many combinations are impossible in crystals because the symmetry elements would react repeatedly on each other to produce an infinite multiplicity of these elements. To illustrate, two 6-fold axes, a_1 and b in Figure 1-8, cannot pass through a common point p in a crystal and be compatible with its symmetry. Thus the operation of axis b on a_1 requires five

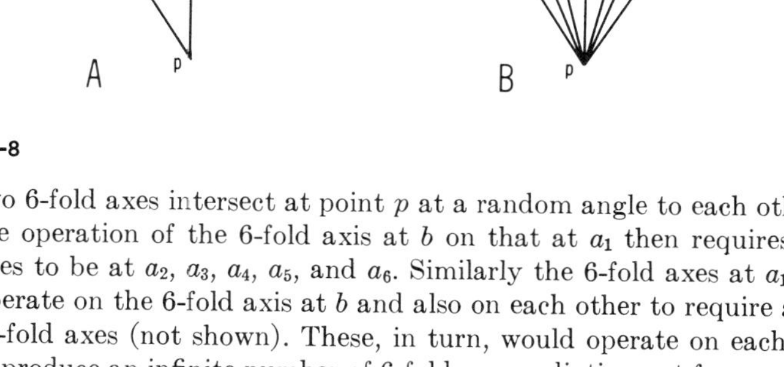

FIGURE 1–8

(A) Two 6-fold axes intersect at point p at a random angle to each other.
(B) The operation of the 6-fold axis at b on that at a_1 then requires additional 6-fold axes to be at a_2, a_3, a_4, a_5, and a_6. Similarly the 6-fold axes at a_1, a_2, . . ., a_6 would operate on the 6-fold axis at b and also on each other to require a multitude of new 6-fold axes (not shown). These, in turn, would operate on each other, and so on, to produce an infinite number of 6-fold axes radiating out from p.

additional 6-fold axes at a_2, a_3, . . ., a_6. These then operate on b, on a_1, and on each other to require still more 6-fold axes which, in turn, interact with each other (and those already established). The process would continue ad infinitum until 6-fold axes extended outward from p in an infinite variety of directions. Such an infinite multiplicity of symmetry axes extending outward from a single point is compatible only with the external symmetry of a perfect sphere, a geometric shape alien to all crystals. (However, all crystals *do* possess an infinite number of 1-fold axes, that is, axes of nonsymmetry; these represent all the directions in the crystal that do not coincide with 2-, 3-, 4-, 6-, $\bar{1}$-, $\bar{2}$-, $\bar{3}$-, $\bar{4}$-, or $\bar{6}$-fold axes.)

The external symmetry of any crystal whatsoever must necessarily correspond to (1) one of the five proper axes, (2) one of the five improper axes, or (3) one of the 22 combinations of these axes that do not result in infinite multiplicities. The nature of (and restrictions governing) these 22 combinations will be discussed next.

Permissible combinations of axes

Two, and only two, symmetry axes[2] intersecting at a point is an impossibility. If two such axes intersect, *at least* a third axis must also pass through their common point. To illustrate, let X and Y in Figure 1-9A represent symmetry axes with repeat angles of θ_x and θ_y, respectively, and with a construction plane (shaded) perpendicular to each. The angle at which axes X and Y intersect, called their *interaxial angle*, is symbolized as $X \wedge Y$. In the lightly shaded plane perpendicular to X, we select a point a_1 such that if rotated θ_x degrees about X, it will coincide with and therefore be equivalent to a_2. Similarly, a rotation of θ_y degrees about Y will relate a_2 to its equivalent a_3. Thus since a_1 is equivalent to a_2, and a_2 to a_3, then a_1 must be equivalent to a_3. Hence a symmetry axis must exist, labeled Z in Figure 1-9B, that relates a_3 to a_1 by a rotation of θ_z.

A simple test exists for determining whether an X-fold, a Y-fold, and a Z-fold symmetry axis can compatibly intersect in a point without producing an infinite multiplicity. Three such intersecting axes, X, Y, Z, define a spherical triangle[3] whose sides (the dashed arcs in Fig. 1-9C) equal the interaxial angles $X \wedge Y$, $Z \wedge X$, and $Z \wedge Y$. The angles of this spherical triangle (enlarged in Fig. 1-9D) can be demonstrated through Euler's or Rodrigues' construction (Buerger, 1956, p. 37) to equal $\theta_x/2$, $\theta_y/2$, and $\theta_z/2$, that is, half the repeat angles associated with X, Y, and Z. Because the sum of the angles of a spherical triangle must exceed 180° but be less than 540°, then

$$180^\circ < \frac{\theta_x}{2} + \frac{\theta_y}{2} + \frac{\theta_z}{2} < 540^\circ \qquad \text{(Ineq. 1-2)}$$

[2] The term *symmetry axis* is now used in *sensu stricto* so as to exclude from consideration 1-fold axes, that is, axes of nonsymmetry.

[3] Spherical triangles are described and discussed in Figure 13–36.

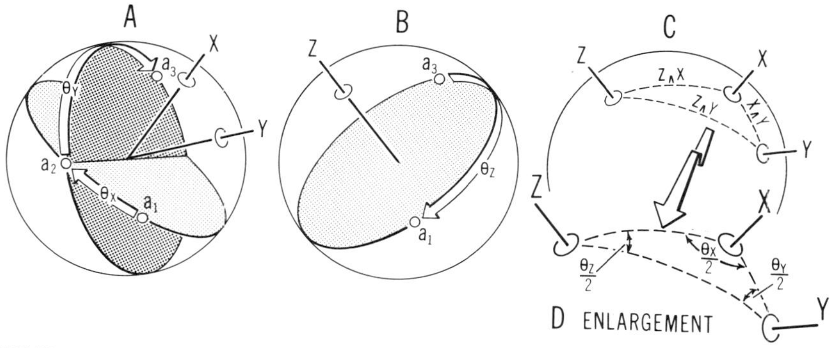

FIGURE 1–9

(A) A sphere has been drawn around the point of intersection between two axes of symmetry, X and Y. The arcs of intersection between this sphere and the planes perpendicular to X and to Y are also shown. Let a_1 be a point in the plane perpendicular to X such that a rotation of θ_x, the repeat angle for axis X, brings a_1 to coincidence with a_2, a point common to both of the shaded planes. A rotation of θ_y about Y then relates a_2 to a_3. Thus a_1 equals a_2 equals a_3; in short, a_3 equals a_1.
(B) Because a_3 equals a_1, a third symmetry axis Z must thus exist with a repeat angle corresponding to θ_z.
(C) The three symmetry axes, X, Y, and Z, emerging from the sphere define a spherical triangle whose sides (dashed) correspond to the angles between these axes, that is, to the interaxial angles $Z \wedge X$, $Z \wedge Y$, and $X \wedge Y$.
(D) Enlargement of the spherical triangle in (C) to show its angles. These Buerger (1956) shows to be equal to one-half of the repeat angle, that is, $\theta_z/2$, $\theta_x/2$, or $\theta_y/2$, for each symmetry axis.

From Equation (1-1),

$$\theta_x = \frac{360°}{X} \qquad \theta_y = \frac{360°}{Y} \qquad \theta_z = \frac{360°}{Z}$$

Substituting these into Inequality (1-2) and dividing by 180°, we obtain

$$1 < \frac{1}{X} + \frac{1}{Y} + \frac{1}{Z} < 3 \qquad \text{(Ineq. 1-3)}$$

To test, for example, whether a 4-, 3-, and 2-fold axis can compatibly pass through a common point, we substitute 4, 3, and 2 for X, Y, and Z in Inequality (1-3). If Inequality (1-3) holds true after this substitution (as it does), the combination is compatible. If Inequality (1-3) does not hold, the three symmetry axes being tested cannot intersect compatibly, unless they happen to be one of the five exceptional cases wherein X, Y, Z equal 2, 4, 4; 3, 3, 3; 3, 3, 4; 3, 4, 4; or 4, 4, 4. These exceptions *can* compatibly intersect, but do not fulfill Inequality (1-3) because for them the Euler triangle (Fig. 1-9C) is indeterminate. In applying Inequality (1-3), X, Y, or Z must always represent symmetry axes in *sensu stricto*; in other words, X, Y, or Z can never be 1-fold axes.

Three symmetry axes, even if indicated by Inequality (1-3) as able to intersect compatibly, can do so only at certain angles. At random angles they too would interact to produce an infinite number of axes, all emanating outward from their point of intersection. The angles at which three symmetry axes, X, Y, and Z, can compatibly intersect can be calculated by applying the law of cosines to the spherical triangle in Figure 1-9D. These are

$$\cos X \wedge Y = \frac{\cos \theta_z/2 + (\cos \theta_x/2)(\cos \theta_y/2)}{(\sin \theta_x/2)(\sin \theta_y/2)} \qquad \text{(Eq. 1-4)}$$

$$\cos Y \wedge Z = \frac{\cos \theta_x/2 + (\cos \theta_y/2)(\cos \theta_z/2)}{(\sin \theta_y/2)(\sin \theta_z/2)} \qquad \text{(Eq. 1-5)}$$

$$\cos X \wedge Z = \frac{\cos \theta_y/2 + (\cos \theta_x/2)(\cos \theta_z/2)}{(\sin \theta_x/2)(\sin \theta_z/2)} \qquad \text{(Eq. 1-6)}$$

Depending on whether X, Y, and Z represent 2-, 3-, 4-, or 6-fold axes—or the improper axes $\bar{2}$, $\bar{3}$, $\bar{4}$, or $\bar{6}$—the values of $\theta/2$ to be inserted into Equations (1-4) to (1-6) are, respectively, 180°/2, 120°/2, 90°/2 or 60°/2, that is, 90°, 60°, 45°, or 30°. Table 1-1 summarizes the compatible trios of intersecting symmetry axes as well as the angles, computed from Equations (1-4) to (1-6), at which they must intersect.

In the first four axial combinations of Table 1-1 (illustrated as A to D in Fig. 1-10) Y and Z represent 2-fold axes, each at 90° to X and thus lying in a plane (shaded in Fig. 1-10) perpendicular to X. In Figure 1-10A, X is only a 2-fold axis and consequently its operation on Y or Z "generates" no additional 2-fold axes. In Figures 1-10B, C, or D, however, where X is, respectively, a 3-, 4-, or 6-fold axis, X operates on Y (now labeled Y_1) to relate it to an additional 2-fold axis at Y_2 and, in some cases, at Y_3 as well. It similarly relates the 2-fold axis at Z (now labeled Z_1) to identical 2-fold axes at Z_2 and, in some cases, at Z_3. Consequently, Y_1, Y_2, and, if present, Y_3 constitute a set of symmetrically related 2-fold axes in Figures 1-10B, C, D, as do Z_1, Z_2, and Z_3. In Figure 1-10B the 3-fold axis operation relates the 2-fold axes at Y_1, Y_2, and Y_3 to the opposite ends of the 2-fold axes at Z_1, Z_2, and Z_3; in short, the 2-fold axes labeled Y and Z constitute a *single* set of symmetrically related 2-fold axes. In Figures 1-10A, C,

TABLE 1-1

Compatible Trios of Symmetry Axes and Their Required Interaxial Angles

X	Y	Z	$X \wedge Y$	$X \wedge Z$	$Y \wedge Z$
2	2	2	90°	90°	90°
3	2	2	90°	90°	60°
4	2	2	90°	90°	45°
6	2	2	90°	90°	30°
4	3	2	54°44′	45°	35°16′
2	3	3	54°44′	54°44′	70°32′

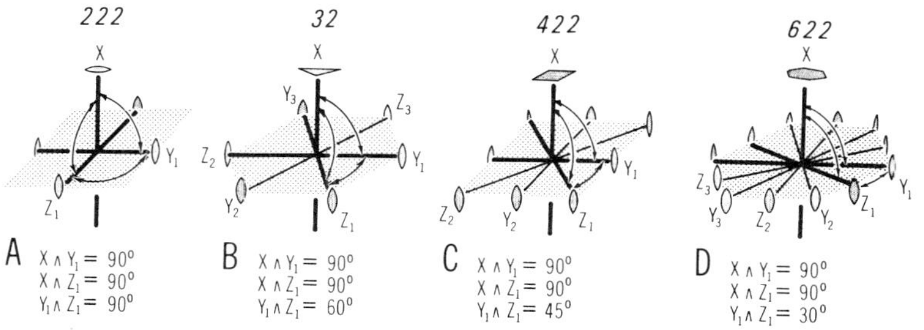

FIGURE 1–10

The first four trios of axes listed in Table 1–1 are shown, respectively, by the three bold lines in (A) to (D). Lighter lines denote the additional 2-fold axes that arise in the shaded plane perpendicular to X because axis X operates to repeat the original (bold line) 2-fold axes. In every case note that $Y_1 \wedge Z_1$, the angle between neighboring 2-fold axes in the shaded plane, always equals $\theta_x/2$, that is, one-half of the repeat angle inherent to X.

and D, however, *two* sets of 2-fold axes exist because no symmetry operation is present that can relate a 2-fold axis in a Y position to one in a Z position; thus the set of symmetrically related 2-fold axes labeled with Y's is symmetrically independent of the set labeled with Z's.

Crystallographers often use the symbol $X2$ to indicate an unspecified symmetry axis X plus a set of 2-fold axes related to each other through X's operation, and lying in the plane perpendicular to X. If within this plane there is a second such set of 2-fold axes that are not related by X to those of the first set, the symbol $X22$ may be used. Thus the specific symbols for the four combinations shown in Figure 1-10 are written 222, 32, 422, or 622 and pronounced "two-two-two, three-two, . . . ," and so on. In these specific symbols the first digit describes the X axis; the second digit indicates the 2-fold axes in the Y sites, and the third digit indicates the 2-fold axes in the Z sites. The angle between neighboring 2-fold axes in the plane perpendicular to X is always $\theta_x/2$, where θ_x is the repeat angle for X. Hence, from the symbols 222, 32, 422, and 622 the angle between the closest 2-fold axes in the plane perpendicular to X is easily deduced as 180°/2, 120°/2, 90°/2, or 60°/2, respectively, that is, 90°, 60°, 45°, or 30°.

For each of the trios of axes listed on the bottom two lines in Table 1-1, $X \wedge Y$ equals 54°44′. Because 54°44′ equals the angle between the perpendicular to a cube's face and its body diagonal,[4] the geometry for each of the trios is conveniently illustrated relative to an imaginary cube (here called the "guide cube"). Thus the 233 trio is represented by the heavy lines, X_1, Y_1, and Z_1 in Figure 1-11A; the 432 trio, by the heavy lines X_1, Y_1, and Z_1 in Figure 1-11B. In each case axes X_1, Y_1, and Z_1 then operate on each other to produce equivalent axes in identical geometrical attitudes to the guide cube. For example, the 3-fold axis at Y_1 in Figure 1-11A operates on the 2-fold axis at X_1 to require

[4] This diagonal is a line from the cube's center to its corner.

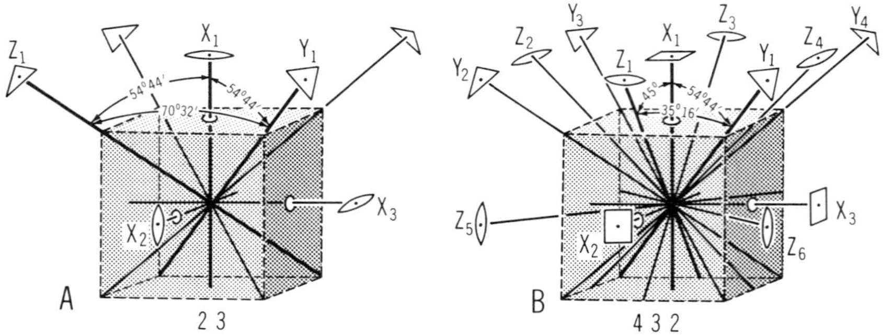

FIGURE 1–11

(A) To intersect compatibly, a 2-, 3-, and another 3-fold axis must be angularly disposed as are X_1, Y_1, and Z_1. The imaginary cube simply emphasizes that X_1 bears the same angular relationship to Y_1 and Z_1 that a perpendicular to a cube's face bears to two of its body diagonals. The operation of Y_1 on X_1 requires additional 2-fold axes at X_2 and X_3. Similarly, X_1 operates on Y_1 and on Z_1 to require the two unlettered 3-fold axes where shown. (The guide cube is purely imaginary and its symmetry is not to be considered; if it were considered, X_1, X_2, and X_3 would have to be 4-fold axes.)
(B) To intersect compatibly, a 4-, 3-, and 2-fold axis must be disposed as are X_1, Y_1, and Z_1, their angular relationship being analogous to that between the perpendicular to a cube's face, its body diagonal, and a line joining midpoints of opposite edges. The operation of Y_1 on the 4-fold axis at X_1 requires 4-fold axes to be at X_2 and X_3 as well. The operation of Y_1 on Z_1 requires additional 2-fold axes at Z_4 and Z_6. The operation of X_1 on Z_1 would require 2-fold axes at Z_2, Z_3, and Z_4. Such interactions would eventually generate all the symmetry shown.

equivalent 2-fold axes at X_2 and X_3, these being, like X_1, perpendicular to the cube's faces. These 2-fold axes then operate on Y_1 and Z_1 to produce two more 3-fold axes (unlabeled in Fig. 1-11A), all four 3-fold axes coinciding with body diagonals of the guide cube.

For a combination of symmetry axes wherein the interaxial angle 54°44′ appears, the Hermann-Mauguin symbol consists of the following parts: (1) the first digit identifies the rotational symmetry of the three equivalent, mutually perpendicular axes (X_1, X_2, and X_3 in Fig. 1-11) oriented perpendicular to the faces of the guide cube; (2) the second digit identifies the four equivalent axes (Y_1, Y_2, Y_3, and Y_4 in Fig. 1-11B) oriented along the body diagonals; (3) the third digit identifies the six equivalent axes (Z_1, Z_2, . . . , Z_6 in Fig. 1-11B) that join the midpoints of opposite edges of the guide cube. For example, 432 symbolizes the combination of axes in Figure 1-11B. For some of the symmetry combinations which involve the interaxial angle 54°44′, the lines which join the midpoints of opposite edges of the guide cube are trivial 1-fold (or $\bar{1}$-fold) axes and the third digit is then omitted from the symbol. For example, the symbol 23 denotes the combination in Figure 1-11A. In this figure the 3-fold axes labeled Y_1 and Z_1 are equivalent, because the 2-fold at X_3 relates Y_1 to the unlabeled end of Z_1.

To emphasize this equivalency, it might now be better to relabel Z_1 as Y_2 in Figure 1-11A.

For a *set of compatibly intersecting axes referrable to a guide cube*, the *Hermann-Mauguin symbol always has a 3 as its second digit*. Hence 23 indicates the axial combination in Figure 1-11A, whereas the symbol 32 indicates the combination in Figure 1-10B.

Combinations of axes and planes

If a point on a 2-, 4-, or 6-fold axis is a center of symmetry, then a plane of symmetry perpendicular to this axis must also pass through this point. We can demonstrate this by letting the tiny circle on the 2-fold axis in Figure 1-12A represent a center of symmetry. This 2-fold axis and center then operate on spheres a_0 and b_0 as follows: (1) the 2-fold axis relates a_0 to a_2 and b_0 to b_2; (2) the center relates a_0 to $a_{0,\bar{1}}$, b_0 to $b_{0,\bar{1}}$, a_2 to $a_{2,\bar{1}}$, and b_2 to $b_{2,\bar{1}}$. The resultant distribution of these eight spheres shows that a plane of symmetry (shaded) must necessarily exist perpendicular to the 2-fold axis. This could be similarly illustrated for a 4- or 6-fold axis. The standard symbol for an X-fold axis with a plane of symmetry perpendicular to it is X/m (read this as "ex over em" or "ex normal em" where "normal" means "perpendicular to"). In the Hermann-Mauguin notation X/m represents a single, indivisible symbol.

Consequently, if a center of symmetry is added to a single proper axis, that is, to a 1-, 2-, 3-, 4-, or 6-fold axis, then the 2-, 4-, or 6-fold axes become the

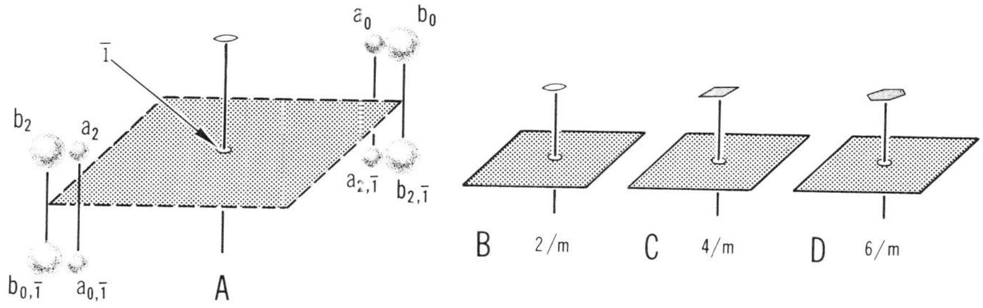

FIGURE 1–12

(A) A center of symmetry (the hollow circle labeled $\bar{1}$) has been added to an axis of 2-fold symmetry. Their successive operations now produce the same effect as a symmetry plane (shaded) perpendicular to the 2-fold axis. For example, atoms a_2 and b_2 are related to a_0 and b_0 by operation of the 2-fold axis, whereas the center of symmetry relates a_0 to $a_{0,\bar{1}}$ and b_0 to $b_{0,\bar{1}}$ and also relates a_2 to $a_{2,\bar{1}}$ and b_2 to $b_{2,\bar{1}}$. The disposition of these atoms thus indicates that the shaded symmetry plane is necessarily present. Such a perpendicular plane of symmetry arises whenever a center is added to an even-fold axis.

(B) (C) (D) The three new compatible X/m combinations obtained by adding a center of symmetry to an even-fold axis such as a 2-fold (B), a 4-fold (C), or a 6-fold axis (D).

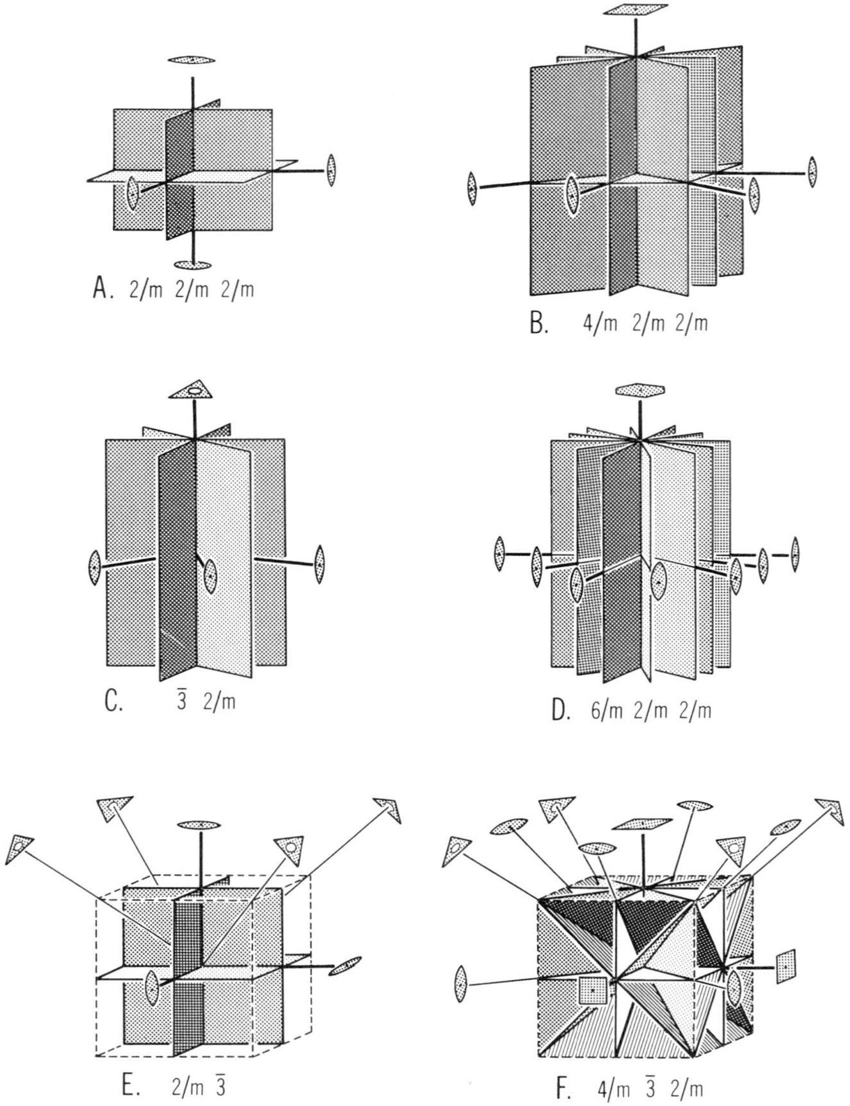

FIGURE 1–13

The six new combinations that result if a center of symmetry is added to the axial combinations shown in Figures 1–10 and 1–11. Thus a center of symmetry, if added to combinations (A) to (D) in Figure 1–10, changes them to the combinations also labeled (A) to (D) in this figure. Similarly, a center added to combinations (A) and (B) in Figure 1–11 converts them, respectively, to the combinations here labeled *E* and *F*.

compatible combinations, $2/m$, $4/m$, or $6/m$ (Fig. 1-12B, C, D), whereas the 1- and 3-fold axes merely become, as already discussed, the improper axes, $\bar{1}$ and $\bar{3}$. Similarly, if a center of symmetry is added to the point of intersection of the axes in the compatible combinations already demonstrated, that is, to 222, 32, 422, 622, 23 or 432, then a symmetry plane must exist perpendicular to every even-fold axis; in addition, all 3-fold axes become $\bar{3}$-fold axes. The six new compatible combinations that then result (Fig. 1-13) are symbolized as $2/m\ 2/m\ 2/m$; $\bar{3}\ 2/m$; $4/m\ 2/m\ 2/m$; $6/m\ 2/m\ 2/m$; $2/m\ \bar{3}$; and $4/m\ \bar{3}\ 2/m$. Regarding X/m as a single digit, we note that $\bar{3}$ occupies the second place in the symbols, $2/m\ \bar{3}$ and $4/m\ \bar{3}\ 2/m$, an indication that each is referable to a guide cube.

The general principle partly underlying the preceding discussion is worth stating. *If an intersection occurs between any two of the following three symmetry*

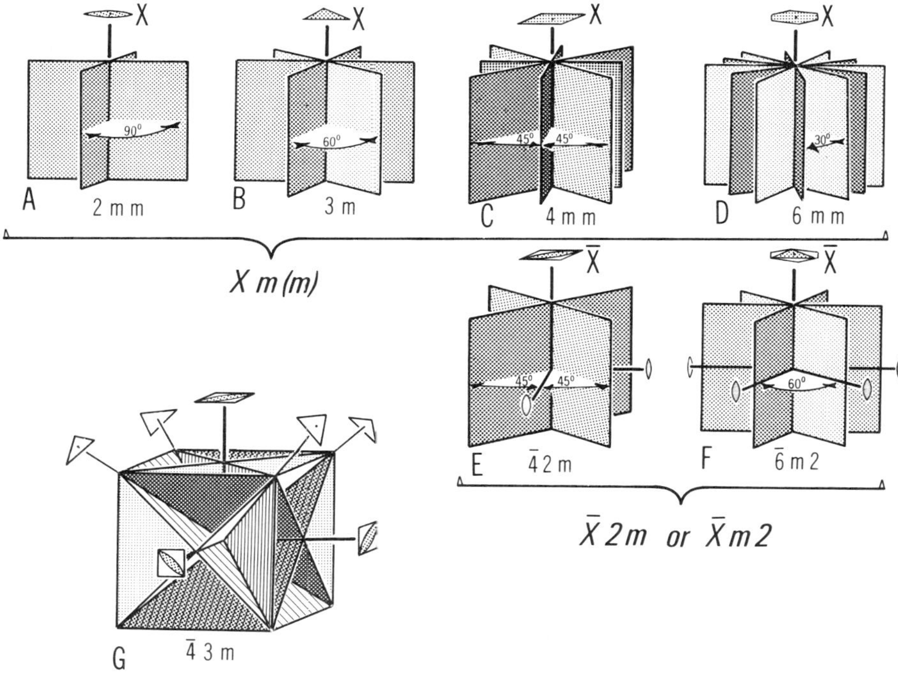

FIGURE 1–14

(A) to (D) $Xm(m)$ combinations, where X equals a 2-, 3-, 4-, or 6-fold axis. The angle between adjacent planes equals $\theta_x/2$ as shown by the arcs.
(E) (F) $\bar{X}2(m)$ or $\bar{X}m(2)$ combinations. The angle between adjacent 2-fold axes and planes of symmetry as measured in the plane perpendicular to the $\bar{X}$ axis equals $\theta_x/2$ as shown by the arcs.
(G) The combination $\bar{4}3m$ obtained by changing the two even-fold axes in combination 432 to improper axes is shown here in reference to a guide cube.

elements—(1) an even-fold axis such as a 2-, 4-, or 6-fold axis, (2) a symmetry plane perpendicular to this axis, or (3) a center of symmetry—*then the third member of the trio must also be present at that point.* We could verify this using an approach like that used in Figure 1-12A.

Combinations of proper and improper axes

For the first five of the six compatible trios of axes listed in Table 1-1, we can replace *two* of their *even-fold axes* with their improper counterparts to develop seven new compatible sets in which both proper and improper axes occur.[5] Thus 222 becomes $2\bar{2}\bar{2}$; 32(2)[6] becomes $3\bar{2}(\bar{2})$; 422 becomes either $4\bar{2}\bar{2}$ or $\bar{4}2\bar{2}$; 622 becomes $6\bar{2}\bar{2}$ or $\bar{6}\bar{2}2$; and 432 becomes $\bar{4}3\bar{2}$. In these, every $\bar{2}$ axis may be replaced by its synonymous operation, namely, a symmetry plane perpendicular to it. After this is done (Fig. 1-14), their symbols become $2mm$, $3m(m)$[6], $4mm$, $\bar{4}2m$, $6mm$, $\bar{6}m2$, and $\bar{4}3m$. Except for $\bar{4}3m$ (Fig. 1-14G), these new combinations consist of either (1) a single X axis, where X equals 2, 3, 4, or 6 (Figs. 1-14A to D), which coincides with the line of intersection of all the symmetry planes present, the angle between adjacent planes being $\theta_x/2$; or (2) a single $\bar{X}$ axis, where $\bar{X}$ is $\bar{4}$ or $\bar{6}$ (Figs. 1-14E, F), which also coincides with the intersection of all the symmetry planes present. In Figures 1-14E, F 2-fold axes are also present that bisect the angles between the symmetry planes; the angle between each such 2-fold axis and adjacent plane is $\theta_{\bar{x}}/2$ where $\theta_{\bar{x}}$, the repeat angle, equals 90° for $\bar{4}$ or 60° for $\bar{6}$.

The general symbols for the combinations in Figures 1-14A to D are either Xm or Xmm. The symbol Xmm stresses that *two sets* of symmetry planes, no plane in the first being relatable by symmetry to one in the second, intersect in a common line, the X axis. For the combinations in Figures 1-14E, F the general symbol often used is $\bar{X}m$ (or $\bar{X}2$), although a more complete symbol such as $\bar{X}m2$ (or $\bar{X}2m$) would emphasize that 2-fold axes bisect the angle between the symmetry planes (or vice versa).

[5] At this point the reader may ask (1) why replace only even-fold axes, why not change 3 to $\bar{3}$? Or (2) why replace precisely *two* of the trio, why not one or all three? To answer the first question, the change of 3 to $\bar{3}$ is equivalent to adding a center of symmetry to the combination and results will be as already discussed for such addition. To answer the second, three axes, X, Y, and Z, if they are to intersect compatibly, must be able to operate successively on any given object and, after a single operation by each, restore it to its original position and shape. However, a single operation by an improper axis changes an object's shape; for example, if the original object was a disembodied right hand, an improper axis operates on it to require a left hand. Now a second improper axis will be necessary to change this left hand back to a right hand once more. The third axis, a proper one, can then relate this right hand, by a simple rotation, back to the original right hand. For a combination $\bar{X}$, Y, Z where only one of the trio is improper, the original right hand is changed by the single $\bar{X}$ operation into a left hand and successive operations by Y and Z maintain it as a left hand, being unable to restore it to its original right-handed shape.

[6] The parentheses indicate that this symbol is usually omitted because this 2-fold axis is related by operation of the 3-fold axis (Fig. 1-10B) to the 2-fold axis whose symbol is not in parentheses. For similar reasons, $3m(m)$ is usually written as $3m$.

TABLE 1–2

The Subgroups and Supergroups among the 32 Point Groups[a]

$4/m\ \bar{3}\ 2/m$	*																															
$\bar{4}3m$	x	*																														
432	x		*																													
$2/m\ \bar{3}$	x			*																												
23	x	x	x	x	*																											
$6/m\ 2/m\ 2/m$						*																										
$\bar{6}m2$						x	*																									
$6mm$						x		*																								
622						x			*																							
$6/m$						x				*																						
$\bar{6}$						x	x			x	*																					
6						x		x	x	x		*																				
$\bar{3}\ 2/m$	x					x							*																			
$3m$	x	x				x	x	x					x	*																		
$3\ 2$	x		x			x	x		x				x		*																	
$\bar{3}$	x			x		x				x			x			*																
3	x	x	x	x	x	x	x	x	x	x	x	x	x	x	x	x	*															
$4/m\ 2/m\ 2/m$	x																	*														
$\bar{4}2m$	x	x																x	*													
$4mm$	x																	x		*												
422	x		x															x			*											
$4/m$	x																	x				*										
$\bar{4}$	x	x																x	x			x	*									
4	x		x															x		x	x	x		*								
$2/m\ 2/m\ 2/m$	x			x		x												x							*							
$2mm$	x	x		x		x	x	x										x	x	x					x	*						
222	x	x	x	x	x	x			x									x	x		x				x		*					
$2/m$	x			x		x				x			x					x				x			x			*				
m	x	x		x		x	x	x		x	x		x	x				x	x	x		x			x	x		x	*			
2	x	x	x	x	x	x	x	x	x	x		x	x		x			x	x	x	x	x	x	x	x	x	x	x		*		
$\bar{1}$	x			x		x				x			x			x		x				x			x			x			*	
1	x	x	x	x	x	x	x	x	x	x	x	x	x	x	x	x	x	x	x	x	x	x	x	x	x	x	x	x	x	x	x	*

[a] The asterisk at the top of each vertical column indicates the supergroup; the x's vertically below it indicate the subgroups which belong to this supergroup. Figure 1–16 illustrates supergroup $2/m\ 2/m\ 2/m$ and its various subgroups.

Unlike the Xm or $\bar{X}m$ combinations, combination $\bar{4}3m$ (Fig. 1-14G) is referable to a guide cube as implied by the presence of a 3 in the second position. Hence $\bar{4}$ indicates the three $\bar{4}$-fold axes normal to the cube faces; 3 indicates the four 3-fold axes along the cube's body diagonals; and m indicates the six planes that cut diagonally through the cube at its center, these planes being perpendicular to the directions labeled Z in Figure 1-11B.

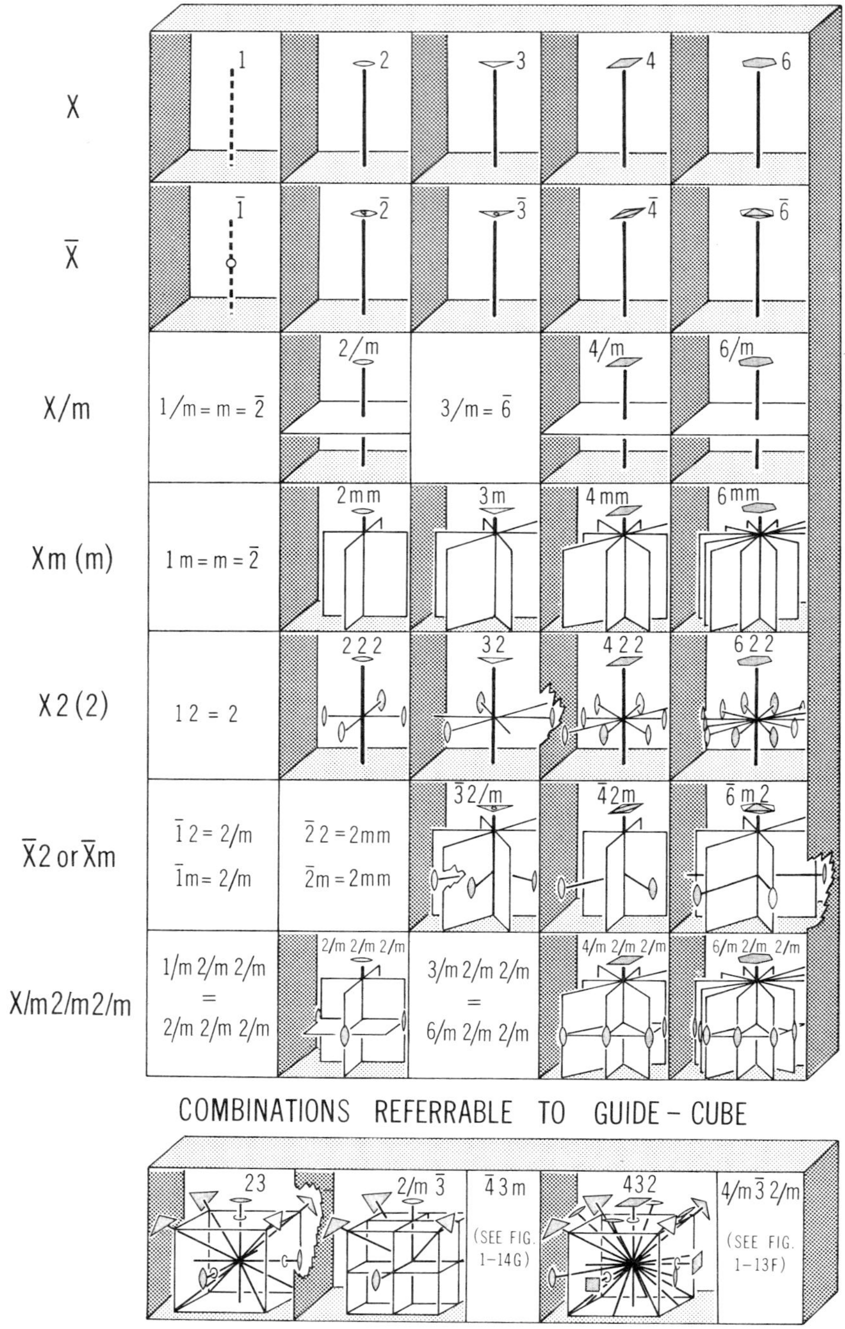

FIGURE 1–15

Graphic portrayal of the geometric relationships between the symmetry elements present in each of the 32 point groups. The tiers labeled X and $\bar{X}$ represent the ten

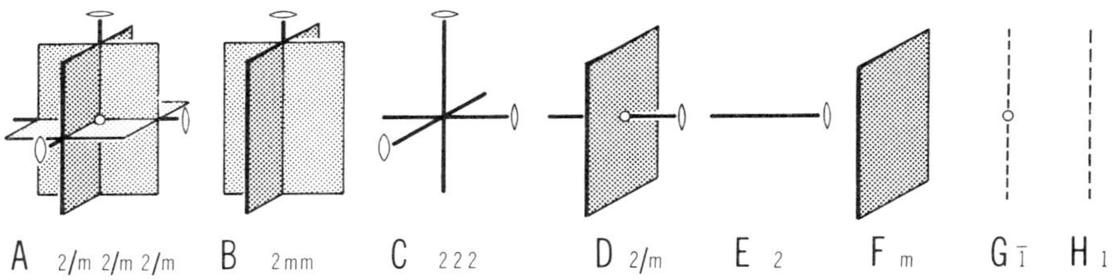

FIGURE 1–16

(A) Point group $2/m\ 2/m\ 2/m$ and (B to H) its various subgroups. In these drawings the small circle represents a center of symmetry. The heavy, dashed axis in (H), which represents a 1-fold axis, could have been drawn in any direction.

The thirty-two point groups

Combinations of symmetry elements that can compatibly pass through a common point are called *point groups.* In summary, only 32 nonidentical point groups are possible. These include the 10 uncombined X or $\bar{X}$ axes (Fig. 1-15, upper two tiers) plus the 22 compatible combinations just discussed (Fig. 1-15, remaining tiers). No $\bar{X}/m$ combinations are drawn in Figure 1-15 because such combinations would be identical to point groups in the top three tiers; thus (Fig. 1-7), $\bar{2}/m$ equals m, $\bar{3}/m$ equals $6/m$, $\bar{4}/m$ equals $4/m$, $\bar{6}/m$ equals $\bar{6}$. In the $\bar{X}2$ tier we note that if $\bar{X}$ equals $\bar{3}$, then a center of symmetry must also be present, whereupon 2 becomes $2/m$.

Subgroups and supergroups

If a single symmetry element (or compatible combination) is a component of a more complex symmetry element (or combination), it is called a *subgroup* of the more complex group, the latter being termed the *supergroup.* For example, the symmetry operation of the most complex single element, a 6-fold axis, involves a 60° rotation. Two such rotations equal 120°, the rotation associated with a 3-fold axis. Thus if we choose to ignore or destroy every other item related by a 6-fold axis, a 3-fold axis remains. Since its operation is enveloped by that of the 6-fold axis, the 3-fold is a subgroup of the 6-fold axis. Similarly, the 2-fold and 1-fold operations are enveloped by that of the 6-fold axis. In a like manner if we subtract the appropriate symmetry elements from the point group $2/m\ 2/m\ 2/m$, (Fig. 1-16A to H), we see that it contains the subgroups $2mm$, 222, $2/m$, 2, m, $\bar{1}$, and 1. The distribution of the subgroups for the 32 point groups is summarized in Table 1-2.

---

basic operations uncombined. Tiers X/m, $Xm(m)$, and $X2(2)$ represent the combination of a proper axis X with a mirror plane perpendicular to it, parallel to it, or with a 2-fold axis perpendicular to it. Tier $\bar{X}2$ or $\bar{X}m$ represents the combination of an improper $\bar{X}$ axis with a 2-fold axis perpendicular to it or a mirror plane parallel to it. Note that for $\bar{6}m2$ the 2-fold axes lie in the mirror planes and that no 2-fold axis is perpendicular to a mirror plane.

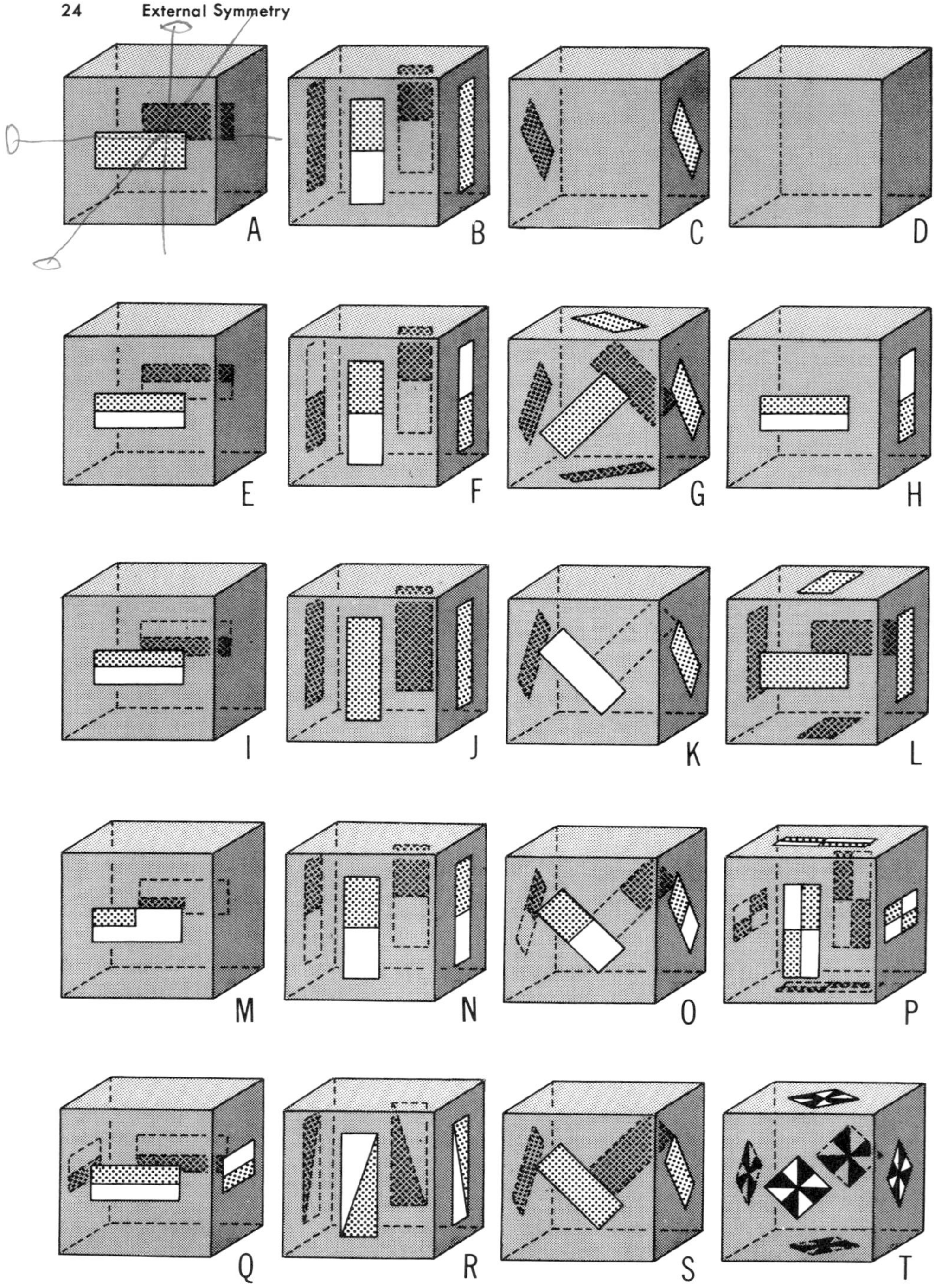

FIGURE 1–17

Cubes with patterns on their faces whose point groups are to be determined by the reader.

REFERENCE AND RECOMMENDED READING

Buerger, M. J. (1956) *Elementary Crystallography*, New York: Wiley.

QUESTIONS AND PROBLEMS

1-1 (a) What point group is developed through the interaction of two symmetry planes at 60° to each other? (b) at 30°? (c) at 45°? (d) at 90°?

1-2 (a) What point group is developed through the interaction of two 2-fold axes at 60° to each other? (b) at 30°? (c) at 45°? (d) at 90°?

1-3 State the point group that results if a center of symmetry is added to a point occupied by (a) point group 1; (b) point group 2; (c) 3; (d) $\bar{4}$; (e) 4; (f) $\bar{6}$; (g) 6; (h) point group X, where X is an even number; (i) 222; (j) m; (k) $2mm$; (l) 432; (m) 23; (n) $\bar{4}3m$; (o) 422; (p) 32.

1-4 For the cubes in Figure 1-17 consider the patterns on their faces to change their symmetry from that of the unmodified cube in Figure 1-17D. Determine the point-group symmetry for each cube.

1-5 For which supergroups is point group $\bar{3}\ 2/m$ a subgroup?

ANSWERS

1-1 (a) $3m$; (b) $6mm$; (c) $4mm$; (d) $2mm$.

1-2 (a) 32; (b) 622; (c) 422; (d) 222.

1-3 (a) $\bar{1}$; (b) $2/m$; (c) $\bar{3}$; (d) $4/m$; (e) $4/m$; (f) $6/m$; (g) $6/m$; (h) X/m; (i) $2/m\,2/m\,2/m$; (j) $2/m$; (k) $2/m\,2/m\,2/m$; (l) $4/m\,\bar{3}\,2/m$; (m) $2/m\ \bar{3}$; (n) $4/m\ \bar{3}\ 2/m$; (o) $4/m\ 2/m\ 2/m$; (p) $\bar{3}\ 2/m$.

1-4 (a) $2/m\ 2/m\ 2/m$; (b) $2mm$; (c) $2/m\ 2/m\ 2/m$; (d) $4/m\ \bar{3}\ 2/m$; (e) $2mm$; (f) $\bar{4}2m$; (g) $\bar{4}3m$; (h) 1; (i) $2/m$; (j) $4/m\ 2/m\ 2/m$; (k) 222; (l) $2/m\,\bar{3}$; (m) 2; (n) $4mm$; (o) 4; (p) 23; (q) m; (r) $\bar{4}$; (s) 422; (t) 432.

1-5 $4/m\ \bar{3}\ 2/m$ and $6/m\ 2/m\ 2/m$ (as read from Table 1-2).

2 Crystal Classes, Axes, and Systems

The Thirty-Two Crystal Classes

A crystal is a solid formed by the systematic arrangement or packing of atoms, ions, or molecules (see Fig. 6-6). The repetitive nature of this packing, to be discussed later, may cause smooth surfaces called *faces* to develop on the crystal during the course of its unobstructed growth. A crystal is called *anhedral*, *subhedral*, or *euhedral* if it, respectively, lacks such faces (Fig. 2-1A), is only partly bounded by them (Fig. 2-1B), or is completely bounded (Fig. 2-1C). These terms originate from the Greek word *hedra* (seat or base and hence a crystal plane) in combination with the prefixes *an* (not or without), *sub* (under or below and hence incomplete), and *eu* (well, that is, complete).

Solids which consist of atoms, ions, or molecules packed together randomly rather than systematically are incapable of developing faces during their unobstructed growth and are hence said to be *amorphous* (Greek *a*, "without," and *morph*, "form"). They are *noncrystalline* solids, which by definition lack the systematic internal structure characteristic of the crystalline state. Lack of faces, however, is not proof of noncrystallinity. For example, anhedral crystals, although they lack faces because neighboring crystals obstructed their growth or because these faces were later broken off, are crystalline because of their systematic internal structure.

In an ideally developed euhedral crystal, the faces may be so disposed as to impart to the crystal an external symmetry identical with that of one of the 32 point groups. For example, the idealized crystal of the mineral zincite (ZnO) in Figure 2-2A possesses symmetry identical to that for point group $6mm$. Moreover careful examination shows that *all* crystals of zincite possess the symmetry $6mm$. Not only zincite, but also the minerals bromellite (BeO), greenockite (CdS),

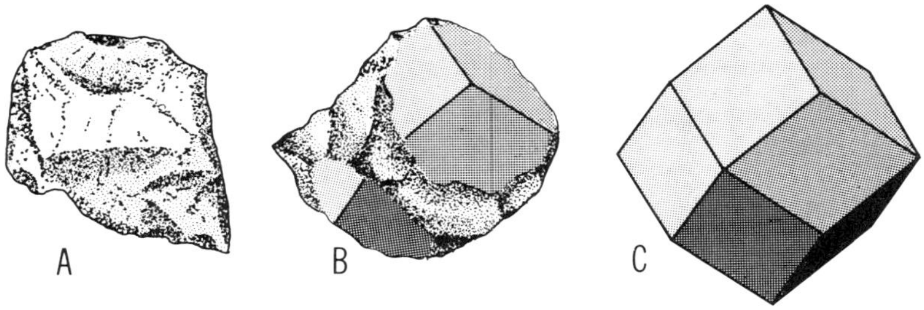

FIGURE 2–1

Crystals classified according to the degree to which they possess smooth faces: (A) An *anhedral* crystal, that is, one which lacks faces; (B) a *subhedral* crystal, that is, one partly bounded by faces; (C) A *euhedral* crystal, that is, one entirely bounded by faces.

wurtzite (ZnS), and swedenborgite ($NaBe_4SbO_7$) are represented by crystals having external symmetry $6mm$; for example, a wurtzite crystal is shown in Figure 2-2B.

All minerals (and artificial compounds) whose crystals possess external symmetry identical to the same point group are said to belong to the same *crystal class*. Thus zincite, bromellite, greenockite, wurtzite, swedenborgite, and many artificial compounds all belong to crystal class $6mm$. Since there are only 32 distinct point groups, crystalline materials the world over may be classified into one or the other of 32 crystal classes. For example, crystals of chalcanthite ($CuSO_4 \cdot 5H_2O$), albite ($NaAlSi_3O_8$), and anorthite ($CaAl_2Si_2O_8$), plus all other minerals and artificial compounds that possess only a center of symmetry, all belong to crystal class $\bar{1}$.

The number of minerals and artificial compounds falling within each of the 32 crystal classes varies greatly (Table 2-1). From the mineralogic view, crystal classes are important if they contain either a large number of minerals or a

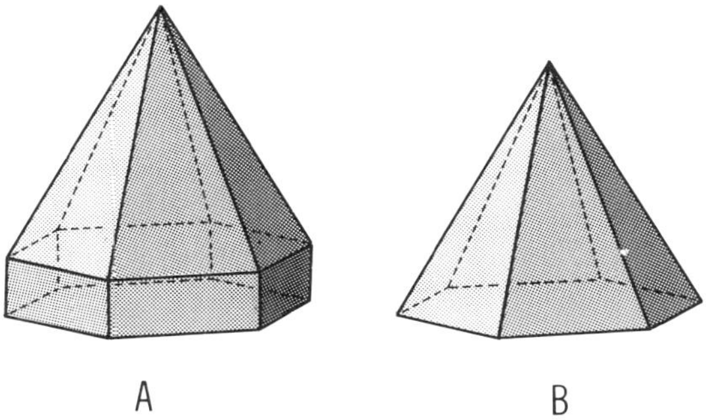

FIGURE 2–2

Crystals of minerals that belong to crystal class $6mm$: (A) zincite (ZnO) and (B) wurtzite (ZnS).

TABLE 2–1

Distribution of 3837 Crystalline Compounds (Mineral and Synthetic) among the 32 Crystal Classes[a]

	Number of Compounds Represented		
Crystal Class	Minerals	Synthetic	Total
1	1	19	20
$\bar{1}$	18	51	69
m	9	23	32
2	8	159	167
$2/m$	131	466	597
$mm2$	18	82	100
222	14	187	201
$2/m\ 2/m\ 2/m\ (mmm)$	121	338	459
$\bar{4}$	2	12	14
4	1	5	6
$4/m$	12	61	73
$\bar{4}2m$	8	54	62
$4mm$	1	8	9
422	3	18	21
$4/m\ 2/m\ 2/m\ (4/mmm)$	37	258	295
3	1	11	12
$\bar{3}$	11	61	72
$3m$	13	23	36
32	6	46	52
$\bar{3}\ 2/m$	38	194	232
$\bar{6}$	0	0	0
6	3	17	20
$6/m$	19	31	50
$\bar{6}m2$–$\bar{6}2m$	2	10	12
$6mm$	5	22	27
622	4	14	18
$6/m\ 2/m\ 2/m\ (6/mmm)$	20	175	195
23	6	33	39
$2/m\ \bar{3}$	14	126	140
$43m$	22	105	127
432	0	6	6
$4/m\ \bar{3}\ 2/m\ (m3m)$	61	613	674

[a] Data compiled from Donnay and Nowacki (1954), *Geol. Soc. Am. Memoir 60.* More recent data show no significant changes in trend.

mineral of widespread geologic importance. For example, class 32 is important because it contains the mineral quartz. In contrast, class $\bar{6}$ is of academic importance only.

The correct determination of a crystal's external symmetry is seldom accomplished by visual inspection. The difficulty becomes apparent if an *idealized*

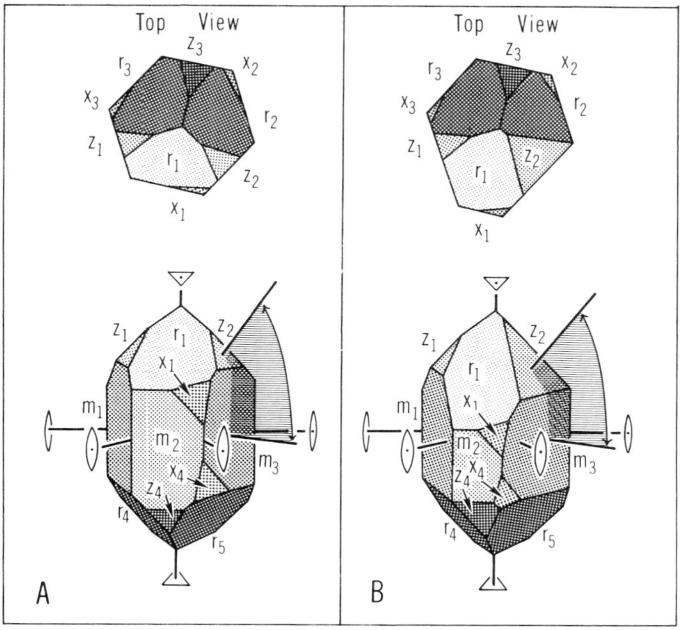

FIGURE 2–3

(A) An idealized crystal of quartz (SiO_2) wherein all the faces labeled *m*, each of which is parallel to the 3-fold axis and to one of the 2-fold axes, are drawn at the same distance from the crystal's center. Similarly, all faces labeled *z* are equidistant from the crystal's center. So are all the faces labeled *r*, and so on.
(B) A malformed crystal of quartz. The various faces labeled *m* are now at unequal distances from the crystal's center. This is also true for the set of faces labeled *z*, and those labeled *x*. Interfacial angles, by definition the angle between the perpendiculars to the two faces in question, are unaffected by distance of the face from the center. For example, the interfacial angle between faces z_2 and m_3, that is, $z_2 \wedge m_3$, has the same value whether measured on the ideal quartz crystal in (A) or on the malformed one in (B).

crystal of, for example, the mineral quartz (Fig. 2-3A) is compared with the somewhat *misshaped* or *malformed* crystals usually found (Fig. 2-3B). In idealized crystals all faces that bear the same *angular relationship* to the crystal's *symmetry axes* or *planes* (or both) are also at *equal distances* from the crystal's center; in misshaped or malformed crystals they are *not* at equal distances. For example, faces m_1, m_2, m_3, ..., each of which is parallel to the 3-fold axis, are at equal distances from the crystal's center in Figure 2-3A but not in Figure 2-3B. As a consequence the presence of the 3-fold axis and of the three 2-fold axes is easily seen for the idealized crystal in Figure 2-3A but not for its misshaped equivalent in Figure 2-3B.

Misshaped or malformed crystals are common because some faces are more accessible than others to the nutrient solutions causing crystal growth. Consequently these grow outward from the crystal's center more rapidly than do the less favored faces. Thus in Figure 2-3B face m_2 grew outward more rapidly than did m_1 or m_3. Consequently, it is (1) more distant from the crystal's center and (2) of lesser area than those other faces, both features being characteristic of rapidly grown faces.

The unequal areal development of similar faces in misshaped crystals may obscure the identity of the symmetry elements that are present. Moreover it may cause crystals of the same substance to be rather diverse in appearance. However, in 1669 the pioneer crystallographer Niels Stensen (Nicolaus Steno) observed that regardless of the size and shape of quartz crystals, the angles between any given pair of faces remained constant in value. For example, the *interfacial angle between two faces,* m_3 and z_2 (symbolized as $m_3 \wedge z_2$), is defined as the *angle between* the *perpendiculars or normals to these faces* and has the same

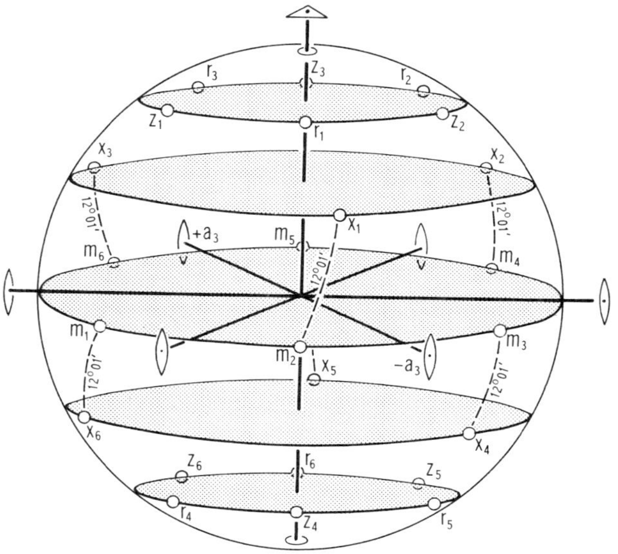

FIGURE 2–4

If a sphere was placed around *either* crystal in Figure 2–3 and if from this sphere's center a line perpendicular to each face was extended outward, these lines would intersect the sphere at the encircled points, called *face poles.* Note that if any face pole is reconnected to the sphere's center, the resultant line represents the perpendicular to the corresponding face. Interfacial angles are now represented by the arcs joining the face poles on the sphere. The distribution on the sphere of the arcs corresponding to the interfacial angles 12° 01′ conforms to the presence of a 3-fold axis and of three 2-fold axes where shown. For example, a 120° rotation about the 3-fold axis superimposes one 12° 01′ arc onto another. For illustrative purposes the distance of face poles x, r, and z (from the equatorial plane containing face poles m) was exaggerated; consequently the arcs representing interfacial angles of 12° 01′ are also exaggerated.

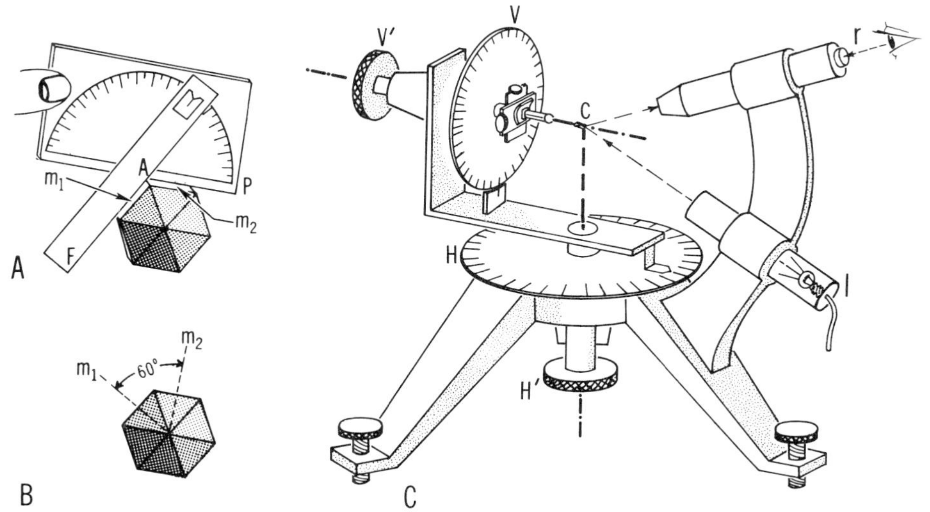

FIGURE 2–5

(A) A contact goniometer of the type presently available from Ward's Scientific Establishment. The interfacial angle, $m_1 \wedge m_2$, is seen to be 60° for the crystal shown as is also seen in the cross-sectional view of the crystal in (B).
(C) A reflecting goniometer of the two-circle type. The crystal at c can be rotated about the dash-dotted axis by turning knurled knob V' and therefore vertical circle V; it is rotatable about the dashed axis by turning knob H' and therefore horizontal circle H. Each crystal face is successively oriented until a line perpendicular to it bisects the angle between two fixed lines, lc and cr, whereupon a tiny, incident lightbeam, which travels along lc, will be reflected as beam cr into the telescope.

value regardless of which quartz crystal in Figure 2-3 is measured. This property also holds true for the angles $m_1 \wedge m_2$, $r_4 \wedge r_5$, $r_4 \wedge z_5$, and so on.

Other crystallographers, particularly Romé de l'Isle during the period 1772–1783, extended Steno's work to crystals of substances other than quartz. Their work established *Steno's law* or the *law of constancy of interfacial angles.* This law states that *the angle measured between a given pair of crystal faces will be equal for all crystals of this substance.* This law, also known as the *Steno-de l'Isle law,* implies that the measurement of a crystal's interfacial angles may aid in its identification.

Regardless of whether a crystal is misshapen or ideal, the geometric distribution of the normals to its faces will be the same (Fig. 2-4). Moreover, this distribution may define the crystal's symmetry and, therefore, its crystal class. For example, the spatial distribution of the recurrent interfacial angle 12°01′, the value of $m_1 \wedge x_6$, $m_2 \wedge x_1$, $m_3 \wedge x_4$, $m_4 \wedge x_2$, $m_5 \wedge x_5$, and $m_6 \wedge x_3$, indicates the presence of the 3-fold axis and the three 2-fold axes where shown. Thus a 120° clockwise rotation about the 3-fold axis, if performed on interfacial angle $m_2 \wedge x_1$, superimposes it on the identical angle, $m_6 \wedge x_3$. Therefore, the symmetry of distribution of quartz's interfacial angles, even though measured for a malformed crystal like that of Figure 2-3B, indicates quartz to belong to crystal class 32.

Thus because the spatial distribution of identically valued interfacial angles often disclosed the external symmetry present in a crystal, the early crystallographers developed devices called "goniometers" (after the Greek *gonia* meaning "angle" and *metron* meaning "measure") to measure the interfacial angles on euhedral or subhedral crystals. Carangeot in 1780 developed a simple contact goniometer (Fig. 2-5A). In use, its protractor is held in a plane perpendicular to the two faces, for example, m_1 and m_2, whose interfacial angle is desired; then with protractor edge AP held parallel to one of these faces, arm AF is rotated until its edge is parallel to the second face, whereupon the value of interfacial angle $m_1 \wedge m_2$ can be read from the protractor's scale (that is, 60° in Fig. 2-5B). At present, however, contact goniometers are relegated to student exercises largely because accurate measurements of interfacial angles, that is, to within a few minutes, are possible only with a *reflecting goniometer* (Fig. 2-5C) like that devised by W. H. Wollaston in 1809. Modern versions secure increased accuracy by rotating the crystal on a vertical and a horizontal axis, successively, until each crystal face in turn intercepts a tiny beam of light from a collimator and reflects it into an eyepiece (Fig. 2-5C). The reflecting position for each face of the crystal is then recorded in terms of the reading on the horizontal and vertical circular scales (the H and V angles, respectively). From the H and V readings for each face on the crystal, precise interfacial angles are readily calculated.

Crystallographic Axes

For convenience, a set of reference axes called *crystallographic axes* is always assigned to a crystal; such axes are indispensable for labeling the crystal's faces. So that crystallographers, even though separated by space and time, will assign the same crystallographic axes to the different crystals of a given substance, certain rules for choosing crystallographic axes have evolved. In general, crystallographic axes are chosen within any given crystal so that the following requirements are met.

1. They are three in number (occasionally four are employed).
2. They are mutually perpendicular, if possible.
3. They coincide with the crystal's symmetry axes or, lacking these, with the normals to its symmetry planes.
4. In crystals that lack sufficient symmetry axes or planes to guide the selection of their crystallographic axes, the axes are selected to coincide with the lines of intersection between faces of greatest area (either with actual intersections or with possible intersections obtained by extending these faces until they do meet).

X-ray studies have necessitated some modifications and additions to these rules. Moreover, in the choice of crystallographic axes for a given crystal, the rules may not all be fulfilled. The extent of fulfillment will vary according to the crystal class to which the crystal belongs. Crystal classes with insufficient symmetry so that rule 4 must be observed will usually possess crystallographic axes that will not be perpendicular to each other. Thus adhering as closely as possible to the four rules, crystals in crystal classes 1 or $\bar{1}$ are assigned three, nonperpendicular axes and are grouped together into the *triclinic system*, so-called because all three crystallographic axes are almost invariably inclined to each other at angles other than 90°. On the other hand, crystals in crystal classes $2mm$, 222, or $2/m\ 2/m\ 2/m$ will always be assigned three mutually perpendicular axes (since the symmetry directions chosen as crystallographic axes are themselves mutually perpendicular). Consequently crystal classes $2mm$, 222, or $2/m\ 2/m\ 2/m$ are grouped together within the *orthorhombic system*, the prefix *ortho* alluding to the right-angle relations between the crystallographic axes. All 32 crystal classes may thus be organized into six (or seven) crystal systems on the basis of similar crystallographic axes. A brief discussion of these crystal systems (and of the crystal classes grouped within each system) follows.

The Triclinic System

Crystals that belong to classes 1 or $\bar{1}$ lack symmetry axes or planes to guide the selection of their crystallographic axes. Thus directions parallel to the edges (or possible edges) between the three faces of largest area were usually chosen to be axes a, b, and c (Fig. 2-6A). Prior to X-ray techniques, the longest such edge was assumed to be parallel to the c axis; the intermediate, to the a axis; and the shortest, to the b axis. Almost invariably the three axes so chosen are obliquely inclined to each other, being at 90° only by coincidence. Consequently crystals of classes 1 and $\bar{1}$ are assigned to the *triclinic system*.

By custom the crystallographic axes are assumed to intersect at the crystal's center (Fig. 2-6). Furthermore, a crystal is conventionally oriented so that the positive ends of its a, b, and c axes point, respectively, (1) toward the observer, (2) to the observer's right, and (3) vertically upward. The Greek equivalents of a, b, and c, that is, α, β, and γ, serve as symbols for the interaxial angles. For example, $a \wedge b$ (read this as "the angle between the a and b axes") is labeled γ; similarly, $b \wedge c$ equals α and $a \wedge c$ equals β (Fig. 2-6A). Note that the symbols for any two axes plus the angle between them always include the first three letters of the alphabet with the angle symbol being in Greek.

For triclinic crystals the interaxial angles α, β, and γ are measured between the positive ends of the crystallographic axes. However, ascribing a positive or negative sign to the end of an axis is purely arbitrary. The convention adopted states that positive ends should be assigned to axes a, b, and c in a way that will cause angles α and β to be obtuse in value. Under this convention γ may be either acute or obtuse, depending on the angular relations between the three triclinic axes under consideration. In Figure 2-6, for example, the left end of the b axis could not have been taken as the positive end because this would make α, that is, $+b \wedge +c$, acute.

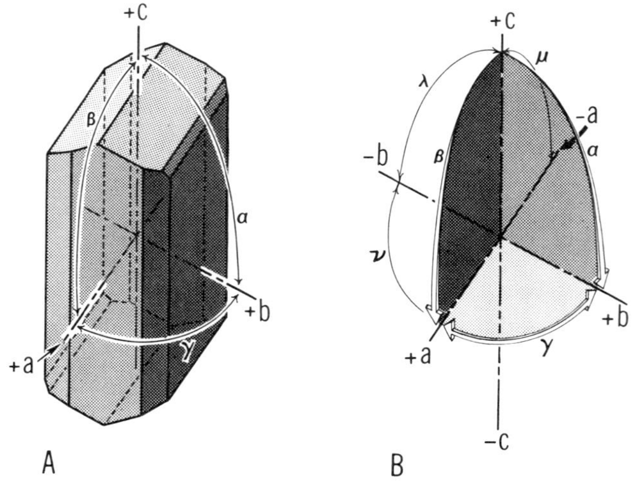

FIGURE 2–6

(A) A triclinic crystal whose a, b, and c axes were chosen to be parallel to the lines of intersection between the crystal's largest faces.
(B) The interaxial angles α, β, and γ in relation to the crystallographic axes in the triclinic system. Angles λ, μ, and υ are, respectively, the supplements of α, β, and γ.

For the time being the foregoing instructions will serve to guide the selection of crystallographic axes in triclinic crystals. Actually an unambiguous choice of crystallographic axes is possible only by utilizing X-ray methods and the conventions formalized by Donnay (p. 155).

The angles supplementary to α, β, and γ are called λ (lambda), μ (mu), and ν (nu), respectively. They are indicated on Figure 2-6B but, in general, the beginning student will seldom need to consider them.

The Monoclinic System

In crystals that belong to classes 2, m, or $2/m$, the b crystallographic axis is usually assumed to coincide with the 2-fold axis (Fig. 2-7A) or with the normal to the symmetry plane, in other words with the $\bar{2}$ axis (Fig. 2-7B) or with both (Fig. 2-7C). The two remaining crystallographic axes, a and c, are invariably located within the plane perpendicular to axis b (Fig. 2-7D) and at the same time are usually parallel to the intersections between prominent faces (or to such intersections as would occur if these prominent faces were extended). Because a and c lie in the plane perpendicular to the b axis, each is at 90° to b and angles α and γ thus equal 90°. However, axes a and c are inclined to each other, the angle β equaling 90° only by coincidence. Because only one pair of axes are inclined to each other, this set of axes is called a *monoclinic set.* Crystals belonging to classes 2, m, and $2/m$ thus comprise the *monoclinic system.*

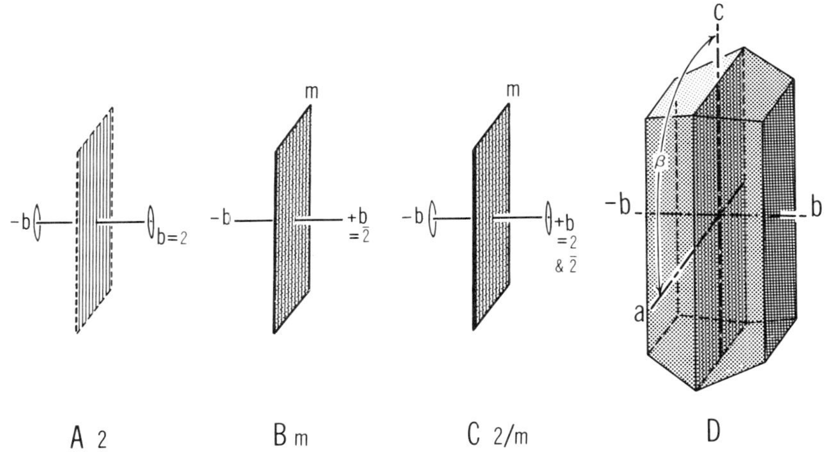

FIGURE 2–7

Choice of crystallographic axes in monoclinic crystals. (A) For crystals in class 2, the 2-fold axis is selected to be the b axis. The ruled plane perpendicular to this b axis will always contain axes a and c. To indicate that for class 2 it is *not* a mirror plane, it has been outlined with dashed lines. (B) For crystals in class m, the normal to the mirror plane m, that is, the $\bar{2}$ axis, is selected to be the b axis. (C) For crystals in class $2/m$, the b axis coincides with both the 2 and $\bar{2}$ axis present. (D) The a and c axes are chosen so that both lie in the (vertically ruled) plane perpendicular to the b axis. Within this plane, a and c are selected to coincide in direction with the lines of intersection between prominent crystal faces.

Monoclinic crystals are conventionally oriented (Fig. 2-7D) with their c axis vertical, their b axis horizontal with its plus end to the observer's right, and their a axis sloping downward toward the observer. Positive ends are assigned to the a and c axes so that the angle β $(+a \wedge +c)$ exceeds 90°. The supplement of β is labeled μ as for triclinic crystals.

The Orthorhombic System

For crystals with symmetry $2mm$, 222, or $2/m\ 2/m\ 2/m$, the 2-fold axes and/or perpendiculars to symmetry planes are chosen to be crystallographic axes a, b, and c (Fig. 2-8). Consequently a, b, and c are mutually perpendicular and no symmetry operation present in the crystal can relate a to b, a to c, or b to c. Crystal classes $2mm$, 222, or $2/m\ 2/m\ 2/m$ are thus grouped together into the *orthorhombic system*; *ortho*rhombic signifies that their crystallographic axes are symmetrically nonequivalent and at right angles to each other. The question as to *which* of the 2-fold axes should be called the a axis, which the b, and which the c is governed by convention. Prior to X-ray methods, the convention was based on the crystal's dimensions. For example, the 2-fold axis along which crystals of a given substance usually grew to be longest was customarily chosen

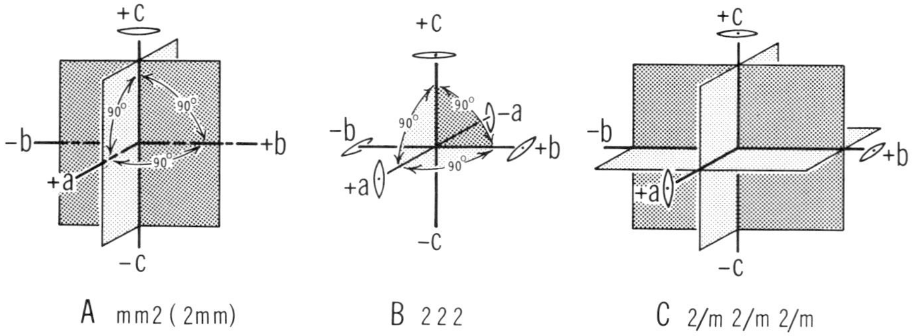

FIGURE 2–8

Choice of the crystallographic axes a, b, and c for crystals in the orthorhombic system according to whether they belong to crystal class (A) $2mm$; (B) 222; (C) $2/m\ 2/m\ 2/m$. In all cases a, b, and c are mutually perpendicular.

to be the c axis; the 2-fold axis along which they were usually shortest was called the b axis. The remaining 2-fold axis, corresponding to an intermediate dimension, was then called the a axis.

By convention the three parts of the symbol for an orthorhombic crystal class are always listed in the order to which they refer to the a, b, and c axes. Thus for the orthorhombic crystal class that contains just one 2-fold axis (Fig. 2-8A), this convention can indicate whether the single 2-fold axis coincides with the a axis ($2mm$), the b axis ($m2m$), or the c axis ($mm2$). Crystallographers conventionally select this single 2-fold axis to coincide with the c axis. Consequently the correct symbol for this crystal class is $mm2$ rather than the $2mm$ previously used.

Orthorhombic crystals, if correctly oriented, have their $+a$ axis directly toward the observer, the b axis horizontal with its plus end to the observer's right, and the c axis vertical with its plus end up (Fig. 2-8).

The Tetragonal System

For crystals belonging to crystal classes 4, $\bar{4}$, $4/m$, $4mm$, $\bar{4}2m$, 422, or $4/m\ 2/m\ 2/m$, the single 4 or $\bar{4}$ axis present is always selected to be the c axis (Fig. 2-9). The two remaining mutually perpendicular crystallographic axes, a_1 and a_2, must lie in the plane perpendicular to the c axis. Within this plane they are selected to coincide with 2-fold axes or, lacking these, with normals to symmetry planes. If these symmetry elements are lacking, as in crystal classes 4, $\bar{4}$, and $4/m$ (Fig. 2-9A, B, and C), the a_1 and a_2 axes are chosen to coincide with the edges between prominent faces. In crystal classes $4mm$, 422, or $4/m\ 2/m\ 2/m$, there are *two* pairs of mutually perpendicular 2-fold axes and/or symmetry planes (Fig. 2-9D, F, and G). Consequently ambiguity arises as to which pair should be chosen as the a_1 and a_2 axes; such ambiguity is best settled by X-ray studies and use of Donnay's rules (p. 155).

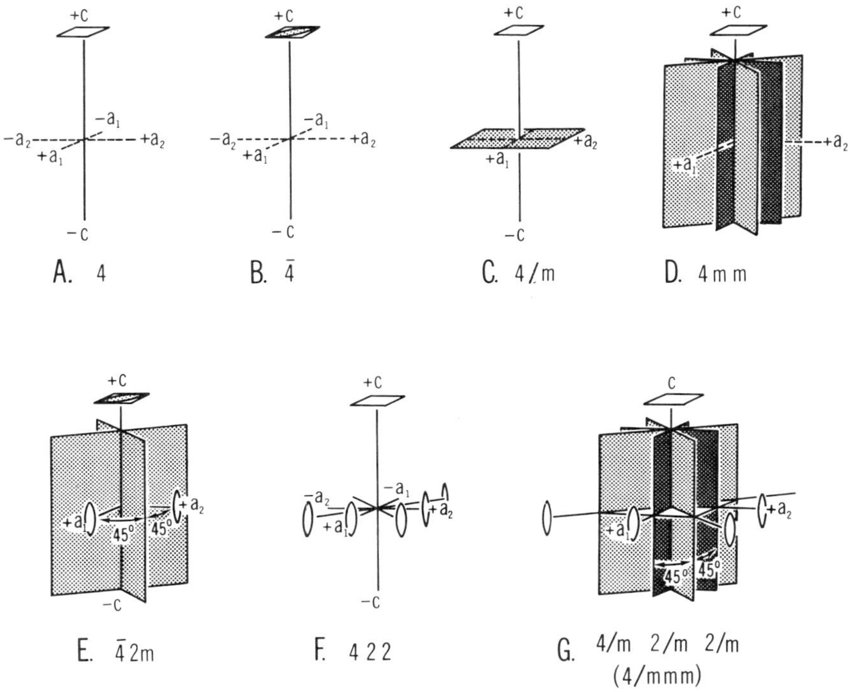

FIGURE 2–9

Relation of the crystallographic axes to the symmetry elements in the seven crystal classes grouped into the tetragonal system. The tetragonal crystallographic axes consist of an axis labeled c which is symmetrically unique, being the only 4 or $\bar{4}$ axis present in the crystal, plus two mutually perpendicular axes, a_1 and a_2, in the plane perpendicular to c axis. The symmetry operation associated with the c axis (4 or $\bar{4}$ axis) indicates that axes a_1 and a_2 are interchangeable.

Any two mutually perpendicular directions within a plane normal to a 4-fold or $\bar{4}$-fold axis are symmetrically related to each other by the operation of this axis and are hence equivalent to each other. For example, the 4-fold axis at c in Figure 2-9A requires the presence at a_2 of a direction identical to a_1. Therefore a_1 and a_2 are interchangeable, one being identical to the other. Such identical (and therefore interchangeable) crystallographic axes are by custom designated with the letter a (plus a numerical subscript chiefly added for convenience of reference). This symbolism indicates at a glance that axes a_1 and a_2 are symmetry related and, consequently, identical to each other in all respects. On the other hand, the c axis is unique; it is not symmetry-related to any other direction within the crystal. Axes such as a_1, a_2, and c constitute a set of *tetragonal* crystallographic axes, the word "tetragonal" being derived from the *tetrad* (4-fold or $\bar{4}$-fold) nature of the c axis. Crystals of crystal classes 4, $\bar{4}$, $4/m$, $4mm$, $\bar{4}2m$, 422, and $4/m\ 2/m\ 2/m$ constitute the *tetragonal system*.

By convention the first portion of the symbol for a tetragonal crystal class applies to the crystal's c axis, the second to its a_1 and a_2 axes, the third to the directions bisecting the (right) angles between a_1 and a_2. Thus the crystal-class symbol $4mm$ indicates that (1) the c axis is a 4-fold axis; (2) the a_1 and a_2 axes are normals to symmetry planes; and (3) the bisectors of the angles between a_1 and a_2 are also normals to symmetry planes. Similarly, $\bar{4}2m$ indicates that (1) the c axis equals a $\bar{4}$-fold axis; (2) the a_1 and a_2 axes coincide with 2-fold axes; and (3) the bisectors of the angles between a_1 and a_2 are normals to symmetry planes. The crystal-class symbol $4/m$ indicates that the c axis is a 4-fold axis with a symmetry plane perpendicular to it and that *neither* the a_1 and a_2 axes nor their bisectors coincide with symmetry axes or normals to symmetry planes.

The Hexagonal System

Four rather than three crystallographic axes are usually chosen for crystals possessing external symmetry corresponding to one of the following point groups: 3, $\bar{3}$, $3m$, $\bar{3}\ 2/m$, 32, 6, $\bar{6}$, $6/m$, $6mm$, $\bar{6}m2$, 622, or $6/m\ 2/m\ 2/m$. Their single 3, $\bar{3}$, 6, or $\bar{6}$ axis is always selected to be the c axis. Three additional axes (a_1, a_2, and a_3 in Fig. 2-10) are chosen from directions within the plane perpendicular to c such that the positive ends of a_1, a_2, and a_3 are 120° apart (although the axes themselves are at 60° to each other). Whenever possible, a_1, a_2, and a_3 are selected to coincide with a trio of 2-fold axes or normals to symmetry planes, that is, $\bar{2}$ axes.

The 3-, $\bar{3}$-, 6-, or $\bar{6}$-fold axis selected as the c axis always symmetrically relates axis a_1 to a_2 to a_3. Thus they are given the same letter symbol and differentiated only by subscripts. The standard orientation for these axes is to have the

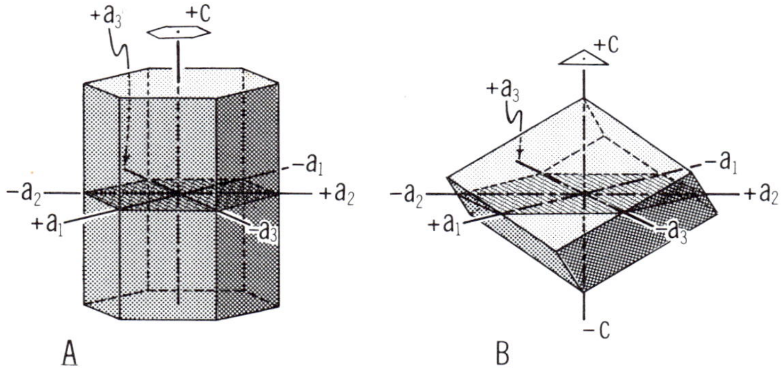

FIGURE 2–10

Location of the crystallographic axes for crystals of the hexagonal system: (A) in a crystal belonging to the hexagonal subsystem; (B) in a crystal belonging to the trigonal subsystem.

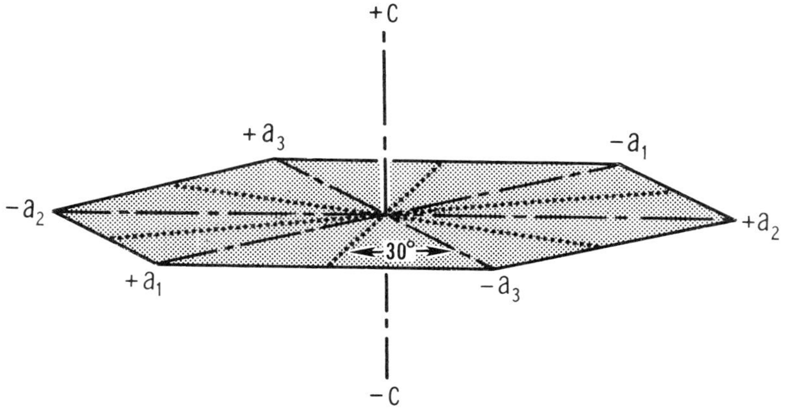

FIGURE 2–11

A set of hexagonal axes wherein the dotted lines bisect the angles between the *a* axes. The international symbol for a hexagonal crystal class customarily identifies first the symmetry associated with the *c* axis, then that associated with the *a* axes, and last that associated with the dotted bisectors of the *a* axes. (See text.)

c axis vertical (plus end up), whereupon the plane containing a_1, a_2, and a_3 (heavily shaded in Fig. 2-10) will be horizontal; axis a_2 should be normal to the observer's line of sight with its plus end to his right. This orientation aligns the plus a_1 axis to the observer's left and negative a_3 axis to his right. Note that the negative end of an axis always bisects the angle between the plus ends of the other two axes.

By convention the first portion of the symbol for a hexagonal crystal class applies to the crystal's *c* axis, the second to its a_1, a_2, and a_3 axes, and the third to the bisectors (dotted in Fig. 2-11) of the angles between these *a* axes. Thus $6/m\ 2/m\ 2/m$ indicates that (1) the *c* axis is a 6-fold axis with a symmetry plane normal thereto; (2) the *a* axes, and (3) the bisectors of the angles between the *a* axes are 2-fold axes, each with a symmetry plane normal to it. Similarly, symbol $\bar{3}\ 2/m$ indicates that (1) the *c* axis is a $\bar{3}$-fold axis; (2) the *a* axes are 2-fold axes with planes normal thereto; but (3) the bisectors of the angles between the *a* axes, since a third part is missing, are axes of nonsymmetry, that is, 1-fold axes.

American and British crystallographers differ in their definition of the *hexagonal system*. The American crystallographers consider this system to include all crystals for which axes like *c*, a_1, a_2, and a_3 in Figure 2-11 may be assigned. They then subdivide the hexagonal system into a *trigonal subsystem* and a *hexagonal subsystem* depending on whether the crystal's *c* axis is a *triad* (3- or $\bar{3}$-fold) or a *hexad* (6- or $\bar{6}$-fold) axis of symmetry. The trigonal subsystem therefore consists of crystal classes 3, $\bar{3}$, $3m$, 32, $\bar{3}\ 2/m$; the hexagonal subsystem consists of classes 6, $\bar{6}$, $6/m$, $6mm$, $\bar{6}m2$, 622, or $6/m\ 2/m\ 2/m$. British crystallographers, however, elevate both subsystems to system rank.

The Isometric or Cubic System

All crystals that possess *four* 3- or $\bar{3}$-fold axes will possess additional symmetry elements such that their symmetry will be identical to one of the five point groups, 23, $\bar{4}3m$, 432, $2/m\ \bar{3}$, and $4/m\ \bar{3}\ 2/m$. These crystals always possess three interchangeable mutually perpendicular symmetry axes (labeled a_1, a_2, and a_3 in Fig. 2-12A) that are either 4-, $\bar{4}$-, or 2-fold axes and are at an angle of 54°44′ to each of the four 3- or $\bar{3}$-fold axes present. These three axes are always chosen as the crystallographic axes. Their interchangeability may be shown by considering the 3-fold axis labeled d in Figure 2-12A. Its symmetry operation (120° rotation) requires that given a direction at a_1, identical directions must exist at a_2 and a_3.

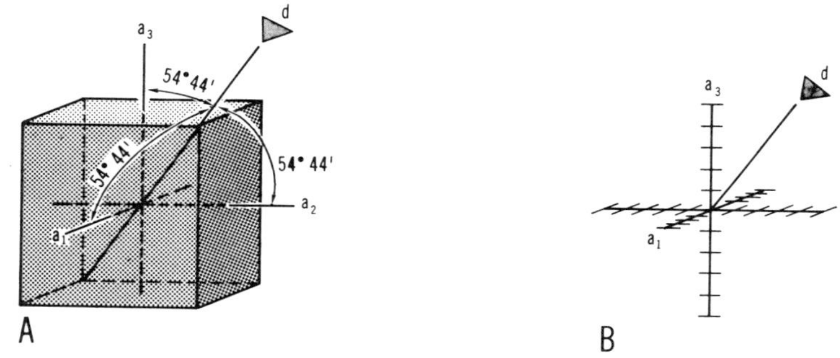

FIGURE 2–12

(A) The three, mutually perpendicular, crystallographic axes, a_1, a_2, and a_3, for isometric crystals. Depending on which isometric crystal class is considered, these will be either all 2-fold axes, all $\bar{4}$-fold axes, or all 4-fold axes. Each of the four body-diagonals of the guide cube, of which line d is one, will be a 3- or $\bar{3}$-fold axis, again depending on the crystal class in question. As shown for axis d, each of these body diagonals is at 54° 44′ to each of the three crystallographic axes. Moreover the 3-fold symmetry operation inherent in d symmetrically relates a_1 to a_2 to a_3. Consequently, as shown in (B), if units are marked off on axis a_1, then equal units must also be marked off along a_2 and a_3 or else d would no longer be a 3-fold axis. This symmetric interchangeability of the three mutually perpendicular crystallographic axes a_1, a_2, and a_3—and therefore the necessity for equal units to occur along them—distinguishes all the crystal classes (23, $\bar{4}3m$, 432, $2/m\ \bar{3}$, and $4/m\ \bar{3}\ 2/m$) that are grouped together into the isometric system.

Just as units are marked off along the axes of a system of Cartesian coordinates, it is also convenient to mark off units along crystallographic axes.[1] Since the crystallographic axes a_1, a_2, and a_3 are interchangeably related by a 3-fold axis, it is logical that equal units be marked off on each (Fig. 2-12B). Consequently these axes are called *isometric*, the word isometric meaning "equal measure."

All crystals that possess four 3- or $\bar{3}$-fold axes, in short those that belong to crystal classes 23, $2/m\ \bar{3}$, 432, $\bar{4}3m$, and $4/m\ \bar{3}\ 2/m$, necessarily require isometric axes and are thus said to crystallize in the *isometric* or *cubic system*. The conventions governing the symbols for these five isometric crystal classes are the same as those already discussed for the comparable point groups. For example, crystal-class symbol $2/m\ \bar{3}$ indicates that (1) the a axes (normals to guide cube's face) are $2/m$ combinations; (2) four $\bar{3}$ axes join opposite corners of the guide cube; and (3) no symmetry elements join the opposite edges of the guide cube. Symbol 432 indicates that (1) the a axes are 4-fold axes; (2) four 3-fold axes join opposite corners of the guide cube; and (3) six 2-fold axes join the midpoints of opposite edges of the guide cube. Thus the crystal-class symbol or point-group symbol identifies the symmetry elements and also their geometrical relations to each other.

Abbreviated Symbols

Crystallographers often use abbreviated symbols for crystal classes or point groups instead of the more descriptive (but lengthier) symbols. For example, in all except the monoclinic system, they may simply write m rather than $2/m$. Consequently, the orthorhombic symbol $2/m\ 2/m\ 2/m$ becomes mmm. Similarly, the tetragonal symbol $4/m\ 2/m\ 2/m$ becomes $4/m\ m\ m$ and the hexagonal symbol $6/m\ 2/m\ 2/m$ becomes $6/m\ m\ m$. In the abbreviated isometric symbols the $2/m$ and $4/m$ combinations are both indicated by m. In addition the $\bar{3}$ axes are indicated as 3 (the bar over the 3 being eliminated). Thus $4/m\ \bar{3}\ 2/m$ becomes simply $m3m$. Similarly $2/m\ \bar{3}$ becomes $m3$.

The reasoning behind these abbreviations is relatively simple. Thus mmm to the crystallographer indicates three mutually perpendicular planes whose order of listing indicates them to be perpendicular to the a, b, and c orthorhombic axes, respectively. The line of intersection between each two of these planes is of necessity a 2-fold axis (Fig. 2-13); moreover, this 2-fold axis is perpendicular to the third plane. Thus mmm is a briefer form of $2/m\ 2/m\ 2/m$. The abbreviated isometric symbols are less obvious. However, considering symbol $m3m$, if there is placed through a guide cube's center (1) a single symmetry plane parallel to a cube face, (2) a 3-fold axis from corner to corner, and (3) a diagonal symmetry plane from edge to edge, their interaction would produce all the symmetry elements more obviously stated in the symbol $4/m\ \bar{3}\ 2/m$. If this diagonal symmetry plane were absent, the abbreviated symbol would be $m3$, and the inter-

[1] The true significance of these units will become apparent only after reading Chapter 7.

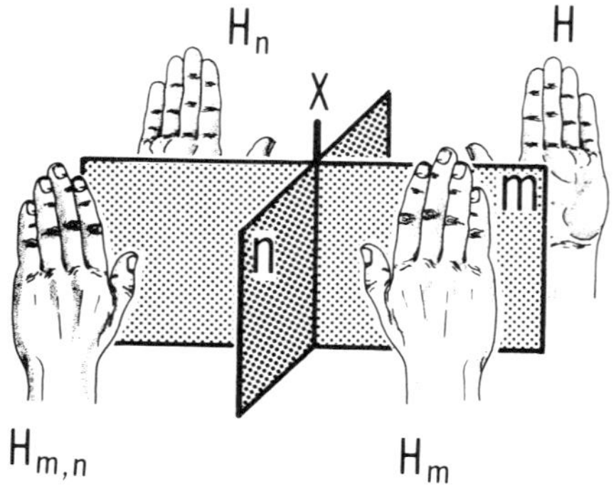

FIGURE 2–13

Demonstration that X, the line of intersection between two mutually perpendicular symmetry planes such as m and n, is necessarily a 2-fold axis. For example, plane m operates on the hand at H to produce another at H_m. Similarly, plane n repeats H at H_n and H_m at $H_{m,n}$. The distribution of the four hands reveals X to be a 2-fold axis since a 180° rotation about X relates H to $H_{m,n}$ and H_m to H_n. It could be similarly shown that if a 2-fold axis lies within a plane of symmetry, then a second mirror plane must exist perpendicular to the first and intersect this first along the line of the 2-fold axis.

action between symmetry (1) and (2) above would yield the symmetry combination more fully expressed by $2/m\ \bar{3}$.

Symbolism for Axial Units

The use of axial units in developing a nomenclature for crystal faces will be discussed in Chapter 3, whereas the true physical significance of these axial units will be discussed in Chapter 7. For the time being, the units marked off along a crystallographic axis will be symbolized by the same letter as the axis except that a small zero subscript will be added. For example, the unit distances along the a, b, and c axes of a crystal would be, respectively, a_0, b_0, and c_0. The units along symmetrically interchangeable axes—that is, along a_1, a_2, and a_3 in either the isometric system (Fig. 2-14A) or hexagonal system (Fig. 2-14B), and along a_1 and a_2 in the tetragonal system (Fig. 2-14C)—are necessarily equal (for reasons of symmetry) and hence are all symbolized as a_0. In the hexagonal or tetragonal system the units along the c axis, that is, c_0, may be larger or smaller than a_0 and are equal to it only by great coincidence. The hexagonal and tetragonal systems involve axial units of two different values, a_0 and c_0, and, consequently, are sometimes grouped together as the *dimetric* systems. As Figures 2-14D to F show, the a, b, and c axes in the orthorhombic, monoclinic, and triclinic systems are not symmetrically related; therefore unequal units, a_0, b_0, and c_0, respectively, occur along each. These three crystal systems are sometimes grouped together as the *trimetric* systems.

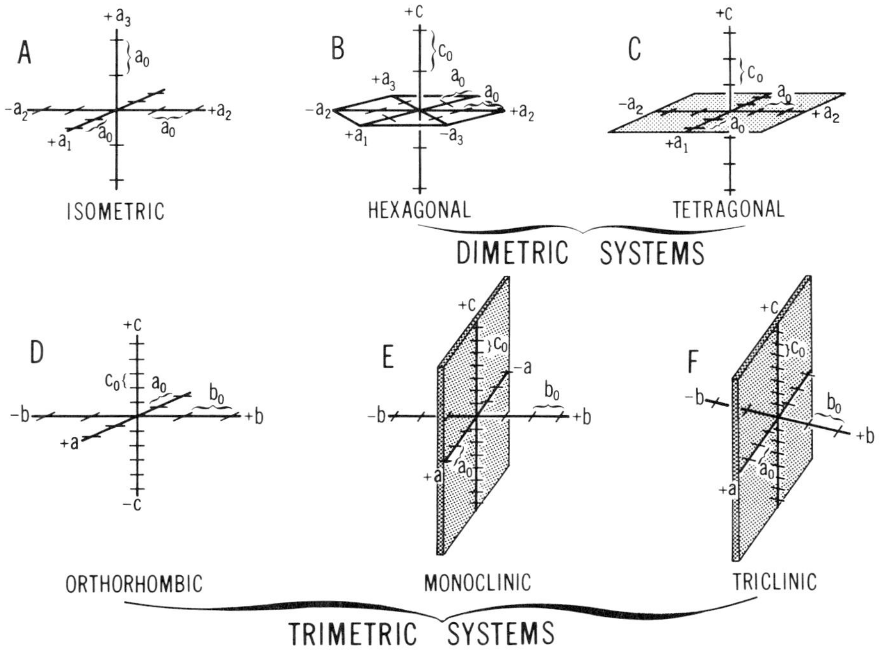

FIGURE 2–14

For those systems wherein the units on two (or more) crystallographic axes must be equal—because the symmetry present in the classes comprising this system demands this—the axes along which these units pertain are all called a axes (but differentiated by subscripts). Note classification of systems into the isometric (equal units a_0 on all axes), dimetric types (two sizes of units: a_0 and c_0), and trimetric types (three sizes: a_0, b_0, and c_0).

Relation of Crystal Classes to Crystal Systems

The grouping of 27 of the 32 crystal classes into the various (nonisometric) crystal systems is summarized in Figure 2-15. The crystal class that possesses more symmetry elements than the other classes in this same system is lightly shaded in Figure 2-15 and is called the *holohedral* class (from the Greek *holos* which means "whole" or "entire"). The less symmetrical (unshaded) classes within each system are called *merohedral* classes (from the Greek *meros* which means "part").

The remaining 5 of the 32 crystal classes (Fig. 2-16) are grouped into the isometric system. Since a $\bar{3}$ axis is identical to a 3-fold axis plus a center of symmetry $\bar{1}$, the lower $X\bar{3}$ tier in Figure 2-16 differs from the upper $X3$ tier only by the addition of a center of symmetry. This, however, has far-reaching consequences in that the addition of a center of symmetry to a class containing

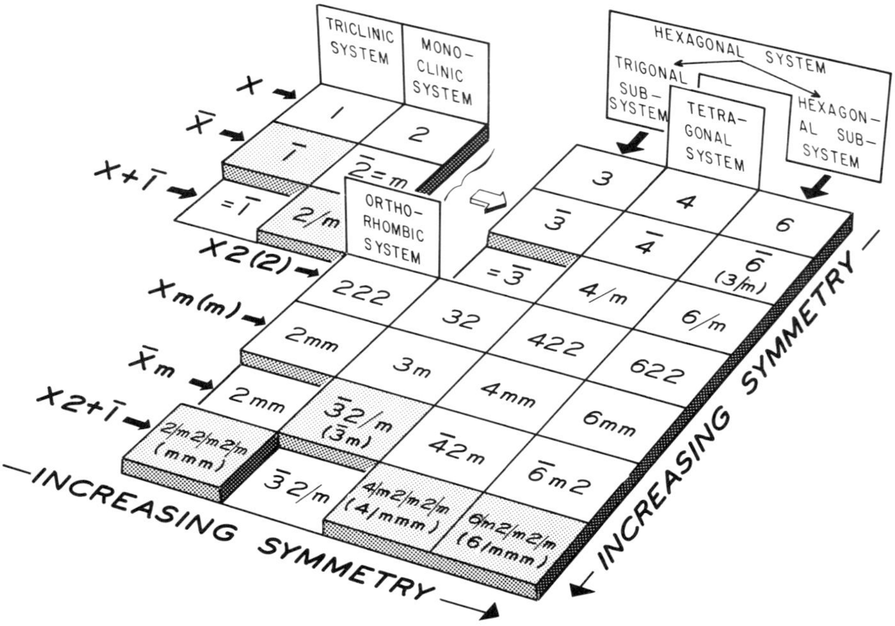

FIGURE 2–15

Distribution of the 27 nonisometric crystal classes among the crystal systems. Note that the column for which X equals 2 is split between the monoclinic and orthorhombic systems, the monoclinic portion being moved to the left. Symbols in parentheses are alternatives to the generally accepted (nonparenthesized) symbols located immediately above them. Within a given crystal system, the class containing the greatest number of symmetry elements, the so-called holosymmetric class, is lightly shaded. If a given symmetry combination is identical to a second, more frequently stated symmetry combination, it is written on a lower surface.

even-fold axes (2, $\bar{4}$, 4, or 6) will cause planes of symmetry to be developed normal to them, whereas its addition to a class containing planes of symmetry will cause 2-fold axes to develop perpendicular to these planes. Thus with the addition of a center of symmetry, class 23 becomes $2/m\,\bar{3}$, class 432 becomes $4/m\,\bar{3}\,2/m$, and class $\bar{4}3m$ would also become $4/m\,\bar{3}\,2/m$.

Schoenflies' Notation (Optional)

An older notation devised for crystal classes by Schoenflies still finds use, principally by spectroscopists and chemists. Schoenflies used C_n, where C stands for cyclic, to indicate a class possessing only a single axis of n-fold symmetry.

Thus C_1, C_2, C_3, C_4, and C_6, respectively, indicate crystal classes 1, 2, 3, 4, and 6. Crystal classes $\bar{1}$ and m were indicated by C_i and C_s, the symbol i indicating a center of symmetry and s standing for *Spiegelung*, which means "reflection" in German. A single cyclic axis with a symmetry plane normal to it, that is, X/m types, was symbolized as C_{nh}. Thus C_{2h} represents $2/m$; C_{3h} is $3/m$ or $\bar{6}$; C_{4n} is $4/m$; and C_{6h} is $6/m$. The letter h signifies a *horizontal* symmetry plane because crystals from X/m classes were conventionally oriented with X vertical and, therefore, m horizontal.

Schoenflies indicated class $\bar{3}$ as C_{3i} because a 3-fold axis plus a center of symmetry equate to $\bar{3}$. Generally, instead of the inversion axes $\bar{1}$, $\bar{2}$, $\bar{3}$, $\bar{4}$, and $\bar{6}$, Schoenflies used the concept of axes of rotatory reflection. The operation of the five possible rotatory reflection axes [modern symbols $\tilde{1}$, $\tilde{2}$, $\tilde{3}$, $\tilde{4}$, and $\tilde{6}$ (pronounced "tilde one, . . ., tilde six") and comparable Schoenflies symbols S_1, S_2, S_3, S_4, and S_6] is illustrated in Figure 2-17. We can generalize by saying that a rotation of $360°/X$ around a $\tilde{X}$ axis plus reflection across a plane normal to this axis brings a point p_1 into coincidence with an equivalent point p_2. Study of Figure 2-17 reveals that S_1 is equivalent to C_s, S_2 to C_i, S_3 to C_{3h}. S_4 is a new Schoenflies class equivalent to $\bar{4}$, and S_6 equals C_{3i}.

An X-fold axis with 2-fold axes at 90° to it—$X2(2)$ combinations in modern terminology—is indicated as D_n, where n indicates both the "fold" of the X axis and the number of 2-fold axes normal to it. Thus D_2 is equivalent to 222, D_3 to 32, D_4 to 422, and D_6 to 622. On adding a horizontal plane of symmetry (which will be perpendicular to X) to the preceding D_n point groups, they become D_{nh} types. Of these D_{2h} is equivalent to $2/m\ 2/m\ 2/m$, D_{3h} to $\bar{6}m2$, D_{4h} to $4/m\ 2/m\ 2/m$, and D_{6h} to $6/m\ 2/m\ 2/m$. If we add vertical symmetry planes that bisect the angles between the 2-fold axes in point groups D_2 and D_3, there result point groups D_{2d} ($\bar{4}2m$) and D_{3d} ($\bar{3}\ 2/m$), d indicating these *diagonal* planes.

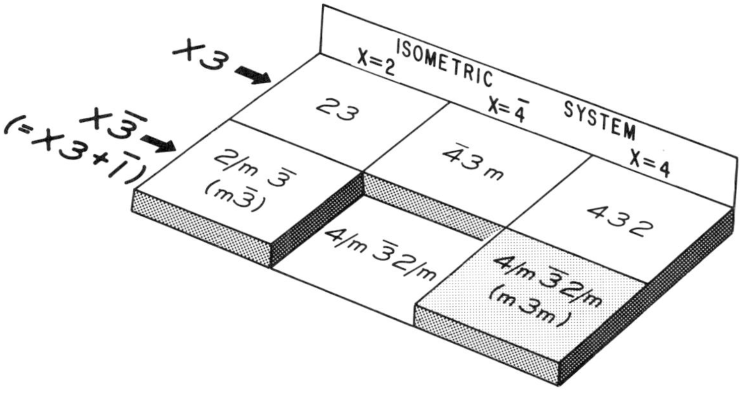

FIGURE 2–16

The five isometric crystal classes.

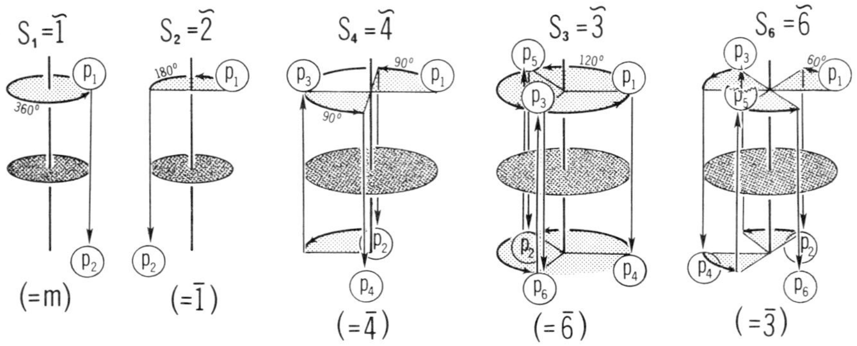

FIGURE 2–17

Operation of each of the five axes of rotatory reflection whose Schoenflies' symbols are S_1, S_2, S_3, S_4, and S_6 but whose modern symbols are $\tilde{1}$, $\tilde{2}$, $\tilde{3}$, $\tilde{4}$, and $\tilde{6}$. The operation for a $\tilde{X}$ axis consists of a rotation of $360°/X$ plus reflection across a plane (stippled) which is normal to the axis.
Taking the $\tilde{3}$ axis as an example, sphere p_1 is related to p_2 by a 120° rotation plus reflection across the stippled plane (as indicated by the downward arrow). Similarly, p_2 is related to p_3 by another 120° rotation plus a reflection as indicated by the upward arrow. Successive application of this operation then relates p_3 to p_4, p_4 to p_5, p_5 to p_6, and p_6 back to the starting point, p_1. The Hermann-Mauguin equivalencies are indicated in parentheses below each $\tilde{X}$ operation. Thus a $\tilde{1}$ axis is seen to be equivalent to a mirror plane m normal to this $\tilde{1}$ axis.

The five point groups comparable to the five isometric crystal classes are symbolized as O if they possess three 4-fold axes, but as T if they do not. Thus T is equivalent to 23, T_h to $2/m\,\bar{3}$, T_d to $\bar{4}3m$, O to 432, and O_h to $4/m\,\bar{3}\,2/m$.

In summary, the equivalencies between the Schoenflies symbols and the Hermann-Mauguin symbols are as follows:

		Hexagonal		
Triclinic	Tetragonal	Trigonal	Hexagonal	Isometric
$C_1 = 1$	$C_4 = 4$	$C_3 = 3$	$C_6 = 6$	$T = 23$
$C_i = \bar{1}$	$S_4 = \bar{4}$	$C_{3i} = \bar{3}$	$C_{3h} = \bar{6}$	$T_h = 2/m\,\bar{3}$
	$C_{4h} = 4/m$	$C_{3v} = 3m$	$C_{6h} = 6/m$	$T_d = \bar{4}3m$
Monoclinic	$C_{4v} = 4mm$	$D_3 = 32$	$C_{6v} = 6mm$	$O = 432$
$C_2 = 2$	$D_{2d} = \bar{4}2m$	$D_{3d} = \bar{3}\,2/m$	$D_6 = 622$	$O_h = 4/m\,\bar{3}\,2/m$
$C_s = m$	$D_4 = 422$		$D_{3h} = \bar{6}m2$	
$C_{2h} = 2/m$	$D_{4h} = 4/m\,2/m\,2/m$		$D_{6h} = 6/m\,2/m\,2/m$	
Orthorhombic				
$C_{2v} = mm2$				
$D_2 = 222$				
$D_{2h} = 2/m\,2/m\,2/m$				

QUESTIONS AND PROBLEMS

2-1 (a) To what crystal system would a crystal belong if its symmetry were the same as for Figure 1-17A? (b) to (t) Same question for Figures 1-17B to 1-17T.

2-2 (a) State the crystal class that results if a center of symmetry is added to the symmetry of a crystal belonging to (a) class $4mm$; (b) class $6mm$; (c) class $\bar{6}$; (d) class $\bar{4}2m$; (e) class $\bar{6}m2$; (f) class 622; (g) class $\bar{3}\ 2/m$.

2-3 For which point group symbols are (a) $\bar{3}m$, (b) $6/mmm$, and (c) $4/mmm$ the shortened forms?

Tiger! Tiger! burning bright
In the forests of the night,
What immortal hand or eye
Could frame thy fearful symmetry?

William Blake

ANSWERS

2-1 (a), (b), (c) Orthorhombic; (d) isometric; (e) orthorhombic; (f) tetragonal; (g) isometric; (h) triclinic; (i) monoclinic; (j) tetragonal; (k) orthorhombic; (l) isometric; (m) monoclinic; (n) and (o) tetragonal; (p) isometric; (q) monoclinic; (r) and (s) tetragonal; (t) isometric.

2-2 (a) $4/m\ 2/m\ 2/m$; (b) $6/m\ 2/m\ 2/m$; (c) $6/m$; (d) $4/m\ 2/m\ 2/m$; (e) $6/m\ 2/m\ 2/m$; (f) $6/m\ 2/m\ 2/m$; (g) $\bar{3}\ 2/m$.

2-3 The reader is required to deduce the presence of a center of symmetry, and therefore of 2-fold axes perpendicular to all planes of symmetry, from the presence of $\bar{3}$, $6/m$, or $4/m$ in the shortened symbols. Thus the answers are (a) $\bar{3}\ 2/m$; (b) $6/m\ 2/m\ 2/m$; and (c) $4/m\ 2/m\ 2/m$.

3 Crystal Nomenclature and Calculations

Nomenclature for Crystal Faces

The unit face

Early in the study of crystallography the need for referring unambiguously to the different faces on a crystal was recognized. To achieve this, crystallographers assigned one of the six sets of references axes (Fig. 2-14) to the crystal, roughly following the conventions previously discussed. They assumed these axes to intersect at a common point, the origin, arbitrarily chosen somewhere within the interior of the crystal. Of those faces on the crystal that cut all three crystallographic axes, they selected the one of greatest area and called it the crystal's *unit face* or *parametral plane*. This unit face, extended if necessary, was assumed to cut each *positive crystallographic axis* at an *equal but unknown number* (n) *of units*.

To illustrate such choice of a unit face, assume that a goniometric study of a sulfur crystal in Figure 3-1A places it in crystal class $2/m\ 2/m\ 2/m$. Its three mutually perpendicular 2-fold axes then serve as its a, b, and c axes. Although neither face s nor p is parallel to these axes, p would be chosen as the unit face because of its greater area. Let us define a face's *intercepts* on a set of crystallographic axes as the distances from the origin at which the face cuts these axes. Thus selection of p as the unit face is the same as assuming that p's intercepts on axes a, b, and c—distances OA_p, OB_p, and OC_p, respectively, in Figure 3-1B—equal na_0, nb_0, and nc_0.

Weiss parameters

Weiss parameters, introduced in 1818 by C. S. Weiss, give an approximate indication of a face's angular attitude to the crystallographic axes and also serve as a symbol for this face. Weiss parameters state the *relative number* of units at

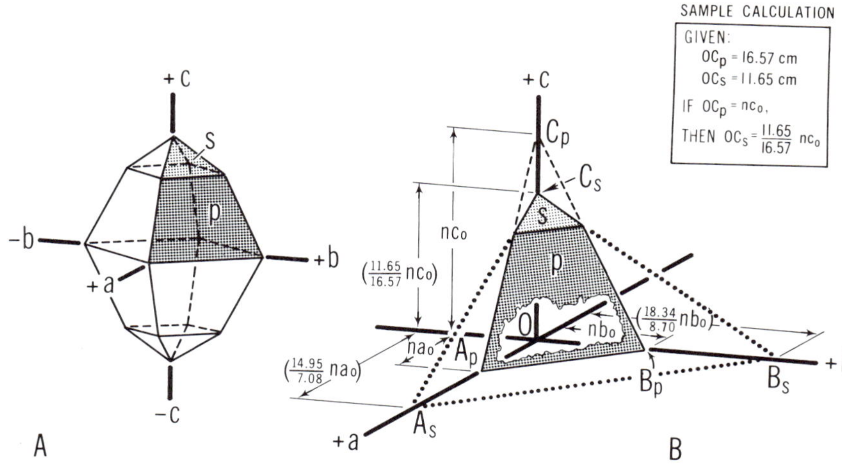

FIGURE 3–1

(A) Faces *s* and *p*, extended as necessary, will each intersect all three crystallographic axes, *a*, *b*, and *c*, of the crystal of orthorhombic sulfur shown here. However, by virtue of its greater area, *p* rather than *s* is selected to be the *unit face*.
(B) Faces *s* and *p*, drawn enlarged and isolated, show *s* to possess the intercepts OA_s, OB_s, and OC_s, whereas for *p* they are OA_p, OB_p, and OC_p. Note that distance OC_s is 11.65/16.57 as long as OC_p and hence would equal only 11.65/16.57 as many c_0 units as OC_p.

which the face cuts each axis; negative signs indicate that the negative end of an axis is intersected. We can illustrate this with the Weiss parameters $2a{:}2b{:}1c$ for a face on an orthorhombic crystal; these indicate that this face intercepts the *a* and *b* axes at twice as many a_0 and b_0 units as it cuts the *c* axis in c_0 units. For a unit face on this same crystal, the Weiss parameters would be $1a{:}1b{:}1c$; these indicate that the number of units at which this face intercepts each of the axes is in the proportion 1:1:1, that is, it cuts the *a*, *b*, and *c* axes at na_0, nb_0, and nc_0 units, respectively.

Once the crystallographic axes are located and a unit face is chosen, the Weiss parameters for other faces may be determined by dividing their intercepts (on each of the axes) by the intercepts of the unit face. For example, let us assume in Figure 3-1B that the intercepts on the *a*, *b*, and *c* axes, respectively, are for face *p*, 7.08:8.70:16.57 cm, and for face *s*, 14.95:18.34:11.65 cm. To obtain the Weiss parameters for *s*, we divide its intercepts by *p*'s, thus

$$\frac{14.95}{7.08}a : \frac{18.34}{8.70}b : \frac{11.65}{16.57}c \qquad \text{or} \qquad 2.111a{:}2.108b{:}0.703c$$

The sample calculation in Figure 3-1B summarizes the reasoning behind these computations by using the intercepts of faces *s* and *p* on the *c* axis as examples. To recapitulate, selection of *p* as the unit face was to assume that *p*'s intercepts,

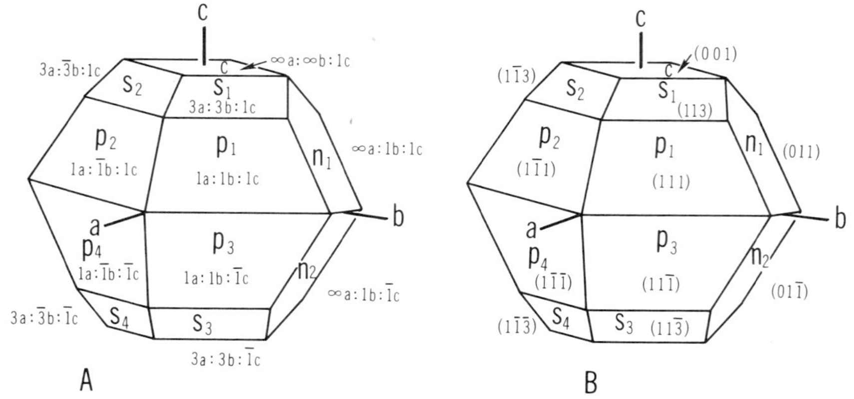

FIGURE 3–2

A crystal of orthorhombic sulfur whose faces are labeled with their Weiss parameters (A) and with their Miller indices (B).

given as 7.08, 8.70, and 16.57 cm, equaled na_0, nb_0, and nc_0, respectively. Consequently because face s's intercepts on axes a, b, and c are 14.95/7.08 (or 2.111), 18.34/8.70 (or 2.108), and 11.65/16.57 (or 0.703) as long as p's, they must equal 2.111 na_0 units, 2.108 nb_0 units, and 0.703 nc_0 units. Face s's Weiss parameters are thus the proportion, $2.111\ na:2.108\ nb:0.703\ nc$. Like any proportion, it remains unchanged if divided by a common factor. Thus attributing to experimental error the slight deviation of 2.111 and 2.108 from 2.109, then 0.703 n can be factored out to simplify the preceding proportion to $3a:3b:1c$. This more acceptable form for the Weiss parameters of face s more readily indicates that face s intercepts the a and b axes at three times as many units as it does the c axis.

TABLE 3–1

Intercepts for Faces in Figure 3-2A

	Intercept on Axis		
Face	*a*	*b*	*c*
p_1	4.60	5.66	10.77
p_2	4.60	−5.66	10.77
p_3	4.60	5.66	−10.77
p_4	4.60	−5.66	−10.77
s_1	8.28	10.18	6.46
s_2	8.28	−10.18	6.46
s_3	8.28	10.18	−6.46
s_4	8.28	−10.18	−6.46
n_1	∞	4.53	8.62
n_2	∞	4.53	−8.62
c	∞	∞	2.98

The general procedure for obtaining an unknown face's Weiss symbol is now apparent. (1) Divide its intercept on a given axis by that of the unit face on this same axis and list the three ratios obtained in the order to which they refer to the a, b, or c axes (or a_1, a_2, and c axes for tetragonal and a_1, a_2, and a_3 axes for isometric). (2) Divide through by the ratio of smallest value; in the preceding orthorhombic example this happened to be 0.703 because face s cut the c axis at fewer units than it did the a or b axis. (3) This division of all three ratios by the smallest one of them always results in relatively simple numbers or fractions—a point to be discussed under the *law of rational indices.*

Figure 3-2A shows a sulfur crystal that is labeled with the Weiss parameters for each face. These could be calculated from the intercepts for these faces (Table 3-1). If a face is parallel to an axis, it is assumed to intercept this axis at an infinite distance (symbol: ∞) from the origin.

Miller indices

In the early nineteenth century the noted Cambridge crystallographer, W. H. Miller, popularized a method of face denotation that has so many advantages over the foregoing Weiss symbols it has won general acceptance. His symbols, called Miller indices, are simply the reciprocals of the Weiss parameters, cleared of fractions, with the letters denoting the axes omitted. To illustrate, face s in Figure 3-1 was shown to have the Weiss symbol $3a{:}3b{:}1c$; thus the reciprocals of these parameters are 1/3, 1/3, 1. Multiplying this set of three numbers by 3, which is the smallest number necessary to clear them of fractions, they become 1, 1, 3. Miller then placed parentheses around this set of three numbers and eliminated the commas to obtain (113), the Miller index for face s. Commas are used in Miller indices only in the rare case when a two-digit number occurs, thus, (12, 2, 1). The Miller index (113) would be read as "one-one-three" but never as "one hundred and thirteen." *Miller indices contain neither fractions nor a common factor.* For example, the Miller index (226) would be divided through by its common factor 2 to become (113). The numbers in a Miller index, like those in Weiss parameters, are listed in the order to which they refer to (1) the a, b, and c axes for triclinic, monoclinic, and orthorhombic; (2) the a_1, a_2, and c axes for tetragonal crystals; (3) the a_1, a_2, and a_3 axes for isometric crystals; or (4) the a_1, a_2, a_3, and c axes for hexagonal crystals. Comparison of Figures 3-2A and 3-2B reveals the reciprocal relationship between Weiss parameters and Miller indices; we note that the reciprocal of infinity is considered to equal zero (and that the reciprocal of zero is considered to be infinity).

From a face's Miller indices we can visualize its approximate angular attitude to the crystallographic axes by considering the reciprocals of these indices, in effect converting them back to Weiss parameters. For example, the face (210), which has reciprocals: $\frac{1}{2}$, 1, ∞, cuts axis a (or a_1) at one-half as many units as it cuts axis b (or a_2) but is parallel to axis c (or a_3). Similarly, the face (001), which has reciprocals: ∞, ∞, 1, is parallel to the a (or a_1) and b (or a_2) axes but cuts the c (or a_3) axis. The Miller indices for a unit face (111), which has reciprocals: 1, 1, 1, indicate that the face cuts all three axes at equal numbers of units.

Determination of Miller indices from intercepts

To determine Miller indices from intercepts, we (1) divide the unit face's intercept on each axis by that of the unknown face on this same axis; we then (2) divide the three resulting intercept ratios by the smallest of them. (3) The resultant numbers may then all be multiplied (or divided) by an appropriate factor, if necessary, to clear them of fractions or to remove any common factor. For example, using the intercepts of faces *s* and *p* as given in Table 3-1, application of step (1) yields the following ratios:

$$\frac{4.60}{8.28} : \frac{5.66}{10.18} : \frac{10.77}{6.46} \qquad \text{or} \qquad 0.556:0.556:1.667$$

Step (2) requires their division by 0.556 to get 1:1:3, which indicates the Miller index for *s* to be (113).[1]

Law of Rational Indices (Haüy's Law). Step (2) in the preceding paragraph, after due allowance for experimental errors, always yields simple rational numbers, that is, simple fractions, or simple integers usually less than six in value. Irrational numbers such as $\sqrt{2}$ never occur. Crystallographers have observed this rational nature of Miller indices (or Weiss parameters) ever since Haüy's work first disclosed it, and the phenomenon is referred to as the *law of rational indices* or *Haüy's law.* For many years this law was simply a statement of observed facts, but its validity becomes apparent once the internal structure of crystals, particularly crystal lattices (Chapter 6), is understood.

The generalized Miller index (hkl) is used to specify a face that may bear any orientation except parallelism to the three crystallographic axes. The indices $(0kl)$, $(h0l)$, and $(hk0)$ signify faces that are parallel to the a, b, or c axes, respectively, but may bear any orientation except parallelism to the remaining two axes; in the special cases where these faces cut the remaining two axes at equal numbers of axial units, they are not usually written as $(0kk)$, $(h0h)$, or $(hh0)$ to indicate this. Rather, because Miller indices should be divided through by any common factor, in this case k or h, they are written as (011), (101), or (110). The general Miller index (hhl) indicates a face that cuts the first two axes at equal numbers of axial units, whereas it cuts the third axis at a different number of units. A face that cuts all three axes at the same number of units

[1] For convenience in discussion, Miller indices (and Weiss parameters) have been here explained on the basis of the relative intercepts of the faces on the crystallographic axes. In actual practice these indices, as well as the crystal's axial ratio and interaxial angles, were determined from the crystal's interfacial angles as measured with a reflecting goniometer. The techniques involved have been largely supplanted by the development of X-ray crystallography, but excellent presentations of these techniques are still available in Tutton (1922) and in Phillips (1963).

would have the index (111) and not (hhh) because (hhh) may be divided through by the common factor h. By the same token, faces parallel to two of three crystallographic axes would have the indices (001), (010), or (100) rather than $(00l)$, $(0k0)$, or $(h00)$.

Miller-Bravais indices

In the hexagonal system, face indices require special consideration because crystals belonging to this system are assigned *four* crystallographic axes. This set of four axes (Fig. 2-11) is often called a *Bravais axial set* because Auguste Bravais, the great French crystallographer, introduced them in 1851. In assigning indices to the faces on these crystals, Bravais followed the Miller method; hence they are called *Miller-Bravais indices.* Where fully expressed, a Miller-Bravais index consists of four numbers cited in the order in which they refer to the a_1, a_2, a_3, and c axes. Thus the Miller-Bravais index $(11\bar{2}1)$, which is comparable to Weiss symbol, $2a_1{:}2a_2{:}-1a_3{:}2c$, will refer to a face that cuts the positive a_1 and a_2 axes at twice as many (a_0) units as it cuts the negative a_3 axis and cuts the c axis at the same number of (c_0) units as it cuts the a_1 and a_2 axes in a_0 units. Similarly, the index $(1\bar{1}01)$ refers to a face that cuts the positive a_1, the negative a_2, and the positive c axis in equal numbers of units, whereas it is parallel to the a_3 axis.

Miller-Bravais indices have one outstanding feature; their first three numbers, which refer to the a_1, a_2, and a_3 axes, invariably total zero. In other words, when $(hkil)$ is the general symbol of a Miller-Bravais index, then

$$h + k + i = 0 \qquad \text{(Eq. 3-1)}$$

Thus Equation (3-1) holds for $(11\bar{2}1)$ and $(1\bar{1}01)$.

The general validity of Equation (3-1) can be demonstrated by assuming that a random face P (shaded in Fig. 3-3) cuts the hexagonal axes a_1, a_2, and $-a_3$ at ma_0, na_0, and xa_0 units, respectively. Its intersection on the c axis need not enter into this discussion. Face P intersects these a axes at points M, X, and T so as to define three triangles related in area as

$$\triangle MOT = \triangle MOX + \triangle XOT$$

Because the area of a triangle equals one-half the product of two of its sides times the sine of the angle between these sides, the foregoing equation can be rewritten as

$$\frac{ma_0 \cdot na_0 \cdot \sin 120°}{2} = \frac{ma_0 \cdot xa_0 \cdot \sin 60°}{2} + \frac{na_0 \cdot xa_0 \cdot \sin 60°}{2}$$

Because sin 120° equals sin 60°, the above equation, if divided through by the common factor $(\frac{1}{2}a_0^2 \sin 60°)$, simplifies to

$$x = \frac{mn}{m + n}$$

As Figure 3-3 illustrates, face P's Weiss parameters in respect to the a_1, a_2, and

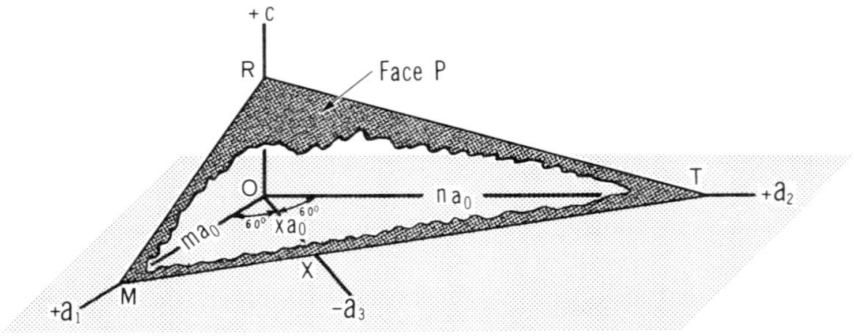

FIGURE 3–3

The plane P intersects the a_1-axis at ma_0 units from the origin, the a_2-axis at na_0 units from the origin, and the $-a_3$ axis at xa_0 units from the origin. The area of triangle MOT equals that of triangle MOX plus triangle XOT as discussed in the text.

$-a_3$ axes are ma_1:na_2:$-xa_3$. Thus the Miller indices in respect to these axes would be

$$h = \frac{1}{m} \qquad k = \frac{1}{n} \qquad i = -\frac{1}{x} = -\left(\frac{m+n}{mn}\right)$$

Clearing of fractions by multiplying throughout by mn, they become

$$h = n \qquad k = m \qquad i = -(m+n) = -(h+k) \qquad \text{QED}$$

Some workers replace the third digit in a Miller-Bravais index by an asterisk or a dot. Thus $(11\bar{2}1)$ becomes $(11{*}1)$. The identity of the number replaced by the asterisk is easily determined because, as just proved, it is equal but opposite in sign to the algebraic sum of the first two digits. Others even eliminate the asterisk so that $(11\bar{2}1)$ becomes (111), but this frequently leads to confusion.

Axial Elements

The axial elements of a crystal consist of that data which, for the purposes of morphologic crystallography, adequately describe its crystallographic axes. The axial elements thus consist of two types of data: (1) data that stipulate the angles between the crystal's crystallographic axes, that is, its interaxial angles, and (2) data that state the relative sizes of the units along the crystallographic axes, that is, the axial ratio.

Interaxial angles

Specific values for the interaxial angles α, β, and γ for triclinic crystals and for β in monoclinic crystals are invariably cited. These values serve as identifying characteristics of the crystal. No values for interaxial angles other than the foregoing need be stated because all the remaining interaxial angles possess the fixed, standard value of 90° (or of 60° between a axes in hexagonal crystals).

Axial ratios

For orthorhombic, monoclinic, or triclinic crystals, the ratio of lengths a_0, b_0, and c_0 relative to b_0, that is,

$$\frac{a_0}{b_0} : \frac{b_0}{b_0} : \frac{c_0}{b_0} \qquad \text{or} \qquad \frac{a_0}{b_0} : 1 : \frac{c_0}{b_0}$$

constitutes their *axial ratio*. Although unable to measure the actual values of a_0, b_0, and c_0, the early crystallographers nevertheless were able to determine their lengths relative to each other and to express this information in an axial ratio. For example, returning to the orthorhombic sulfur crystal of Figure 3-1, we see that the intercepts for unit face p were given as 7.08, 8.70, and 16.57 cm. Acceptance of p as the unit face was to conclude that its intercepts equaled na_0, nb_0, and nc_0 units, respectively, where n is an unknown number. Thus

$$7.08 = na_0 \qquad 8.70 = nb_0 \qquad 16.57 = nc_0$$

From this it is obvious that sulfur's axial ratio is

$$\frac{na_0}{nb_0} : \frac{nb_0}{nb_0} : \frac{nc_0}{nb_0} = \frac{7.08}{8.70} : 1 : \frac{16.57}{8.70} = 0.814:1:1.905$$

Thus for a trimetric crystal, the axial ratio may be calculated from the unit face's intercepts; we merely divide its intercept on the b axis into its intercepts on the a, b, and c axes. Sulfur's axial ratio concisely expresses the fact that the units along its a axis are slightly over 0.8 as long as those along its b axis, and that the units along its c axis are almost twice as long as along its b axis.

For the second crystal of sulfur (Fig. 3-2), the intercepts for unit face p were given as 4.60, 5.65, and 10.77 cm. Dividing them by 5.65 we obtain the axial ratio 0.814:1:1.906 which, allowing for experimental errors, is the same as before. In fact all orthorhombic sulfur crystals would yield the same axial ratio. Interestingly, the X-ray methods now available to crystallographers disclose the *actual* values of the units along the axes in orthorhombic sulfur to be $a_0 = 10.45$ Å, $b_0 = 12.84$ Å, and $c_0 = 24.46$ Å, where one angstrom (1 Å) represents 10^{-8} cm. These actual values lead to the same axial ratio for sulfur as was obtained by the earlier crystallographers.

Prior to X-ray crystallography, axial ratios were an important means of crystal identification because trimetric crystals of differing composition rarely possess identical axial ratios; these ratios thus served as fingerprints. However, the axial ratio is less definitive for tetragonal or hexagonal crystals because axial units of only two sizes occur; consequently, the axial ratio, as defined for dimetric crystals, is simply the ratio in size of the a_0 and c_0 units to that of a_0. In quartz, for example, X-ray measurements indicate that $a_0 = 4.913$ and $c_0 = 5.405$ Å. From these measurements the axial ratio can be computed and stated as

$a{:}c = 1.000{:}1.100$ or, more briefly, as $c/a = 1.100$ (the 1.000 being omitted), or as $c = 1.100$ (the a being now omitted as well). Interestingly, the early crystallographers, working solely from crystal-face data, had already computed quartz's axial ratio at the above value.

For all isometric crystals the axial ratio is of no significance because, regardless of identity, it invariably equals 1.000; that is, equal units occur along all three axes.

Nomenclature for Linear Directions

In Figure 3-4 the three axes of an orthorhombic sulfur crystal are marked with units of no special size except that their lengths are in the same proportion to each other as indicated by the axial ratio 0.814:1:1.905. These axes, so marked, now define a three-dimensional coordinate system that permits a point in space to be specified by three numbers, u, v, and w, which are the point's coordinates in terms of units along the a, b, and c axes, respectively. Several such points, some labeled with their coordinates, are indicated by crosses or ellipses in Figure 3-4.

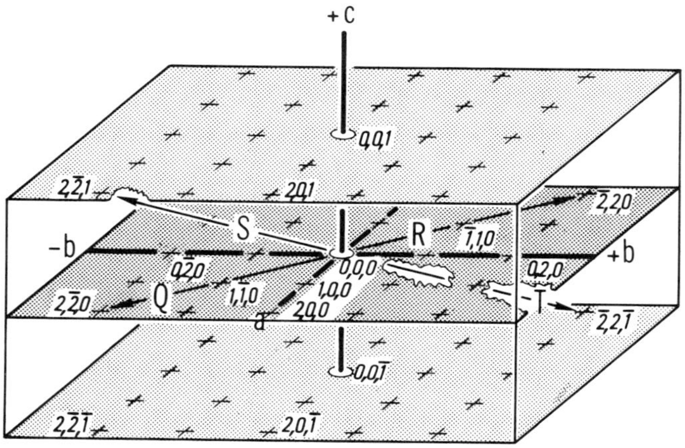

FIGURE 3–4

The three mutually perpendicular crystallographic axes for sulfur, bold lines a, b, and c, if units are marked off along them with lengths respectively proportional to sulfur's axial ratio, 0.814:1:1.905, define a coordinate system. The coordinates in respect to this system are indicated for several points (usually to their lower right). Following custom, the coordinates for a point are listed in the sequence to which they refer to the number of units along the a, b, and c axis.

Any line in a crystal may be assumed to emanate from the origin of the coordinate system. Consequently, the coordinates of the closest point to the origin through which the line must pass will constitute the line's symbol, if enclosed in brackets. Thus line Q has the symbol $[1\bar{1}0]$.

Symbolism for directions

Any direction within a crystal can be assumed to emanate from the origin of the crystallographic axes. For example, we let arrows Q, R, S, and T (Fig. 3-4) represent directions of crystallographic interest in the sulfur crystal. The coordinates of the closest point to the origin through which directions Q, R, S, and T pass will be the symbols for these lines (if enclosed in brackets). Thus the symbols are $[1\bar{1}0]$, $[\bar{1}10]$, $[2\bar{2}1]$, and $[\bar{2}2\bar{1}]$ for Q, R, S, and T, respectively. We note that $[1\bar{1}0]$ and $[2\bar{2}0]$, for example, represent the same line; however, $[1\bar{1}0]$ is the preferred symbol because it contains no common factors. If the symbols for a line are multiplied through by -1, the resultant symbol is for a line precisely opposite in direction. Thus $[1\bar{1}0]$ and $[\bar{1}10]$ represent precisely opposite directions, as do $[2\bar{2}1]$ and $[\bar{2}2\bar{1}]$. Similarly, the symbols for crystallographic axes a, $-a$, b, $-b$, c, and $-c$ in a crystal are $[100]$, $[\bar{1}00]$, $[010]$, $[0\bar{1}0]$, $[001]$, and $[00\bar{1}]$.

Zones, zone axes, zone symbols

A *zone* of faces represents a set of faces parallel to a common line called the *zone axis*. If faces lying in the same zone meet, they intersect in edges that are parallel to their zone axis. For example, faces $(1\bar{1}0)$, (110), and (010) in Figure 3-5 intersect in edges that are parallel to $[001]$, the zone axis for the zone containing these faces. The symbol for a zone axis, for example, $[001]$, is usually called the *zone symbol*. Zone symbol $[uvw]$ indicates a zone axis of unspecified orientation, the letters u, v, and w standing for any digits whatsoever.

In the staurolite crystal in Figure 3-5 the faces are disposed in three main zones whose zone axes are shown by the heavy lines. Thus faces $(1\bar{1}0)$, (110), and (010) lie in the zone $[001]$; faces $(20\bar{1})$, (201), and (001) lie in the zone

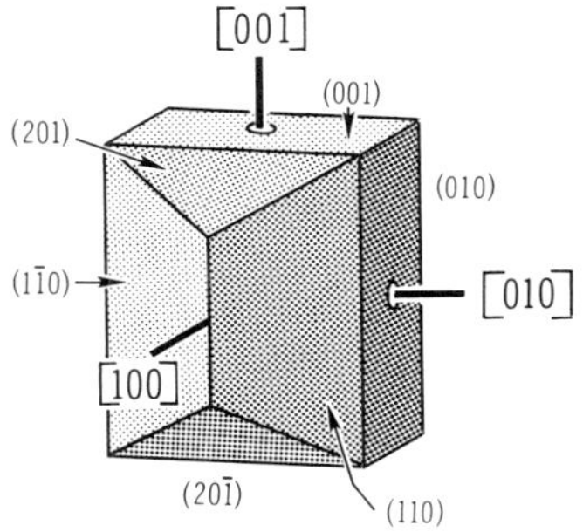

FIGURE 3–5

A crystal of staurolite whose crystallographic axes, a, b, and c, are, respectively, the zone axes [100], [010], and [001]. Faces $(1\bar{1}0)$, (110), and (010) all lie in the zone [001] as evidenced by the fact that their lines of intersection (and they themselves) are all parallel to [001].

[010]; and faces (001) and (010) lie in the zone [100]. The three zones are important because they contain either (1) a large number of faces or (2) the faces of largest area on the crystal. Often the zone axes for the more important zones on a crystal will coincide with its crystallographic axes. In Figure 3-5, for example, zone axes [100], [010], and [001] are actually the crystallographic axes a, b, and c as well. Less important zones in Figure 3-5 would be those containing faces (201) and $(1\bar{1}0)$, faces (201) and (110), faces $(1\bar{1}0)$ and $(20\bar{1})$, or faces (110) and $(20\bar{1})$. The line of intersection between each of these pairs of faces is parallel to the axis of the zone to which the two faces belong.

Crystallographic Calculations

General background

Detailed mathematical treatments of the crystallographic calculations discussed next are given by Terpstra and Codd (1961) and by deJong (1959). For present understanding of the mechanics of these calculations, however, an elementary knowledge of determinants and of matrix multiplication, as provided in this chapter's appendix, should be acquired before reading the following sections.

Finding the zone symbol for two nonparallel faces

Any two nonparallel faces, $(h_1k_1l_1)$ and $(h_2k_2l_2)$, determine a line by their intersection. The indices $[uvw]$ of this line or zone axis, in other words, the zone symbol, can be calculated as follows.

1. Write the Miller indices of each face twice, that is,

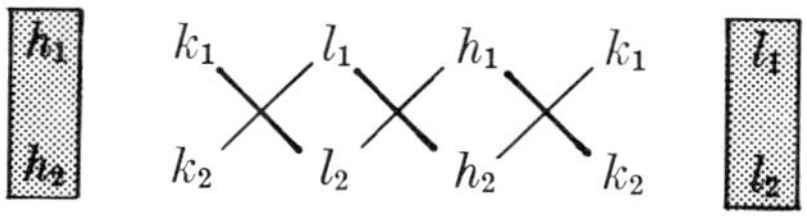

2. Discard the (shaded) indices at both ends.

3. For each of the three crosses above, multiply the two indices at the ends of the bold bar and subtract from this the product of the two indices at the ends of the light bar. For the cross at the left, this operation yields u; for the middle cross, v; for the cross on the right, w. In other words,

$$u = k_1l_2 - k_2l_1 \qquad v = l_1h_2 - l_2h_1 \qquad w = h_1k_2 - h_2k_1$$

As a numerical example consider two faces with Miller indices (345) and (012). The symbol for the line to which both are parallel, that is, their line of intersection, is obtained by writing

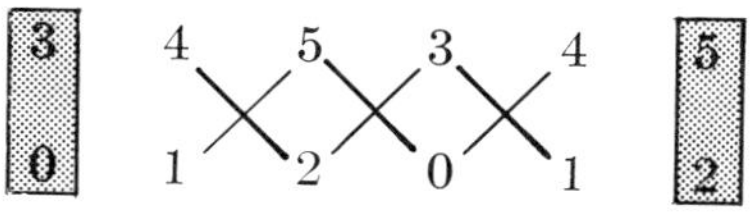

whereupon

$$u = 8 - 5 = 3 \qquad v = 0 - 6 = -6 \qquad w = 3 - 0 = 3$$

The resultant symbol $[3\bar{6}3]$ may be divided by three to become $[1\bar{2}1]$, the more conventional form.

Computing Miller index of a face parallel to two given lines

Any two, nonparallel, intersecting lines, $[u_1v_1w_1]$ and $[u_2v_2w_2]$, determine a plane. The Miller indices (hkl) of this face can be quickly calculated in a manner identical to that just discussed. Thus the line symbols are written twice,

$$\begin{array}{cccccc} u_1 & v_1 & w_1 & u_1 & v_1 & w_1 \\ & \times & \times & \times & & \\ u_2 & v_2 & w_2 & u_2 & v_2 & w_2 \end{array}$$

and the end indices (shaded) are discarded. Taking the differences between the cross products as before, we obtain

$$h = v_1w_2 - v_2w_1 \qquad k = w_1u_2 - w_2u_1 \qquad l = u_1v_2 - u_2v_1$$

As a numerical example let the symbols of the two lines be $[30\bar{1}]$ and $[210]$. To calculate the indices of the face parallel to both these lines, we write

$$\begin{array}{cccccc} 3 & 0 & \bar{1} & 3 & 0 & \bar{1} \\ & \times & \times & \times & & \\ 2 & 1 & 0 & 2 & 1 & 0 \end{array}$$

from which we see that

$$h = 0 - (-1) \qquad k = -2 \qquad l = 3$$

and that the face has Miller indices $(1\bar{2}3)$.

Test for tautozonality of crystal faces

If three given planes $(h_1k_1l_1)$, $(h_2k_2l_2)$, and $(h_3k_3l_3)$ are *tautozonal,* that is, all lie in the same zone and are therefore parallel to the same zone axis, then the determinant formed with their indices must equal zero, thus

$$\begin{vmatrix} h_1 & k_1 & l_1 \\ h_2 & k_2 & l_2 \\ h_3 & k_3 & l_3 \end{vmatrix} = 0$$

Note that the faces (111), (101), and $(1\bar{1}1)$ can be quickly determined to be tautozonal because

$$\begin{vmatrix} 1 & 1 & 1 \\ 1 & 0 & 1 \\ 1 & \bar{1} & 1 \end{vmatrix} = 0$$

Test for coplanarity of directions in a crystal

If the three directions $[u_1v_1w_1]$, $[u_2v_2w_2]$, and $[u_3v_3w_3]$ are coplanar, that is, all lie in (or are parallel to) the same plane in the crystal, then the determinant

$$\begin{vmatrix} u_1 & v_1 & w_1 \\ u_2 & v_2 & w_2 \\ u_3 & v_3 & w_3 \end{vmatrix} = 0$$

Thus the three directions [101], [313], and [111] are disclosed to be coplanar by applying this test.

The Zone Condition. A given face $(h_1k_1l_1)$ lies in a given zone $[u_1v_1w_1]$ only if the following condition is satisfied:

$$u_1h_1 + v_1k_1 + w_1l_1 = 0$$

The foregoing equation is sometimes called the *zone condition* or the *Weiss zone law.* It tests whether a given face $(h_1k_1l_1)$ is parallel to a given line $[u_1v_1w_1]$ or vice versa.

The Addition and Subtraction Rules. The Miller indices of additional faces lying in the same zone as two given faces, $(h_1k_1l_1)$ and $(h_2k_2l_2)$, may be calculated by employing the *addition rule* or the *subtraction rule.* The addition rule states that the symbols of additional faces in the zone are obtainable by adding the indices of the given faces to obtain $(h_1 + h_2, k_1 + k_2, l_1 + l_2)$ or by adding multiples of their indices to obtain $(mh_1 + nh_2, mk_1 + nk_2, ml_1 + nl_2)$, where m and n are integers. For example, face (011) necessarily lies in the same zone as (001) and (010). The subtraction rule is analogous to the addition rule except that differences are taken to obtain $(h_1 - h_2, k_1 - k_2, l_1 - l_2)$ or $(mh_1 - nh_2, mk_1 - nk_2, ml_1 - nl_2)$.

Transformation matrices

Occasionally good reasons arise for selecting crystallographic axes along directions other than those chosen by earlier investigators. If this is done, then the Miller indices that the earlier investigators assigned to the crystal's faces are no longer correct in respect to this new set of axes. Consequently, the investigator who has chosen the new set of axes will generally supply a *transformation matrix,*

$$\begin{bmatrix} u_1 & v_1 & w_1 \\ u_2 & v_2 & w_2 \\ u_3 & v_3 & w_3 \end{bmatrix}$$

This matrix can transform (hkl), the Miller index applied to the face by the earlier investigators, into (HKL), the Miller index for the face with respect to the new axes; capital letters here distinguish the new Miller indices from the old. If symbols (hkl) and (HKL) are each written as a matrix, then

$$\begin{bmatrix} u_1 & v_1 & w_1 \\ u_2 & v_2 & w_2 \\ u_3 & v_3 & w_3 \end{bmatrix} \cdot \begin{bmatrix} h \\ k \\ l \end{bmatrix} = \begin{bmatrix} H \\ K \\ L \end{bmatrix}$$

In other words, if multiplied by a matrix formed from the old index (hkl), the transformation matrix will yield a second matrix whose numbers (HKL) correspond to the new Miller indices for the face. The correct procedure for this matrix multiplication is discussed in the appendix to this chapter. However, to save typographical effort, the transformation matrix is often written as

$$u_1v_1w_1/u_2v_2w_2/u_3v_3w_3$$

This suggests an easily remembered, but unorthodox, way for carrying out this matrix multiplication. Write the old index (hkl) below each of these three, now side-by-side rows of the transformation matrix, then multiply each top digit by that directly below it. Thus

$$\begin{array}{ccccccccc} u_1 & v_1 & w_1\ / & u_2 & v_2 & w_2\ / & u_3 & v_3 & w_3 \\ \times & \times & \times & \times & \times & \times & \times & \times & \times \\ h & k & l & h & k & l & h & k & l \end{array}$$

$$\underbrace{hu_1 + kv_1 + lw_1}_{H}/\underbrace{hu_2 + kv_2 + lw_2}_{K}/\underbrace{hu_3 + kv_3 + lw_3}_{L}$$

Then, as shown, the algebraic sum of the first three products connected by the brace equals H; that of the second three equals K; and so on.

Peacock's (1950) article on hauchecornite furnishes a numerical example. The X-ray evidence led Peacock to choose axes a_1 and a_2 (dashed in Fig. 3-6) at 45° to those (A_1 and A_2) chosen by Scheibe about sixty years earlier. Peacock's new orientation of axes plus his revision of the sizes of the units along them were accompanied by two transformation formulas:

Scheibe to Peacock: $\frac{1}{2}\bar{\frac{1}{2}}0/\frac{1}{2}\frac{1}{2}0/00\frac{1}{2}$

and

Peacock to Scheibe: $110/\bar{1}10/002$

The first formula transforms any Miller index that is correct for Scheibe's choice of axes and axial units into one that is correct for Peacock's choice of axes and units; the second formula performs the reverse transformation. Employing the first transformation formula, we find that a face indexed as $(10\bar{1})$ by Scheibe

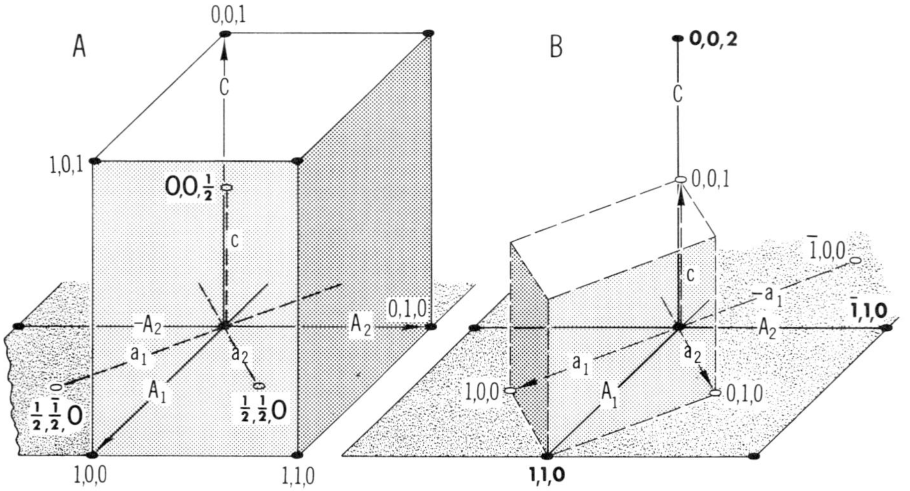

FIGURE 3–6

Comparison of Scheibe's choice of crystal axes and unit lengths for the mineral hauchecornite (solid vectors A_1, A_2, and C) with those chosen by Peacock (dashed vectors a_1, a_2, and c). The coordinates of the points at the ends of these six vectors are given relative to the Scheibe coordinate system in (A) and relative to the Peacock coordinate system in (B).

becomes $(11\bar{1})$ in respect to the axes chosen by Peacock. Mechanical details of the computation follow.

$$\begin{array}{ccc|ccc|ccc}
\frac{1}{2} & \frac{\bar{1}}{2} & 0 & \frac{1}{2} & \frac{1}{2} & 0 & 0 & 0 & \frac{1}{2} \\
\times & \times & \times & \times & \times & \times & \times & \times & \times \\
1 & 0 & \bar{1} & 1 & 0 & \bar{1} & 1 & 0 & \bar{1} \\
\frac{1}{2} & 0 & 0 & \frac{1}{2} & 0 & 0 & 0 & 0 & \frac{\bar{1}}{2} \\
 & \frac{1}{2} & & & \frac{1}{2} & & & \frac{\bar{1}}{2} &
\end{array}$$

Thus the index is $(\frac{1}{2}\frac{1}{2}\frac{\bar{1}}{2})$ which, cleared of fractions, is $(11\bar{1})$. If the Peacock-to-Scheibe matrix is applied to $(11\bar{1})$, it will be transformed back to $(20\bar{2})$ or, dividing through by the common factor, to $(10\bar{1})$, its Scheibe equivalent.

Obtaining the transformation matrix (optional)

Let the solid arrows A_1, A_2, and C in Figure 3-6A represent the directions and lengths of units for the set of tetragonal axes chosen by Scheibe for the mineral hauchecornite; similarly, let the dashed lines a_1, a_2, and c be those chosen by Peacock. The three hollow dots at the ends of Peacock axes a_1, a_2, and c have Scheibe coordinates $\frac{1}{2}$, $\frac{\bar{1}}{2}$, 0; $\frac{1}{2}$, $\frac{1}{2}$, 0; and 0, 0, $\frac{1}{2}$, respectively. These constitute the Scheibe-to-Peacock transformation matrix $\frac{1}{2}\frac{\bar{1}}{2}0/\frac{1}{2}\frac{1}{2}0/00\frac{1}{2}$. In Figure 3-6B all points have coordinates relative to the Peacock axes (dashed arrows); thus the three solid dots marking the tips of Scheibe axes A_1, A_2, and C have Peacock coordinates 1, 1, 0; $\bar{1}$, 1, 0; and 0, 0, 2, respectively. These constitute the Peacock-to-Scheibe transformation matrix 110/$\bar{1}$10/002. In general for a new set of axes, the coordinates of their tips relative to the old set define a matrix that transforms $(hkl)_{\text{old}}$ to $(hkl)_{\text{new}}$.

Buerger (1942, pp. 10-11) provides a clear and more rigorous approach to transformation matrices that requires only a knowledge of the addition of vectors.

Calculations involving Bravais indices

All calculations applicable to three-digit Miller indices become applicable to the four-digit Bravais indices if their third digit, that is, the one applicable to the somewhat superfluous a_3 axis, is replaced by an asterisk. Thus the symbol for the zone containing the (10*1) and the (21*1) face can be determined to be [$\bar{1}$1*1] by using the method discussed on page 58. In effect the a_1, a_2, and c axes of a four-axis Bravais set may be handled as though they constitute a normal three-axis set.

Hexagonal zone symbols

The three-number, hexagonal zone symbols (general form, [uv*w]) are becoming increasingly popular. In these symbols the number replaced by the asterisk is most emphatically *not* a value such that $(u + v + *)$ equals zero. Moreover, these three-number symbols should not be confused with the four-number symbols of the general form [$UVTW$] whose use to describe directions in hexagonal crystals should be discontinued. In *these symbols* $(U + V + T)$ *does* total zero just as $(h + k + i)$ did for the Bravais-Miller indices for a face. The distinction between a direction indicated by the three-number symbol [21*0] and one indicated by the four-number symbol [21$\bar{3}$0] is illustrated in Figure 3-7. We note that [21*0] conforms to the ordinary convention for direction symbols (p. 57), the a_3 axis being ignored. In contrast direction [21$\bar{3}$0] extends outward from the origin toward the point located by going two units along a_1, one unit along a_2, and three units along the negative a_3 axis. For any given direction in a hexagonal crystal, its four-number symbol [$UVTW$] converts to its three-number symbol [uv*w],

$$u = U - T \qquad v = V - T \qquad w = W$$

Similarly, [uv*w] converts to [$UVTW$],

$$U = \tfrac{1}{3}(2u - v) \qquad V = \tfrac{1}{3}(2v - u) \qquad T = -(U + V) \qquad W = w$$

These equations indicate that [21*0] signifies the same direction as [10$\bar{1}$0] and vice versa. We can check this equivalency by locating direction [10$\bar{1}$0] in Figure 3-7A.

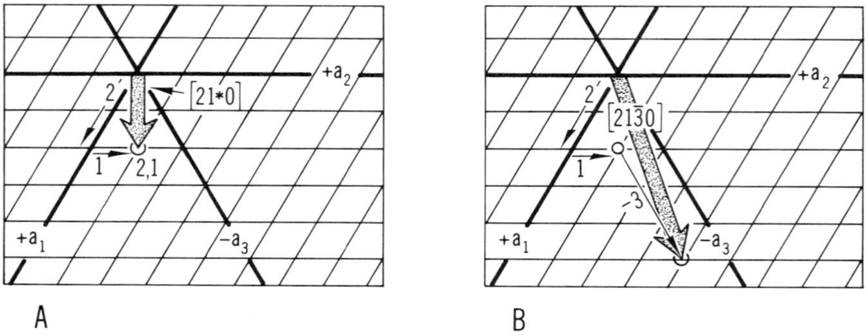

FIGURE 3–7

The coordinate system defined by a set of hexagonal axes is marked onto a plane through the origin and perpendicular to the c axis. In both (A) and (B), therefore, the coordinates of any point relative to the c axis are zero. In (A) the direction of the stippled arrow is symbolized by its three-number, hexagonal zone-symbol [21*0], where 2, 1, and 0 represent the coordinates relative to a_1, a_2, and c of the point closest to the origin through which this direction passes. The asterisk signifies the omission of any reference to axis a_3.
In (B) the stippled arrow is along a direction indicated by the four-number zone-symbol [$21\bar{3}0$], which means that the direction is reproduced by drawing an arrow from the origin to the point located by going two units along a_1, one along a_2, three along the negative direction of a_3, and zero along c. Four-number zone symbols [$UVTW$] are not as suited to crystallographic calculations as are the three-number symbols [$uv*w$] and should be discontinued.

Calculation of interfacial angles

For monoclinic crystals if the axial ratio $a{:}b{:}c$ and angle β are known, then θ, the interfacial angle between faces $(h_1k_1l_1)$ and $(h_2k_2l_2)$, may be calculated from the formula (Terpstra, 1952, p. 19)

$$\cos\theta = \frac{h_1h_2c^2 + k_1k_2a^2c^2\sin^2\beta + l_1l_2a^2 - (h_1l_2 + h_2l_1)ac\cos\beta}{\sqrt{n_1n_2}} \qquad \text{(Eq. 3-2)}$$

where

$$n_1 = h_1^2c^2 + k_1^2a^2c^2\sin^2\beta + l_1^2a^2 - 2h_1l_1ac\cos\beta$$

and

$$n_2 = h_2^2c^2 + k_2^2a^2c^2\sin^2\beta + l_2^2a^2 - 2h_2l_2ac\cos\beta$$

The formula for triclinic crystals is even more complex than the foregoing, but for the other systems the formulas for the interfacial angle θ are simplified cases of the monoclinic formula given above. Such simplification results because (1) for orthorhombic crystals, $\sin\beta = 1$ and $\cos\beta = 0$; (2) for tetragonal crystals, $\sin\beta = 1$, $\cos\beta = 0$, $a = 1$; and (3) for isometric crystals, $\sin\beta = 1$, $\cos\beta = 0$, $a = 1$, $c = 1$.

Appendix

Determinants

A determinant is a *square* array of numbers such as

$$\begin{vmatrix} a_1 & b_1 \\ a_2 & b_2 \end{vmatrix} \quad \text{or} \quad \begin{vmatrix} a_1 & b_1 & c_1 \\ a_2 & b_2 & c_2 \\ a_3 & b_3 & c_3 \end{vmatrix}$$

The two-by-two determinant at the left is called a *second-order* determinant; the one at the right is called a *third-order* determinant. The second-order determinant above is simply a convenient abbreviation for the expression $a_1b_2 - a_2b_1$; in other words, by definition

$$\begin{vmatrix} a_1 & b_1 \\ a_2 & b_2 \end{vmatrix} = a_1b_2 - a_2b_1$$

Thus a second-order determinant is equal to the product of the numbers lying along its *principal diagonal*, which is marked by the heavy line above, minus the products of its numbers along the diagonal marked by the dotted line above. The second-order determinant

$$\begin{vmatrix} -1 & 1 \\ 2 & 1 \end{vmatrix}$$

can be thus evaluated as equal to -3, that is, $(-1 \times 1) - (2 \times 1)$.

Similarly, by mathematical definition the third-order determinant given above is a convenient abbreviation for the expression,

$$a_1b_2c_3 + b_1c_2a_3 + c_1a_2b_3 - c_1b_2a_3 - b_1a_2c_3 - a_1c_2b_3$$

The three algebraically positive terms in this expression are merely the products of the determinant's numbers as they are considered along NW-SE diagonals (heavy lines in Fig. 3-8). We note that b_1 and c_2 are multiplied by a_3, the number that is farthest from them and that can be considered on the bottom of their NW-SE diagonal. Similarly, c_1 is multiplied by a_2 and b_3, which can be thought of as occupying the bottom of c_1's interrupted (heavy line) diagonal. All three negative terms in the expression that the determinant represents are merely the products of the determinant's numbers along NE-SW diagonals (dotted in Fig. 3-8). We also note that when only two numbers appear along a dotted diagonal, for example b_1 and a_2 in Figure 3-8, they are multiplied by the number farthest from their diagonal, for example c_3, which is considered to lie on the diagonal's interrupted end.

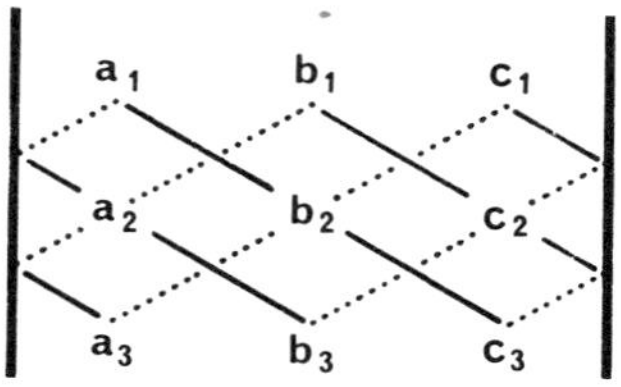

FIGURE 3–8

A 3-by-3 determinant with bold lines along its principal diagonals, each principal diagonal consisting of three digits. The product of the numbers along one principal diagonal is $a_1b_2c_3$, along another $b_1c_2a_3$, and along the third $c_1a_2b_3$. Its remaining three diagonals are marked by dotted lines, the products of the numbers along these being $a_3b_2c_1$, $a_2b_1c_3$, and $a_1b_3c_2$, respectively.

To ensure that he knows how to evaluate a determinant, the reader is invited to verify the values given for the following determinants:

A

$$\begin{vmatrix} 1 & 2 & 1 \\ 0 & 1 & 2 \\ 3 & \bar{1} & 1 \end{vmatrix} = 12$$

B

$$\begin{vmatrix} 1 & 1 & 1 \\ \bar{1} & 0 & 4 \\ 2 & 0 & 1 \end{vmatrix} = 9$$

C

$$\begin{vmatrix} 2 & 2 & 0 \\ 3 & 1 & 1 \\ 0 & 1 & 1 \end{vmatrix} = -6$$

Matrices

A rectangular array of numbers such as

$$\begin{bmatrix} a_1 & b_1 & c_1 & d_1 \\ a_2 & b_2 & c_2 & d_2 \\ a_3 & b_3 & c_3 & d_3 \end{bmatrix}$$

is defined by mathematicians as a *matrix*. This particular matrix consists of three *rows* and *four* columns. A matrix may also have the same number of rows as columns, in which case it is called a *square matrix*. Just as mathematicians have defined the way in which a determinant should be evaluated, they have also formulated the rules for the multiplication of one matrix by a second. Thus the multiplication of a square matrix, that is,

$$\begin{bmatrix} u_1 & v_1 & w_1 \\ u_2 & v_2 & w_2 \\ u_3 & v_3 & w_3 \end{bmatrix}$$

by one consisting of three rows but just a single column, that is,

$$\begin{bmatrix} h \\ k \\ l \end{bmatrix}$$

is defined to be the third matrix

$$\begin{bmatrix} (u_1h + v_1k + w_1l) \\ (u_2h + v_2k + w_2l) \\ (u_3h + v_3k + w_3l) \end{bmatrix}$$

In other words,

$$\begin{bmatrix} u_1 & v_1 & w_1 \\ u_2 & v_2 & w_2 \\ u_3 & v_3 & w_3 \end{bmatrix} \cdot \begin{bmatrix} h \\ k \\ l \end{bmatrix} = \begin{bmatrix} (u_1h + v_1k + w_1l) \\ (u_2h + v_2k + w_2l) \\ (u_3h + v_3k + w_3l) \end{bmatrix}$$

To use a numerical example,

$$\begin{bmatrix} \bar{1} & 0 & 4 \\ 0 & 1 & \bar{1} \\ 3 & 2 & 1 \end{bmatrix} \cdot \begin{bmatrix} \bar{2} \\ 1 \\ 5 \end{bmatrix} = \begin{bmatrix} (\bar{1} \times \bar{2}) + (0 \times 1) + (4 \times 5) \\ (0 \times \bar{2}) + (1 \times 1) + (\bar{1} \times 5) \\ (3 \times \bar{2}) + (2 \times 1) + (1 \times 5) \end{bmatrix}$$

Thus

$$\begin{bmatrix} \bar{1} & 0 & 4 \\ 0 & 1 & \bar{1} \\ 3 & 2 & 1 \end{bmatrix} \cdot \begin{bmatrix} \bar{2} \\ 1 \\ 5 \end{bmatrix} = \begin{bmatrix} 22 \\ \bar{4} \\ 1 \end{bmatrix}$$

The foregoing matrix multiplication operation in which a three-by-three matrix is multiplied by a one-by-three matrix finds frequent use in crystallography and the mechanics of the operation should be mastered. For the time being only the operation of matrix multiplication will be needed. In the Appendix to Chapter 11, additional operations with matrices will be described.

REFERENCES AND RECOMMENDED READING

Buerger, M. J. (1942) *X-ray Crystallography,* New York: Wiley. 10–28.
deJong, W. F. (1959) *General Crystallography,* San Francisco: Freeman.
Peacock, M. A. (1950) "Hauchecornite," *Am. Mineral.,* **35,** 440.

Phillips, F. C. (1963) *An Introduction to Crystallography*, New York: Wiley.
Terpstra, P. (1952) *A Thousand and One Questions on Crystallographic Problems*, Groningen: J. B. Walters.
Terpstra, P., and L. W. Codd (1961) *Crystallometry*, London: Longmans, Green.
Tutton, A. E. H. (1922) *Crystallography*, Vol. I, New York: Macmillan.

QUESTIONS AND PROBLEMS

3-1 The intercepts in centimeters of a unit face (111) and of faces A to E on an orthorhombic crystal are

	Unit Face	Face A	Face B	Face C	Face D	Face E
on a axis	0.393	0.405	0.324	0.340	0.117	0.247
on b axis	0.742	0.765	1.224	∞	0.330	0.234
on c axis	0.353	0.728	0.194	0.611	0.314	∞

(a) What are the Miller indices for faces A to E?
(b) What is the axial ratio?

3-2 What is the Weiss symbol for (a) face (113)? (b) face (120)? (c) face (351)? (d) face (001)?

3-3 What is the Miller index for the face whose Weiss symbol is (a) $4a{:}2b{:}c$? (b) $\infty a{:}1b{:}3c$? (c) $10a{:}3b{:}5c$? (d) $1a{:}\infty b{:}\infty c$?

3-4 What is the symbol for the zone containing faces (112) and (341)?

3-5 Does (112) lie in the [$\bar{2}$01] zone?

3-6 What face is common to zones [$\bar{7}$51] and [20$\bar{1}$]?

3-7 What is the angle between [001] and the normal to a (001) plane (a) in orthorhombic crystals? (b) in a monoclinic crystal for which $\beta = 106°$?

3-8 For which crystal system(s) will the unit face cut (a) three axes at equal angles? (b) two axes at the same angle but the third at a different angle? (c) all three axes at different angles?

3-9 From a copy of Donnay et al. (*Crystal Data, Am. Cryst. Assoc. Monograph* 5, 1963) identify the monoclinic crystal whose axial elements are $0.545{:}1{:}0.292$, $\beta = 104.7°$.

3-10 What would Peacock's indices be for the faces on hauchecornite indexed by Scheibe as (110), (112), (111), and (011)?

3-11 Investigator II, restudying a monoclinic crystal previously studied by Investigator I, relabeled axis a_{I} as b_{II}, b_{I} as c_{II}, c_{I} as a_{II}. (a) What is the I to II transformation matrix? (b) The II to I matrix?

3-12 In Figure 3-7 direction [21$\bar{3}$0] would be described as [54*0] in three-number symbolism. What is the three-number symbol for the direction represented by the four-number symbol [$\bar{2}$4$\bar{2}$3]?

ANSWERS

3-1 (a) *A* (221), *B* (213), *C* (201), *D* (321), *E* (120); (b) 0.530:1:0.476.
3-2 (a) $3a:3b:1c$; (b) $2a:1b:\infty c$; (c) $5a:3b:15c$; (d) $\infty a:\infty b:1c$.
3-3 (a) (124); (b) (031); (c) (3,10,6); (d) (100).
3-4 $[\bar{7}51]$.
3-5 Yes, the *zone condition* (p. 60) is fulfilled.
3-6 (112).
3-7 (a) 0°; (b) 16°.
3-8 (a) Isometric; (b) tetragonal, hexagonal; (c) orthorhombic, monoclinic, or triclinic.
3-9 Tremolite.
3-10 Respectively, (010), (011), (021), and $(\bar{1}11)$.
3-11 (a) 001/100/010; (b) 010/001/100.
3-12 [02*1].

112112
341341

1-8 6-1 4-3

-7 5 1

$\bar{7}51$

4 Crystal Projections

Direct and Reciprocal Projections

Crystallographic problems involving the angles between lines or planes may often be quickly solved by projecting these elements onto a sphere. To project a line or plane onto the surface of a given reference sphere, the line or plane is (1) passed (with its orientation preserved) through the center of the sphere and (2) extended outward until it intersects the sphere. For example, in Figure 4-1 lines *OP* and *OR* project as points *P* and *R*, whereas plane *A* projects as the dashed great circle.[1] Such intersections represent the *direct projections* of the given geometrical elements onto the sphere; the term *direct projection* indicates that the geometrical element itself was extended until it intersected the sphere.

In contrast a given geometric element may also be projected *reciprocally*. By definition this signifies the construction through the sphere's center of a second geometric element that is *uniquely* perpendicular to the given element. The intersection of this perpendicular element with the sphere then represents the *reciprocal projection* of the given element. For example, the geometric element uniquely perpendicular to plane *A* in Figure 4-1 is line *OP*. Consequently, point *P*, which is line *OP*'s point of intersection with the sphere, now represents the *reciprocal projection* of plane *A*; in this event *P* is called the *face pole* of plane *A*. Conversely, the geometric element uniquely perpendicular to line *OP* is plane *A*. Consequently, the intersection of this plane with the sphere, that is, the dashed great circle in Figure 4-1, represents the *reciprocal projection* of line *OP*.

To summarize: (1) A plane or crystal face is projected *directly* as a great

[1] A great circle may be defined as the arc of intersection of a sphere with a plane passing through its center.

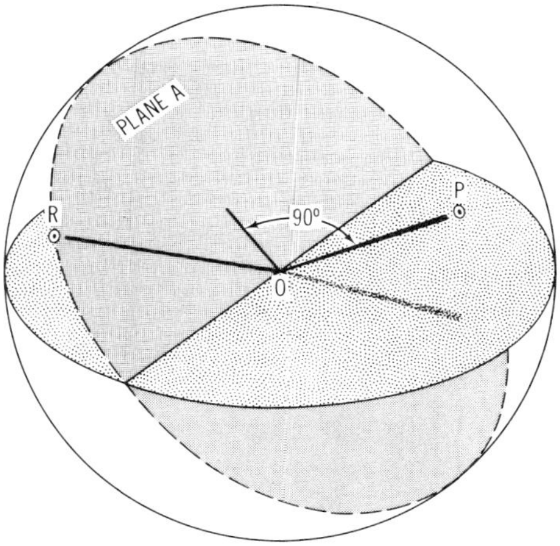

FIGURE 4–1

The dashed great circle and points *R* and *P*, respectively, represent the direct projection of plane *A*, line *OR*, and line *OP* on a reference sphere. Since plane *A* and line *OP* are mutually perpendicular, the dashed great circle could also represent the reciprocal (polar) projection of line *OP*; similarly, point *P* could be the reciprocal projection of plane *A*.

circle but *reciprocally* as a face pole, whereas (2) a line or crystal axis is projected *directly* as a point but *reciprocally* as a great circle. (In either type of projection the geometric element intersecting the sphere must necessarily pass through the sphere's center.) In general, it is usually most convenient to project crystal faces or planes *reciprocally* (as face poles or points) and crystal axes or lines *directly* (as points). Such a procedure avoids construction of great circles. However, it may sometimes be desirable to project a plane directly (or a line reciprocally) if the resultant great circles are necessary to solve a problem.

The Cyclographic and Stereographic Projections

Unfortunately, a sphere cannot be printed on a flat page—a grave disadvantage for communication of ideas. Several methods, however, exist for transferring the points or great circles from a spherical surface to a flat plane. In the case of the cyclographic and stereographic projections, this flat plane is an equatorial plane of the sphere (stippled in Fig. 4-1). It is often called the *plane of projection* and divides the sphere into an upper and lower hemisphere. The line perpendicular to this equatorial plane (*UL* in Fig. 4-2) intersects the sphere in an upper pole (*U*) and a lower pole (*L*).

A point on a hemisphere surface (for example, *P* in Fig. 4-2) may be transferred to the equatorial plane by (1) constructing a line that joins this given point to the pole of the opposite hemisphere (for example, dotted line *LP*) and

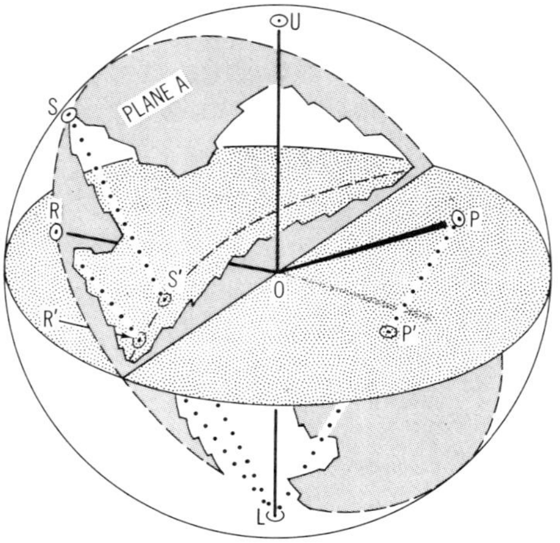

FIGURE 4–2

Transferring points from the surface of the sphere to its equatorial plane (stippled). Each point is joined to the pole of the opposite hemisphere by a dotted line. This dotted line pierces the equatorial plane at the projected position of the point in this plane; R', S', and P' are examples. Point P' is the reciprocal projection of plane A onto the equatorial plane.

(2) locating P', the intersection between this line and the equatorial plane. Point P' now marks the projected position of point P. A great circle on the sphere may be projected onto the equatorial plane by similarly treating all individual points along it (for example, R and S) and obtaining their projected positions (R' and S').

Point P' in Figure 4-2 thus marks the projection in the equatorial plane of point P, which, in turn, may represent (1) the direct projection of line OP or (2) the reciprocal or polar projection of plane A. If (1) is true, point P' represents the *cyclographic* projection of line OP; if (2) is true, P' represents the *stereographic* projection of plane A. The sole difference between these two projections is that the cyclographic stems from a direct projection, whereas the stereographic stems from a reciprocal projection of a plane or line onto the sphere. Similarly, the dashed curve in the stippled equatorial plane of Figure 4-2 may represent the cyclographic projection of plane A or the stereographic projection of line OP.

Coordinates of a point

A face-normal OP (which projects in Fig. 4-3 on the sphere at P and on the equatorial plane at P') becomes oriented in space if two angular coordinates, ϕ (phi) and ρ (rho), are defined for it. The value of ϕ for line OP is ϕ_1, the angle between two planes hinged on axis UL; one is an arbitrarily selected reference

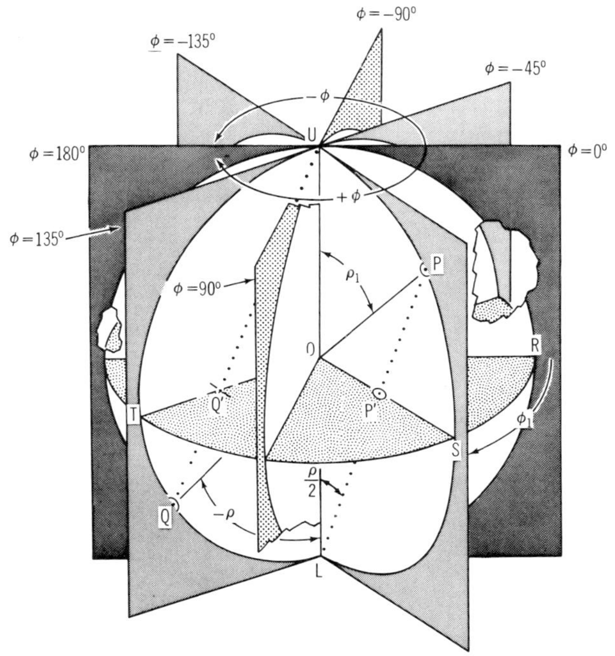

FIGURE 4–3

Spherical coordinate system for a sphere as related to pole *UL*. The shaded planes hinge on *UL* at various angles to the heavily shaded reference plane containing *R*. These angles represent the ϕ (phi) coordinates for these planes; negative ϕ values are measured counterclockwise from the reference plane, positive ϕ values clockwise. A line, for example, *OP*, is fixed in space if (1) the ϕ value of the plane containing it is specified and (2) its angle to pole *UL*, that is, its ρ (rho) angle, is also specified. In other words, if ϕ and ρ represent specific values, line *OP* is fixed in space.

plane (heavily shaded), the second contains line *OP*. The value of ρ for *OP* is the angle between *OP* and the polar axis *UL*, that is, angle ρ_1 in Figure 4-3. The value of ρ is defined as positive if measured from the upper pole (for example, as for *OP*) and negative if measured from the lower pole (for example, as for *OQ*). By way of example (Fig. 4-3), the coordinates of line *OP* are $\phi_1 = 45°$, $\rho_1 = 48°$. Those for line *OQ* are $\phi = 135°$, $\rho = -65°$.

Plotting points

Let Figure 4-4A represent Figure 4-3 as seen by looking down the polar axis *UL*. Any radius of the circle in Figure 4-4A may thus be considered to represent the edge-on view of a plane hinged on axis *UL*. The value of ϕ for such a plane can be measured with a protractor; it is the angle between this radius and

OR, where OR in Figure 4-4A represents the trace of the reference plane for which ϕ was defined to equal zero degrees. Positive ϕ angles are measured clockwise from OR; negative ones are measured counterclockwise. Radius OS in Figure 4-4A, for example, represents the edge-on view of a plane that contains all directions (of which OP in Fig. 4-3 is one) for which ϕ equals 45°. Similarly, radius OT represents the edge-on view of a plane that contains all directions for which ϕ equals 135°. Along such radii (for example, OS in Fig. 4-4) the distance

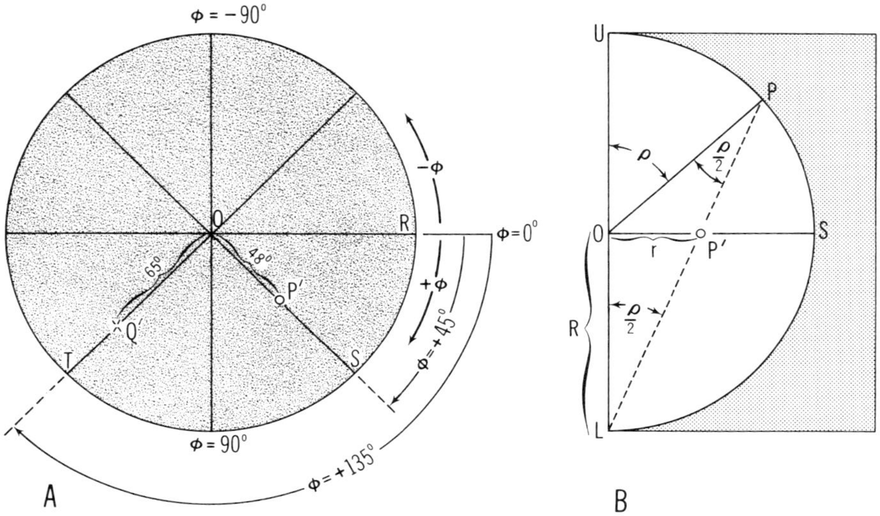

FIGURE 4–4

(A) The stippled plane of projection in Figure 4–3 as seen looking down pole UL. The point P' has the coordinates $\phi = 45°$ and $\rho = 48°$. Line OR represents the edge-view of the heavily shaded reference plane in Figure 4–3.
(B) Plane $UOLPS$ in Figure 4–3 is enlarged to show that projection line LP (dashed) defines an isosceles triangle LOP for which $LO = OP$. Note that

$$\angle OLP + \angle OPL + \angle LOP = 180°$$

and that

$$\rho + \angle LOP = 180°$$

Hence, because

$$\angle OLP = \angle OPL$$

$$\angle OLP = \frac{\rho}{2}$$

Thus r, the distance of projected point P' from O, the center of the stereogram in A, can be calculated from its ρ angle because

$$r = R \tan \frac{\rho}{2}$$

where R is the radius of the sphere of projection.

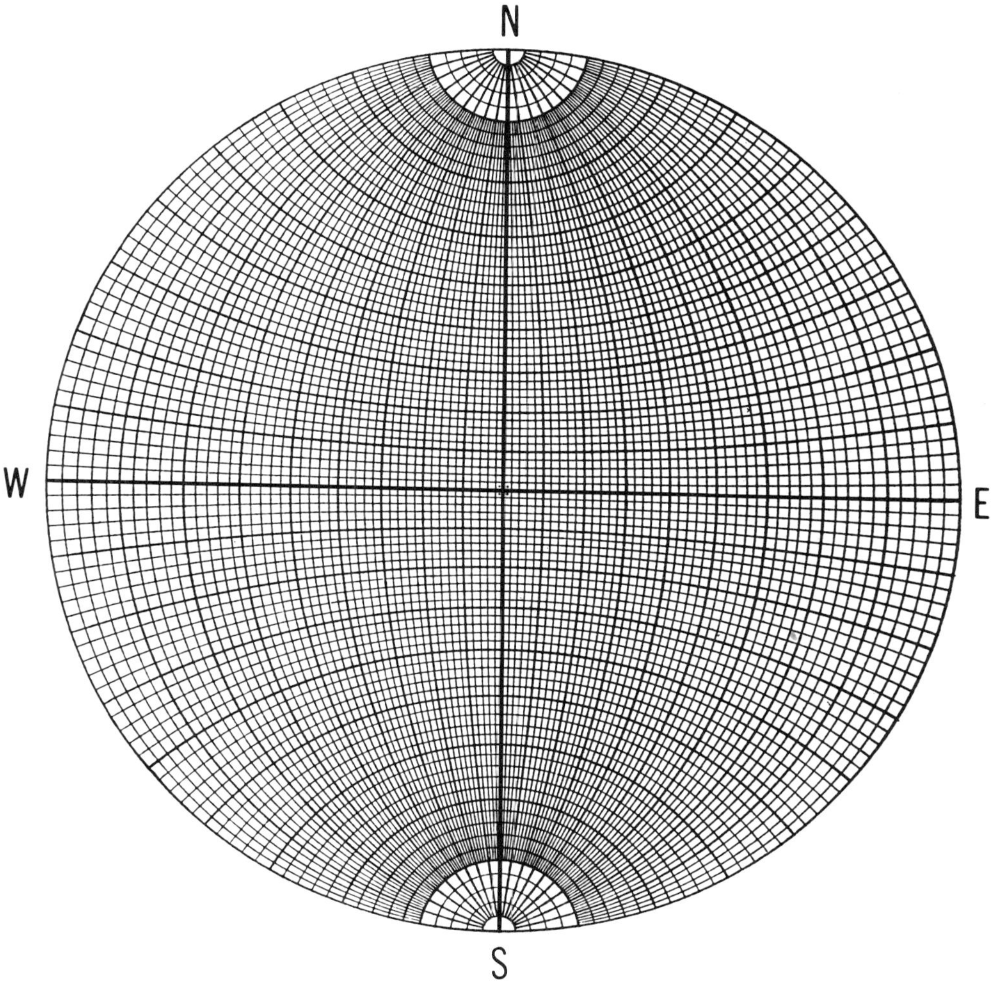

FIGURE 4–5

The stereonet (Wulff net) drawn with 2° intervals. The 45 divisions along each of the four bold radii mark positions where a point would plot (when stereographically projected as in Fig. 4–4B) if its ρ angle were 2°, 4°, 6°, . . . , 88°, 90°, their distances from the stereonet's center equaling $R \tan \rho/2$. The circumference of the stereonet is divided into 180 equal arcs, each of 2°. This permits the stereonet to serve as a protractor when needed.

r at which a point (P') plots outward from the circle's center increases with ρ since as shown in Figure 4-4B,

$$r = R \tan \frac{\rho}{2} \qquad \text{(Eq. 4-1)}$$

where R equals the radius of the reference sphere. If ρ equals zero degrees, P' plots at the circle's center (whereupon any value of ϕ pertains because P' is then common to all radii of the circle). If ρ equals 90°, P' plots on the circumference. For values of ρ intermediate between 0 and 90°, P' would plot somewhere along OS. The distance outward from the circle's center for a particular ρ value can be determined from any of the four radii that terminate at points E, S, W, and N on the stereographic (Wulff) net in Figure 4-5, each radius being divided into 2° intervals of ρ.

The Stereonet

Description and significance

The series of curves joining points N and S of the stereonet (Fig. 4-5) represent great circles, that is, the cyclographic projection of a family of planes that hinge on line NS (Fig. 4-6). The angular interval between them is actually

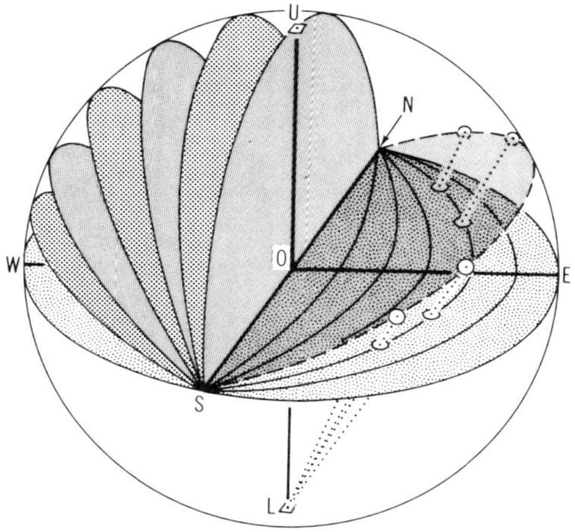

FIGURE 4–6

Significance of the great circles joining points N and S in Figure 4–5. They represent the direct projection of a series of planes hinged on line NS and at varying angles to the (stippled) plane of projection. Four points on the dashed great circle representing the intersection between the gray transparent plane and the sphere of projection have been projected onto the plane of projection to illustrate how the dashed great circle would be transferred to this plane.

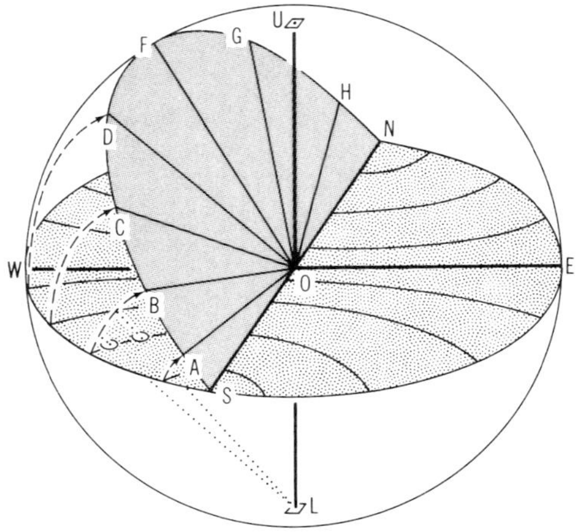

FIGURE 4–7

Significance of the small circles in Figure 4–5. If the shaded plane is rotated on *NS* as a hinge, the lines within it (*OA*, *OB*, . . . , *OH*) intersect the reference sphere so as to describe small circles (dashed). If all points on these small circles are transferred to the equatorial plane by the standard stereographic method, the small circles of the stereonet result.

2°, but they are shown only at 15° intervals in Figure 4-6. With one exception these planes have been omitted from the upper right quadrant of the sphere to show better their cyclographic projections on the equatorial plane. The dashed great circle which represents the transparent plane in Figure 4-6 is projected onto the equatorial plane in detail.

A second set of curves on the stereonet is roughly transverse to the great circles just described. To understand their significance, we shall assume that semicircle *SFN* (Fig. 4-7) contains 90 equally spaced radii, a few of which (for example, *OS*, *OA*, *OB*, *OC*) are drawn in Figure 4-7. If semicircle *SFN* is rotated around hinge line *NS* its radii trace out a series of small circles[2] (dashed) on the sphere surface. If each point on these small circles is projected onto the stippled plane, the solid curves shown in Figure 4-7 result.

Practical use

Remove Figure 4-5 from the book and mount it on a square of heavy cardboard; pierce its precise center with a thumbtack whose point has been ground very sharp and thin (Fig. 4-8A). On a sheet of tracing paper draw two, mutually perpendicular, intersecting lines. Pierce their intersection with a needle,

[2] A small circle may be defined as any circle on a sphere's surface whose radius is less than that of the sphere.

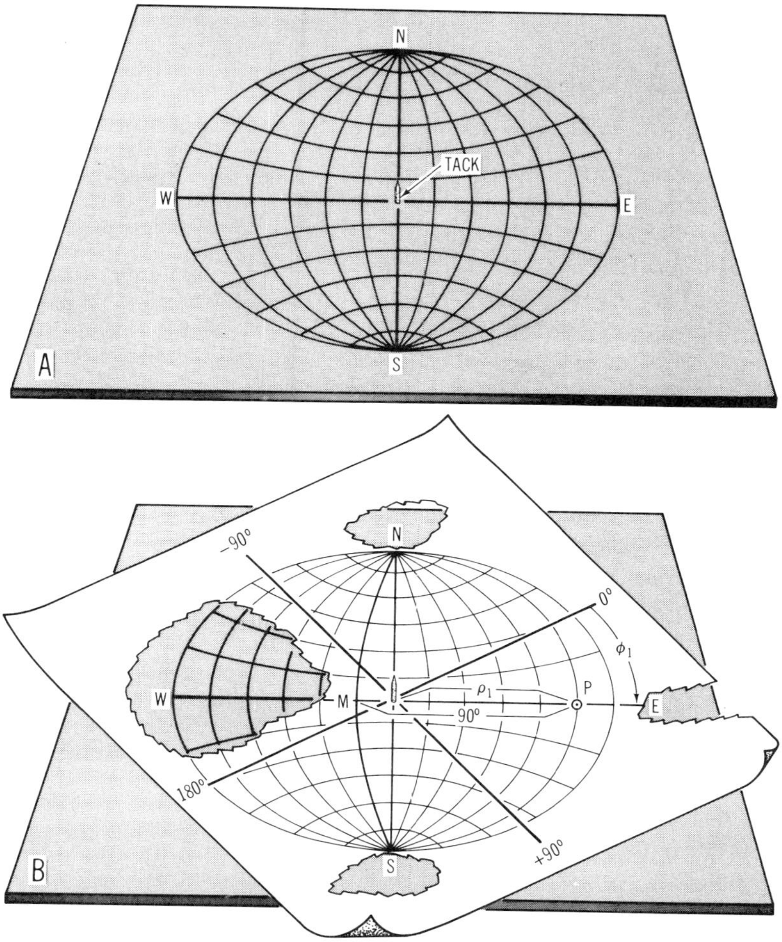

FIGURE 4–8

(A) The stereonet (shaded gray for illustrative purposes) made ready for use by insertion of a fine-pointed tack through its center.
(B) Insertion of the stereonet's tack through the intersection of two perpendicular lines on a sheet of tracing paper. The ends of the lines are labeled with their ϕ values (0°, 90°, 180°, and −90°). The paper has been rotated above the stereonet until diameter *WE* of the underlying stereonet makes an angle of ϕ_1 (=30°) with the paper's zero-degree line, the circumference of the stereonet serving as a protractor. A point *P* whose coordinates are ϕ_1 and ρ_1 may then be plotted on the overlying paper by counting a number of degrees equal to ρ_1 outward from the tack toward *E*. If *P* represents the face pole of a stereographically plotted crystal plane, point *M*, which is 90° from *P* on diameter *WE*, marks the great circle *NMS* that is the cyclographic projection of this crystal plane.

label their ends $\phi = 0°$, $90°$, $-90°$, and $180°$, and impale this tracing paper on the stereonet's tack (Fig. 4-8B) using the previously made needle hole as a guide. To plot a point P ($\phi_1 = 30°$, $\rho_1 = 75°$) on the tracing paper, rotate the paper above the stereonet until the stereonet's radius OE makes an angle of $30°$ ($= \phi_1$) with the $\phi = 0°$ line on the paper. Looking through the tracing paper, count off $75°$ ($= \rho_1$) outward along radius OE and mark the point thus located (that is, point P) with a small pinhole in the tracing paper.

If this point P represents the stereographic projection of a crystal plane whose spherical plot is on the upper hemisphere, draw a tiny circle around the pinhole (both to indicate what P is and to make the pinhole easier to find). If point P had represented a face projected onto the lower hemisphere (in which case its coordinates would be $\phi_1 = 30°$ and $\rho_1 = -75°$), we would draw a small cross over the pinhole instead of encircling it.

To plot the *cyclographic* projection of a crystal plane, if its ϕ and ρ coordinates are known, simply plot its face pole P as just described, whereupon point P will overlie diameter WE of the stereonet (Fig. 4-8B). Now looking through the tracing paper, count off $90°$ from P along stereonet diameter WE to locate point M. The great circle that intersects WE at point M may now be traced onto the tracing paper (see heavy curve NMS in Fig. 4-8B) to mark the cyclographic projection of the plane whose face pole was P.

In Figure 4-8B point P and great circle NMS represent, respectively, the stereographic and the cyclographic projections of one and the same plane. Alternatively point P could represent the cyclographic projection of a line, whereas great circle NMS could represent the stereographic projection of this same line. If P in Figure 4-8B did represent the cyclographic projection of a line (in which case the pinpoint marking P's precise location would be encircled by a tiny square rather than a circle so as to mark it as the cyclographic projection of a line) then this line's stereographic projection (that is, great circle NMS) could be located and drawn on the tracing paper by using the procedure described in the preceding paragraph. For example, we again rotate the tracing paper until P overlies diameter WE, then count $90°$ from P along WE (Fig. 4-8B) to locate great circle NMS. It is instructive to imagine that P represents the cyclographic projection of a steel rod and that welded perpendicular to this rod is a flat plate which projects cyclographically as great circle NMS.

Given the cyclographic projection of two directions within a crystal (for example, Q and R in Fig. 4-9), the cyclographic projection of the crystal plane that contains or is parallel to both these directions is obtained by rotating the tracing paper so that both points fall on the same great circle. If this great circle (curve $NQRS$ in Fig. 4-9) is traced onto the paper, it represents the cyclographic projection of the desired crystal plane. To understand this, we note that the cyclographic projection R' of a line such as OR in Figure 4-2 necessarily falls on the cyclographic projection (dashed great circle, Fig. 4-2) of a plane that contains this line. Thus just as two lines determine a plane, their cyclographic projections fall on the great circle that represents this plane cyclographically.

Similarly, the angle θ between two directions cyclographically projected as points Q and R (in Fig. 4-9) can be measured by rotating the underlying stereonet

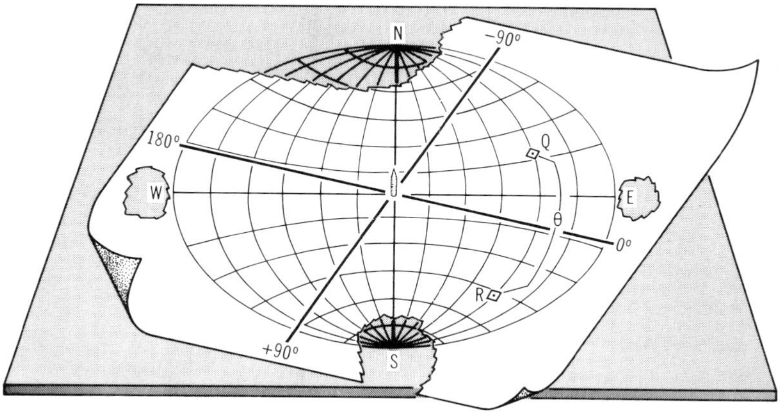

FIGURE 4–9

Use of the stereonet to measure θ, the angle between two points Q and R. The tracing paper has been rotated until the two points Q and R lie on the same great circle of the underlying stereonet. The angle θ is then determined by counting the degree divisions between Q and R along this great circle.

until both points fall along the same great circle. The portion of this great circle lying between points Q and R is divided (by its intersections with the small circles) into 2° intervals on the stereonet. Counting the number of these intervals between points Q and R determines the value of θ. If Q and R had represented face poles, θ would denote the interfacial angle between the crystal faces they represented.

Crystal Projection

Standard orientations

In projecting the faces or axes of a crystal, the crystallographic axes are usually assigned a standard orientation with respect to the sphere of projection. For example, a crystal's vertical axis (a_3 in isometric, c in all other systems) must always coincide with the sphere's polar axis (UL in Fig. 4-3) and will therefore plot at the center of the stereogram (Fig. 4-10). Similarly, its a or a_1 axis (except for hexagonal crystals) is always made to lie within the $\phi = +90°$ plane as Figure 4-10 illustrates. For hexagonal crystals oriented in what is known as the G_2 position, the $+a_1$ axis has a ϕ angle of 120°.[3]

If a crystal's axial ratio and interaxial angles are known, and if, as just discussed, it is oriented in the standard orientation for the system to which it

[3] In the less used G_1 position (not shown in Fig. 4-10) the $+a_1$ axis is at ϕ equal 60°.

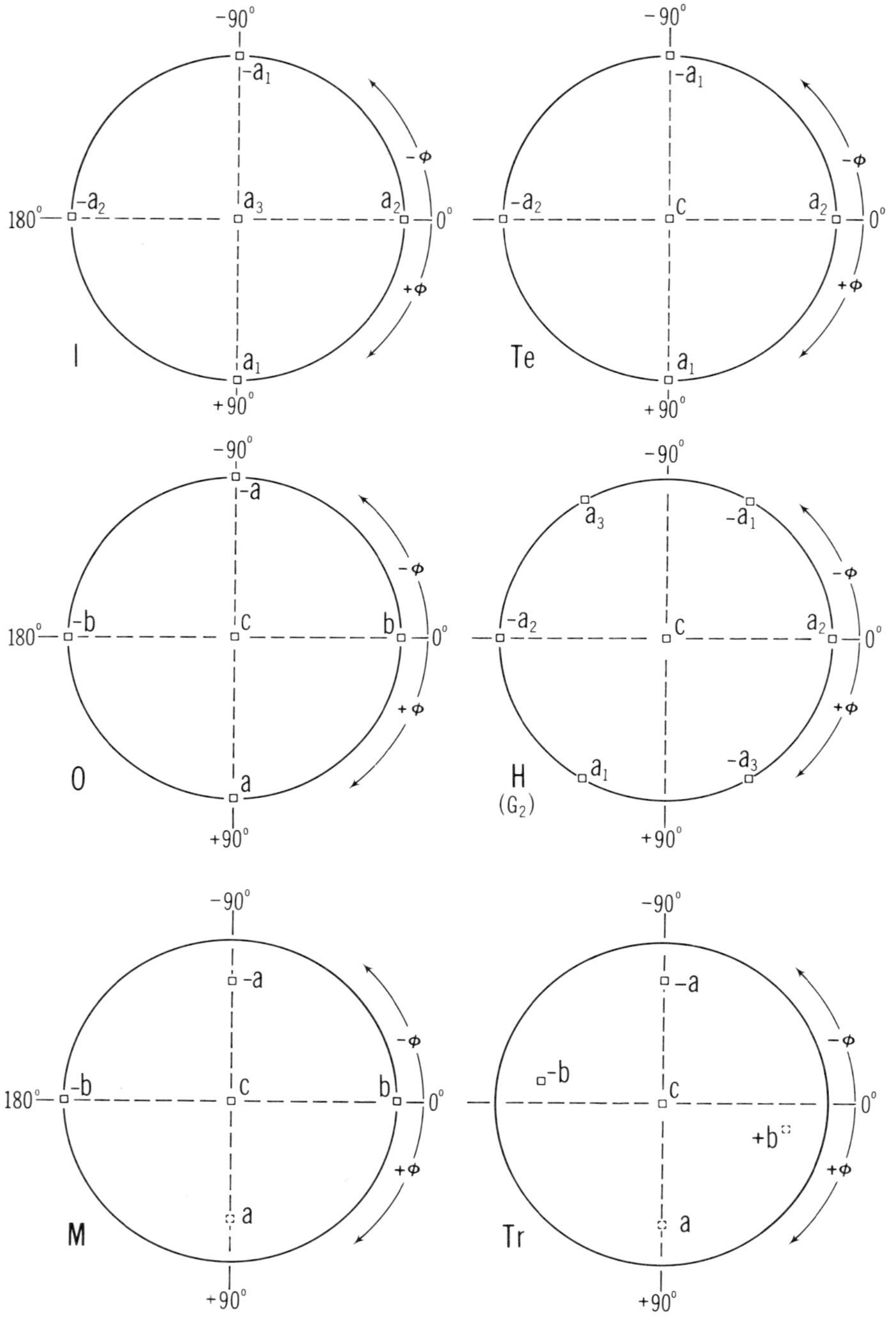

FIGURE 4–10

Standard coordinates for the crystal axes (cyclographically projected) in the isometric system (I), tetragonal system (Te), orthorhombic system (O), hexagonal →

TABLE 4–1

Formulas for Computation of Tan ϕ and ρ for a Face from Its Miller Indices (*hkl*) and from the Axial Elements

Tan ϕ	Tan ρ	
	$k \neq 0$	$k = 0$
	Monoclinic	
$\dfrac{hc/a + l\sin(\beta - 90)}{kc \sin\beta}$	$\dfrac{kc}{l\cos\phi}$	$\dfrac{hc/a + l\sin(\beta - 90)}{l\sin\phi\cos(\beta - 90)}$
	Orthorhombic	
$\dfrac{h}{ak}$	$\dfrac{kc}{l\cos\phi}$	$\dfrac{hc}{la\sin\phi}$
	Tetragonal	
$\dfrac{h}{k}$	$\dfrac{c}{l}\sqrt{h^2 + k^2}$	
	Hexagonal (G_2)	
$1.1547\left(\dfrac{h}{k} + 0.5\right)$	$\dfrac{kc}{l\cos\phi}$	$\dfrac{hc}{l\sin(\phi - 30)}$
	Isometric	
$\dfrac{h}{k}$	$\sqrt{\dfrac{h^2 + k^2}{l^2}}$	

belongs, then the ϕ and ρ coordinates can be calculated for any face whose Miller indices (hkl) are known. The formulas expressing these relationships are given in Table 4-1 for all crystal systems except the rather complex triclinic case. Thus the ϕ and ρ coordinates for a (111) face, for example, would differ on monoclinic crystals of different compositions because their axial elements, $a{:}b{:}c$ and β, would probably differ in value. Note, however, that for isometric crystals, no axial elements exist or are involved in the computations. Thus the equations in Table 4-1

$$\tan\phi = \frac{h}{k} \qquad \tan\rho = \sqrt{\frac{h^2 + k^2}{l^2}}$$

indicate, for example, that the (111) face on all isometric crystals, regardless of composition, invariably has the spherical coordinates $\phi = 45°$, $\rho = 54°44'$.

system (H), monoclinic system (M), and triclinic system (Tr). For monoclinic or triclinic crystals the coordinates of the positive end of the a axis are $\phi = 90°$ and $\rho = \beta$. Because β is almost always an obtuse angle, this positive end therefore projects upon the lower hemisphere and is consequently indicated here by a dashed square rather than a solid square.

Orientations alternative to those above are sometimes used. Thus the above orientation for hexagonal crystals is the so-called G_2 position (c at $\rho = 0°$; a_2 at $\phi = 0°$), but the G_1 position (c at $\rho = 0°$; a_2 at $\phi = -60°$) is also used.

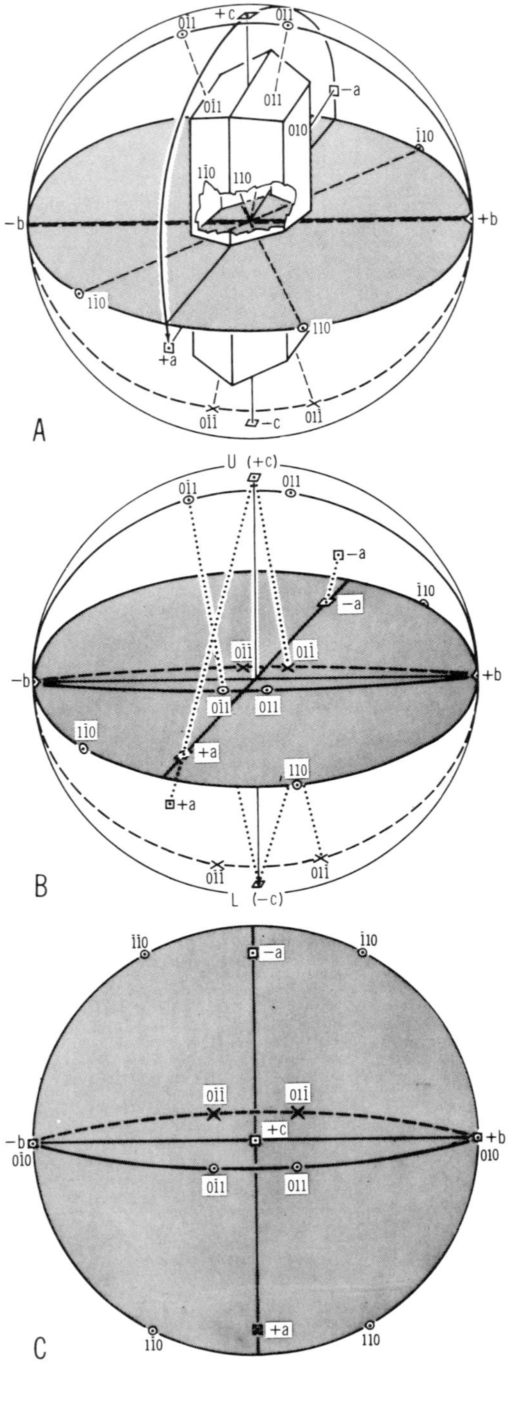
+c
−a
−b
+b
+a
−c
A
U (+c)
L (−c)
B
C

Monoclinic example

Figure 4-11 illustrates the projection on the fundamental sphere of the axes and faces of a monoclinic crystal; the faces are projected reciprocally by the dashed lines (normal to them) that emanate from the sphere's center and intersect the sphere's surface. In Figure 4-11B these intersections have been transferred to the equatorial plane by joining them to the pole of the opposite hemisphere, *U* or *L*, as the case may be. This method of transferral is shown in detail for face poles $(0\bar{1}1)$ and $(01\bar{1})$ and for both ends of the *a* axis, with the imaginary lines joining them to the opposite pole drawn as dotted lines. Faces or lines that were originally projected on the upper hemisphere are symbolized, respectively, with tiny circles or squares; faces or lines originally on the lower hemisphere are symbolized, respectively, as tiny crosses or dashed squares. Similarly, a great circle on the upper hemisphere such as the one containing face poles $(0\bar{1}1)$ and (011) in Figure 4-11 will be shown by a solid line, whereas one on the lower hemisphere such as that containing face poles $(0\bar{1}\bar{1})$ and $(01\bar{1})$ will be shown by a dashed line. Figure 4-11C represents the equatorial plane of Figure 4-11B tilted into the plane of the paper; it is therefore the *stereogram* of the crystal shown in Figure 4-11A. Note that the *b* axis and the normals to faces (110), $(1\bar{1}0)$, $(0\bar{1}0)$, $(\bar{1}\bar{1}0)$, and $(\bar{1}10)$ all lie within the equatorial plane (in Fig. 4-11A) and thus intersect the sphere at points along its equator. Consequently, these points retain their same positions after transferral to the equatorial plane (Fig. 4-11B, C).

Zone planes and zone circles

Six faces, (011), $(0\bar{1}1)$, $(0\bar{1}0)$, $(0\bar{1}\bar{1})$, $(01\bar{1})$, and (010), from the crystal in Figure 4-11A are redrawn in isolation in Figure 4-12A. These six faces all lie in the same zone, their zone axis being [100], that is, the *a* axis. Because all six are parallel to the same line [100], their normals, if extended outward from the crystal's center, lie in a plane (stippled in Fig. 4-12A) called the *zone plane*. *This zone plane is always perpendicular to its corresponding zone axis*, the *a* axis in this case, because it contains lines that are perpendicular to this zone axis. A zone plane intersects the sphere of projection in a great circle called the *zone circle*. Thus the six face poles in Figure 4-12B all plot on the same zone circle.

← FIGURE 4-11

(A) Spherical projection of a monoclinic crystal. The dashed lines normal to the various faces emanate from the sphere's center and intersect the sphere's surface where shown.
(B) Transferral of the spherically projected points to the equatorial plane is shown in detail for planes $(0\bar{1}1)$ and $(01\bar{1})$ and for both ends of the *a* axis.
(C) View of the equatorial plane as seen looking down *UL*; this is the "stereogram" of the crystal.

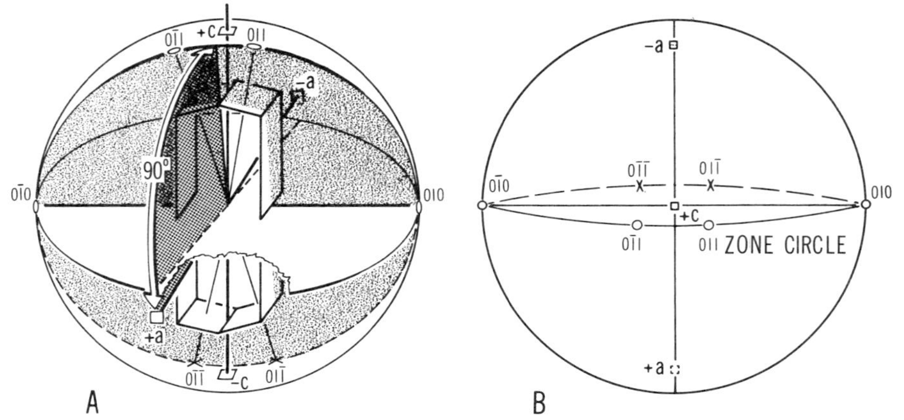

FIGURE 4–12

(A) The six faces, each parallel to the *a* axis in Figure 4–11A and thus all lying in the same zone, are here drawn by themselves. The normals to these faces lie in (and define) a plane called the *zone plane* (stippled), which is perpendicular to the *a* axis and may thus be regarded as the reciprocal projection of this axis.
(B) The poles for the six faces projected stereographically fall along a great circle called their *zone circle,* because it represents the direct projection of their zone plane or, alternatively, the reciprocal projection of axis *a,* their zone axis. The dashed great circle was projected from the lower hemisphere.

Plotting zone circles and zone poles

The zone circle can be alternatively regarded as (1) the direct projection of the zone plane or (2) the reciprocal projection of the zone axis perpendicular to this zone plane. Thus in Figure 4-12B the *direct* projection of zone axis [100] onto the upper hemisphere is marked by the small square denoting the negative *a* axis, whereas the *reciprocal* projection of [100] is the zone circle shown. Given a zone axis's direct projection, that is, its *zone pole,* the corresponding zone circle is readily located. We first rotate the tracing paper until the given zone pole lies above the *E-W* diameter of the underlying stereonet (Fig. 4-13). From the given zone pole we now count 90° along *E-W* to locate, transverse to *E-W*, the desired zone circle, tracing it onto the paper as in Figure 4-13. Conversely, to locate the zone pole if given the zone circle, we rotate the tracing paper until this zone circle coincides with a great circle on the underlying stereonet. Having done this, the situation is as in Figure 4-13 provided the zone pole were erased. From where the given zone circle intersects the stereonet's *E-W* diameter we count 90° along *E-W* to locate the desired zone pole, which would thus plot at the site shown in Figure 4-13.

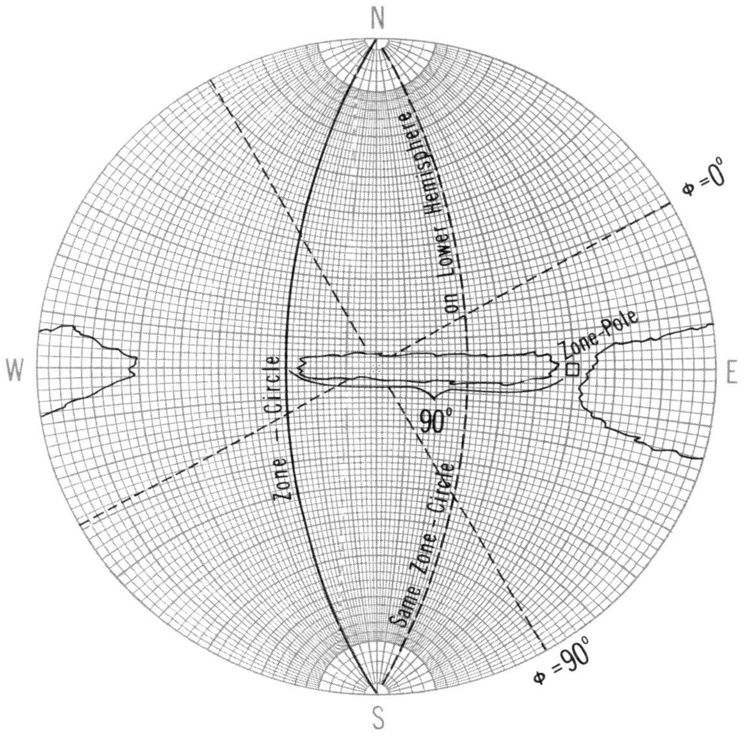

FIGURE 4–13

To locate a zone circle corresponding to a given zone pole, rotate the paper above the stereonet until the given zone pole falls on the stereonet's *E–W* diameter. Now count 90° along *E–W* toward (and beyond) the stereonet's center to locate the point of intersection between *E–W* and that great circle of the underlying stereonet that represents the desired zone circle; this great circle can then be traced onto the overlying paper (as done here).
If we equate the direction represented by the zone pole to line *OP* in Figure 4–1, we note that the plane normal to it (= plane *A* in Fig. 4–1) will also intersect the lower hemisphere of the sphere of projection. To locate the zone circle on the lower hemisphere, count off 90° along diameter *E–W*, starting again at the zone pole but this time proceeding *away* from the stereonet's center. Upon reaching the stereonet's circumference at *E*, the rest of the 90° is counted off from *E* back toward the stereonet's center. The great circle (dashed) that passes through the point so located represents the zone circle on the lower hemisphere.

Triclinic example

If the interaxial angles are known for a crystal, its crystallographic axes may be plotted either directly as zone poles, or reciprocally as zone circles. The face pole for any face parallel to a given axis must plot along the zone circle that is the reciprocal projection of this axis. Construction of the stereographic pro-

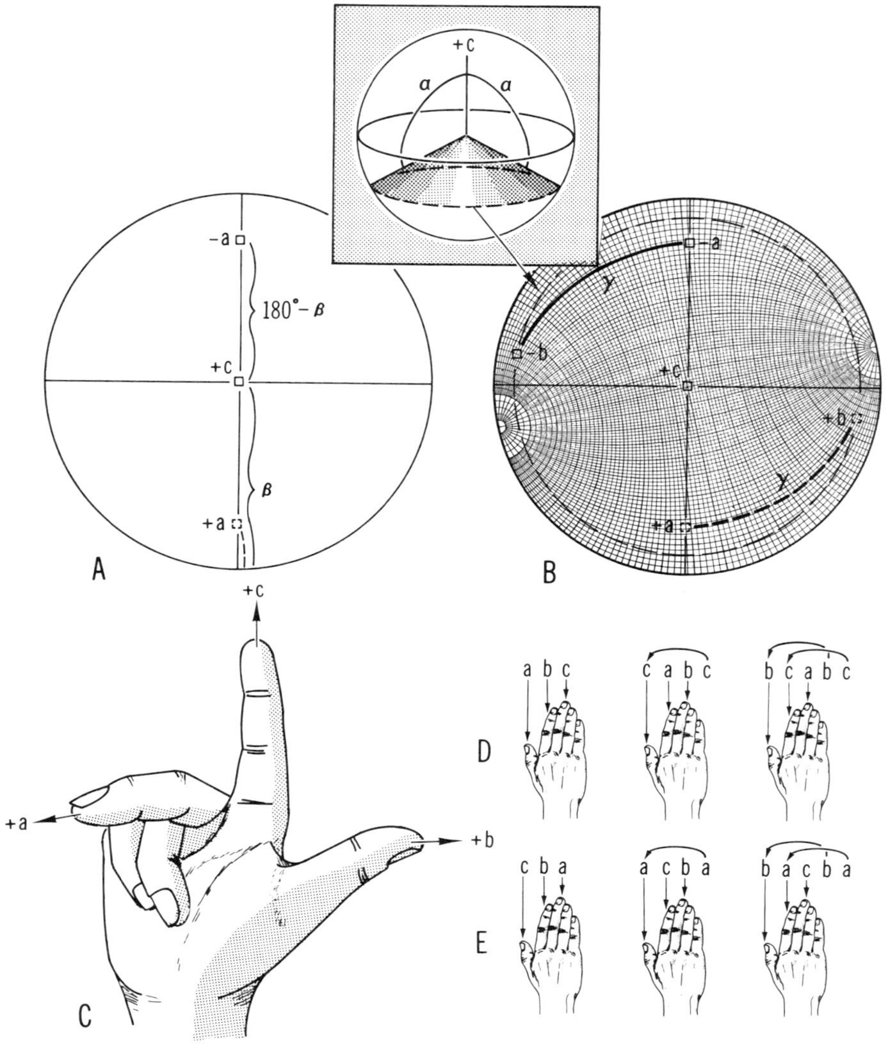

FIGURE 4–14

Plotting the direct projections of a triclinic crystal's axes provided the interaxial angles are known.

(A) The c and a axes are easily plotted because by convention the c axis plots at $\rho = 0°$ and the a axis plots with its positive and negative ends at $\phi = 90°$ and $\phi = -90°$, respectively. The ρ coordinate for the $+a$ axis is β; for $-a$ it is $180° - \beta$.

(B) The cone illustrated in the square inset contains all lines that are at an angle α to the $+c$ axis. Because α exceeds 90° (the usual case), this cone projects cyclographically onto the lower hemisphere and is hence represented by the small circle (dashed) centered around $+c$. The $+b$ axis necessarily plots on this small circle (because $+b \wedge +c = \alpha$). We can locate the specific position of $+b$ on this small jection (*stereogram*) of a crystal of chalcanthite, $CuSO_4 \cdot 5H_2O$ (triclinic: α, 97.5°; β, 107°; γ, 77.5°) will clarify the foregoing concepts. Following the standard orientation for a triclinic crystal, chalcanthite's c axis is plotted at $\rho = 0°$, whereas its a axis is plotted along the $\phi = 90°$ line at a point where it makes an angle of 107°, that is, β, with the c axis (Fig. 4-14A). Because the angle β exceeds 90° for chalcanthite, the $+a$ axis plots on the lower hemisphere, whereas the $-a$ axis plots on the upper hemisphere (at $\phi = -90°$ and at an angle of 180°-β from the c axis). Locating the cyclographic projection of the b axis is more difficult. However, the $+b$ axis is at an angle α to the c axis; thus a small circle can be drawn around the c axis so that it (the dashed circle in Fig. 4-14B) contains the zone poles for all lines that make an angle of α with the c axis. If $\alpha = 90°$, this circle would coincide with the circumference of the stereonet. However, α actually equaled 97.5°, so that the circle corresponding to $\rho = 97.5°$ falls 7.5° to the inside of the stereonet's circumference as measured along its *E-W* diameter. In short this circle plotted on the lower hemisphere and for that reason was indicated by a dashed line. The $+b$ axis must plot somewhere on this dashed circle or else α cannot equal 97.5°. The $+b$ axis also makes an angle γ (77.5° for chalcanthite) with the $+a$ axis. Thus to locate where the $+b$ axis plots on the dashed circle, we rotate the tracing paper until a great circle of the underlying stereonet (1) passes through the $+a$ zone pole and (2) intersects the lower right quadrant of the dashed circle at an angle of 77.5°, as counted along this great circle (the bold dashed line in Fig. 4-14B). The point of intersection then represents the zone pole for the $+b$ axis, which here plots on the lower hemisphere.

The zone poles corresponding to the a, b, and c axes of chalcanthite are replotted in Figure 4-15A. The zone circles corresponding to each of these zone poles have been added using the technique already discussed. The zone circle corresponding to the a axis thus contains all face poles of the type $(0kl)$ (that is, of all faces parallel to the a axis) and may be called the $(0kl)$ zone circle. Similarly, the zone circle corresponding to the b axis contains all $(h0l)$ face poles,

---

circle by rotating the underlying stereonet (as shown) until one of its great circles (here copied onto the paper as a bold, dashed line) simultaneously (1) passes through $+a$ and (2) intersects the small circle at a point whose angle from $+a$ as counted along this great circle equals γ. This latter point locates $+b$. Diametrically opposite from $+b$ but equidistant from the stereonet's center locate $-b$, which will be on the upper hemisphere because $+b$ was on the lower.

The foregoing location of $+b$ at $\phi = 10°$ (ca) yields a *right-handed set of axes* as is demanded by convention. To test a set of axes for right handedness, point the thumb and adjacent two fingers of your right hand as nearly as possible along the positive directions for these axes (as in C). These fingers, labeled like these axes and then unbent, will correspond to one of the three permutations in (D) for a right-handed set, but to (E) if left handed. Although a construction like that which located $+b$ at $\phi = 10°$ would also locate $+b$ at $\phi = 170°$ on the small circle in (B), this second solution is rejected because the resultant set of axes would not be right handed.

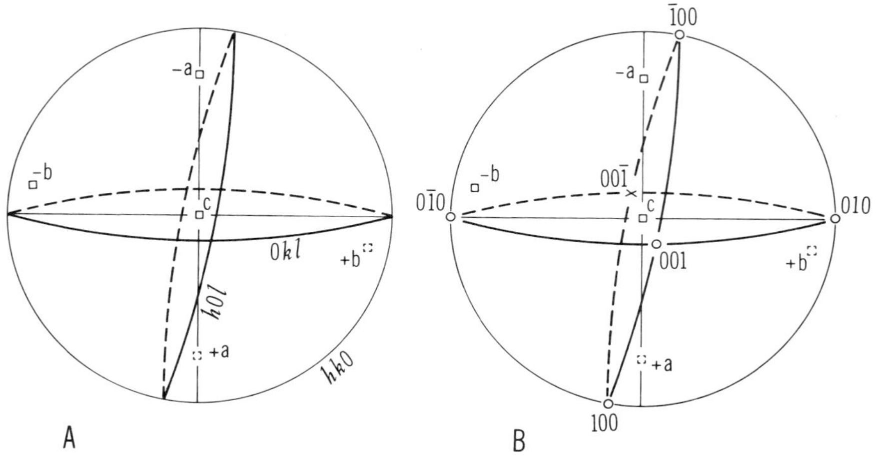

FIGURE 4–15

(A) The great circles labeled $(0kl)$, $(h0l)$, and $(hk0)$ represent, respectively, the zone circles corresponding to axes a, b, and c. The face poles for all faces parallel to a will thus fall along great circle $(0kl)$ or, if their poles fall on the lower hemisphere, along its dashed continuation on the lower hemisphere.
(B) The face poles for faces parallel to *two* crystallographic axes, that is, (100), (010), (001), and so on, must plot at the intersection between the *two* zone circles that correspond to the two axes to which the face is parallel.

whereas the zone circle corresponding to the c axis, which always coincides with the circumference of the stereonet, contains all $(hk0)$ face poles. The intersections between the zone circles corresponding to the a, b, and c axes (Fig. 4-15B) then mark the locations of face poles which, because they are common to *two* zone circles, are parallel to *two* crystallographic axes. Hence they have indices of the type $(h00)$, $(\bar{h}00)$, $(0k0)$, $(0\bar{k}0)$, $(00l)$, or $(00\bar{l})$. Dividing through by the common factor, h, k or l, as the case may be, these indices could be better written (100), ($\bar{1}$00), (010) (0$\bar{1}$0) (001) and (00$\bar{1}$). Note that face poles of all faces of the type $(0\bar{k}l)$ will plot along the great circle connecting (0$\bar{1}$0) to (001) in Figure 4-15B (see Question 4-2).

FIGURE 4–16

(A) Perspective view of three directions of interest in a crystal, that is, heavy lines, OX, OY, OZ; direction OZ is at 90° to the b axis. If this crystal, which itself has not been drawn, were rotated around its b axis until OZ became perpendicular to the (stippled) plane of projection, OZ would locate at OZ_r, OY at OY_r, and OX at OX_r. The opposite ends of lines OX and OX_r are labeled OX' and OX'_r.
(B) Projected positions of OX, OY, OZ, and the b axis in their original orientation in respect to the plane of projection.
(C) The stereonet located in position below the tracing paper—that is, with its

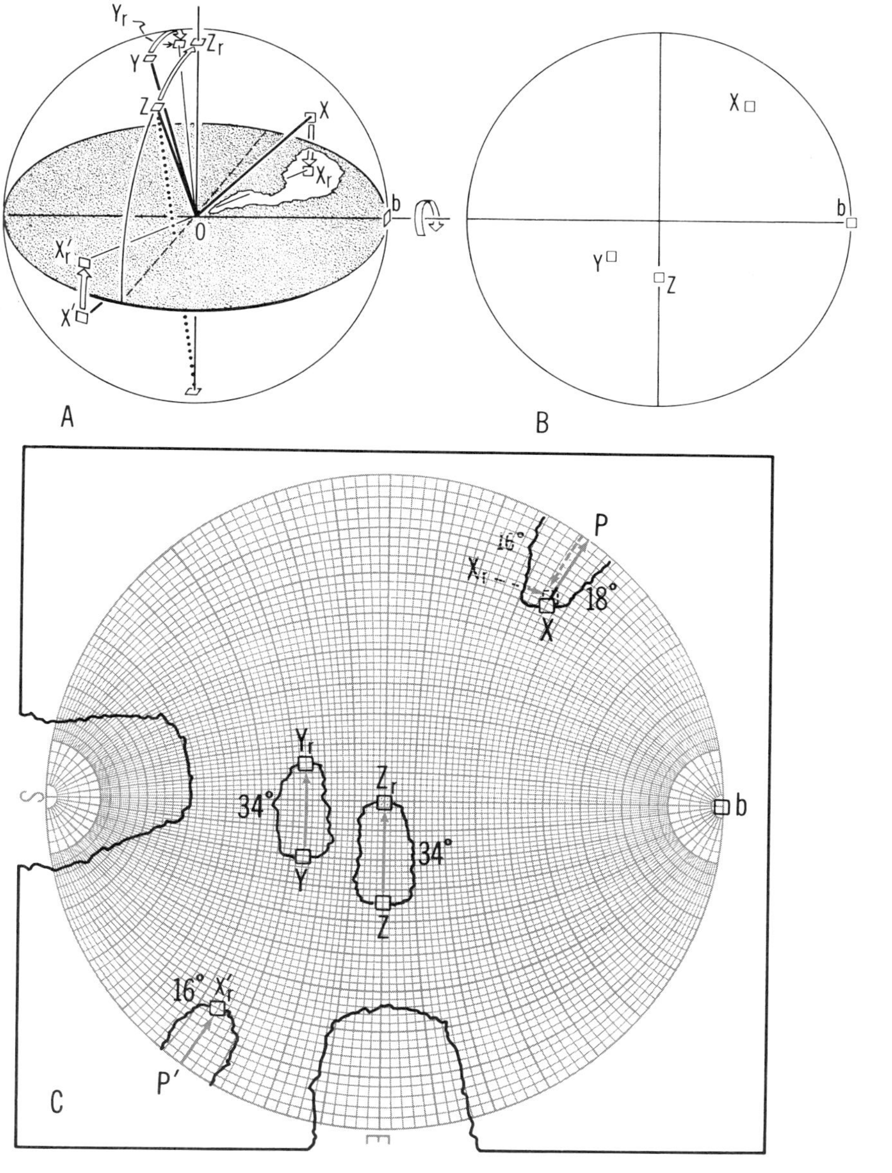

WE diameter below point Z—to permit Z to be rotated 34° to Z_r whereupon Y and X also rotate 34°, each along the small circle of the stereonet below it, to occupy sites Y_r and X_r. X_r is on the lower hemisphere; however, its opposite end X_r' is on the upper hemisphere. Points P and P' mark the ends of OX after only 18° of the 34° rotation had been effected at which instant line OX lay in the plane of projection.

Further Operations with the Stereonet

Rotation of points

The b axis and a few additional directions OX, OY, and OZ for an invisible crystal are shown in perspective in Figure 4-16A and in projection in Figure 4-16B. However, if the crystal were rotated around its b axis to make OZ normal to the plane of projection, and thus at $\rho = 0°$, the ends of its axes would describe arcs on the sphere as illustrated by the arrows in Figure 4-16A. Only the b axis would be unchanged in position because it happens to be the axis about which the crystal was rotated. The stereogram in Figure 4-16B can be converted to the rotated position by revolving it on the tack until Z, the direction to be made normal to the projection plane, overlies the E-W diameter of the underlying stereonet (Fig. 4-16C). Thus as Z rotates 34° to its central position Z_r, the other directions also rotate 34° along the small circles on which they happen to fall. The direction X, after only 18° of this 34° rotation, is horizontal, one end being at P and its opposite end X' being at P'. After the full 34° rotation, therefore, its X' end travels the remaining 16° along a small circle on the upper hemisphere to occupy the site X_r'. At the same time its X end travels this remaining 16° along a small circle on the lower hemisphere that happens to coincide with the small circle on the upper hemisphere already labeled 18°.

Construction of small circles that represent the locus of all face poles or lines at an angle θ to a given direction

Line OY (coordinates ρ_Y and ϕ_Y in Fig. 4-17A) has a cone centered about it that contains all lines at an angle θ to OY. Of these lines, OX and OZ have the same ϕ coordinate ϕ_Y as OY, but their ρ coordinates are $(\rho_Y - \theta)$ and $(\rho_Y + \theta)$, respectively. If the circle in which this cone intercepts the sphere is projected stereographically point by point onto the (stippled) plane of projection in Figure 4-17B, a true circle results. This fundamental property of the stereographic projection, that circles on the sphere are also true circles in projection, is ably demonstrated by Phillips (1963, pp. 24–25). Points X and Z in Figure 4-17B thus represent two points on the projection of the small circle that contains all directions at an angle θ to OY. However, the center of this small circle is at the midpoint between X and Z (black dot in Fig. 4-17B); it is *not* at Y, the projected position of OY on the equatorial plane. This is proved by Phillips (1963).

In practice such a small circle in stereographic projection may be constructed by plotting Y (coordinates, ϕ_Y and ρ_Y), X (coordinates, ϕ_Y and $\rho_Y - \theta$), and Z (ϕ_Y and $\rho_Y + \theta$) as shown in Figure 4-18 (where ρ_Y is drawn as 55° and θ is taken to equal 25°). Next bisect the distance between X and Z to locate M, the midpoint of line XZ. The desired small circle can now be drawn with a compass because (1) its radius r equals the half-distance MX and (2) the circle is

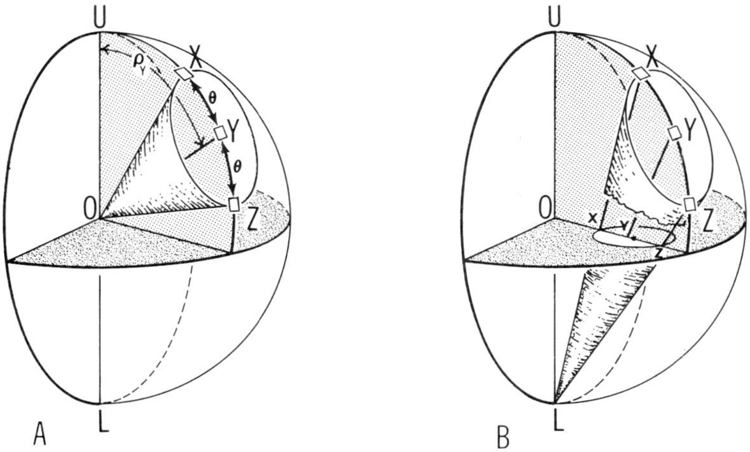

FIGURE 4–17

(A) Direction OY and the cone that contains all lines at an angle θ to it, two of which, OX and OZ, have the same ϕ coordinate as OY. The cone intersects the sphere of projection in a circle.
(B) This circle stereographically projects as a true circle in the plane of projection (stippled); however, Y, the projected site of OY, does not coincide with the black dot marking the center of this projected circle.

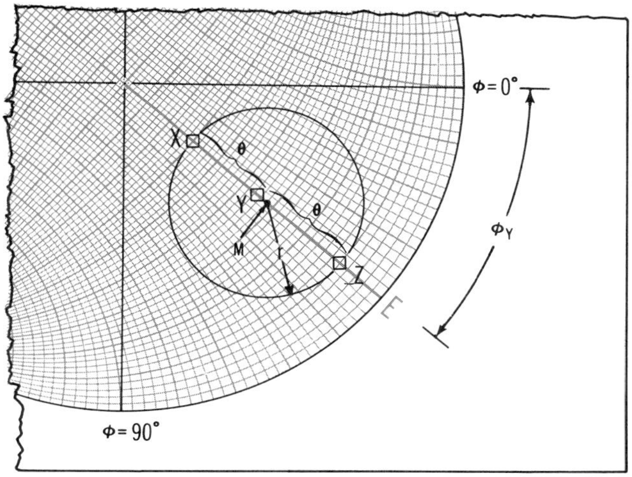

FIGURE 4–18

Construction of the small circle that contains all directions at an angle θ, in this case 25°, to the direction whose stereographic projection occurs at Y. The radius r of this small circle equals one-half of the distance between X and Z.

centered on M. Note that Y, which is at a 25° angle to X (and to Z), does not coincide with midpoint M (except for the special cases when Y falls on the stereonet's center or on its circumference).

Depiction of symmetry relationships

The ten basic symmetry operations, except for 1 and $\bar{1}$ which are directionless, may be plotted cyclographically in stereograms and labeled with their usual graphic symbols. Those great circles that represent symmetry planes in stereograms will be drawn with bolder lines than those that do not represent such planes. With a crystal's symmetry elements and face poles plotted, a stereogram may illustrate (1) the angular relationship of the crystal's faces to its symmetry elements and (2) which of its faces are related to each other by these symmetry elements. In respect to point (1), if a face pole plots at the same site as the cyclographic projection of a symmetry axis, the face must be perpendicular to this axis because the face normal and this axis necessarily coincide. As previously discussed in respect to zones, if a face pole plots on the great circle that represents the reciprocal projection of a symmetry axis, the face must be parallel to this axis. Furthermore, if a face pole plots on *neither* the direct nor reciprocal projection of an axis, the face is at an oblique angle to this axis.

Recognition of symmetry-related face poles

If they represent faces related by a plane of symmetry, two face poles, p and p', may be joined by a great circle (Fig. 4-19) that will intersect the symmetry plane's direct projection in a point I. Measured along this great circle,

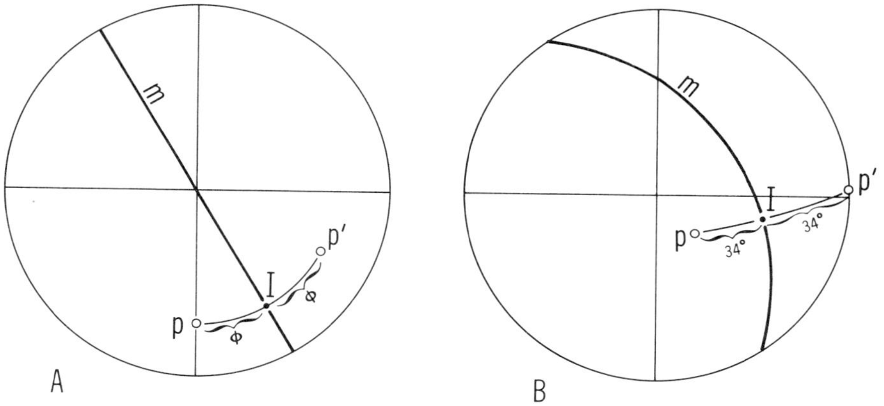

FIGURE 4–19

Two faces p and p' that are related by a plane of symmetry m. Plane m is perpendicular to the plane of projection in (A) but oblique to it in (B). In both cases the great circle joining face pole p to p' intercepts plane m at a point I that is at the same angle to p as to p'.

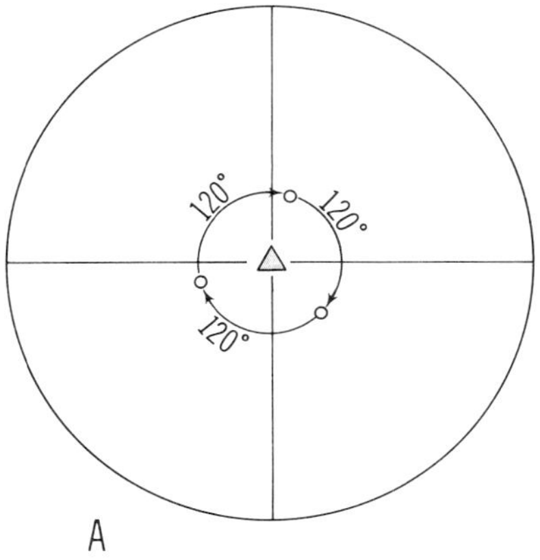

FIGURE 4–20

The face poles for three faces related by a 3-fold axis that is (A) perpendicular and (B) oblique to the plane of projection. In (B) the black dot at the center of the small circle containing the face poles does not quite coincide with the location of the 3-fold axis (for reasons discussed in Figure 4-17B).

point I will be the same number of degrees away from p as from p'. This is true because just as with any two faces related to each other by a plane of symmetry, faces p and p' must each be at the same angle to this plane.

A group of n faces, each related to the other by an n-fold axis of symmetry and therefore each at the same angle θ to this axis, will plot stereographically along the small circle that represents the locus of all face poles at an angle θ to

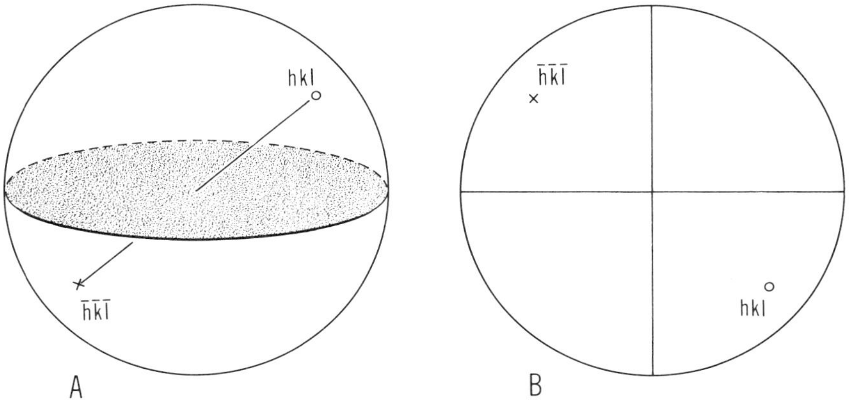

FIGURE 4–21

The normals to two faces that are symmetrically related by a center of symmetry are shown (A) in perspective in respect to the (stippled) plane of projection and (B) in stereographic projection.

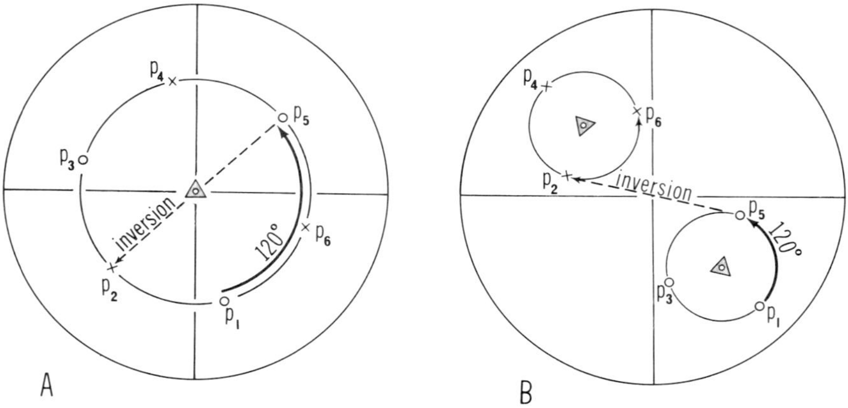

FIGURE 4–22

The face poles for six faces related by a $\bar{3}$ axis that is (A) perpendicular and (B) oblique to the plane of projection. The $\bar{3}$ operation whereby p_1 relates to p_2 combines a rotation of 120° coupled with inversion through the center of the sphere. This same operation relates p_2 to p_3, p_3 to p_4, p_4 to p_5, and so on. In (B) the $\bar{3}$ axis at $\phi = 45°$ has been extended downward so that it intersects the lower hemisphere and consequently also plots at $\phi = -135°$. Thus the $\bar{3}$ symbol in the upper left quadrant of (B) should be construed as being on the lower hemisphere.

this axis. More specifically, their face poles will be spaced along this small circle (Fig. 4-20) so as to conform to the repeat angle inherent to this symmetry axis.

If a center of symmetry $\bar{1}$ is present in a crystal, a face (hkl) will be related by it to a second face $(\bar{h}\bar{k}\bar{l})$ which will be parallel to the first but on the opposite side of the crystal. The normals to these two faces thus pierce the sphere of projection at diametrically opposite points (Fig. 4-21A). Consequently, two face poles related by a center of symmetry will have ϕ coordinates that differ by 180° (Fig. 4-21B) and ρ coordinates that are equal but opposite in sign.

Inversion axes relate one face to another by an operation that combines both rotation and inversion. Thus the $\bar{3}$ axis shown perpendicular to the plane of projection in Figure 4-22A and oblique to it in Figure 4-22B will relate face p_1 to p_2 by a 120° rotation about this axis combined with inversion through the sphere. Face p_2 is related to p_3 by a similar 120° rotation plus inversion through the sphere. This combined operation of rotation plus inversion thus relates p_1 to p_2, p_2 to p_3, p_3 to p_4, p_4 to p_5, p_5 to p_6, and p_6 to p_1.

The Gnomonic Projection

For the gnomonic projection the plane of projection, instead of passing through the center of the sphere of projection (as in the stereographic projection), is tangent to the sphere's upper pole (Fig. 4-23). Thus each face is projected by

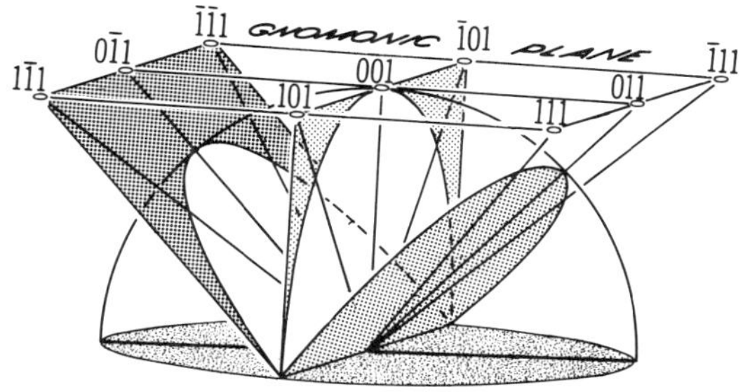

FIGURE 4–23

The normals to nine faces, the faces themselves not being shown, are drawn extended so as to intersect not only the sphere of projection but also a plane tangent to this sphere at its upper pole *U*. Their points of intersection with this tangent plane mark their *gnomonic* projection. The face poles for faces that lie in the same zone, for example, faces (111), (011), and ($\bar{1}11$) whose normals all lie in the same zone plane, plot in gnomonic projection along the same line. In other words, the zone circles of the stereographic projection (Fig. 4–12B) have become straight lines in the gnomonic projection.

extending its normal until it intersects the plane of projection. In Figure 4-23 the normals to the three faces (111), (011), and ($\bar{1}11$) lie in the shaded plane which thus constitutes a zone plane for the zone containing these three faces. We observe that the gnomonic projections of three or more faces lying in the same zone, for example, (111), (011), and ($\bar{1}11$), all lie along the same straight line. In other words, the zone circles of the stereographic projection have become straight lines in the gnomonic projection. In spite of this advantage the gnomonic projection is difficult to work with because small circles on the sphere do not project as circles, as they do in stereographic projection, but rather as relatively complex conic sections. Consequently, for beginning work the stereographic projection is usually favored.

REFERENCE AND RECOMMENDED READING

Phillips, F. C. (1963) *An Introduction to Crystallography*, 3d ed., New York: Wiley.

QUESTIONS AND PROBLEMS

4-1 What are the ϕ and the ρ coordinates of the zone pole in Figure 4-13?

4-2 Any face lying on the great circle joining (100) to (001) in Figure 4-15 would have a generalized Miller index of the type (*h*0*l*). (a) What would be the generalized index for a face plotting on the great circle joining (001) to ($\bar{1}00$)? (b) joining ($\bar{1}00$) to ($00\bar{1}$)? (c) ($00\bar{1}$) to (100)? (d) ($0\bar{1}0$) to ($00\bar{1}$)? (e) ($\bar{1}00$) to (010)?

4-3 Two face poles plot at $\phi = 68°$, $\rho = 60°$ and $\phi = -80°$, $\rho = 40°$, respectively. (a) What is the interfacial angle between these two faces? (b)

What are the ϕ and ρ coordinates for the line of intersection, that is, zone axis, between these two faces?

4-4 Two face poles plot at $\phi = 110°$, $\rho = 50°$ and at $\phi = 70°$, $\rho = -80°$. What is the interfacial angle between them? *Hint*: Observe from Figure 4-15 how a given plane's intersection with the sphere of projection results in a solid-line great circle on the upper hemisphere and a dotted-line great circle on the lower hemisphere.

4-5 Two lines, A and B, have the coordinates $\phi = -45°$, $\rho = 50°$ and $\phi = 10°$, $\rho = 90°$, respectively. (a) What is the angle between them? (b) What are the coordinates of the face pole for a face parallel to both lines?

4-6 The two lines cited in Question 4-5 are rotated bodily by the minimum amount necessary to cause line A to move to the center of the projection (at $\rho = 0°$). After this rotation what would be the new coordinates of (a) line B? (b) the face pole for the plane parallel to both lines?

4-7 In Figure 4-19 what is the angle between the plane of symmetry and the plane that contains the normals to faces p and p'?

4-8 Given a face p whose face pole plots at $\phi = 52°$, $\rho = 30°$. Give the ϕ and ρ coordinates for the face (or faces) symmetrically related to p by (a) a 2-fold axis at $\phi = 180°$, $\rho = 70°$; (b) a $\bar{4}$-fold axis at $\phi = 180°$, $\rho = 70°$.

ANSWERS

4-1 $\phi = 30°$, $\rho = 60°$.

4-2 (a) $(\bar{h}0l)$; (b) $(\bar{h}0\bar{l})$; (c) $(h0\bar{l})$; (d) $(0\bar{k}\bar{l})$; (e) $(\bar{h}k0)$.

4-3 (a) 96°; (b) $\phi = 168°$, $\rho = 73°$.

4-4 63°.

4-5 (a) 64°; (b) $\phi = 100°$, $\rho = 46°$.

4-6 (a) $\phi = 21°$, $\rho = 64°$; (b) $\phi = 111°$, $\rho = 90°$.

4-7 90°.

4-8 (a) $\phi = -128°$, $\rho = -30°$; (b) $\phi = -82°$, $\rho = 68°$; $\phi = 98°$, $\rho = -68°$; $\phi = -128°$, $\rho = -30°$.

5 Crystal Forms and Class Determination

Crystal Forms

Definition

A face on a crystal may be related through operation of the crystal's symmetry axes and/or planes to other faces on this crystal; such an assemblage of symmetry-related faces is called a *crystal form*. The number of faces belonging to a crystal form depends on two factors: (a) the crystal class to which the crystal belongs and (b) the orientation of the original face to the crystal's symmetry axes or planes.

Influence of the crystal class

The influence of the crystal class is seen by ignoring all but a single face (a_1 in Fig. 5-1) on a representative crystal from each of the crystal classes, 1, $\bar{1}$, 2, m, and $2/m$. That from class 1 has no symmetry to operate on a_1; thus face a_1 by itself (Fig. 5-1A) constitutes a form, such a single-faced form being called a *pedion*. The crystal from class $\bar{1}$ (Fig. 5-1B) has a center of symmetry that operates on a_1 to relate it to a parallel face a_2; faces a_1 and a_2 thus constitute a *pinacoid*, which is by definition a form consisting of *two parallel faces*. For the crystal from class 2 (Fig. 5-1C) the 2-fold axis operates on a_1 to relate it to a_2; faces a_1 and a_2 constitute a *sphenoid*, which is by definition a form consisting of *two nonparallel faces related by a 2-fold axis*. For the crystal from class m (Fig. 5-1D) the symmetry plane relates a_1 to a_2; although the resultant form actually resembles a sphenoid, it is called a *dome*, which by definition consists of *two*

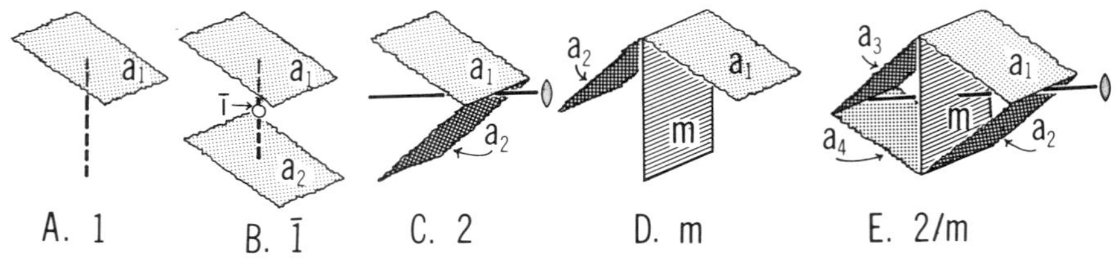

FIGURE 5–1

Repetition of a randomly oriented face a_1 according to whether it is present in (A) class 1, (B) class $\bar{1}$, (C) class 2, (D) class m, or (E) class $2/m$. The resultant collections of symmetry-related faces constitute the following forms: (A) pedion, (B) pinacoid, (C) sphenoid, (D) dome, (E) prism.

nonparallel faces related by a symmetry plane. For the crystal from class $2/m$ (Fig. 5-1E) the 2-fold axis relates a_1 to a_2 and the symmetry plane relates a_1 to a_3 and a_2 to a_4; the resultant form, consisting of the faces a_1, a_2, a_3, and a_4, is called a *prism*, a form defined to consist of *three or more faces all lying in the same zone*; that is, all are parallel to a given line.

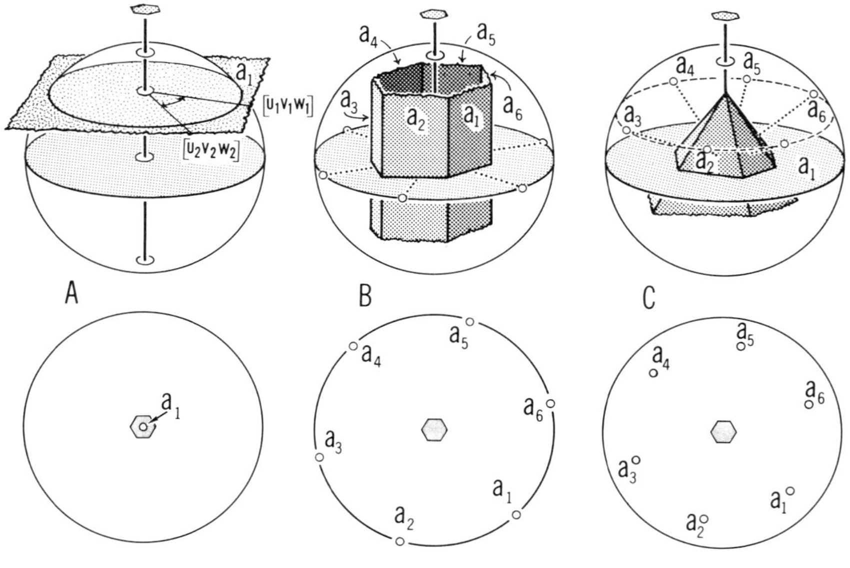

FIGURE 5–2

Comparison of the repetition of a face a_1 according to whether it is (A) perpendicular, (B) parallel, or (C) at a random angle to a 6-fold axis. The resultant forms are a *pedion* (A), a *prism* (B), or a *pyramid* (C); the stereographic projection for each, made with the 6-fold axis normal to the plane of projection, is shown below.
In (A), although it does not repeat face a_1, the 6-fold axis would relate line $[u_1v_1w_1]$ in plane a_1 to five additional lines at 60° intervals in this plane, $[u_2v_2w_2]$ being the only one of the five here drawn.

Influence of orientation

If a face is perpendicular to a proper symmetry axis or to a symmetry plane, the operation of the axis or the plane on the face will not require any additional faces to be present. For example, a face perpendicular to the 6-fold axis in a crystal from crystal-class 6 (Fig. 5-2A) would not be repeated by the axis's operation, face a_1 thus being a pedion. On the other hand, however, if face a_1 is parallel to the 6-fold axis, the 6-fold axis' operation relates a_1 to additional faces $a_2 \ldots a_6$ to produce a prism (Fig. 5-2B) or if a_1 had been oblique to the 6-fold axis, a *pyramid* (Fig. 5-2C). By definition *a pyramid is any form consisting of* 3 *or more nonparallel faces, all capable of intersecting at a common point.* If the 6-fold axis in Figure 5-2C were replaced by a 4-fold axis, this 4-fold axis' operation on a_1 would "generate" a 4-faced pyramid just like one of the great pyramids of Egypt.

As demonstrated in Figure 5-2A, if a face is perpendicular to a proper axis, this axis can relate it to no other faces. Similarly, if a face a_1 is perpendicular to a mirror plane as in Figures 5-3A, B, the mirror plane cannot relate this face to

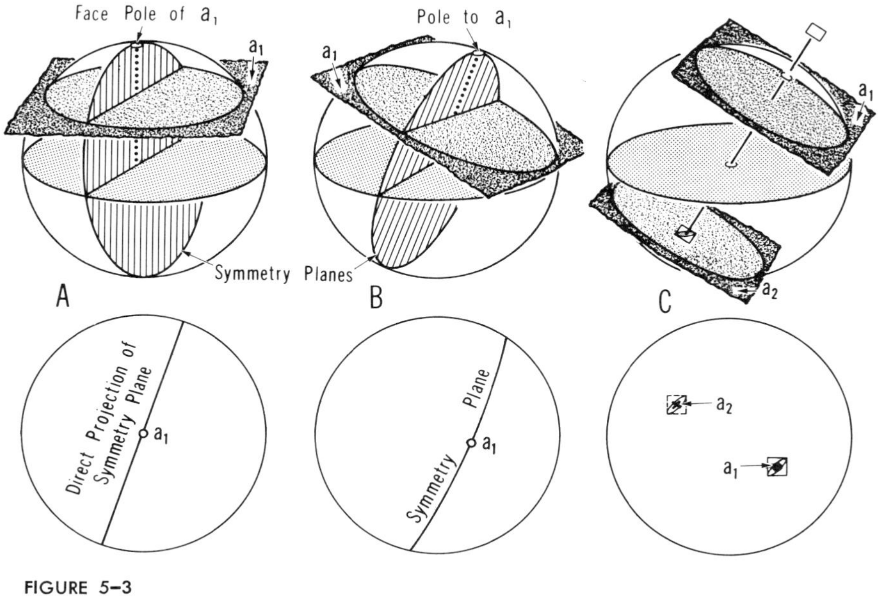

FIGURE 5–3

The stippled face a_1 is perpendicular to, and therefore not repeated by, the ruled symmetry plane in (A) and (B). This symmetry plane is drawn normal to the plane of projection for (A) but not for (B). Their stereographic representation is shown below (A) and (B). However, if a face a_1 is normal to an improper axis such as the $\bar{4}$ axis in (C), this axis's operation requires the presence of a second and parallel face a_2. Note that a face's stereographic projection falls on the cyclographic projection of any symmetry plane or axis to which it is perpendicular (as shown in C).

any other faces. In contrast, if a face a_1 is perpendicular to an improper axis, the inversion operation inherent to this improper axis will relate a_1 to a second face a_2 which will be parallel to a_1 (Fig. 5-3C).

In stereographic projections (Fig. 5-3, lower) *if a face is perpendicular to a symmetry axis or plane, its face pole always falls on the direct projection of this axis or plane.* An unfortunate side effect of this is that the face pole for a face normal to a 3-fold axis will plot at the center of the triangle that represents this 3-fold axis, the resultant combination closely resembling the standard graphic symbol for a $\bar{3}$ axis. Consequently, to avoid such ambiguity a triangle containing a bold minus sign, either at its center or above it, is substituted as the graphic symbol for a $\bar{3}$ axis in this chapter. A triangle with a circle at its center will thus represent a 3-fold axis plus the pole for a face that is perpendicular to it.

Special and general forms

For each crystal class, starting with the least symmetric, we shall consider the interaction of the symmetry elements for this class with the faces (001), then (010), then (100), (011), (0*kl*), (101), (*h*0*l*), (110), (*hk*0), (111), (*hhl*), (*hll*), and, finally, (*hkl*), in that order. This particular order is such that a face cited earlier in the list will likely bear the special relationship of perpendicularity or parallelism to a greater number of the symmetry elements of the class than a face cited later in the list. By comparison, the last face in the list (*hkl*) bears the most general angular relationship to the class's symmetry elements. In other words, an (*hkl*) face will not be perpendicular or parallel to a single symmetry element regardless of which crystal class is being considered. Consequently, face (*hkl*) and the faces that are related to it by the symmetry present in the crystal class constitute a form called the *general form*. On the other hand, a face perpendicular to one or more of the symmetry elements present in the crystal class plus the faces related to it by these symmetry elements constitute a *special form*. For most crystal classes the general form consists of a larger number of faces than do any of the several special forms possible for this class.

Form symbols

When the indices for one face of a form are enclosed in curly braces, they symbolize the entire form. For example, {111} represents face (111) plus all other faces symmetrically related to (111) by the symmetry axes or planes inherent in the crystal class being discussed. In crystal class 1, for example, {111} stands for a form consisting of the single face (111), whereas in crystal class $\bar{1}$ {111} stands for faces (111) *and* ($\bar{1}\bar{1}\bar{1}$), the two being related through operation of the center of symmetry now present. It becomes apparent, therefore, that a form {*hkl*} may embrace differing numbers of faces depending on the symmetry present in the crystal class involved. In crystal drawings, to emphasize that the faces of a given form are symmetry-related to each other, all faces that belong to the same form are often labeled with the same letter symbol. For example, if the faces (111), ($\bar{1}$11), (1$\bar{1}$1), and ($\bar{1}\bar{1}$1) all belonged to the form

{111}, they might all be labeled with the letter *p*. Indeed the form {111} might be accorded the letter symbol *p*. Certain conventions for assigning letter symbols to forms are followed. Thus forms that are parallel to two crystallographic axes are usually symbolized with the same letter used to designate the third crystallographic axis, that is, the one they intersect. Thus in trimetric crystals *c* refers to the {001} form, *b* to {010}, and *a* to {100}.

Triclinic Forms

Class 1: the pedial class

Any crystal belonging to class 1 lacks symmetry of any type. Consequently, no face on it can be symmetrically related to another face and hence each face present constitutes a *pedion,* which is by definition a form consisting of just a single face. Thus the crystal in Figure 5-4A is bounded by eight pedions. Those pedions that are parallel to any *two* crystallographic axes but cut the positive end of the third axis are called the *positive front pedion* {100}, the *positive side pedion* {010}, and the *positive basal pedion* {001}. Similarly, pedions {$\bar{1}00$}, {$0\bar{1}0$}, and {$00\bar{1}$} represent the *negative* front, side, and basal pedions, respectively.

Class $\bar{1}$: the pinacoidal class

The center of symmetry present in this class relates any face (*hkl*) to a second ($\bar{h}\bar{k}\bar{l}$), the two faces so related always having the same Miller indices but with opposite signs. Thus for rhodonite, whose crystals belong to class $\bar{1}$ (Fig. 5-4B), the center relates (100) to ($\bar{1}00$) but to no other faces, these two thus

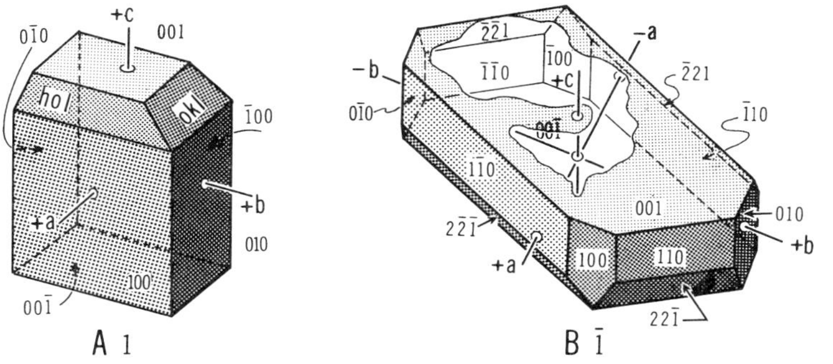

FIGURE 5–4

(A) A triclinic crystal that belongs to class 1 and for which each single face, of and by itself, constitutes the one-faced form known as a *pedion.*
(B) A triclinic crystal that belongs to class $\bar{1}$. Each pair of parallel faces is symmetrically related by the $\bar{1}$ axis and constitutes a pinacoid. A part of the (001) face has been cut away for illustrative purposes only.

constituting the {100} pinacoid or the *front pinacoid*. Similarly, the center relates (001) to (00$\bar{1}$), these composing the {001} or *basal pinacoid*; (010) to (0$\bar{1}$0), these being the {010} or *side pinacoid*; and, in general, (*hkl*) to ($\bar{h}\bar{k}\bar{l}$) to form an {*hkl*} pinacoid. Because even the general form {*hkl*} would be a pinacoid, this class is called the *pinacoidal class*.

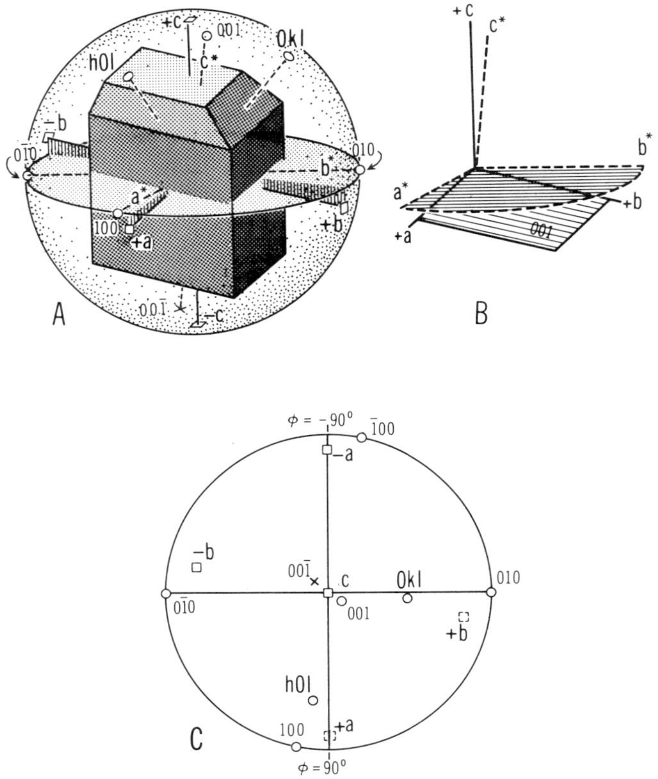

FIGURE 5–5

(A) The triclinic crystal of Figure 5-4A is now drawn at the center of the sphere of projection.
(B) Its direct axes *a*, *b*, and *c* and its reciprocal axes a^*, b^*, and c^* are drawn relative to its (001) plane (widely ruled) and to the plane (finely ruled) perpendicular to the *c* axis.
(C) The stereographic projection of its faces and axes if the crystal is oriented, as shown in (A), that is, with its *c* axis perpendicular to the plane of projection and its (010) face pole at $\phi = 0°$. On this stereogram and most subsequent ones, mutually perpendicular lines are drawn through its center at ϕ equal 0°, 90°, 180°, and −90°. All stereograms will be oriented with the $\phi = 0°$ line to the reader's right and with the $\phi = 90°$ line pointing toward the bottom of the page.

Reciprocal and direct triclinic axes

Only if three axes are mutually perpendicular can a plane parallel to two of them be simultaneously perpendicular to the third. Thus because the *a*, *b*, and *c* axes for triclinic crystals are in general *not* mutually perpendicular, a face parallel to two of these axes (for example, (100), (010), or (001)) *cannot* be perpendicular to the third. In other words, the normals to faces (100), (010), and (001)—labeled a^*, b^*, c^* and drawn with dashed lines in Figure 5-5A—cannot coincide with axes *a*, *b*, or *c*, respectively. Directions a^*, b^*, and c^* (called "*a* star, *b* star, *c* star") are designated as the crystal's *reciprocal axes* because they are the directions used in reciprocally projecting faces (100), (010), and (001) onto the sphere of projection. By convention asterisks differentiate the symbols for reciprocal axes from axes *a*, *b*, and *c*, these latter often being called *direct axes.*

The following geometric relations pertain to triclinic and, indeed, to all trimetric crystals: (1) *A direct and a reciprocal axis, if their letter symbols are unlike, will inevitably be at 90° to each other.* Thus, as Figure 5-5B illustrates, c^* is at 90° to axis *a* and axis *b* because it was defined as the normal to (001). Similarly, *c* is at 90° to both a^* and b^*. In like manner, a^* could be shown to be at 90° to axes *b* and *c*; b^*, to *a* and *c*; *a*, to b^* and c^*; and *b*, to a^* and c^*. (2) *A direct axis will be parallel to two of the three faces (100), (010), and (001) and thus to the line of intersection between these two faces.*

Stereographic projection

In stereographically projecting a triclinic crystal, it is oriented so that its *c* axis is perpendicular to the plane of projection (lightly shaded in Fig. 5-5A) and so that its *a* axis is at $\phi = 90°$. Consequently, the *c* axis projects cyclographically at the center of the stereogram (Fig. 5-5C) and face poles (010) and (100) plot on the stereogram's circumference, face pole (010) being at $\phi = 0°$. Face poles (100), (010), and (001) may also be regarded as marking the projected positions of reciprocal axes a^*, b^*, and c^*, respectively. In contrast, because direct axes *a*, *b*, and *c* are not perpendicular to faces (100), (010), and (001), respectively, they project at different sites from face poles (100), (010), and (001).

Monoclinic Forms

In classes *m*, 2, or 2/*m* of the monoclinic system, the 2-fold axis and/or the direction normal to the plane of symmetry is usually assumed to be the *b* axis. Such an orientation of the crystallographic axes to the symmetry directions is known as the "second setting." Some crystallographers, however, prefer the "first setting" in which the 2-fold axis and/or the direction normal to the symmetry plane is chosen to be the *c* axis instead of the *b* axis. If desired, the nomenclature that follows can be applied to this "first setting" by reversing the positions of the second and third digits in any Miller indices mentioned and by reading "*c* axis" for "*b* axis" in the ensuing discussions.

Class m: the domatic class

Since the second setting is adopted here, the direction normal to plane of symmetry *m* is specified as the *b* axis (Fig. 5-6A). All faces parallel to this axis are hence distinguished by indices of the type $(h0l)$; that is, their second index always equals zero. Such faces [(001), (100), $(00\bar{1})$, and $(\bar{1}00)$ in Figure 5-6A] are perpendicular to the plane of symmetry and not repeated by it. Consequently, each, by itself, constitutes a pedion. Face (010), which is parallel to the symmetry plane, is related by it to face $(0\bar{1}0)$, the two constituting a side pinacoid {010}. The general face (hkl) is oblique to the symmetry plane and is related by it to $(h\bar{k}l)$, these two nonparallel faces constituting the $\{hkl\}$ dome from which the class gets the name *domatic class.* In stereographic projections of the crystal under discussion (Fig. 5-6B), the heavy line represents the cyclographic projection of the single symmetry plane.

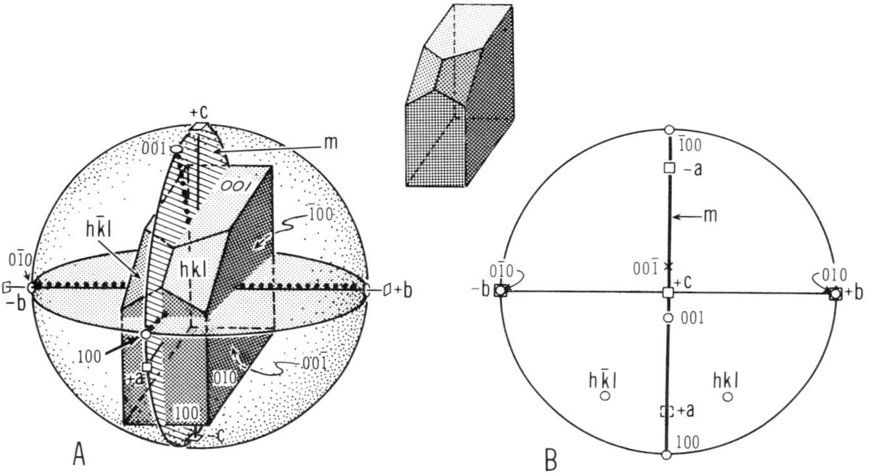

FIGURE 5–6

(A) A crystal that belongs to class *m* is oriented here in the second setting; that is, the direction normal to its (ruled) symmetry plane is chosen to be its *b* axis. Note that the dotted line normal to (010), which is reciprocal axis b^*, coincides with the *b* axis. The crystal is redrawn on a smaller scale (upper right) so as to be unobscured by labels.

(B) Stereogram of crystal at left. The heavy line represents the cyclographic projection of the ruled symmetry plane. For faces (100), (001), $(\bar{1}00)$, and $(00\bar{1})$, all of which are perpendicular to the symmetry plane, the poles plot on this heavy great circle. Such heavy great circles will be used to depict the cyclographic projections of mirror planes in all stereograms which follow.

A useful generalization now becomes apparent, namely, *if a symmetry plane is perpendicular to a crystallographic axis, then the Miller indices for any two faces symmetrically related by this plane will differ only in respect to the sign of the index referring to this axis.* For example, the symmetry plane perpendicular to the *b* axis in Figure 5-6 symmetrically relates (hkl) to $(h\bar{k}l)$ and (010) to $(0\bar{1}0)$.

Class 2: the sphenoidal class

Using the second setting, the single 2-fold axis in a crystal of class 2 is denoted as the *b* axis. In Figure 5-7A such a crystal is correctly oriented so that (1) its *c* axis is normal to the plane of stereographic projection and (2) the normal to its (010) face, which coincides with the *b* axis, intercepts the sphere of projection at $\phi = 0°$, $\rho = 90°$. All faces parallel to the *b* axis, that is, faces of type $(h0l)$, will be related by the *b* axis' 2-fold operation to a second face $(\bar{h}0\bar{l})$ parallel

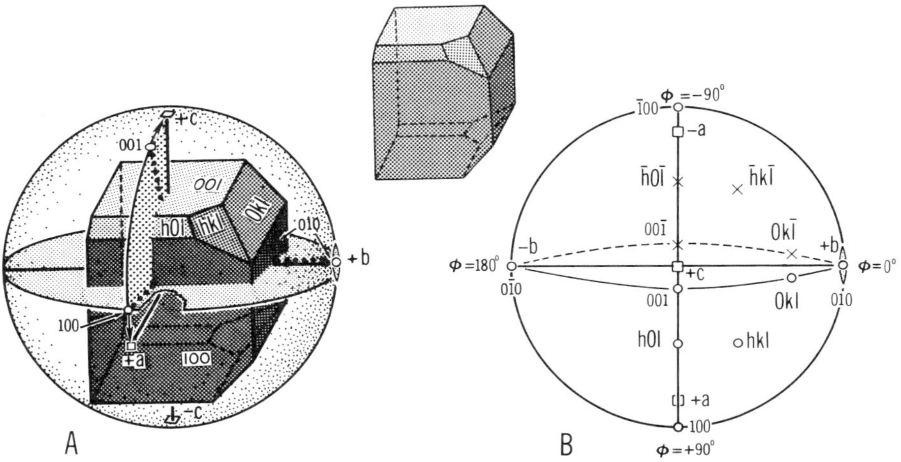

FIGURE 5–7

(A) An idealized crystal belonging to class 2. The 2-fold relates $(h0l)$ to a similarly shaped face $(\bar{h}0\bar{l})$; (hkl), to a similarly shaped face $(\bar{h}k\bar{l})$; and so on. However, (010), which is perpendicular to the 2-fold axis, is not related by it to any additional faces. The crystal is redrawn on a smaller scale (upper right).
(B) Stereogram for the crystal at left. The straight line between face poles (100) and $(\bar{1}00)$ can be regarded as the great circle that represents the stereographic (reciprocal) projection of the *b* axis. All faces parallel to the *b* axis have face poles that lie on this great circle. Similarly, the circumference represents the great circle that is the stereographic projection of the *c* axis. Thus all face poles on the circumference represent faces that are parallel to the *c* axis. The stereographic projection of the *a* axis is the great circle through face poles $(0\bar{1}0)$, (001), $(0kl)$, and (010) and, where dashed because it is on the lower hemisphere, through poles $(0k\bar{l})$ and $(00\bar{1})$; these faces are each parallel to the *a* axis as evidenced by the fact that their poles plot on its stereographic projection.

to this first. Thus the faces labeled $(h0l)$ and $(\bar{h}0\bar{l})$ in Figure 5-7B constitute the $\{h0l\}$ pinacoid. Similarly, (001), which is actually a special case of $(h0l)$ wherein h also equals zero, is related to $(00\bar{1})$ to form the basal pinacoid $\{001\}$. In like manner (100), which is a special case of $(h0l)$ wherein l equals zero, is related by the 2-fold axis to $(\bar{1}00)$ to form the front pinacoid $\{100\}$. On the other hand, faces (010) and $(0\bar{1}0)$ are each perpendicular to the 2-fold axis and thus not related to another face, (010) constituting the positive side-pedion $\{010\}$, and $(0\bar{1}0)$ constituting the negative side-pedion $\{0\bar{1}0\}$. Faces that intersect the b axis at an oblique angle will be related to a second face that intersects the b axis at this same angle. Thus $(0kl)$ is related to $(0k\bar{l})$, these two 2-fold, related, non-parallel faces constituting a sphenoid. Similarly, the general face (hkl) is related by the 2-fold axis to $(\bar{h}k\bar{l})$, these two nonparallel faces constituting another sphenoid $\{hkl\}$. A second useful generalization may be added to that in the preceding paragraph, namely, if a 2-fold axis coincides with a symmetry axis, then the Miller indices for two faces symmetrically related by this axis will differ only in respect to the signs of the two indices not referring to this axis. Thus (hkl) was related to $(\bar{h}k\bar{l})$ by the 2-fold axis coincident with the b axis in Figure 5-7. This second generalization applies in all except the hexagonal system.

Class 2/m: the prismatic class

A crystal from this class possesses a 2-fold axis (which for the second setting is chosen to be the b axis), a plane of symmetry (ruled in Fig. 5-8A) normal to it, and at their intersection, a center of symmetry. In Figure 5-8A the crystal is correctly oriented with its c axis perpendicular to the lightly shaded plane of stereographic projection and Figure 5-8B represents the stereogram of this crystal. Any face parallel to the b axis, and hence with indices of the type $(h0l)$, will be perpendicular to the symmetry plane and thus not "repeated" by it; however, because the b axis is a 2-fold axis, it relates $(h0l)$ to a second parallel face, $(\bar{h}0\bar{l})$. The center of symmetry also relates $(h0l)$ to $(\bar{h}0\bar{l})$ and the form $\{h0l\}$ is thus a pinacoid consisting of two parallel faces. Similarly, $\{001\}$ is the basal pinacoid and $\{100\}$ the front pinacoid, these being special cases of $\{h0l\}$ for which h or l equals zero. Face (010) is perpendicular to the 2-fold axis and not repeated by it, but the plane of symmetry and the center relate it to $(0\bar{1}0)$, these two constituting the side pinacoid $\{010\}$.

A face of type (hkl), or of type $(hk0)$ or $(0kl)$, will obliquely intersect both the 2-fold axis and the plane of symmetry and consequently will be "repeated" by each. For example, $(0kl)$ is related to $(0k\bar{l})$ by the 2-fold axis and to $(0\bar{k}l)$ by the symmetry plane (Fig. 5-8); similarly, $(0\bar{k}l)$ is related to $(0\bar{k}\bar{l})$ by the 2-fold axis' operation. These four faces, $(0kl)$, $(0k\bar{l})$, $(0\bar{k}l)$, and $(0\bar{k}\bar{l})$, lie in a zone parallel to [100], the a axis, and hence constitute the $\{0kl\}$ prism. The $\{hkl\}$ form, not shown in Figure 5-8, would be a four-faced prism very similar to the $\{0kl\}$ prism.

We observe from Figures 5-6 to 5-8 that *the Miller indices are identical but opposite in sign for two mutually parallel faces.* Thus (312) and $(\bar{3}\bar{1}\bar{2})$ would be mutually parallel. In stereographic projection two such parallel faces, for ex-

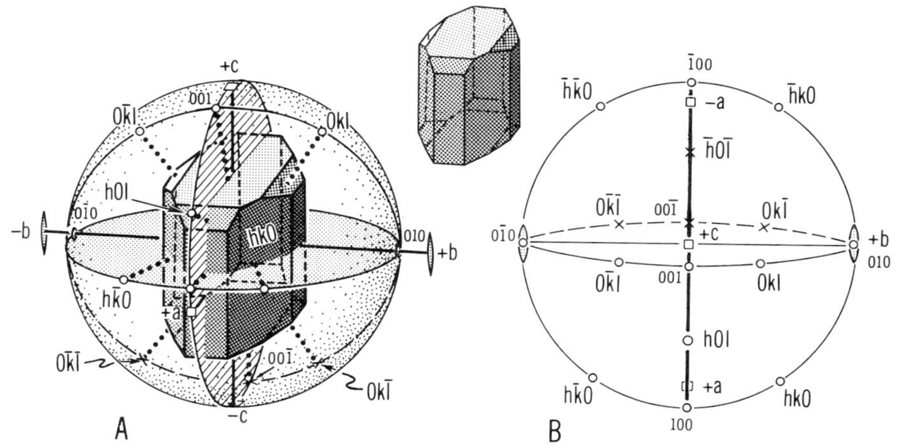

FIGURE 5–8

A monoclinic crystal that belongs to class $2/m$ is shown in spherical projection (A) and in stereographic projection (B). In (B) the stereographic projection of the a axis is represented on the lower hemisphere by the dashed great circle and on the upper hemisphere by the comparable solid-line great circle. The face poles for all faces parallel to the a axis, for example, $(0\bar{k}\bar{l})$, $(0\bar{k}l)$, and $(0kl)$, fall on these great circles. The heavy vertical line represents (a) the cyclographic projection of the symmetry plane and also (b) the stereographic projection of the b axis, which is normal to this plane.

ample, $(h0l)$ and $(\bar{h}0\bar{l})$ in Figure 5-8B, have their face poles at equal but opposite distances from the circle's center, the face pole on the upper hemisphere being denoted by a tiny circle and the face pole on the lower hemisphere by a tiny cross.

Orthorhombic Forms

For the three crystal classes $mm2$, 222, and $2/m\ 2/m\ 2/m$ that constitute the orthorhombic system, the choice of axes has already been discussed (page 35). Thus the class symbols are listed in the order with which they describe the symmetry of the a, b, and c axes. In review, symbol $mm2$ indicates a symmetry plane to be perpendicular to the a axis and to the b axis, whereas the solitary 2-fold axis coincides with the c axis. Symbol 222 indicates that the a, b, and c axes are 2-fold axes while symbol $2/m\ 2/m\ 2/m$ indicates that they also have planes of symmetry normal to them.

Class mm2: the orthorhombic pyramidal class

In crystals of this class the (001) face, being perpendicular to the 2-fold axis and to both symmetry planes m_1 and m_2 in Figure 5-9A, is related to no other faces and constitutes the positive basal pedion {001}. The negative basal

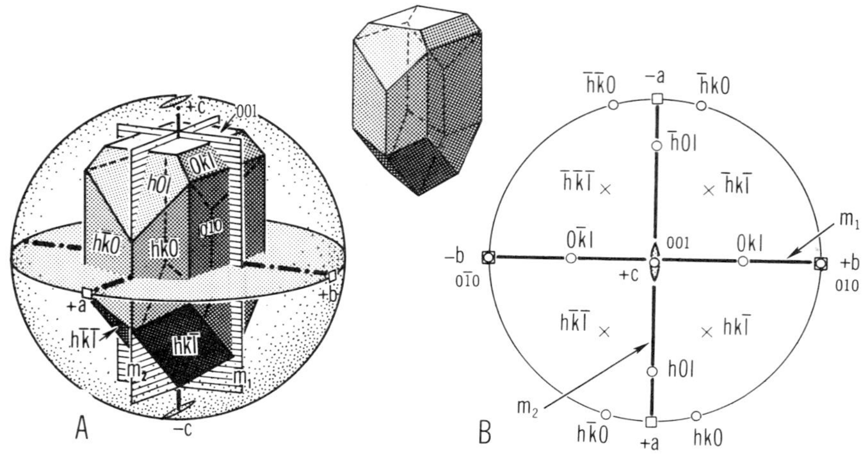

FIGURE 5–9

An orthorhombic crystal from class $mm2$ is shown in spherical projection (A) and in stereographic projection (B).

pedion $\{00\bar{1}\}$, the only other one-faced form in this class, is not shown in Figure 5-9A. The 2-fold axis and symmetry plane m_2 each relate (010) to $(0\bar{1}0)$ to form the side pinacoid {010} and they relate $(0kl)$ to the nonparallel face $(0\bar{k}l)$ to form a dome $\{0kl\}$. Because the 2-fold axis as well as m_2 relate $(0kl)$ to $(0\bar{k}l)$, form $\{0kl\}$ could also be considered to be a sphenoid. Similarly, $(h0l)$ and $(\bar{h}0l)$ are related by m_1 (or the 2-fold axis) to form another dome $\{h0l\}$. In like manner if (100) were present in Figure 5-9A, it would be related by m_1 and by the 2-fold axis to $(\bar{1}00)$, these two parallel faces constituting the front pinacoid {100}. Face $(hk0)$ is related by m_2 to $(h\bar{k}0)$ and by m_1 to $(\bar{h}k0)$; furthermore, m_2 relates $(\bar{h}k0)$ to $(\bar{h}\bar{k}0)$ so that the form consists of four faces, all parallel to the same direction [001] and thus constituting the prism $\{hk0\}$. The general face $(hk\bar{l})$, which cuts the 2-fold axis and both symmetry planes obliquely, is related by m_2 to $(h\bar{k}\bar{l})$ and both are related by m_1 to $(\bar{h}k\bar{l})$ and $(\bar{h}\bar{k}\bar{l})$, the two other, similarly shaped, lower faces in Figure 5-9A. The resultant form $\{hk\bar{l}\}$ consists of four nonparallel faces which meet at a point; it is thus an *orthorhombic pyramid* and resembles an up-side-down Egyptian pyramid. If the pyramid $\{hkl\}$ had been illustrated, it would resemble a right-side-up Egyptian pyramid.

In the orthorhombic system the direct axes a, b, and c are all mutually perpendicular. Hence faces (100), (010), and (001), which are parallel to two of these axes, are perpendicular to the third direct axis. In other words, the normals to (100), (010), and (001) (that is, the reciprocal axes a^*, b^*, and c^*) coincide with direct axes a, b, and c, respectively. Consequently, as illustrated for (010) and (001) in Figure 5-9B, if the face poles for (100), (010), and (001) are present, they always plot at the sites of axes a, b, and c in stereograms of orthorhombic crystals. For a monoclinic crystal *only one* direct and reciprocal axis coincide, these being b and b^* if the second setting is adopted. In triclinic crystals no coincidence between direct and reciprocal axes usually occurs.

Class 222: the orthorhombic disphenoidal class

In this class any face that is parallel to two crystallographic axes is related by the 2-fold operation of each to a second and parallel face. Thus in Figure 5-10A the (100) face is related by the 2-fold nature of the b axis and by that of the c axis to, and only to, a second, parallel face $(\bar{1}00)$, the two faces constituting the front pinacoid {100}. Similarly, {010} and {001} would be pinacoids. A face parallel to just one crystallographic axis (for example, $(hk0)$ in Fig. 5-10A) is related by the three 2-fold axes to three others also parallel to this crystallographic axis; hence it forms a prism. Thus axis a's 2-fold operation relates $(hk0)$ to $(h\bar{k}0)$; axis b's relates $(hk0)$ to $(\bar{h}k0)$ and $(h\bar{k}0)$ to $(\bar{h}\bar{k}0)$; axis c's relates $(hk0)$ to $(\bar{h}\bar{k}0)$ and $(h\bar{k}0)$ to $(\bar{h}k0)$ but adds no new faces to the form. These four faces thus constitute the $\{hk0\}$ prism. Similarly, $\{0kl\}$ and $\{h0l\}$ could be demonstrated to be four-faced prisms with faces parallel, respectively, to the a and to the b axis. The (hkl) face, which obliquely intersects the three crystallographic axes, is related to $(h\bar{k}\bar{l})$ by the 180° (2-fold) rotation about a as shown in the stereogram for the crystal (Fig. 5-10B). Similarly, axis c's 2-fold operation symmetrically relates (hkl) to $(\bar{h}\bar{k}l)$ and $(h\bar{k}\bar{l})$ to $(\bar{h}k\bar{l})$. These four faces constitute the $\{hkl\}$ form because the 2-fold operation of the b axis relates them

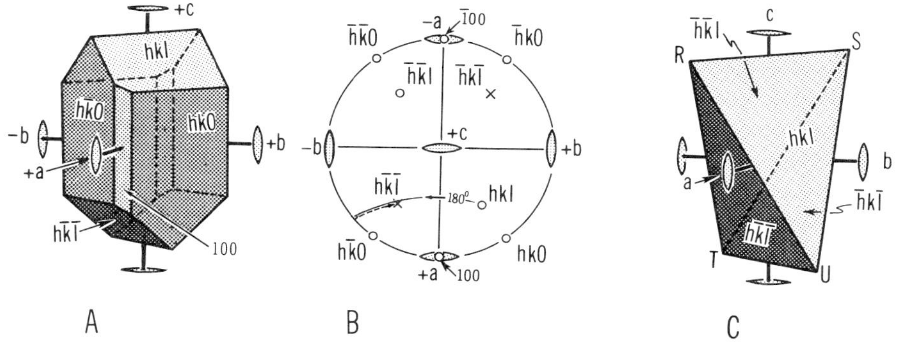

FIGURE 5–10

(A) An orthorhombic crystal from class 222 with its 2-fold axes drawn and labeled.

(B) Its stereogram showing how face pole (hkl) would migrate to the site of $(h\bar{k}\bar{l})$ during a 180° rotation about the a axis. The dashed small circle that represents its path on the bottom hemisphere has been slightly displaced for illustrative purposes.

(C) Drawing of the $\{hkl\}$ orthorhombic disphenoid in isolation, the 2-fold axes again being added.

only to each other, not to any additional faces. This form, if present alone on a crystal (Fig. 5-10C), can be likened to two sphenoids, (hkl) and $(\bar{h}\bar{k}l)$ being one and $(h\bar{k}\bar{l})$ plus $(\bar{h}k\bar{l})$ being the other; hence it is called the *orthorhombic disphenoid* $\{hkl\}$. No two edges of intersection between two faces of an orthorhombic sphenoid will ordinarily be at 90° to each other. For example, line RS in Figure 5-10C will not be at 90° to TU.

Class 2/m 2/m 2/m: the orthorhombic dipyramidal class

In the holohedral class of the orthorhombic system, axes a, b, and c not only coincide with 2-fold axes but also with perpendiculars to symmetry planes. Any face parallel to two of these axes will be related to a second parallel face to form a pinacoid. For example, (010) in Figure 5-11A is related to $(0\bar{1}0)$ by the 2-fold operation of axes a and c and by reflection across plane m_2. Thus {010} is a side pinacoid. Although not shown in Figure 5-11A, the front pinacoid {100} and the basal pinacoid {001} are also possible. A face parallel to only one crystallographic axis will be symmetrically related to three other faces, also parallel to this axis, to form a four-faced prism. Thus in Figure 5-11A and B, plane m_2 relates $(hk0)$ to $(h\bar{k}0)$ and plane m_1 relates $(hk0)$ to $(\bar{h}k0)$ and $(h\bar{k}0)$ to $(\bar{h}\bar{k}0)$. The remaining symmetry plane and the three 2-fold axes relate no new faces to these. Thus form $\{hk0\}$ is a four-faced prism. Similarly, forms $\{h0l\}$ and $\{0kl\}$ would be four-faced prisms. The general face (hkl) is related by the 2-fold axes and symmetry planes to seven additional faces, all eight faces being labeled with u in Figure 5-11B and in Figure 5-11C, where the form is drawn as it would look if it alone were present on the crystal. This eight-faced form is called an *orthorhombic dipyramid* because its upper half resembles a pyramid with apex up and its lower half a second pyramid with its apex pointing down. We note that if any face on

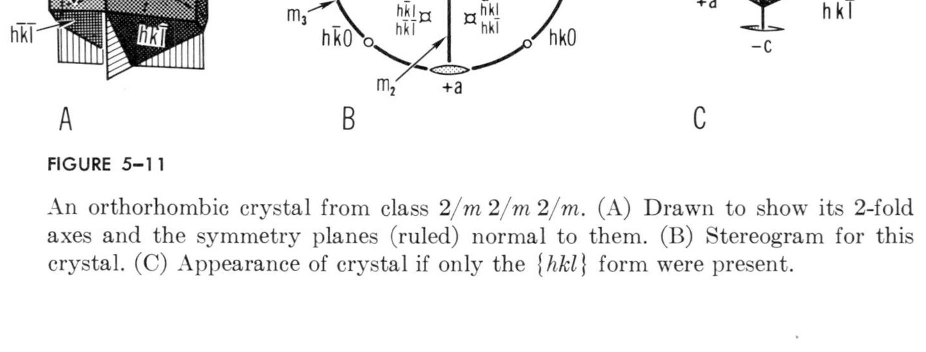

FIGURE 5–11

An orthorhombic crystal from class $2/m\ 2/m\ 2/m$. (A) Drawn to show its 2-fold axes and the symmetry planes (ruled) normal to them. (B) Stereogram for this crystal. (C) Appearance of crystal if only the $\{hkl\}$ form were present.

the lower half of a dipyramid is compared with the corresponding face on the upper half, it has the same ϕ and ρ angles in the stereographic projection (Fig. 5-11B) except that the ρ angle is negative in sign. Consequently, these two faces (for example, (hkl) and $(hk\bar{l})$) plot at the same point. Similarly, the Miller indices of such a pair of faces differ only in respect to the sign of l.

Tetragonal Forms

The single 4 or $\bar{4}$ axis serves as the c axis in tetragonal crystal classes 4, $\bar{4}$, $4/m$, $4mm$, 422, $\bar{4}2m$, or $4/m\ 2/m\ 2/m$, whereas the mutually perpendicular a_1 and a_2 axes in the plane normal to the c axis are chosen to coincide with 2-fold axes in classes 422, $\bar{4}2m$, or $4/m\ 2/m\ 2/m$, with normals to symmetry planes in $4mm$ or with no directions of symmetry in classes 4, $\bar{4}$, and $4/m$. Thus the seven faces (001), (100), (110), $(hk0)$, $(h0l)$, (hhl), and (hkl) are listed not only in order of their increasingly random orientation to the c, a_1, and a_2 axes but also to the symmetry elements which these axes may represent. The forms developed by action of each crystal class's symmetry on these faces will next be considered.

The {001} form

The (001) face is related to the $(00\bar{1})$ face (1) by the inversion operation inherent in the $\bar{4}$ axis of classes $\bar{4}$ or $\bar{4}2m$; (2) by the symmetry plane normal to the 4-fold axis, that is, to the c axis, in classes $4/m$ or $4/m\ 2/m\ 2/m$; or (3) by the 2-fold axes normal to the 4-fold axes in classes 422, $\bar{4}2m$, or $4/m\ 2/m\ 2/m$. Hence in classes $\bar{4}$, $\bar{4}2m$, $4/m$, 422, or $4/m\ 2/m\ 2/m$ the {001} form is a basal pinacoid consisting of faces (001) and $(00\bar{1})$. In classes 4 or $4mm$, however, there is no symmetry capable of relating (001) to $(00\bar{1})$. Thus they represent the positive and the negative basal pedions {001} and $\{00\bar{1}\}$, respectively.

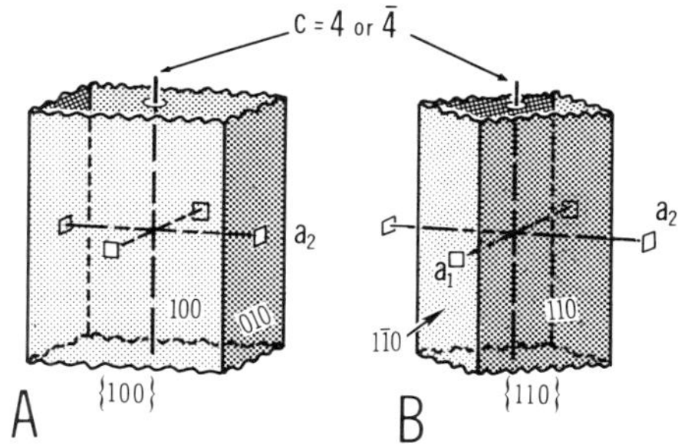

FIGURE 5–12

The {100} and {110} prisms that can occur in any tetragonal crystal class.

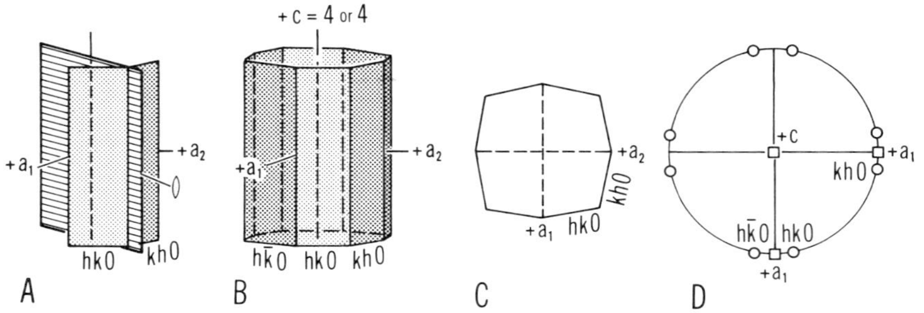

FIGURE 5–13

In classes $4mm$, $\bar{4}2m$, or $4/m\ 2/m\ 2/m$, face $(hk0)$ is related to $(kh0)$ by operation of the 2-fold axis and/or symmetry plane (ruled) as shown in (A). The operation of the c axis on this pair of faces produces the ditetragonal prism $\{hk0\}$ shown in perspective (B), in top view (C), and in stereographic projection (D).

Forms {100}, {110}, and {hk0}

The faces (100), (110), and $(hk0)$ are each parallel to the c axis, but (100) is the most specially oriented of them because it is also parallel to the a_2 axis. Face $(hk0)$ is the most randomly oriented, cutting the a_1 and a_2 axes at unequal angles in contrast to (110) which cuts each at 45°. In all seven tetragonal classes, face (100) is related to (010), $(\bar{1}00)$, and $(0\bar{1}0)$ by operation of the 4- or $\bar{4}$-fold axis (Fig. 5-12A) that serves as its c axis. Regardless of which tetragonal crystal class is involved, the class's remaining symmetry will relate these four faces only to each other, not to any additional faces. Thus they constitute the tetragonal prism {100}. In like manner for all the crystal classes, (110) is related to $(1\bar{1}0)$, $(\bar{1}\bar{1}0)$, and $(\bar{1}10)$ by the c axis's 4- or $\bar{4}$-fold operation (Figs. 5-12B) and the four faces constitute the tetragonal prism {110}.

The number of faces in form $\{hk0\}$ depends on whether or not the crystal class has a plane of symmetry or a 2-fold axis that bisects the angle between its a_1 and a_2 axes. If such a plane or 2-fold axis (or both) is present (Fig. 5-13A), its operation relates face $(hk0)$ to a second face $(kh0)$ whereupon the 4- or $\bar{4}$-fold axis that serves as the c axis will relate this pair of faces to six others (Fig. 5-13B to D). The resultant form, called a *ditetragonal prism*, resembles an ordinary tetragonal prism with a slight "crease" down the center of each face. The $\{hk0\}$ form is a ditetragonal prism in classes $4mm$, 422, $\bar{4}2m$, and $4/m\ 2/m\ 2/m$, the third part of each class's symbol indicating that a symmetry plane and/or

FIGURE 5–14→

The $\{h0l\}$, $\{hhl\}$, and $\{hkl\}$ forms are shown in relation to the crystallographic axes and the symmetry elements for each of the seven tetragonal crystal classes. For classes possessing symmetry planes, these planes (ruled) are sometimes shown near the bottom of the crystal; in class $4/m\ 2/m\ 2/m$ all its symmetry planes have been omitted in order to simplify the drawing.

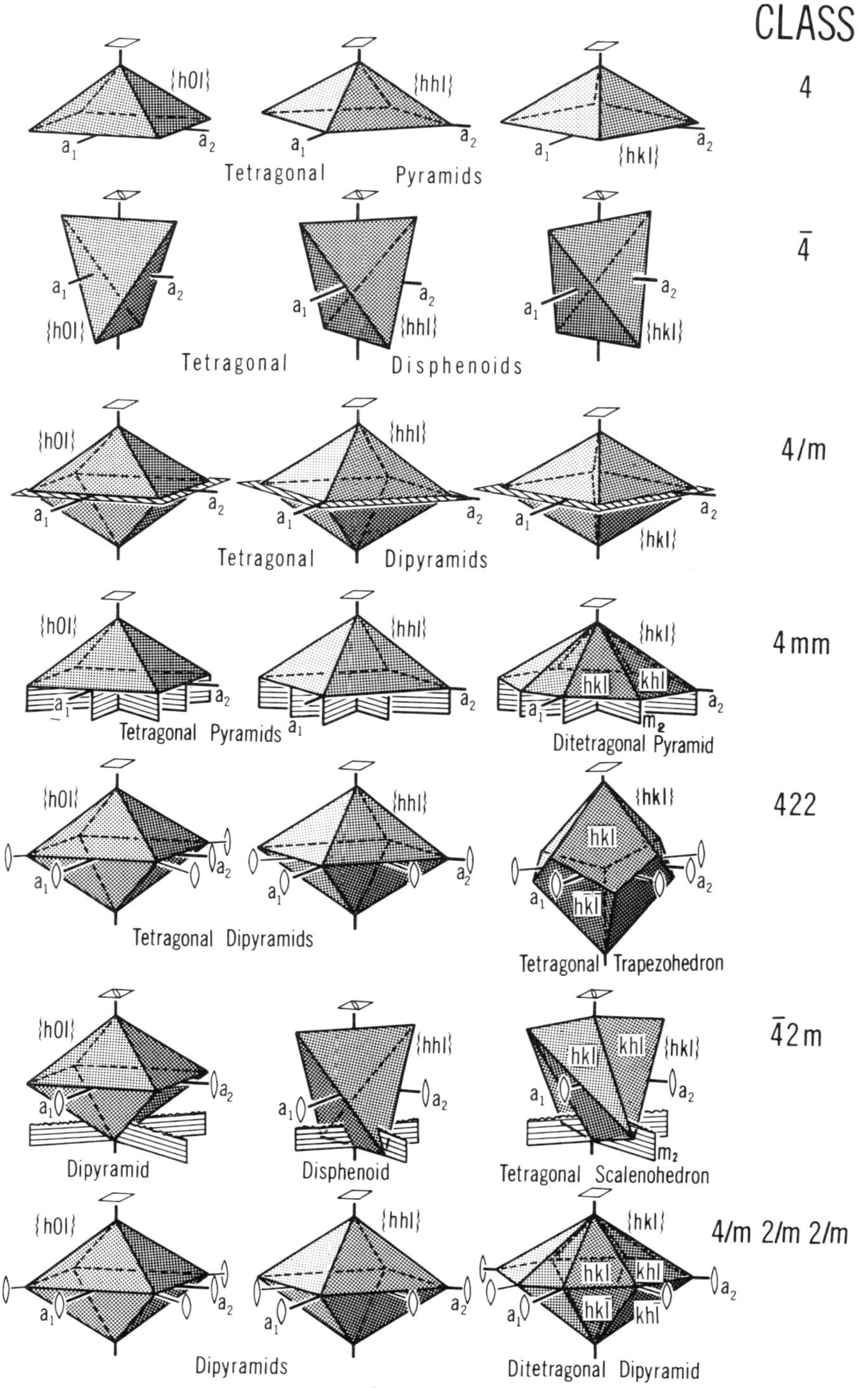
CLASS
4
{h0l}
{hhl}
{hkl}
a1
a2
Tetragonal Pyramids
4̄
Tetragonal Disphenoids
4/m
Tetragonal Dipyramids
4mm
hkl
khl
m2
Tetragonal Pyramids
Ditetragonal Pyramid
422
hk̄l̄
Tetragonal Dipyramids
Tetragonal Trapezohedron
4̄2m
Dipyramid
Disphenoid
Tetragonal Scalenohedron
4/m 2/m 2/m
hkl̄
khl̄
Dipyramids
Ditetragonal Dipyramid

2-fold bisects the angle between axes a_1 and a_2. However, in the three remaining tetragonal classes, 4, $\bar{4}$, and $4/m$, such symmetry is lacking and $\{hk0\}$ is an ordinary, four-faced tetragonal prism.

Forms {h0l}, {hhl}, and {hkl}

Faces $(h0l)$, (hhl), and (hkl) intersect the c axis obliquely; however, (hkl) is the most randomly oriented of the trio, cutting all three axes at unequal angles. Actually, since face $(h0l)$ cuts a_1 obliquely but is parallel to a_2, it bears the same orientation to the a_1 and a_2 axes as face (hhl) bears to directions [110] and [1$\bar{1}$0], face (hhl) cutting [110] obliquely but being parallel to [1$\bar{1}$0]. Thus if directions [110] and [1$\bar{1}$0] possess the same symmetry as do the a_1 and a_2 axes (as is true for all tetragonal classes except $\bar{4}2m$), then forms $\{h0l\}$ and $\{hhl\}$ will look alike. However, for class $\bar{4}2m$ ($=\bar{4}2\bar{2}$), in which the a axes are 2 folds but [110] and [1$\bar{1}$0] are $\bar{2}$ fold, form $\{h0l\}$ is a dipyramid, whereas $\{hhl\}$ is a disphenoid (Fig. 5-14). Figure 5-14 also shows the forms "generated" by operation of each class's symmetry on an (hkl) face. The names of these general $\{hkl\}$ forms are sometimes used for the class as well. For example, class 4 is called the *tetragonal-pyramidal* class; class $\bar{4}$, the *tetragonal-disphenoidal* class; class $4/m$, the *tetragonal-dipyramidal* class; class $4mm$, the *ditetragonal-pyramidal* class; and so on. Note that in class $4mm$ (Figure 5-14), face (hkl) is not perpendicular to any symmetry planes whereas $(h0l)$ and (hhl) are; thus symmetry plane m_2 relates (hkl) to (khl) and these two faces are each repeated by the 4-fold axis to generate a *ditetragonal pyramid*. This form resembles a tetragonal pyramid whose faces have been creased down the center and therefore doubled in number.

In class 422, face (hkl) is related to $(h\bar{k}\bar{l})$ by the 2-fold axis labeled a_1 in Figure 5-14. Each of these faces is repeated by the 4-fold axis to generate a

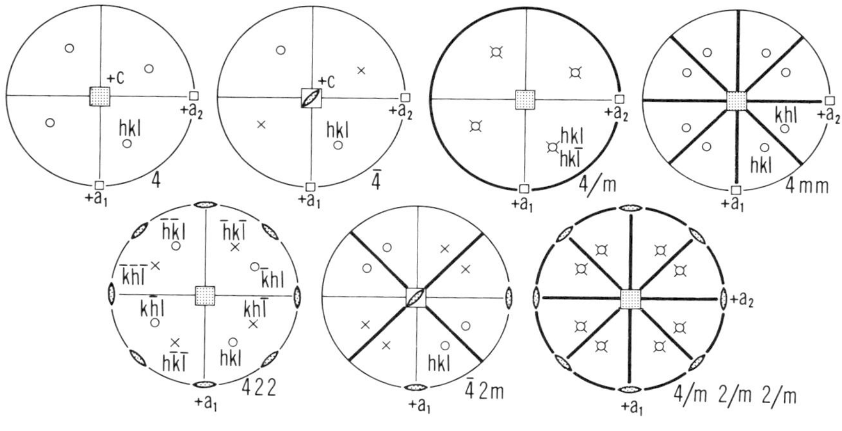

FIGURE 5–15

Stereograms of the $\{hkl\}$ forms for the seven tetragonal crystal classes.

tetragonal trapezohedron $\{hkl\}$, so-called because each individual face is a trapezium, a four-sided figure with nonparallel sides. Class 422 is thus called the *tetragonal trapezohedral* class. In class $\bar{4}2m$, face (hkl) is symmetrically related to face (khl) by a mirror plane (labeled m_2 in Fig. 5-14). The $\bar{4}$ axis, in turn, operates on these two faces to produce a *tetragonal scalenohedron* $\{hkl\}$, so named because its individual faces are scalene triangles. Class $\bar{4}2m$ is hence called the *tetragonal scalenohedral* class. With some justice a tetragonal disphenoid could be called a ditetragonal disphenoid because it resembles a tetragonal disphenoid whose faces are creased down the center. In Fig. 5-14 the mirror planes are drawn for classes $4mm$, $4/m$, and $\bar{4}2m$ but not for $4/m\ 2/m\ 2/m$ to simplify this lowest tier of drawings. Because of their presence in class $4/m\ 2/m\ 2/m$, face (hkl) relates to (khl) through a mirror comparable to m_2. The mirror normal to the 4-fold axis relates to (hkl) to $(hk\bar{l})$ and also (khl) to $(kh\bar{l})$. The 4-fold then operates on each of these faces to develop a *ditetragonal dipyramid* $\{hkl\}$, the general form with 16 faces from which class $4/m\ 2/m\ 2/m$ derives the name *ditetragonal dipyramidal class.*

The stereograms of the general $\{hkl\}$ form for each tetragonal class (Fig. 5-15) should be compared with the drawings of these forms in Figure 5-14.

Hexagonal Forms (Trigonal Subsystem)

The single 3 or $\bar{3}$ axis serves as the c axis in the trigonal crystal classes 3, $\bar{3}$, 32, $3m$, and $\bar{3}\ 2/m$. In the plane normal to the c axis, the a_1, a_2, and a_3 axes, whose positive ends are at 120° to each other, are selected to coincide with 2-fold axes or with normals to symmetry planes in classes 32, $3m$, and $\bar{3}\ 2/m$, but with 1-fold axes in classes 3 and $\bar{3}$. As for tetragonal crystals, the seven hexagonal faces (00∗1), (10∗0), (11∗0), $(hk{*}0)$, $(h0{*}l)$, $(hh{*}l)$, and $(hk{*}l)$ bear increasingly random orientations to the crystallographic axes and thus to any symmetry directions with which these axes coincide. The indices for these seven faces with the third index explicitly stated are (0001), $(10\bar{1}0)$, $(11\bar{2}0)$, $(hki0)$, $(h0\bar{h}l)$, $(hh\overline{2h}l)$, and $(hkil)$, where i, as demonstrated earlier (Fig. 3-3), equals $-(h+k)$.

Form {0001}

Faces (0001) and $(000\bar{1})$ are each perpendicular to the c axis and as a consequence classes 3 and $3m$ possess no symmetry capable of relating (0001) to $(000\bar{1})$ or to any other face. Thus in classes 3 and $3m$ each of these faces constitutes a one-faced form, the *positive basal pedion* {0001} and the *negative basal pedion* $\{000\bar{1}\}$. In classes $\bar{3}$, 32, or $\bar{3}\ 2/m$, however, face (0001) relates to $(000\bar{1})$ by the inversion operation inherent in the $\bar{3}$ axis and/or by any 2-fold axes that are perpendicular to the c axis. These two faces thus constitute the *basal pinacoid* {0001} in classes $\bar{3}$, 32, or $\bar{3}\ 2/m$.

Forms $\{10\bar{1}0\}$, $\{11\bar{2}0\}$, *and* $\{hki0\}$

The faces $(10\bar{1}0)$, $(11\bar{2}0)$, and $(hki0)$ are each parallel to the c axis, but $(10\bar{1}0)$ is more specially oriented, being also parallel to the a_2 axis. Face $(hki0)$ is the most randomly oriented of the trio, cutting the a_1 and a_2 axes at unequal angles, whereas $(11\bar{2}0)$ intersects both at 30°. In classes 3 and $3m$, the 3-fold axis relates face $(10\bar{1}0)$ to $(\bar{1}100)$ and to $(0\bar{1}10)$, these three faces constituting form $\{10\bar{1}0\}$, a *trigonal prism* (Fig. 5-16A). The stereogram of this trigonal prism for class $3m$ indicates that the prism's faces are perpendicular to the symmetry planes and hence are not repeated by them. In classes $\bar{3}$, $\bar{3}\ 2/m$, and 32, face $(10\bar{1}0)$ is related to faces $(01\bar{1}0)$, $(0\bar{1}10)$, $(\bar{1}100)$, $(\bar{1}010)$, $(1\bar{1}00)$; these six related faces constitute a *hexagonal prism* (Fig. 5-16B) whose cross section (normal to the c axis) is a hexagon. The stereograms of this hexagonal prism indicate how its faces are related by the symmetry elements for classes $\bar{3}$, $\bar{3}\ 2/m$, and 32. In class $\bar{3}$, for example, a 120° clockwise rotation plus inversion through the center relates $(10\bar{1}0)$ to $(01\bar{1}0)$, $(01\bar{1}0)$ to $(\bar{1}100)$, and so on. In class 32 a 180° rotation about the 2-fold axis that serves as the a_3 axis relates $(10\bar{1}0)$ to $(01\bar{1}0)$ as shown by the dashed arrow in the stereogram; these two faces, repeated by the 3-fold axis, generate the remaining four faces of the hexagonal prism.

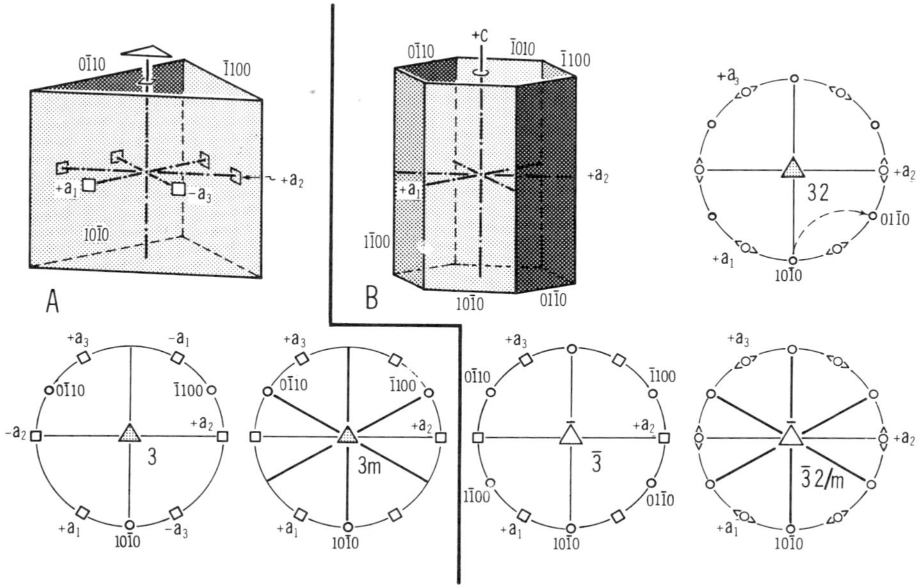

FIGURE 5–16

(A) The $\{10\bar{1}0\}$ trigonal prism shown in perspective and (below) projected stereographically for the two classes in which it is possible: 3 and $3m$.
(B) The $\{10\bar{1}0\}$ hexagonal prism shown in perspective as well as in stereograms (to the right and below) for the three trigonal classes in which it can occur.

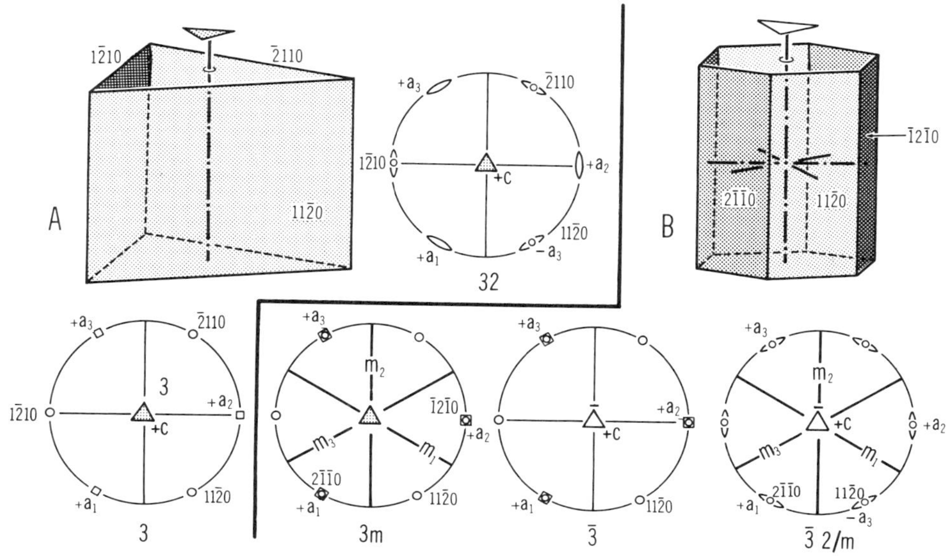

FIGURE 5–17

(A) The $\{11\bar{2}0\}$ trigonal prism and its stereograms for the two classes, 32 and 3, in which it occurs.
(B) The $\{11\bar{2}0\}$ hexagonal prism and its stereograms for the three trigonal classes $3m$, $\bar{3}$, and $\bar{3}\ 2/m$ in which it occurs.

In class 3 the a axes are not directions of symmetry; thus only its 3-fold or c axis relates face $(11\bar{2}0)$ to $(1\bar{2}10)$ and $(\bar{2}110)$, the three constituting the $\{11\bar{2}0\}$ *trigonal prism* (Fig. 5-17A). In class 32 even though the a axes are 2 folds, $\{11\bar{2}0\}$ is again a trigonal prism because $(11\bar{2}0)$ is perpendicular to the a_3 axis and hence not repeated by it, being related to $(1\bar{2}10)$ and $(\bar{2}110)$ by operations of the 3-fold and of the 2-fold axes at a_1 and a_2. In classes $3m$, $\bar{3}$, and $\bar{3}\ 2/m$, however, $(11\bar{2}0)$ is related to five more faces to produce the *hexagonal prism* $\{11\bar{2}0\}$ (Fig. 5-17B). As the stereograms of this form for classes $3m$ and $\bar{3}\ 2/m$ show, symmetry plane m_2, which is perpendicular to axis a_2, relates $(11\bar{2}0)$ to $(2\bar{1}\bar{1}0)$ and this pair of faces is related by the c axis' symmetry to the remaining faces in the form. In class $\bar{3}$ although the a axes are only 1 fold, the operation of the $\bar{3}$ axis relates $(11\bar{2}0)$ to the remaining five faces in the $\{11\bar{2}0\}$ hexagonal prism.

Although the $(hki0)$ face is at a random angle to the a axes, these can relate it to no additional faces if they are merely 1-fold axes. Consequently, in classes 3 and $\bar{3}$, where such is the case, only the symmetry inherent in the c axis relates $(hki0)$ to other faces; hence $\{hki0\}$ is a trigonal prism in class 3 (Fig. 5-18A), a hexagonal prism in class $\bar{3}$ (Fig. 5-18C). However, in classes 32, $3m$, or $\bar{3}\ 2/m$, the a axes are, respectively, 2-fold axes, $\bar{2}$ axes (or both) and therefore relate $(hki0)$ to a second face; for example, the 2-fold axis labeled $-a_3$ in Figures 5-18B and D relates $(hki0)$ to $(khi0)$. If c is a 3-fold axis as in classes 32 or $3m$, this

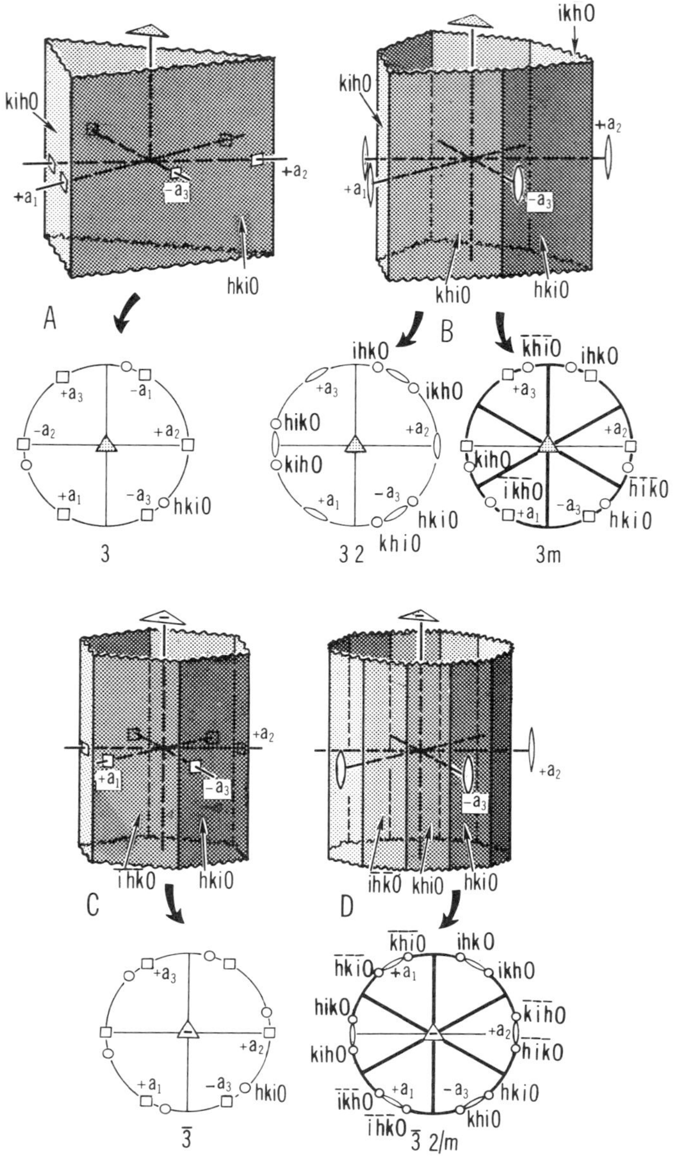

FIGURE 5–18

The $\{hki0\}$ forms for the five crystal classes of the trigonal subsystem. (A) In class 3, $\{hki0\}$ is a *trigonal prism* as shown in perspective and in stereogram. The small squares in the stereogram merely mark the cyclographic projections of the a axes and are not the symbols for 4-fold axes. (B) In class 32, the 2-fold at $-a_3$ relates $(hki0)$ to $(khi0)$, both faces being then repeated by the 3-fold axis to produce the *ditrigonal prism* shown in perspective and (below, left) in stereogram. In class $3m$, pair of faces relates to two other pairs, thereby constituting a *ditrigonal prism* (Fig. 5-18B); if, however, c is a $\bar{3}$-fold axis as in class $\bar{3}\ 2/m$, the pair relates to five other pairs to constitute a *dihexagonal prism* (Fig. 5-18D). In summary, $\{hki0\}$ is a trigonal prism in class 3, a ditrigonal prism in classes 32 or $3m$, a hexagonal prism in class $\bar{3}$, and a dihexagonal prism in class $\bar{3}\ 2/m$. We note that if *di* is prefixed to the adjectival form of the system's or subsystem's name, as in *ditrigonal* or *dihexagonal prism*, it indicates a trigonal or hexagonal prism whose faces are doubled as if by a "crease down their center." Whereas all interfacial angles equal 60° for a hexagonal prism, those around a ditrigonal prism are alternately greater or less than 60° (Fig. 5-18B).

Forms $\{h0\bar{h}l\}$, $\{hh\overline{2h}l\}$, and $\{hkil\}$

Forms $\{h0\bar{h}l\}$, $\{hh\overline{2h}l\}$, and $\{hkil\}$ are trigonal pyramids in class 3 (Fig. 5-19A). In class $\bar{3}$, however, the center of symmetry present in addition to the 3-fold axis would require each face of a trigonal pyramid to have a second face parallel to it. The resultant form, called a *rhombohedron* (Fig. 5-19B), thus consists of three pairs of parallel faces, three of these intersecting the c axis at its positive end and the others intersecting at its negative end. Consequently, $\{h0\bar{h}l\}$, $\{hh\overline{2h}l\}$, and $\{hkil\}$ are each a rhombohedron in class $\bar{3}$. In class 32 the a axes are 2 fold and hence the differing orientations to these axes of faces $(h0\bar{h}l)$, $(hh\overline{2h}l)$, and $(hkil)$ cause differences in the forms $\{h0\bar{h}l\}$, $\{hh\overline{2h}l\}$, and $\{hkil\}$. In Figure 5-19C, for example, the 2-fold axis labeled $-a_3$ relates $(h0\bar{h}l)$ to $(0h\bar{h}\bar{l})$, $(hh\overline{2h}l)$ to $(hh\overline{2h}\bar{l})$, and $(hkil)$ to $(khi\bar{l})$ and the 3-fold axis operates on the resultant pairs of faces to generate the others in the form. In class 32, therefore, $\{h0\bar{h}l\}$ is a *rhombohedron*, $\{hh\overline{2h}l\}$ is a *trigonal dipyramid*, and $\{hkil\}$ is a *trigonal trapezohedron*. As the stereograms for these forms illustrate, the lower faces of a rhombohedron have ϕ angles midway between those for its upper faces; the lower faces of a dipyramid have the same ϕ angles as its upper ones; but the lower faces of a trapezohedron have ϕ angles neither midway nor equal to those of its upper faces.

---

the mirror plane between the a_3 and a_2 axes would relate $(hki0)$ to $(\bar{h}ik0)$ a face shown in the stereogram for class $3m$ but not in the perspective drawings. The 3-fold axis then operates on these two faces to produce another ditrigonal prism which would look like the one drawn in perspective, if this perspective drawing were rotated 30° counterclockwise around its 3-fold axis (and relabeled with indices to conform to those in the stereogram for class $3m$). (C) In class $\bar{3}$, the symmetry operates on $(hki0)$ to produce the *hexagonal prism* $\{hki0\}$. (D) In class $\bar{3}\ 2/m$, the symmetry operates on $(hki0)$ to produce the dihexagonal prism $\{hki0\}$.

For Miller-Bravais indices, i equals $-(h+k)$. Hence i is by definition a negative index which denotes a face that cuts a negative axis. In contrast $\bar{i}$ is positive and indicates the face to cut a positive axis. For example, face $(\bar{i}\bar{h}\bar{k}0)$ in (D) intercepts the $+a_1$, $-a_2$, and $-a_3$ axes at distances proportional to $1/(h+k)$ on a_1, $1/h$ on $-a_2$, and $1/k$ on $-a_3$. The $\{hki0\}$ forms in this figure have been drawn as if $h < k$, that is, as if the $\{hki0\}$ face is more nearly perpendicular to the a_2 axis (and thus plots closer to it) than to the a_1 axis.

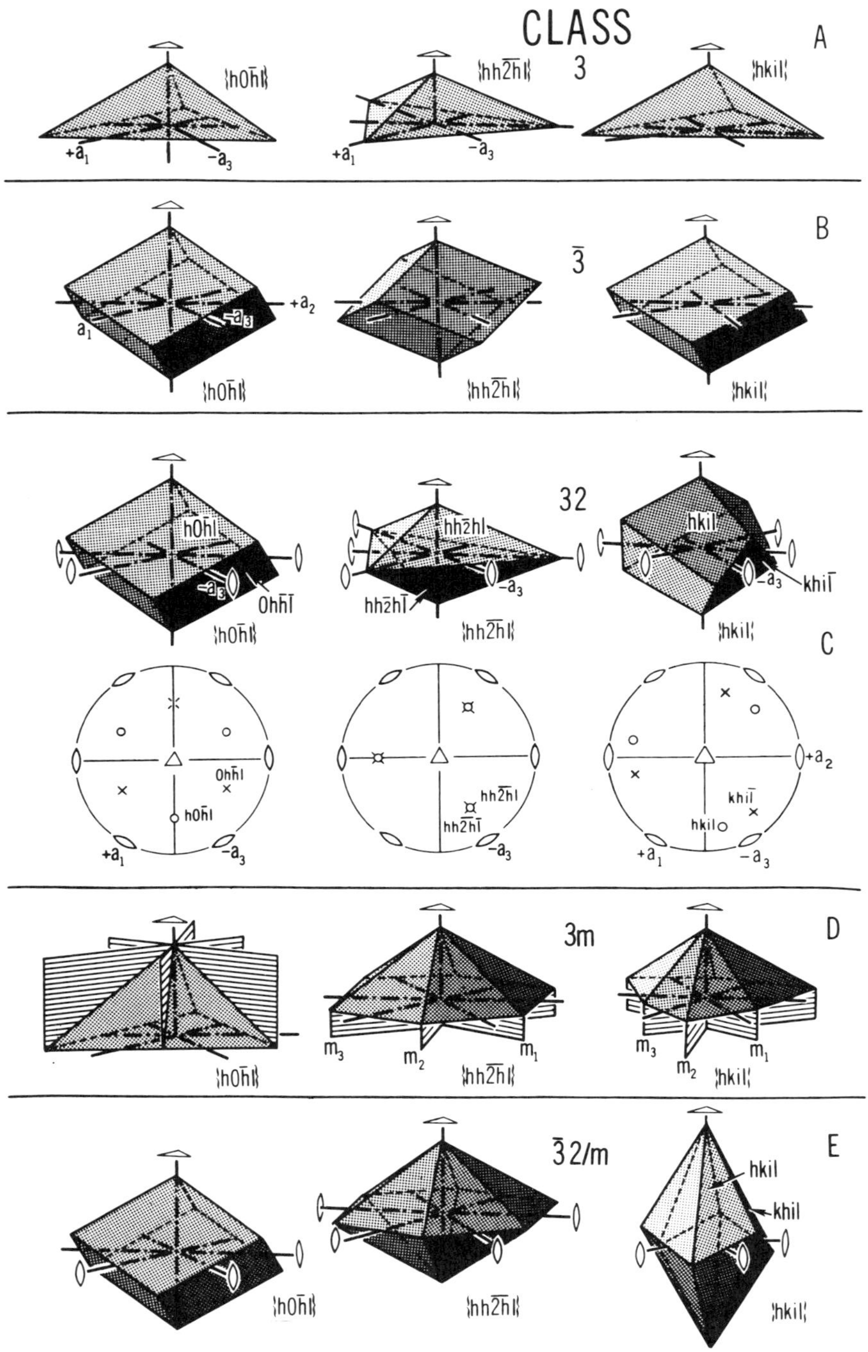
CLASS
A
3
{h0hl}
{hh2hl}
{hkil}
+a1
-a3
B
3
+a2
a1
-a3
32
h0hl
0hhl
hh2hl
hh2hl
-a3
hkil
khil
C
0hhl
h0hl
hh2hl
hh2hl
khil
hkil
+a1
-a3
+a2
D
3m
m3
m2
m1
E
3 2/m
hkil
khil

In class $3m$ the three a axes are each normal to a symmetry plane (ruled in Fig. 5-19D). Thus $\{h0\bar{h}l\}$ is a *trigonal pyramid* each face of which is perpendicular to one symmetry plane but related by the other two symmetry planes (and by the 3-fold axis) to the form's remaining two faces. In contrast $\{hh\overline{2h}l\}$ is a *hexagonal pyramid,* but $\{hk\bar{i}l\}$ is a *ditrigonal pyramid,* the difference between them arising because face $(hh\overline{2h}l)$ intersects symmetry planes m_1 and m_2 at equal angles (Fig. 5-19D), whereas $(hk\bar{i}l)$ does not.

In class $\bar{3}\ 2/m$ the a axes are 2-fold axes and also normals to symmetry planes. The operation of these a axes and of the c axis, which is a $\bar{3}$ axis, on $(h0\bar{h}l)$ results in the $\{h0\bar{h}l\}$ *rhombohedron* (Fig. 5-19E). Their operation on $(hh\overline{2h}l)$ and $(hk\bar{i}l)$ results, respectively, in the $\{hh\overline{2h}l\}$ *hexagonal dipyramid* and the $\{hk\bar{i}l\}$ *trigonal scalenohedron.* Symmetry planes m_1, m_2, and m_3 are present in class $\bar{3}\ 2/m$ just as in class $3m$ but were omitted from Figure 5-19E for simplicity. If plane m_1 were drawn, it would be normal to axis a_1 but bisect the angle between axes $-a_3$ and $+a_2$ and would thus symmetrically relate face $(hk\bar{i}l)$ to $(kh\bar{i}l)$. The $\bar{3}$ operation on these two faces then generates the remaining faces of trigonal scalenohedron $\{hk\bar{i}l\}$.

Hexagonal Forms (Hexagonal Subsystem)

For hexagonal crystal classes 6, $\bar{6}$, $6/m$, $\bar{6}m2$, $6mm$, 622, and $6/m\ 2/m\ 2/m$, the forms developed as their symmetry operates on faces (0001), $(10\bar{1}0)$, $(11\bar{2}0)$, $(hki0)$, $(h0\bar{h}l)$, $(hh\overline{2h}l)$, and $(hkil)$ are similar to those just described for the trigonal crystal classes. They differ chiefly because now c is a 6 or $\bar{6}$ axis rather than a 3 or $\bar{3}$ axis. Where c is a $\bar{6}$ axis, because $\bar{6}$ equals $3/m$, the resemblance to trigonal forms is particularly close.

Form {0001}

In classes $\bar{6}$ $(3/m)$, $6/m$, $\bar{6}m2$, 622, and $6/m\ 2/m\ 2/m$, a plane of symmetry or a 2-fold axis (or both) is normal to the c axis and consequently relates (0001) to $(000\bar{1})$, these two faces constituting the *basal pinacoid* $\{0001\}$. In classes 6 and

---

← FIGURE 5–19

The $\{h0\bar{h}l\}$, $\{hh\overline{2h}l\}$, and $\{hk\bar{i}l\}$ forms for the five crystal classes of the trigonal subsystem. (A) For class 3 each is a trigonal pyramid as shown. (B) For class $\bar{3}$ each is a rhombohedron. (C) For class 32, as shown with stereograms as well, $\{h0\bar{h}l\}$ is a rhombohedron, $\{hh\overline{2h}l\}$ is a trigonal dipyramid, and $\{hk\bar{i}l\}$ is a trigonal trapezohedron. (D) For class $3m$, $\{h0\bar{h}l\}$ is a trigonal pyramid, $\{hh\overline{2h}l\}$ is a hexagonal pyramid, and $\{hk\bar{i}l\}$ is a ditrigonal pyramid. (E) For class $\bar{3}\ 2/m$, $\{h0\bar{h}l\}$ is a rhombohedron, $\{hh\overline{2h}l\}$ a hexagonal dipyramid, and $\{hk\bar{i}l\}$ a trigonal scalenohedron. In all cases the class is named after its $\{hk\bar{i}l\}$ form. Thus class 3 is called the trigonal pyramidal class; $\bar{3}$, the rhombohedral class; 32, the trigonal trapezohedral class; $3m$, the ditrigonal pyramidal class; and $\bar{3}\ 2/m$, the trigonal scalenohedral class. In each of these classes the $\{hk\bar{i}l\}$ form was drawn as if $h > k$.

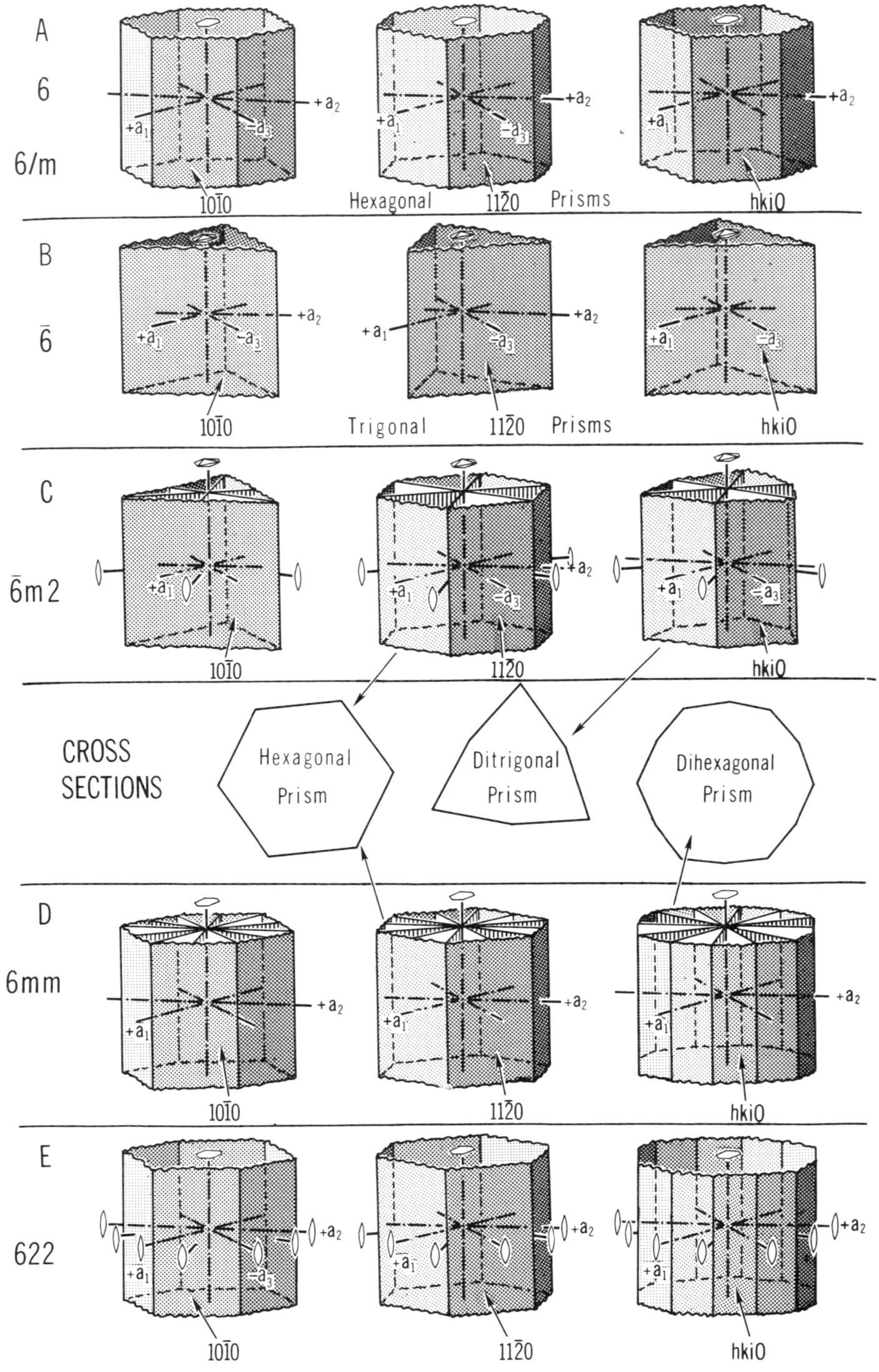

FIGURE 5–20

The $\{10\bar{1}0\}$, $\{11\bar{2}0\}$, and $\{hki0\}$ forms as they occur in the various classes of the hexagonal subsystem. For class $6/m\,2/m\,2/m$ these forms would be as shown in (E) for class 622.

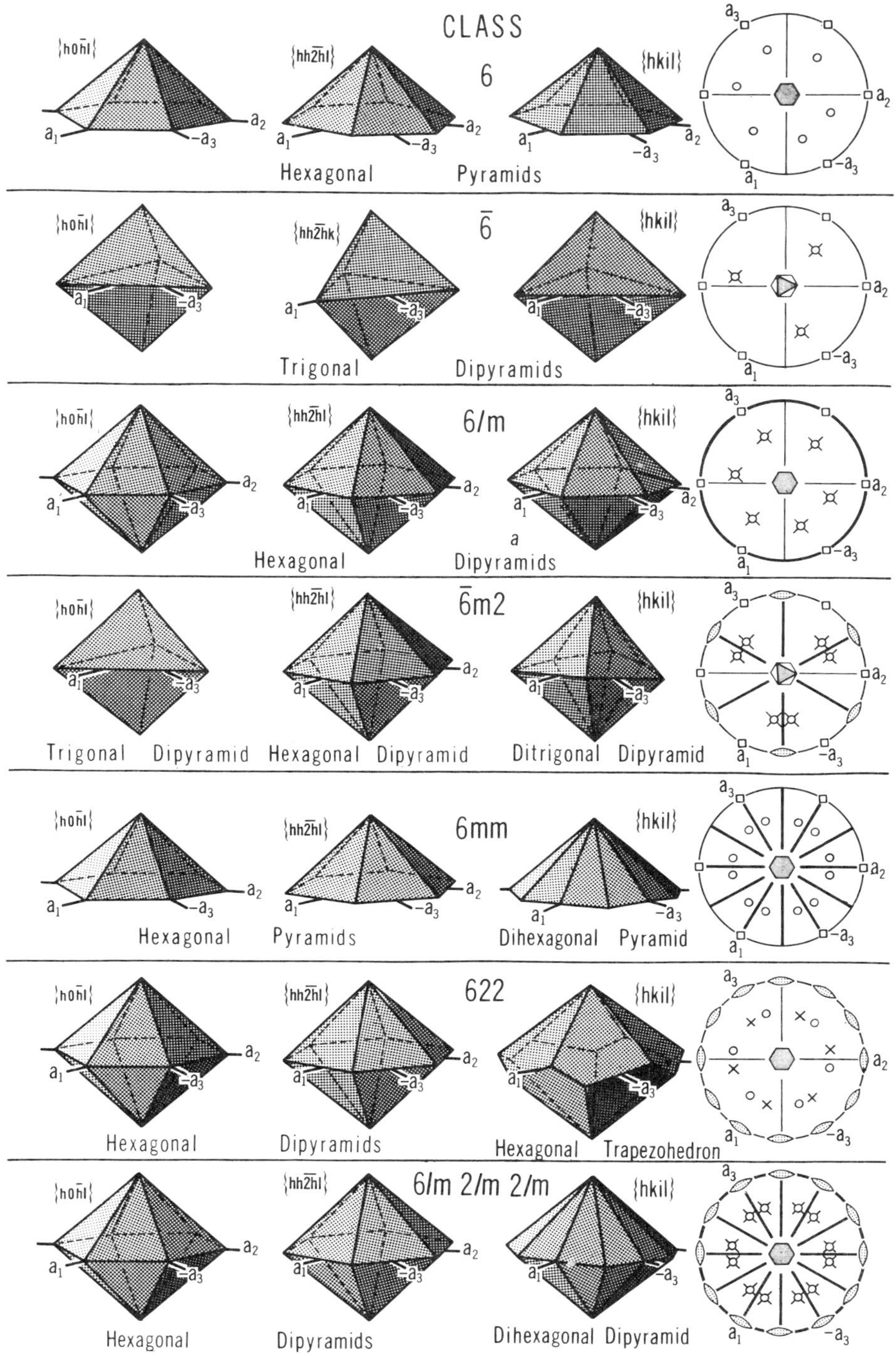

FIGURE 5–21

The $\{h0\bar{h}l\}$, $\{hh\overline{2h}l\}$, and $\{hkil\}$ forms that occur in the seven classes of the hexagonal subsystem. The symmetry elements are not drawn in respect to the crystals (except in the stereogram of each $\{hkil\}$ form shown to the right).

$6mm$, however, which lack such symmetry perpendicular to the c axis, the two faces are unrelated by symmetry and individually constitute the *positive basal pedion* $\{0001\}$ and the negative one $\{000\bar{1}\}$.

Forms $\{10\bar{1}0\}$, $\{11\bar{2}0\}$, and $\{hki0\}$

In classes 6, $\bar{6}$, and $6/m$ the a axes are merely 1 folds and, consequently, only the symmetry of the c axis operates on face $(10\bar{1}0)$, on $(11\bar{2}0)$, or on $(hki0)$. Thus in spite of the more random orientations of $(hki0)$ in respect to the a axes, form $\{hki0\}$ looks the same as forms $\{11\bar{2}0\}$ or $\{10\bar{1}0\}$, these three forms being hexagonal prisms for classes 6 and $6/m$ (Figure 5-20A) but trigonal prisms for class $\bar{6}$ (Figure 5-20B). In classes $\bar{6}m2$, $6mm$, 622, and $6/m\ 2/m\ 2/m$, in which the a axes are either $\bar{2}$ folds, 2 folds, or both, the differing angular attitudes to the a axes of faces $(10\bar{1}0)$, $(11\bar{2}0)$, and $(hki0)$ may cause them to develop into different forms. Thus for class $\bar{6}m2$ (Fig. 5-20C), form $\{10\bar{1}0\}$ is a trigonal prism; $\{11\bar{2}0\}$ is a hexagonal prism; and $\{hki0\}$ is a ditrigonal prism. Forms $\{10\bar{1}0\}$ and $\{11\bar{2}0\}$ are hexagonal prisms, whereas $\{hki0\}$ is a dihexagonal prism for classes $6mm$ (Fig. 5-20D), 622 (Fig. 5-20E), and, although not illustrated, for $6/m\ 2/m\ 2/m$.

Forms $\{h0\bar{h}l\}$, $\{hh\overline{2h}l\}$, and $\{hkil\}$

Figure 5-21 shows forms $\{h0\bar{h}l\}$, $\{hh\overline{2h}l\}$, and $\{hkil\}$ for each crystal class in the hexagonal subsystem. Except for class $6/m\ 2/m\ 2/m$ from which the planes of symmetry normal to the 2-fold axes were omitted in Figure 5-21, the symmetry elements in each crystal class are added to the form in order to show how the individual faces of each form are related by this symmetry to each other. The general form $\{hkil\}$ is a *hexagonal pyramid* for class 6, a *trigonal dipyramid* for class $\bar{6}$, a *hexagonal dipyramid* for class $6/m$, a *ditrigonal dipyramid* for class $\bar{6}m2$, a *dihexagonal pyramid* for class $6mm$, a *hexagonal trapezohedron* for class 622, and a *dihexagonal dipyramid* for class $6/m\ 2/m\ 2/m$. The adjectival form of the name of these general forms serves as the class's name. Thus class $\bar{6}$ is the *trigonal dipyramidal* class; 622 is the *hexagonal trapezohedral* class, and so on.

Isometric Forms

The five isometric crystal classes, 23, $2/m\ \bar{3}$, $\bar{4}3m$, 432, and $4/m\ \bar{3}\ 2/m$, are characterized by four 3 or $\bar{3}$ axes located along the four body diagonals of the guide cube. Their crystallographic axes, the three a axes, coincide with 2-fold axes (classes 23, $2/m\ \bar{3}$), $\bar{4}$ axes ($\bar{4}3m$), or with 4-fold axes ($4/m\ \bar{3}\ 2/m$); hence each is normal to a face of the guide cube. The forms resulting from operation of each crystal class's symmetry on faces (100), (110), $(hk0)$, (111), (hhl), and, finally, (hkl) will be considered next. We shall start with face (100) because of its very special orientation to the a axes and their associated symmetry, then proceed through the less special faces, and finally consider the general forms developed from the randomly oriented face (hkl).

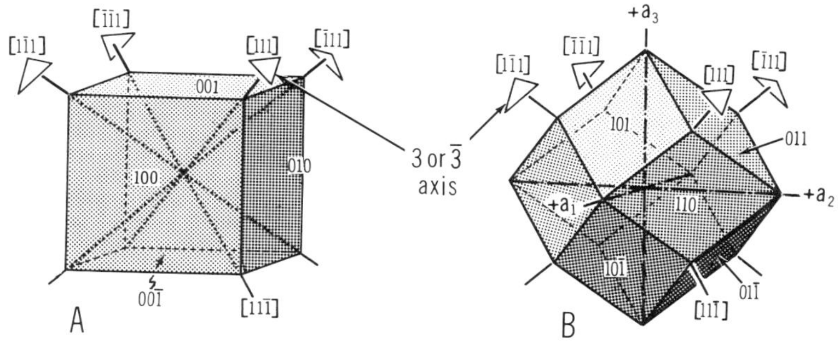

FIGURE 5–22

The two forms that can occur in all five isometric classes. (A) The cube or hexahedron, {100}, and (B) The dodecahedron, {110}. In respect to the four 3- or $\bar{3}$-fold axes invariably present in an isometric crystal class, observe that 120° rotations about the axis labeled [111] symmetrically relates (100) to (001) to (010) in (A) or (110) to (101) to (011) in (B). The 3-fold axis whose end is labeled [$11\bar{1}$] similarly relates (100) to (010) to ($00\bar{1}$) or (110) to ($01\bar{1}$) to ($10\bar{1}$), and so on.

Form {100}

The four 3 or $\bar{3}$ axes present in every isometric crystal class relate face (100) to the remaining five faces of a cube (Fig. 5-22A). Thus a 120° rotation about the 3-fold axis whose direction is [111] relates (100) to (001) and to (010). Similarly, rotation about the 3-fold axis at [$\bar{1}\bar{1}1$] relates (001) to ($0\bar{1}0$) and ($\bar{1}00$); also, as can best be seen from the other end of this 3-fold axis, which is labeled [$11\bar{1}$] in Figure 5-22, the rotation relates (010) to (100) and to ($00\bar{1}$). In each crystal class the symmetry elements additional to the 3 or $\bar{3}$ axes merely relate these faces to each other, adding no more faces to the form. As a consequence {100} is a *cube* in each of the five isometric crystal classes. Often it is called a *hexahedron* because it consists of six faces.

Forms {110} and {hk0}

Face (110) is related to eleven additional faces by operation of the four 3 or $\bar{3}$ axes for each crystal class, their remaining symmetry elements adding no new faces to the form. Form {110} consists of twelve faces (Fig. 5-22B) and is therefore, called a *dodecahedron* or, more specifically, a *rhombic dodecahedron* because its individual faces are rhomb-shaped.

Although parallel to one a axis, as is face (110), face ($hk0$) differs because it intercepts the remaining two a axes at unequal angles whereas (110) intercepted each at 45°. As a consequence, form {$hk0$} is markedly different from the {110} dodecahedron. In classes 23 and $2/m\ \bar{3}$, {$hk0$} is a twelve-faced form known as a *pyritohedron* (Fig. 5-23A, B) because crystals of the mineral pyrite often have this form. In classes $\bar{4}3m$, 432, and $4/m\ \bar{3}\ 2/m$, however, {$hk0$} consists of twenty-four faces (Fig. 5-23C); it is called a *tetrahexahedron* because it frequently

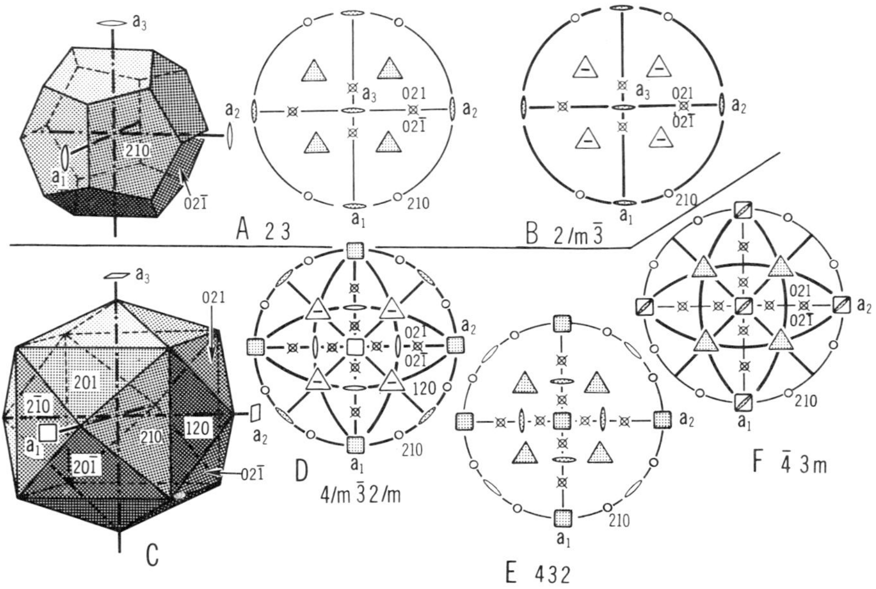

FIGURE 5–23

The two types of $\{hk0\}$ forms possible in the isometric system as exemplified by (A) the pyritohedron $\{210\}$ and (C) the tetrahexahedron $\{210\}$. The two stereograms to the right of the pyritohedron are shown for the two classes in which it occurs. The three stereograms to the right of the tetrahexahedron represent it in the three crystal classes in which it occurs.

resembles the cube or hexahedron in overall shape except that each face of the hexahedron is now replaced by four (hence the prefix, *tetra*), these four intersecting in a point somewhat above the center of each cube face. The three stereograms to the right of the tetrahexahedron in Figure 5-23C disclose how the faces in the form are related by symmetry in each of the three crystal classes in which this form may occur. In class $4/m\,\bar{3}\,2/m$, for example, the 4-fold axis at a_1 (Fig. 5-23C) relates (210) to (201), $(\bar{2}\bar{1}0)$ and $(20\bar{1})$; moreover the $\bar{3}$-fold axes operate on this group of four faces just as they did on an individual face of a cube, so that four faces now occur in place of each single cube face. In summary, whereas $\{110\}$ is a dodecahedron for all five isometric crystal classes, $\{hk0\}$ is a pyritohedron for those classes in which the a axes are 2 fold (23, $2/m\,\bar{3}$) but a tetrahexahedron for those in which they are 4 or $\bar{4}$ fold ($\bar{4}3m$, 432, $4/m\,\bar{3}\,2/m$).

Form {111}

Face (111) is normal to one of the 3-fold axes and hence not repeated by it. The symmetry inherent in the a axes relates (111) to seven additional faces (Fig. 5-24A) in classes $4/m\,\bar{3}\,2/m$, 432, and $2/m\,\bar{3}$, each face being perpendicular to a 3- or $\bar{3}$-fold axis and thus, in the stereograms, plotting at the sites of these axes. The resultant form, because it possesses eight faces, is called an *octahedron*. In classes 23 and $\bar{4}3m$, however, the diminished symmetry of the a axes can only relate (111) to three other faces (Fig. 5-24B), these four faces constituting the *positive tetrahedron* {111}. If these a axes operated on $(11\bar{1})$ in classes 23 or $\bar{4}3m$, the *negative tetrahedron* $\{11\bar{1}\}$ would result, as illustrated for class 23 (Fig. 5-24C). The two tetrahedra, {111} and $\{11\bar{1}\}$, together contain all the faces present in

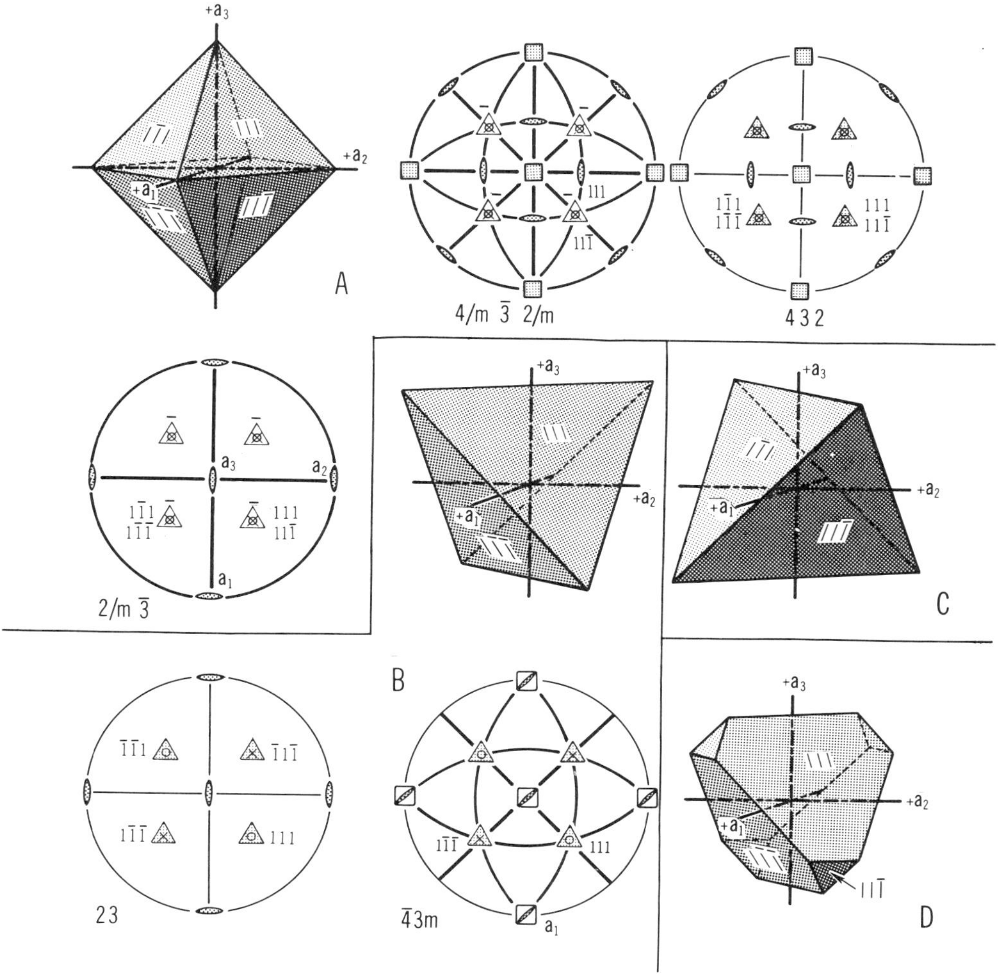

FIGURE 5–24

The two types of {111} forms and their stereograms for the classes in which they can occur. (A) The octahedron; (B) the positive tetrahedron {111}; (C) the negative tetrahedron $\{11\bar{1}\}$ which may also be present in classes 23 or $\bar{4}3m$. If both the positive and negative tetrahedra are present on the same crystal as in (D), the faces of one form will often be larger (better developed) than those of the other.

the octahedron and are thus known as *correlate forms.* Both forms may be developed on the same crystal, either (1) unequally as in Figure 5-24D or (2) equally so that the $\{11\bar{1}\}$ faces equal the $\{111\}$ faces in area (and the crystal then mimics a true octahedron). If either form is the only one present on a crystal, $\{111\}$ is indistinguishable from $\{11\bar{1}\}$ just as the models in Figures 5-24B and C would be indistinguishable were their a axes erased.

The {hhl} forms

In isometric classes the $\{hhl\}$ forms divide into two distinct types according to whether h is less than or exceeds l in value. Because face (hhl) cuts the a_1 and a_2 axes at equal angles, all faces of this type plot along the $\phi = 45°$ line in any stereogram (Fig. 5-25A). More precisely, if h is less than l, they plot between the (001) and (111) face poles, whereas if h exceeds l, they plot between (111) and (110). Face pole (111) marks the point of emergence of one of the four 3- or $\bar{3}$-fold axes characteristic of isometric crystals. This 3- or $\bar{3}$-fold axis then operates on (hhl) and repeats it if h is less than l, as shown in Figure 5-25B, whereas if h exceeded l, (hhl) is repeated as shown in Figure 5-25C. The basic difference is that in Figure 5-25B *one* face (that is, (hhl)) plots near axis a_3 and therefore more nearly approaches perpendicularity to this axis, whereas in Figure 5-25C the *two* faces (hlh) and (lhh) plot equally near to axis a_3 and therefore one is as near perpendicularity to it as the other.

For classes 23 and $\bar{4}3m$, whose $\{111\}$ forms were tetrahedra, the three faces whose face poles were shown in Figures 5-25B and C are then each repeated by the remaining symmetry elements of the class; the resultant $\{hhl\}$ form is a

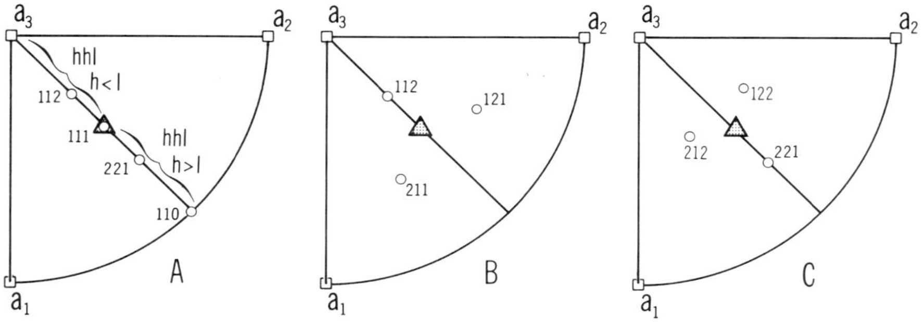

FIGURE 5–25

(A) Different locations along the $\phi = 45°$ line in one quadrant of a stereogram of face poles of the type (hhl) compared to the projection of the 3-fold axis (at $\rho = 54°44'$) for isometric crystals in standard orientation.
(B) Operation of the 3-fold axis upon (112), an (hhl) face for which $h < l$, requires the presence of (211) and (121) with face poles where shown.
(C) Operation of the 3-fold axis upon (221) requires (212) and (122) to be present with face poles where shown.

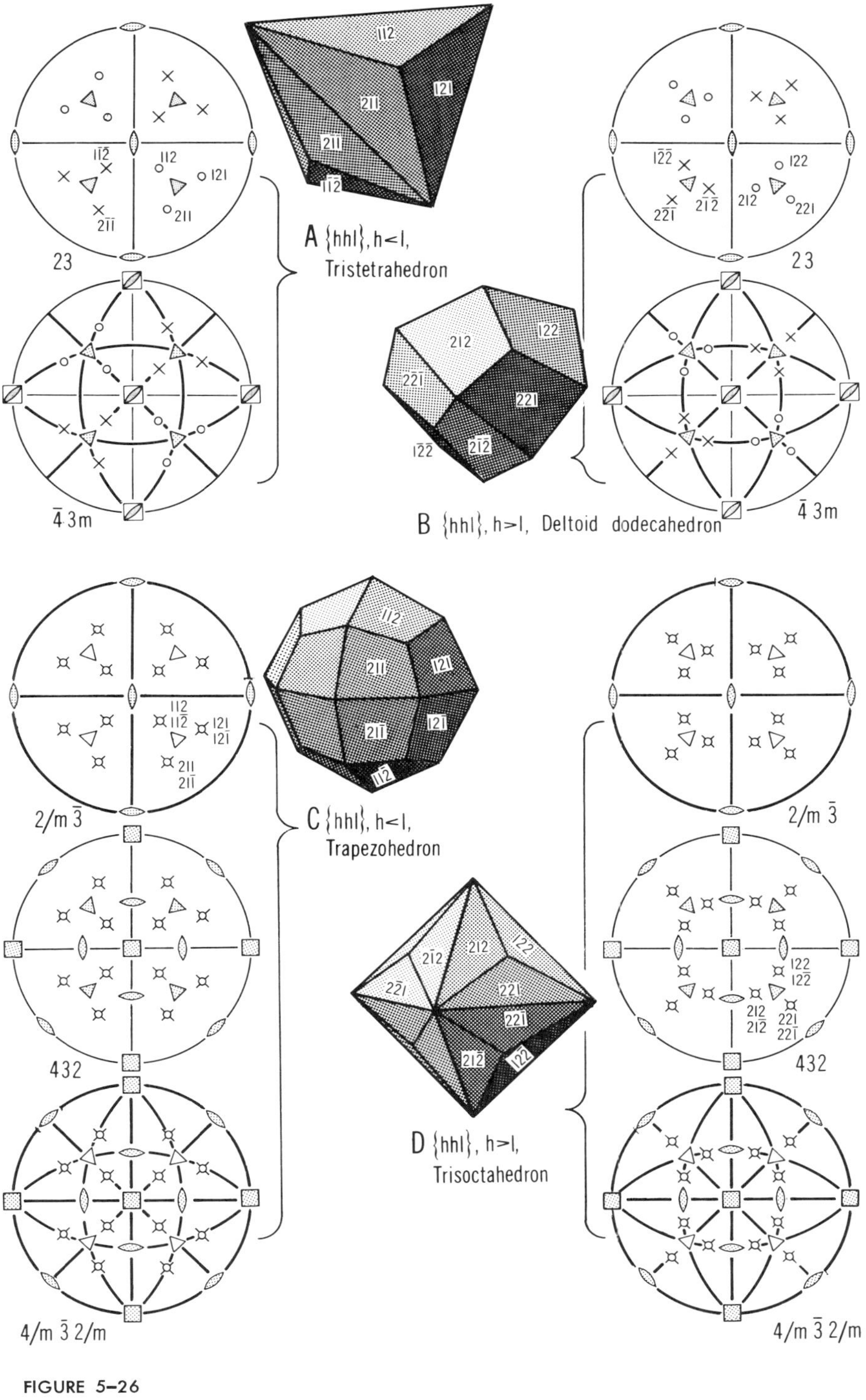

FIGURE 5–26

The types of $\{hhl\}$ forms and their stereograms that are possible in the isometric system.

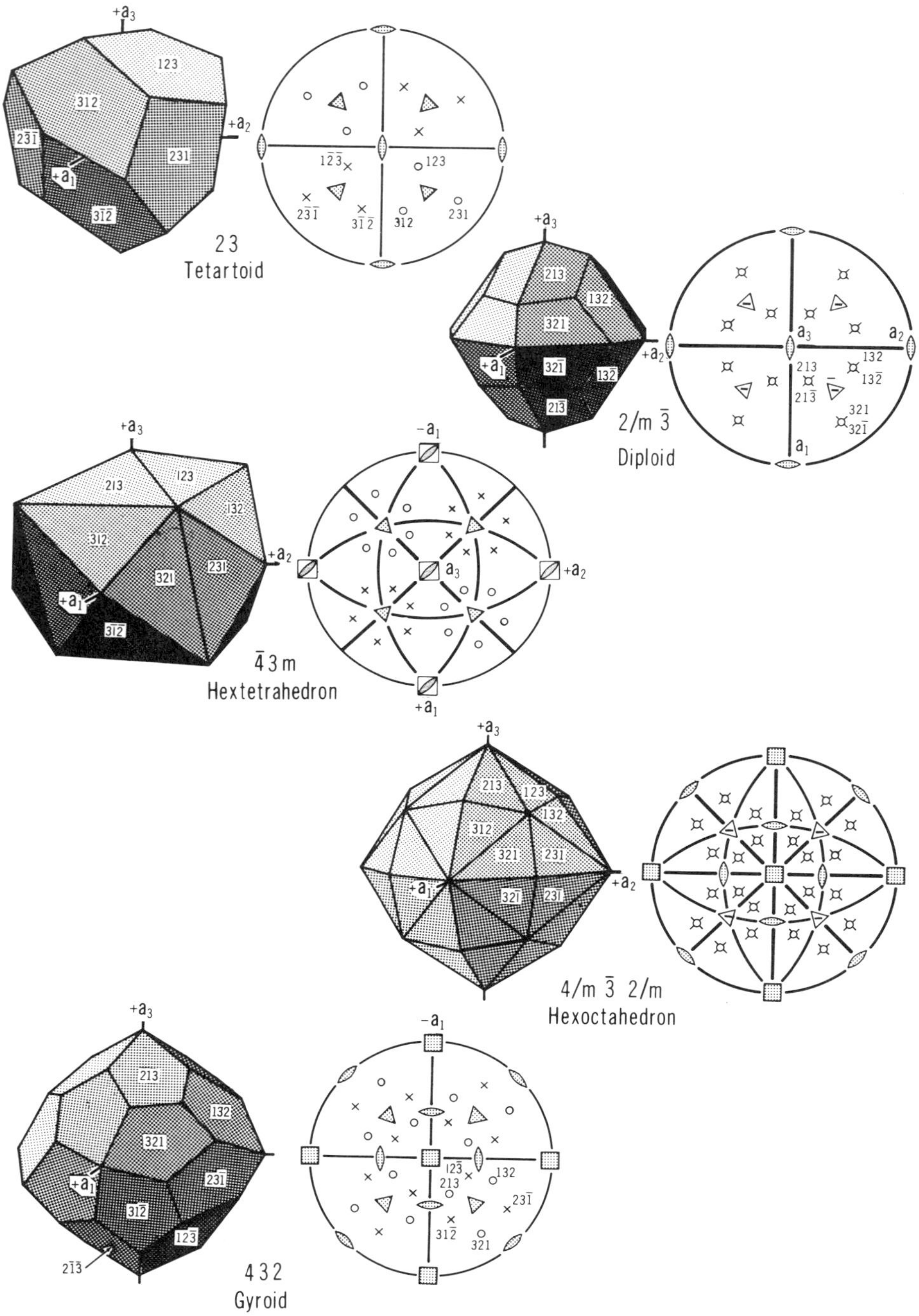

FIGURE 5–27

The general $\{hkl\}$ forms and their stereograms for the five isometric crystal classes.

tristetrahedron (Fig. 5-26A) if h is less than l, but a *deltoid dodecahedron* (Fig. 5-26B) if h exceeds l. The tristetrahedron is so named because it resembles a tetrahedron each of whose faces has been displaced by three. Like the deltoid dodecahedron, it consists of twelve faces. For class 23 the stereogram of each form indicates how the three faces in the lower-right quadrant are related to the three in the upper-left quadrant by a 180° rotation about the 2-fold axis at a_3. Rotation of 180° about the 2-fold axis at either a_1 or a_2 then relates these six faces to those shown to plot on the lower hemisphere.

For classes $2/m\ \bar{3}$, 432, and $4/m\ \bar{3}\ 2/m$, whose {111} forms were octahedra, the three faces whose face poles were shown in Figures 5-25B and C, are repeated by their symmetry. For each class the resultant $\{hhl\}$ form is seen to be a *trapezohedron* (Fig. 5-26C) if h is less than l, but a *trisoctahedron* (Fig. 5-26D) if h exceeds l. The trapezohedron consists of twenty-four trapezium-shaped faces, whereas the trisoctahedron consists of twenty-four faces, each an isosceles triangle. Both the trapezohedron and the trisoctahedron may be likened to an octahedron, each face of which is replaced by three, the trisoctahedron being so named because of this. Note that the trisoctahedron $\{hhl\}$ is a form for which h exceeds l, whereas for the tristetrahedron $\{hhl\}$, h is less than l.

The {hkl} forms

The $\{hkl\}$ forms are general forms and, consequently, unlike the special forms, differ for each of the five isometric crystal classes (Fig. 5-27). The names of these general forms are often applied to the crystal class. Thus 23 is called the tetartoidal class; $\bar{4}3m$, the hextetrahedral class; $2/m\ \bar{3}$, the diploidal class; 432, the gyroidal class; and $4/m\ \bar{3}\ 2/m$, the hexoctahedral class. In class $\bar{4}3m$, for example, face (hkl) is repeated five more times in the lower-right quadrant of the stereogram, merely through operation of the nearby 3-fold axis and the three symmetry planes that intersect at this 3-fold axis. The remaining symmetry then operates on these six faces to generate a hextetrahedron, which resembles a tetrahedron, each face of which is replaced by six. Similarly, the hexoctahedron may be likened to an octahedron, each face of which is replaced by six.

Enantiomorphous Forms and Crystals

The general form, and occasionally others, for crystals that lack a plane of symmetry, inversion axes, and a center of symmetry—that is, belong to classes 1, 2, 222, 4, 422, 3, 32, 6, 622, 23, or 432—occur as either of two nonsuperposable types, each the other's mirror image and therefore said to be *enantiomorphous*. For example, the two trigonal trapezohedrons, x_L $\{6\bar{1}\bar{5}1\}$ and x_R $\{51\bar{6}1\}$ (Fig. 5-28A), which can occur on crystals of quartz (trigonal class 32), are enantiomorphous. Like right- and left-handed gloves, two enantiomorphous forms are never superposable and, by analogy, one is called right handed and the other left handed. On *individual crystals* only right-handed or only left-handed forms, but never both, may occur. Thus crystals from the previously cited eleven crystal

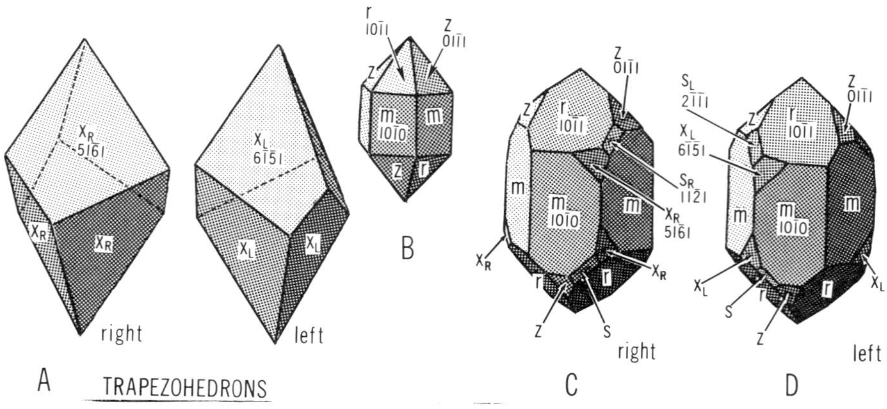

FIGURE 5–28

(A) The two trigonal trapezohedra $\{51\bar{6}1\}$ and $\{6\bar{1}\bar{5}1\}$ that may occur in quartz (class 32) are called *enantiomorphs,* a name applied to pairs of nonsuperimposable objects that bear a mirror-image relationship to each other. By analogy to right and left hands, which are also enantiomorphs, $\{51\bar{6}1\}$ is called a "right-handed" and $\{6\bar{1}\bar{5}1\}$ a "left-handed" trapezohedron.
(B) A quartz crystal that lacks any enantiomorphous forms and possesses only the three special forms, the two rhombohedra r $\{10\bar{1}1\}$ and z $\{01\bar{1}1\}$ and the hexagonal prism m $\{10\bar{1}1\}$.
(C) and (D) Quartz crystals that exhibit enantiomorphous forms in addition to the special forms; (C) contains only the right-handed forms, x_R and s_R, and is called a right-handed crystal; (D) contains only left-handed forms and is called a left-handed crystal. As a mnemonic device, regard the $(10\bar{1}0)$ face as the back of a man. For right-handed quartz (C), faces x and s resemble an upraised right arm; for left-handed quartz (D), they resemble a left arm.

classes are themselves enantiomorphous and classifiable as right- or left-handed individuals whose internal structures are frequently related as are the threads of right- and left-handed screws. For example, throughout the world all individual quartz crystals are either left handed or right handed. If only special forms are present such as the hexagonal prism m $\{10\bar{1}0\}$, the positive rhombohedron r $\{10\bar{1}1\}$, and the negative rhombohedron z $\{01\bar{1}\bar{1}\}$[1] the "handedness" of such a crystal (Fig. 5-28B) is not determinable from mere inspection. Such visual determination is possible only if an enantiomorphic form is also present (Figs. 5-28C, D). Thus for a quartz crystal held in standard orientation, that is, with c axis vertical and with a well-developed rhombohedral face r $(10\bar{1}1)$ toward the observer, the right trigonal trapezohedron x_L $\{51\bar{6}1\}$, would occur on right-handed crystals (Fig. 5-28C) as an upper and lower face immediately to the observer's *right* of the forward prism face $(10\bar{1}0)$; on left-handed crystals (Fig. 5-28D), the left trigonal trapezohedron x_R $\{6\bar{1}\bar{5}1\}$ would occur as two faces

[1] For quartz, the z faces are usually of smaller area than r and thereby distinguished from them.

immediately to the observer's left of $(10\bar{1}0)$. Determination of the "handedness" of a crystal from its optical activity will be discussed later.

A clear distinction should be made in respect to right- and left-handed forms as opposed to positive and negative forms. Like a right and left shoe, right- and left-handed forms cannot be superimposed, whereas the positive and negative versions of a particular form could be (Figs. 5-24B and C). Moreover, positive and negative forms can simultaneously occur on an *individual* crystal (Fig. 5-24D), whereas right- and left-handed forms cannot.

Class Determination

Morphologic determination

Unequivocal designation of the crystal class for an unknown crystal by inspection of the forms present is often difficult and sometimes impossible. This is true even if precise measurements of its interfacial angles are enlisted to sort out and recognize the forms that are present on malformed crystals. One obvious difficulty occurs if the only forms present are special forms. For example, the cube $\{100\}$ is possible in isometric classes 23, $2/m\,\bar{3}$, $\bar{4}3m$, 432, or $4/m\,\bar{3}\,2/m$; hence a crystal that is a cube might belong to any of these five classes. Another difficulty occurs if the general form for one class resembles the special form in others. In class 4, for instance, the general $\{hkl\}$ form is a tetragonal pyramid of the same appearance as the $\{hhl\}$ or $\{h0l\}$ tetragonal pyramids that occur as special forms in class $4mm$ (Fig. 5-14). Consequently, the tests that follow may aid in the determination of the crystal class of an unknown crystal.

Etch tests

If a crystal is immersed in (or lightly coated with) a fluid in which it is soluble, solution initially occurs at isolated spots so that minute depressions called *etch pits* develop on its faces. The frequently used etchants, that is, water, dilute acids, acetone, alcohol, benzene, carbon disulfide, are usually applied by a light touch with a cloth or by causing their vapor to condense on and lightly coat the crystal's faces; for less soluble crystals a period of immersion may be needed. From isolated spots on the crystal's faces, solution will then proceed radially outward at different rates for different directions, the rates being identical for symmetrically equivalent directions. To illustrate, if a 4-fold axis is perpendicular to the face, the rate of solution will be identical along directions in the face that are at 90° to each other; hence the resultant etch pits will possess 4-fold symmetry. As a consequence the shapes of the etch pits, these being best observed microscopically, will often identify those symmetry elements that are normal to the face.

The shapes of all the etch pits on a given face, or on faces belonging to the same form, will be similar although between two faces of a form they may differ in orientation so as to conform to the crystal's point-group symmetry. This is

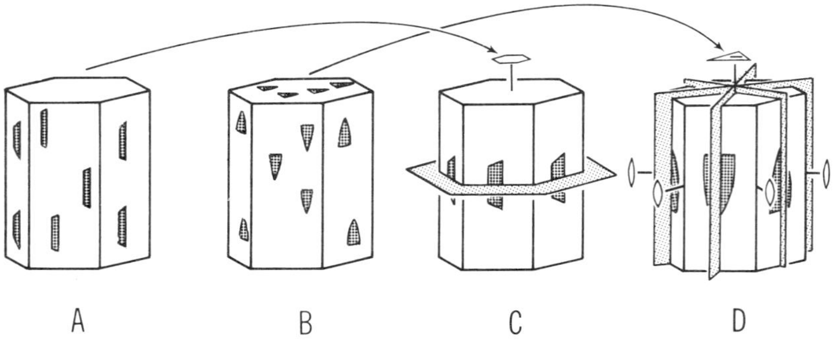

FIGURE 5–29

Crystals of (A) apatite and (B) calcite if etched with HCl. The shapes of the etch pits, drawn enlarged and in outline form, conform to class $6/m$ for apatite (A) and to class $\bar{3}\ 2/m$ for calcite (B). This is most easily verified by considering the etch pits, which are actually tiny and randomly distributed on the above crystal faces, to be replaced by a single pit of like shape at each face's center (C, D).

illustrated by the etch figures on a hexagonal prism of apatite (Fig. 5-29A) or on a hexagonal prism of calcite (Fig. 5-29B). Although a hexagonal prism may be present in several crystal classes, the etch figures indicate apatite to belong to $6/m$ (Fig. 5-29C) and calcite to $\bar{3}\ 2/m$ (Fig. 5-29D). Similarly, the shape and orientation of the etch pits on the faces of a cube might reveal the specific isometric crystal class to which a particular cube belonged (whereas the cube's shape alone would invariably indicate $4/m\ \bar{3}\ 2/m$).

Etch figures may be immature and misleading if too little solvent or too short an etch time is used. In view of this, X-ray techniques (Chapter 13) are now used almost exclusively for the determination of crystal class. However, X-ray techniques possess an outstanding weakness: they cannot determine whether or not a center of symmetry is present because all crystals react to X-rays as if a center of symmetry were present, even if none is. For example, crystals from classes $mm2$ or 222 yield similar X-ray results to those belonging to class $2/m\ 2/m\ 2/m$. In other words, X-ray techniques cannot usually determine whether a crystal is *centrosymmetric,* that is, possesses a center of symmetry, or *noncentrosymmetric,* that is, lacks one. Hence they often need to be supplemented by the tests for noncentrosymmetry that are described next.

Pyroelectric and piezoelectric tests for noncentrosymmetry

A crystallographic direction $[uvw]$ in a crystal is called *nonpolar* if it is related by symmetry to the diametrically opposite direction $[\bar{u}\bar{v}\bar{w}]$; if it is not so related, it is *polar.* Thus for a crystal of class m (Fig. 5-30A), direction $[010]$ is nonpolar because the mirror plane relates it to $[0\bar{1}0]$; in this same class, however, general directions such as $[uvw]$ are polar, the mirror plane being incapable of relating $[uvw]$ to the opposite direction $[\bar{u}\bar{v}\bar{w}]$. In crystals that possess a center

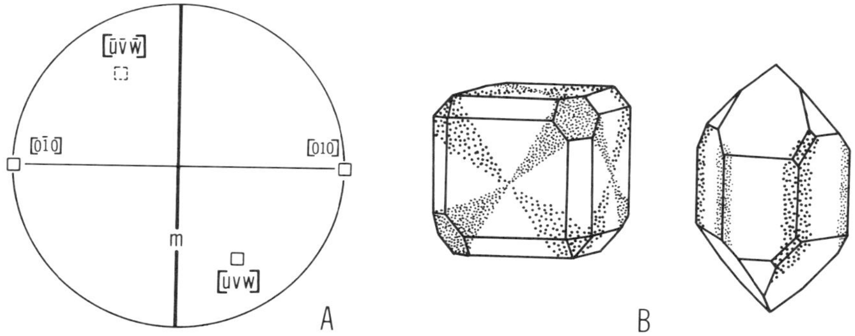

FIGURE 5–30

(A) Stereogram for a crystal that belongs to class *m* and for which only direction [010] is nonpolar, the mirror plane *m* relating [010] to [01̄0]. Other directions such as [*uvw*] are polar because plane *m* cannot relate them to [*ūv̄w̄*], their opposite direction.
(B) Kundt's test as applied to a crystal of boracite (left) and quartz (right) discloses their noncentricity because sulfur particles adhere to the positively charged (finely stippled) areas and red lead to the negatively charged (coarsely stippled) areas (after Dana).
The end of a polar axis is called the *analogous pole,* if it becomes positively charged on heating (= negative on cooling) but the *antilogous pole* if it becomes negative on heating (= positive on cooling).

of symmetry, all directions are necessarily nonpolar because the center always relates [*uvw*] to [*ūv̄w̄*]. Thus only noncentrosymmetric crystals can possess polar axes and therefore to demonstrate the existence of polar axes for a crystal is to demonstrate noncentrosymmetry.

If a crystal is heated, cooled, or subjected to a directed pressure, the positively charged ions composing it may be slightly displaced relative to its negative ions. For a nonpolar direction [*uvw*] in a crystal, this charge displacement along [*uvw*] will be the same as along the opposite direction [*ūv̄w̄*], the symmetrical equivalence of [*uvw*] to [*ūv̄w̄*] for a nonpolar direction requiring this. However, if [*uvw*] is a polar direction, the charge displacement along [*uvw*] need no longer be balanced by an equal charge displacement along [*ūv̄w̄*]. Consequently, a polar axis [*uvw*] in a crystal may develop on its opposite ends static electric charges of opposite signs in response to heating or cooling (=*pyroelectricity*) or to application or release of a directed stress (=*piezoelectricity*). If this pyroelectric or piezoelectric effect is strong, the presence of polar axes and therefore the noncentrosymmetry of the crystal may be demonstrated by *Kundt's method.* Thus if a finely pulverized mixture of red lead (PbO) and of sulfur is sieved (through fine muslin) onto a noncentrosymmetric crystal after it has been cooled, heated, or released from a directed stress, then the fine red particles of PbO, having become positively charged during the sieving, will adhere to negatively charged areas on the crystal, whereas the negatively charged sulfur particles will adhere to the positively charged surfaces on the crystal (Fig. 5-30B). Even though some crystals are noncentrosymmetric, the charges developed in them may be so weak as to escape detection by the Kundt method. In this event more sensitive detectors of pyroelectricity or piezoelectricity must be utilized, often at temperatures from room temperature down to the boiling temperature of liquid nitrogen. If such sensitive methods yield negative results and thus disclose no evidence for noncentrosymmetry, the crystal is likely to be centrosymmetric. Unfortunately, however, the possibility always remains that the crystal is noncentrosymmetric *but so weakly pyro- or piezoelectric that this escapes detection.*

Form Symbols for Directions

Just as $\{hkl\}$ represents the form of faces related by the symmetry operations of the crystal class, the symbol $\langle uvw \rangle$ represents the "form" of lines or directions so related. Thus in crystal class $\bar{1}$ the symbol $\langle 112 \rangle$ represents directions [112] plus $[\bar{1}\bar{1}\bar{2}]$.

QUESTIONS AND PROBLEMS[2]

5-1 The number of faces in a given form constitutes its *multiplicity factor.* For example, the multiplicity factor for a cube is six; for an octahedron it is eight. (a) What is the multiplicity factor for the $\{432\}$ form in class 1? (b) in class $2/m$? (c) in 222? (d) in 422? (e) in $\bar{4}2m$? (f) in 23? (g) $2/m\,\bar{3}$? (h) 432? (i) $4/m\,\bar{3}\,2/m$?

5-2 Name the $\{432\}$ form for (a) class 1; (b) class $2/m$; (c) 222; (d) 422; (e) $\bar{4}2m$; (f) 23; (g) $2/m\,\bar{3}$; (h) 432; (i) $4/m\,\bar{3}\,2/m$.

5-3 (a) What is the multiplicity factor for $\{201\}$ in class $2/m$? (b) in $mm2$? (c) in m? (d) in 222? (e) in $4/m$? (f) in 422? (g) in 4? (h) in $\bar{4}2m$? (i) in 23? (j) in $\bar{4}3m$? (k) in 432? (l) in $4/m\,\bar{3}\,2/m$?

5-4 Name the $\{201\}$ form for the classes cited in (a) to (l) of Question 5-3.

5-5 (a) What is the multiplicity factor for $\{10\bar{1}1\}$ in class 32? (b) $\bar{3}\,2/m$? (c) 6? (d) $\bar{6}$? (e) $6/m$? (f) 622? (g) $\bar{6}m2$?

5-6 Name the $\{10\bar{1}1\}$ form for the classes cited in Question 5-5.

5-7 (a) For what crystal system(s) is [111] perpendicular to face (111)? (b) [110] to (110)? (c) [001] to (001)? (d) [100] to (100)? (e) [010] to (010)?

5-8 What direction(s), if any, will be symmetrically related to [001]: (a) in crystal class $\bar{4}$? (b) in class 4?

5-9 What direction(s), if any, will be symmetrically related to [110]: (a) in

[2] Assume all monoclinic crystals to be in the second setting.

class 4? (b) in $\bar{4}$? (c) in $2/m$? (d) in 222? (e) in m? (f) in $\bar{1}$? (g) in class 1?

5-10 What direction(s), if any, will be symmetrically related to $[uvw]$ in (a) class $\bar{1}$? (b) class m? (c) class $2/m$? (d) class 1? (e) class 222?

ANSWERS

5-1 (a) 1; (b) 4; (c) 4; (d) 8; (e) 8; (f) 12; (g) 24; (h) 24; (i) 48.

5-2 (a) Pedion; (b) prism; (c) orthorhombic disphenoid; (d) tetragonal trapezohedron; (e) tetragonal scalenohedron; (f) tetartoid; (g) hextetrahedron; (h) gyroid; (i) hexoctahedron.

5-3 (a) 2; (b) 2; (c) 1; (d) 4; (e) 8; (f) 8; (g) 4; (h) 8; (i) 12; (j), (k), (l) 24.

5-4 (a) Pinacoid; (b) dome; (c) pedion; (d) prism; (e) tetragonal dipyramid; (f) tetragonal dipyramid; (g) tetragonal pyramid; (h) tetragonal dipyramid; (i) pyritohedron; (j), (k), (l), tetrahexahedrons.

5-5 (a), (b) 6; (c) 6; (d) 6; (e), (f) 12; (g) 6.

5-6 (a), (b) Rhombohedron; (c) hexagonal pyramid; (d) trigonal dipyramid; (e), (f) hexagonal dipyramid; (g) trigonal dipyramid.

5-7 (a) Isometric only; (b) isometric and tetragonal; (c) isometric, tetragonal, and orthorhombic; (d) isometric, tetragonal, and orthorhombic; (e) isometric, tetragonal, orthorhombic, and monoclinic. (See Fig. 5-31 for explanation.)

5-8 (a) $[00\bar{1}]$; (b) none.

5-9 (a), (b), (c), (d) $[\bar{1}10]$, $[\bar{1}\bar{1}0]$, $[1\bar{1}0]$; (e) $[1\bar{1}0]$; (f) $[\bar{1}\bar{1}0]$; (g) none.

5-10 (a) $[\bar{u}\bar{v}\bar{w}]$; (b) $[u\bar{v}w]$; (c) $[u\bar{v}w]$, $[\bar{u}\bar{v}\bar{w}]$, $[\bar{u}v\bar{w}]$; (d) none; (e) $[u\bar{v}\bar{w}]$, $[\bar{u}\bar{v}w]$, $[\bar{u}v\bar{w}]$.

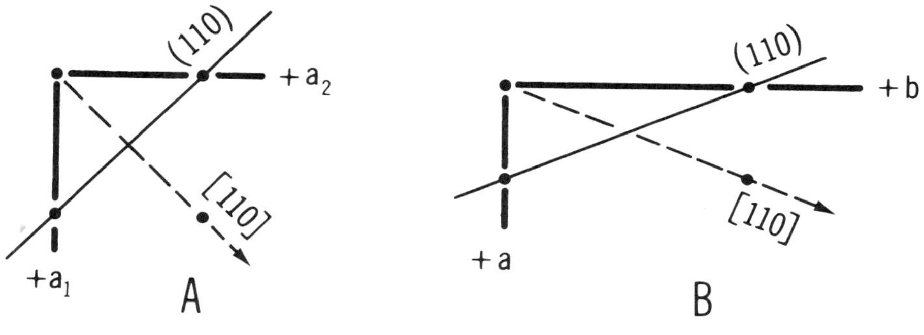

FIGURE 5–31

Illustration of the principle that a face and a direction with identical indices, for example, (110) and [110], will be mutually perpendicular *only if the face intercepts just those axes whose units are equal.* Thus drawings with the *c* axis perpendicular to the page disclose (110) and [110] to be mutually perpendicular for (A) the tetragonal system but not for (B) the orthorhombic system.

6 Translational Symmetry: Lattices

Arrays and Identipoints

The regular spacing of identical objects, each being identically oriented in space, may be called an *array* of such objects. To illustrate, the chessmen in Figure 6-1A represent a *linear* or *one-dimensional array* because they are regularly spaced along a line. Those in Figure 6-1B represent a *planar* or *two-dimensional array,* whereas those in Figure 6-1C represent a *three-dimensional array.* These arrays represent a type of symmetry known as *translational symmetry.* Thus if chessman 1 in Figure 6-1A were moved translationally[1] along (and for distances equal to) vector **b**, it successively coincides with chessmen 2, 3, 4, . . . , and vector **b** is called the *unit of translation* along this line. Similarly, a plane array (Fig. 6-1B) may be considered to result from the successive translation of an entire linear array, or of each object composing this linear array, along (and for a distance equal to) a second vector such as **a**. We could as well consider the translation of the linear array or its objects to occur along a vector such as **a′**, but **a** is conventionally preferred because it is the shortest of the possibilities. In like manner a three-dimensional array may be regarded as resulting from the repeated translation of a planar array, for example, along vector **c** in Figure 6-1C.

Such arrays are most easily described by specifying (1) the nature of the object being repeated in space, for example, the chessman in Figure 6-1, and (2) the scheme of such repetition. The scheme of repetition underlying an array is laid bare if an identical point, here called an *identipoint,* is located on each object

[1] Translational motion is fulfilled only if all points on the object trace out straight-line paths during such motion. Thus a jet airplane on perfectly even keel (in straight-line flight) illustrates translational motion, whereas if it goes into a barrel roll its motion ceases to be translational.

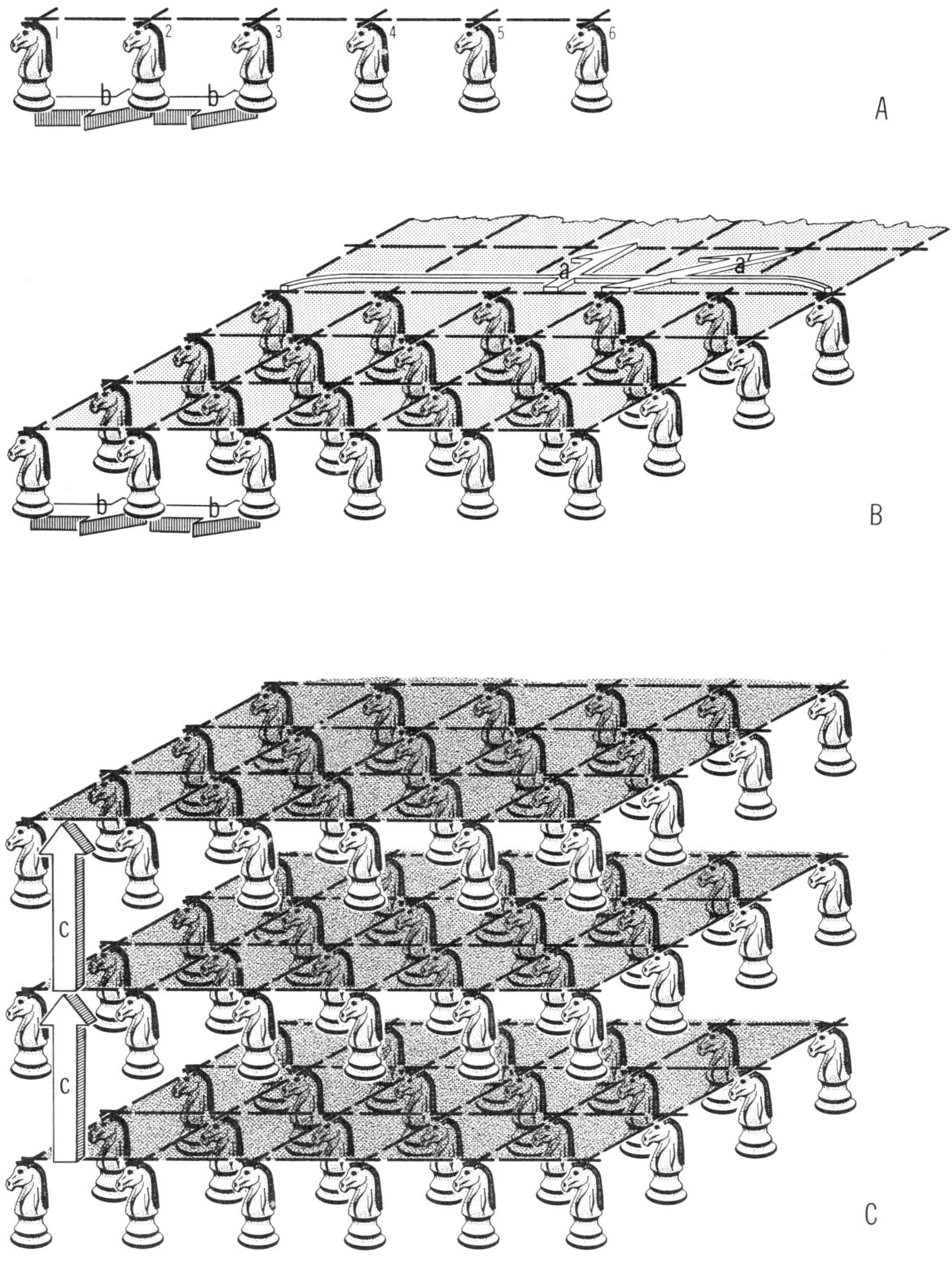

FIGURE 6–1

Examples of one-, two-, and three-dimensional arrays, *A*, *B*, and *C*, respectively. Vectors **a**, **b**, and **c** indicate the three shortest amounts (and directions) that a chessman need be moved to coincide in position with a neighbor.

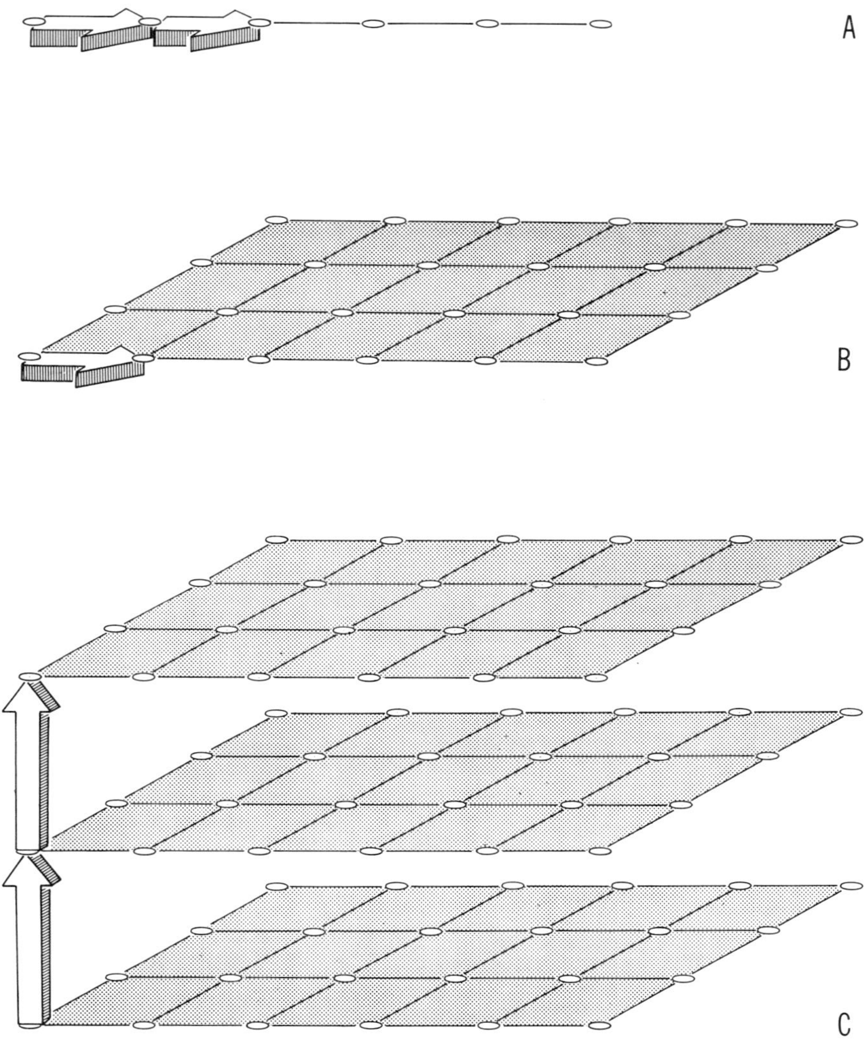

FIGURE 6-2

An identical point on each chessman in Figure 6-1, for example, the left ear tip, has been selected and the rest of the chessman faded away. The location of each such identipoint is indicated by a small circle that appears elliptical in perspective. Three types of point lattices thus result: (A) a *line lattice* or *lattice row*; (B) a *plane lattice* or *net*; (C) a *space lattice*.

of the array. In Figure 6-1, for example, the left ear tip of each chessman is marked with a cross. Now if each chessman were faded away until only this identipoint remained, an array of identipoints called a *point lattice* would result. Depending on whether a point lattice summarizes the scheme of repetition of a

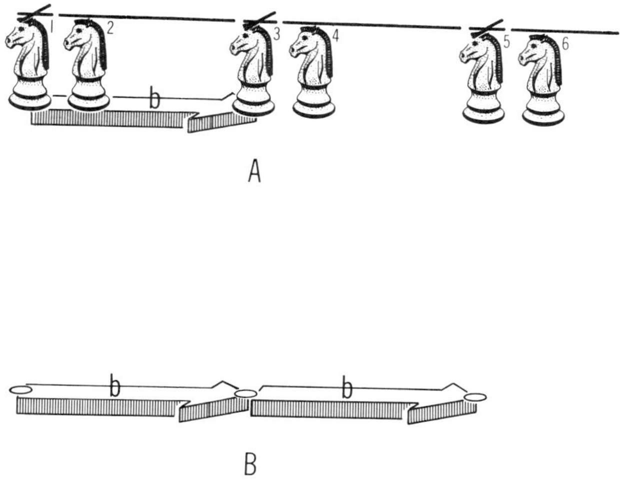

FIGURE 6–3

(A) A pair of chessmen translationally repeated according to a line lattice.
(B) The line lattice that summarizes the nature of the translational repetition of the pair of chessmen shown.

one-, two-, or three-dimensional array of objects, it is called a *line lattice* or *lattice row* (Fig. 6-2A), a *plane lattice or net* (Fig. 6-2B), or a *space lattice* (Fig. 6-2C). For any given array of objects, the same point lattice will be obtained regardless of which set of identipoints is singled out to represent this lattice. We can verify this by imagining each chessman in Figure 6-1 as being replaced by a point at its center of gravity or at the corner of its left eye, and so on. The point lattice only shifts bodily up or down, or left or right, if a different set of identipoints is chosen.

Occasionally caution must be exercised in locating a set of identipoints in an array. In Figure 6-3A, for example, only every second chessman is identical. Thus chessmen 3 and 4 differ because the closest neighbor is *behind* for 3 but *in front* for 4. Consequently, *all* left ear tips do not comprise a single set of identipoints, only those of every second horse. Therefore the crosses comprise one set of identipoints; the centers of the bases of horses 1, 3, and 5 comprise a second set. The line lattice that summarizes the scheme of repetition is drawn in Figure 6-3B. The object to be translationally repeated by vector **b** is now a *pair* of chessmen; for example, if chessmen 1 and 2 are successively repeated along vector **b**, they generate the array in Figure 6-3A.

Types of Line Lattices and Nets

A line lattice consists of identipoints equally spaced along a line. Hence only one basic type of line lattice is possible because there is only one way of spacing points equally along a line. Thus line lattices can differ only in respect

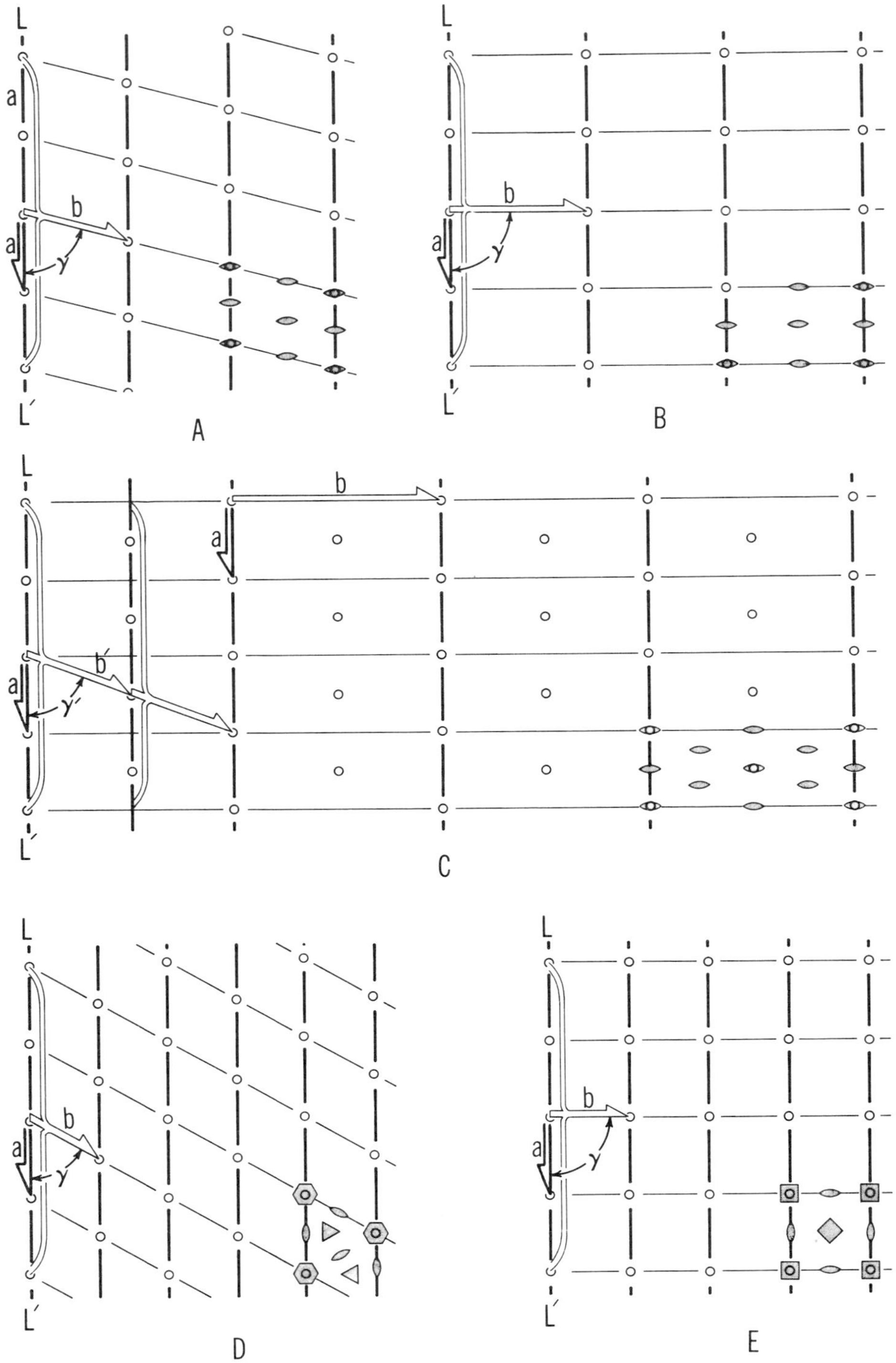
L
a
b
a
γ
L′
A
L
b
a
γ
L′
B
L
b
a
b′
a
γ′
L′
C
L
b
a
γ
L′
D
L
b
a
γ
L′
E

to the distance between their points. This distance is called their *unit of translation* and is represented by the length of vector **b** in Figure 6-3 (or by the length of vector **a** for line LL' in Figure 6-4).

The five types of nets or plane lattices may be developed by considering a line lattice LL' (Fig. 6-4), whose points are translationally repeated by a vector **a**, as being itself bodily translated by a second vector **b** at an angle γ to **a**. The types differ according to whether length **a** equals **b** and whether γ equals 90°, 60°, or an arbitrary angle. The five types are thus as follows:

1. *Clinonet* (Fig. 6-4A): $\mathbf{a} \neq \mathbf{b}$; $\gamma \neq 90°$.
2. *Orthonet* (Fig. 6-4B): $\mathbf{a} \neq \mathbf{b}$; $\gamma = 90°$.
3. *Centered orthonet* (Fig. 6-4C): To obtain this, line LL' in Figure 6-4C was translated by vector $\mathbf{b}'$ which is at an angle (γ') to vector **a** such that $\cos \gamma' = \mathbf{a}/2\mathbf{b}'$. The resultant net is conventionally regarded to possess two orthogonal vectors **a** and **b**, which define rectangles that now have lattice points at their centers as well as at their corners.
4. *Hexanet* (Fig. 6-4D): $\mathbf{a} = \mathbf{b}$; $\gamma = 60°$.
5. *Tetranet* (Fig. 6-4E): $\mathbf{a} = \mathbf{b}$; $\gamma = 90°$.

Page-size limitations permitted only small areas of each net to appear in Figure 6-4; actually each should be regarded as extending indefinitely and thus containing innumerable identipoints. When considered this way, there is a symmetry to the distribution of these identipoints as follows: Perpendicular to each net, at each lattice point, there are either 2-fold axes (Figs. 6-4A, B, C), 6-fold axes (Fig. 6-4D), or 4-fold axes (Fig. 6-4E), whereas midway between lattice points 2-fold axes exist for all five nets. For each net the symmetry axes are indicated only in its lower right portion in Figure 6-4, but the planes of symmetry perpendicular to these nets are not indicated at all. We note that 3-fold as well as 6-fold axes occur in hexanets.

Lattice-Imposed Restrictions on Symmetry Axes

As just discussed, only 1-, 2-, 3-, 4-, or 6-fold axes, but not 5-, 7-, 9-fold axes, and so on, are perpendicular to the plane lattices in Figure 6-4. This restriction of symmetry axes to those whose repeat angles are, respectively, 360°, 180°, 120°, 90°, or 60° is a fundamental property of plane lattices and no mere chance event. To demonstrate the nature of this restriction, we let A, B, and C represent lattice points, each separated by the translation unit **a**, along a lattice row LL' of a

---

← FIGURE 6–4

The five basic types of plane lattices or nets as developed by translation of line lattice LL' by vector **b**: (A) clinonet, (B) orthonet, (C) centered orthonet, (D) hexanet, and (E) tetranet. The locations and types of symmetry axes normal to each net are shown only for its lower right corner. The planes of symmetry normal to these nets are not indicated.

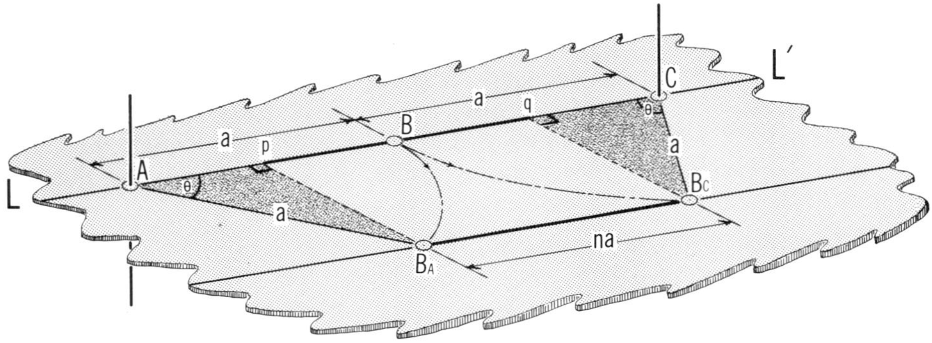

FIGURE 6–5

A perspective view of a small area from a generalized net that shows three neighboring identipoints, A, B, and C, separated by a single unit of translation **a** along line LL'. Two more identipoints, B_A and B_C, are shown to be separated by an integral number of units of translation, that is, by $n\mathbf{a}$, along a line parallel to LL'. An indeterminate symmetry axis at A is presumed to relate B to B_A by a rotation of $\theta°$; an identical axis at C similarly relates B to B_C. The dash-dot arcs thus represent sectors of a circle with radius **a**. Perpendiculars dropped from B_A and B_C to LL', respectively, locate p and q and the heavy line pq necessarily equals $n\mathbf{a}$ because pqB_CB_A is a rectangle.

generalized plane lattice (Fig. 6-5). Any symmetry axis perpendicular to this plane, if it occurs at one lattice point, must be present at *all* lattice points. For simplicity the generalized symmetry axis with repeat angle θ is drawn only at points A and C, although it exists at all other points. Let B_A and B_C represent two lattice points (on a row parallel to LL') such that a rotation of θ about the axis at A relates B to B_A and of θ about the axis at C relates B to B_C. As for any two lattice points on a row parallel to LL' in the net in Figure 6-5, the distance between points B_A and B_C must equal $n\mathbf{a}$, where n is an integer (0, 1, 2, 3, . . .). (Figure 6-5 is drawn as if n equals 1.) The perpendiculars from B_A and B_C to LL' define (1) a rectangle whose boldly drawn sides equal $n\mathbf{a}$ in length and (2) two right triangles (stippled) whose sides Ap and Cq each equal $\mathbf{a}\cos\theta$ in length. The distance between points A and C equals 2**a**, but it also equals the rectangle's bold edge $n\mathbf{a}$ plus the sum of Ap and Cq. Each of these latter equals $\mathbf{a}\cos\theta$, hence

$$2\mathbf{a} = 2\mathbf{a}\cos\theta + n\mathbf{a}$$

an equation that reduces to

$$\cos\theta = 1 - n/2 \qquad \text{(Eq. 6-1)}$$

Because the cosine is restricted to values from 1 to -1, only values of n equal to 0, 1, 2, 3, or 4 offer valid solutions of Eq. (6-1). Thus from Eq. (6-1) the value of the repeat angle θ (and the identity of the symmetry axis that this value of θ

defines) is related to the value of n as follows:

n	0	1	2	3	4
θ	360°	60°	90°	120°	180°
Rotation axis	1 or $\bar{1}$	6 or $\bar{6}$	4 or $\bar{4}$	3 or $\bar{3}$	2 or $\bar{2}$

The condition of periodic repetition of points in a plane lattice thus restricts its axes of symmetry to one of the ten rotation axes just cited. Because crystals, as we shall next see, have internal structures based on such lattices, their possible axes of symmetry are also restricted to these ten types and can never be, for example, axes of 5-fold, 7-fold, 8-fold, or 9-fold symmetry.

The Fourteen Types of Space Lattices

If a crystal could be magnified until its component atoms or ions became distinguishable, it would be seen to consist of a three-dimensional array of these atoms or ions. The purely translational symmetry inherent in such an array—in other words, the scheme of repetition of the atoms, groups of ions, or molecules in forming the crystal's internal structure—is summarized by the space lattice obtained through isolating any single set of identipoints in the array. Figure 6-6 illustrates this; the identipoints located where a chloride ion touches the sodium ion above it define the space lattice, that is, scheme of translational repetition, inherent in the sodium chloride structure. To emphasize this the ions are completely faded away to the right in Figure 6-6 to leave, like the Cheshire cat's smile, only this set of identipoints. Any other set of identipoints, such as the centers of the chloride ions or the center of the sodium ions, would yield an identical space lattice.

The great French crystallographer, Auguste Bravais, first demonstrated that there could be only 14 different types of space lattices. This means that there are only 14 basically different types of three-dimensional arrays whereby atoms, ions, or molecules may pack together when forming a crystal. Because of Bravais' contributions to their study, space lattices are sometimes called *Bravais lattices*.

In essence a space lattice consists of a stack of equidistant parallel nets or, alternatively, of a net that has been successively translated along a vector, here called the *stacking vector* and shown by the hollow arrows in the examples (Fig. 6-7). The 14 possible types of space lattices differ with respect to (1) which of the five basic nets is being stacked or translated and (2) the length of the stacking vector as well as its angle to the net (and to rows of points in the net).

Isometric (cubic) space lattices

Three types of space lattices, each characterized by point-group symmetry $4/m\ \bar{3}\ 2/m$, result from stacking tetranets in the three ways shown in Figure 6-7. These lattices represent the only three schemes of repetition that can possibly

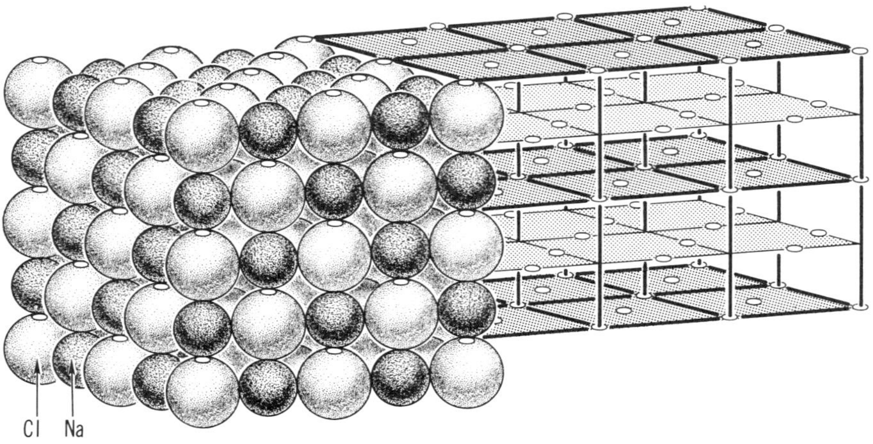

FIGURE 6–6

The arrangement of sodium and chlorine ions in the crystal structure of common table salt, sodium chloride. The identipoint at the contact of each chlorine ion with the smaller sodium ion directly above it is shown as a circle that appears elliptical in perspective. To the right the ions are faded away so as to leave just this set of identipoints. These define the space lattice that summarizes the scheme of repetition inherent in the sodium chloride structure. Thus considering the repeated object to be a chlorine with a tangent sodium directly overlying it, the translational repetition of this object by the lattice at right would reproduce the structure at the left.
Note the difference between the meanings of the two words, "structure" and "lattice." The distribution of atoms at left represents the *structure* for sodium chloride; the distribution of identipoints at right represents the *lattice.*

underlie the atomic structures of isometric crystals. For each, a single *unit cell* is drawn at the right, this usually being defined as the *smallest portion of a lattice with identipoints at its corners* which still *retains the same point-group symmetry* as the entire lattice. Customarily, only a single unit cell is drawn to represent the entire lattice, this single unit cell being considered as coming from deep within the lattice and therefore sharing its corners, edges, and faces with the identical unit cells that surround it in the lattice. Thus if an identipoint is at a corner, it is shared by eight unit cells; if at an edge, it is shared by four; and if at a face (like some in Fig. 6-7C), it is shared by two. In Figure 6-7A the unit cell "owns" only one identipoint, that is, one-eighth of each at its corners, and hence is called a *primitive* unit cell. By contrast the cell in Figure 6-7B is *doubly primitive* "owning" the identipoint at its center plus one-eighth of each at its corners. Similarly, the cell in Figure 6-7C is *quadruply primitive* "owning" four identipoints. Thus in respect to the object being repeated to form the three-dimensional array, a unit cell will contain one such object if it is a primitive cell, two such objects if doubly primitive, and four if quadruply primitive.

The names and symbols of the three isometric lattices are (1) the *isometric P* or *primitive cubic lattice* (Fig. 6-7A); (2) the *isometric I* or *body-centered cubic*

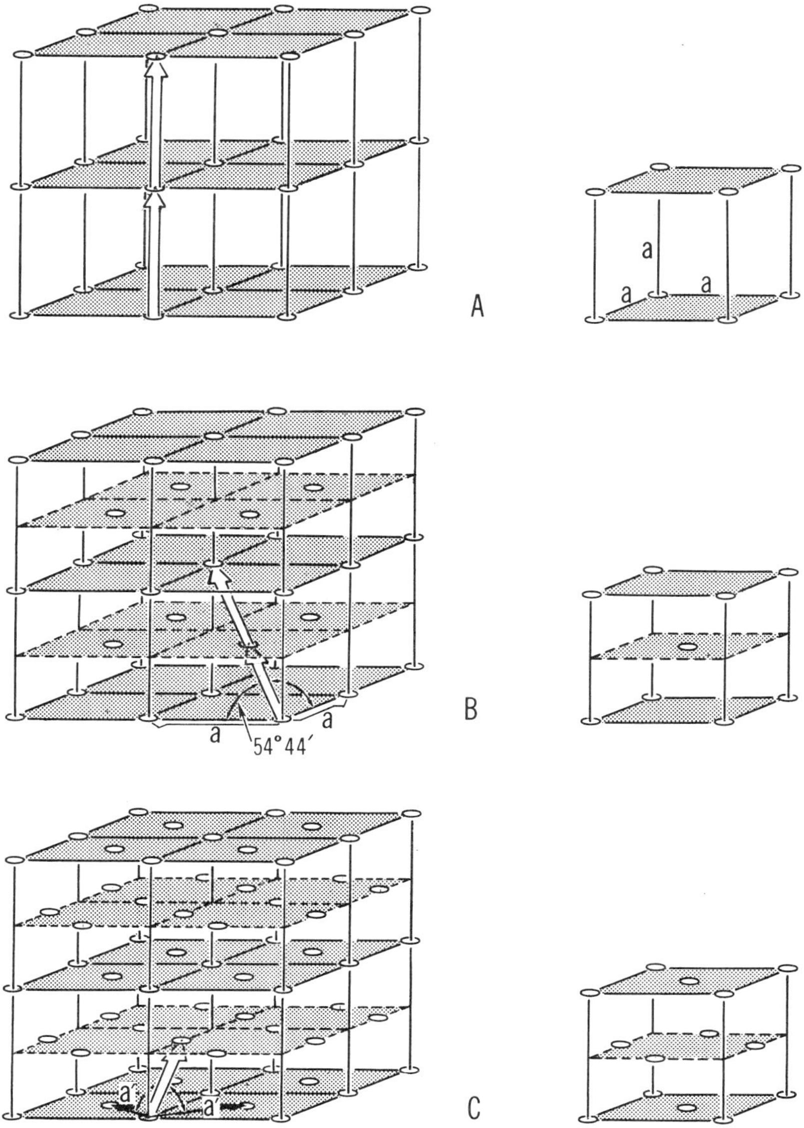

FIGURE 6–7

The three types of isometric lattices: (A) primitive P, (B) body centered I, (C) face centered F. These lattices result if tetranets are translated by a stacking vector (hollow arrow) which is (A) at 90° to the tetranet and equal in length to **a**, the shortest unit of translation *within* the plane of the tetranet; (B) at 54°44′ to each **a** vector but equal in length to 0.866 **a**; (C) at 45° to the tetranet but at 60° to each of the within-net vectors labeled **a′**, the stacking vector now equaling **a′** in length. A single unit cell is shown at the right for each lattice.

lattice (Fig. 6-7B) whose symbol I stands for *innenzentrierte* (the German word for "body centered"); and (3) the *isometric F* or *face-centered cubic lattice* (Fig. 6-7C), so-called because points are centered on all faces of its cubic unit cell as well as being at its corners. In this text the symbols P $(4/m\ \bar{3}\ 2/m)$, I $(4/m\ \bar{3}\ 2/m)$, and F $(4/m\ \bar{3}\ 2/m)$ are coined for these lattices to emphasize that the symmetry elements of point group $4/m\ \bar{3}\ 2/m$ occur through *each point* of these indefinitely extended lattices. The parentheses distinguish these symbols from the space-group symbols that will be introduced in the next chapter.

The vectors that outline the edges of the unit cells are called the *unit cell edges* or *unit cell vectors*. In isometric lattices they are cube edges and consequently are three mutually perpendicular vectors of equal length, each commonly labeled **a**.

Hexagonal space lattices

Hexanets may be stacked to form lattices in two different ways. For one, the stacking vector **c** is at 90° to the hexanet and the resultant lattice (Fig. 6-8A) has point-group symmetry $6/m\ 2/m\ 2/m$. Although Figure 6-8A represents the smallest portion of this lattice that can retain this symmetry, it is not usually considered to be the unit cell because it is triply primitive "owning" one-sixth of each corner point and one-half of those on its faces. Instead, a primitive unit cell which is one-third its size is chosen (Fig. 6-8B) even though its symmetry is not $6/m\ 2/m\ 2/m$. This *primitive hexagonal lattice*, P $(6/m\ 2/m\ 2/m)$, represents the scheme of repetition underlying the atomic structures of all crystals of the hexagonal subsystem and of some that crystallize in the trigonal subsystem.

Figure 6-8C is a plan view of part of a hexanet and shows the sites of the perpendicular 6- and 3-fold axes in relation to the identipoints. For discussion purposes only, the 3-fold axes are divided into an r set and an s set. In Figure 6-8A the 6-fold axes of one net directly overlie (and thus coincide with) those of the net below. This is not the case in the second possible method of stacking hexanets to form a lattice (Fig. 6-8D). Instead, as illustrated by the small segments of the hexanets at levels 1, 2, 3, and 4, net 2's identipoints (the sites of its 6-fold axes) directly overlie the 3-fold sites marked r in net 1. Similarly, net 3's identipoints overlie the 3-fold sites marked s in net 1. The resultant lattice has symmetry $\bar{3}\ 2/m$ and the smallest primitive cell that retains this symmetry is the rhombohedron outlined in Figure 6-8D. This lattice, called the *rhombohedral* or *R lattice* and symbolized here by R $(\bar{3}\ 2/m)$, summarizes the type of schemes of repetition that underlie the atomic structures of many, but not all, crystals of the trigonal subsystem; lattice P $(6/m\ 2/m\ 2/m)$ accounts for the remainder as well as for all crystals of the hexagonal subsystem.

The unit cell conventionally chosen for a primitive hexagonal lattice is a right prism of height **c** (Fig. 6-8B) whose base is a 60° rhombus with edges **a**. For a rhombohedral lattice (Fig. 6-8D) the unit cell may be selected to be a rhombohedron with edges $\mathbf{a}_R$ and with an angle between these edges of α_R (Fig. 6-8E), this angle being measured between two edges that intersect at the $\bar{3}$ axis

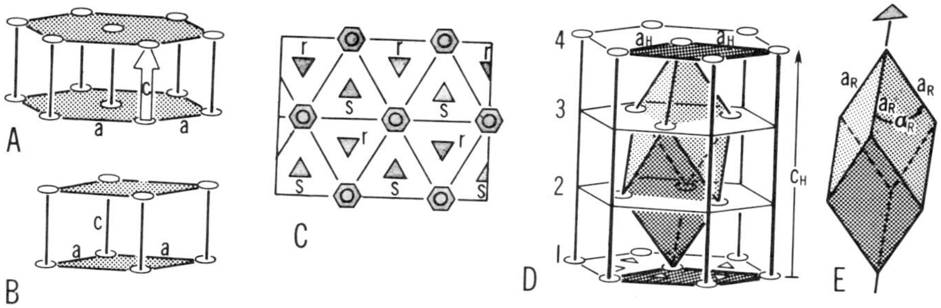

FIGURE 6–8

The types of space lattices that underlie structures crystallizing in the hexagonal system.
(A) The repetition of hexanets by a perpendicular stacking vector to form the primitive hexagonal lattice P $(6/m\,2/m\,2/m)$ that underlies all crystals in the hexagonal subsystem as well as some in the trigonal subsystem.
(B) A single unit cell for the lattice P $(6/m\,2/m\,2/m)$.
(C) A plan view of a hexanet for which the 3-fold sites have been classed into two groups, those at s and those at r.
(D) The repetition of hexanets by a stacking vector (not drawn) that coincides with one of the edges of the shaded rhombohedron. As thus stacked, the identipoints of net 2 lie directly above the r sites of net 1, these r sites being indicated on net 1 by small triangles with apices toward the reader. The identipoints of net 3, however, lie above the s sites of net 1, these being indicated by small triangles with their bases toward the reader. The resultant rhombohedral lattice R $(\bar{3}\,2/m)$ underlies the atomic structure of many crystals of the trigonal subsystem.
(E) The rhombohedral cell from (D) shown in relation to its $\bar{3}$ axis. The two parameters that define the shape and size of a rhombohedral cell are $\mathbf{a}_R$ (the length of its edges) and α_R (the angle between any two edges that intersect at the $\bar{3}$ axis).

of the rhombohedron. For this same rhombohedral lattice, however, many investigators prefer to choose a unit cell that is a right-prism of the height $\mathbf{c}_H$ in Figure 6-8D with a base that is a 60° rhombus with edges $\mathbf{a}_H$ (like the dark-shaded area on plane 4 in Fig. 6-8D). This latter cell is triply primitive. Equations relating the dimensions of these two alternative cells are

$$\mathbf{a}_H = 2\mathbf{a}_R \sin\frac{\alpha_R}{2} \qquad \text{(Eq. 6-2)}$$

$$\mathbf{c}_H = 3\mathbf{a}_R\sqrt{1-\frac{4}{3}\left(\sin^2\frac{\alpha_R}{2}\right)} = \sqrt{9\mathbf{a}_R{}^2 - 3\mathbf{a}_H{}^2} \qquad \text{(Eq. 6-3)}$$

$$\mathbf{a}_R = \tfrac{1}{3}\sqrt{3\mathbf{a}_H{}^2 + \mathbf{c}_H{}^2} \qquad \text{(Eq. 6-4)}$$

$$\sin\frac{\alpha_R}{2} = \frac{3\mathbf{a}_H}{2\sqrt{3\mathbf{a}_H{}^2 + \mathbf{c}_H{}^2}} \qquad \text{(Eq. 6-5)}$$

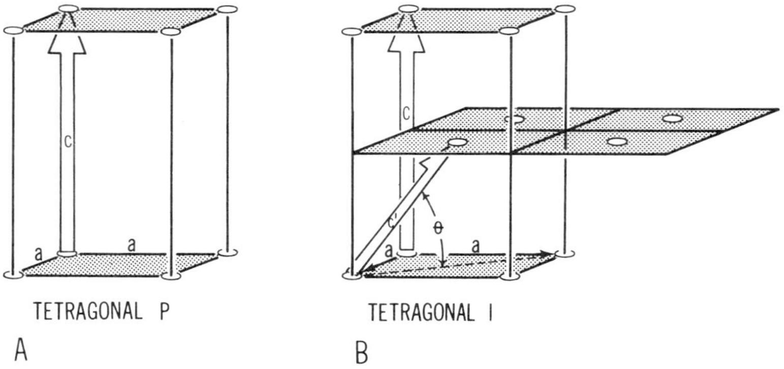

FIGURE 6–9

The two basic types of tetragonal lattices.
(A) The primitive lattice P ($4/m\ 2/m\ 2/m$) resultant from translating tetranets according to a stacking vector **c** which is at 90° to them but not equal to vector **a** in length.
(B) The body-centered lattice I ($4/m\ 2/m\ 2/m$) for which tetranets are translated as by stacking vector **c**′. The angle θ has a cosine equal to 0.707 **a**/**c**′. The edges of its unit cell are **a** and **c**.

Tetragonal space lattices

Tetranets may be stacked in two ways to form the two types of space lattices that, one or the other, invariably underlie the structure of any tetragonal crystal. If the stacking vector **c** is normal to the tetranet but not equal in magnitude to the within-net vectors **a** (Fig. 6-9A), the tetragonal primitive lattice results. Alternatively, the stacking vector now labeled **c**′ in Figure 6-9B, may (1) vertically overlie a diagonal of the tetranet, one such diagonal being dashed in Figure 6-9B and (2) be of a length **c**′ and at an angle θ to the tetranet such that **c**′ cos θ equals $0.707a$. In this case a body-centered tetragonal lattice (Fig. 6-9B) results with mutually perpendicular cell edges, **a**, **a**, and **c**. Both of these lattices have point-group symmetry $4/m\ 2/m\ 2/m$ and can thus be respectively symbolized P ($4/m\ 2/m\ 2/m$) and I ($4/m\ 2/m\ 2/m$), where P signifies primitive and I stands for body centered.

Orthorhombic space lattices

Four types of orthorhombic space lattices occur, each with symmetry $2/m\ 2/m\ 2/m$. Orthonets stacked directly over each other, that is, with stacking vector **c** at 90° to each orthonet, produce a primitive orthorhombic lattice (Fig. 6-10A) provided **c** equals neither **a** nor **b**, the within-net vectors. Alternatively, orthonets can be stacked to yield the body-centered orthorhombic I lattice (Fig. 6-10B). Centered orthonets, stacked as in Figure 6-10D, yield the all-face-centered orthorhombic F lattice. Centered orthonets stacked directly over each other (Fig. 6-10C) yield a space lattice with unit cell edges **a**, **b**, and **c** but with

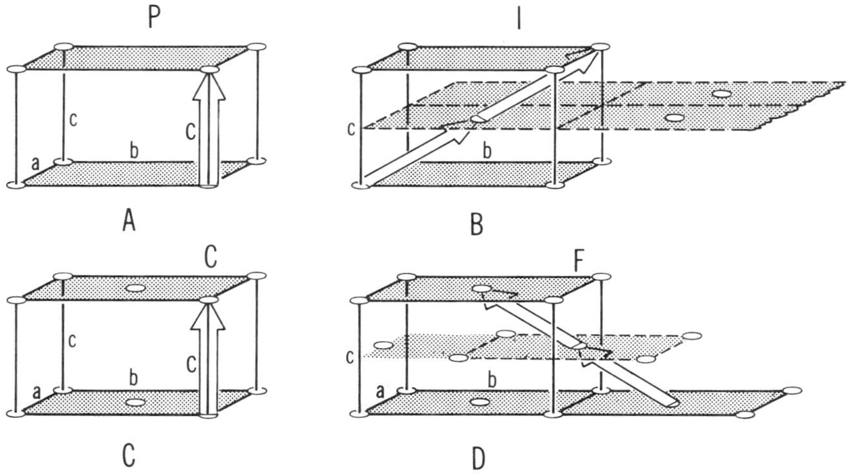

FIGURE 6–10

The four types of orthorhombic space lattices.
(A) The primitive lattice *P*.
(B) The body-centered lattice *I*. Both (A) and (B) result from stacking of orthonets as shown by the hollow arrows.
(C) The *C*-face centered lattice.
(D) The (all) face-centered lattice *F*. Both (C) and (D) result from stacking of centered orthonets as shown by the hollow arrows.

identipoints centered on *one* face only. In Figure 6-10C the centering occurs on the cell face parallel to the (001) or *c*-pinacoidal face and the lattice is consequently called a *C*-face-centered orthorhombic lattice. Let us suppose that for the crystal whose lattice is shown in Figure 6-10C the present *c* axis were labeled *a* and the present *a* axis were labeled *c*. The lattice in Figure 6-10C would then need to be rotated 90° around its *b* axis to make its new *c* axis vertical. So relabeled and oriented, the centering would be on (100), the *a* pinacoid, and this same lattice would then be called an *A*-face-centered lattice. After relabeling of its axes so that centering would occur on the (010) or *b* pinacoid, this same lattice would be a *B*-face-centered orthorhombic lattice. Thus the orthorhombic *A*, *B*, and *C* lattices represent the same lattice type and differ only in respect to whether the axis normal to the centered face is labeled, *a*, *b*, or *c*, respectively.

Our impromptu symbols for the four possible orthorhombic lattices would thus be $P\ (2/m\,2/m\,2/m)$, $I\ (2/m\,2/m\,2/m)$, $F\ (2/m\,2/m\,2/m)$, and C, A, or $B\ (2/m\,2/m\,2/m)$.

Monoclinic space lattices

The distribution of identipoints within the atomic structure of a monoclinic crystal conforms to one or the other of two lattice types, each with point-group

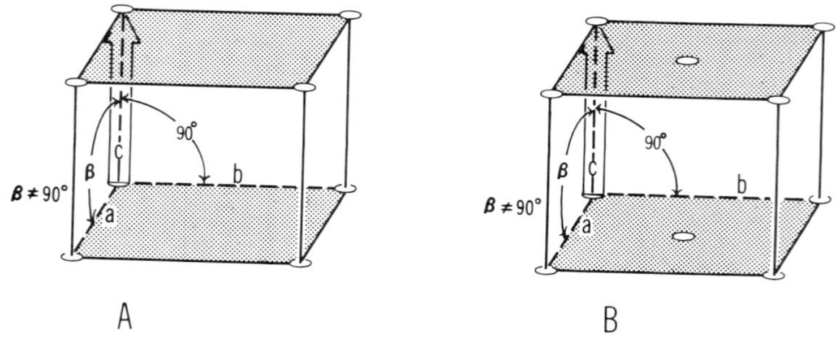

FIGURE 6–11

The two possible types of monoclinic space lattices.
(A) The primitive P lattice, P $(2/m)$, resultant from stacking orthonets as shown ($\beta \neq 90°$).
(B) The C-face centered lattice, C $(2/m)$, resultant from stacking centered orthonets as shown ($\beta \neq 90°$).

symmetry $2/m$. One, the primitive or monoclinic P lattice (Fig. 6-11A), can be regarded as consisting of orthonets stacked with their stacking vector **c** at 90° to *one* of the within-net vectors, for example, to **b** in Figure 6-11A, but not to the other. The second, the C-face centered or monoclinic C lattice (Fig. 6-11B), can be regarded as consisting of a similar stacking of *centered* orthonets. If the labels on the a and c axes were interchanged in Figure 6-11B, this lattice would be called an A lattice because the centered face would then be the a pinacoid.

Triclinic space lattices

The single lattice type that resembles the spatial distribution of identipoints within the atomic structure of a triclinic crystal has the point-group symmetry $\bar{1}$. This lattice can be regarded as consisting of clinonets stacked so that the stacking vector **c** is *not* at 90° to any within-net vector (Fig. 6-12). For this lattice type there are many different choices for the unit cell, each of which would be primitive and have symmetry $\bar{1}$ (Fig. 6-12). Of these, however, the cell with the three shortest edges, whose sides are stippled in Figure 6-12, is preferably selected and is called the *reduced cell* because of its short edges.

Choice and Orientation of Unit Cell

Donnay (1943) formalized the rules to be used for choosing a unit cell once the distribution of identipoints in a crystal's atomic structure, that is, its space lattice, has been determined by X-ray diffraction techniques (Chapter 13). These rules, to be applied in the order given, are the following:

1. The cell edges selected should, whenever possible, coincide with any symmetry axes present in the lattice.
2. If possible, these edges should be related to each other by the lattice's symmetry.
3. The smallest possible cell, that is, *reduced cell,* which fulfills conditions (1) and (2) should be chosen (see Azaroff and Buerger, 1958).

If a crystal structure lacks symmetry axes along three noncoplanar directions, then two (or three) cell edges must be chosen that do not fulfill rule 1. These two (or three) cell edges must then be located in conformity to rule 3 along the shortest, next shortest, and so on, translations in the crystal's lattice. During our discussions of space lattices, rules 1 to 3 were fulfilled by choosing the smallest unit cell having the same point-group symmetry as the entire lattice.

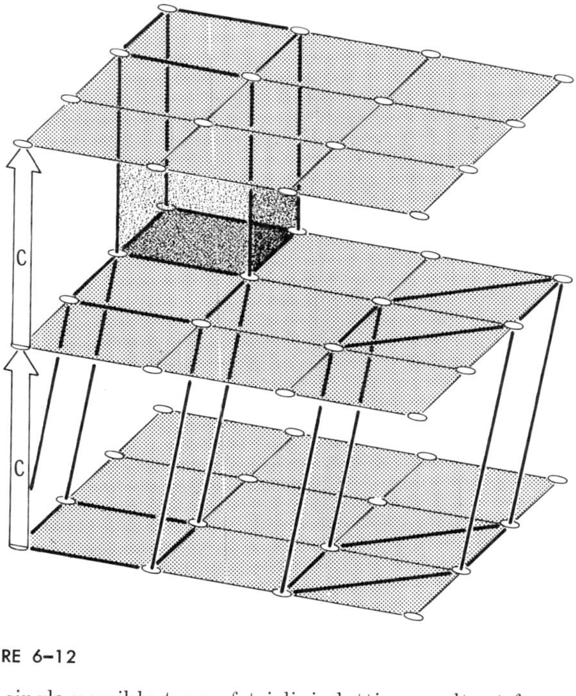

FIGURE 6–12

The single possible type of triclinic lattice resultant from stacking clinonets at an arbitrary angle. Various possible primitive unit cells, whose edges would thus define the direction of the crystallographic axes, a, b, and c, and the repeat distances along them, are drawn with heavy lines. Each of these cells possesses $\bar{1}$ symmetry identical to the entire lattice; however, the cell with the three shortest edges, usually called the *reduced cell,* is generally selected as the unit cell, its edges thereby defining axes a, b, and c. To distinguish the reduced cell in this figure, its faces have been stippled.

For triclinic crystals the unit cell edges are chosen to coincide with the three shortest noncoplanar translations of the lattice because the triclinic space lattice lacks symmetry axes. Donnay recommends that the shortest edge of the trio be named **c**; the next shortest, **a**; and the third, **b**. The three, mutually perpendicular, cell edges for orthorhombic crystals are also conventionally labeled to observe this inequality, that is, $\mathbf{c} < \mathbf{a} < \mathbf{b}$, except that for class $mm2$ the cell edge coinciding in direction with the single 2-fold axis is conventionally labeled **c** regardless of whether or not its length is less than **a** or **b**. In monoclinic crystals the cell edge which coincides with the lattice's 2-fold axis is named either **c** (first setting) or **b** (second setting); for the remaining two edges, which are the two shortest lattice translations in the plane normal to this 2-fold axis, the convention is $\mathbf{a} < \mathbf{b}$ (first setting) or $\mathbf{c} < \mathbf{a}$ (second setting). For tetragonal, hexagonal, or isometric cells, the two or more cell edges that are symmetrically equivalent and therefore equal in length are usually labeled **a**.

Observance of the $\mathbf{c} < \mathbf{a} < \mathbf{b}$ rule is not unanimous. Some crystallographers use the rule $\mathbf{a} < \mathbf{b} < \mathbf{c}$ in labeling cell edges for triclinic and orthorhombic crystals. Similarly, although the *second setting* for monoclinic crystals is most used, the earlier-used *first setting* is still preferred by some. It has the advantage of preserving the convention that any *unique* symmetry axis—for example, the single 4 or $\bar{4}$ in tetragonal crystals, the single 3, $\bar{3}$, 6, or $\bar{6}$ in hexagonal crystals, and the single 2-fold axis in class $mm2$—will be named **c**.

Summary

Any single set of identipoints within a crystal's atomic structure summarizes the translational periodicity inherent in it. For any crystal whatsoever the distribution of these identipoints in space will conform to one of the 14 possible space lattices (Fig. 6-13). The unit cells for these lattices usually represent their smallest segment that retains the same point-group symmetry as the entire lattice. Exceptions to this are the cells usually chosen for hexagonal or rhombohedral lattices (Fig. 6-13), unless, as is frequently done, a rhombohedron-shaped unit cell is chosen for the rhombohedral lattice. The unit cell's edges by their directions define the crystallographic axes for the crystal and by their lengths define the units, that is, repeat distances, along these axes. The actual lengths of these units, for example, **a**, **b**, and **c** for triclinic crystals, became measurable only with the advent of X-ray crystallography.[2] These units are submicroscopically short, usually being less than 20 angstroms[3] in length.

For each crystal system Table 6-1 summarizes (1) the types of lattices possible for crystals of this system and (2) the parameters whose values must be specified to describe completely an individual lattice of the type. The less sym-

[2] These lengths are also symbolized as a_0, b_0, and c_0, a practice frequently followed in this text to acquaint the reader with this alternative symbolism.

[3] One angstrom (symbol: Å or A) equals 10^{-8} cm, that is, 100 million angstroms equal 1 cm.

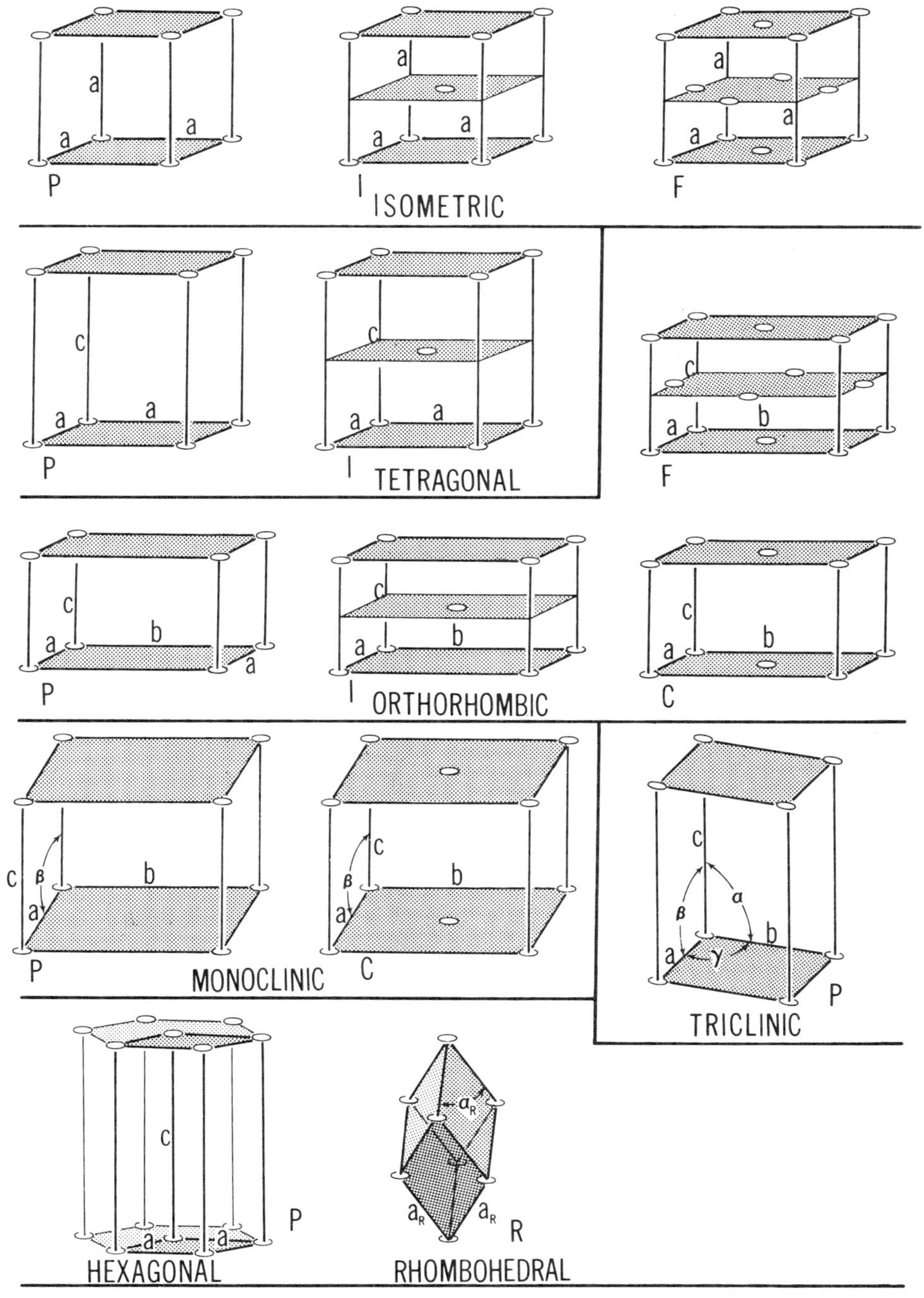

FIGURE 6–13

The unit cells for the 14 possible types of space lattices.

TABLE 6–1

Lattice Types and the Parameters to be Specified in Describing Them

System or Subsystem	Possible Lattices: Type P	I	F	C	R	Symmetry	Parameters a	b	c	α	β	γ
Trimetric Systems												
Triclinic	*					$\bar{1}$	*	*	*	*	*	*
Monoclinic												
2d setting	*			*		$2/m$	*	*	*		*	
Orthorhombic	*	*	*	*		$2/m\ 2/m\ 2/m$	*	*	*			
Dimetric Systems												
Tetragonal	*	*				$4/m\ 2/m\ 2/m$	*		*			
Hexagonal	*					$6/m\ 2/m\ 2/m$	*		*			
Trigonal[a]					*	$\bar{3}\ 2/m$	*		*			
Monometric System												
Isometric	*	*	*			$4/m\ \bar{3}\ 2/m$	*					

[a] Some trigonal crystals possess the hexagonal *P* lattice but others the rhombohedral *R* lattice indicated here. For the *R* lattice, if a rhombohedral unit cell is chosen, the parameters $\mathbf{a}_R$ and α_R (Fig. 6–8E) will be specified.

metric lattices require more parameters to describe them and are hence more individualistic. Present-day X-ray techniques permit measurement for a crystal of any of its lattice parameters listed in Table 6-1. In contrast the earlier studies, which consisted chiefly of measurement of a crystal's interfacial angles with an optical goniometer, permitted measurement of the interaxial angles, for example, α, β, and γ for triclinic crystals; it was not, however, the actual values of **a**, **b**, and **c**, that were measured but rather of the axial ratio $a{:}b{:}c$ or, in other words, of $\mathbf{a}/\mathbf{b}{:}\mathbf{b}/\mathbf{b}{:}\mathbf{c}/\mathbf{b}$.

REFERENCES AND RECOMMENDED READING

Azaroff, L. V., and M. J. Buerger (1958) *The Powder Method*, New York: Wiley, p. 124.

Donnay, J. D. H. (1943) "Rules for the Conventional Orientation of Crystals," *Am. Mineral.*, 313–328, 470.

QUESTIONS AND PROBLEMS

6-1 Let the point at the lower left rear corner of each unit cell in Figure 6-13 represent the origin of the lattice and therefore have coordinates 0, 0, 0 with respect to axes a_1, a_2, a_3; a_1, a_2, c or a, b, c, as the case may be. Coordinates of other lattice points can then be expressed in terms of units of translation along these axes. Thus the right rear lower corner has coordinates 0, 1, 0; the left front lower corner has coordinates 1, 0, 0; and so on. (a) State the coordinates for the center point in I lattices; (b) for the face-centered points in a C lattice; (c) for the six face-centered points in an F lattice; (d) for the corner points of a rhombohedral unit cell; (e) for the points of the cell (edges: a_H, a_H, and c_H) used for the rhombohedral lattice in Figure 6-8D.

6-2 Why cannot a valid space lattice be formed by interstacking two different types of nets?

6-3 Given a point with coordinates 0.2, 0.3, 0.4, give the coordinates of a second point in a body-centered cell that is related to this first solely by the body-centered translation inherent in the lattice.

6-4 (a) Demonstrate by drawings (or otherwise) that a tetragonal C lattice is the same as tetragonal P; (b) that tetragonal F equals tetragonal I; (c) that monoclinic B (second setting) equals monoclinic P; (d) that monoclinic I equals C.

ANSWERS

6-1 (a) $\frac{1}{2}, \frac{1}{2}, \frac{1}{2}$. (b) $\frac{1}{2}, \frac{1}{2}, 0; \frac{1}{2}, \frac{1}{2}, 1$. (c) $\frac{1}{2}, \frac{1}{2}, 0; \frac{1}{2}, \frac{1}{2}, 1; \frac{1}{2}, 0, \frac{1}{2}; \frac{1}{2}, 1, \frac{1}{2}; 0, \frac{1}{2}, \frac{1}{2}; 1, \frac{1}{2}, \frac{1}{2}$. (d) 0, 0, 0; 0, 0, 1; 0, 1, 0; 0, 1, 1; 1, 0, 0; 1, 0, 1; 1, 1, 0; 1, 1, 1. (e) Corner points: 0, 0, 0; 1, 0, 0; 0, 1, 0; 1, 1, 0; 0, 0, 1; 1, 0, 1; 0, 1, 1; 1, 1, 1. For points within the body of this unit cell, the coordinates are $\frac{1}{3}, \frac{2}{3}, \frac{1}{3}$ for the point on plane 2 and $\frac{2}{3}, \frac{1}{3}, \frac{2}{3}$ for the point on plane 3.

6-2 The points in a space lattice are identipoints, that is, identical to each other in every respect except location in space. For each and every point in the lattice, therefore, the environment of fellow lattice points surrounding it must be identical. However, a point in one type of net would have a different environment of points around it than a point in a second type of net. Thus the points of the one net type could not be identical to those of a second type.

6-3 The body-centered translation of $\frac{1}{2}, \frac{1}{2}, \frac{1}{2}$ or 0.5, 0.5, 0.5 would require an identical point at coordinates 0.7, 0.8, 0.9, that is, $0.2 + 0.5$, $0.3 + 0.5$, $0.4 + 0.5$.

7 Internal Symmetry: Space Groups

For each crystal system the possible space lattices possess the symmetry of the holosymmetric class. Yet in each system crystals belonging to less symmetric classes occur. For example, all four possible orthorhombic lattices possess the symmetry $2/m\ 2/m\ 2/m$, yet orthorhombic crystals belong to classes 222 or $mm2$ as well. The explanation lies in the symmetry possessed by the object, usually a cluster of atoms or ions, which is translationally repeated by the lattice. If this object's symmetry is the same as (or a supergroup of) that of the lattice, the resultant array will retain the lattice's symmetry, provided that the repeated object has its symmetry axes and planes oriented so as to coincide in direction with the comparable ones for the lattice. However, if the symmetry of the repeated object merely conforms to a subgroup of the lattice's symmetry, then the array (and crystal structure that it represents) will conform to this subgroup's symmetry.

For example, the structure of hemimorphite, $Zn_4Si_2O_7(OH)_2 \cdot H_2O$, is based on a body-centered orthorhombic lattice I (Fig. 7-1A) of symmetry $2/m\ 2/m\ 2/m$. Viewing the structure with its c axis normal to the page, the centers of its component ions occur where shown in Figure 7-1B and the near-by numbers indicate the height of each ion's center above an arbitrary (001) plane in terms of percent of the **c** translation of 5.12 Å. Thus 50 would indicate a height of 2.56 Å above this plane. Actually, in Figure 7-1B the ions are indicated by circles smaller than their true size at that scale; if drawn at true scale, neighboring ions would probably be tangent spheres. A group of ions that can be regarded as the object being repeated by the lattice is set off by a light gray background. This group possesses symmetry $mm2$, a subgroup of $2/m\ 2/m\ 2/m$. This object, translated by the **a** vector or the **b** vector (or both), would then coincide with identical groups of ions at the unit cell's corners. If translated as demanded by the body

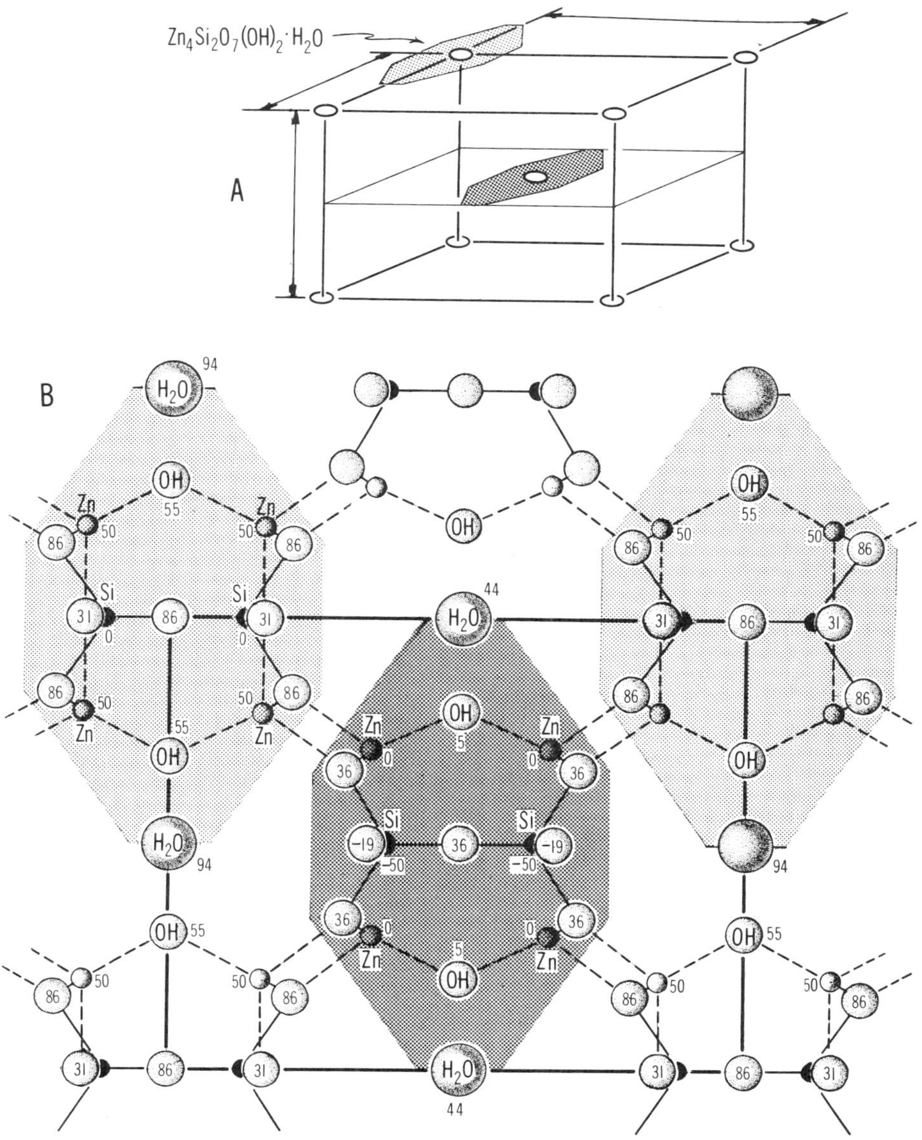

FIGURE 7–1

(A) The body-centered, orthorhombic cell that summarizes the translational symmetry inherent to the structure of hemimorphite, $Zn_4Si_2O_7(OH)_2 \cdot H_2O$, is shown in standard orientation (edge **c**, vertical, plus end up; edge **b**, the longest edge, horizontal with its plus end to the reader's right; edge **a** with its plus end toward the reader).

(B) The hemimorphite structure with the component atoms shrunken but their center-to-center distances maintained to scale as viewed along its *c* axis, that is, projected on (001). The numbers by each atom indicate the height of its center in terms of percent of the **c** translation of 5.12 Å, above the plane of projection. The **a** and **b** cell edges, drawn with bold lines, are now parallel to the plane of the paper.

centering, that is, successively by **a**/2, **b**/2, and **c**/2, it coincides with an identical group set off by a dark gray background. The heights of ions on dark gray differ by 50, which represents **c**/2, from comparable ions on the light gray background.

Simple Space Groups

The symmetry inherent in an array comprises a *space group*. The space-group symmetry is described by citing (1) the space lattice or scheme of repetition used for the array plus (2) the symmetry of the object so arrayed. The following space groups result if a lattice typical of the crystal system repeats an object whose symmetry conforms to one of the point groups for that system.

Triclinic:	$P\,1; P\,\bar{1}$
Monoclinic:	$P\,2; P\,m; P\,2/m$
	$C\,2; C\,m; \mathrm{C}\,2/m$
Orthorhombic:	$P\,222; P\,mm2; P\,2/m\,2/m\,2/m$
	$C\,222; C\,mm2; A\,mm2; C\,2/m\,2/m\,2/m$
	$F\,222; F\,mm2; F\,2/m\,2/m\,2/m$
	$I\,222; I\,mm2; I\,2/m\,2/m\,2/m$
Tetragonal:	$P\,4; P\,\bar{4}; P\,4/m; P\,422; P\,4mm; P\,\bar{4}2m;$
	$P\,\bar{4}m2; P\,4/m\,2/m\,2/m$
	$I\,4; I\,\bar{4}; I\,4/m; I\,422; I\,4mm; I\,\bar{4}2m$
	$I\,\bar{4}m2; I\,4/m\,2/m\,2/m$
Trigonal:	$P\,3; P\,\bar{3}; P\,312; P\,321; P\,3m1; P\,31m;$
	$P\,\bar{3}1\,2/m; P\,\bar{3}\,2/m\,1$
	$R\,3; R\,\bar{3}; R\,32; R\,3m; R\,\bar{3}\,2/m$
Hexagonal:	$P\,6; P\,\bar{6}; P\,6/m; P\,622; P\,6mm; P\,\bar{6}m2$
	$P\,\bar{6}2m; P\,6/m\,2/m\,2/m$
Isometric:	$P\,23; P\,2/m\,\bar{3}; P\,432; P\,\bar{4}3m; P\,4/m\,\bar{3}\,2/m$
	$I\,23; I\,2/m\,\bar{3}; I\,432; I\,\bar{4}3m; I\,4/m\,\bar{3}\,2/m$
	$F\,23; F\,2/m\,\bar{3}; F\,432; F\,\bar{4}3m; F\,4/m\,\bar{3}\,2/m$

Points of difference between some of the space groups listed above merit explanation. For example, $C\,mm2$ and $A\,mm2$ differ because the 2-fold axis of the object being repeated is perpendicular to the centered face of the lattice in $C\,mm2$ (Fig. 7-2A) but parallel to it in $A\,mm2$ (Fig. 7-2B). Space groups $P\,\bar{4}2m$ and $P\,\bar{4}m2$ differ because in $P\,\bar{4}2m$ the repeated object's 2-fold axes coincide with translation vectors **a** (Fig. 7-2C), whereas in $P\,\bar{4}m2$ these 2-fold axes are at 45° to vectors **a** (Fig. 7-2D). Similarly, space groups $P\,3m1$ and $P\,31m$ are based on repetition of an object with symmetry $3m$ by a primitive hexagonal lattice. For $P\,3m1$ the object has its planes of symmetry *normal* to the lattice's **a** vectors (Fig. 7-2E); for $P\,31m$ the object has its planes of symmetry *parallel* to the lattice's **a** vectors (Fig. 7-2F). The conventions underlying the "point-group portion" of symbols $P\,\bar{4}2m$ and $P\,\bar{4}m2$ or $P\,3m1$ and $P\,31m$ are the same as

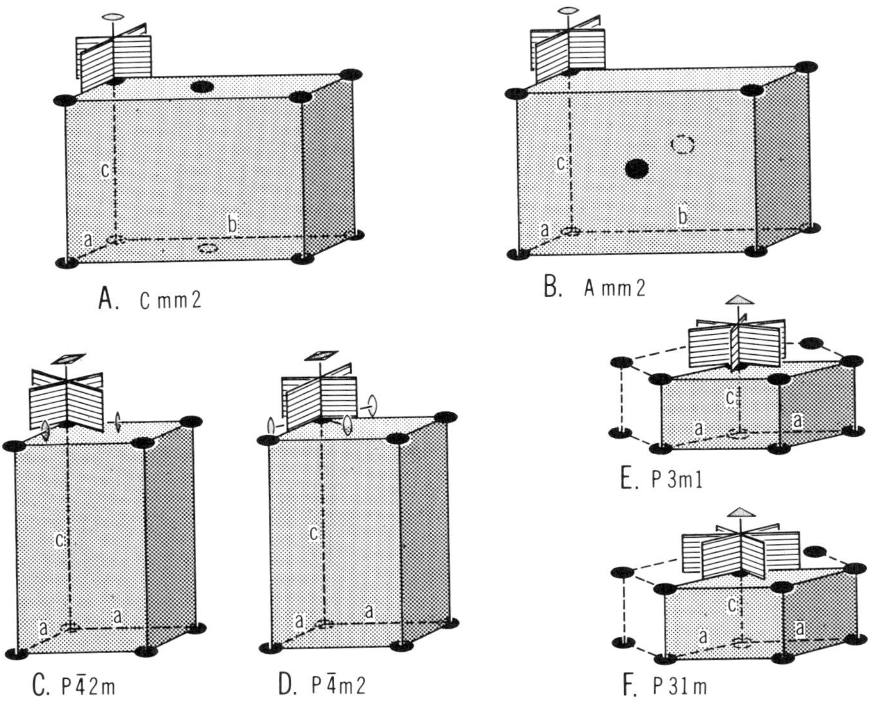

FIGURE 7–2

Differences produced in space groups because the lattice may be differently oriented with respect to the symmetry of the object being repeated, for crystals this object being a cluster of either atoms or ions. The symmetry axes and planes shown at the corner of each unit cell represent the symmetry (and its orientation) for the object being repeated by the lattice, the object itself not being drawn.
Thus an orthorhombic lattice, one face of which is centered, repeats an object whose unique 2-fold axis is perpendicular to the centered face in (A) but parallel to it in (B). By convention the resultant space groups are usually oriented so that the unique 2-fold axis is parallel to the *c* axis. Thus (A) represents space group *C mm*2, whereas (B) represents *A mm*2. If for the cell in (B), the labels on its *a* and *b* axes are interchanged, then *A mm*2 becomes *B mm*2, but these are identical space groups that differ only as to the labeling of the *a* and *b* axes.
The tetragonal unit cells (C) and (D) give rise to different space groups because in (C) the planes of symmetry of the repeated object are at 45° to the {100} faces of the unit cell, whereas in (D) these mirror planes are parallel to {100}. Their different symbols, $P\,\bar{4}2m$ and $P\,\bar{4}m2$, indicate that for (C) the crystal's *a* axes are 2 fold, whereas in (D) the *a* axes are normal to the planes of symmetry. Hexagonal unit cells (E) and (F) represent a situation similar to that which exists between (C) and (D).

discussed in respect to point groups (pp. 38, 39). In this list of symbols the numeral 1 indicates that the crystal direction indicated by that "place" in the symbol is merely a 1-fold axis. Thus *P* 121, if used instead of *P* 2, indicates a monoclinic crystal in the second setting; that is, it indicates the *a* axis to be a 1 fold, the *b* axis to be a 2 fold, and the *c* axis to be a 1 fold. The symbol *P* 112 would indicate this same crystal in the first setting, that is, the 2 fold is now labeled as the *c* axis.

On rare occasions a group of ions possessing symmetry corresponding to a crystal class in one system may be repeated according to a lattice whose symmetry is typical of another. To illustrate this, let the chessman in Figure 7-3 represent a group of ions whose symmetry corresponds to monoclinic class *m*; in Figure 7-3 this group is translationally repeated according to a primitive orthorhombic lattice. As a whole the resultant array, that is, the crystal structure if atoms replace the chessman, has the symmetry *m* and X-ray studies of the crystal would disclose that it possessed only a single plane of symmetry. On the other hand, this crystal would be said to have "orthorhombic geometry" because β, the angle between its *a* and *c* axes, would equal 90°. Thus the fact that the interaxial angles α, β, and γ are all 90° is not quite an infallible guarantee that a crystal is orthorhombic. X-ray studies might disclose it to be monoclinic or, if the repeated object only possessed symmetry 1 or $\bar{1}$, triclinic.

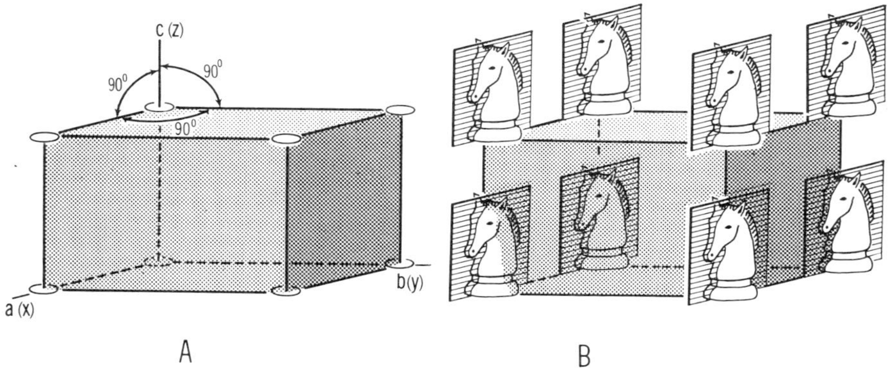

FIGURE 7–3

(A) A primitive orthorhombic space lattice that in itself possesses point group symmetry $2/m\ 2/m\ 2/m$. All interaxial angles equal 90°.
(B) A chessman, used here as symbolic of a cluster of atoms that possesses only a plane of symmetry (ruled), repeated according to the space group to the left. The resultant array only possesses planes of symmetry parallel to (010) Considering the chessmen as clusters of atoms, the array would represent the structure of a monoclinic crystal belonging to the point group *m*. Interestingly, the *a*, *b*, and *c* axes for this crystal would be mutually perpendicular; that is, this monoclinic crystal would possess "orthorhombic geometry."

Rotational-Translational Symmetry

The 10 symmetry elements basic to the 32 point groups represent pure rotational symmetry. On the other hand, pure translational symmetry, a second major type of symmetry, is summarized for three dimensions by the space lattices of which there are 14 types. Within the internal structure of crystals, however, the atoms are sometimes so distributed as to conform to a symmetry operation that combines rotation *plus* translation. These operations are the *screw axes* and *glide planes* to be discussed next.

Screw axes

Within crystalline structures, a set of equivalent atoms $a_1, a_2, a_3, a_4, \ldots$ may be related by an operation that combines (1) a rotation corresponding to a 2-, 3-, 4-, or 6-fold axis with (2) a translation parallel to this axis. In Figure 7-4A, for example, atom a_1 is related to a_2 by the 180° rotation inherent in a 2-fold axis plus a translation corresponding to the heavy dashed arrow. Similarly, a_2 is so related to a_3. Such an operation is called a "2-fold screw axis" and given the written symbol 2_1. Its graphic symbol is the usual 2-fold symbol but with "tails" added. Note that atoms a_1 and a_3 (Fig. 7-4A) are related by a pure translation parallel to the screw axis. This translation, labeled **t** in Figure 7-4A, represents

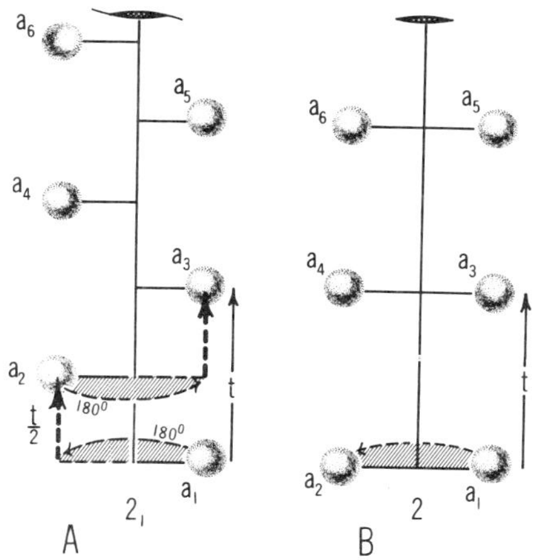

FIGURE 7–4

Comparison of the two types of internal symmetry that produce a 2-fold axis in the crystal's external form. The 2-fold axis at the right operates as in morphological symmetry. Thus it relates a_1 to a_2 by a 180° rotation. These two are then related by the lattice translation **t** to those at a_3 and a_4. The axis labeled 2_1 at left is a 2-fold screw axis. It relates atom a_1 to a_2 or a_2 to a_3 through a 180° rotation in the shaded plane normal to this axis combined with a translation (heavy dashed arrow **t**/2) parallel to this axis.

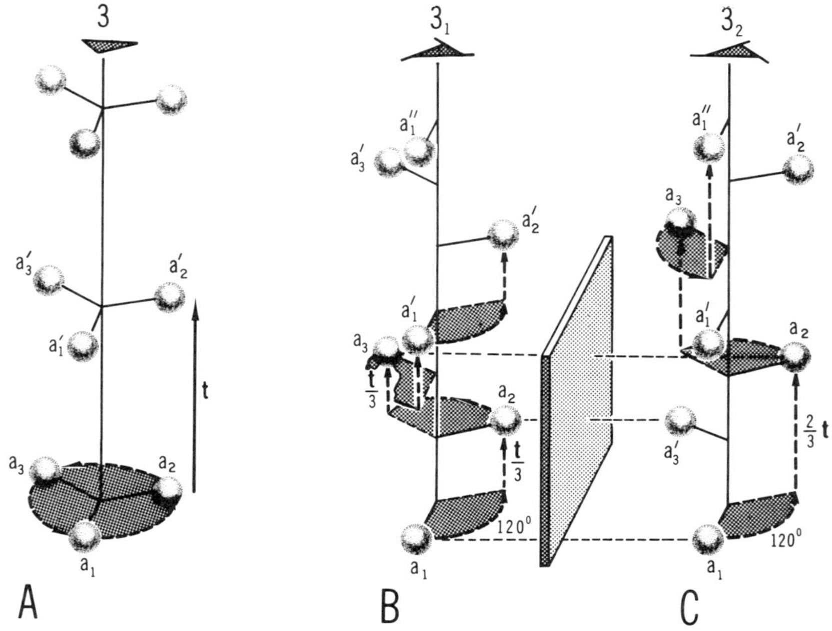

FIGURE 7–5

The types of axes of internal symmetry that can produce a 3-fold axis in the external form of a crystal. (A) A 3-fold axis that involves just rotations of 120° to relate a_1 to a_2 to a_3. These atoms are related to those above them by **t**, the repeat distance along this axis within the structure, that is, the lattice translation parallel to the axis. (B) In contrast, a 3_1 axis relates a_1 to a_2 to a_3 by a 120° anticlockwise rotation plus a translation of **t**/3 (the heavy dashed arrow). (C) A 3_2 axis relates a_1 to a_2 to a_3 by a 120° anticlockwise rotation plus a translation of $\frac{2}{3}$**t**. The lattice translation **t** then requires that a_1' directly overlie a_1 by the distance **t**, that a_3' underlie a_3 by **t**, and so on. The distribution of symmetrically equivalent atoms around the 3_2 axis would also be attained by relating a_1 to a_3' by a 120° *clockwise* rotation plus a translation of **t**/3; a_3' then relates to a_2 by this same operation.
The two axes 3_1 and 3_2 bear a mirror-image (enantiomorphic) relationship. For example, spheres a_1, a_2, and a_3 in (B) are related to spheres a_1, a_3', and a_2 in (C) by reflection across the mirror plane drawn between (B) and (C).

two of the dashed vectors laid end-to-end. Hence each dashed vector equals **t**/2. The written symbol for a screw axis, and this applies to all screw axes, discloses the fraction of **t** that corresponds to the translation inherent in a single operation of the screw axis. Thus 2_1, after its subscript is placed over the main symbol, becomes $\frac{1}{2}$, which is the fraction of **t** equal to the translation involved in a single 2_1 operation.

The 2_1 axis and a "normal" 2-fold axis (Fig. 7-4B) are the only kinds of 2-fold axes possible in the internal structure of a crystal. Either may give rise to the 2-fold axes observed from the crystal's external symmetry.

An externally observed 3-fold axis may result from the presence in the crystal's internal structure of a "normal" 3-fold axis (Fig. 7-5A) or of either of two kinds of 3-fold screw axes, 3_1 (Fig. 7-5B) or 3_2 (Fig. 7-5C). Screw axis 3_1 relates an atom a_1 to a second, a_2, by a 120° anticlockwise rotation plus a translation (equals **t**/3 in Fig. 7-5B). Repeating this operation, a_2 relates to a_3, a_3 to a_1', and so on. The pure translation **t** relates a_1 to a_1', a_2 to a_2', and so on. In other words, parallel to the screw axis within a crystal's internal structure, identipoints are spaced at intervals equal to **t**, the lattice translation. The screw axis 3_2 relates a_1 to a_2, a_2 to a_3, and a_3 to a_1'' by a 120° anticlockwise rotation plus a translation of $\frac{2}{3}$ of **t** (Fig. 7-5C), whereas the lattice translation **t** relates a_1 to

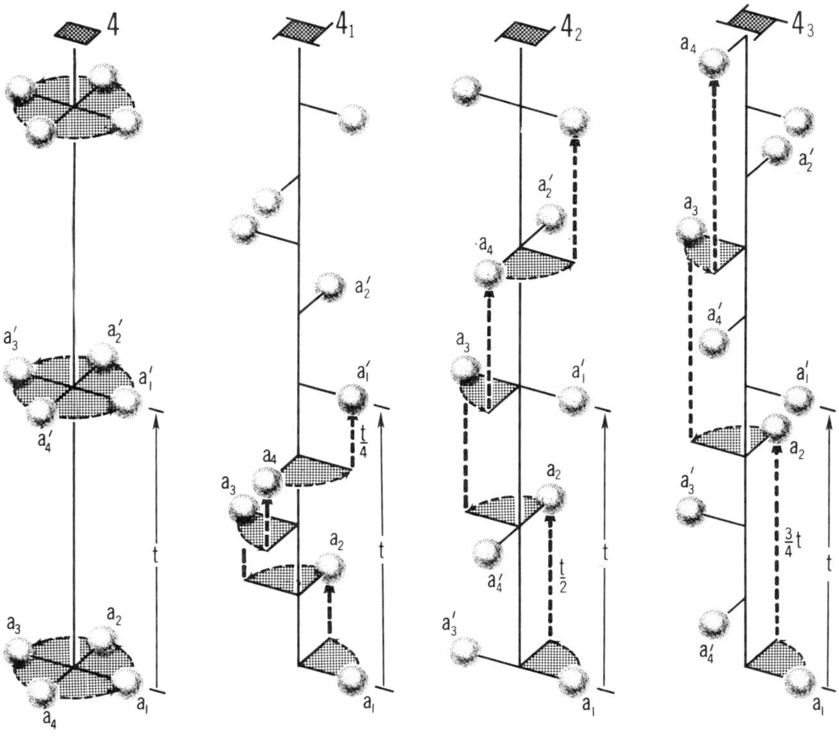

FIGURE 7–6

The four types of internal axes of symmetry that produce a 4-fold axis in the crystal's external shape. The lattice translation parallel to the axis is shown as **t** in all cases. Atoms a_1, a_2, a_3, and a_4 are shown to be related by the operation inherent to the axis. Atoms indicated by primes are related by translations of **t** or 2**t** to atoms similarly subscripted (but unprimed). By convention, screw axes combine *anticlockwise* rotations with *upward* translations (bold dashed arrows). However, axis 4_3 can be also considered to relate atom a_1 to a_4' (and a_4' to a_3') by a 90° *clockwise* rotation plus an upward translation of **t**/4. Thus 4_3 differs from 4_1 only in its direction of rotation, the two bearing an enantiomorphous relationship. For example, a mirror plane drawn through the 4_3 axis so as to pass through atoms a_1, a_1', a_3, and a_3' as well would operate on 4_3 so as to make it identical to 4_1.

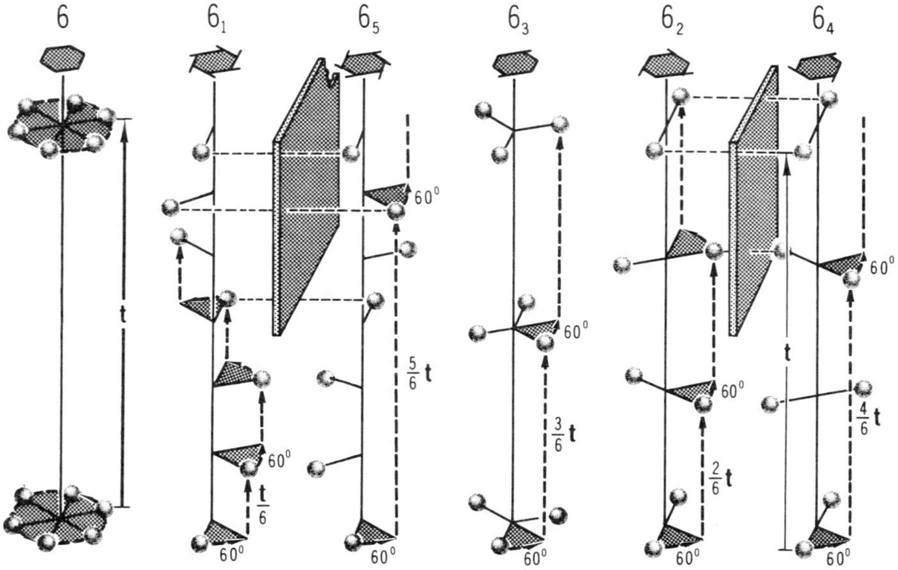

FIGURE 7–7

The six types of internal axes of symmetry that produce a 6-fold axis in the crystal's external shape. The lattice translation parallel to the axis, that is, **t**, has been given the same length for each type of axis. Screw axes 6_1 and 6_5 are enantiomorphous as are 6_2 and 6_4. To illustrate this, a small part of the mirror plane which relates them is drawn between each pair. Atoms related by reflection across this plane are connected by horizontal dashed lines.

a_1' and a_3 to a_3'. Alternatively, the 3_2 operation may be considered to relate a_1 to a_3', a_3' to a_2, a_2 to a_1', and so on by a 120° *clockwise* rotation plus a screw translation of $\frac{1}{3}$ of **t**. When it is considered in this way, 3_2 differs from 3_1 only in its sense of rotation. Axes 3_1 and 3_2 differ only in being mirror images of each other. Consequently, a crystal whose internal structure contains 3_1 axes may with equal facility crystallize in a mirror-image structure that thus contains 3_2 axes. In this event the two crystals will bear an enantiomorphous relationship as discussed for quartz (p. 133) whose two possible mirror-image structures are based on space groups $P3_12$ or $P3_22$.

The remaining screw axes require little additional discussion. Thus, externally expressed 4-fold axes on crystals may result from internal symmetry axes that relate atoms by the operation of a normal 4-fold axis or of screw axes 4_1, 4_2, or 4_3 (Fig. 7-6). Axes 4_1 and 4_3 are enantiomorphous. Successive operations of 4_2 relate atom a_1 to a_2, a_2 to a_3, a_3 to a_4, whereas the lattice translation of **t** along this axis relates these atoms to their primed counterparts, for example, a_1 to a_1', a_2 to a_2'. Externally expressed 6-fold axes result from internal axes 6, 6_1, 6_2, 6_3, 6_4, or 6_5 (Fig. 7-7). Note the enantiomorphism of 6_1 to 6_5 and of 6_2 to 6_4.

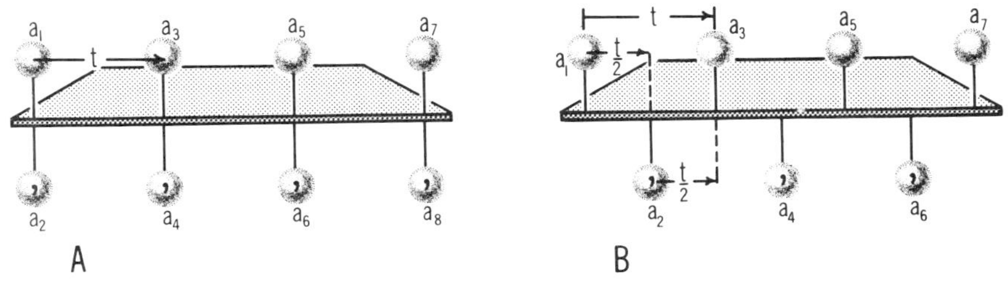

FIGURE 7–8

The two types of internal symmetry planes.
(A) The ordinary mirror plane is shown to relate by reflection a_1 to a_2, a_3 to a_4, and so on. Commas are used to indicate that a_2, a_4, a_6, and a_8 are mirror images of a_1, a_3, a_5, and a_7. This would be apparent if a_1, a_3, a_5, and a_7 were shapes less symmetric than a sphere (for example, a left hand), then a_2, a_4, a_6, and a_8 would be right hands. A lattice translation **t** relates a_1 to a_3 to a_5 to a_7 or a_2 to a_4 to a_6 to a_8.
(B) The *glide plane* relates a_1 to a_2 by a glide translation **t**/2, which is parallel to the plane, plus reflection across the plane. Continued operation of the glide plane relates a_2 to a_3, a_3 to a_4, and so on. The purely translational repeat parallel to the plane is **t**.

Glide planes

Although the symmetry planes observed in crystals may result from the presence of ordinary mirror planes within the crystal's internal structure (Fig. 7-8A), in many instances they result from the presence of *glide planes*. A glide plane's operation involves (1) reflection perpendicularly across the plane *plus* (2) translation parallel to this plane for a distance and direction equal to the *glide component*. For example, the glide plane in Figure 7-8B relates atoms a_1 to a_2, a_2 to a_3, a_3 to a_4 by a translation equal to the glide component **t**/2 plus a reflection across the plane.

The nomenclature for glide planes is illustrated in Figure 7-9 for an orthorhombic crystal whose unit cell alone is shown. *Axial glides* are planes whose glide component is parallel to a crystallographic axis and equal in length to one-half the repeat unit along this axis. Such axial glides are symbolized as *a*, *b*, or *c* according to whether their glide component was along the *a*, *b*, or *c* axis and was equal to, respectively, **a**/2, **b**/2, or **c**/2, where **a**, **b**, and **c** represent the lattice vectors or unit cell edges. *Diagonal glides*, whose general symbol is *n*, are glide planes whose glide component represents the vector sum of any two of the following vectors: **a**/2, **b**/2, **c**/2.[1] For *diamond glides*, which are so named because they occur in the structure of diamond, the general symbol is *d* and their glide component represents the vector sum of any two of the following: **a**/4, **b**/4, **c**/4.[1]

[1] For glide planes parallel to {110}, which occur only in tetragonal or isometric crystals, their glide component is the vector sum of all three, where **a**, **b**, and **c** now indicate lattice vectors along a_1, a_2, and *c* or along a_1, a_2, and a_3. (Only if this glide component is the vector sum of all three, will it be a direction parallel to a {110} face.)

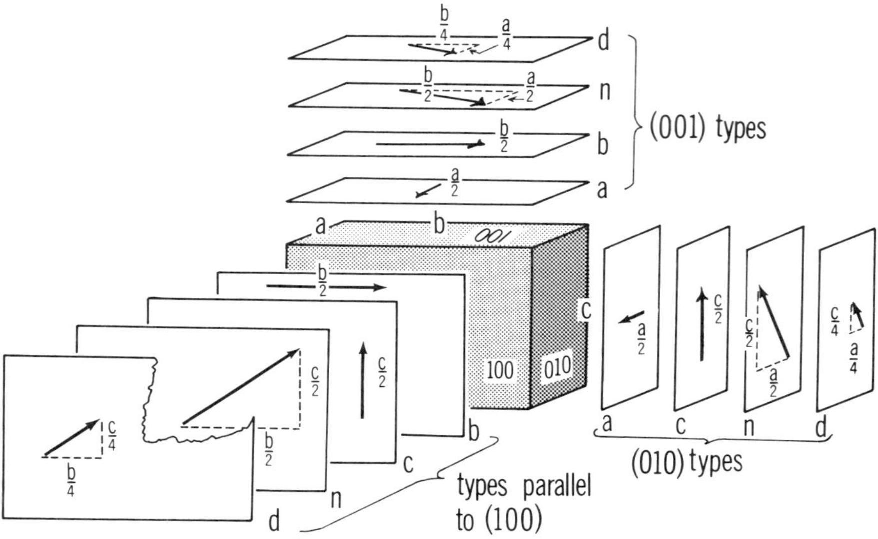

FIGURE 7–9

The possible types of glide planes parallel to the (100), (010), and (001) faces are drawn for an orthorhombic crystal, one unit cell of which is shown. For each possible glide plane the direction and length of the glide component is shown by an arrow. The dashed lines represent the vectors parallel to the crystallographic axes into which the glide component can be resolved. Note the impossibility of an *a* glide parallel to (100), a *b* glide parallel to (010), or a *c* glide parallel to (001) because these call for a glide component that is not parallel to the glide plane.

Let us consider an orthorhombic crystal from class $2/m\ 2/m\ 2/m$ whose internal structure possesses the following characteristics: (1) The space lattice is primitive; (2) the *a* axis, which shows up externally as a 2 fold, is actually a 2_1-screw axis with a *b*-glide plane normal to it; (3) the *b* axis is also a 2_1 screw with an *a*-glide plane normal to it; (4) the *c* axis is an ordinary 2 fold, even internally, with a mirror plane normal to it. These data are concisely summarized in the crystal structure's space-group symbol, $P\ 2_1/b\ 2_1/a\ 2/m$. The shortened form, $P\ b\ a\ m$, is less informative but often used.

The 230 Space Groups

In the 73 simple space groups already cited, the cluster of atoms being repeated by the space lattice was presumed to possess only symmetry corresponding to one of the 32 point groups. Actually, however, the cluster's symmetry might be such that screw axes rather than the comparable purely rotational axes may be present: that is, 2_1 instead of 2; 3_1 or 3_2 instead of 3; 4_1, 4_2, or 4_3 instead of 4. Or the cluster might possess glide planes rather than mirror planes, that is,

TABLE 7–1 The Space Groups of the Triclinic, Monoclinic, and Orthorhombic Systems

System	Triclinic		Monoclinic (Second Setting: b = 2 and/or $\bar{2}$)			Orthorhombic		
Point Group	1	$\bar{1}$	2	m	$2/m$	222	$mm2$	mmm or $2/m\ 2/m\ 2/m$
Space Groups	(1)$P1$	(2)$P\bar{1}$	(3)$P2$	(6)Pm	(10)$P2/m$	(16)$P222$	(25)$Pmm2$	(47)$Pmmm$ or $P2/m\ 2/m\ 2/m$
			$P121$	$P1m1$	$P1\ 2/m\ 1$	(17)$P222_1$	(26)$Pmc2_1$	(48)$Pnnn$ or $P2/n\ 2/n\ 2/n$
			(4)$P2_1$	(7)Pc	(11)$P2_1/m$	(18)$P2_12_12$	(27)$Pcc2$	(49)$Pccm$ or $P2/c\ 2/c\ 2/m$
			$P12_11$	$P1c1$	$P1\ 2_1/m\ 1$	(19)$P2_12_12_1$	(28)$Pma2$	(50)$Pban$ or $P2/b\ 2/a\ 2/n$
			(5)$C2$	(8)Cm	(12)$C2/m$	(20)$C222_1$	(29)$Pca2_1$	(51)$Pmma$ or $P2_1/m\ 2/m\ 2/a$
			$C121$	$C1m1$	$C1\ 2/m\ 1$	(21)$C222$	(30)$Pnc2$	(52)$Pnna$ or $P2/n\ 2_1/n\ 2/a$
				(9)Cc	(13)$P2/c$	(22)$F222$	(31)$Pmn2_1$	(53)$Pmna$ or $P2/m\ 2/n\ 2_1/a$
				$C1c1$	$P1\ 2/c\ 1$	(23)$I222$	(32)$Pba2$	(54)$Pcca$ or $P2_1/c\ 2/c\ 2/a$
					(14)$P2_1/c$	(24)$I2_12_12_1$	(33)$Pna2_1$	(55)$Pbam$ or $P2_1/b\ 2_1/a\ 2/m$
					$P1\ 2_1/c\ 1$		(34)$Pnn2$	(56)$Pccn$ or $P2_1/c\ 2_1/c\ 2/n$
					(15)$C2/c$		(35)$Cmm2$	(57)$Pbcm$ or $P2/b\ 2_1/c\ 2_1/m$
					$C1\ 2/c\ 1$		(36)$Cmc2_1$	(58)$Pnnm$ or $P2_1/n\ 2_1/n\ 2/m$
							(37)$Ccc2$	(59)$Pmmn$ or $P2_1/m\ 2_1/m\ 2/n$
			Monoclinic (First Setting: c = 2 and/or $\bar{2}$)				(38)$Amm2$	(60)$Pbcn$ or $P2_1/b\ 2/c\ 2_1/n$
							(39)$Abm2$	(61)$Pbca$ or $P2_1/b\ 2_1/c\ 2_1/a$
							(40)$Ama2$	(62)$Pnma$ or $P2_1/n\ 2_1/m\ 2_1/a$
			2	m	$2/m$		(41)$Aba2$	(63)$Cmcm$ or $C2/m\ 2/c\ 2_1/m$
							(42)$Fmm2$	(64)$Cmca$ or $C2/m\ 2/c\ 2_1/a$
			(3)$P2$	(6)Pm	(10)$P2/m$		(43)$Fdd2$	(65)$Cmmm$ or $C2/m\ 2/m\ 2/m$
			$P112$	$P11m$	$P1\ 1\ 2/m$		(44)$Imm2$	(66)$Cccm$ or $C2/c\ 2/c\ 2/m$
			(4)$P2_1$	(7)Pb	(11)$P2_1/m$		(45)$Iba2$	(67)$Cmma$ or $C2/m\ 2/m\ 2/a$
			$P112_1$	$P11b$	$P1\ 1\ 2_1/m$		(46)$Ima2$	(68)$Ccca$ or $C2/c\ 2/c\ 2/a$
			(5)$B2$	(8)Bm	(12)$B2/m$			(69)$Fmmm$ or $F2/m\ 2/m\ 2/m$
			$B112$	$B11m$	$B1\ 1\ 2/m$			(70)$Fddd$ or $F2/d\ 2/d\ 2/d$
				(9)Bb	(13)$P2/b$			(71)$Immm$ or $I2/m\ 2/m\ 2/m$
				$B11b$	$P1\ 1\ 2/b$			(72)$Ibam$ or $I2/b\ 2/a\ 2/m$
					(14)$P2_1/b$			(73)$Ibca$ or $I2/b\ 2/c\ 2/a$
					$P1\ 1\ 2_1/b$			(74)$Imma$ or $I2/m\ 2/m\ 2/a$
					(15)$B2/b$			
					$B1\ 1\ 2/b$			
Laue Group Symmetry[a]	$\bar{1}$		$2/m$			$m\ m\ m$ or $2/m\ 2/m\ 2/m$		

[a] See Figure 11-4 for a discussion of Laue classes.

TABLE 7–2

The Tetragonal Space Groups

System	Tetragonal						
Point Group	4	$\bar{4}$	$4/m$	422	$4mm$	$\bar{4}2m$	$4/mmm$ or $4/m\ 2/m\ 2/m$
Space Groups	(75)$P4$	(81)$P\bar{4}$	(83)$P4/m$	(89)$P422$	(99)$P4mm$	(111)$P\bar{4}2m$	(123)$P4/mmm$ or $P4/m\ 2/m\ 2/m$
	(76)$P4_1$	(82)$I\bar{4}$	(84)$P4_2/m$	(90)$P42_12$	(100)$P4bm$	(112)$P\bar{4}2c$	(124)$P4/mcc$ or $P4/m\ 2/c\ 2/c$
	(77)$P4_2$		(85)$P4/n$	(91)$P4_122$	(101)$P4_2cm$	(113)$P\bar{4}2_1m$	(125)$P4/nbm$ or $P4/n\ 2/b\ 2/m$
	(78)$P4_3$		(86)$P4_2/n$	(92)$P4_12_12$	(102)$P4_2nm$	(114)$P\bar{4}2_1c$	(126)$P4/nnc$ or $P4/n\ 2/n\ 2/c$
	(79)$I4$		(87)$I4/m$	(93)$P4_222$	(103)$P4cc$	(115)$P\bar{4}m2$	(127)$P4/mbm$ or $P4/m\ 2_1/b\ 2/m$
	(80)$I4_1$		(88)$I4_1/a$	(94)$P4_22_12$	(104)$P4nc$	(116)$P\bar{4}c2$	(128)$P4/mnc$ or $P4/m\ 2_1/n\ 2/c$
				(95)$P4_322$	(105)$P4_2mc$	(117)$P\bar{4}b2$	(129)$P4/nmm$ or $P4/n\ 2_1/m\ 2/m$
				(96)$P4_32_12$	(106)$P4_2bc$	(118)$P\bar{4}n2$	(130)$P4/ncc$ or $P4/n\ 2_1/c\ 2/c$
				(97)$I422$	(107)$I4mm$	(119)$I\bar{4}m2$	(131)$P4_2/mmc$ or $P4_2/m\ 2/m\ 2/c$
				(98)$I4_122$	(108)$I4cm$	(120)$I\bar{4}c2$	(132)$P4_2/mcm$ or $P4_2/m\ 2/c\ 2/m$
					(109)$I4_1md$	(121)$I\bar{4}2m$	(133)$P4_2/nbc$ or $P4_2/n\ 2/b\ 2/c$
					(110)$I4_1cd$	(122)$I\bar{4}2d$	(134)$P4_2/nnm$ or $P4_2/n\ 2/n\ 2/m$
							(135)$P4_2/mbc$ or $P4_2/m\ 2_1/b\ 2/c$
							(136)$P4_2/mnm$ or $P4_2/m\ 2_1/n\ 2/m$
							(137)$P4_2/nmc$ or $P4_2/n\ 2_1/m\ 2/c$
							(138)$P4_2/ncm$ or $P4_2/n\ 2_1/c\ 2/m$
							(139)$I4/mmm$ or $I4/m\ 2/m\ 2/m$
							(140)$I4/mcm$ or $I4/m\ 2/c\ 2/m$
							(141)$I4_1/amd$ or $I4_1/a\ 2/m\ 2/d$
							(142)$I4_1/acd$ or $I4_1/a\ 2/c\ 2/d$
Laue Group Symmetry		$4/m$			$4/m\ m\ m$ or $4/m\ 2/m\ 2/m$		

TABLE 7–3

The Space Groups of the Hexagonal System

System	Trigonal Subsystem				
Point Group	3	$\bar{3}$	32	$3m$	$\bar{3}m$ or $\bar{3}\ 2/m$
Space Groups	(143)$P3$ (144)$P3_1$ (145)$P3_2$ (146)$R3$	(147)$P\bar{3}$ (148)$R\bar{3}$	(149)$P312$ (150)$P321$ (151)$P3_112$ (152)$P3_121$ (153)$P3_212$ (154)$P3_221$ (155)$R32$	(156)$P3m1$ (157)$P31m$ (158)$P3c1$ (159)$P31c$ (160)$R3m$ (161)$R3c$	(162)$P\bar{3}1m$ or $P\bar{3}\ 1\ 2/m$ (163)$P\bar{3}1c$ or $P\bar{3}\ 1\ 2/c$ (164)$P\bar{3}m1$ or $P\bar{3}\ 2/m\ 1$ (165)$P\bar{3}c1$ or $P\bar{3}\ 2/c\ 1$ (166)$R\bar{3}m$ or $R\bar{3}\ 2/m$ (167)$R\bar{3}c$ or $R\bar{3}\ 2/c$
Laue Group Symmetry	$\bar{3}$		$\bar{3}\ 2/m$		

Hexagonal Subsystem

Point Group	6	$\bar{6}$	$6/m$	622	$6mm$	$\bar{6}m2$	$6/mmm$ or $6/m\ 2/m\ 2/m$
Space Groups	(168)$P6$ (169)$P6_1$ (170)$P6_5$ (171)$P6_2$ (172)$P6_4$ (173)$P6_3$	(174)$P\bar{6}$	(175)$P6/m$ (176)$P6_3/m$	(177)$P622$ (178)$P6_122$ (179)$P6_522$ (180)$P6_222$ (181)$P6_422$ (182)$P6_322$	(183)$P6mm$ (184)$P6cc$ (185)$P6_3cm$ (186)$P6_3mc$	(187)$P\bar{6}m2$ (188)$P\bar{6}c2$ (189)$P\bar{6}2m$ (190)$P\bar{6}2c$	(191)$P6/mmm$ or $P6/m\ 2/m\ 2/m$ (192)$P6/mcc$ or $P6/m\ 2/c\ 2/c$ (193)$P6_3/mcm$ or $P6_3/m\ 2/c\ 2/m$ (194)$P6_3/mmc$ or $P6_3/m\ 2/m\ 2/c$
Laue Group Symmetry	$6/m$			$6/mmm$ or $6/m\ 2/m\ 2/m$			

TABLE 7–4

The Space Groups of the Isometric System

System	Isometric System				
Point Group	23	$m3$ or $2/m\,\bar{3}$	432	$\bar{4}3m$	$m3m$ or $4/m\,\bar{3}\,2/m$
Space Groups	(195)$P23$	(200)$Pm3$ or $P2/m\,\bar{3}$	(207)$P432$	(215)$P\bar{4}3m$	(221)$Pm3m$ or $P4/m\,\bar{3}\,2/m$
	(196)$F23$	(201)$Pn3$ or $P2/n\,\bar{3}$	(208)$P4_232$	(216)$F\bar{4}3m$	(222)$Pn3n$ or $P4/n\,\bar{3}\,2/n$
	(197)$I23$	(202)$Fm3$ or $F2/m\,\bar{3}$	(209)$F432$	(217)$I\bar{4}3m$	(223)$Pm3n$ or $P4_2/m\,\bar{3}\,2/n$
	(198)$P2_13$	(203)$Fd3$ or $F2/d\,\bar{3}$	(210)$F4_132$	(218)$P\bar{4}3n$	(224)$Pn3m$ or $P4_2/n\,\bar{3}\,2/m$
	(199)$I2_13$	(204)$Im3$ or $I2/m\,\bar{3}$	(211)$I432$	(219)$F\bar{4}3c$	(225)$Fm3m$ or $F4/m\,\bar{3}\,2/m$
		(205)$Pa3$ or $P2_1/a\,\bar{3}$	(212)$P4_332$	(220)$I\bar{4}3d$	(226)$Fm3c$ or $F4/m\,\bar{3}\,2/c$
		(206)$Ia3$ or $I2_1/a\,\bar{3}$	(213)$P4_132$		(227)$Fd3m$ or $F4_1/d\,\bar{3}\,2/m$
			(214)$I4_132$		(228)$Fd3c$ or $F4_1/d\,\bar{3}\,2/c$
					(229)$Im3m$ or $I4/m\,\bar{3}\,2/m$
					(230)$Ia3d$ or $I4_1/a\,\bar{3}\,2/d$
Laue Group Symmetry	$m3$ or $2/m\,\bar{3}$		$m3m$ or $4/m\,\bar{3}\,2/m$		

a, b, c, n, or d glides instead of m. Because of these additional possibilities, there are 230 possible types of space groups in all. In the summaries of their distribution among the different crystal systems and crystal classes (Tables 7-1 to 7-4), the parenthesized number preceding each space group's symbol represents their order of listing in Volume 1 of *International Tables for X-ray Crystallography*, the authoritative, modern compilation of space-group descriptions.

Each space group represents a scaffoldlike array of symmetry elements. To illustrate, Figure 7-10A is a representative portion of space group $P\ 2_1/b\ 2_1/a\ 2/m$. Alternatively, Figure 7-10A may be regarded to represent a single unit cell for a crystal of space group $P\ 2_1/b\ 2_1/a\ 2/m$ as it would appear if the atoms or ions encompassed by this cell were invisible, whereas the symmetry operations relating them were not. In depicting each space group the *International Tables for X-ray Crystallography* customarily draw its appearance as sighted along a line normal to the (001) plane that is, projected on (001). Thus $P\ 2_1/b\ 2_1/a\ 2/m$ is depicted as in Figure 7-10B rather than three dimensionally as in Figure 7-10A. In these (001) projections, structural crystallographers generally employ the following conventions. (1) The letters x, y, and z denote axes a, b, and c as well as, on occasion, one translational repeat along them; for more symmetric systems, they denote a_1, a_2, c or a_1, a_2, a_3. (2) In (001) projections like Figure 7-10B, for

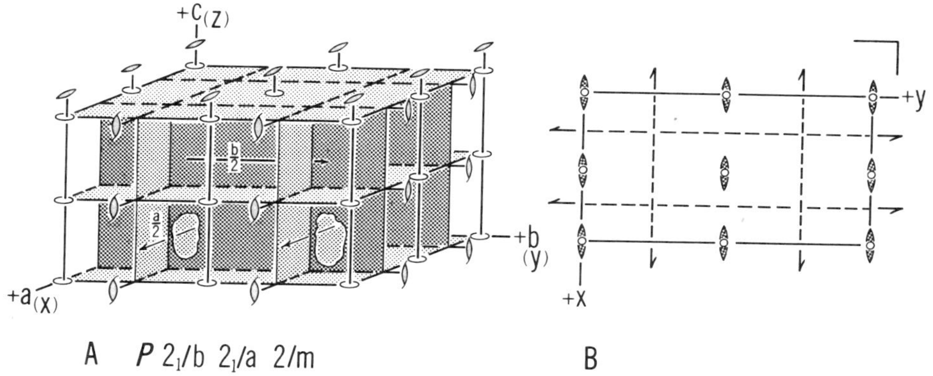

FIGURE 7–10

(A) The operations of internal symmetry present in the unit cell of a crystal whose structure conforms to space group $P\ 2_1/b\ 2_1/a\ 2/m$. The horizontal planes are mirror planes; the vertical planes are glide planes whose glide components equal one-half the cell's edge along axes b and a respectively. The circles, shown in perspective as ellipses, represent the centers of symmetry that occur where the 2 folds intersect the horizontal mirror planes.

(B) Top view of the unit cell at left, that is, its appearance as projected onto the (001) plane. The single-barbed arrows represent the 2_1 axes that are now parallel to the page; the right angle symbol at the upper right indicates the planes of symmetry parallel to the page at heights of $z = 0$ and $z = \frac{1}{2}$, where these fractions refer to the fraction of the cell's edge along the c axis.

which necessarily x and y lie in the plane of the page, y is oriented horizontally (plus end to the reader's right), whereas the plus x axis extends down the page; the value of the angle between x and y, that is, γ, depends on the crystal system involved and, for monoclinic crystals, on whether the first or second setting was chosen. (3) Different graphic symbols are used for symmetry operations that are parallel to the plane of projection (Fig. 7-11A) as opposed to those that are perpendicular to it (Fig. 7-11B). Symmetry planes parallel to the (001) plane projection are thus denoted by two short intersecting lines that are parallel to the y and z axes; if these symmetry planes are glide planes, an arrow indicates the glide direction. A double-barbed arrow represents a 2-fold axis that is parallel to the plane of projection; a single-barbed arrow similarly indicates a parallel 2_1 axis. Glide planes perpendicular to the plane of projection are represented by (1) dashed, (2) dotted, (3) dash dotted, or (4) a dash-arrow-dot line depending on whether, in respect to the plane of projection, the glide translation is (1) parallel, (2) perpendicular, or (3) and (4) is resolvable into both parallel and perpendicular components. The dash-dotted line indicates a diagonal glide n, whereas a dash-arrow-dotted line indicates a diamond glide d (Fig. 7-11B).

For each space group the *International Tables for X-ray Crystallography* stipulate the conventional choice of origin for the space lattice, that is, where the corners of the unit cell are considered to be located. Thus for space group $P\ 2_1/b\ 2_1/a\ 2/m$ the statement reads, "Origin at center $(2/m)$." In other words, for crystal structures of this symmetry the unit cell defined by lattice translations **a**, **b**, and **c** should have its corners at the set of identipoints located at centers of symmetry in the structure, these centers being at the points of intersection

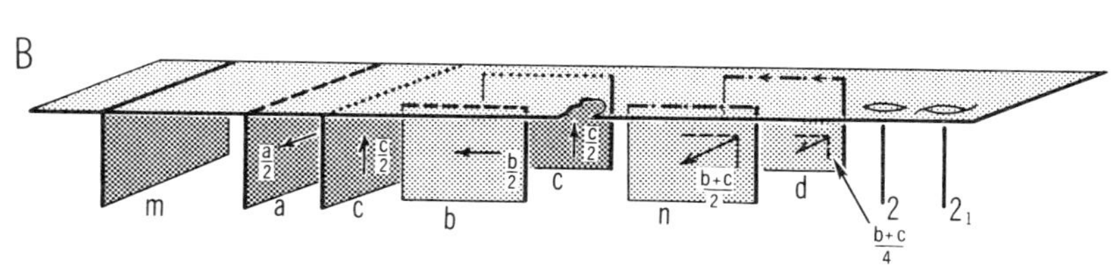

FIGURE 7–11

The symbols used for several of the operations of internal symmetry if they are projected onto a plane.

(A) The symbols used to indicate a mirror plane m, glide planes a, b, n, or d, and axes 2 or 2_1 if these are parallel to the plane of projection.

(B) The symbols for these if they are perpendicular to the plane of projection.

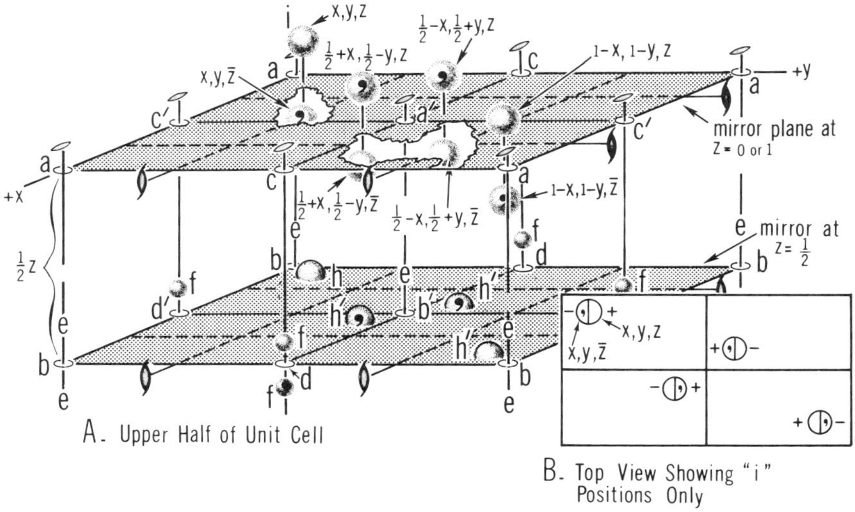

FIGURE 7–12

(A) The upper half of the unit cell for space group $P\,2_1/b\,2_1/a\,2/m$. The missing lower half could be generated from this upper half by reflection across the (shaded) mirror plane at $z = \frac{1}{2}$. If two atoms are related by mirror reflection, one will be marked with a comma. Thus the atoms at x, y, z and $x, y, \bar{z}$, which are in the positions designated as i, are related by the mirror at $z = 1$.
(B) A top view of (A) at reduced scale with the general positions i indicated. The split circles indicate two superimposed i positions related by the mirror plane between them. For example, the atom at x, y, z in (A) is now represented by the circle's right half, the adjacent plus sign indicating that this atom lies above the (mirror) plane at $z = 0$. The atom at $x, y, \bar{z}$ in (A) is represented by the left half of the circle, the adjacent negative sign indicating that this site lies *below* the mirror at $z = 0$.

between the 2-fold axes and mirror planes. Consequently, the origin for this space group must be located at one of the eight ellipses marked a, a', c, c', b, b', d, or d' in Figure 7-12A. Each is at a center of symmetry in the crystal structure and each is relatable to identical points in neighboring unit cells by the translational symmetry expressed by the cell-edge vectors **a**, **b**, and **c**.

Equipoints

Any given point or position in a unit cell *plus* all points in the cell symmetrically related to it through operation of the structure's symmetry axes, centers, planes, screw axes, and glide planes constitute a set of symmetrically *equivalent positions* or *equipoints*. Thus in Figure 7-12A point a is related to a' through operation of either the b glide that is parallel to (100) at $x = \frac{1}{4}$, the a glide parallel to (010) at $y = \frac{1}{4}$, or one of the 2_1 axes lying in the upper mirror plane.

The number of equipoints or equivalent positions in a single unit cell depends on (1) the nature and number of symmetry operations inherent to the space group and (2) the location of the initial equipoint relative to these operations. A point that falls on an axis of rotational symmetry ($2, 3, 4, 6, \bar{1}, \bar{3}, \bar{4}, \bar{6}$) or a mirror plane ($m$) is in a *special position* because it will not then be repeated by these. This is not true for screw axes or glide planes whose translational component *does* repeat even atoms centered on them. In contrast to such special positions an atom whose center is at coordinates x, y, and z in a unit cell, where x, y, and z represent random fractions of the cell edge along each of these axes, is conventionally regarded as not centered on any symmetry axis or plane; that is, it is in a *general position*. As shown for atom i at x, y, z (Fig. 7-12A), a general position within the unit cell for space group $P\,2_1/b\,2_1/a\,2/m$ is symmetrically equivalent to seven more with coordinates as given in Figure 7-12A, the eight constituting a set of equivalent points. Commas denote atoms of the set that are enantiomorphously related to i; the environment around these atoms would be the mirror image of that surrounding i.

For each space group the *International Tables for X-ray Crystallography* summarize for the general site, and for each of the different special sites, the following: (1) the number of equivalent positions in the set; (2) the letter, *a, b, c, d, e, f, g, h, i*, and so on, which Wyckoff used to denote this site, those letters that are later in the alphabet denoting sites through which fewer symmetry elements pass; (3) the point symmetry existent at this site; and (4) the coordinates of the equivalent positions. Thus for $P\,2_1/b\,2_1/a\,2/m$ the data are as follows:

Number of Positions, Wyckoff Notation, and Point Symmetry			Coordinates of Equivalent Positions
8	i	1	$x, y, z;\ \ \bar{x}, \bar{y}, z;\ \ \frac{1}{2}+x, \frac{1}{2}-y, \bar{z};\ \ \frac{1}{2}-x, \frac{1}{2}+y, \bar{z};$ $\bar{x}, \bar{y}, \bar{z};\ \ x, y, \bar{z};\ \ \frac{1}{2}-x, \frac{1}{2}+y, z;\ \ \frac{1}{2}+x, \frac{1}{2}-y, z.$
4	h	m	$x, y, \frac{1}{2};\ \ \bar{x}, \bar{y}, \frac{1}{2};\ \ \frac{1}{2}+x, \frac{1}{2}-y, \frac{1}{2};\ \ \frac{1}{2}-x, \frac{1}{2}+y, \frac{1}{2}.$
4	g	m	$x, y, 0;\ \ \bar{x}, \bar{y}, 0;\ \ \frac{1}{2}+x, \frac{1}{2}-y, 0;\ \ \frac{1}{2}-x, \frac{1}{2}+y, 0.$
4	f	2	$0, \frac{1}{2}, z;\ \ 0, \frac{1}{2}, \bar{z};\ \ \frac{1}{2}, 0, z;\ \ \frac{1}{2}, 0, \bar{z}.$
4	e	2	$0, 0, z;\ \ 0, 0, \bar{z};\ \ \frac{1}{2}, \frac{1}{2}, z;\ \ \frac{1}{2}, \frac{1}{2}, \bar{z}.$
2	d	$2/m$	$0, \frac{1}{2}, \frac{1}{2};\ \ \frac{1}{2}, 0, \frac{1}{2}.$
2	c	$2/m$	$0, \frac{1}{2}, 0;\ \ \frac{1}{2}, 0, 0.$
2	b	$2/m$	$0, 0, \frac{1}{2};\ \ \frac{1}{2}, \frac{1}{2}, \frac{1}{2}.$
2	a	$2/m$	$0, 0, 0;\ \ \frac{1}{2}, \frac{1}{2}, 0.$

In Figure 7-12A the coordinates of the i positions correspond to those listed above except that $1 - x, 1 - y, z$ and $1 - x, 1 - y, \bar{z}$ are given instead of $\bar{x}, \bar{y}, z$ and $\bar{x}, \bar{y}, \bar{z}$. The difference is trivial, however, because to add or subtract 1.0 from one or more of the coordinates of a point is simply to translate it into a translationally identical site in the next cell. Thus if 1.0 is added to the x and y coordinates of points $\bar{x}, \bar{y}, z$ and $\bar{x}, \bar{y}, \bar{z}$, these become $1 - x, 1 - y, z$ and $1 - x, 1 - y, \bar{z}$ and the points are translated into the cell illustrated in Figure 7-12. All the Wyckoff sites except g are located in Figure 7-12A. The atoms shown to occupy site h and its three equivalent positions h' are outlined in black so they will not be confused with the similarly sized atoms in the i sites. Small atoms are shown in the f sites.

In determining the number of positions equivalent to a given Wyckoff site, positions centered on the unit cell's face are shared equally with the neighboring unit cell, each "owning" one-half. Similarly, a cell "owns" only one-fourth of each position centered on its edge and one-eighth of those on its corners. For example, the unit cell in Figure 7-12A owns four e positions (one-fourth of each of the eight e sites on the cell's vertical edges plus both of the e sites in the cell's interior, the mirror at $z = \frac{1}{2}$ hiding the lower of these).

Rather than a three-dimensional view like Figure 7-12A, the *International Tables for X-ray Crystallography* substitute a (001) projection (Fig. 7-12B) in which the equivalent atoms for a general position are shown. By convention the plane of the paper contains the origin, and atoms labeled + or − are, respectively, considered to be centered slightly above or below this plane. The symbol $\frac{1}{2}+$ would indicate an atom above the plane by an amount slightly more than one-half of that unit cell edge that is normal to the page. A pair of atoms such as the two at x, y, z and $x, y, \bar{z}$ in Figure 7-12A will plot at the same position in Figure 7-12B. The circle representing them is thus divided by a vertical line, and the circle's left half represents the one atom, that at $x, y, \bar{z}$ in this case, whereas its right half represents the atom at x, y, z. Whenever a plane of symmetry is parallel to the plane of projection, this convention will be employed.

Schoenflies notation

The Schoenflies symbolism for space groups is rapidly becoming obsolete. Because it may be encountered in older references, however, Table 7-5 is included here. It lists the Schoenflies symbols in the same order as their listing in Tables 7-1 to 7-4, thereby permitting the transliteration of a Schoenflies symbol to its corresponding international symbol. For example, the 200th space group is symbolized in Schoenflies notation (Table 7-5) as T_h^1 but in international notation (Table 7-4) as $P\,2/m\,\bar{3}$. The Schoenflies space group symbol consists of the point group symbol plus an arbitrarily assigned superscript which serves to distinguish this space group from other space groups belonging to this same point group. For example, point group T_h ($= 2/m\,\bar{3}$) embraces the seven space groups T_h^1 ($P\,2/m\,\bar{3}$), T_h^2 ($P\,2/n\,\bar{3}$), T_h^3 ($F\,2/m\,\bar{3}$), T_h^4 ($F\,2/d\,\bar{3}$), T_h^5 ($I\,2/m\,\bar{3}$), T_h^6 ($P\,2_1/a\,\bar{3}$), and T_h^7 ($I\,2_1/a\,\bar{3}$). Note the more explicit nature of the inter-

TABLE 7–5

Schoenflies Notation for Space Groups

No.	Sch.	No.	Sch.	No.	Sch.	No.	Sch.	No.	Sch.
(1)	C_1^1	(47)	D_{2h}^1	(93)	D_4^5	(139)	D_{4h}^{17}	(185)	C_{6v}^3
(2)	C_i^1	(48)	D_{2h}^2	(94)	D_4^6	(140)	D_{4h}^{18}	(186)	C_{6v}^4
(3)	C_2^1	(49)	D_{2h}^3	(95)	D_4^7	(141)	D_{4h}^{19}	(187)	D_{3h}^1
(4)	C_2^2	(50)	D_{2h}^4	(96)	D_4^8	(142)	D_{4h}^{20}	(188)	D_{3h}^2
(5)	C_2^3	(51)	D_{2h}^5	(97)	D_4^9	(143)	C_3^1	(189)	D_{3h}^3
(6)	C_s^1	(52)	D_{2h}^6	(98)	D_4^{10}	(144)	C_3^2	(190)	D_{3h}^4
(7)	C_s^2	(53)	D_{2h}^7	(99)	C_{4v}^1	(145)	C_3^3	(191)	D_{6h}^1
(8)	C_s^3	(54)	D_{2h}^8	(100)	C_{4v}^2	(146)	C_3^4	(192)	D_{6h}^2
(9)	C_s^4	(55)	D_{2h}^9	(101)	C_{4v}^3	(147)	C_{3i}^1	(193)	D_{6h}^3
(10)	C_{2h}^1	(56)	D_{2h}^{10}	(102)	C_{4v}^4	(148)	C_{3i}^2	(194)	D_{6h}^4
(11)	C_{2h}^2	(57)	D_{2h}^{11}	(103)	C_{4v}^5	(149)	D_3^1	(195)	T^1
(12)	C_{2h}^3	(58)	D_{2h}^{12}	(104)	C_{4v}^6	(150)	D_3^2	(196)	T^2
(13)	C_{2h}^4	(59)	D_{2h}^{13}	(105)	C_{4v}^7	(151)	D_3^3	(197)	T^3
(14)	C_{2h}^5	(60)	D_{2h}^{14}	(106)	C_{4v}^8	(152)	D_3^4	(198)	T^4
(15)	C_{2h}^6	(61)	D_{2h}^{15}	(107)	C_{4v}^9	(153)	D_3^5	(199)	T^5
(16)	D_2^1	(62)	D_{2h}^{16}	(108)	C_{4v}^{10}	(154)	D_3^6	(200)	T_h^1
(17)	D_2^2	(63)	D_{2h}^{17}	(109)	C_{4v}^{11}	(155)	D_3^7	(201)	T_h^2
(18)	D_2^3	(64)	D_{2h}^{18}	(110)	C_{4v}^{12}	(156)	C_{3v}^1	(202)	T_h^3
(19)	D_2^4	(65)	D_{2h}^{19}	(111)	D_{2d}^1	(157)	C_{3v}^2	(203)	T_h^4
(20)	D_2^5	(66)	D_{2h}^{20}	(112)	D_{2d}^2	(158)	C_{3v}^3	(204)	T_h^5
(21)	D_2^6	(67)	D_{2h}^{21}	(113)	D_{2d}^3	(159)	C_{3v}^4	(205)	T_h^6
(22)	D_2^7	(68)	D_{2h}^{22}	(114)	D_{2d}^4	(160)	C_{3v}^5	(206)	T_h^7
(23)	D_2^8	(69)	D_{2h}^{23}	(115)	D_{2d}^5	(161)	C_{3v}^6	(207)	O^1
(24)	D_2^9	(70)	D_{2h}^{24}	(116)	D_{2d}^6	(162)	D_{3d}^1	(208)	O^2
(25)	C_{2v}^1	(71)	D_{2h}^{25}	(117)	D_{2d}^7	(163)	D_{3d}^2	(209)	O^3
(26)	C_{2v}^2	(72)	D_{2h}^{26}	(118)	D_{2d}^8	(164)	D_{3d}^3	(210)	O^4
(27)	C_{2v}^3	(73)	D_{2h}^{27}	(119)	D_{2d}^9	(165)	D_{3d}^4	(211)	O^5
(28)	C_{2v}^4	(74)	D_{2h}^{28}	(120)	D_{2d}^{10}	(166)	D_{3d}^5	(212)	O^6
(29)	C_{2v}^5	(75)	C_4^1	(121)	D_{2d}^{11}	(167)	D_{3d}^6	(213)	O^7
(30)	C_{2v}^6	(76)	C_4^2	(122)	D_{2d}^{12}	(168)	C_6^1	(214)	O^8
(31)	C_{2v}^7	(77)	C_4^3	(123)	D_{4h}^1	(169)	C_6^2	(215)	T_d^1
(32)	C_{2v}^8	(78)	C_4^4	(124)	D_{4h}^2	(170)	C_6^3	(216)	T_d^2
(33)	C_{2v}^9	(79)	C_4^5	(125)	D_{4h}^3	(171)	C_6^4	(217)	T_d^3
(34)	C_{2v}^{10}	(80)	C_4^6	(126)	D_{4h}^4	(172)	C_6^5	(218)	T_d^4
(35)	C_{2v}^{11}	(81)	S_4^1	(127)	D_{4h}^5	(173)	C_6^6	(219)	T_d^5
(36)	C_{2v}^{12}	(82)	S_4^2	(128)	D_{4h}^6	(174)	C_{3h}^1	(220)	T_d^6
(37)	C_{2v}^{13}	(83)	C_{4h}^1	(129)	D_{4h}^7	(175)	C_{6h}^1	(221)	O_h^1
(38)	C_{2v}^{14}	(84)	C_{4h}^2	(130)	D_{4h}^8	(176)	C_{6h}^2	(222)	O_h^2
(39)	C_{2v}^{15}	(85)	C_{4h}^3	(131)	D_{4h}^9	(177)	D_6^1	(223)	O_h^3
(40)	C_{2v}^{16}	(86)	C_{4h}^4	(132)	D_{4h}^{10}	(178)	D_6^2	(224)	O_h^4
(41)	C_{2v}^{17}	(87)	C_{4h}^5	(133)	D_{4h}^{11}	(179)	D_6^3	(225)	O_h^5
(42)	C_{2v}^{18}	(88)	C_{4h}^6	(134)	D_{4h}^{12}	(180)	D_6^4	(226)	O_h^6
(43)	C_{2v}^{19}	(89)	D_4^1	(135)	D_{4h}^{13}	(181)	D_6^5	(227)	O_h^7
(44)	C_{2v}^{20}	(90)	D_4^2	(136)	D_{4h}^{14}	(182)	D_6^6	(228)	O_h^8
(45)	C_{2v}^{21}	(91)	D_4^3	(137)	D_{4h}^{15}	(183)	C_{6v}^1	(229)	O_h^9
(46)	C_{2v}^{22}	(92)	D_4^4	(138)	D_{4h}^{16}	(184)	C_{6v}^2	(230)	O_h^{10}

national symbol (parenthesized) in describing a space group. Usually Schoenflies space group symbols are set with the superscript directly above the subscript, thus T_h^1 rather than $T_h{}^1$. However, the form $T_h{}^1$ shows more clearly that these space group symbols consist of the point group symbol with a superscript added.

REFERENCE AND RECOMMENDED READING

Henry, N. F. M., and K. Lonsdale, eds. (1952), *International Tables for X-ray Crystallography*. Birmingham, England: Kynoch Press.

QUESTIONS AND PROBLEMS

7-1 For space group *Pcc*2 (Fig. 7-13A) give the coordinates of positions equivalent to (a) the general position at x, y, z; (b) the special position at $\frac{1}{2}, 0, z$; (c) the special position at $0, 0, z$. Also state the point-group symmetry existent at each of these three points.

7-2 For space group $P2_1/m$ in the second setting (Fig. 7-13B) give the co-

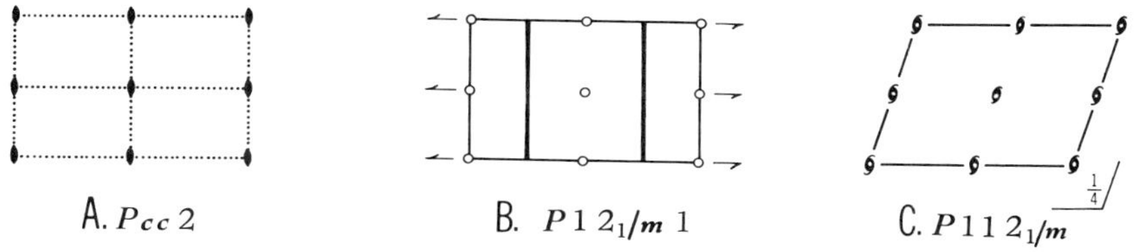

FIGURE 7–13

The symmetry elements, projected along the c axis, for (A) space group *Pcc*2 and for space group $P\,2_1/m$ in the second setting (B) and in the first setting (C). In all three cases the c axis is perpendicular to the page. A projection along the c axis represents a view of the lattice (or of the crystal structure) as sighted along the c axis, *if drawn on a plane perpendicular to this c axis.* For an orthorhombic crystal as in (A) or for a monoclinic crystal in the first setting as in (C), the plane perpendicular to the c axis is also (001). However, for a monoclinic crystal in the second setting, for example (B), the plane perpendicular to the c axis is at an angle $\beta - 90°$ to (001); these two planes have the b-axis, that is, the top edge of the rectangle in (B), as their common line or hinge. Consequently, whereas the plane perpendicular to the c axis lies in the plane of the page in (B), the (001) plane tilts so that a ball placed on it would roll directly toward the reader. The short edge of the rectangle in (B) consequently equals $\mathbf{a}\cos(\beta - 90°)$, the projected length of the a axis on the plane perpendicular to the c-axis.

In (A), (B), and (C), the origin of the lattice is by convention the upper left corner of the figure. Thus for *Pcc*2 the origin is located on a 2-fold axis. For $P\,2_1/m$, the origin coincides with a center of symmetry which is indicated by a circle in (B) and by a white dot on the 2_1 symbol in (C). Centers of symmetry at $z = \frac{1}{2}$ occur below the centers indicated in (B) and (C). In (C) a plane of symmetry is indicated at $z = \frac{1}{4}$, but there is also one at $z = \frac{3}{4}$, the centers of symmetry at $z = \frac{1}{2}$ requiring this.

ordinates of the positions equivalent to (a) x, y, z; (b) $x, \frac{1}{4}, z$; (c) $\frac{1}{2}, 0, 0$; and (d) $0, 0, 0$. State the point-group symmetry at each of these points.

7-3 For space group $P2_1/m$ in the first setting (Fig. 7-13C) give coordinates of the positions equivalent to (a) x, y, z; (b) $x, y, \frac{1}{4}$; (c) $0, 0, 0$. State the point-group symmetry at each of these points.

7-4 Let $\langle uvw \rangle$ represent a form of symmetrically equivalent *directions* just as $\{hkl\}$ represented a form of symmetrically equivalent planes for a given point group. Describe the symmetry associated with directions $\langle 001 \rangle$, $\langle 100 \rangle$, and $\langle 110 \rangle$ in space group $I4_1/a\ 2/c\ 2/d$.

ANSWERS

7-1 (a) x, y, z; $\bar{x}, \bar{y}, z$; $\bar{x}, y, \frac{1}{2} + z$; $x, \bar{y}, \frac{1}{2} + z$. Point symmetry at x, y, z: 1 fold. (b) $\frac{1}{2}, 0, z$; $\frac{1}{2}, 0, \frac{1}{2} + z$. Point symmetry at $\frac{1}{2}, 0, z$: a 2 fold, but no mirror planes (only c glides). (c) $0, 0, z$; $0, 0, \frac{1}{2} + z$. Point symmetry at $0, 0, z$: 2 fold.

7-2 (a) x, y, z; $\bar{x}, \bar{y}, \bar{z}$; $\bar{x}, \frac{1}{2} + y, \bar{z}$; $x, \frac{1}{2} - y, z$. Point symmetry: 1. (b) $x, \frac{1}{4}, z$; $\bar{x}, \frac{3}{4}, \bar{z}$ (the 2_1 relates these). Point symmetry: m. (c) $\frac{1}{2}, 0, 0$; $\frac{1}{2}, \frac{1}{2}, 0$. Point symmetry: $\bar{1}$. (d) $0, 0, 0$; $0, \frac{1}{2}, 0$ (the 2_1 relates these). Point symmetry: $\bar{1}$.

7-3 (a) x, y, z; $\bar{x}, \bar{y}, \bar{z}$; $\bar{x}, \bar{y}, \frac{1}{2} + z$; $x, y, \frac{1}{2} - z$. Point symmetry: 1. (b) $x, y, \frac{1}{4}$; $\bar{x}, \bar{y}, \frac{3}{4}$. Point symmetry: m. (c) $0, 0, 0$; $0, 0, \frac{1}{2}$. Point symmetry: $\bar{1}$.

7-4 $\langle 001 \rangle$ equals $4_1/a$; $\langle 100 \rangle$ equals $2/c$; $\langle 110 \rangle$ equals $2/d$. More specifically, $\langle 001 \rangle$ is parallel to a 4_1 screw axis but perpendicular to an a glide.

8 Crystal Chemistry

Atoms and Their Structures

The fundamental particles

All crystals consist of atoms, each basic chemical element having its own characteristic atom. Atoms may be regarded as spheres so small, usually between 2 to 5 Å in diameter, that present-day electron microscopes cannot produce a definitive picture of one. However, physical experiments reveal that atoms consist primarily of three fundamental particles, *protons, electrons,* and *neutrons,* with the properties given in Table 8-1. Because a neutron consists of a proton combined with an electron, their equal-but-opposite charges nullify each other to produce an electrically neutral particle, hence the name "neutron." A neutron's mass exceeds the sum of those for a proton and electron (Table 8-1) because, in combining, the proton and electron release energy that is converted into mass; the Einstein equation describes such interconversion between energy and matter.

The nucleus of an atom

An atom possesses a dense, positively charged nucleus consisting of protons and neutrons clumped together in response to strong attractive forces of an unknown nature.[1] The number of protons in an atom's nucleus is called its *atomic number* and determines the chemical properties of the element that the atom represents. Thus far 103 different types of atoms or elements, representing atomic numbers 1 to 103, are known. However, the elements with atomic numbers 43, 61, 85, and 93–103 have not been observed to occur naturally. So far as we know today, less than 90 different types of elements comprise the earth's bulk. Element 104 may have been recently discovered, but is subject to confirmation.

[1] Normally, like-charged particles such as protons would repel rather than attract each other.

TABLE 8–1

	Electric Charge[a]	Atomic Mass Units[b] (u)
Proton	+1	1.00728
Electron	−1	0.00055
Neutron	0	1.00867

[a] The charge on an electron is defined to be one negative unit; actually it equals 1.6×10^{-19} ampere-seconds or coulombs.

[b] One atomic mass unit (u) equals one-twelfth of the weight of a (^{12}C) carbon atom.

Orbital electrons: energy levels

Electrons, equal in number to the protons in an atom's nucleus, orbit about the nucleus; this numerical equality between orbital electrons and nuclear protons causes each atom to be electrically neutral. Speaking approximately, if an atom were enlarged so that its nucleus equaled the size of a basketball, its orbital electrons, insofar as a size can be assigned to an electron, would be about the size of ping-pong balls and would at times occupy orbits up to 2 miles from the nucleus. Resorting to analogy, bricks at different heights above the ground represent different levels of potential energy; thus the higher bricks occupy higher energy levels because, if dropped, they release more energy, that is, deliver more telling blows, on striking the ground. In like manner, orbital electrons would release energy if they succumbed to the positive attraction of the nucleus and plunged into it rather than orbiting around it. The electrons that would release the most energy on doing so are presumed to occupy shells farthest out from the nucleus; those releasing lesser energies are considered to occupy the smaller shells closer to the nucleus. In "falling" from an outer shell to an inner shell, an orbital electron releases a "packet" of energy known as a *quantum* or *photon*; the atom then emits either light or a nonvisible radiation such as X-rays or ultraviolet light.

Shells and quantum numbers

Starting with the inner shells, that is, those with lower energy quantum levels, and proceeding to those farther from the nucleus, the shells are conventionally known as the K shell, L shell, M shell, . . . , Q shell; for these shells the assigned quantum number n is, respectively, 1, 2, 3, . . . , 7. Electrons repel each other and hence cannot be packed together too closely in a given area or volume. Thus the outer shells, in effect being more spacious, can contain more electrons than can the inner shells. Actually the maximum number of electrons a shell can contain is equal to $2n^2$, where n is the shell's quantum number. The maximum number of electrons each shell can contain is given in Table 8-2.

The electrons within a given shell are confined to particular regions known

TABLE 8-2

Shell	Principal Quantum Number (n)	Maximum Electron Occupancy ($2n^2$)
K	1	2
L	2	8
M	3	18
N	4	32
O	5	50
P	6	72
Q	7	98

as the *s*, *p*, *d*, and *f* orbitals.[2] The orbital belonging to a given shell is designated by the principal quantum number n for that shell. For example, the *s* orbital for the *K* shell is denoted as 1*s*; the *p* orbital of the *M* shell as 3*p*, and so on. The nature of these orbitals can be described in terms of the variations in the probability of encountering an electron (1) along different directions, that is, angles, outward from the nucleus or (2) at different distances, that is, radial lengths, outward from the nucleus. These probabilities are called (1) the *angular probability* and (2) the *radial probability.*

The s Orbitals. For *s* orbitals the probability of encountering an electron is equal for all directions outward from the nucleus. This can be expressed graphically by drawing a sphere (Fig. 8-1A) and letting the length of any direction that intercepts the sphere's surface, for example, radii *OA* and *OB*, correspond to the probability of encountering *s*-orbital electrons along these directions. In contrast the probability of encountering such *s*-orbital electrons *does* vary with radial distance from the nucleus as is shown schematically for an *s* orbital of the *K* shell (Fig. 8-1B), of the *L* shell (Fig. 8-1C), or of the *M* shell (Fig. 8-1D). Thus within these *s* orbitals, the maximum probability of encountering an electron occurs closer to the nucleus for the *K* shell, more distant for the *L* shell, and most distant for the *M* shell.

The p Orbitals. For an atom in a magnetic field three mutually perpendicular axes, x, y, z, may be defined to extend outward from this atom's center. Relative to these axes, three different *p* orbitals, p_x, p_y, and p_z, may be described. Taking the p_x orbital as an example, the probability of encountering an electron in it at any particular instant in time is greatest along the direction of the x axis but decreases rapidly for directions at increasingly large angles to this x axis. This angular probability can thus no longer be expressed graphically by the surface of a sphere but rather by the dumbbell-like surface symmetrically disposed about the x axis in Figure 8-2A. The probability of encountering an electron in the p_x

[2] The letters *s*, *p*, *d*, and *f* are abbreviations of spectroscopic terms and stand, respectively, for *sharp, principal, diffuse,* and *fundamental.* The phrase, "subdivided pretty darn fine," may aid in recall of these letters.

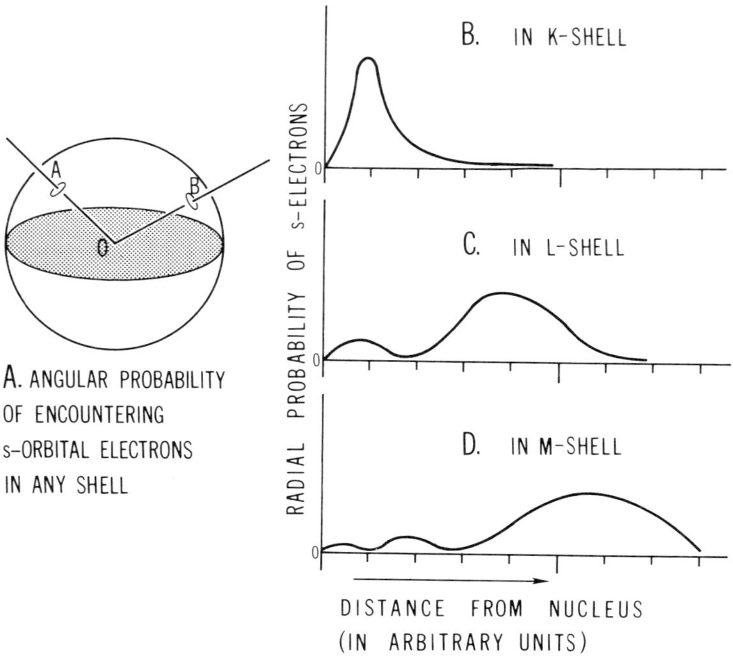

FIGURE 8–1

Graphical depiction of the probability of finding an *s*-orbital electron at a particular instant in time.

(A) *The angular probability.* The intercept on a sphere of any direction outward from an atom's nucleus centered at *O* is proportional in length to the probability of finding an *s*-orbital electron along that direction. In other words the angular probability is equal along *any* direction outward from the nucleus, for example, along *OA* or *OB*.

(B)(C)(D) *Radial probabilities* of encountering *s*-orbital electrons at distances (in arbitrary units) outward from an atom's nucleus.

orbital along a particular direction *OP* is then proportional to the length *OP′*, where *P′* is the intercept of *OP* on the dumbbell's surface; this probability is nil along directions in the *yz* plane. Similar dumbbell-like surfaces serve to express graphically the angular probability of encountering electrons in a p_y orbital (Fig. 8-2B) or a p_z orbital (Fig. 8-2C).

The radial probability of encountering an electron in a *p* orbital, be it p_x, p_y, or p_z, is nil at the nucleus, but increases with distance from the nucleus until a maximum is reached as shown for the *L* shell (Fig. 8-3A) and the *M* shell (Fig. 8-3B) of a highly simple atom. For the more external *M* shell the maximum radial probability occurs farther from the nucleus than for the *L* shell.

For any particular orbital the probability of encountering an electron along a particular direction and at a given distance from the nucleus is proportional to the product of the angular probability for this direction times the radial probability for the given distance.

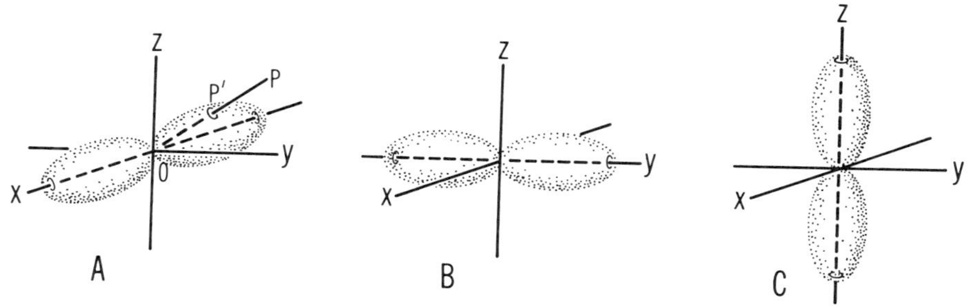

FIGURE 8–2

The angular probability of encountering an electron in (A) a p_x orbital, (B) a p_y orbital or (C) a p_z orbital. For any given direction outward from the nucleus such as *OP* this probability is proportional to the length of its intercept on the dumbbell-like figure. Axes *x*, *y*, and *z* are mutually perpendicular.

The d Orbitals. The angular probabilities for the five possible *d* orbitals (symbols: d_{z^2}, $d_{x^2-y^2}$, d_{xy}, d_{xz}, d_{yz}) are shown in Figure 8-4. The radial probability for each of these *d* orbitals would resemble those in Figure 8-3. The *f* orbitals, seven in number, are usually less important in chemical bonding and will not be discussed.

In a single shell there can only be one *s* orbital, three *p* orbitals (p_x, p_y, p_z), five *d* orbitals, and seven *f* orbitals. Moreover, a single *s*, *p*, *d*, or *f* orbital can contain at most two electrons, these electrons spinning in opposite directions as required by the Pauli exclusion principle. The maximum number of *s*, *p*, *d*, and *f* electrons that can be present in a single shell is therefore

s	$1 \times 2 = 2$
p	$3 \times 2 = 6$
d	$5 \times 2 = 10$
f	$7 \times 2 = 14$

An atom's *K* shell consists of a single *s* orbital and hence has a maximum capacity of two electrons. Its *L* shell consists of one *s* and three *p* orbitals and has a

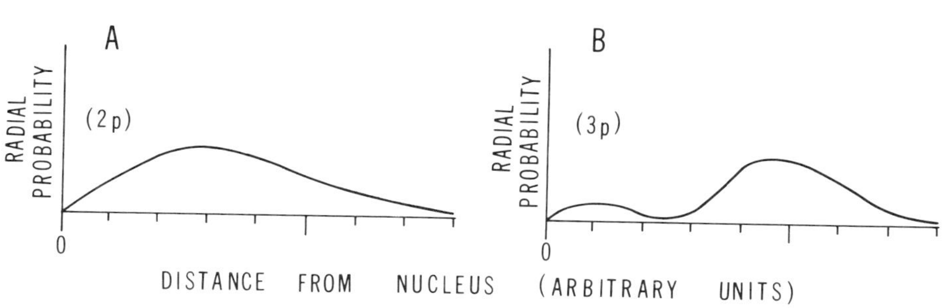

FIGURE 8–3

The radial probability for a *p* orbital in the *L* shell (A) and in the *M* shell (B).

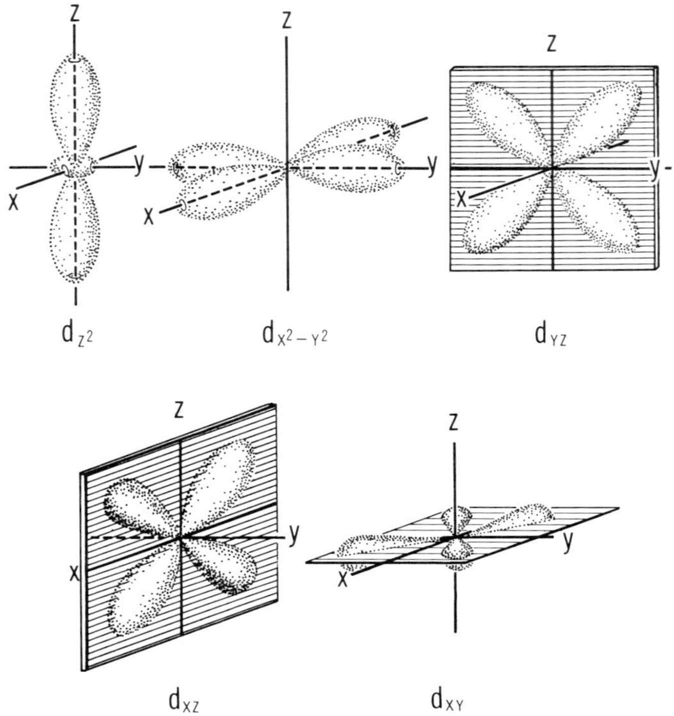

FIGURE 8–4

The angular probability for the five d orbitals. (Their radial probability for the M shell resembles the distribution in Figure 8–3B except that the maximum occurs farther from the nucleus.) The ruled planes bisect the lobes as is best seen for d_{xy}.

maximum capacity of eight electrons. These maximum capacity values conform to $2n^2$, where n is the shell's principal quantum number. The M shell with a maximum capacity of 18 electrons consists of one s, three p, and five d orbitals. Shell N with a maximum capacity of 32 electrons consists of one s, three p, five d, and seven f orbitals. The several orbitals are (somewhat loosely) regarded to represent various energy levels as roughly indicated in Figure 8-5. Thus $1s$ is the orbital of lowest energy, then in order of increasing energy $2s$, $2p$, $3s$, $3p$, $4s$, $3d$, $4p$, $5s$, $4d$, $5p$, $6s$, $4f$, $5d$, $6p$, $7s$, $6d \cong 5f$. In Figure 8-5 slight differences in the energy levels of like-labeled orbitals are ignored; for example, the $3d$ orbitals would not necessarily all plot at the same energy level. Moreover, an orbital's energy level, for example, that of $3p$, will be different for atoms of different elements (or even for atoms of the same element if in different environments).

Electron configurations of atoms

The number of electrons that orbit around the nucleus of an atom is equal to the number of protons in this nucleus, this number corresponding to the atomic number of the atom. In general these orbital electrons rarely occupy a higher

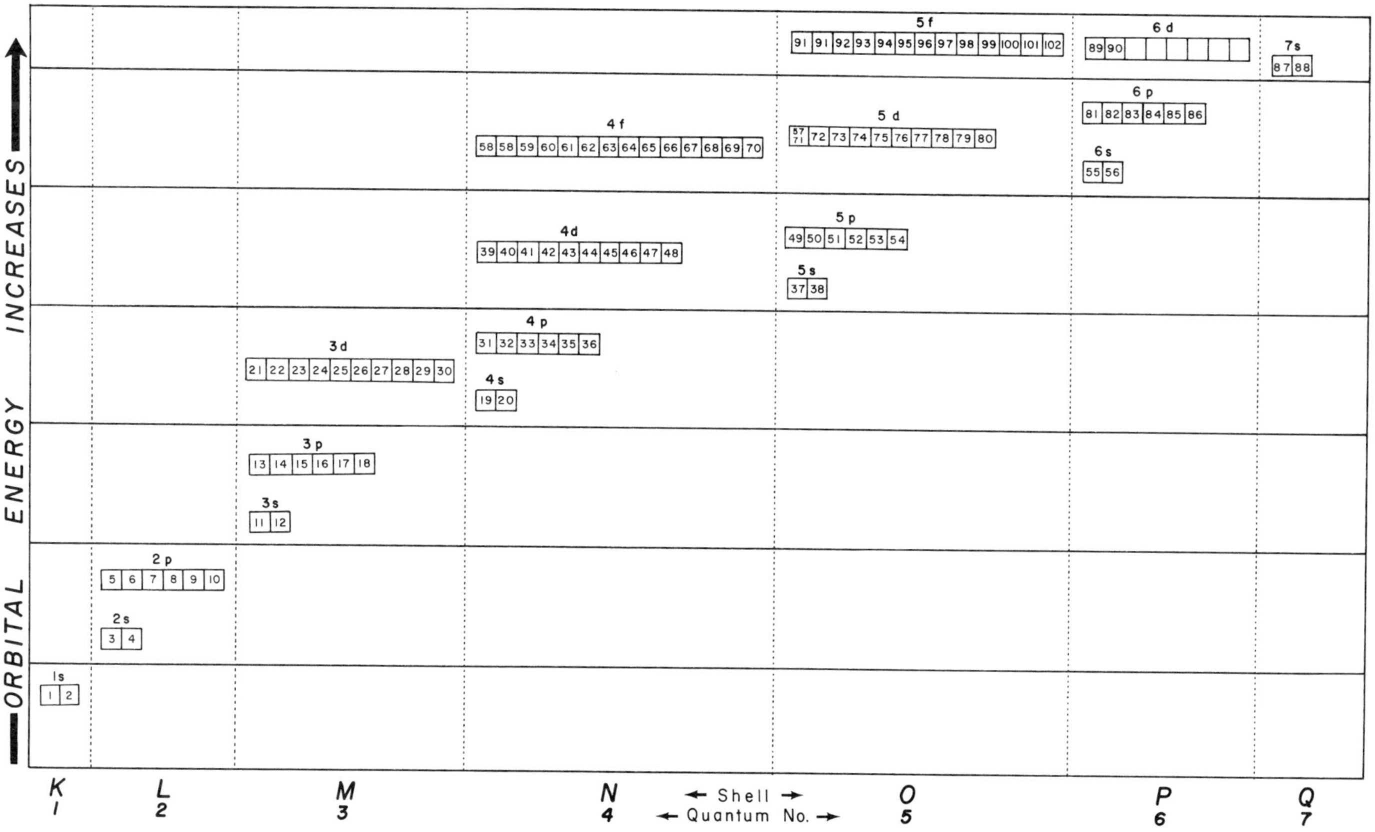

FIGURE 8–5 Schematic representation of the relative energy levels for the various orbital electrons in the various shells. For each orbital the number of squares indicates the number of electrons necessary to fill that orbital. The number within each square indicates the atomic number of the element whose atoms have an electron in that site (and in all marked with lower atomic numbers).

energy orbital until all lower energy orbitals are completely occupied (the *aufbau* principle). In Figure 8-5 the number of electrons necessary to fill each orbital to capacity is schematically indicated with squares. Thus the *s*, *p*, *d*, and *f* orbitals are assigned 2, 6, 10 or 14 squares, respectively. The orbital of an outer shell is occasionally lower in energy than an orbital of an inner shell, for example, the 4*s* orbital in comparison to the 3*d* orbital.

Figure 8-5 permits the electron configuration of an atom, that is, the distribution of its electrons in the various orbitals, to be determined if its atomic number is known. Thus the electron configuration for fluorine with atomic number 9 is indicated by recognition that all squares with numbers 9 or less represent orbital electrons for fluorine, whereas squares with numbers greater than 9 represent unfilled orbitals or portions of them. A few exceptions in the use of Figure 8-5 arise. Elements 24 and 29, chromium and copper, possess one less 4*s* electron and one more 3*d* electron than the diagram would indicate. A similar situation arises for elements 41, 42, 44, 45, and 47 which have one less 5*s* electron and one more 4*d* electron than the diagram would indicate. Lanthanum, atomic number 57, is exceptional in that its 57th electron fills a 5*d* orbital rather than beginning the lower energy 4*f* orbital. For elements 58-70, however, the 5*d* orbital is vacant; element 58 starts the filling of the 4*f* orbital with *two* electrons. To indicate this, there are two squares numbered 58 in Figure 8-5. A similar situation exists for elements 91-102; element 91 begins filling the 5*f* orbital with two electrons.

The symbol for each orbital, if followed by an exponent indicating the number of electrons occupying this orbital, can summarize an atom's electron configuration. Thus $1s^2\ 2s^2\ 2p^6\ 3s^2\ 3p^1$ indicates the configuration for aluminum, which has two electrons in its 1*s* orbital, two in its 2*s* orbital, six in its 2*p* orbital, two in its 3*s*, and one electron in its 3*p* orbital. Because $1s^2$ and $2s^2\ 2p^6$, respectively, represent a completely filled *K* and *L* shell, aluminum's configuration can be more briefly written as $KL\ 3s^2\ 3p^1$.

The periodic table (Fig. 8-6) is a sequential ordering of elements by atomic number and by similarity of chemical properties. It is organized so that those elements in one of the vertical columns headed I, II, III, . . . , or VIII have atoms whose outer shells contain electrons equal in number to the Roman numeral designating the group. Thus group I elements, with atomic numbers 1, 3, 11, 19, 37, 55, and 87, have just one electron in their outermost shell, this occupying an *s* orbital. Similarly, group II elements have two *s* orbital electrons in their outer shell; group III atoms possess three electrons (two *s*, one *p*); group IV atoms possess four electrons (two *s*, two *p*); group VIII possess eight outer-shell electrons (two *s*, six *p*) except for helium whose outer shell consists of a completely filled *K* shell that is simply a filled *s* orbital. Group VIII elements are distinguished by outer shells that are filled to capacity.

In Figure 8-6 the elements are organized into horizontal tiers known as *periods*, these periods being numbered from 1 to 7. From left to right in any given period, the outer shell is progressively filled until for the element at the far right, the *s* and ultimately the *p* orbitals are filled to capacity, that is, contain two and

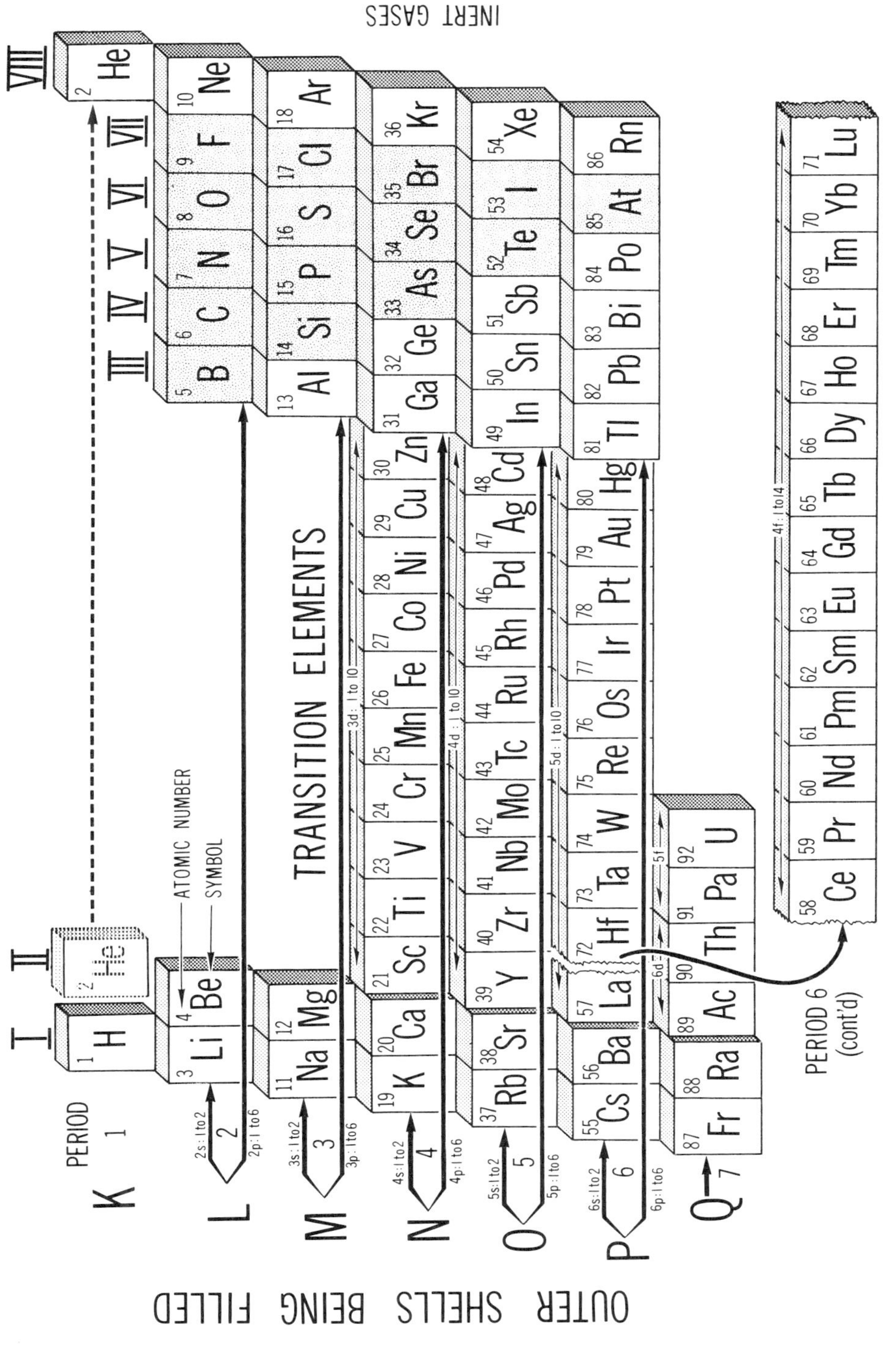

FIGURE 8–6 The periodic table. The lightly shaded nonmetallic elements separate the metals (at left) from the inert gases (on the right).

six electrons, respectively. The outer shells being filled are the K shell for period 1, this shell having only a capacity of two electrons; the L shell for period 2; M for period 3; N for period 4; and so on. Period 4 contains elements 21 to 30; these are called *transition* elements because the orbital electrons by which they exceed calcium occupy the $3d$ orbitals of the more interior M shell. Only after this more interior shell is filled to its capacity of ten, as in zinc, is there a resumption of filling of the outer shell's p orbitals (elements 31 to 36). Period 6 contains 24 transition elements because elements 57 to 80 not only digress to fill the inner $5d$ orbital but also to fill the $4f$ orbital to its capacity of 14 electrons; with these filled, elements 81 to 86 then return to building up the outer shell's $6p$ orbital to complete the period.

The chemical properties of an atom depend on (1) whether it loses, gains, or shares electrons (called *valence electrons*) from its outermost or *valence shell*; (2) the actual number of such valence electrons customarily lost, gained, or shared; and (3) the atom's alacrity in doing so. Elements belonging to the same group have, to a first approximation, similar chemical properties because their valence shells are alike and it is these that largely control factors (1) to (3). The completely filled outer shells of group VIII elements are evidently very stable configurations because their atoms tend neither to lose, gain, nor share electrons and hence are chemically inert; consequently, group VIII elements are called the *inert gases*. The elements in groups V, VI, and VII, on the other hand, frequently add three, two, or one valence electrons, respectively, to complete their outer shell and attain the stable, inert-gas configuration. In doing so they become *anions*, that is, negatively charged atoms, with charges of -3 (group V anions), -2 (group VI), or -1 (group VII). Similarly, elements from groups I, II, or III may lose, respectively, one, two, or three orbital electrons from their outer shells to become *cations*, that is, positively charged atoms, with charges of $+1$ (group I), $+2$ (group II), or $+3$ (group III); when stripped of valence electrons, they attain the inert-gas configuration except in the case of hydrogen which becomes H^+, that is, a single proton. Thus through loss of valence electrons sodium becomes Na^+, calcium becomes Ca^{2+}, whereas through gain of valence electrons fluorine becomes F^- and oxygen become O^{2-}. Group IV atoms, rather than losing or gaining, will often share orbital electrons so as to attain the inert-gas configuration.

Ionization potentials

The amount of energy required to withdraw successively a first, a second, or a third electron from an atom in the gaseous state is called the atom's *first*, *second*, or *third ionization potential*, respectively. An electron can be withdrawn from an atom more readily if (1) no others have been previously withdrawn; (2) such withdrawal will produce an ion with an inert-gas configuration (or close thereto) for its outer shell; (3) its distance from the nucleus is large; (4) the charge on the nucleus is small; or if (5) there exist numerous electrons in interior orbitals that act as a screen and prevent the full attraction of the nucleus from being exerted on the electrons in the outer shell.

TABLE 8-3

Ionization Potentials and Electron Configurations for Elements of the Third Period

	Na(I)	Mg(II)	Al(III)	Si(IV)	P(V)	S(VI)	Cl(VII)	Ar(VIII)
				Ionization Potentials[a] (after Addison, 1961)				
1st	5.1	7.6	6.0	8.1	11.0	10.4	13.0	15.0
2nd	**47.3**	15.0	18.8	16.3	19.7	23.4	23.8	27.6
3rd		**80.1**	28.4	33.5	30.2	35.0	39.9	
4th			**120.0**	45.1	51.4	47.2	53.5	
5th				**166.7**	65.0	72.5	67.8	
6th					**220.4**	88.0	96.7	
				Electron Configuration				
K				Completely Full				
L				Completely Full				
M *s*	1	2	2	2	2	2	2	2
M *p*			1	2	3	4	5	6

Elements (and Their Group in Periodic Table)

[a] Values are in electron volts (eV) where one eV equals the energy acquired by an electron when it falls through a potential difference of one volt; 1 eV $= 1.602 \times 10^{-12}$ erg $= 3.83 \times 10^{-20}$ cal. Boldfaced numbers represent the energies required to withdraw an electron from an ion that has achieved the inert-gas configuration. Note their relatively large increase over the value just above them.

In Table 8-3 the first ionization potential for the elements of the third period increases fairly steadily from left to right. This is as expected because nuclear charge also increases from left to right and the atoms to the right therefore hold their orbital electrons more tightly. For magnesium and phosphorus the first ionization potential is above the general trend. For magnesium this is attributed to the greater difficulty of withdrawing an electron from a filled *s* orbital; for phosphorus, it is due to the fact that three *p* orbital electrons, distributed one each to the p_x, p_y, and p_z orbitals, constitute a relatively stable configuration. In Table 8-3 the successive increase in the first, second, and third ionization potentials for any given atom indicates that as electrons are removed the nucleus exerts greater attraction for those remaining. Thus they are attracted closer to the nucleus and the effective radius of the atoms is thereby decreased. As a consequence a cation is always smaller than the comparable atom.

After the electrons of an atom's outer shell have been stripped away, additional electrons are difficult to remove because these must be withdrawn from a tightly held, inner shell. Thus each dark, boldfaced ionization potential in

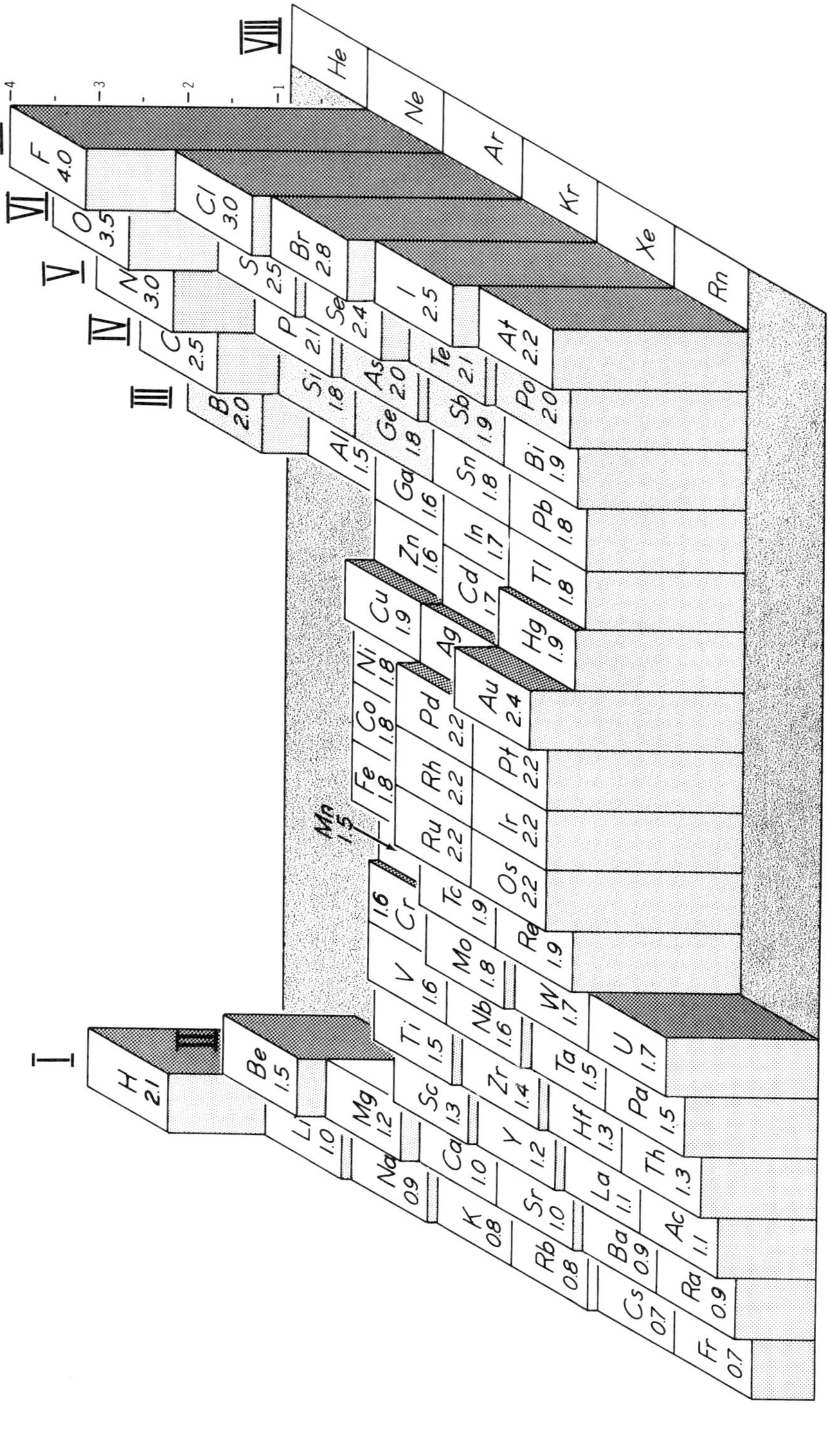
I
II
III
IV
V
VI
VII
VIII
−4
−3
−2
−1
H 2.1
Li 1.0
Be 1.5
Na 0.9
Mg 1.2
K 0.8
Ca 1.0
Sc 1.3
Ti 1.5
V 1.6
1.6 Cr
Mn 1.5
Fe 1.8
Co 1.8
Ni 1.8
Cu 1.9
Zn 1.6
Ga 1.6
Ge 1.8
As 2.0
Se 2.4
Br 2.8
Rb 0.8
Sr 1.0
Y 1.2
Zr 1.4
Nb 1.6
Mo 1.8
Tc 1.9
Ru 2.2
Rh 2.2
Pd 2.2
Ag
Cd 1.7
In 1.7
Sn 1.8
Sb 1.9
Te 2.1
I 2.5
Cs 0.7
Ba 0.9
La 1.1
Hf 1.3
Ta 1.5
W 1.7
Re 1.9
Os 2.2
Ir 2.2
Pt 2.2
Au 2.4
Hg 1.9
Tl 1.8
Pb 1.8
Bi 1.9
Po 2.0
At 2.2
Fr 0.7
Ra 0.9
Ac 1.1
Th 1.3
Pa 1.5
U 1.7
B 2.0
Al 1.5
C 2.5
Si 1.8
N 3.0
P 2.1
O 3.5
S 2.5
F 4.0
Cl 3.0
He
Ne
Ar
Kr
Xe
Rn

Table 8-3 represents such a difficult withdrawal from an inner shell and consequently greatly exceeds the value cited immediately above it.

On the average, readiness to give up electrons is greatest for group I elements, somewhat less for group II, and increasingly less as groups III to VIII are considered. Comparison of the first ionization potentials for the third period's elements (Table 8-3) confirms this if allowance is made for magnesium and phosphorus which, for reasons already discussed, hold their electrons more tightly than expected (and thus fall above the general trend). Within a single group or vertical column in the periodic table (Fig. 8-6), ability to lose electrons increases for elements successively lower in the column, that is, for those with higher atomic numbers. Evidently, as atomic number increases in a given group, the ability to retain outer electrons diminishes because these electrons are (1) increasingly distant from the nucleus and (2) screened from the nucleus's attraction by an increasing number of shells of inner orbital electrons.

Electronegativity

The ability of an atom *in a crystal structure or molecule* to attract electrons into its outer shell is called its *electronegativity* and varies in value with position in the periodic table (Fig. 8-7). Except for hydrogen, electronegativity values are low for group I elements but successively increase for groups II to VII. The maximum electronegativity values in group VII result because atoms of these elements will strongly attract an electron in order to complete their *p* orbital and thereby attain an inert-gas configuration. For group VIII elements, which already possess this configuration, the electronegativity values are minimal. Within any particular group as atomic number increases, the ability of the atom's nucleus to attract electrons, that is, its electronegativity value, decreases, chiefly because factors (1) and (2), as cited in the previous paragraph, come into play.

Elements are broadly classified as (1) *metals*, these generally having electronegativity values below 1.9 and therefore tending to give up electrons to become cations during chemical reactions; (2) *nonmetals*, these generally having electronegativity values above 2.1 and therefore tending to gain electrons to become anions during chemical reactions; and (3) *metalloids* or semimetals, these having electronegativity values and therefore chemical properties that are intermediate between the metals and nonmetals. In Figure 8-7 the metalloids are

---

← FIGURE 8–7

The electronegativities for the more common elements (after Pauling). Each element is plotted above the stippled reference plane in proportion to its electronegativity value, that is, its ability to capture electrons during chemical reactions. The lightly shaded elements, which are intermediate in their ability to capture electrons, are called metalloids or semimetals. Extending above them are the nonmetals whose tendency to capture electrons is strong, fluorine being the strongest of all. At the left of the metalloids, and generally characterized by lower electronegativity values, are the metals.

shaded to distinguish them from the metals (to the left in Fig. 8-7) and the nonmetals (to the right).

Chemical Bonds

Any force of attraction between two or more atoms or ions is known as a *chemical bond.* The more important types of chemical bonds include (1) *ionic bonds,* (2) *covalent bonds,* (3) *metallic bonds,* (4) *van der Waals bonds,* and (5) *hydrogen bonds.* In minerals and in inorganic crystals in general, ionic bonds and covalent bonds are of major importance. The metallic bond is operative in relatively few minerals (gold, silver, platinum, iron, and some sulfides), but it is important in metallurgy; van der Waals bonds are chiefly operative as a force of attraction between molecules and thus tend to occur in organic rather than in inorganic crystals.

Ionic bonds

An atom whose electronegativity exceeds that of another by 2.0 or more will generally be able to attract the valency electrons of this other atom into its own outer shell. As a consequence the highly electronegative atom becomes a negatively charged ion, whereas the loser becomes positively charged. The electrostatic attraction that will now exist between these two oppositely charged ions constitutes an *ionic bond.* In crystals formed through the reaction of elements that differ by 2.0 or more on the electronegativity scale (Fig. 8-7), the ionic bond will be the principal cohesive force.

Just as the earth's gravitational attraction radiates out from it in all directions, so does the attractive force of an ion for oppositely charged ions. In other words the ionic bond is *nondirected*; this is an important difference between it and the covalent bond to be described next.

Covalent bonds

Two atoms with high, but nearly equal, electronegativity values will each be unable to capture the other's outer electrons. Consequently, an ionic bond between them is not possible. Both atoms can, however, attain or approach an inert-gas configuration for their outer shells by drawing near each other until an incompletely filled orbital from one, that is, an orbital that contains only one electron instead of its full capacity of two, partly overlaps a similarly incomplete orbital from the other. During the times that both electrons, one from each orbital, are in the region of overlap, both orbitals in effect possess the inert-gas configuration of two oppositely spinning electrons. The compulsion of two such atoms to stay close together and thereby maintain overlap between their electron-sharing orbitals produces a strong attraction between them which is known as the *covalent bond* or the *shared-electron bond.* A *double bond* occurs if *two* orbitals from one atom overlap *two* from the other so that *two* pairs of electrons are shared between these atoms. A *triple bond* occurs if *three* orbitals from one atom overlap *three* from another so that *three* pairs of electrons are shared. The center-to-center distance between these atoms would be least for triple bonding, more for double bond, and most for single bonding because bonding strength decreases from the triple, to the double, to the single bond.

Hybrid orbitals

The directed nature of covalent bonds as well as the concept of *hybrid orbitals* may be illustrated by comparing carbon atoms in their uncombined "ground" state with those participating in the diamond and graphite structures. A carbon atom's electronic configuration is believed to change from its ground state to its state in diamond as follows:

↑↓	↑↓	↑	↑	
$1s$	$2s$	$2p$	$2p$	$2p$

Ground state

$\longrightarrow$

↑↓	↑	↑	↑	↑
$1s$	$2(sp^3)$	$2(sp^3)$	$2(sp^3)$	$2(sp^3)$

State in diamond

In the foregoing, a pair of oppositely barbed arrows signifies that the orbital designated below it is filled by two oppositely spinning electrons. Absence of an arrow indicates a vacant orbital; for example, one of the three $2p$ orbitals is vacant in carbon's ground state; vertical lines separate orbitals of different energy levels, higher levels being to the right. If a carbon atom enters into the structure of diamond, its two $2s$ and two $2p$ electrons redistribute into four new equal-energy-level orbitals called $2(sp^3)$ orbitals because their properties are mixtures of the original properties of the *one* $2s$ and the *three* $2p$ orbitals. The use of exponents denotes this in the symbol $2(sp^3)$. The total of these exponents always equals the number of participating orbitals (or of resultant hybrid orbitals). The angular probability for these four $2(sp^3)$ orbitals can be illustrated by drawing four lobes whose axes are at 109°28′ to each other, the axes of these lobes thus extending toward the corners of an imaginary tetrahedron centered around the carbon atom (Fig. 8-8A). Although the radial probability also enters in, for purposes of visualization we regard the lobes in Figure 8-8A as the regions where electrons in the $2(sp^3)$ orbitals are likely to be, maximum probability of encountering an electron occurring along the central axes within these lobes.

Quantum-mechanical calculations indicate that greater overlap between orbitals results in a stronger covalent bond. In diamond (Fig. 8-8B) the carbon atoms are located tetrahedrally about each other. Each of the four $2(sp^3)$ orbitals of any given carbon is hence strongly overlapped by similar orbitals from its neighbors; to do so, the C—C—C angles, the angle formed by lines from a carbon atom's center to the center of two of its neighbors, must be 109°28′. The directed nature of covalent bonds thus emerges from the directed nature of the bonding orbitals. The diamond structure represents a three-dimensional network of strong covalent bonds which explains why diamond is so hard. To scratch a diamond crystal, it is necessary to break some of these strong bonds.

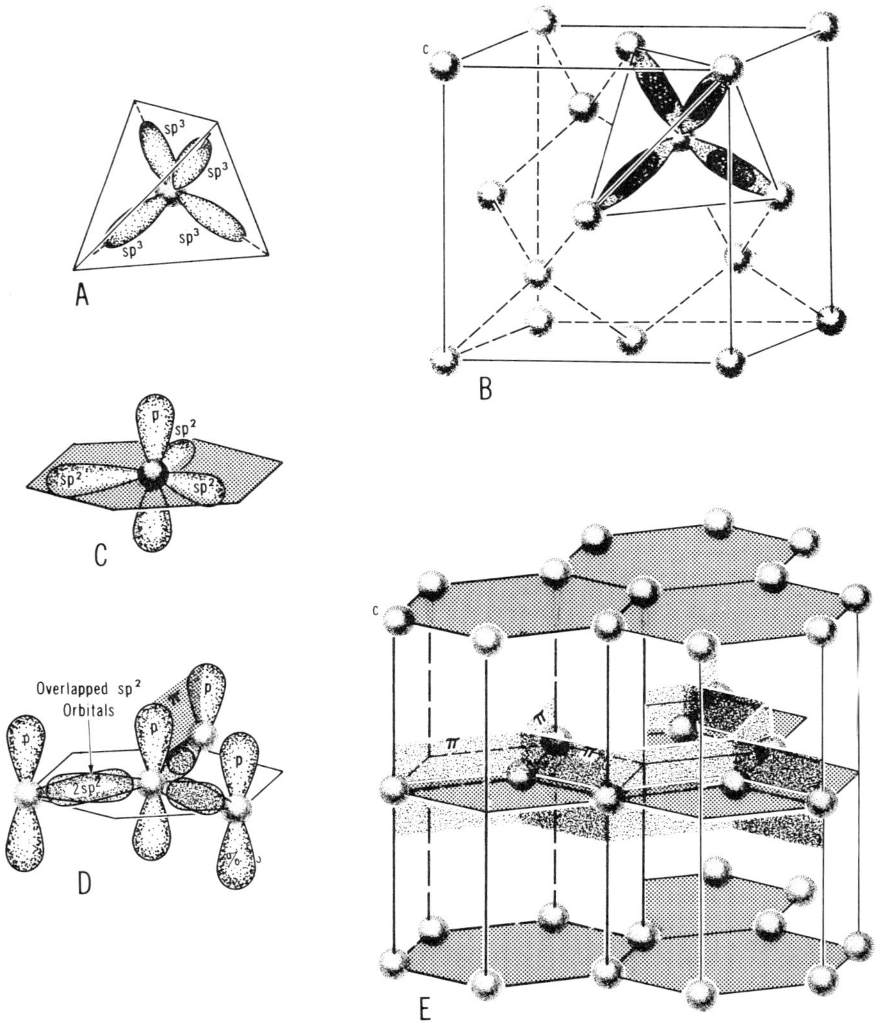

FIGURE 8–8

(A) Schematic illustration of the hybrid sp^3 orbitals emanating outward from a carbon atom. The central axis of each lobe (dashed line) extends outward toward the corners of an imaginary tetrahedron drawn around the atom.
(B) Carbon atoms disposed relative to each other as in diamond. Each carbon is bonded to four others distributed tetrahedrally about it, that is, centered on the corners of an imaginary tetrahedron surrounding it. The overlap of its sp^3 orbitals with those of its four neighbors is shown for just one carbon atom. For the remaining carbon atoms the dashed lines joining neighbors schematically represent their overlapped sp^3 orbitals.

In going from its ground state to the graphite structure, a carbon atom's electronic configuration is believed to change as follows:

$$\begin{array}{c|c|ccc} \uparrow\downarrow & \uparrow\downarrow & \uparrow & \uparrow & \_ \\ 1s & 2s & 2p & 2p & 2p \end{array} \longrightarrow \begin{array}{c|ccc|c} \uparrow\downarrow & \uparrow & \uparrow & \uparrow & \uparrow \\ 1s & 2(sp^2) & 2(sp^2) & 2(sp^2) & 2p \end{array}$$

Three of the two $2s$ and two $2p$ electrons in carbon's ground state redistribute into three hybrid $2(sp^2)$ orbitals which are a mathematical mixing of the s orbital with two of the three p orbitals. The angular probabilities for these $2(sp^2)$ orbitals can be represented by three coplanar lobes at 120° to each other (Fig. 8-8C). The fourth electron of the original two $2s$ and two $2p$ electrons fills that p orbital which does not participate in the $2(sp^2)$ hybrid, the lobe for this p orbital being perpendicular to the plane defined by the three $2(sp^2)$ orbitals. In the graphite structure, overlap occurs between the $2(sp^2)$ orbitals of neighboring atoms in the same plane (Fig. 8-8D). For such neighbors a side-to-side overlap also occurs between their unhybridized p orbitals. A resultant side-to-side bonding known as "π bonding" results between these neighbors. The electrons participating in this π bonding seem able to move across these π bonds from one atom to the next, that is, they become "delocalized." This feature explains graphite's ability to conduct electricity along the sheets of carbon atoms parallel to the (0001) direction in its structure (Fig. 8-8E). Normal to (0001), adjacent sheets of carbon atoms are held together by the weak van der Waals bonds, which will be discussed later.

When some chemical elements bond covalently, they adopt hybrid orbitals involving d-orbital electrons as well as s and p. For example, copper in CuO crystals develops dsp^2 hybrids; here the exponents again indicate the number of orbitals reorganized into the hybrids. The four dsp^2 hybrid bonds for copper all lie in the same plane and are directed outward toward the corners of a square.

A covalent bond will be partly ionic in nature to the extent that the pair of shared electrons can occasionally leave the area of overlap and occupy only the

---

(C) Distribution of the three hybrid sp^2 orbitals of a carbon atom relative to its lone p orbital.
(D) Four carbon atoms disposed as in a layer of the graphite structure. A strong bond occurs between neighboring carbon atoms in the same layer primarily because of their overlapped sp^2 orbitals. A side-to-side overlap (π bonding) also occurs between the lone p orbitals of these neighbors. However, distances separating these neighbors have been so exaggerated that this overlap could not be drawn and it is instead schematically portrayed by the fencelike shaded area labeled π.
(E) A larger portion showing three layers of carbon atoms parallel to {0001} in the graphite structure. The sp^2 and lone p orbitals are here omitted. However, the π bonding between adjacent carbons in the same layer is drawn schematically to show how electrons, which appear to be able to pass freely along these π bonds, can be easily conducted parallel to (0001) in graphite, but not perpendicular to (0001).

orbital of the more electronegative atom of the bonded pair. At such times the more electronegative atom has a negative charge and the less negative one has a positive charge; and the bond between them is ionic in nature. The degree or percentage of time that a covalent bond will be ionic depends partly on the difference in electronegativity between the two bonded atoms. According to Pauling (1959, p. 100) if the difference is about 1.7, the bond between the atoms will be about 50 percent ionic. The relationship between the difference in electronegativity Δ and the percent of ionic character of the bond formed is roughly as follows:

Δ	Percent	Δ	Percent	Δ	Percent	Δ	Percent
0.1	0.05	0.9	19	1.7	51	2.5	79
0.2	1	1.0	22	1.8	55	2.6	82
0.3	2	1.1	26	1.9	59	2.7	84
0.4	4	1.2	30	2.0	63	2.8	86
0.5	6	1.3	34	2.1	67	2.9	88
0.6	9	1.4	39	2.2	70	3.0	89
0.7	12	1.5	43	2.3	74	3.1	91
0.8	15	1.6	47	2.4	76	3.2	92

The rapid fluctuation of the electron configuration of an atom between two or more types is called *resonance*. Thus atoms that bond partly ionically and partly covalently are resonating between two configurations.

Metallic bonds

The outer shell of a metal atom usually consists of loosely held valence electrons. If metal atoms are closely packed, each atom may share its valence electrons with its near neighbors, these pairs of electrons forming, in effect, covalent bonds. They are very weak covalent bonds, however, because the valence electron from a given atom "takes turns" in forming an electron-pair bond with the adjacent atoms. If the given atom is surrounded by 12 others, it forms an electron-pair bond with each neighbor only during one-twelfth of the time. As a consequence the metallic bond is much weaker than the covalent or ionic bonds. In a packing of metal atoms the energy level of the outer p orbitals becomes so close to that of the outer s orbitals that electrons can freely transfer from the s orbital in one atom to the s or p in another, or vice versa. Therefore the crystal or melt of a metal may be regarded as an array of positively charged spheres (namely, the metal atoms minus their valence electrons) immersed in a sea of mobile electrons.

Between elements of low but nearly equal electronegativity values, a metallic bond will usually arise.

Van der Waals bonds

Van der Waals bonds, the weakest of the chemical bonds, are best observed between electrically neutral particles such as molecules. These molecules, which usually consist of several atoms bonded together covalently, may develop a slight concentration of negative charge at one end, this becoming its negative pole, and a corresponding dearth of negative charge at the other end, its positive pole. These molecules thus become very weak dipoles, oppositely charged at two ends. In forming crystals these molecules pack so that the negative pole for one lies against the positive pole for its neighbor. The extremely weak attraction between the oppositely charged poles of these neighboring molecules bonds them together in the crystal and the bond is called a *van der Waals bond.*

In crystals consisting of ionically bonded ions or of covalently bonded atoms, van der Waals attractions may also be present, but their contribution to the overall bonding is small by comparison.

Hydrogen bonds

If hydrogen is covalently bonded to a highly electronegative element such as fluorine, oxygen, chlorine, or nitrogen, its single orbital electron will be confined largely to the zone of orbital overlap. As a consequence the part of the hydrogen atom farthest from the overlap consists of an exposed proton because, unlike other atoms, once hydrogen shares its single electron, no inner electron shells remain to shield its exposed nucleus. As a result the exposed hydrogen nucleus attracts negative ions and the force of this attraction is known as the *hydrogen bond.*

Atomic and Ionic Radii

The radius of an atom or ion cannot be rigorously defined because the radial probability of encountering an orbital electron (Figs. 8-1B, C, D; Fig. 8-3) never quite becomes zero, even at large distances from the nucleus. On the other hand, two atoms or oppositely charged ions cannot be brought indefinitely close together because the mutual interpenetration of their inner shells produces a strong repulsion which ultimately equals the force of attraction between them and prevents closer approach. Within the internal structure of a crystal, the center-to-center distance of two atoms or ions that are bonded together constitutes the *bond length* between them (L in Figure 8-9A, B) and is accepted to equal the sum of the radii for the two participating atoms or ions.

Measurement of radii

If two identical atoms are bonded together, as in the crystal of a pure metal, their bond length equals twice the radius of an individual atom. Thus if bond lengths can be determined in crystals of pure elements, the radii for the individual atoms become known. For example, crystals of pure metals very often adopt a face-centered cubic (FCC) or body-centered cubic (BCC) structure; this is to say that if the tangent atoms are replaced by points at their centers, the underlying space lattice will be either face-centered cubic or body-centered cubic (Fig. 8-9C, D). The edge **a** of these cubic cells can be readily measured by X-ray techniques (Chapter 13); then the bond length L and therefore r, the radius of the component atoms, is easily computed (see Problem 8-1).

As shown in Figure 8-9B the bond length L between two oppositely charged ions consists of the sum of two *different* radii, that is,

$$L = r_C + r_A$$

where r_C and r_A, respectively, represent the radii of the bonded cation and anion. Consequently, although X-ray techniques permit measurement of the lengths of the various bonds in a crystal structure, we do not know how much of a given bond L represents r_A and how much r_C for the two ions being bonded. However, in 1922 Wasastjerna, on the basis of molecular refraction, calculated the ionic radii for F^- and O^{2-} to be 1.33 and 1.32 Å, respectively. Consequently, if we could measure the bond length from a given cation to either F^- or O^{2-} in a

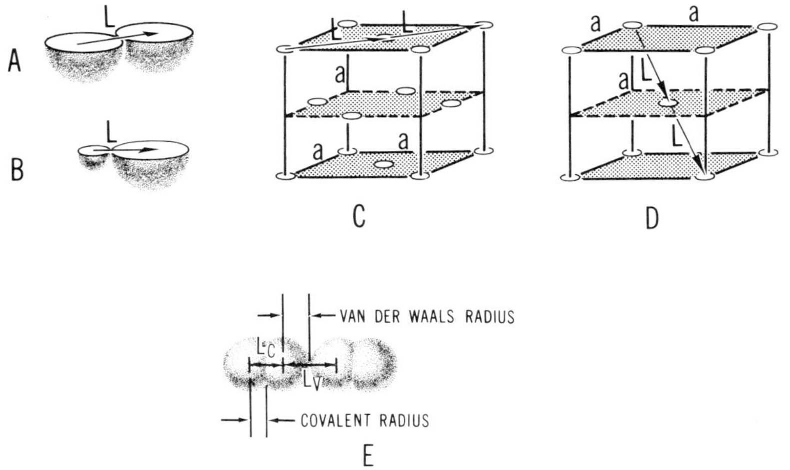

FIGURE 8–9

The bond length L is illustrated for (A) two atoms with identical radii and (B) two ions with unlike radii. For identical atoms packed so that their centers occur at the points of (C) a face-centered cubic lattice and (D) a body-centered cubic lattice, the bond length L between nearest neighbors is as indicated. (E) Two neighboring diatomic molecules isolated from a hypothetical crystal structure in which they were packed as shown. Two bond lengths are illustrated: L_C, which is between two atoms in the same molecule, and L_V, which is between two atoms in adjacent molecules.

variety of crystal structures, then subtraction of the radius for F^- or O^{2-} from this bond length would yield the radius of the given cation. Once the radii of several cations were known, their bond lengths with anions other than F^- or O^{2-} could be measured to disclose the radii of anions other than F^- and O^{2-}. Thus by subtracting Wasastjerna's radius for O^{2-} from bond lengths measured principally in ionically bonded oxide crystals, V. M. Goldschmidt and co-workers in 1926 determined an empirical set of ionic radii as well as of atomic radii. In 1927 Linus Pauling recalculated the radius of O^{2-} to be 1.40 Å and, interjecting some theoretical considerations, calculated an alternative set. L. H. Ahrens in 1952 suggested revisions to the Pauling set.

Causes of variations

An atom or ion may vary in size from one crystal structure to another, particularly if (1) the relative percentages of the ionic, covalent, or metallic characteristics of its bonding to its neighbors differ for these structures, or (2) the number of closest neighbors tangent to it, *this number being known as its coordination number,* changes from one structure to the other. As a consequence, because the values cited in tables of atomic and ionic radii are average values, they may differ somewhat from an individual value determined for a specific crystalline material. In the case of covalent radii, Schomaker and Stevenson suggest that if r_A and r_B are, respectively, the covalent radii of atoms A and B, and if x_A and x_B are their electronegativities, then their center-to-center distance d_{AB} is related to these as follows:

$$d_{AB} = r_A + r_B - 0.09 \mid x_A - x_B \mid \qquad \text{(Eq. 8–1)}$$

The absolute value $\mid x_A - x_B \mid$ is in effect an index of the ionic character of this (dominantly) covalent bond.

For crystals consisting of diatomic molecules, the shortest bond length (L_C) will be between two atoms in the same molecule (Fig. 8-9E) and will equal twice the atom's covalent radius. The bond length (L_V) between atoms from two adjacent molecules will equal twice the atom's van der Waal radius.

In Figure 8-10 the atoms and ions of the earth's more common elements are drawn approximately to scale, the ionic radii after Ahrens (1952), the atomic radii after Evans (1964). The ionic radii cited in Figure 8-10 are for the ion in 6-fold coordination, that is, tangent to six oppositely charged ions, and to modify them to apply to 4-, 8-, or 12-fold coordination requires their multiplication by 94, 103, or 112 percent, respectively. Whereas a 6-fold coordination is common for ions, a 12-fold coordination commonly occurs for atoms metallically bonded to each other. Accordingly, the atomic radii cited in Figure 8-10 for all elements except C, N, O, F, Si, P, S, and Cl represent the atom size if it is metallically bonded to 12 neighbors. To correct these values to 4-, 6-, or 8-fold coordination requires their multiplication by 88, 96, or 98 percent, respectively.

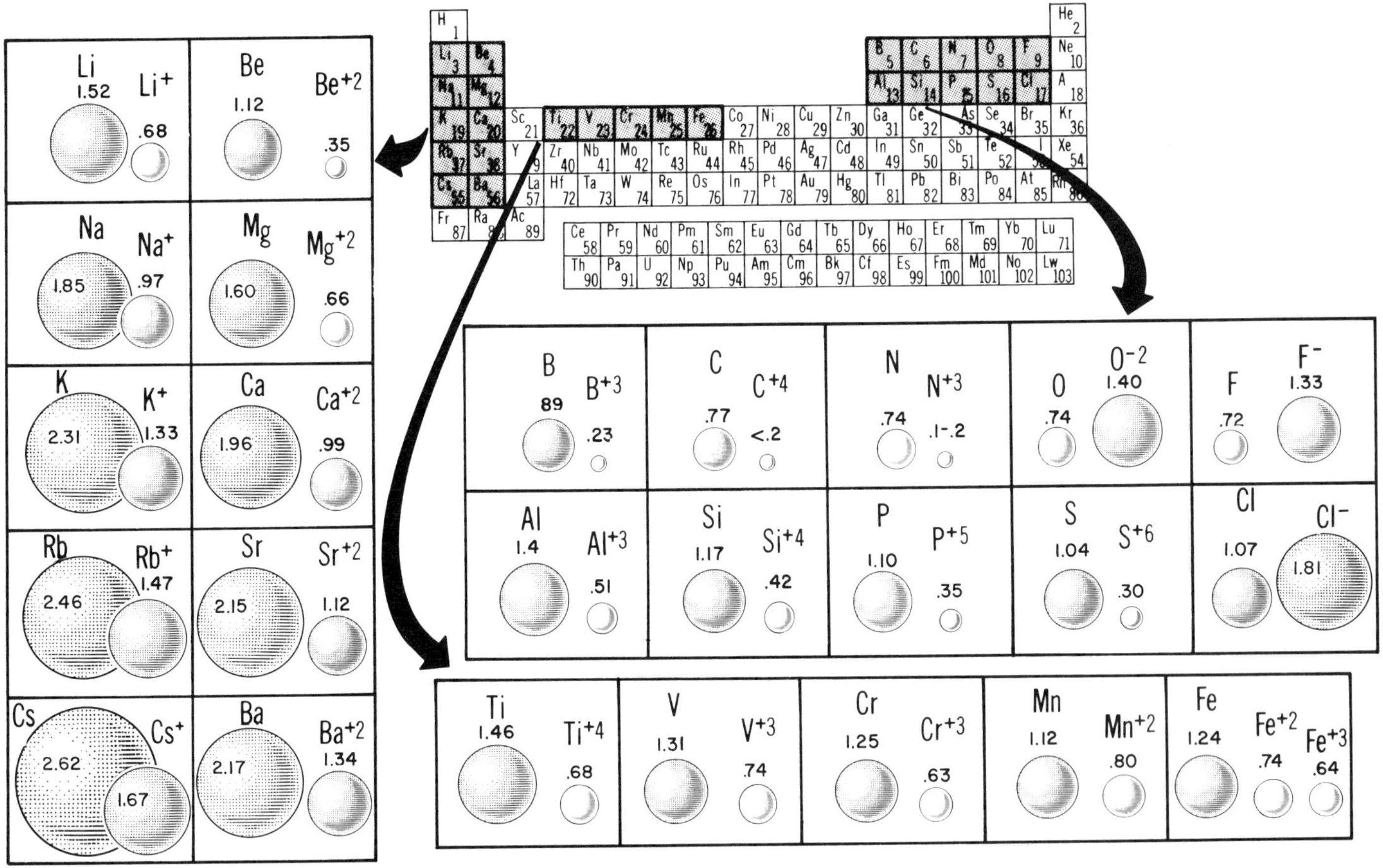

FIGURE 8–10 The more important atoms and ions drawn to scale. The values cited for the ions represent their Pauling-Ahrens radii in angstroms.

The elements C, N, O, F, Si, P, S, and Cl are so electronegative that their atoms usually form covalent rather than metallic bonds. Hence the atomic radii for these eight elements (in Fig. 8-10) represent their covalent radii for single bonding. If doubly or triply bonded, the atom's single-bond radius must be multiplied by 86 or 78 percent, respectively. The stronger the bond between two atoms, and this applies to all bond types, not just the covalent, the closer they approach each other and therefore the smaller their effective radii. Usually the covalent radii measured from individual molecules or crystals for C, N, O, F, Si, P, S, and Cl will be within 0.1 Å of those predicted by Figure 8-10.

Tables of reported radii

A recent table of atomic and ionic radii (Zhdanov, 1965, pp. 192–193) follows Bokii and Belov who assumed the radius of O^{2-} to be 1.36 Å, that is, the average of the Pauling value, 1.40 Å, and that used by Goldschmidt, 1.32 Å; they calculated the remaining ionic radii (for 6-fold coordination) on this basis. This table, reproduced here as Table 8-4, also cites the atomic radii for (1) metallic bonding and 12-fold coordination, (2) covalent bonding and single bonding, and (3) van der Waals bonding. The radius for a given element, as measured in a crystal structure, may fall between the element's ionic radius and its atomic radius because its bonding is hybrid between two or more bond types. In some covalently bonded materials the bond may fluctuate with lightninglike rapidity, at times being single, at other times double. Such resonance of a covalent bond between single and double bonding frequently causes empirically measured covalent radii to differ from those predicted by the table.

According to Ahrens (1952, p. 153), who reviews the problems of measuring ionic radii, the discrepancies between the various published sets of ionic radii are chiefly attributable to (1) the primary difficulty of whether to accept as the starting point a radius for O^{2-} of 1.32 Å, 1.40 Å, or some other value and (2) the ever-present possibility that the center-to-center distance between two ions in a crystal structure is *unequal* to the sum of their standard ionic radii.

The two standard sets of ionic radii, Pauling's and Goldschmidt's, are listed in alphabetical order in Table 8-5. The Pauling set includes revisions made by L. H. Ahrens, one of which was to accept 0.68 Å rather than 0.60 Å as the radius of Li^{+}. This 14 percent increase then required a similar 14 percent increase in Pauling's radii for Be^{2+}, B^{3+}, C^{4+}, and N^{5+}. More recently Donnay and Gryder (1963, p. 239) cited reasons for accepting 0.76 Å as the radius for Li^{+}. In this event the radii of Be^{2+}, B^{3+}, C^{4+}, and N^{5+} must be further increased to 0.39, 0.26, 0.18, and 0.15 Å, respectively, and these are inserted as alternatives in Table 8-5.

Shannon and Prewitt (1969) have determined a set of "empirical" ionic radii (column *S-P* in Table 8-5) as determined from oxides and fluorides. They confirm the suggestion by Jeffrey and Slaughter (1963) that an anion's radius will increase as the number of cations coordinating to it, that is, its coordination number (*CN*), increases. For example, they find the radius of O^{2-} in oxides to increase with its coordination number (in Roman numerals) as follows: (II)

TABLE 8–4

Crystal-Chemical Radii (Å)

Correction for Coordination No.

C. N.	Radius, percent: Ionic	Radius, percent: Metallic
12	112	100
8	103	98
6	100	96
4	94	88

Correction for Multiplicity of the Covalent Bond

Multiplicity of bond	Bond length, percent
1	100
2	86
3	78

I A	II A	III A	IV A	V A	VI A	VII A	VIII A	VIII A
Li 1^+ 0.68 K 1.33 M 1.55	Be 2^+ 0.34 K (1.00) M 1.13							
Na 1^+ 0.98 K 1.54 M 1.89	Mg 2^+ 0.74 K (1.38) M 1.60							
K 1^+ 1.33 M 2.36	Ca 2^+ 1.04 M 1.97	Sc 3^+ 0.83 M 1.64	Ti 4^+ 0.64 3^+ 0.69 2^+ 0.78 M 1.46	V 5^+ 0.4 4^+ 0.61 3^+ 0.67 2^+ 0.72 M 1.34	Cr 6^+ 0.35 3^+ 0.64 2^+ 0.83 M 1.27	Mn 7^+ (0.46) 4^+ 0.52 3^+ 0.70 2^+ 0.91 M 1.30	Fe 3^+ 0.67 2^+ 0.80 M 1.26	Co 3^+ 0.64 2^+ 0.78 M 1.25
Rb 1^+ 1.49 M 2.48	Sr 2^+ 1.20 M 2.15	Y 3^+ 0.97 M 1.81	Zr 4^+ 0.82 M 1.60	Nb 5^+ 0.66 4^+ 0.67 M 1.45	Mo 6^+ 0.65 4^+ 0.68 M 1.39	Tc M 1.36	Ru 4^+ 0.62 M 1.34	Rh 4^+ 0.65 3^+ 0.75 M 1.34
Cs 1^+ 1.65 M 2.68	Ba 2^+ 1.38 M 2.21	La 4^+ 0.90 3^+ 1.04 M 1.87	Hf 4^+ 0.82 M 1.59	Ta 5^+ (0.66) M 1.46	W 6^+ 0.65 4^+ 0.68 M 1.40	Re 6^+ 0.52 M 1.37	Os 4^+ 0.65 M 1.35	Ir 4^+ 0.65 M 1.35
Fr	Ra 2^+ 1.44	Ac 3^+ 1.11						

Lanthanides

Ce	Pr	Nd	Pm	Sm
4^+ 0.88 3^+ 1.02 M 1.83	3^+ 1.00 M 1.82	3^+ 0.99 M 1.82	3^+ (0.98) M	3^+ 0.97 M 1.81

Actinides

Th	Pa	U	Np	Pu
4^+ 0.95 3^+ 1.08 M 1.80	4^+ 0.91 3^+ 1.06 M 1.62	4^+ 0.89 3^+ 1.04 M 1.53	4^+ 0.88 3^+ 1.02 M 1.50	4^+ 0.88 3^+ 1.01 M

SOURCE: Zhdanov, *Crystal Physics,* New York: Academic Press, 1965.

Ionic radii are for C.N. 6
M = metallic radii for C.N. 12
K = covalent radii for single bond
V_{12} = van der Waals radii for C.N. 12

VIII A	I B	II B	III B	IVB	V B	VI B	VII B	VIII B
							H 1^- 1.36 K 1^+ 0.00	He V_{12} 1.22
			B 3^+ (0.20) K 0.83 M 0.89	C K 0.77	N 5^+ 0.15 3^+ K 0.74 3^- 1.48	O K 0.73 2^- 1.36	F K 0.71 1^- 1.33	Ne V_{12} 1.60
			Al 3^+ 0.57 K 1.26 M 1.43	Si 4^+ 0.39 K 1.17	P 5^+ 0.35 3^+ K 1.10 3^- 1.86	S 6^+ (0.29) K 1.04 2^- 1.82	Cl 7^+ (0.26) K 0.99 1^- 1.81	Ar V_{12} 1.92
Ni 2^+ 0.74 M 1.24	Cu 2^+ 0.80 1^+ 0.98 M 1.28	Zn 2^+ 0.83 K 1.31 M 1.39	Ga 3^+ 0.62 K 1.27 M 1.39	Ge 4^+ 0.44 2^+ 0.65 K 1.22	As 5^+ (0.47) 3^+ 0.69 K 1.21 3^- 1.91	Se 6^+ 0.35 4^+ 0.69 K 1.17 2^- 1.93	Br 7^+ (0.39) K 1.14 1^- 1.96	Kr V_{12} 1.98
Pd 4^+ 0.64 M 1.37	Ag 1^+ 1.13 M 1.44	Cd 2^+ 0.99 K 1.48 M 1.56	In 3^+ 0.92 K 1.44 M 1.66	Sn 4^+ 0.67 2^+ 1.02 K 1.40 M 1.58	Sb 5^+ 0.62 3^+ 0.90 K 1.41 M 1.61 3^- 2.08	Te 6^+ (0.56) 4^+ 0.89 K 1.37 2^- 2.11	I 7^+ (0.50) K 1.33 1^- 2.20	Xe V_{12} 2.18
Pt 4^+ 0.64 M 1.38	Au 1^+ (1.37) M 1.44	Hg 2^+ 1.12 M 1.60	Tl 3^+ 1.05 1^+ 1.49 M 1.71	Pb 4^+ 0.76 2^+ 1.26 M 1.75	Bi 5^+ (0.74) 3^+ 1.20 K M 1.82 3^- 2.13	Po	At	Rn

Eu 3^+ 0.97 M 2.02	Gd 3^+ 0.94 M 1.79	Tb 3^+ 0.89 M 1.77	Dy 3^+ 0.88 M 1.77	Ho 3^+ 0.86 M 1.76	Er 3^+ 0.85 M 1.75	Tu 3^+ 0.85 M 1.74	Yb 3^+ 0.81 M 1.93	Lu 3^+ 0.80 M 1.74
Am 4^+ 0.85 3^+ 1.00 M								

1.349 ± 0.005 Å; (III) 1.357 ± 0.010; (IV) 1.378 ± 0.004; (VI) 1.396 ± 0.009; (VIII) 1.42 Å. Gibbs and Brown (1968) independently obtained like values for O^{2-} in silicates. For the radius of F^- in fluorides, Shannon and Prewitt observe: (II) 1.285 ± 0.009; (III) 1.300 ± 0.012; (IV) 1.31; (VI) 1.33 Å. The Shannon and Prewitt radii for cations are adjusted for these changes in size of O^{2-} and F^-, whereas radii of cations in the standard sets are not.

Shannon and Prewitt give an ion's radius for the several coordination numbers observed for it. Of these, some generalized examples of the geometric distribution of neighbors X around a central ion A are illustrated for coordination number II (Fig. 9-11C), III (Fig. 9-11B), IV (Fig. 9-11A), VI (Fig. 9-10A, B), VIII (Fig. 9-9), and XII (Fig. 9-8A, D). If the centers of these neighbors are joined by lines, a polyhedron known as the *coordination polyhedron* results (see Fig. 9-8B, E and Fig. 9-10B). In addition, Shannon and Prewitt give radii for coordination numbers V, VII, IX, and X. Radii are also cited for two additional types of IV coordination. Thus IV *SQ* indicates the central ion A to be at the center of a square with its four X neighbors at the corners, whereas IV *PY* indicates the central A ion to be located above the center of this square (but still

TABLE 8–5

Sets of Ionic Radii in Angstroms[a]

Ion	Coordination Number	*G*	*P-A*	*S-P*
Ac^{3+}	VI		1.18	
Ag^+	II			0.67
	IV *SQ*			1.02
	V			1.12
	VI	1.13	1.26	1.15
	VII			1.24
	VIII			1.30
Ag^{2+}	VI		0.89	
Ag^{3+}	IV *SQ*			0.65

[a] All radii are in angstroms. Column *G* is from Goldschmidt (1954). Column *P-A* represents Pauling radii as revised and supplemented by Ahrens (1952). Several alternative radii suggested by Donnay and Gryder (1963) are also entered in column *P-A* and distinguished by the superscript *D*.

Columns *S-P* are from Shannon and Prewitt (1969). The superscripts *LS* and *HS*, respectively, denote *low-spin* and *high-spin* radii (p. 217). Negative values for radii are to be subtracted from the radius of the neighboring anion to obtain the interatomic distance. Radii that differed by no more than ±0.01 Å for at least five structural determinations are in boldface type. Question marks follow those radii in column *S-P* that are in doubt because of uncertainty about (1) the ion's oxidation state, (2) its coordination number, or (3) the precise chemistry of the compound in which it occurred.

If we let *P* represent the coordination number for cation A, *Q* that for cation B, and *R* that for anion X in the compound $A_pB_qX_r$, these coordination numbers are related to the atomic frequencies *p*, *q*, and *r* for the compound as follows:

$$pP + qQ = rR$$

Shannon and Prewitt utilize this equation, when necessary, to determine *R*, the anion coordination, as well as to check the coordination numbers stated in the original papers.

TABLE 8–5 (*Continued*)[a]

Ion	Coordination Number	*G*	*P-A*	*S-P*
Al^{3+}	IV			**0.39**
	V			0.48
	VI	0.57	0.51	**0.53**
Am^{3+}	VI		1.07	1.01
Am^{4+}	VI		0.92	
	VIII			0.95
As^{3+}	VI		0.58	
As^{5+}	IV			**0.335**
	VI		0.46	0.50
As^{3-}	VI ?	2.0	2.22	
At^{7+}	VI		0.62	
Au^{+}	VI		1.37	
Au^{3+}	IV *SQ*			0.70
	VI		0.85	
B^{3+}	III			0.02
	IV			**0.12**
	VI		0.23	
			0.26^D	
Ba^{2+}	VI	1.43	1.34	1.36
	VII			1.39
	VIII			1.42
	IX			1.47
	X			1.52 ?
	XII			1.60
Be^{2+}	III			0.17
	IV			**0.27**
	VI	0.34	0.35	
	VI		0.39^D	
Bi^{3+}	V			0.99
	VI		0.96	1.02
	VIII			1.11
Bi^{5+}	VI		0.74	
Bk^{3+}	VI			0.96
Bk^{5+}	VIII			0.93
Br^{-}	VI	1.96	1.95	
Br^{7+}	VI		0.39	
C^{4+}	III			−0.08
	IV	0.2	0.15	
	VI		0.16	
	VI		0.18^D	

TABLE 8–5 (*Continued*)[a]

Ion	Coordination Number	*G*	*P-A*	*S-P*
Ca^{2+}	VI	1.06	0.99	1.00
	VII			1.07
	VIII			1.12
	IX			1.18
	X			1.28
	XII			1.35
Cd^{2+}	IV			0.84
	V			0.87
	VI	1.03	0.97	0.95
	VII			1.00
	VIII			1.07
	XII			1.31
Ce^{3+}	VI	1.18	1.07	1.034
	VIII			1.14
	XII			1.29
Ce^{4+}	VI	1.02	0.94	0.80 ?
	VIII			0.97
Cf^{3+}	VI			0.95
Cl^{-}	VI	1.81	1.81	
Cl^{5+}	III			0.12
Cl^{7+}	IV			0.20
	VI		0.27	
Cm^{3+}	VI			0.98
Cm^{4+}	VIII			0.95
Co^{2+}	VI			0.65^{LS}
		0.82	0.72	0.735^{HS}
Co^{3+}	VI			0.525^{LS}
			0.63	0.61^{HS}
Cr^{2+}	VI			0.73^{LS}
				0.82^{HS}
Cr^{3+}	VI	0.64	0.63	0.615
Cr^{4+}	IV			0.44
	VI			0.55
Cr^{5+}	IV			0.35
	VIII			0.57
Cr^{6+}	IV			0.30
	VI	0.3–0.4	0.52	
Cs^{+}	VI	1.65	1.67	1.70
	IX			1.78
	X			1.81
	XII			1.88

TABLE 8–5 (*Continued*)[a]

Ion	Coordination Number	*G*	*P-A*	*S-P*
Cu^{+}	II			0.46
	VI		0.96	
Cu^{2+}	IV *SQ*			0.62
	V			0.65
	VI	0.8	0.72	0.73
Dy^{3+}	VI	1.07	0.92	0.908
	VIII			1.03
Er^{3+}	VI	1.03	0.89	0.881
	VIII			1.00
Eu^{2+}	VI	1.24		1.17
	VIII			1.25
Eu^{3+}	VI	1.12	0.98	0.95
	VIII			1.07
F^{-}	II			1.285
	III			1.30
	IV			1.31
	VI	1.33	1.33	1.33
F^{7+}	VI		0.08	
Fe^{2+}	IV			0.63^{HS}
	VI			0.61^{LS}
		0.83	0.74	0.77^{HS}
Fe^{3+}	IV			0.49^{HS}
	VI			0.55^{LS}
		0.67	0.64	0.645^{HS}
Fr^{+}	VI		1.80	
Ga^{3+}	IV			0.47
	V			0.55
	VI	0.62	0.62	0.620
Gd^{3+}	VI	1.11	0.97	0.938
	VIII			1.06
Ge^{2+}	VI	0.78	0.73	
Ge^{4+}	IV			0.40
	VI	0.44	0.53	0.540
Hf^{4+}	VI	0.84	0.78	0.71
	VIII			0.83
Hg^{+}	III			0.97
Hg^{2+}	II			0.69
	IV			0.96
	VI	1.12	1.10	1.02
	VIII			1.14
Ho^{3+}	VI	1.05	0.91	0.894
	VIII			1.02

TABLE 8–5 (*Continued*)[a]

Ion	Coordination Number	*G*	*P-A*	*S-P*
I^-	VI	2.20	2.20	
I^{5+}	VI		0.62	0.95 ?
I^{7+}	VI		0.50	
In^{3+}	VI	0.92	0.81	0.790
	VIII			0.923
Ir^{3+}	VI			0.73 ?
Ir^{4+}	VI	0.66	0.68	0.63
K^+	VI	1.33	1.33	1.38
	VII			1.46 ?
	VIII			1.51 ?
	IX			1.55 ?
	X			1.59 ?
	XII			1.60 ?
La^{3+}	VI	1.22	1.14	1.061
	VII			1.10
	VIII			1.18
	IX			1.20 ?
	X			1.28
	XII			1.32 ?
Li^+	IV			0.59
	VI		0.68	0.74
	VI		0.76^D	
Lu^{3+}	VI	0.99	0.85	0.848
	VIII			0.97
Mg^{2+}	IV			0.49
	VI	0.78	0.66	0.720
	VIII			0.89
Mn^{2+}	VI			0.67^{LS}
		0.91	0.80	0.820^{HS}
	VIII			0.93
Mn^{3+}	V			0.58
	VI			0.58^{LS}
		0.70	0.66	0.65^{HS}
Mn^{4+}	VI	0.52	0.60	0.54
Mn^{6+}	IV			0.27
Mn^{7+}	IV			0.26
	VI		0.46	
Mo^{3+}	VI			0.67
Mo^{4+}	VI	0.68	0.70	0.650
Mo^{5+}	VI			0.63
Mo^{6+}	IV			0.42
	V			0.50
	VI	0.69	0.62	0.60
	VII			0.71

TABLE 8–5 (*Continued*)[a]

Ion	Coordination Number	*G*	*P-A*	*S-P*
N^{3+}	VI		0.16	
N^{5+}	III			−0.12
	VI	0.1–0.2	0.13 0.15[D]	
Na^{+}	IV			0.99 ?
	V			1.00 ?
	VI	0.98	0.97	1.02
	VII			1.13 ?
	VIII			1.16 ?
	IX			1.32 ?
Nb^{2+}	VI			0.71 ?
Nb^{3+}	VI			0.70
Nb^{4+}	VI	0.69	0.74	0.69
Nb^{5+}	IV			0.32 ?
	VI	0.69	0.69	0.64
	VII			0.66
Nd^{3+}	VI	1.15	1.04	0.995
	VIII			1.12
	IX			1.09 ?
Ni^{2+}	VI	0.78	0.69	0.700
Ni^{3+}	VI			0.56[LS] 0.60[HS]
Np^{2+}	VI			1.10
Np^{3+}	VI		1.10	1.04
Np^{4+}	VI		0.95	
	VIII			0.98
Np^{7+}	VI		0.71	
O^{2-}	II			1.35
	III			1.36
	IV			1.38
	VI	1.32	1.40	1.40
	VIII			1.42
Os^{4+}	VI		0.69	0.630
P^{3+}	VI		0.44	
P^{5+}	IV			0.17
	VI	0.3–0.4	0.35	
P^{3-}		1.8	2.12	
Pa^{3+}	VI		1.13	
Pa^{4+}	VI		0.98	
	VIII			1.01
Pa^{5+}	VI		0.89	
	IX			0.95

TABLE 8–5 (*Continued*)[a]

Ion	Coordination Number	*G*	*P-A*	*S-P*
Pb^{2+}	IV *PY*			0.94
	VI	1.32	1.20	1.18
	VIII			1.29
	IX			1.33
	XI			1.39
	XII			1.49
Pb^{4+}	VI	0.84	0.84	0.775
	VIII			0.94
Pd^{+}	II			0.59
Pd^{2+}	IV *SQ*			0.64
	VI	0.89	0.80	0.86
Pd^{3+}	VI			0.76
Pd^{4+}	VI		0.65	0.62
Pm^{3+}	VI		1.06	0.979
Po^{4+}	VIII			1.10
Po^{6+}	VI		0.67	
Pr^{3+}	VI	1.16	1.06	1.013
	VIII			1.14
Pr^{4+}	VI	1.00	0.92	0.78 ?
	VIII			0.99
Pt^{2+}	VI		0.80	
Pt^{4+}	VI		0.65	0.63
Pu^{3+}	VI		1.08	1.00
Pu^{4+}	VI		0.93	0.80 ?
	VIII			0.96
Ra^{2+}	VI	1.52	1.43	
Rb^{+}	VI	1.49	1.47	1.49
	VII			1.56 ?
	VIII			1.60
	XII			1.73
Re^{4+}	VI		0.72	0.63
Re^{5+}	VI			0.52 ?
Re^{6+}	VI			0.52
Re^{7+}	IV			0.40
	VI		0.56	0.57
Rh^{3+}	VI	0.68	0.68	0.665
Rh^{4+}	VI			0.615
Ru^{3+}	VI			0.68
Ru^{4+}	VI	0.65	0.67	0.620
S^{2-}	VI	1.74	1.84	
S^{4+}	VI		0.37	
S^{6+}	IV			0.12
S^{6+}	VI	0.34	0.30	

TABLE 8–5 (Continued)[a]

Ion	Coordination Number	*G*	*P-A*	*S-P*
Sb^{3+}	IV *PY*			0.77
	V			0.80
	VI		0.76	
Sb^{5+}	VI		0.62	0.61
Sb^{3-}	VI	2.1	2.45	
Sc^{3+}	VI	0.83	0.81	0.730
	VIII			0.87
Se^{2-}	VI	1.91	1.98	
Se^{4+}	VI		0.50	
Se^{6+}	IV			0.29
	VI		0.42	
Si^{4+}	IV			0.26
	VI	0.39	0.42	0.400
Si^{4-}	VI	2.0	2.71	
Sm^{3+}	VI	1.13	1.00	0.964
	VIII			1.09
Sn^{2+}	VI		0.93	
	VIII			1.22
Sn^{4+}	VI	0.74	0.71	0.690
Sr^{2+}	VI	1.27	1.12	1.16
	VII			1.21
	VIII			1.25
	X			1.32
	XII			1.44
Ta^{3+}	VI			0.67
Ta^{4+}	VI			0.66
Ta^{5+}	VI	0.71	0.68	0.64
	VIII			0.69
Tb^{3+}	VI	1.09	0.93	0.923
	VIII			1.04
Tb^{4+}	VI	0.89	0.81	0.76
	VIII			0.88
Tc^{4+}	VI			0.64
Tc^{7+}	VI		0.56	
Te^{2-}	VI	2.11	2.21	
Te^{4+}	III			0.52
	IV			
	VI	0.89	0.70	
	VI		0.91[D]	
Te^{6+}	VI		0.56	
Th^{4+}	VI	1.10	1.02	1.00
	VIII			1.06
	IX			1.09

TABLE 8–5 *(Concluded)*[a]

Ion	Coordination Number	*G*	*P-A*	*S-P*
Ti^{2+}	VI			0.86
Ti^{3+}	VI	0.69	0.76	0.67
Ti^{4+}	V			0.53
	VI	0.64	0.68	0.605
Tl^{+}	VI	1.49	1.47	1.50
	VIII			1.60
	XII			1.76
Tl^{3+}	VI	1.05	0.95	0.880
	VIII			1.00
Tm^{3+}	VI	1.01	.87	0.869
	VIII			0.99
U^{3+}	VI			1.06
U^{4+}	VI	1.05	0.97	
	VII			0.98
	VIII			1.00
	IX			1.05
U^{5+}	VI			0.76
	VII			0.96
U^{6+}	II			0.45
	IV			0.48
	VI		0.80	0.75
	VII			0.88
V^{2+}	VI		0.88	0.79
V^{3+}	VI	0.65	0.74	0.640
V^{4+}	VI	0.61	0.63	0.59
V^{5+}	IV	0.4 ?		0.355
	V			0.46
	VI		0.59	0.54
W^{4+}	VI	0.68	0.70	0.650
W^{6+}	IV			0.41
	VI	0.69 ?	0.62	0.58
Y^{3+}	VI	1.06	0.92	0.892
	VIII			1.015
	IX			1.10
Yb^{3+}	VI	1.00	0.86	0.858
	VIII			0.98
Zn^{2+}	IV			0.60
	V			0.68
	VI	0.83	0.74	0.745
Zr^{4+}	VI	0.87	0.79	0.72
	VII			0.78
	VIII			0.84

on the square's 4-fold axis of symmetry) so that the centers of the A and the four X ions, if joined by lines, define a tetragonal pyramid.

For certain transitional metals in octahedral coordination (as in Figure 9-10A), the ion may be larger if its outer *d*-orbital electrons are in a high-spin state (*HS*) as compared to a low-spin state (*LS*). To understand spin states, segregate the outer *d* orbitals of a transition metal ion into a set that is called t_{2g} orbitals (d_{xy}, d_{xz}, d_{yz}) and a set called e_g orbitals ($d_{x^2-y^2}$, d_{z^2}). An energy difference Δ may arise between electrons in the t_{2g} set and those in the e_g set according to the nature of the transition metal ion's closest neighbors and the shape of the coordination polyhedron they form around it (see Figures 12-18 and 12-19). If Δ is large, the outer *d* electrons fill the metal's t_{2g} orbitals to capacity before any enter the higher energy e_g orbitals (the *aufbau principle*). If Δ is small, occupancy of the e_g orbitals can begin after each of the t_{2g} orbitals is only half-filled. Keeping this in mind and accepting the *spin quantum number* of an unpaired electron in an orbital to equal one-half, whereas for an oppositely spinning pair it equals zero (because they nullify each other), we see that transition metal ions with four to seven electrons in their *d* orbitals, that is, cases d^4 to d^7 in Table 8-6, may adopt a high-spin state if Δ is small but a low-spin state if Δ is large. The Shannon and Prewitt radii (column *S-P* in Table 8-5) show that ions of transition metals are larger for the high-spin state but smaller for the low-spin state.

Beginning with W. L. Bragg in 1920, crystal chemists have attempted to treat atoms and ions as hard spheres, to determine their radii from the interatomic (center-to-center) distances in known crystal structures, and to use these radii to predict interatomic distances in newly studied crystal structures. However,

TABLE 8–6

NO. OF d ELECTRONS	LOW Δ						HIGH Δ					
	t_{2g}			e_g		SPIN	t_{2g}			e_g		SPIN
	EQUAL SPINS											
1 (=d^1)	↑					$\frac{1}{2}$	↑					$\frac{1}{2}$
2 (=d^2)	↑	↑				1	↑	↑				1
3 (=d^3)	↑	↑	↑			$1\frac{1}{2}$	↑	↑	↑			$1\frac{1}{2}$
	HIGH SPIN CASES						LOW SPIN CASES					
4 (=d^4)	↑	↑	↑	↑		2	↑↓	↑	↑			1
5 (=d^5)	↑	↑	↑	↑	↑	$2\frac{1}{2}$	↑↓	↑↓	↑			$\frac{1}{2}$
6 (=d^6)	↑↓	↑	↑	↑	↑	2	↑↓	↑↓	↑↓			0
7 (=d^7)	↑↓	↑↓	↑	↑	↑	$1\frac{1}{2}$	↑↓	↑↓	↑↓	↑		$\frac{1}{2}$
	EQUAL SPINS											
8 (=d^8)	↑↓	↑↓	↑↓	↑	↑	1	↑↓	↑↓	↑↓	↑	↑	1
9 (=d^9)	↑↓	↑↓	↑↓	↑↓	↑	$\frac{1}{2}$	↑↓	↑↓	↑↓	↑↓	↑	$\frac{1}{2}$
10 (=d^{10})	↑↓	↑↓	↑↓	↑↓	↑↓	0	↑↓	↑↓	↑↓	↑↓	↑↓	0

Shannon and Prewitt note that such predicted values will likely deviate from the actual values because a simple table of radii cannot incorporate such effects as (1) anion-anion repulsion, (2) metal-metal bonding, and (3) irregular coordination polyhedra (wherein all X neighbors are not equidistant from the central ion A).

Several sets of ionic radii have been here presented: (1) the Bokii-Belov radii (Table 8-4) and in Table 8-5 the (2) Goldschmidt, (3) Pauling-Ahrens, and (4) Shannon-Prewitt radii. Of these the Shannon-Prewitt radii are recommended because they incorporate the greater number of contingencies and are based on 1000 experimentally determined interatomic distances of superior accuracy.

Polarization

The distortion of an ion's charge distribution by an oppositely charged neighbor is called *polarization.* The phenomenon introduces difficulties in determining ionic radii because the center-to-center distance between an adjacent anion and cation in a crystal structure can equal the sum of their ionic radii only if polarization is absent (Fig. 8-11A). This will not be true if the cation is small and highly charged so that the high positive charge of its nucleus is only sparsely shielded by its orbital electrons. In this case considerable attractive force "leaks through" the shield to attract and distort the electron cloud of any adjacent anion. As a result the anion's orbital electrons "spend more time" near this cation, for example, in the region where negative signs are concentrated in Figure 8-11B, but less time in regions such as that lightly shaded in Figure 8-11B. Consequently, the anion possesses a slight negative pole (where the negative signs are concentrated) and a relatively positive pole (the shaded area) where the negative charge is less strong. The anion is now said to be *polarized* by the cation and, in effect, begins to share electrons with the cation. To the extent this occurs, the bond between them develops covalent character and the two approach closer than the sum of their ionic radii.

A complex anion such as $(OH)^-$ may be polar by its very nature (Fig. 8-11C).

Fajans' rules state that polarization or covalency increases to the extent that

1. The cation is small and/or highly charged.
2. The anion is large and/or highly charged.
3. The cation deviates from an inert gas configuration.

The greater polarizability of anions as size increases reflects a large anion's inability to hold its outer electrons as tightly because (1) they are more distant from its nucleus and (2) they are screened from the nucleus by a larger number of electrons in inner orbitals. In polarizability, therefore, $F^- < Cl^- < Br^- < I^-$. In general, a large, highly charged anion will be strongly polarized if adjacent to a small, highly charged cation. Cations with an inert-gas configuration (eight

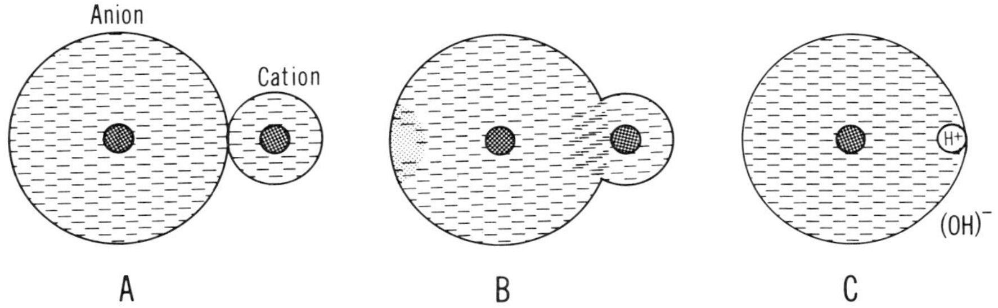

FIGURE 8–11

(A) Hypothetical case of a neighboring anion and cation, neither of which affects the other's homogeneous distribution of negative charge about its nucleus.
(B) Hypothetical case in which the cation attracts electrons of the cloud around the anion's nucleus. As a result the anion develops polarity; that is, near the cation it develops a relative concentration of negative charge, whereas distant from the cation it develops an area (shaded) in which the negative charge is somewhat depleted. The ion is thus slightly more negative at one end or pole and, in comparison, slightly more positive, at the other. Such dipolarity is further discussed under the topic of ferroelectricity (Chapter 11).
(C) The $(OH)^-$ ion is a natural dipole. Its H^+ proton marks its positive pole, the site opposite this proton being, by contrast, its negative pole.

electrons in their outer orbital) cause less polarization and undergo less themselves than those with pseudo inert-gas configurations (18 outer electrons). Thus Cu^+ polarizes an adjacent anion more strongly than would Na^+.

The polarizing power of a cation is proportional to the ratio of its charge divided by its radius (rule 1). The *polarizability* of an anion is proportional to its charge times its radius (rule 2). Whereas polarization of an anion by a cation may be appreciable in ionic crystals in which cations are usually small and anions large, polarization of a cation by an anion is usually negligible in comparison. However, polarization of a cation by an anion might become appreciable if: (1) both are univalent; (2) the cation is relatively large; and (3) the anion is relatively small.

REFERENCES AND RECOMMENDED READING

Addison, W. E. (1961) *Structural Principles in Inorganic Compounds.* New York: Wiley.

Ahrens, L. H. (1952) "The use of ionization potentials," *Geochimica et Cosmochimica Acta,* 155–169.

Donnay, G., and J. W. Gryder (1963) "The ionic radius of lithium," *Annual Report of the Director of the Geophysical Laboratory, 1963–1964,* Washington, D. C., 238.

Evans, R. C. (1964) *Crystal Chemistry.* Cambridge, England: Cambridge Univ. Press.

Gibbs, G. V., and G. E. Brown (1968) "Oxygen coordination and the Si—O bond." (abstr.) *Program, Ann. Meet., S. E. Sect., Geol. Soc. Amer.,* Durham, North Carolina; also see Brown and Gibbs (1969) *Amer. Mineral.,* **54,** 1528–1539.

Goldschmidt, V. M. (1954) *Geochemistry* (Alex Muir, ed.). London: Oxford Univ. Press.

Jeffrey, G. A., and M. Slaughter (1963) "The structure of aluminum tetroxycarbide," *Acta Cryst.*, 177–184.
Pauling, L. (1960) *The Nature of the Chemical Bond*, 3d ed. Ithaca, New York: Cornell Univ. Press.
Schomaker, V., and D. P. Stevenson (1941) "Some revisions of the covalent radii and the additivity rule for the lengths of partially ionic single covalent bonds," *J. Am. Chem. Soc.*, 37–40.
Shannon, R. D., and C. T. Prewitt (1969) "Effective ionic radii in oxides and fluorides," *Acta Cryst.*, 925–946.
Zhdanov, G. S. (1965) *Crystal Physics* (A. F. Brown, translator). New York: Academic Press.

QUESTIONS AND PROBLEMS

8-1 Find the atomic radius r in terms of the edge **a** of the cubic unit cell if metal atoms are packed according to (a) a face-centered lattice (Fig. 8-9C) and (b) a body-centered lattice (Fig. 8-9D).

8-2 The radius of a gold atom is 1.44 Å. What is the length of the cell edge for gold crystals whose atoms are packed according to a face-centered lattice?

8-3 Crystals of platinum ($F\ 4/m\ \bar{3}\ 2/m$) exhibit a unit cell for which **a** equals 3.91 Å and for which $Z = 4$. What is the radius of a platinum atom?

ANSWERS

8-1 (a) $r = 0.3535\mathbf{a}$; (b) $r = 0.433\mathbf{a}$.

8-2 4.08 Å.

8-3 1.38 Å.

9 Principles of Crystal Structures

Broadly speaking, crystals are divisible into two major types: (1) *molecular* and (2) *nonmolecular* crystals. The nonmolecular type can then be further subdivided into (a) *metallic,* (b) *covalent,* and (c) *ionic* crystals, depending on the dominant bonding force to which the crystal owes its coherence. This classification of nonmolecular crystals is largely for convenience of discussion because, actually, the bonding between the atoms in a crystal's structure is seldom solely metallic, covalent, or ionic in nature; it is more likely to be a hybrid between these. Moreover, demonstration that a crystal's structure conforms to one of the prototypes that we are now going to describe does not necessarily indicate that the dominant bonding is that under which the prototype is classified.

Molecular Crystals

Molecular crystals consist of a regular packing of molecules, a molecule being analogous to a "fraternity of atoms" because their bonding to each other, which is ordinarily covalent, is much stronger than any bond they can form to an atom outside the group. For example, the crystal structure of cyanuric triazide, C_3N_{12}, consists of C_3N_{12} molecules (Fig. 9-1A) systematically packed together (Fig. 9-1B). In molecular crystals the bonding of one molecule to the next is by relatively weak van der Waals forces. As a consequence molecular crystals are usually soft and weak, being easily scratched or broken. As its temperature is increased and the thermal vibration of its component molecules steadily increases, a molecular crystal will increase measurably in volume, that is, it generally has a large *coefficient of expansion,* because the weak van der Waals forces can no longer keep the molecules as close together in face of their increasing thermal agitation.

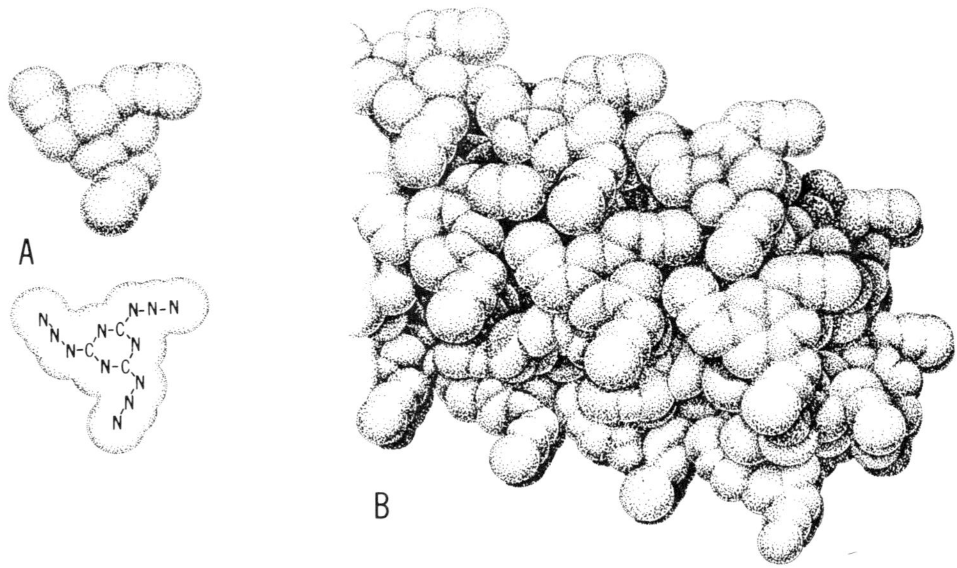

FIGURE 9–1

(A) Individual molecule of cyanuric triazide, C_3N_{12}. (B) Systematic packing of C_3N_{12} molecules in the crystal structure of cyanuric triazide. (From *General Chemistry*, Second Edition, by Linus Pauling, W. H. Freeman and Company. Copyright © 1953.)

Moreover, a molecule on the crystal's surface may become so agitated it breaks its van der Waals bonds with its neighbors and escapes into space. If the crystal is confined in a sealed, evacuated container, the molecules escaping from the crystal's surface now constitute a vapor which, like any gas, exerts a measurable pressure. This, its *vapor pressure*, is relatively high for molecular crystals and increases as temperature increases. As its temperature is elevated the thermal vibration will ultimately become violent enough to disrupt all bonds between molecules, even those deep in the interior of the crystal. At this point the crystal melts, the regularity of packing of its molecules being destroyed and their mobility increased. Because molecular crystals depend on weak van der Waals forces for their coherence, they characteristically melt or decompose at relatively low temperatures and conduct electricity poorly; that is, they act as insulators.

Metallic Crystals

The bond between atoms of low, but equal (or nearly equal) electronegativities is usually metallic in nature. The metallic bond, unlike the covalent bond, extends outward in all directions from a given central atom. In consequence a metallic bond can form at *any* point of contact between neighboring metal atoms.

Thus pure metals frequently have crystal structures in which their component atoms are packed together as closely as possible so that (1) each atom touches 12 neighbors, as will be demonstrated later, and (2) the total pore space between atoms is minimized. Conditions (1) and (2) are synonymous; that is, if (1) is true for an array of equally sized spheres, then (2) must also be true, or vice versa. To the extent that the coordination number is less than 12 in an array of like-sized atoms, total pore space increases and the atoms no longer represent a closest-packed array.

Closest packings

The distribution of atoms in the structures of crystals of pure metallic elements frequently conforms to a closest-packed array that is either the type called *cubic closest packing* (CCP) or that called *hexagonal closest packing* (HCP). Each may be regarded as consisting of single layers of closest-packed atoms, like the layer formed by the atoms labeled *a* at the left in Figure 9-2A, with additional closest-packed layers like layers *b* and *c* stacked atop. Only the manner of layer stacking distinguishes cubic closest packing from hexagonal closest packing. To clarify this, the interstices between the *a* atoms (Fig. 9-2A, left) are divided

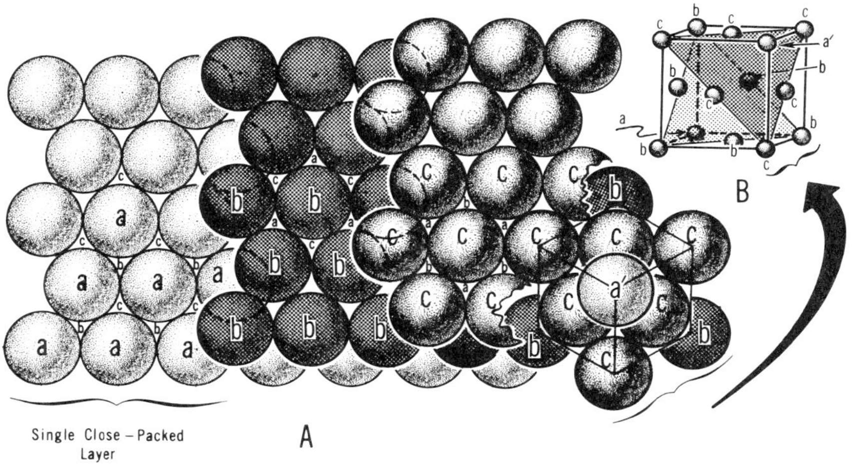

FIGURE 9–2

(A) Stacking of spheres in a cubic closest packing. The spheres in the different layers are shaded differently for illustrative purposes only. The spheres labeled *a* represent a single closest-packed layer, those labeled *b* and *c* represent two more such layers, each stacked on the layer below it. The centering of the spheres on the points of a face-centered cubic lattice is illustrated by drawing the cube whose uppermost corner is at *a′*. Sphere *a′* represents the start of another *a* layer that will overlie the *c* layer.

(B) This same cube, re-oriented, and with the spheres reduced in size for better illustration of its face centering. In this cube the dark plane marks the layer of closest-packed *c* spheres, the lighter plane that of *b* spheres.

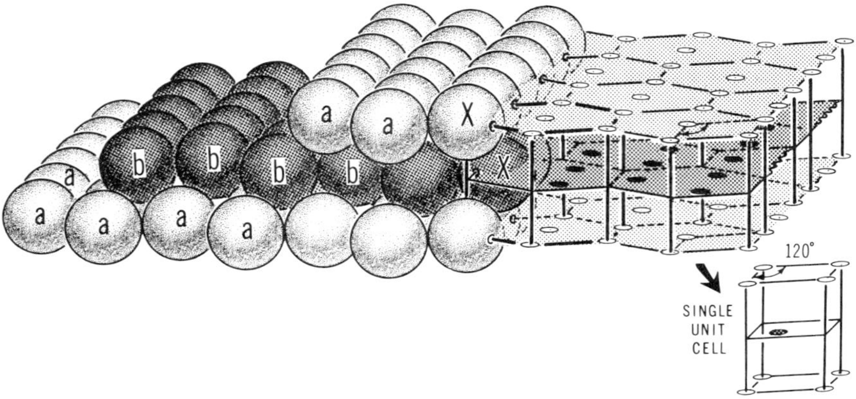

FIGURE 9–3

Hexagonal closest packing of spheres. The spheres labeled *a* and *b* are identical and have been shaded differently for illustrative purposes only. To the right, the centers of the spheres are replaced by circles (seen as ellipses in perspective) so as to reveal the hexagonal space lattice (white ellipses) that underlies hexagonal closest packing. The two spheres marked X constitute the unit of pattern being repeated.

into a set *b* and a second set *c*. The heavily shaded atoms of the closest-packed layer atop this first *a* layer are labeled *b* because these *b* atoms directly overlie, and in fact nestle down into, the indentations in the *a* layer formed by the *b* set of interstices. In Figure 9-2A the next layer of closest-packed atoms are labeled *c* because they directly overlie the *c* set of interstices in both the *a* and *b* layers, nestling into the indentations on the *b* layer caused by these interstices. The fourth closest-packed layer in Figure 9-2A is represented by only the atom labeled *a′*. This atom, and all others in this fourth layer had they been drawn, directly overlie the atoms in the original *a* layer and this fourth layer is simply another *a* layer. Using a left-to-right sequence to summarize the sequence in which the *a, b,* and *c* layers are stacked atop each other, the stacking in Figure 9-2A could be symbolized as *abca*. In cubic closest packing this sequence of stacking continues indefinitely, thus *abcabcabc. . . abc*.

Replacing each atom in Figure 9-2A by a point at its center, a face-centered cubic lattice results. To emphasize this, those atoms that define a face-centered cube are heavily outlined at the lower right in Figure 9-2A, the cube being viewed down one of its $\bar{3}$-fold axes; these same atoms are shrunk in Figure 9-2B and the cube is reoriented to a standard position. In cubic closest packing a single sphere constitutes the unit of pattern being repeated by the face-centered cubic lattice. Consequently, a cubic closest-packing array of spheres belongs to the space group, $F\,4/m\,\bar{3}\,2/m$, and thus to the isometric or cubic system, hence the name *cubic closest packing*. Alternatively, this cubic closest-packing structure is called a *face-centered cubic* structure (abbreviated FCC) in allusion to the lattice underlying it.

If the stacking of the closest-packed layers in Figure 9-2A had been *ababab...*, or *bcbcbc...*, or *acacac...*, the three-dimensional closest packing of spheres known as *hexagonal closest packing* (HCP) would result. In this type of packing (Fig. 9-3) the spheres of every other closest-packed layer, for example, the two layers of unshaded spheres, lie vertically above (or below) each other. If the spheres are shrunk until only points at their centers remain (Fig. 9-3, right side), *two* sets of identipoints (one set unshaded, the other shaded) result. This is true because the shaded and unshaded spheres, although representing the same type of atom, differ in respect to the directions from them to their nearest neighbors. The set of white identipoints thus define the hexagonal lattice underlying hexagonal closest packing. The two spheres marked X in Figure 9-3 *together* represent the unit of pattern that this lattice repeats and the resultant space group is $P\,6_3/m\;2/m\;2/c$ ($P\,6_3/m\,mc = D_{6h}^4$).

Body-centered cubic structures ($I\;4/m\;\bar{3}\;2/m = Im3m = O_h^5$)

The two types of closest packing (CCP and HCP) are commonly adopted in metal crystals. A third frequently observed structural type is the body-centered cubic structure (Fig. 9-4) in which the metal atoms are repeated by a body-centered cubic lattice. In this structure each metal atom coordinates to eight neighbors, not to twelve as in closest packing. Consequently, the body-centered cubic structure (abbreviated BCC) is not quite so compact a packing, having a slightly higher percentage of "pore space."

Of the crystal structures adopted by pure elements when solidified (Table 9-1), metallic elements favor the CCP, HCP, and BCC types. As a general rule, because of the closeness of packing and relatively high atomic weights of their constituent atoms, metal crystals have high densities. They conduct electricity readily both as solids and when melted. If the same metallic element can occur with different structures, the crystals with the cubic closest-packed structure will frequently be softer and more malleable than those with the hexagonal closest-packed or body-centered cubic structures for reasons discussed later (pp. 336–337, 359).

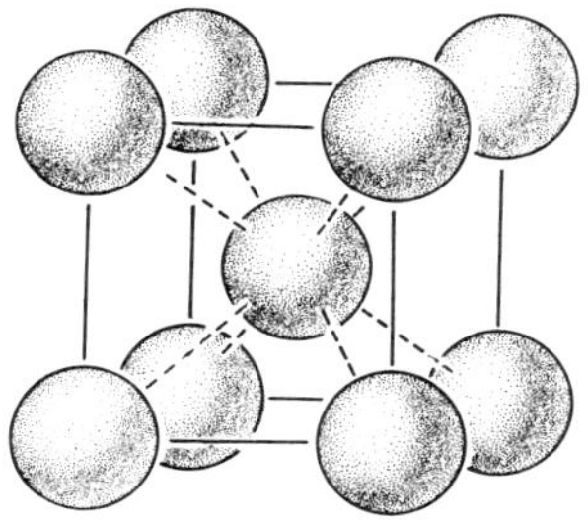

FIGURE 9–4

Body-centered cubic packing of spheres illustrated by a single unit cell. The spheres have been reduced in size for illustrative purposes. If drawn to true scale, the central sphere would be tangent to each corner sphere.

TABLE 9.1 Crystal Structures of the Elements

I	II
1 H ⬡ (as H_2 molecular crystal)	
3 Li ⊠	4 Be ⬡ ↓ 630° hexagonal
11 Na ⊠	12 Mg ⬡

⬡ -h.c.p. = hexagonal close-packed

□ -c.c.p. or f.c.c. = cubic close-packed or face-centered cubic

⊠ -b.c.c. = body-centered cubic

β–W–cubic intermediate between c.c.p. and b.c.c.

α and β Mn–very complex cubic structures

I_a	II_a	111_a	IV_a	V_a	VI_a	VII_a	VIII	VIII	VIII	I_b	II_b
19 K ⊠	20 Ca α-□ 250° → β-⬡ 450° → γ-⊠	21 Sc □ ⬡	22 Ti ⬡ ↓ 882° ⊠	23 V ⊠	24 Cr ⊠ ⬡ (electrolytic form)	25 Mn α-cubic-complex ↓ 727° β-cubic-complex ↓ 1095° γ-□ → 1133° δ-⊠	26 Fe ⊠ 909° → □ 1403° → ⊠	27 Co ⬡ 420° → □	28 Ni □ ⬡ (electrolytic form)	29 Cu □	30 Zn hexagonal (distorted ⬡)
37 Rb ⊠	38 Sr α-□ 248° → β-⬡ 614° → γ-⊠	39 Y ⬡	40 Zr ⬡ ↓ 862° ⊠	41 Nb ⊠	42 Mo ⊠	43 Tc ⬡	44 Ru ⬡	45 Rh □ cubic-unique type (electrolytic form)	46 Pd □	47 Ag □	48 Cd hexagonal (distorted ⬡)
55 Cs ⊠	56 Ba ⊠	57 La double ⬡ □	58 Ce ⬡ □	59 Pr double ⬡ □	60 Nd double ⬡	61 Pm ?	62 Sm rhombohedral	63 Eu ⊠	64 Gd ⬡	65 Tb ⬡	66 Dy ⬡
		LANTHANIDE									
			72 Hf ⬡ ↓ 1950° ⊠	73 Ta ⊠	74 W α-⊠ β-cubic (electrolytic form)	75 Re ⬡	76 Os ⬡	77 Ir □	78 Pt □	79 Au □	80 Hg rhombohedral
87 Fr ?	88 Ra ?	89 Ac ?	90 Th □ ↓ 1400° ⊠	91 Pa tetragonal	92 U α-orthorhombic ↓ 665° β-tetragonal ↓ 775° γ-⊠	93 Np α-orthorhombic ↓ 278° β- tetragonal ↓ 570° γ-⊠	94 Pu 6 phases *	95 Am double ⬡ and other phases	96 Cm ?	97 Bk ?	98 Cf ?
		ACTINIDE									

* α-monoclinic
↓ 120°C
β-(?)
↓ 203°
γ-f.c. orthorhombic
↓ 319°
δ-c.c.p.
↓ 451°
δ′-b.c. tetragonal
↓ 476°
α-b.c.c.

III	IV	V	VI	VII	0
					2 He ⬡ (molecular crystal)
5 B tetragonal, rhombohedral, and other crystal forms	6 C diamond–cubic graphite– α–hexagonal β–rhombohedral	7 N □ ↓ −238° ⬡ } as N_2 molecular crystal	8 O α–orthorhombic ↓−249° β–rhombohedral ↓−229.8° γ–cubic (as pairs of O_2) } as O_2 molecular crystal	9 F ?	10 Ne □ (molecular crystal)
13 Al □	14 Si cubic–diamond	15 P white–cubic, hexagonal violet–rhombohedral black–orthorhombic also amorphous forms	16 S rhombic, three monoclinic, rhombohedral, and other crystal forms also amorphous and plastic forms	17 Cl orthorhombic (as Cl_2 molecular crystal)	18 A □ (molecular crystal)
III_b	IV_b	V_b	VI_b	VII_b	
31 Ga orthorhombic	32 Ge cubic–diamond	33 As yellow—cubic grey —rhombohedral black —amorphous	34 Se hexagonal, two monoclinic forms, also amorphous forms	35 Br orthorhombic (as Br_2 molecular crystal)	36 Kr □ (molecular crystal)
49 In f.c. tetragonal (distorted □)	50 Sn α–cubic–diamond ↓18° β–tetragonal ↓161° γ–⬡	51 Sb yellow—cubic grey —rhombohedral black —amorphous	52 Te hexagonal, also amorphous forms	53 I orthorhombic (as I_2 molecular crystal)	54 Xe □ (molecular crystal)
67 Ho ⬡	68 Er ⬡	69 Tm ⬡	70 Yb □	71 Lu ⬡	
81 Tl ⬡ ↓ 231° ⊠	82 Pb □	83 Bi rhombohedral	84 Po cubic–simple ↓ 75° rhombohedral-simple	85 At ?	86 Rn ?
99 E ?	100 Fm ?	101 Mv ?	102 No ?	103 Lw ?	

From *Molecular Science and Molecular Engineering*, R. von Hippel et al., John Wiley and Sons, 1959.

Polymorphism

The ability of a given substance, be it pure element or complex compound, to crystallize in more than one structure is known as *polymorphism*; accordingly, the different possible structures of the substance are called its *polymorphs* or *polymorphic modifications*.[1] For example, pure iron adopts a body-centered cubic structure for temperatures up to 909°C but a cubic closest-packed, (that is, FCC) structure from 909 to 1403°C. From 1403 to 1537°C it adopts a body-centered cubic structure again. The specific polymorphic modification adopted by a substance depends chiefly on the temperature, pressure, and chemistry of its environment. Polymorphism is frequently observed not only among metallic crystals but also among the covalent and ionic crystals we shall discuss next.

Covalent Structural Types

In molecular crystals covalent bonds hold the atoms together to form molecules, but relatively weak van der Waals bonds link molecule to molecule to form soft crystals with low melting and boiling points. In covalent crystals, however, dominantly covalent bonds link each atom to its neighbor and no weak

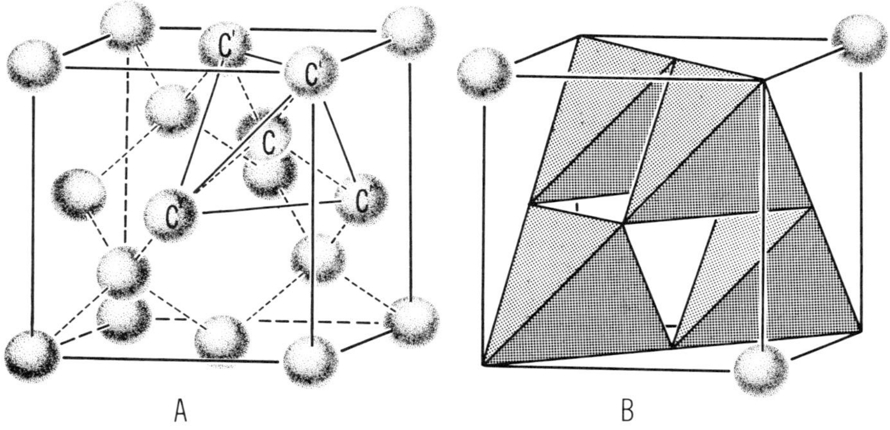

FIGURE 9–5

(A) Diamond structure showing the distribution of the carbon atoms in a unit cell. Each carbon atom, drawn to a reduced scale, is tetrahedrally coordinated by four others.
(B) The diamond structure shown chiefly by means of the tetrahedral coordination polyhedra. For each tetrahedron, carbon atoms would be centered on its corners and at its center.

[1] The polymorphism of a pure element is sometimes called *allotropism* and the polymorphs are then called *allotropes*.

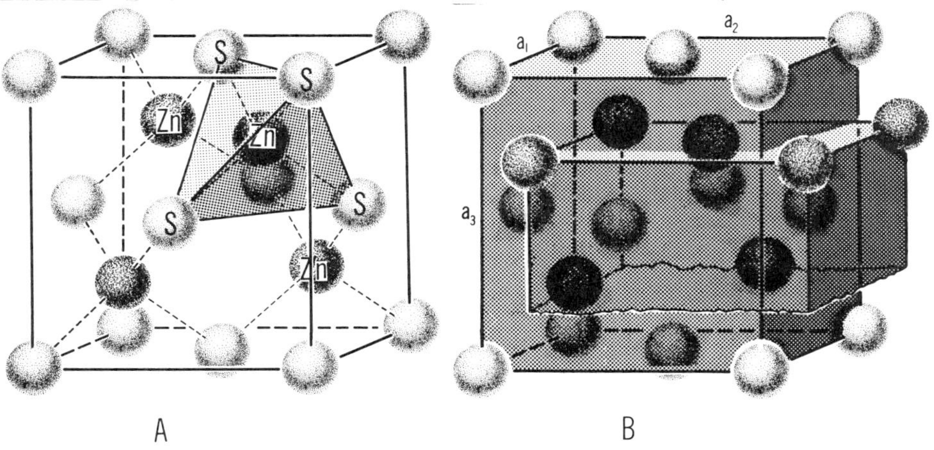

FIGURE 9–6

(A) A single unit cell in the structure of sphalerite, ZnS.
(B) Similarity of the sphalerite structure to a face-centered cubic lattice of sulfur atoms interpenetrating another of zinc atoms.

links exist. Consequently, the entire crystal, in effect, represents a single, gigantic molecule. Therefore covalent crystals may be hard and possess high melting points. The directed nature of the covalent bond (p. 197) tends to favor structures whose atoms have low coordination numbers and are thus loosely packed with considerable pore space. Consequently, covalent crystals usually possess low densities.

The diamond structure ($F\ 4_1/d\ \bar{3}\ 2/m = Fd3m = O_h^7$)

In the diamond structure (Fig. 9-5A) each carbon atom is covalently bonded to four other carbons, the dashed lines in Figure 9-5A indicating such bonds. For example, four carbon atoms (c') bond to carbon atom c in Figure 9-5A. Lines joining the centers of these four atoms define a tetrahedron that contains c at its center. Consequently atom c is said to be *tetrahedrally coordinated* by its four neighbors. The polyhedron (in this case a tetrahedron) defined by lines joining the centers of those atoms that surround a given atom is called a *coordination polyhedron.* Sometimes structures can be more simply visualized in terms of their coordination polyhedra (Fig. 9-5B). This necessity of one carbon atom to be tetrahedrally coordinated by four others reflects the spatial distribution of the hybrid sp^3 orbitals of carbon (Fig. 8-8A).

The sphalerite (zincblende) structure ($F\ \bar{4}3m = T_d^2$)

Two elements, for convenience called A and X, each of which has a high electronegativity and each able to form tetrahedrally distributed orbitals of the hybrid sp^3 type, may combine into a crystal structure like that for sphalerite (ZnS), which is also known as zincblende (Fig. 9-6A). The structure resembles that for diamond except that now each atom of the one element is coordinated by four of the other. The sphalerite structure can be likened to two interpenetrating face-centered cubic lattices, sulfurs being centered on the points of one, zinc atoms on the other (Fig. 9-6B). When so considered, one lattice is displaced from the other by one quarter of the cell edge along a_1, a_2, and a_3. The diamond structure could also have been regarded in this manner but in its case the carbons would be centered at the points of *both* interpenetrating lattices.

The wurtzite structure ($P\ 6_3mc = C_{6v}{}^4$)

Alternative to the isometric sphalerite structure, the compound ZnS may adopt the hexagonal structure of the mineral wurtzite (Fig. 9-7A). In it each zinc atom is coordinated by four sulfurs and, conversely, each sulfur is bonded to four zinc atoms. Redrawing Figure 9-7A with the zinc atoms omitted, the result (Fig. 9-7B) indicates the same distribution of sulfur atoms in the unit cell as for a hexagonally close-packed array (Fig. 9-3). Figure 9-7B would also have resulted if only the zinc atoms in Figure 9-7A had been drawn. Consequently, the wurtzite structure can be likened to two interpenetrating arrays, one offset directly above the other (Fig. 9-7C) and each having the centers of its atoms spatially distributed as in hexagonal closest packing.

Ionic Crystals and Packing

Potential energy and tangency

To a first approximation, ionic crystal structures consist of relatively large ions (usually anions) in a closest-packed arrangement with small ions (usually cations) filling some of the interstices between the large ions. Such a structure will generally only accept cations that are sufficiently large so as not to "rattle" in these interstices. This "nonrattle" requirement of tangency between the cation and its neighboring anions is another way of saying that ions pack so that the potential energy of the resultant crystal is minimized. In contrast if a neighboring anion and cation are not tangent in a structure, then the two will possess greater potential energy in respect to each other (like a magnet and iron bar being held apart). Consequently, in ionic crystal structures neighboring anions and cations will most generally be tangent to each other so as to contribute least to the crystal's overall potential energy, thereby making it more likely to form. In summary, as first said by the great Norwegian geochemist, V. M. Goldschmidt, a cation cannot "rattle" within its niche among the anions in a structure.

Radius ratios and coordination polyhedra

The maximum number of large ions (X) that can surround and touch a second ion (A) depends on their size relative to each other, as summarized by their radius ratio R_A/R_X. If ions A and X are equal in size, that is, R_A/R_X equals

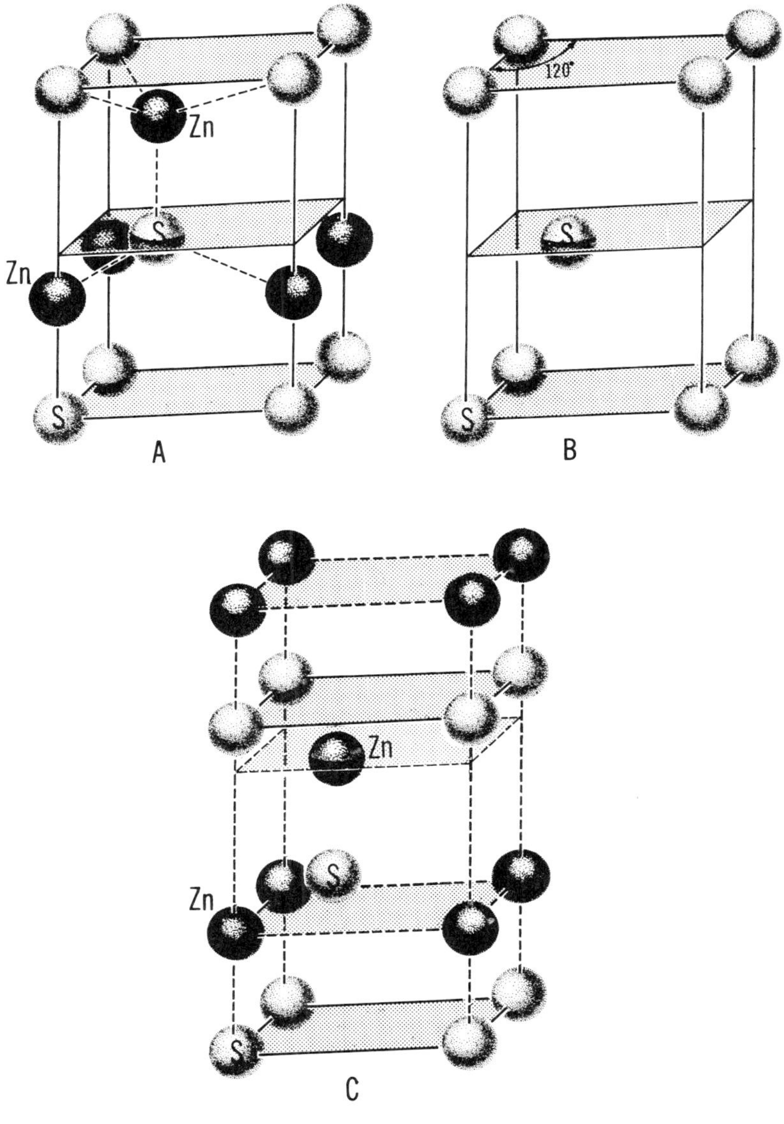

FIGURE 9–7

(A) A single unit cell of the wurtzite, ZnS, structure. Each zinc atom is coordinated by four sulfur atoms and vice versa.
(B) Redrawing of the wurtzite structure with zinc atoms omitted to emphasize that the sulfur atoms are hexagonally closest packed (HCP).
(C) Similarity of the wurtzite structure to two interpenetrating hexagonal closest-packing-like arrays, one of sulfur, the second of zinc atoms.

1.0, then 12 X ions at most can surround A so as to be tangent to it and to each other. The 12 X ions may be arranged about A either as in Figure 9-8A or as in Figure 9-8D. If the A ion were displaced by an X ion, Figure 9-8A would correspond to hexagonal closest packing, Figure 9-8D to cubic closest packing. The coordination polyhedron of X ions around A thus differs (Figs. 9-8B, C vs. 9-8E, F) although the coordination number (CN) for ion A is 12 in either case.

If the A ion in Figure 9-8A and D were smaller, it would "rattle" in its site. Thus as R_A/R_X decreases below 1.0 in value, only eight X ions can surround

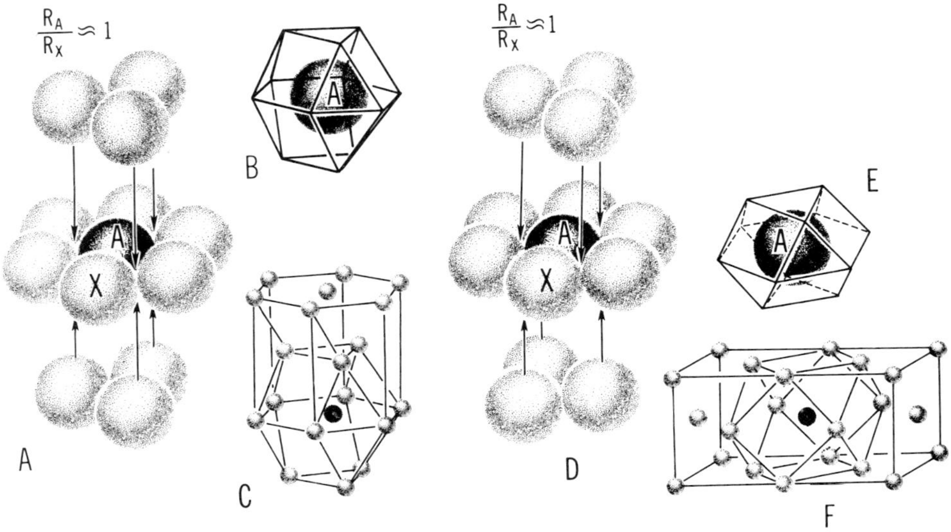

FIGURE 9–8

Two types of 12-fold coordination of X ions (unshaded) about an A ion (shaded) of similar radius.

(A) Hexagonal closest-packing-like arrangement wherein the A ion touches the six X ions in its same plane and also touches the three X ions above it and the three X ions below it which nestle into the arrow-indicated interstices.

(B) Hexagonal closest-packing-like arrangement showing the centers of the X ions as corners of a coordinating polyhedron about A.

(C) The coordination polyhedron in (B) is shown relative to a larger number of the hexagonally closest-packed spheres, these spheres being reduced below their true scale size (as is the dark-shaded A sphere).

(D) Cubic closest-packing-like arrangement that differs from the hexagonal closest-packing type because the three X ions above nestle into interstices that are not directly above the insertion points of the three lower X ions.

(E) Coordination polyhedron for this cubic closest-packing-like coordination is shown to be a cubo-octahedron.

(F) Re-orientation of the cubic closest-packed spheres in (C), with additional spheres added so as to show better their relationship to the cubo-octahedron surrounding A. The cubo-octahedron is now oriented with a 4-fold axis vertical, whereas in (E) its $\overline{3}$-fold axis is vertical.

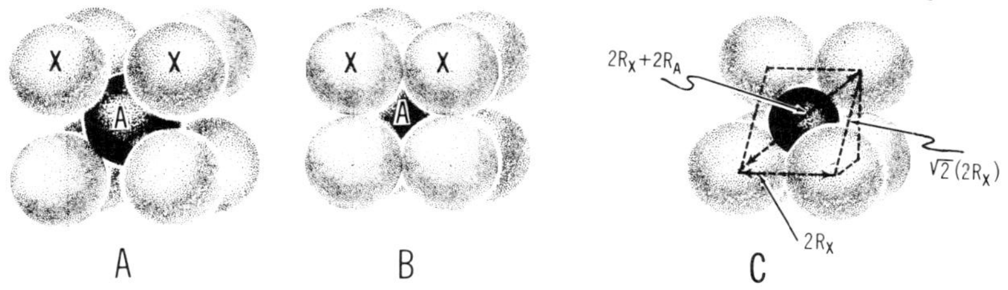

FIGURE 9–9

Eightfold coordination of eight X ions around an A ion as favored for radius ratios R_A/R_X in the range 0.73 to 1.0. In (A) the radius ratio R_A/R_X exceeds 0.73 and consequently the X ions do not touch each other.
(B) This ratio equals 0.73 and they *do* touch.
(C) The cube in (B) after removal of two of the ions previously labeled X. This cube's body diagonal (labeled $2R_X + 2R_A$) equals 1.732 times the cube edge labeled $2R_X$.

and touch ion A (Fig. 9-9A). As R_A/R_X further decreases in value, that is, smaller A ions are packed among the X ions, the X ions at the corners of the coordination cube approach each other until *they* become tangent to each other (Fig. 9-9B). Their center-to-center distance $2R_X$ now represents the cube's edge (Fig. 9-9C); furthermore the cube's body diagonal (the longest dashed arrow) represents $2R_X + 2R_A$. For any cube the body diagonal equals $\sqrt{3}$ times its edge, thus

$$2R_X + 2R_A = \sqrt{3}(2R_X)$$

Dividing both sides by $2R_X$, and solving for R_A/R_X,

$$\frac{R_A}{R_X} = \sqrt{3} - 1 = 0.732$$

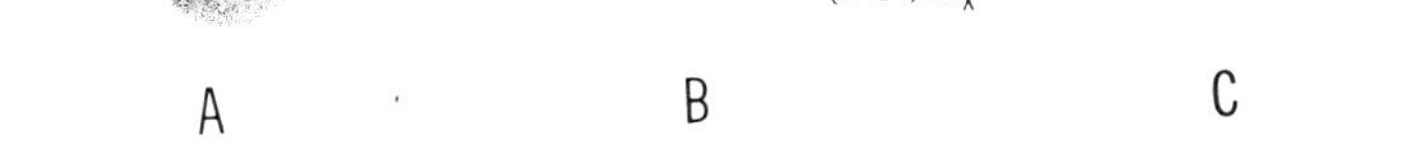

FIGURE 9–10

(A) Octahedral coordination of six X ions about an A ion as favored for radius ratios R_A/R_X in the range 0.73 to 0.41.
(B) Replacement of the X ions by points at their centers which, joined by lines, define an octahedron centered about the A ion.
(C) The coordination of four X ions about A if the radius ratio is 0.41. Note that the square defined by the centers of the X ions has a side equal to $2R_X$ and therefore a diagonal equal to 1.414 times $2R_X$ (and to $2R_X + 2R_A$).

Thus if the radius ratio falls below 0.732, the A ion will "rattle" in its site; that is, the eightfold coordination of X ions around A is no longer the arrangement of lowest potential energy.

For a radius ratio below 0.732 and down to the "no-rattle" limit of 0.414 only six of the larger X ions can surround ion A (Fig. 9-10A). The coordination polyhedron they define about A is an octahedron (Fig. 9-10B) and hence this type of coordination is called *octahedral coordination*. The lower "no-rattle" limit (radius ratio: 0.414) for octahedral coordination is the same as when four large X ions coordinate about A at the corners of a square (Fig. 9-10C). Just as the limit 0.732 represented $\sqrt{3} - 1$, the limit 0.414 represents $\sqrt{2} - 1$.

A tetrahedral coordination of X around A (Fig. 9-11A) is favored for R_A/R_X values from 0.414 down to the limit below which "rattling" would occur, that is, 0.225. Below 0.225 and down to a "no-rattle" limit of 0.155, three X ions centered at the corners of an equilateral triangle surround and touch the A ion (Fig. 9-11B). For radius ratios below 0.155 the disparity in size between the X and A ions is so great that only two X ions on opposite sides of the A ion can simultaneously touch it (Fig. 9-11C).

So far we have been discussing *regular* coordination polyhedra, the adjective *regular* signifying that all the large anions are equidistant from the smaller, central cation. As techniques for determining crystal structures improve, however, increasing numbers of *irregular coordination polyhedra* (in which all anions are *not* equidistant from the central cation) are being discovered. Some of these are merely slight distortions of the regular polyhedra and thus involve coordina-

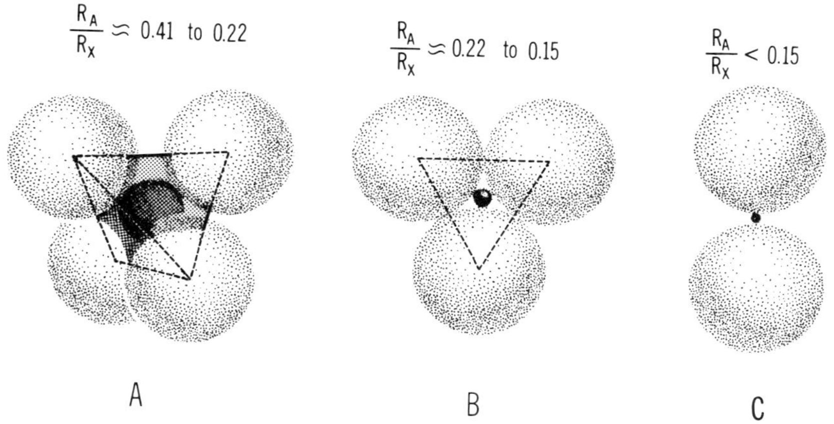

FIGURE 9-11

(A) Tetrahedral coordination of four X ions about the smaller, shaded A ion as favored for radius ratios in the range 0.41 to 0.22. Here, R_A/R_X exceeds 0.22 and consequently the X ions do not touch each other.
(B) Threefold (triangular) coordination of three X ions about an A ion as favored for the R_A/R_X range 0.22 to 0.15. For this drawing R_A/R_X exactly equals 0.155.
(C) Twofold (linear) coordination of two X ions about A as favored for $R_A/R_X <$ 0.15.

tion numbers of 2, 3, 4, 6, 8, or 12. Others are irregularly shaped polyhedra in which 5, 7, 9, or 10 anions are at differing distances from the central cation. The atoms (or ions) tangent to a central atom (or ion) are called its *first nearest neighbors* and together constitute its *first coordination shell*. In turn those atoms tangent to these first nearest neighbors are called the central atom's (or ion's) *second nearest neighbors* and together constitute its *second coordination shell*. Forces between a central atom (or ion) and those in its first coordination shell are relatively strong. Between the central atom and successively outer shells, the forces decrease rapidly.

Pauling's rules

In crystals of most minerals and inorganic compounds the bonding is appreciably ionic. Consequently, their structures largely follow the principles governing the structures of ionic crystals as elucidated by Linus Pauling. These principles, known as *Pauling's rules*, describe features which, if present in an ionic structure, tend to minimize its over-all potential energy. These rules, slightly rephrased, will be discussed at strategic points in the remainder of this chapter.

Pauling's first rule

A coordination polyhedron of anions is formed about each cation, the cation-anion distance equaling the sum of their characteristic packing radii and their radius ratio determining both the nature of the coordination polyhedron and therefore the coordination number of the cation. In other words, as we just discussed in terms of "no-rattle" limits, the maximum number of large ions X that can be simultaneously tangent to a smaller ion A is geometrically dependent on the value of R_A/R_X, their radius ratio. We can summarize for the regular coordination polyhedra as follows:

R_A/R_X	Maximum Number of X Ions around A	Coordination of X around A
~1.00	12	As for CCP (Fig. 9-8A) or HCP (Fig. 9-8D)
1.00–0.73	8	At corners of a cube (Fig. 9-9A)
0.73–0.41	6	At corners of an octahedron (Fig. 9-10A)
	4	At corners of a square (Fig. 9-10C)
0.41–0.22	4	At corners of a tetrahedron (Fig. 9-11A)
0.22–0.15	3	At corners of a triangle (Fig. 9-11B)
<0.15	2	Linear, on opposite sides of A (Fig. 9-11C)

Interstices among closest packings of spheres

Many inorganic crystal structures can be regarded as closest packings of large ions with smaller ions in the interstices. Usually the larger ions are anions that are either in cubic or hexagonal closest packing. Such closest packings were earlier discussed as consisting of close-packed layers of spheres with the spheres of the upper layer nestling into the indentations on the lower ones. Such imposition of two or more close-packed layers (Fig. 9-12) creates two types of interstices: (1) octahedral interstices (labeled *O* in Fig. 9-12), which can accommodate a sphere capable of octahedral coordination to six spheres of the closest-packed array, and (2) tetrahedral interstices (labeled *T* or *T′*), which can accommodate a sphere that can be tetrahedrally coordinated by four of the spheres of the close-packed array.

In a cubic closest-packed array of spheres the unit cell contains eight *T* sites (Fig. 9-13A) and four *O* sites (Fig. 9-13B). The total of four *O* sites includes the *O* site at the cell's center plus one-fourth of each *O* site on its 12 edges, each edge being shared by four unit cells. The unit cell also "owns" four of the closest-packed spheres, that is, one-eighth of each at its eight corners plus one-half of each centered on its six faces. In a cubic closest-packed array of spheres, therefore, the proportion of *T* sites to *O* sites to closest-packed spheres is 8:4:4 or 2:1:1.

In a hexagonal closest-packed array of spheres the proportion of *T* sites to *O* sites to closest-packed spheres is again 2:1:1. For example, a unit cell representative of such an array (Fig. 9-14A) "owns" *six* spheres, namely, the one at its center *plus* one-half of each on its upper and lower faces *plus* one-third of each on its vertical edges. (If the adjacent unit cells had been drawn in Figure 9-14A, each vertical cell edge would be seen to be shared by three such cells, each face by two.) The cell also "owns" *twelve* *T* sites. Six of these are explicitly

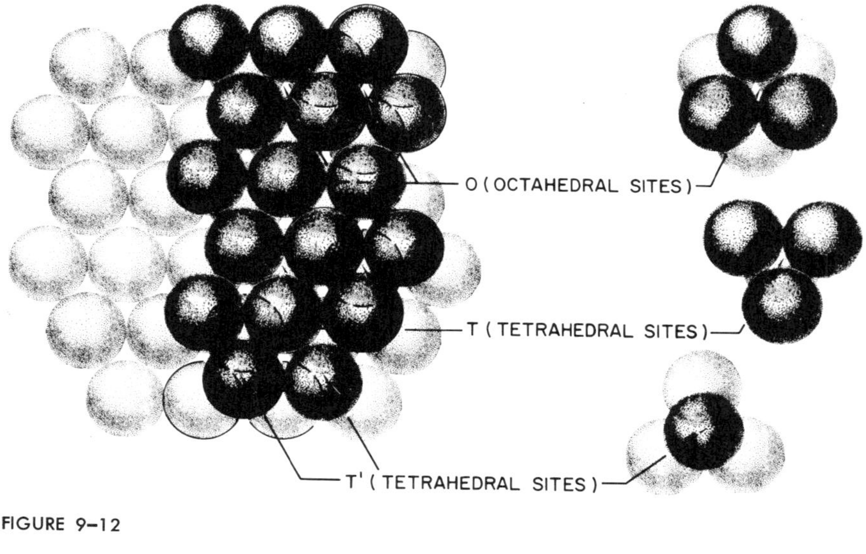

FIGURE 9–12

One closest-packed layer of X ions with a second (shaded) closest-packed layer of X ions stacked upon it. The octahedral interstices, marked *O*, are centered on a plane midway between these two layers. The tetrahedral interstices labeled *T* are centered slightly above this midway plane, whereas those labeled *T′* are slightly below it.

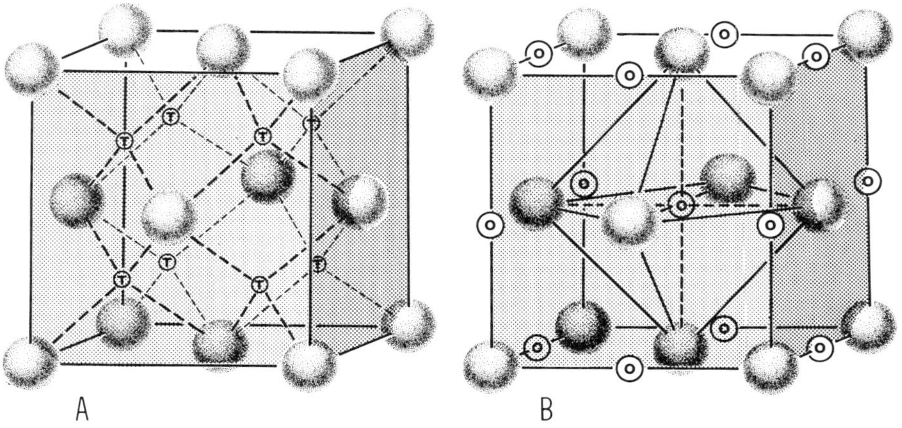

FIGURE 9–13

Cubic closest-packed arrays of spheres with the spheres shrunken so that their face-centered cubic distribution in space is apparent.
(A) Eight tetrahedral interstices such as described in Figure 9–12 are seen to be located at points *T* within a single, face-centered cubic unit cell.
(B) The circles labeled *O* mark the centers of the octahedral interstices in the unit cell. The cell thus "owns" four octahedral sites, one-quarter of each at its 12 edges plus the one at its center.

shown (as circles) in Figure 9-14A. Of the remaining six *T* sites, one lies directly above and the other directly below the sphere at the center, which together with the three spheres on the upper surface (or on the lower surface) defines a tetrahedron. The other *T* sites lie on each of the six vertical edges (at the places marked *T* on just one of these edges). The unit cell thus owns one-third of each of the twelve *T* sites located on its vertical edges plus the *T* sites directly above and below the central sphere. This unit cell also owns *six O* sites of which only two are explicitly drawn in Figure 9-14B.

Rather than have their large X ions in a cubic or hexagonal closest-packed array, some structures have them in a *simple cubic packing* (SCP) (Fig. 9-15A) with the smaller A ions located in the interstices at the center of the cube so as to be coordinated by the eight corner spheres. Thus, as already discussed in respect to Figure 9-9, such structures would be favored for R_A/R_X values of 0.73 to 1.0. The proportion of central interstices to X ions is 1:1 in this structure.

Lima-de-Faria's classification of structures

J. Lima-de-Faria (1965) classifies the classic inorganic structures according to the type of packing of the large X ions and the filling of their interstices. Thus the letter X is superscripted as X^c, X^h, X^{sc} to indicate that the X ions are in cubic closest packing, hexagonal closest packing, or simple cubic packing, respectively. Similarly, the smaller A ions are superscripted as A^t, A^o, or A^{cb} to

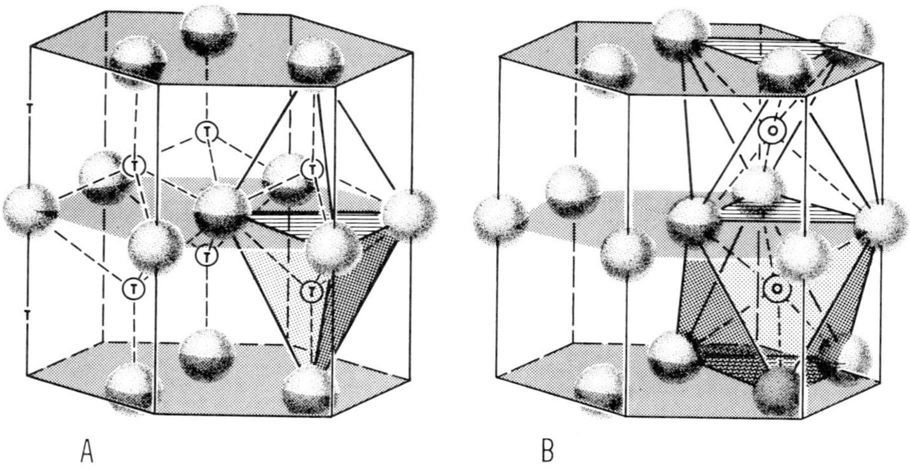

FIGURE 9–14

A representative portion of a hexagonal closest-packed array of spheres is drawn in relation to a nonprimitive unit cell which is actually a hexagonal prism. For illustrative purposes, the spheres are shrunken to less than true scale.
(A) Six of the twelve tetrahedral interstices "owned" by the unit cell are shown by circles marked *T*. The six remaining *T* sites are not drawn specifically but lie (1) on the six vertical edges at the sites marked by the uncircled *T*'s on one of these edges and (2) on a line parallel to these vertical edges but through the center of the prism. The prism thus "owns" one-third of each of the twelve *T* sites on its vertical edges plus both of the *T* sites on its vertical central line. The two *T* sites on this central line result because the sphere at the prism's center plus the three spheres on either the upper or the lower surface define two tetrahedra. The *T* sites on the vertical edges would be similarly defined if we were to draw the six neighboring unit cells which share prism faces with the cell illustrated. Then each vertical cell edge could also be seen to be shared by three unit cells. All tetrahedra will share faces as illustrated by the ruled face shared between the two tetrahedra drawn in outline; this sharing of a face brings their *T* sites close together.
(B) In this same unit cell, the octahedral sites are also illustrated. Within this unit cell there would be six *O* sites, but only two have been specifically drawn. Note that these octahedra share faces (ruled) both with each other and with octahedra in the unit cells above and below the one drawn. Figure 9–21 also shows the octahedral sites in a hexagonal closest-packed array.

indicate that they are surrounded by X ions disposed tetrahedrally, octahedrally, or at the corners of a cube.

The ratio of tetrahedral to octahedral sites to X ions in a cubic or a hexagonal closest-packed array, previously discussed to be 2:1:1, means that complete filling of these sites would yield the formulas, $A_2{}^tA^oX^c$ or $A_2{}^tA^oX^h$. A similar formula for complete filling of the eightfold sites in a simple cubic-packing array is $A^{cb}X^{sc}$. For the equally but oppositely charged A and X ions in an AX structure, depending upon whether the R_A/R_X radius ratio favors (1) tetrahedral, (2) octahedral, or (3) an eightfold coordination of X ions about A, the structure frequently adopted will be, respectively (1) A^tX^c or A^tX^h with 50 percent of the tetrahedral sites occupied, (2) A^oX^c or A^oX^h with 100 percent of the octahedral sites occupied, or (3) $A^{cb}X^{sc}$ with 100 percent of the eightfold sites occupied. In like manner an AX_2 compound, in which the A cations have twice the charge of the X anions, might adopt a structure, again depending on the radius ratio R_A/R_X, such as (1) $A^tX_2{}^c$ or $A^tX_2{}^h$ with 25 percent of the tetrahedral sites occupied, (2) $A^oX_2{}^c$ or $A^oX_2{}^h$ with 50 percent of the octahedral sites occupied, or (3) $A^{cb}X_2{}^{sc}$ with 50 percent of the eightfold sites occupied. Many of the classic structure types can be classified in this way.

AX Structures

CsCl or $A^{cb}X^{sc}$ structures ($P\,4/m\,\bar{3}\,2/m = Pm3m = O_h{}^1$)

Crystals in which the A and X ions are equally but oppositely charged will contain an equal number of each (to achieve neutrality) and therefore have the generalized formula AX. If R_A/R_X exceeds 0.73, a coordination of eight X ions about the A ion is favored. This is attained if the X ions adopt a simple cubic packing and the A ions fill the relatively large interstices between them as in the CsCl structures (Fig. 9-15A) for which Lima-de-Faria's general symbol is

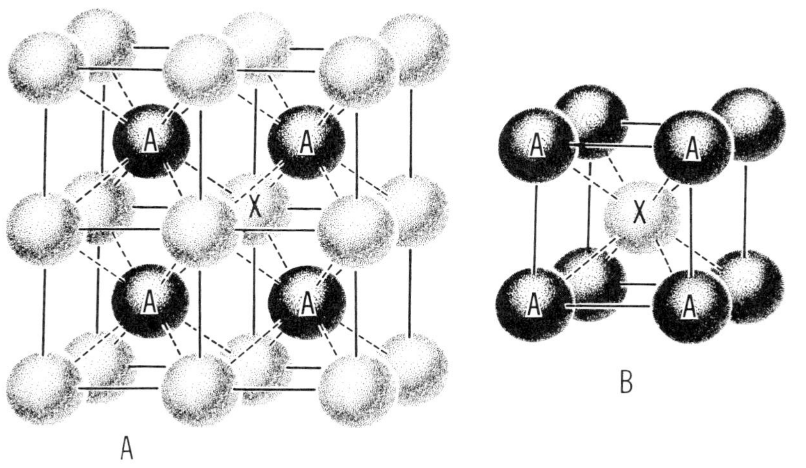

FIGURE 9–15

(A) Four unit cells of the CsCl or $A^{cb}X^{sc}$ structure, which is favored by AX compounds for which R_A/R_X lies between 0.73 and 1.0, particularly if the A cations are large and univalent. If the X anions were by themselves and tangent to each other, their packing would be of the type called *simple cubic*, each X ion being centered on the corner of a cube. Each cube shares a face with six others. This feature makes the CsCl structure unattractive to highly charged cations for reasons discussed under Pauling's third rule.
(B) Coordination of eight A ions around an X ion in the CsCl structure.

$A^{cb}X^{sc}$. If the neighboring unit cells had been drawn behind those in Figure 9-15A, eight A ions would be seen to touch each X ion (Fig. 9-15B). Thus the X ion also has a coordination number of eight. The coordination for $A^{cb}X^{sc}$ compounds can be summarized as CN 8:8, the first eight representing the coordination number for A, the second that for X. The CsCl structure is adopted by the following compounds, the parenthesized numbers representing their R_A/R_X ratio: CsCl (0.92), NH_4Cl (0.79), CsBr (0.86), and CsI (0.76). Pauling-Ahrens' radii (Table 8-5) are used here and in subsequent computations.

NaCl or A^oX^c Structures ($F\,4/m\,\bar{3}\,2/m = Fm3m = O_h^5$)

Values of R_A/R_X in (or near) the range 0.73-0.41 favor an octahedral coordination of six X ions about the A ion. Ionically bonded AX structures attain this coordination through a cubic closest packing of X ions with the smaller A ions occupying the octahedral sites labeled *O* in Figures 9-12 and 9-13B. Sodium chloride or A^oX^c structures thus result (Fig. 9-16). For these, if R_A/R_X exceeds 0.41, the A ions are large enough to prevent the X ions from touching each other and attaining true closest packing. Thus for NaCl itself (Fig. 9-16) the Cl^- ions only approximate cubic closest packing. In such structures each A is touched by six X ions and each X ion is touched by six A ions (CN: 6:6), the coordination polyhedron being an octahedron in both cases.

The NaCl structure is adopted by the following compounds, the parenthesized numbers representing their R_A/R_X ratio:

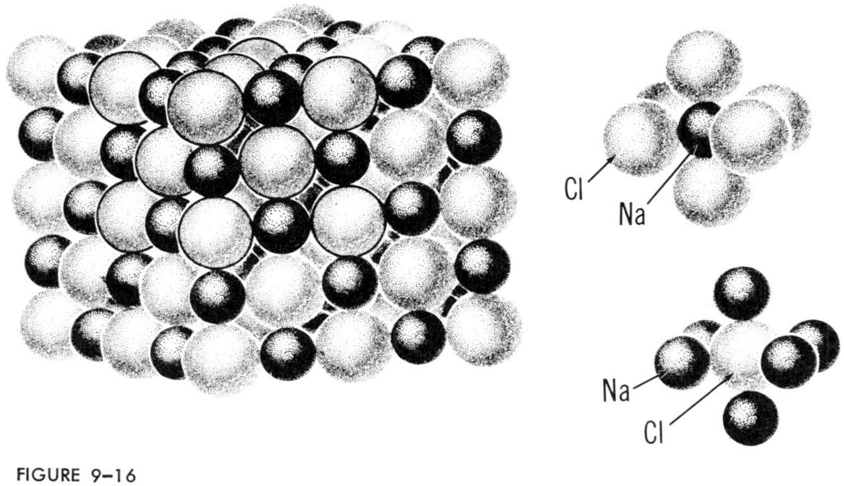

FIGURE 9–16

The NaCl or A^oX^c structure, which is favored by AX compounds in which R_A/R_X lies between 0.41 and 0.73. The octahedral coordination of Cl^- around Na^+ and of Na^+ around Cl^- in the NaCl structure is shown at the right. The ions defining one single unit cell have been outlined in black.

Halides: LiF^2 (0.57), $LiCl^2$ (0.42), $LiBr^2$ (0.39), LiI^2 (0.35), NaF (0.73), NaCl (0.54), NaBr (0.50), NaI (0.44), KF (1.00), KCl (0.73), KBr (0.68), KI (0.60), RbF (1.11), RbCl (0.81), RbBr (0.75), RbI (0.67), CsF (1.26), AgF (0.95), AgCl (0.70), AgBr (0.65).

Oxides: MgO (0.47), CaO (0.71), SrO (0.80), BaO (0.96), VO (0.63), MnO (0.57), FeO (0.53), CoO (0.51), NiO (0.49).

Sulfides: MgS (0.36), CaS (0.54), SrS (0.61), BaS (0.73), MnS (0.43), PbS (0.65).

In this list several of the radius ratios fall outside the range of 0.73 to 0.41. Thus factors in addition to the radius ratio must operate to determine the type of structure adopted. Moreover the bonding in some of these compounds, particularly the sulfides, might be appreciably covalent or metallic whereupon the radius ratio for their ions becomes less pertinent.

Pauling's second rule

The strength of an *electrostatic bond* may be defined as a cation's valence charge divided by its coordination number. In NaCl the Na^+ bond thus has a strength of $+\frac{1}{6}$. Pauling's second rule states that *an ionic structure will be stable to the extent that the sum of the strengths of the electrostatic bonds that reach an anion* (from adjacent cations) *equals the charge on that anion.* In NaCl, for example, each Cl^- anion is surrounded by six Na^+ neighbors, each of which directs a bond of strength $+\frac{1}{6}$ toward it. Consequently, the single negative unit of charge on a Cl^- anion is completely neutralized by the six $(+\frac{1}{6})$ bonds from the adjacent Na^+ ions; no portion of the positive charge from more distant cations is required to neutralize the Cl^- anion's charge. Pauling's second rule thus calls for the neutralization of charge around an anion in an ionic structure to be sharply localized; hence the rule is often known as the *electrostatic valence principle.*[3]

NiAs or A^oX^h structures ($P\,6_3/m\,2/m\,2/c = P6_3/mmc = D_{6h}{}^4$)

Those AX compounds in which the bonding is appreciably metallic may adopt an A^oX^h structure. For example, niccolite (NiAs) consists of nickel atoms filling the octahedral interstices between hexagonal closest-packed arsenic atoms (Fig. 9-17A). Consequently, six arsenic atoms octahedrally surround each nickel

[2] Donnay radius for Li^+ used.

[3] The Verwey-Heilmann *principle of maximal charge neutralization,* which extends and applies Pauling's second rule to cations, states that an ionic structure will be stable to the extent that the charge on any cation is neutralized by adjacent anions, that is, by the cation's "first coordination neighbors," rather than by the next closest anions or by more distant anions. If such neutralization of charge is to be sharply localized around cations, however, then cations with large charges must have high coordination numbers; this is the same as saying that highly charged cations must occupy the large holes in a structure because usually only these will be bounded by a large enough number of anions to neutralize locally the cation's charge. Interestingly, highly charged cations are usually small in radius and thus, on the basis of radius ratio, "seek" to occupy small holes. The two effects are at cross purposes, therefore, and either may dominate depending upon the structure.

atom; conversely, six nickels centered on the corners of a trigonal prism (Fig. 9-17B) surround each arsenic atom. Other compounds adopting the A^oX^h structure are NiS (breithauptite), FeS (pyrrhotite), CoS, VS, and compounds in which a transition metal is combined with Sn, Pb, As, Sb, Bi, Se, or Te (Evans, 1964, p. 142). Purely ionic AX compounds in which R_A/R_X equals 0.73 to 0.41 usually adopt the NaCl (A^oX^c) structure, not the NiAs (A^oX^h) structure. The reason is that in the NiAs structure the octahedra of X ions about A share faces (Fig. 9-19B), whereas in the NaCl structure these octahedra share only edges. Consequently, as will be apparent after discussion of Pauling's third rule, the NaCl structure is favored over the NiAs structure for ionic crystals.

Pauling's third rule

Pauling's third rule states that *the sharing of edges and particularly of faces by two anion polyhedra decreases the stability of an ionic crystal structure.* This is particularly true if the cation within each anion polyhedron (1) is highly charged, (2) has a small coordination number, and (3) has a radius ratio in respect to the anions that is near the lowest limit tolerated by the anion polyhedron. This third rule might be acceptably restated: *Highly charged cations "prefer"* (1) *to maintain as large a separation as possible within a structure and* (2) *to have anions inter-*

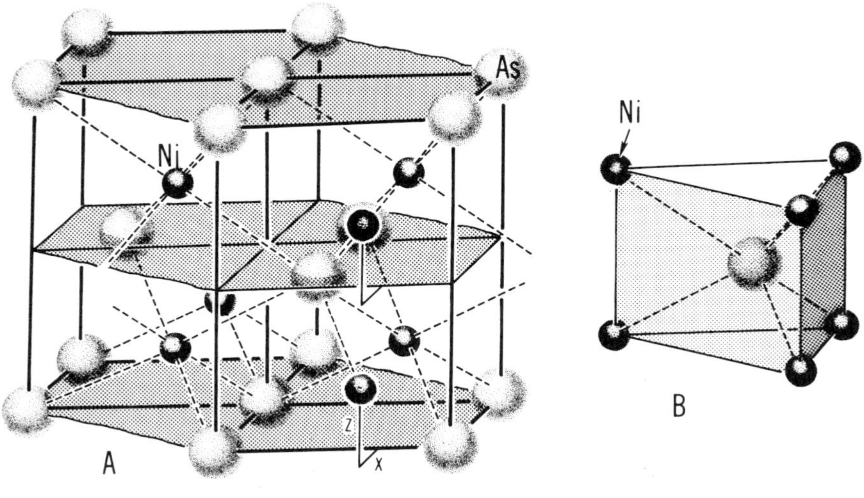

FIGURE 9–17

(A) The NiAs or A^oX^h structure as represented by two unit cells and parts of others. The As (X) atoms are hexagonally closest packed. The Ni (A) atoms fill the octahedral sites in this closest-packed array. Hence each Ni atom is centered within an octahedron which shares a face (see Fig. 9–19B) with the octahedron immediately above it and below it. Two additional Ni atoms from a unit cell closer to the reader are located relative to their x and y coordinates in this cell.
(B) The six Ni (A) atoms which surround each As (X) atom are seen to define a trigonal prism which surrounds the As (X) atom.

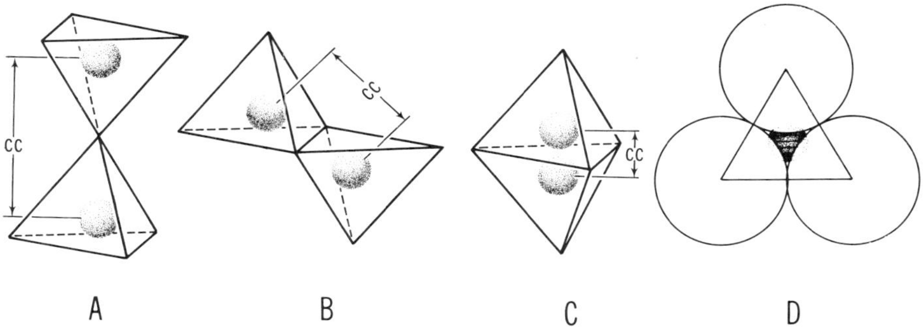

FIGURE 9–18

In (A), (B), and (C) the corners of each tetrahedron represent the centers of O^{2-} ions that are tetrahedrally coordinated to the Si^{4+} cation (shown as a small sphere) at the center of the tetrahedron. The center-to-center distance *CC* between Si^{4+} cations is greatest for corner sharing (A), less for edge sharing (B), and least for face sharing (C).
(D) View of shared face if eye were located at upper Si^{4+} ion in (C). Repulsive force of lower Si^{4+} ion for upper one "shines" directly through the ruled "window" left unscreened by the three large anions.

vening between them so as to screen one from the other. Either feature decreases a crystal's potential energy by minimizing the repulsive forces existing between nearby cations.

To illustrate Pauling's third rule more graphically, we consider the tetrahedron that four O^{2-} ions typically form around a Si^{4+} ion in silicate crystals. In such crystal structures two tetrahedra often share a corner (Fig. 9-18A) but not an edge (Fig. 9-18B) or a face (Fig. 9-18C). Two reasons dictate this: (1) the cation-to-cation distance *CC* is greatest for corner sharing (Fig. 9-18A) and least for face sharing (Fig. 9-18C), and (2) the O^{2-} ions intervene between and thus screen one cation from the other most effectively for corner sharing and least for face sharing. To elaborate on this second point, in Figure 9-18A the shared O^{2-} ion, which the shared corner actually represents, lies directly between the two cations and thereby prevents each from "shining" its repulsive force directly on the other. In Figure 9-18B, however, the two shared O^{2-} ions, which a shared edge actually represents, screen the repulsive forces between cations less effectively. Poorest screening occurs if a face is shared by two coordination polyhedra; thus Figure 9-18D is a drawing of the three O^{2-} ions that the shared face in Figure 9-18C represents. The reader will note the "window" (ruled) through which each cation "shines" its repulsive force on the other. Actually, of course, the "window" area offers a slight degree of screening because an orbital electron's radial probability density never quite drops off to zero as distance from the atom's center increases.

The distance from the center of a polyhedron to its edges or faces increases as we successively consider a tetrahedron, octahedron, cube, and cubo-octahedron, all with identical edge lengths. Consequently, the sharing of edges or faces brings the cations at their centers particularly close together for polyhedra representing low coordination numbers. In other words two tetrahedra are less likely to share a face or edge than are two octahedra, two octahedra less likely than two cubes, and so on.

The principles underlying Pauling's third rule explain why ionic crystals of type AX, where the radius ratio R_A/R_X is favorable for octahedral coordination, usually adopt the A^oX^c (NaCl) structure for which neighboring octahedra merely share edges (Fig. 9-19A) rather than A^oX^h structures for which they share faces (Fig. 9-19B). It also explains why the CsCl structure, in which the anion polyhedra are face-sharing cubes, appears largely confined to halides of the large univalent cation, cesium. Cations, which represent greater concentrations of positive charge (for example, multivalent cations or even univalent cations smaller than cesium) tend to avoid the face-sharing CsCl structure.

Pauling's fourth rule

Pauling's fourth rule states that *in a crystal structure containing different cations, those of high valency and small coordination number tend not to share polyhedron elements with each other.* This fourth rule is merely an extension of the third and the considerations that justified the third rule also justify the fourth.

Pauling's fifth rule

Pauling's fifth rule states that *the number of essentially different kinds of constituents in a crystal tends to be small.* If the simplifying assumption is adopted

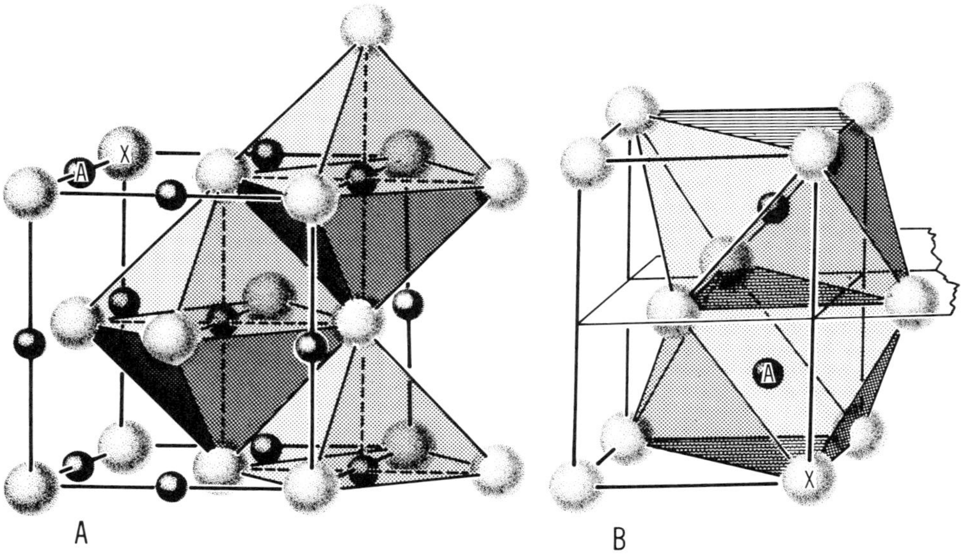

FIGURE 9–19

The octahedra formed by the large atoms around the smaller atoms share corners in the NaCl structure in (A) but share faces in the NiAs structure shown in (B).

that most crystal structures consist of a closest packing of the large ions, or a variation thereof, with the smaller ions in the interstices, then the fifth rule may be restated as: *the number of types of interstitial sites present within a periodically regular packing of anions tends to be small.* For example, in structures based on a closest packing of anions, four types of sites or interstitial voids exist: (1) Very small 3-fold coordinated voids as created when three anions touch each other, (2) the *T* sites, (3) the *O* sites, and (4) the 12-fold sites possible for a cation large enough to fill the place of an anion in the closest-packed array.

A^tX^c and A^tX^h structures

For radius ratios of about 0.22 to 0.41, an AX compound may adopt the A^tX^c ($F\bar{4}3m = T_d^2$) or the A^tX^h ($P6_3mc = C_{6v}^4$) structure. In these the A ions occupy 50 percent of the tetrahedral interstices in a cubic or a hexagonal closest-packed array of the larger X ions. The A^tX^c structure is actually that of sphalerite (Fig. 9-6); the A^tX^h structure is that of wurtzite (Fig. 9-7). Both of these structures were discussed under covalent crystals and, indeed, for ionic compounds adopting these structures the bonding is likely to be appreciably covalent. The nature of the coordination (CN 4:4) in both structures is such that the X atoms are located at the corners of a tetrahedron centered about A, as are the A atoms about X. Such tetrahedral coordination is especially conducive to covalent bonding that involves overlap of orbitals with tetrahedral configurations such as the hybrid sp^3 orbitals. The A^tX^c or sphalerite structure is adopted by a large number of AX compounds, namely: CuCl, CuBr, CuI, BeS, BeSe, BeTe, ZnS, ZnSe, ZnTe, CdS, CdSe, CdTe, AlP, AlAs, AlSb, GaP, GaAs, and GaSb. The A^tX^h or wurtzite structure is adopted by BeO (bromellite), ZnO (zincite), CdS (greenockite), CdSe, AlN, GaN, InN, and MgTe. In these compounds bonding other than ionic is appreciable and, in some cases, dominant. Thus solely on the basis of the radius ratio for Zn^{2+} and O^{2-} (namely, 0.53), ZnO would expectably adopt the A^oX^c (NaCl) structure; evidently the degree of covalency of its bonds disposes it toward the wurtzite structure.

AX_2 Structures

For ionic compounds in which the charge on the A ion is double that of the oppositely charged X ion, the number of X ions in their structure must be double that of A if electrostatic neutrality is to be achieved. Such a proportion of X to A ions results if the A ions fill only 50 percent of the octahedral interstices or 25 percent of the tetrahedral interstices among the X ions of either a cubic or a hexagonal closest-packed array or a modification thereof. The Lima-de-Faria symbolism for these structures is thus $A^oX_2^c$, $A^oX_2^h$, $A^tX_2^c$, or $A^tX_2^h$, respectively. If we remember that the proportion of tetrahedral to octahedral interstices to X atoms in a closest-packed array is 2:1:1, the subscripts in the preceding generalized formulas indicate percentage occupancy of the *O* sites as 50 percent and of the *T* sites as 25 percent.

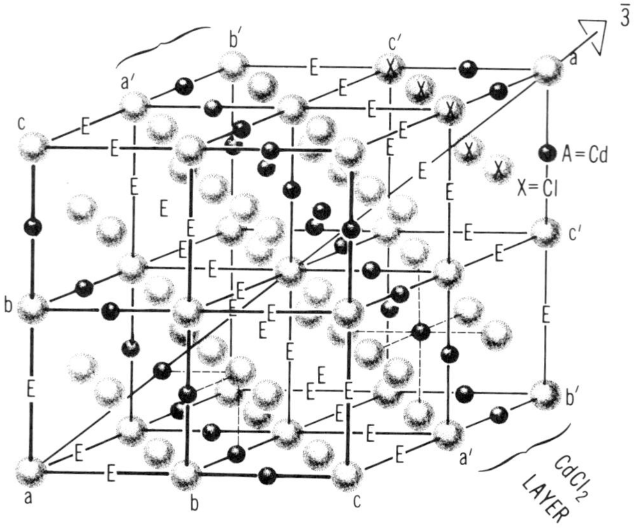

FIGURE 9–20

The $CdCl_2$ or $A^oX_2{}^c$ structure in which the A cations occupy only one-half of the octahedral sites in a cubic closest-packed array of X anions. The empty octahedral sites (labeled *E*) are distributed so that, of the four $\bar{3}$ axes that would exist if all octahedral sites were occupied (as in the NaCl or A^oX^c structure), only one $\bar{3}$ axis remains as shown. The X anions are distributed in closest-packed planes that are perpendicular to the $\bar{3}$ axis. Consequently, the XAX "sandwiches" or $CdCl_2$ layers that compose the structure are in effect giant planar molecules normal to the $\bar{3}$ axis and van der Waals bonds link one planar molecule to the other.

The $A^oX^c{}_2$ or $CdCl_2$ structure ($R\,\bar{3}\,2/m = R\bar{3}m = D_{3d}{}^5$)

In this structure the larger X or Cl^- ions are in a cubic closest-packed array but are drawn to reduced size in Figure 9-20 for illustrative purposes. The smaller A or Cd^+ ions (small, dark spheres) occupy only 50 percent of the octahedral sites, being arranged in layers normal to a $\bar{3}$ axis of the face-centered cube that summarizes the cubic closest-packing distribution of the X ions. Parallel to these layers of Cd^{2+} ions are layers of empty octahedral sites (labeled *E* in Fig. 9-20). The coordination is 6:3, six X ions touching each A ion and three A ions touching each X ion. The Cd—Cl bond represents a resonance between ionic and covalent bonding. Consequently, the Cd—Cl distance is less than the sum of their ionic radii. Van der Waals bonding occurs between the chlorine atoms of one $CdCl_2$ layer (Fig. 9-20) and the chlorine atoms of the neighboring $CdCl_2$ layer. The weakness of this interlayer bonding results in an easy breakage (cleavage) between layers as well as a much larger coefficient of thermal expansion normal to the layers as compared to parallel to the layers.

The $A^oX^h_2$ or CdI_2 structure ($P\,\bar{3}\,2/m\,1 = P\bar{3}m1 = D_{3d}^3$)

In this structure the X or I^- ions are in a hexagonal closest-packed array with only one of the two octahedral sites in the unit cell occupied by the A or Cd^{2+} ions (Fig. 9-21). Like the $A^oX_2^c$ structure, this structure consists of stacked XAX "sandwiches," each consisting of a layer of A ions with a layer of X ions on either side. Whereas the bonding between the A and X ions is ionic, that between one XAX sandwich and another is a weak van der Waals attraction. Consequently, crystals with the structures $A^oX_2^c$ or $A^oX_2^h$ tend to break along the contact planes between adjacent "sandwiches" and to grow as flakes parallel to these planes.

$A^{cb}X_2^{sc}$ or CaF_2 (fluorite) structure ($F\,4/m\,\bar{3}\,2/m = Fm3m = O_h^5$)

If the radius ratio R_A/R_X exceeds 0.73, a coordination of eight X atoms around one A is favored, thus such an AX_2 compound may adopt a structure resembling that of CsCl except that only every other cube formed by the simple

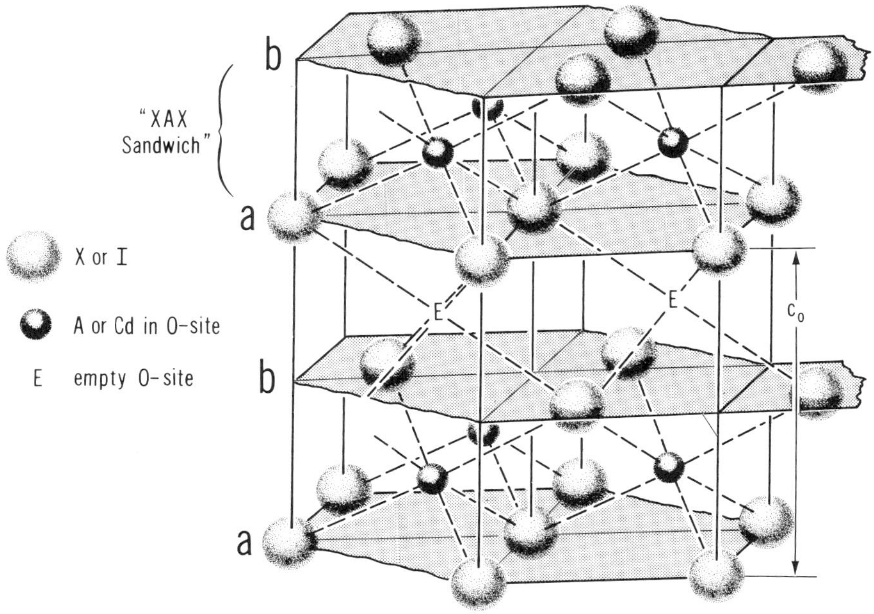

FIGURE 9–21

The CdI_2 or $A^oX_2^h$ structure in which the A ions occupy one-half of the octahedral sites in an hexagonal closest-packed array of X ions. Such 50 percent occupancy eliminates the sharing of faces between occupied octahedra such as occurs in the NiAs structure where Ni occupies *all* the octahedral sites.

cubic packing of the X ions has an A ion at its center. In this structure (Fig. 9-22A), which is actually that of CaF_2 (fluorite), a cube with an A (Ca^{2+}) cation at its center shares a face only with a vacant cube; thus cation-cation repulsion between these divalent cations is minimized. The ions labeled in Figure 9-22A are redrawn (with others added) in Figure 9-22B to show fluorite's conventionally drawn unit cell.

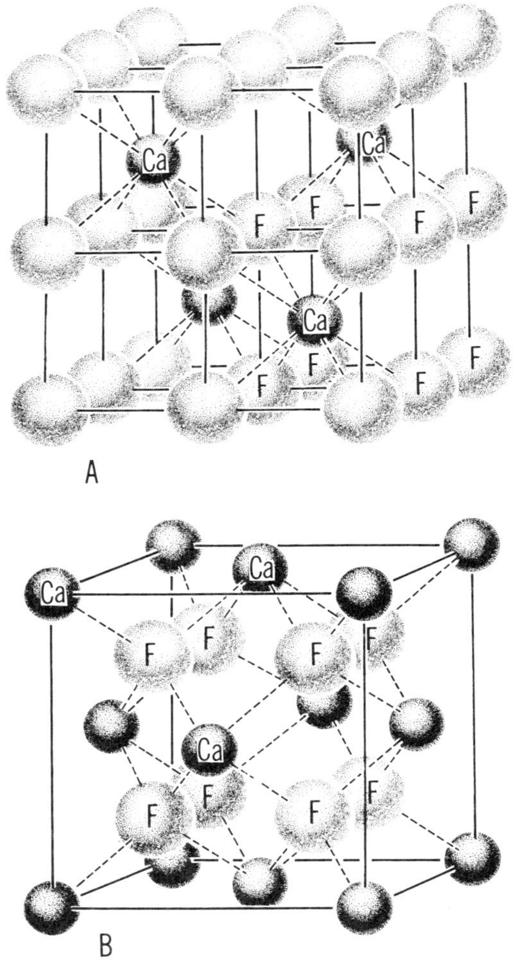

FIGURE 9–22

(A) Structure of fluorite, CaF_2, in which the F^- ions are in simple cubic packing and the Ca^{2+} ions occupy the interstices at the center of every other cube.
(B) Standard unit cell for fluorite in which the origin is chosen at the center of a Ca^{2+} ion. The eight F^- ions correspond to those eight F^- ions that were labeled in (A). Note the face centering of the Ca^{2+} ions.

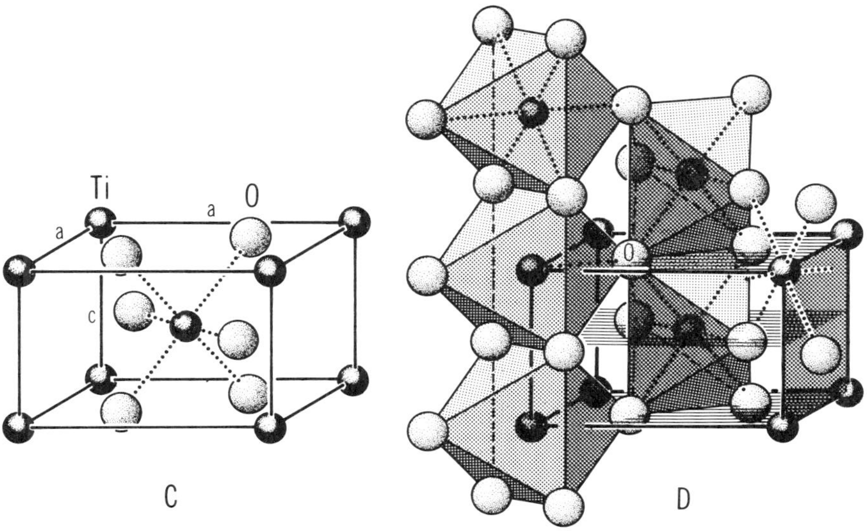

FIGURE 9-22 Continued

(C) Structure of rutile TiO_2 which is frequently adopted by AX_2 compounds for which the ratio R_A/R_X lies between 0.41 and 0.73.
(D) Two of the edge-sharing chains of octahedra in rutile are shown running parallel to the c axis. The oxygen atom labeled O is coordinated by (and coplanar with) the three titaniums that are joined to it by dotted lines. The Ti–O–Ti angle is 120° or close thereto.

Rutile (TiO_2) structure ($P\ 4_2/m\ 2_1/n\ 2/m = P4_2/mnm = D_{4h}^{14}$)

For radius ratios of about 0.73 to 0.41 an AX_2 compound may adopt the rutile structure (Fig. 9-22C) in which A (Ti^{4+}) is octahedrally coordinated by six X (O^{2-}) ions. As the space group indicates, such structures are tetragonal. The unit cell for rutile itself has the dimensions, $a_0 = 4.59$, $c_0 = 2.96$ Å. As drawn in Figure 9-22C, a titanium atom occurs at the corners and center of the unit cell. Six oxygens coordinate octahedrally around each titanium and each oxygen is bonded to three titaniums. For AX_2 compounds with the rutile structure the coordination is 6:3, whereas for fluorite it is 8:4. The rutile structure consists of edge-sharing chains of octahedra (Fig. 9-22D) that run parallel to the c axis; octahedra of neighboring chains share corners.

Isostructuralism

Two or more compounds whose atoms are arranged in the same type of crystal structure are said to be *isostructural.* Thus CaF_2 (fluorite) is isostructural with $BaCl_2$ because if each calcium in Figure 9-22A were labeled barium and each fluorine were labeled chlorine, Figure 9-22A would then represent the $BaCl_2$ structure. Isostructuralism is occasionally called *isotypism* or, less desirably, *isomorphism*. A large number of compounds are isostructural with fluorite and with rutile, these being listed below with their Pauling cation-to-anion radius ratio in parentheses.

Fluorite Structures		Rutile Structures	
HfO_2(0.56)	CaF_2(0.71)	GeO_2(0.39)	SnO_2(0.51)
ZrO_2(0.56)	CdF_2(0.73)	MnO_2(0.43)	NiF_2(0.52)
HfF_2(0.59)	ThO_2(0.73)	VO_2(0.45)	NbO_2(0.53)
ZrF_2(0.59)	$BaCl_2$(0.74)	RuO_2(0.48)	CoO_2(0.54)
$SrCl_2$(0.62)	HgF_2(0.83)	TiO_2(0.49)	FeF_2(0.56)
PrO_2(0.66)	SrF_2(0.84)	IrO_2(0.49)	ZnF_2(0.56)
CeO_2(0.67)	PbF_2(0.90)	OsO_2(0.49)	PbO_2(0.60)
UO_2(0.69)	BaF_2(1.01)	WO_2(0.50)	MnF_2(0.60)
		MoO_2(0.50)	TeO_2(0.65)
		MgF_2(0.50)	

Many of the fluorite structures have radius ratios that are less than 0.73, the "nonrattle" limit for 8-fold coordination. Thus factors in addition to radius ratio must be operative in determining which structure is adopted. Except for GeO_2, the radius ratios of the rutile structures fall in the range 0.73 to 0.41, the range theoretically expected for octahedral coordination.

Model structures

Two isostructural compounds are called *model structures* if the valences or charges on the atoms in the one structure are a given multiple of the comparable atoms in the second structure. The isostructures $Ca^{2+}F_2^{-}$ and $U^{4+}O_2^{2-}$ thus constitute a *model pair*.

A_2X Structures

If the X ions are closest packed, A_2X structures might be of the type $A_2{}^tX^c$ ($F\,4/m\,\bar{3}\,2/m$) or $A_2{}^tX^h$ ($P6_3/m\,2/m\,2/c$) in which the tetrahedral interstices are completely filled by the A ions. Actually the $A_2{}^tX^h$ structure is both unknown and unlikely among ionic compounds because if the X ions are in a hexagonal closest-packed array, the tetrahedra share faces (Fig. 9-14A); therefore electrostatic repulsion between the A cations would contribute significantly to the crystal's potential energy. However, the $A_2{}^tX^c$ structure (Fig. 9-23) does not possess this handicap, the tetrahedra merely sharing edges, and is adopted by a number of appreciably ionic compounds including Li_2O, Li_2S, and Na_2O as well as other oxides, sulfides, selenides, and tellurides of alkali metals.

Antistructuralism

Comparison of the $A_2{}^tX^c$ structure (Fig. 9-23) with the fluorite structure (Fig. 9-22B) reveals their structures to be opposites; that is, at the sites where one has cations, the other has anions, and vice versa. Structures so related may

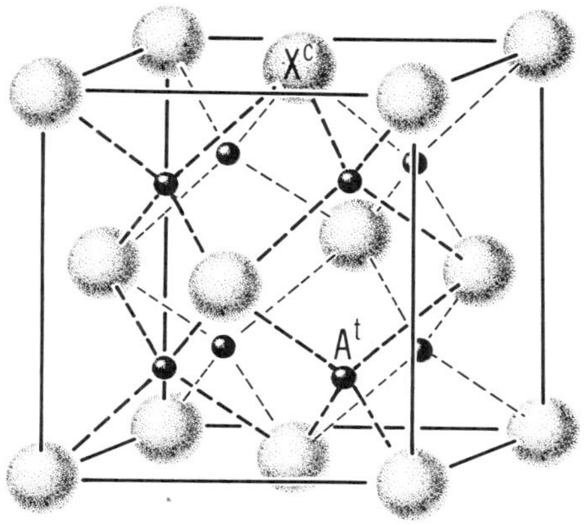

FIGURE 9–23

Antifluorite or $A_2{}^tX^c$ structure that is adopted by Li_2O, Li_2S, and Na_2O. Compare with the fluorite structure (Fig. 9–22B).

be called *antistructures* or *antitypes*. Thus the $A_2{}^tX^c$ structure in Figure 9-23 is usually called the *antifluorite structure.*

A_2X_3 Structures

Corundum (Al_2O_3) structure ($R\ \bar{3}\ 2/c = R\bar{3}c = D_{3d}{}^6$)

Typical of several A_2X_3 structures is that of corundum, Al_2O_3. Let Figure 9-24 represent a section from this structure cut parallel to $(2\bar{1}\bar{1}0)$. Normal to the page, the section is 2.38 Å thick, whereas parallel to the page its repeat distances are 13.00 Å and 8.24 Å as shown. The corundum structure repeats every 2.38 Å along the direction normal to $(2\bar{1}\bar{1}0)$ and the number by each atom's position in Figure 9-24 denotes the percentage of this repeat distance at which each atom is centered above (plus values) or below (negative values) the plane containing the centers of the aluminum atoms. Thus an oxygen marked $\overline{35}$ is vertically below the plane of aluminums by a distance equal to 35 percent of 2.38 Å.

The oxygens in the corundum structure are approximately in hexagonal closest packing. Thus the oxygens at the same level as those labeled *a* in Figure 9-24 represent a single closest-packed layer, the layer atop this is represented by those labeled *b*, the next layer by those labeled *a′*, and so on. The sequence of stacking of essentially closest-packed layers in Figure 9-24 is thus *a, b, a′, b′, a″, b″*, and *a*. But the *a′* layer does not directly overlie the *a* layer of oxygens because atom *a* has a height below the reference plane of $\overline{15}$ whereas the comparable atom *a′* has a height of $\overline{19}$. The aluminum atoms fill two-thirds of the octahedral interstices in the structure, the *E*'s indicating empty octahedral sites. The dashed rectangle in Figure 9-24 contains two octahedra that share a face, that is, hold three *a′* oxygens (heights: $\overline{15}$, 31, $\overline{15}$) in common. The aluminum

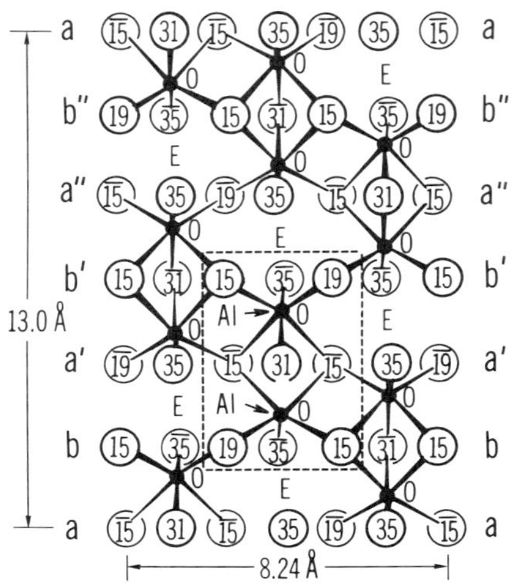

FIGURE 9–24

Structure of corundum, Al_2O_3, projected on $(2\bar{1}\bar{1}0)$. The plane of the paper is presumed to pass through the centers of the aluminums, and the heights of the oxygens (the large open circles) above or below this plane are cited in terms of percentages of the repeat translation normal to $(2\bar{1}\bar{1}0)$ in the structure. The aluminums are approximately in octahedral coordination. Where two aluminums are in octahedra which share a face, for example, the two labeled aluminums above and below the shared octahedral face defined by the oxygens labeled $\overline{15}$, 31, and $\overline{15}$, they mutually repel each other so that neither is entered within its octahedron.

ions are not centered in the octahedra because they repel each other and thereby increase the distance between themselves.

Isostructural with corundum are hematite (Fe_2O_3), eskolaite (Cr_2O_3), karelianite (V_2O_3), and the compounds Ti_2O_3, Rh_2O_3, and Ga_2O_3. All possess a true rhombohedral lattice.

$A_mB_nX_z$ Structures

Space permits the description of only a few of the structures possible for compounds of the composition $A_mB_nX_z$, where A and B represent metallic elements and X a nonmetal.

Perovskite structure ($P\ 2_1/c\ 2_1/m\ 2_1/n = Pcmn = D_{2h}^{16} = V_h^{16}$)

Many ABX_3 compounds adopt the structure of perovskite, $CaTiO_3$ (Fig. 9-25A), which, to a first approximation, consists of cubic closest-packed oxygen ions with, however, calcium ions in the place of one-fourth of these oxygens. Titanium ions then occupy one-fourth of the octahedral interstices in this cubic closest-packed array of oxygens and calciums, filling those octahedral sites not adjacent to calcium. The conventional unit cell choice for perovskite, generally selected so that the origin of the lattice is centered on titanium ions (Fig. 9-25B), shows less clearly that the oxygens and calciums approximate a cubic closest-packing, that is, a face-centered cubic, array. The calciums, being closest packed with the oxygens, are coordinated by twelve oxygens; titanium is coordinated by six oxygens. Each oxygen touches four calcium and two titanium ions.

If calcium were as large as oxygen and thus precisely fit its 12-fold site, then the line joining the centers of oxygens O_1 and O_2 in Figure 9-25B would equal $(2R_O + 2R_{Ca})$ or 1.414 times the cube's edge. The cube's edge in turn equals $(2R_O + 2R_{Ti})$. And thus

$$R_O + R_{Ca} = 1.414(R_O + R_{Ti}) \qquad \text{(Eq. 9-1)}$$

indicates the ideal relationship between radii for ions in the perovskite structure. However, the structure tolerates a degree of variation so that Equation (9-1) may be modified to read

$$R_O + R_{Ca} = t \cdot 1.414(R_O + R_{Ti}) \qquad \text{(Eq. 9-2)}$$

where t is the so-called *tolerance factor* and usually lies between 0.8 and 1.0 in value. If X represents O^{2-} ions in this ABX_3 structure, the valences of the A and B ions must total six. Possible values are thus 1 + 5, 2 + 4, or 3 + 3. If X represents F^- ions, the valences of A plus B must total three or 1 + 2.

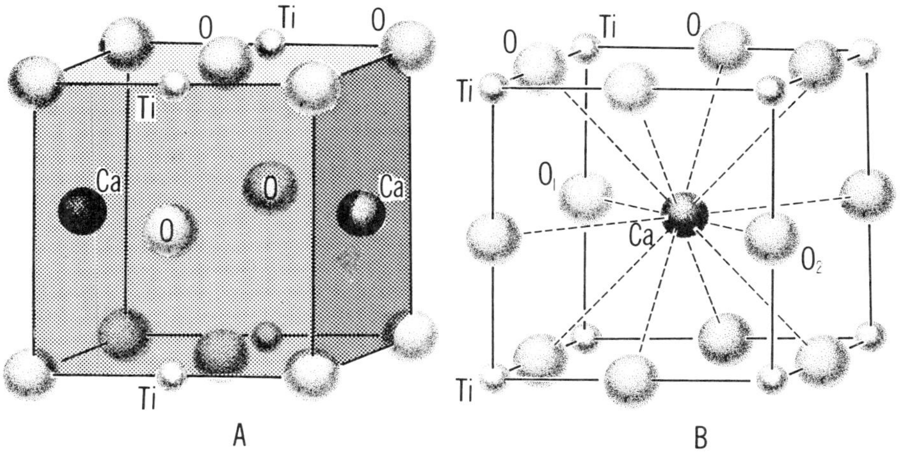

FIGURE 9–25

(A) Structure of perovskite, $CaTiO_3$, adopted by many ABX_3 compounds. In it the calciums and oxygens *together* approximate a cubic closest-packed array as witnessed by their face-centered cubic distribution. The titaniums then occupy those octahedral interstices in the array that do not neighbor the calciums.
(B) Unit cell for perovskite drawn with the origin centered on the titanium atoms. The 12-fold coordination of Ca^{2+} is now apparent.

The ideal isometric structure for perovskite illustrated in Figure 9-25 seems possible only in those few compounds for which the tolerance factor t lies between 0.90 and 1.0. For t less than 0.90, deformed perovskite structures are favored in which the atoms are slightly displaced from their ideal positions, the symmetry being thereby reduced. For example, perovskite itself ($t = 0.81$) appears to be either orthorhombic or monoclinic and thus only pseudoisometric. Some of the compounds adopting the perovskite structure, either ideal or deformed, are listed below; the brackets enclose the valences of the A and B ions in the compounds that are listed after these brackets. The parenthesized numbers represent the tolerance factor t for the compound as computed by solving Equation (9-2) for t and inserting the revised Pauling radii for the ions.

[1 + 5]: $NaNbO_3$ (0.80), $KNbO_3$ (0.92).

[2 + 4]: $CdTiO_3$ (0.81), $CaTiO_3$ (0.81), $SrTiO_3$ (0.86), $PbTiO_3$ (0.88), $BaTiO_3$ (0.93), $CaZrO_3$ (0.77), $SrZrO_3$ (0.81), $PbZrO_3$ (0.84), $BaZrO_3$ (0.88), $CaSnO_3$ (0.80), $SrSnO_3$ (0.84), $BaSnO_3$ (0.92), $CdCeO_3$ (0.72), $CaCeO_3$ (0.72), $SrCeO_3$ (0.76), $PbCeO_3$ (0.79), $BaCeO_3$ (0.83), $SrHfO_3$ (0.82), $BaHfO_3$ (0.89), $BaPrO_3$ (0.84), $BaThO_3$ (0.80).

[3 + 3]: $YAlO_3$ (0.86), $LaMnO_3$ (0.87), $LaFeO_3$ (0.88), $LaCrO_3$ (0.88), $LaAlO_3$ (0.94).

[1 + 2]: $KMgF_3$ (0.95), $PbMgF_3$ (0.90), $KZnF_3$ (0.91), $KNiF_3$ (0.93).

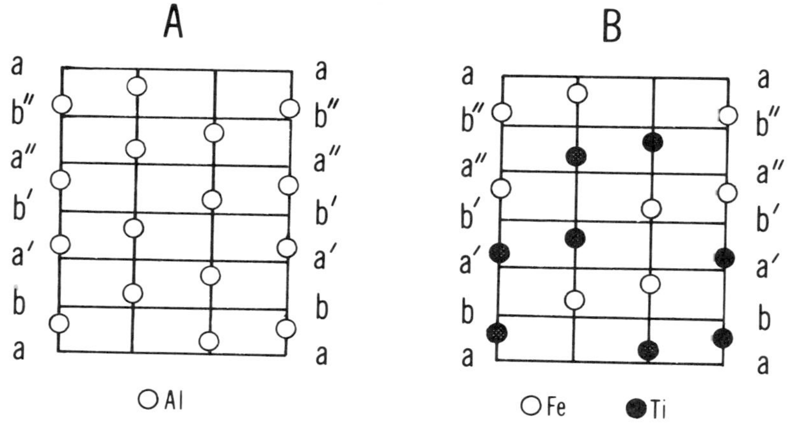

FIGURE 9–26

(A) Schematic representation of the projection on $(2\bar{1}\bar{1}0)$ of the corundum structure in Figure 9–24. Lines *aa*, *bb*, *a′a′*, and so on, represent the similarly labeled lines of oxygen ions in Figure 9–24, each line being actually an edge view of a single, closest-packed layer of oxygens. The circles schematically represent the Al^{3+} sites at 0 heights in Figure 9–24.

(B) Similar schematic representation of the structure of ilmenite, $FeTiO_3$, in which half the aluminum sites are now occupied by Fe^{2+}, the other half by Ti^{4+}.

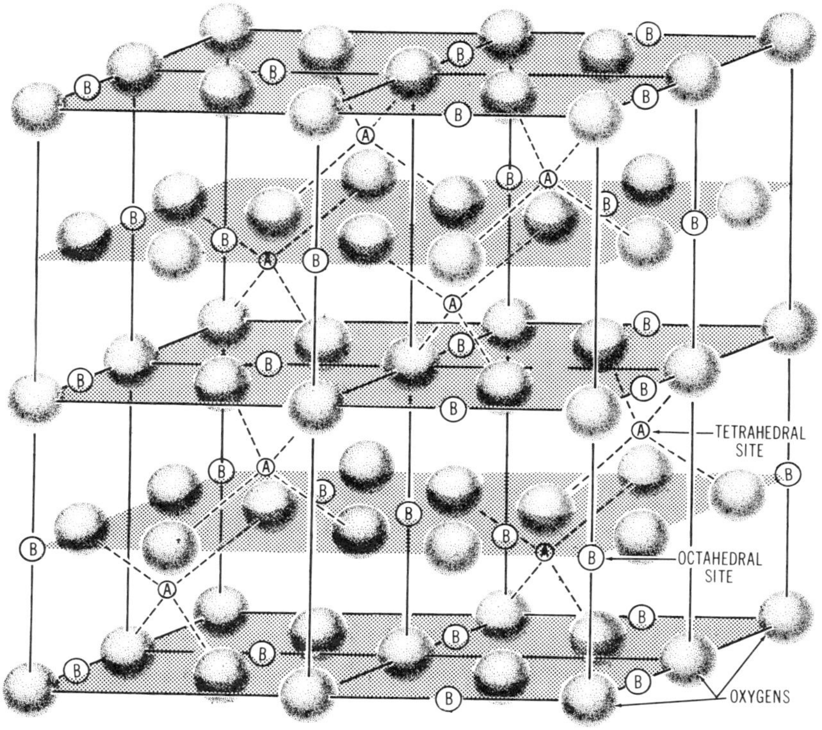

FIGURE 9–27

Idealized spinel structure showing the oxygen ions in a cubic closest-packed array. Those tetrahedral interstices that are occupied by divalent metal ions A are so indicated as are those octahedral interstices normally occupied by trivalent ions B. In *inverse spinels* the tetrahedral sites are occupied by trivalent ions; then half of the octahedral sites labeled B are occupied by divalent ions and the remaining half by the same trivalent ions as occupy the tetrahedral sites.

Ilmenite structure ($R\,\overline{3} = C_{3i}^{\,2}$)

ABX_3 compounds for which the tolerance factor t is below 0.75 are likely to adopt the ilmenite $FeTiO_3$ structure. This structure is closely related to that of corundum. However, in place of aluminum in the octahedral sites (Fig. 9-26A), ilmenite contains iron and titanium in alternating layers (Fig. 9-26B). This reduces the space-group symmetry from $R\,\overline{3}\,2/c$ to $R\,\overline{3}$. The ilmenite structure is adopted by the following compounds, the valences being bracketed and the tolerance factor t being parenthesized.

[2 + 4]: $MgTiO_3$ (0.70), $NiTiO_3$ (0.71), $CoTiO_3$ (0.72), $FeTiO_3$ (0.73), $MnTiO_3$ (0.75).

[1 + 5]: $LiNbO_3$ (0.70).

[3 + 3]: Al_2O_3, Ti_2O_3, V_2O_3, Fe_2O_3, Rh_2O_3, Ga_2O_3 ($t = 0.71$ for all these, which are corundum structures, because the same cation occupies both the A and the B sites).

Spinel structures ($F\,4_1/d\,\bar{3}\,2/m = Fd3m = O_h^7$)

The spinel structure consists of an essentially cubic closest-packed array of oxygen ions with metal cations occupying one-eighth of the tetrahedral interstices and one half of the octahedral interstices. In *normal spinels* (Fig. 9-27) divalent metal ions A occupy the tetrahedral sites and trivalent ions B occupy the octahedral sites. The general formula is thus $A^tB_2^oO_4^c$ and a unit cell ($a_0 = 8.0 - 8.5$ Å) "owns" 32 oxygens, 8 A atoms, and 16 B atoms. *Inverse spinels* are unusual in that the divalent A ions switch places with half of the trivalent B ions; their general formula is thus $B^t[AB]^oO_4^c$. For either spinel structure the space group is $F\,4_1/d\,\bar{3}\,2/m$ ($Fd3m$). Natural and synthetic compounds with the spinel structure are $MgAl_2O_4$ (spinel), $Fe^{2+}Al_2O_4$ (hercynite), $ZnAl_2O_4$ (gahnite), $MnAl_2O_4$ (galaxite), $Fe^{2+}Cr_2O_4$ (chromite), $NiFe_2^{3+}O_4$ (trevorite), $MnFe_2^{3+}O_4$ (jacobsite), $MnCr_2O_4$, and numerous others. Those with inverse spinel structures are $[Fe^{3+}]^t[MgFe^{3+}]^oO_4^c$ (magnesioferrite), $[Fe^{3+}]^t[Fe^{2+}Fe^{3+}]^oO_4^c$ (magnetite), and so on. Magnetite, Fe_3O_4, is interesting because the same element is involved in two valence states. A degree of covalency of bonding in spinel structures may be one of the factors that determines whether the normal or inverse structure is adopted but this is not fully understood. R. C. Evans (1964, p. 173) notes that Zn^{2+} and Cd^{2+} as well as Fe^{3+}, In^{3+}, and Ga^{3+} prefer tetrahedral coordination but that other trivalent or quadrivalent ions prefer the octahedral sites. The presence of Fe^{3+}, In^{3+}, and Ga^{3+} is therefore likely to produce inverse spinels (unless Zn^{2+} and/or Cd^{2+} are also present to pre-empt the tetrahedral sites and confine the trivalent ions to the octahedral sites).

Derivative Structures

M. J. Buerger defined a derivative structure as being any structure related to a basic structure by the suppression of one or more of the symmetry operations in this basic structure's space group. Consequently, the derivative structure's space group will be a subgroup of that of the basic structure. The derivative structure's point group may be the same as for the basic structure, as, for example, if a mirror plane in the basic structure simply becomes a glide plane, or an X-fold axis becomes the comparable X_n screw axis. However, if a mirror plane or glide plane is completely suppressed or an X-fold axis is drastically altered or eliminated, the derivative structure's point group will be a subgroup (see Table 1-2) of the basic structure's point group.

In one type of derivative structure, called by Buerger a *substitution structure*, two or more different chemical elements may substitute for the one present in the basic structure. We can illustrate this with the structure of chalcopyrite, $CuFeS_2$,

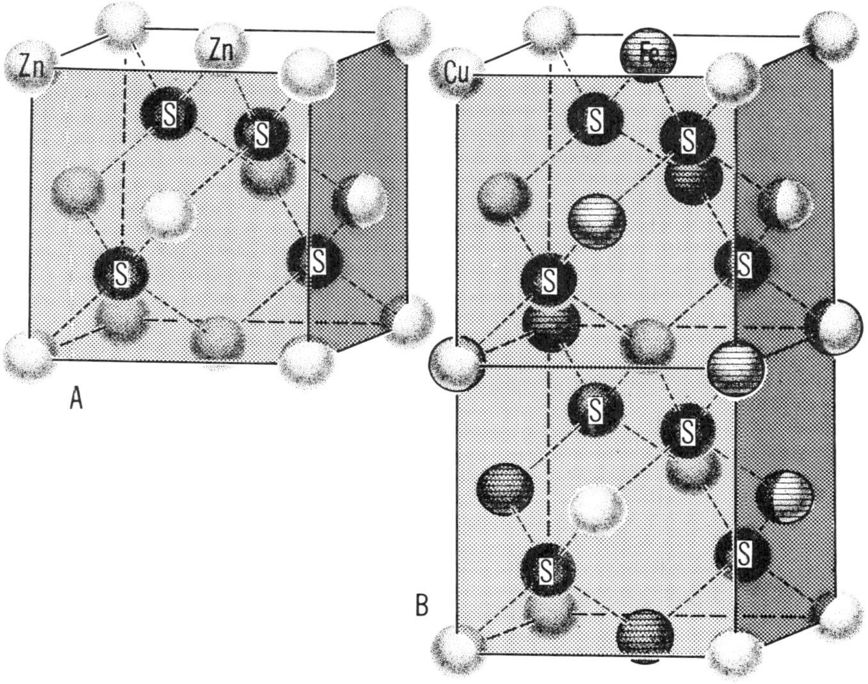

FIGURE 9–28

Comparison of the structure of (A) sphalerite, ZnS (isometric, $F\bar{4}3m$, $a = 5.42$ Å), with that of (B) chalcopyrite, $CuFeS_2$ (tetragonal, $I\bar{4}2d$, $a = 5.24$, $c = 10.30$ Å). Chalcopyrite's structure is like sphalerite's except that Cu and Fe each occupy one-half of the Zn sites. As a consequence, two of sphalerite's $\bar{4}$ axes become 2-fold axes, its 3-fold axes cease to exist, and two of sphalerite's six mirror planes survive as diamond glide planes in chalcopyrite.

which may be regarded as a derivative of the basic sphalerite structure. In Figure 9-28A sphalerite's unit cell is drawn with the origin centered on the tetrahedrally coordinated zinc atoms, whereas an earlier illustration (Fig. 9-6A) showed it centered on the sulfur atoms. If copper and iron atoms are alternately substituted for the zinc atoms, the structure of chalcopyrite is derived (Fig. 9-28B). Whereas sphalerite is isometric ($F\bar{4}3m$; $a_0 = 5.42$ Å), the derivative chalcopyrite structure is tetragonal ($I\bar{4}2d$; $a_0 = 5.24$, $c_0 = 10.30$ Å). Note the doubling along c of chalcopyrite's unit cell as compared with sphalerite's.

A derivative structure may be compositionally identical with a basic structure yet differ structurally because certain atoms have shifted slightly in position and thereby reduced the symmetry. For example, right-handed high-quartz (SiO_2) belongs to space group $P6_422$, whereas low-quartz, the form of SiO_2 stable below 573°C, belongs to space group $P3_121$. In low quartz the silicon atoms have shifted slightly from their high-quartz positions and thereby destroyed one set of 2-fold axes and altered the 6-fold axes to 3-fold axes (Fig. 10-18A, B). Buerger called such derivative structures *distortion structures.* However, a basic structure and the comparable distortion structure need not be identical in composition. Thus Buerger regards cooperite (PtS) as the basic structure and tenorite (CuO) as the distortion structure. Both these AX structures have four X atoms centered on the corners of a square with A at the square's center. However, slight shifts in the positions of the A and X atoms reduce the symmetry from $P\ 4_2/m\ 2/m\ 2/c$ for PtS to $C\ 2/c$ for CuO.

Silicate Structures

Compounds containing silicon and oxygen, that is, the silicates, are major constituents of the earth's crust. Their structures consist of oxygen ions sometimes in closest packing but sometimes less tightly packed. The cations that most frequently occupy the interstices between these oxygens are grouped in Table 9-2 according to whether their size favors 12-, 8-, 6-, or 4-fold coordination to the oxygen ions. Pauling's radii (Table 8-5), corrected by the factors cited in Table 8-4 if the coordination differed from 6 fold, were used in the calculations. Consequently, for example, the radius ratio of Al^{3+} in 4-fold coordination to O^{2-} is 94 percent of that for 6-fold coordination. The ratio R_{Si}/R_O equals 0.28 and favors tetrahedral coordination as does the partly covalent nature of the bond between silicon and oxygen. Thus the silicon ion and the four oxygens tangent to it in silicate structures form a tightly bonded SiO_4^{4-} group (Fig. 9-29A) which is frequently depicted as an ordinary tetrahedron (Fig. 9-29B), its corners marking

TABLE 9–2

Ions Commonly Found in Silicate Structures

Cation	R_c/R_o[a]	CN[b]	Anion	Radius
Si^{4+}	0.28	4	O^{2-}	1.40
Al^{3+}	0.34, 0.36	4, 6	F^-	1.33
Fe^{3+}	0.46	6	$(OH)^-$	1.33
Ti^{4+}	0.49	6		
Mg^{2+}	0.47	6		
Fe^{2+}	0.52	6		
Li^+	0.54	6		
Mn^{2+}	0.57	6		
Na^+	0.69, 0.71	6, 8		
Ca^{2+}	0.73	8		
K^+	0.98, 1.06	8, 12		

[a] Radius ratio of cation to O^{2-} calculated using Pauling radii (Table 8-5), corrected by the factors cited in Table 8-4 for coordinations other than 6 fold.

[b] Coordination often adopted in respect to O^{2-} by these cations in silicate structures. However, coordinations other than those cited here are not infrequent. Occasionally unusual coordinations such as 5, 7, or 9 may even be adopted in conformance with the peculiarities of the structure.

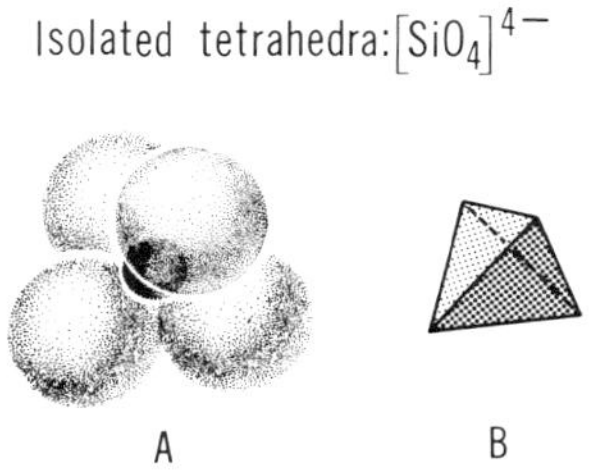

FIGURE 9–29

(A) The $[SiO_4]^{4-}$ group that is the strongest structurally cohesive unit in the *island silicates* or *nesosilicates*.
(B) Conventional depiction of an $[SiO_4]^{4-}$ group as a tetrahedron. The tetrahedron's corners represent the centers of the oxygens and its center marks the site of the Si^{4+} ion.

the centers of the oxygen ions. Centered within this tetrahedron, but not generally shown, is the Si^{4+} ion. In silicate structures these SiO_4^{4-} tetrahedra are nearly ideal in form, their Si—O (center-to-corner) distance averaging 1.60 to 1.64 Å and their edge lengths or O—O distances ranging from 2.5 to 2.7 Å. However, Al^{3+} is small enough so that in some silicates it may partly replace Si^{4+} in this tetrahedral niche, but the resultant AlO_4^{5-} tetrahedra are significantly larger than the SiO_4^{4-} tetrahedra.

Island silicates

In a cooling, natural, subsurface rock melt called a *magma*, the first silicate crystals to form usually have the Si^{4+} ions as widely separated as possible in the structure, thereby following Pauling's third rule as restated on page 242. Consequently, an oxygen ion touches at most one Si^{4+} ion. In other words, if coordination tetrahedra are drawn about each Si^{4+} in the crystal structure, no two tetrahedra share a corner, that is, an oxygen ion. Such crystal structures in which each $[SiO_4]^{4-}$ tetrahedron is isolated from all others are called *island silicates*; they are also called *nesosilicates* after the Greek word *nesogaean* meaning "island." Included among the island silicates are the olivine group of minerals X_2SiO_4 and the garnet group $X_3Y_2(SiO_4)_3$, where X is generally a divalent cation and Y is trivalent. These cations coordinate to the oxygens of neighboring $[SiO_4]^{4-}$ groups and thus link them all together. Island silicates were once believed to be the salts of a hypothetical acid, orthosilicic acid (H_4SiO_4), and island silicates are sometimes called *orthosilicates* even yet.

Double-island silicates

As island silicates crystallize from a magma, the fraction that is still liquid becomes enriched in Si^{4+} with respect to oxygen. Ultimately this enrichment permits silicates with a higher proportion of Si^{4+} to O^{2-} to form. In their structures, therefore, the Si^{4+} ions cannot be so widely spaced among the oxygens,

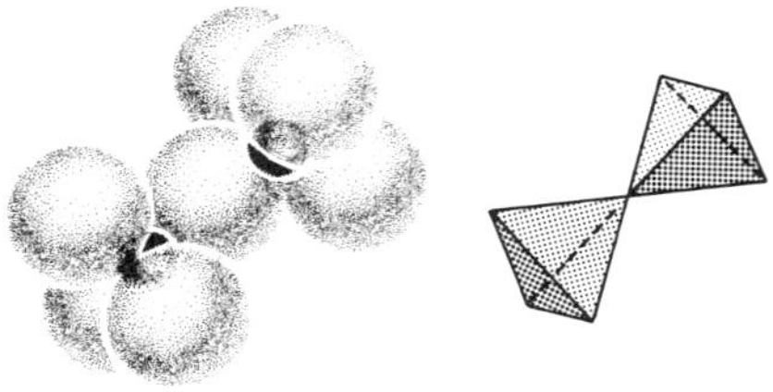

FIGURE 9–30

Representation of the $[Si_2O_7]^{6-}$ group, the strongest structurally cohesive unit in the double-island silicates or sorosilicates. (A) Packing model. (B) Two corner-sharing tetrahedra.

some of the oxygens now touching two Si^{4+} ions. In some relatively rare minerals the tightly knit group is no longer a single tetrahedron but an $Si_2O_7^{6-}$ group or *double island* produced by the sharing of a corner (Fig. 9-30). The mineral, thortveitite ($Sc_2Si_2O_7 \cdot H_2O$) contains such $[Si_2O_7]^{6-}$ groups. Double-island silicates are also called *sorosilicates,* from a Greek word meaning "group," or *orthodisilicates,* in allusion to the hypothetical orthodisilicic acid $H_6Si_2O_7$. Some silicates contain both single islands and double islands in their structures.

Single-chain silicates

With progressive crystallization from the magma, the proportion of Si^{4+} to O^{2-} in the residual liquid increases and, accordingly, this proportion in the crystallizing silicate crystal must also increase. Consequently, after the magma cools to a low enough temperature, island silicates, particularly those containing magnesium, will no longer form. Instead the silicates that do form have Si^{4+} in so many interstices that each tetrahedron shares a corner with two others to form $[SiO_3]^{2-}$ chains (Fig. 9-31A). These chains run parallel to each other in the crystal's structure. Examples of such single-chain structures are the pyroxenes whose generalized composition may be written as $XSiO_3$ or $WY(SiO_3)_2$, where W, X, and Y, respectively, represent uni-, di-, and trivalent cations that bond to the oxygens in two or more chains thereby cross-linking them. In some pyroxenes, notably *augite,* Al^{3+} may occupy a small proportion of the Si^{4+} sites in the chains. The single-chain silicates are sometimes called *metasilicates,* a carry-over from the time they were thought to be salts of the hypothetical metasilicic acid H_2SiO_3.

Double-chain silicates

With increased cooling, crystallization of single-chain silicates will be succeeded by that of double-chain silicates. In these, the double chains (Fig. 9-31B) consist of equal numbers of tetrahedra of two types: (1) tetrahedra like that labeled t_1 which share corners with two neighboring tetrahedra, each such tetrahedron thus "owning" two oxygens outright plus halves of two others, and (2) tetrahedra like t_2 which share corners with *three* neighboring tetrahedra, thus

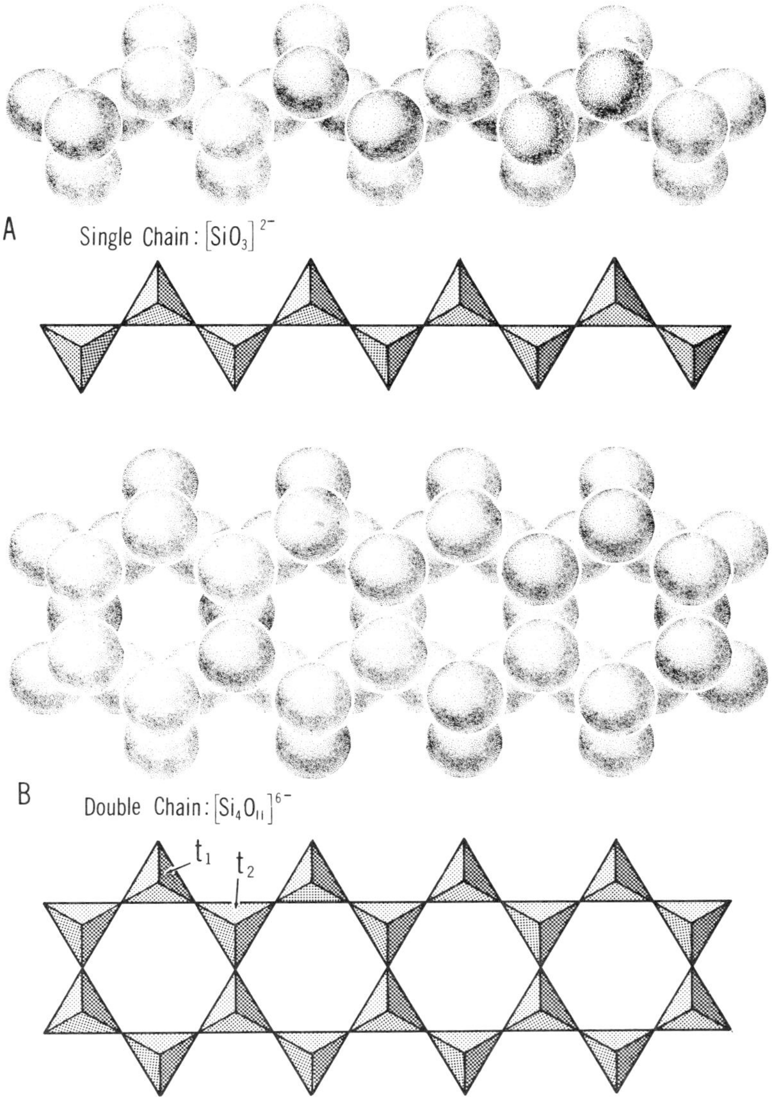

FIGURE 9–31

(A) Representation of the $[SiO_3]^{2-}$ chains, the strongest structurally cohesive units in the single-chain silicates. To simplify this and subsequent figures, the Si^{4+} ions that occupy the tetrahedral sites are not drawn.
(B) Similar representation of the $[Si_4O_{11}]^{6-}$ double chains that are the strongest cohesive units in the double-chain silicates, for example, in the amphiboles.

"owning" one oxygen outright plus halves of three others. Thus a double chain corresponds in composition to $[Si_4O_{11}]^{6-}$. In the double-chain silicates known as *amphiboles* the double chains run parallel to the *c* axis. Oxygen ions from different but neighboring chains will bond to the same uni-, di-, or trivalent cations so as to cross-link the chains. The natural double-chain silicates include $(OH)^-$ ions in their structures and thus can form only if (1) H_2O is present as a source of (OH) ions and (2) the temperature of formation is below that at which this (OH) would be driven out of the structure (as H_2O). A few examples of amphiboles are $Ca_2Mg_5Si_8O_{22}(OH)_2$ (tremolite) and $Ca_2Fe_5Si_8O_{22}(OH)_2$ (ferro-actinolite).

The single- and double-chain silicates are sometimes grouped together as chain silicates or *inosilicates,* the Greek *ino* signifying "thread" or "chain."

Layer silicates

At even lower temperatures than those necessary for amphibole formation, minerals of the layer-silicate type may crystallize. Compared to amphiboles they contain an even higher proportion of Si^{4+} to O^{2-} and of $(OH)^-$ to O^{2-} because at these lower temperatures the residual magma has become increasingly enriched in silicon and water content. The basic structure in the layer silicates is the $[Z_4O_{10}]$ sheet (Fig. 9-32) where Z represents Si^{4+} or Al^{3+} tetrahedrally coordinated to O^{2-}. Each tetrahedron in a sheet has one unshared corner (the *apical oxygen*) and three corners (the *basal oxygens*) shared with other tetrahedra. Depending upon whether their component $[Z_4O_{10}]$ sheets are most nearly $[Si_4O_{10}]^{4-}$, $[AlSi_3O_{10}]^{5-}$, or $[Al_2Si_2O_{10}]^{6-}$, layer silicates are sometimes classified as *tetrasilicic, trisilicic,* or *disilicic,* respectively. Representative layer silicates include most clay minerals; talc, $Mg_3[Si_4O_{10}](OH)_2$; pyrophyllite, $Al_2[Si_4O_{10}](OH)_2$; and all micas, for example, muscovite, $KAl_2[AlSi_3O_{10}](OH)_2$, and margarite, $CaAl_2[Al_2Si_2O_{10}](OH)_2$. In the foregoing, brackets are used to enclose the constituents of the sheets. The hydroxyl ions in layer silicates, for which F^- may substitute, are co-planar with (and centered between) the apical oxygens. Layers of cations between the $[Z_4O_{10}]$ sheets bond to the oxygen ions of the sheets immediately above and below this cation layer, thus linking one sheet to the other. These bonds between sheets, however, are weak in comparison with the Z—O bonds between the ions composing the sheets. Consequently layer silicates are characterized by their ease of fracture and their flaky growth parallel to the $[Z_4O_{10}]$ sheets. The layer silicates are thus often called *phyllosilicates* from the Greek *phyllo* which means "leaf" or "sheet."

Ring silicates

In the structures of the ring silicates the $[SiO_4]^{4-}$ tetrahedra share corners to form three-, four-, or six-membered rings (Fig. 9-33) of the general formula $[Si_xO_{3x}]$. Thus benitoite, $BaTi[Si_3O_9]$, contains three-membered rings; axinite, $Ca_3Al_2(BO_3)[Si_4O_{12}](OH)$, contains four-membered Si_4O_{12} rings; tourmaline, $NaMg_3Al_6(BO_3)_3[Si_6O_{18}](OH)_4$, contains six-membered rings. Various cations,

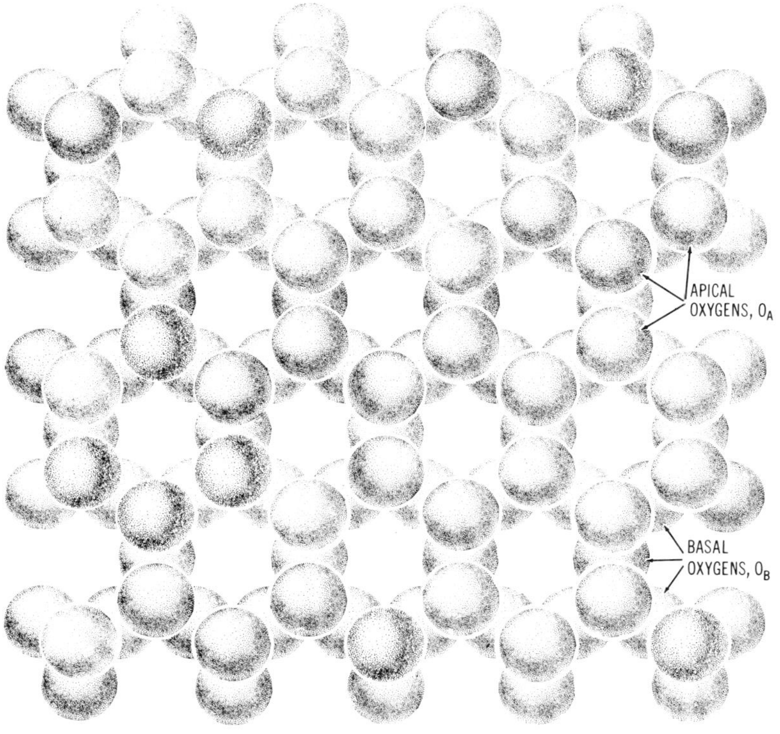

FIGURE 9–32

Packing representation of the $[Z_4O_{10}]$ sheets, the strongest, structurally cohesive units in the layer silicates. In layer silicates Z is generally either Si^{4+} or Al^{3+}. If Si^{+4} occupies all, three-quarters, or one-half of the Z sites (and Al^{3+} the remainder), the $[Z_4O_{10}]$ sheet is called *tetrasilicic* $[Si_4O_{10}]^{4-}$, *trisilicic* $[AlSi_3O_{10}]^{5-}$, or *disilicic* $[Al_2Si_2O_{10}]^{6-}$, respectively.

for example, Ba^{2+}, Ti^{4+}, Mg^{2+}, Al^{3+}, and others, bond to the oxygens from neighboring rings thus linking the rings into a coherent structure. Those ring silicates that crystallize from a magma's residual liquid do so at relatively low temperatures. This is true because their structures require elements, for example, scandium or boron, that are present only in minute amounts in the original magma. As crystallization from the magma progresses, these minor elements, because their size and/or valence does not permit them to incorporate in the early crystallizing silicates, ultimately become sufficiently concentrated in the liquid to promote crystallization of certain ring silicates. Ring silicates are sometimes called *cyclosilicates,* the Greek word *cyclo* meaning "circle" or "ring." They have the same Si:O ratio (1:3) as for pyroxenes and, in effect, represent a chain that closes on itself.

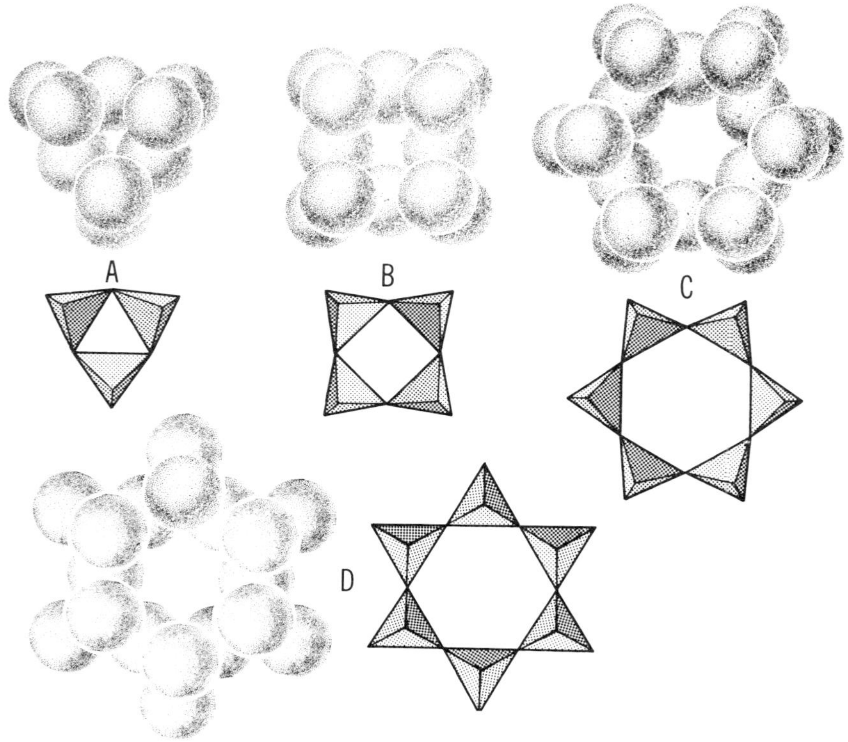

FIGURE 9–33

Three-, four-, and six-membered rings of [SiO_4] tetrahedra that are encountered in the ring silicates. The six-membered ring commonly ascribed to beryl (C) differs from that in tourmaline (D). For illustrative purposes, the three, four, or six uppermost oxygens in (A), (B), and (C) were drawn closer to the center of the ring than they actually are. Simultaneously, the three, four, or six lowermost oxygens, which are 80 percent hidden by these upper spheres, were drawn farther from the center of the ring than they should be. If (A), (B), and (C) were drawn without this distortion, (1) the uppermost oxygens would be centered directly over the lowermost ones and completely hide them and (2) the edges of the shaded tetrahedra, here drawn as 2 mm long (*ca.*), would become points, because these edges would be perpendicular to the plane of the stage.

Framework silicates

In some structures the proportion of Si^{4+} to O^{2-} is so high that *all* tetrahedra share *all* their corners with others. An individual tetrahedron therefore shares each of its four corners with its four neighboring tetrahedra (Fig. 9-34). Each tetrahedron thus "owns" the Si^{4+} ion at its center plus one-half of each O^{2-} ion at its corners. Consequently, if no ion substitutes for silicon, the entire framework or network has the composition SiO_2. For a framework structure in general if Z represents ions tetrahedrally coordinated by oxygen, the compositional formula is ZO_2. The various crystalline forms of SiO_2 are framework structures in which the Z ions are almost exclusively Si^{4+}. The feldspar structures are also frameworks; in these 25 to 50 percent of the tetrahedra are occupied by Al^{3+} instead of Si^{4+}. Thus using brackets to enclose the constituents of the tetrahedral frameworks, the feldspar named albite, $Na[AlSi_3O_8]$, is seen to have 25 percent of its tetrahedra filled by Al^{3+}, whereas for anorthite, $Ca[Al_2Si_2O_8]$, it is 50 percent. The plagioclase feldspars represent a range of compositions from pure albite to pure anorthite. For example, plagioclase which is halfway between albite and anorthite would have 37.5 percent of its tetrahedra filled by Al^{3+}. The charges remain balanced because as Ca^{2+} replaces Na^+, a corresponding number of Al^{3+} ions replace Si^{4+}.

Framework silicates may crystallize from a magma at all stages of its cooling history. Crystallizing early, along with the olivines, are the plagioclases close to anorthite, $Ca[Al_2Si_2O_8]$, in composition. In them the proportion of Si^{4+} to O^{2-} is close to 1 to 4 (as for the olivines). As cooling and crystallization continue,

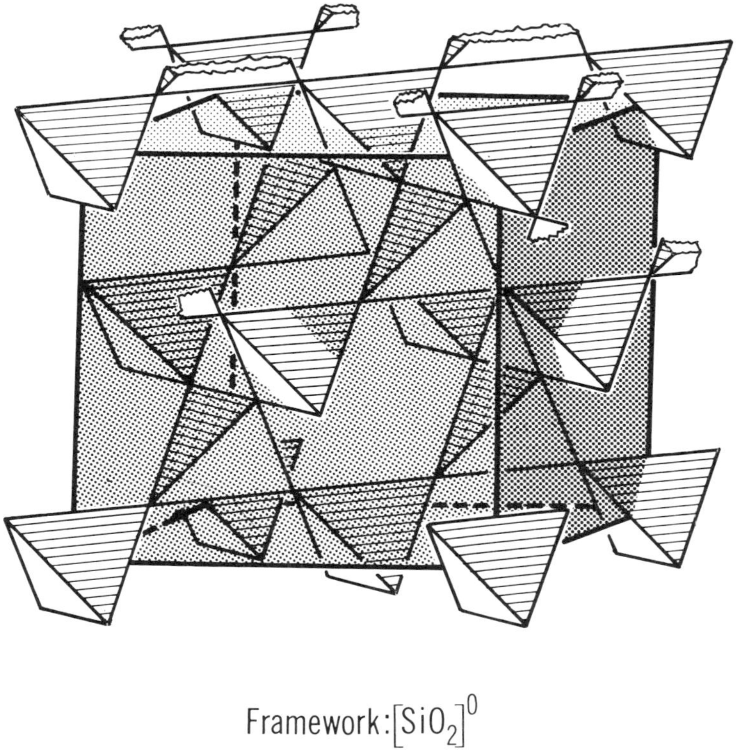

FIGURE 9–34

The structure of β-cristobalite, SiO_2. This is a typical framework structure in which no corners of any tetrahedron remain unshared. If more unit cells were drawn, every tetrahedron would be seen to share each of its four corners with its four neighbors. As the superscript indicates, an $[SiO_2]^0$ framework bears no net charge.

these calcic plagioclases are succeeded by plagioclases successively closer to albite, Na[$AlSi_3O_8$], in composition. The proportion of Si^{4+} to O^{2-} thus rises steadily in the successively crystallizing plagioclases although it always remains lower than in the magma. With successive crystallization, therefore, the magma is steadily enriched in Si^{4+} as opposed to O^{2-}. Ultimately, the ratio of Si^{4+} to O^{2-} in the magma becomes high enough to permit a framework mineral of composition SiO_2, that is, quartz, tridymite, or cristobalite, to crystallize simultaneously with the albitic plagioclases and with orthoclase, K[$AlSi_3O_8$], a potassium-rich feldspar.

Bridging oxygens

In silicate structures an oxygen that is bonded to two silicon atoms, that is, an oxygen that represents a corner shared between two [SiO_4] tetrahedra, is called a *bridging* oxygen. Those representing unshared tetrahedral corners are *nonbridging* oxygens. In the classic Si—O groups just described, the percent of oxygens that are bridging oxygens is zero for isolated tetrahedra (Fig. 9-29), 14.3 for double tetrahedra (Fig. 9-30), 33.3 for single chains (Fig. 9-31A) and for rings (Fig. 9-33), 45.5 for double chains (Fig. 9-31B), 60 for sheets (Fig. 9-32), and 100 for frameworks like that in Figure 9-34. Thus the percent of bridging oxygens increases from isolated tetrahedra to frameworks.

Classification of Silicates

Liebau's classification

Machatschki (1928) conceived the modern basis for classifying silicates into island, double-island, single-chain, double-chain, ring, sheet, or framework silicates according to the linkage of [ZO_4] tetrahedra, where Z was Si^{4+}, Al^{3+}, Be^{2+} or perhaps B^{3+}. More recently, Liebau (1962) described and divided into subtypes the different single chains, double chains, sheets, and frameworks that occur in silicate structures.

Single Chain Varieties. Liebau describes six variations of single chains (Fig. 9-35) that he calls 1-, 2-, 3-, 4-, 5- or 7-unit single chains, respectively, depending on whether the number of tetrahedra in the repeat unit along the chain's length is 1, 2, 3, 4, 5, or 7.[4] The 2-unit single chain, the classic (and most common) type, occurs in pyroxenes ($MgSiO_3$, $CaMgSi_2O_6$, and so on). A 3-unit single chain occurs in wollastonite ($CaSiO_3$) and pectolite ($Ca_2NaHSi_3O_9$). Recently, 5- and 7-unit single chains were discovered in rhodonite ($MnSiO_3$) and pyroxmangite, respectively.

[4] In the original German, the word *einfachkelte* is used for "single chain" and prefixed with *einer, zweier, dreier, vierer, fünfer,* or *siebener* to indicate that the chain's repeat unit consists of 1, 2, 3, 4, 5, or 7 tetrahedra, respectively. Thus *einereinfachkelte* and *zweiereinfachkelte,* respectively, mean a 1-unit single chain and a 2-unit single chain.

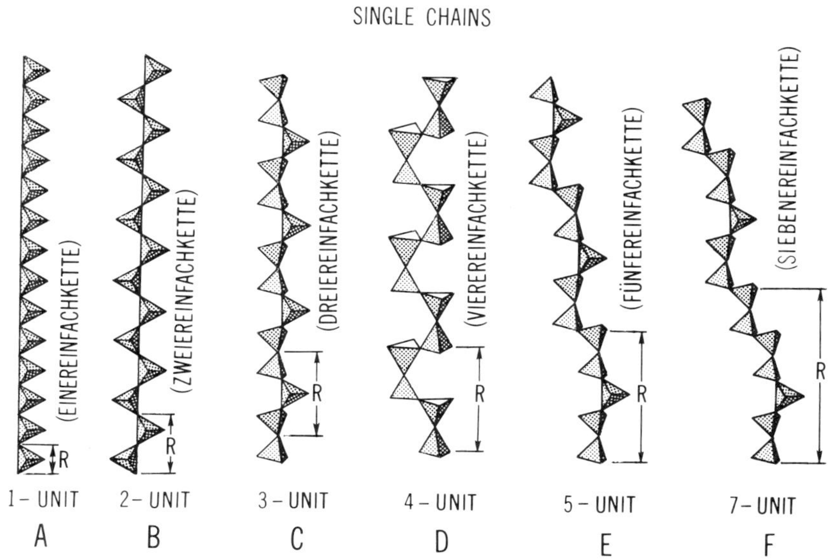

FIGURE 9–35

Types of single chains of [SiO_4] tetrahedra that may occur in the single-chain silicates. *R* represents the repeat unit along each chain. (A) The 1-unit single chain as yet only observed for [GeO_4] tetrahedra in the $CuGeO_3$ structure. (B) The 2-unit single chain that is actually the classic type occurring in pyroxenes. (C) The 3-unit single chain present in wollastonite, $CaSiO_3$. (D) A hypothetical 4-unit single chain that may occur in neptunite, $Na_2FeTi(SiO_3)_4$. (E) The 5-unit single chain found in rhodonite, $MnSiO_3$. (F) The 7-unit single chain found in pyroxmangite, (Mn, Fe)SiO_3.

The absence of 1-unit chains in silicate structures perhaps reflects the appreciable covalency of the Si—O bond. This bond's directional nature apparently constrains two silicon atoms, if bonded to the same oxygen, to be so disposed about this oxygen that the Si—O—Si bond angle is usually in the range of 130 to 140° as noted by Liebau (1961). However, the geometry of 1-unit chains would require Si—O—Si bond angles of 117°, this being apparently too far below 130° to be permissible. In contrast the Ge—O—Ge bond angle does not appear so constrained to the 130 to 140° range because in $CuGeO_3$ the [GeO_4] tetrahedra form 1-unit single chains.

Double-Chain Varieties. Two single chains of a particular type may share corners to form a double chain (German, *doppelkette*). Thus there result 1-unit, 2-unit, 3-unit, and 4-unit double chains (*einerdoppelkette, zweierdoppelkette, drierdoppelkette,* and *viererdoppelkette*) as shown in Figure 9-36. A 1-unit double chain consisting only of [SiO_4] tetrahedra is as unlikely in silicate structures as a 1-unit single chain. However, in sillimanite (Al_2SiO_5) 1-unit double chains occur that consist alternately of [SiO_4] and [AlO_4] tetrahedra, the [AlO_4] tetrahedra (outlined in bold lines in Fig. 9-36A) being slightly larger than the [SiO_4]

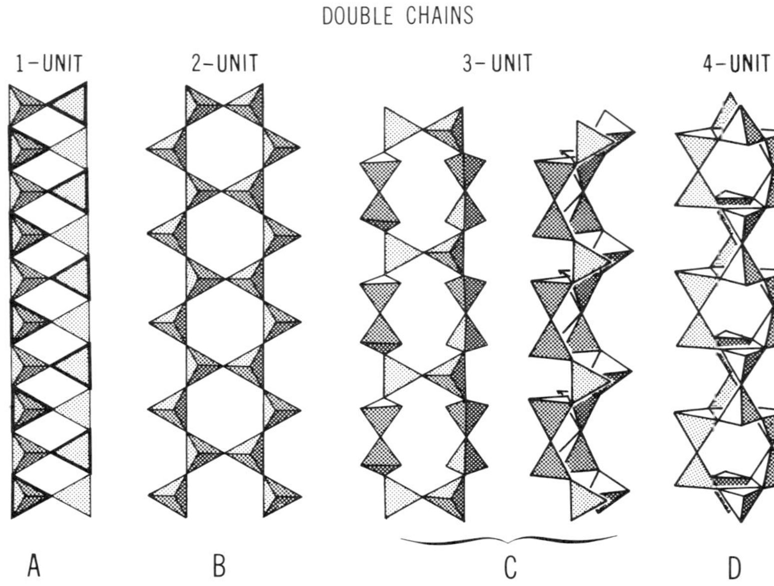

FIGURE 9–36

Types of double chains. (A) The 1-unit double chain whose tetrahedral sites are randomly occupied by Si^{4+} and Al^{3+} in mullite, Al_4SiO_8, and if alternate tetrahedra (boldly outlined) are $[AlO_4]$ and the others $[SiO_4]$, as found in sillimanite, Al_2SiO_5. (B) The 2-unit double chain that is the classic type long observed in amphiboles. (C) Two types of 3-unit double chains, the one $[Si_6O_{17}]$ found in xonotlite $Ca_6Si_6O_{17}(OH)_2$ (at left) and the one $[Si_6O_{15}]$ found in epididymite $Na_2Be_2Si_6O_{15} \cdot H_2O$(?) (at right). (D) The 4-unit double-chain $[Si_4O_{10}]$ found in narsarsukite.

tetrahedra. Evidently the Al—O—Si bond angle is not limited to the range of 130 to 140°.

The 2-unit double chain (Fig. 9-36B), which is actually the classic double chain used in the original Machatschki classification, occurs abundantly in nature as a constituent of the amphiboles. Only recently 3-unit double chains of two types have been found (Fig. 9-36C) as well as a 4-unit double chain (Fig. 9-36D).

Varieties of Sheets. Sheets can be regarded as consisting of single chains. This is emphasized in Figure 9-37 by the alternate single chains being drawn with bolder lines in the three different types of sheet structures. Sheets may thus be called 2-unit, 3-unit, or 4-unit sheets if the single chain composing them is of this type. The 2-unit sheet in Figure 9-37A is the classic variety and is by far the most abundant, occurring in the structures of micas. The rare 3-unit sheet (Fig. 9-37B) occurs in okenite, $Ca_2Si_4O_{10} \cdot 4H_2O$. The 4-unit single-sheet structure (Fig. 9-37C) is found in apophyllite, $KCa_4(Si_4O_{10})_2F \cdot 8H_2O$. Silicate sheets are sometimes classified according to the number of tetrahedra in the rings discernible in them. Thus the 2-unit sheet (Fig. 9-37A) contains just 6-membered rings, whereas the

3-unit sheet in Figure 9-37B contains both 5- and 8-membered rings. The apophyllite sheet (Fig. 9-37C) contains both 4- and 8-membered rings.

Returning to the classic 2-unit sheet (Fig. 9-37A) we find its apical oxygens (unshared corners) are all located on the same side of the sheet. However, additional varieties of 2-unit sheets occur in which this is not the case (Fig. 9-38). Thus in Figure 9-38A two single chains with their apical oxygens pointing up alternate with two whose apical oxygens point down. If we let **⊥** and **T** indicate single chains, the first with its apical oxygens pointing up, the second with them point down, the sheet in Figure 9-38A can be symbolized as . . . **⊥⊥TT**. . . and that in Figure 9-38B as . . . **⊥⊥⊥TTT**. . . . Figure 9-38C shows the so-called double

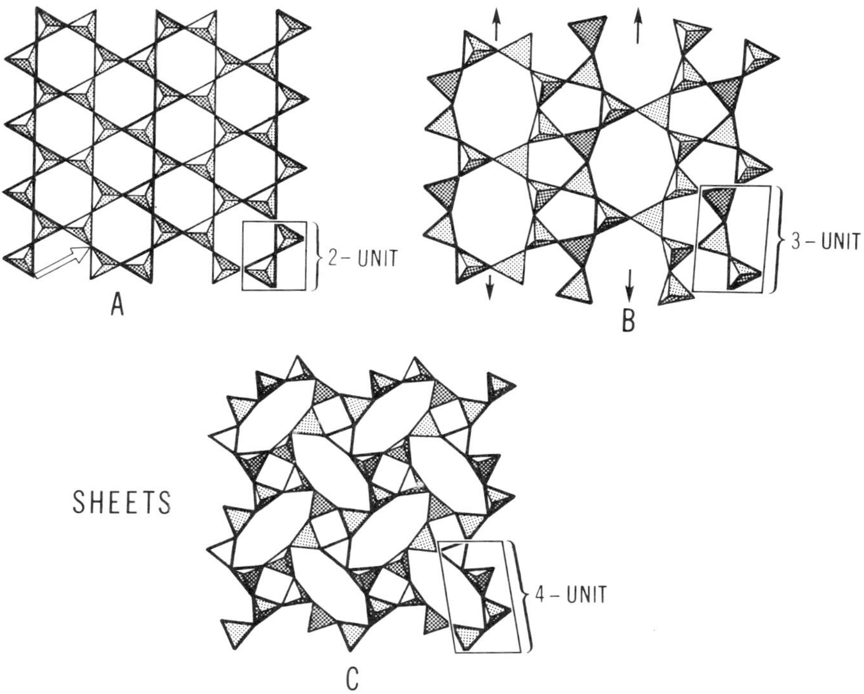

FIGURE 9–37

Three varieties of sheet structures.

(A) The classic 2-unit sheet found in the mica structures, each with alternate chains of tetrahedra boldly outlined to indicate Liebau's concept that sheets can be regarded as consisting of single chains joined together. The hollow arrow indicates the translation that would cause the one single chain to coincide with its neighbor.

(B) A 3-unit sheet such as is possibly present in okenite, $Ca_2Si_4O_{10}\cdot 4H_2O$. The arrows indicate the 2-fold axes that pass through juncture points of the chains and relate the one chain to its neighbor.

(C) The 4-unit sheet occurring in apophyllite, $KCa_4(Si_4O_{10})_2F\cdot 8H_2O$. Neighboring single chains, although joined, are not symmetrically related to each other.

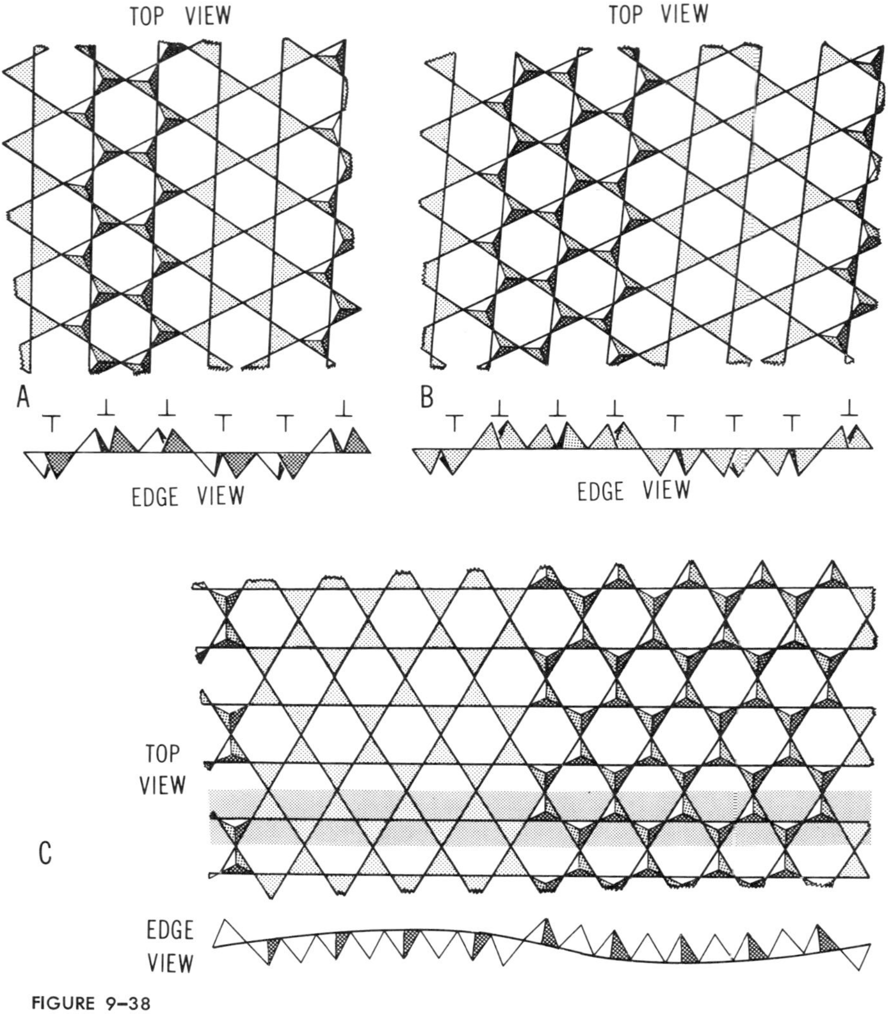

FIGURE 9–38

Three variations of 2-unit sheets in which the apical oxygens sometimes point up, sometimes down. (A) The type ···**⊥⊥TT**··· (**⊥** symbolizes a single chain with apical oxygens pointing up, **T** one with them pointing down) as found in attapulgite. (B) The type ···**⊥⊥⊥TTT**··· which occurs in sepiolite. (C) The so-called double half-wave structure found in antigorite, a serpentine mineral of composition $Mg_3Si_2O_5(OH_4)$. One of its "single chain units" is shown on a gray background. The curvature of the sheets, illustrated by the edge view, decreases the distance between apical oxygens.

half-wave structures, a wavelike variation of a 2-unit sheet as found by Kunze in the structure of antigorite. If we attempt to trace in it a 2-unit single chain (shaded background), the chain has some apical oxygens pointing up, others pointing down.

Double-Sheet Structures in which two tetrahedral sheets join through sharing of their apical oxygens are rare. Two 2-unit sheets, so joined into a double sheet (Fig. 9-39A), occur in the structure of the synthetic hexagonal polymorphs of $CaAl_2Si_2O_8$ and of $BaAl_2Si_2O_8$ (α-celsian).

Rings and Frameworks. Ring structures are classified by Liebau according to the number of [SiO_4] tetrahedra in the ring. Those in Figure 9-33 are thus 3-, 4-, and 6-unit single rings, respectively. Just as there can be double sheets, so can there be double-ring structures but they too are rare. Figure 9-39B illustrates the 6-unit double ring of [SiO_4] tetrahedra observed in milarite, $KCa_2AlBe_2[Si_{12}O_{30}]\cdot\frac{1}{2}H_2O$; however, if the [$AlO_4$] and [$BeO_4$] tetrahedra are also considered (see Zoltai's reclassification), milarite becomes a framework structure.

Framework Structures are difficult to classify into subtypes. Although they can be regarded as consisting of joined single chains, selection of the chain to be so considered is frequently ambiguous. Moreover it is not always possible to build up a given framework through the repeated joining of chains.

Zoltai's reclassification

Zoltai (1960) notes inconsistencies in the present classification of silicates. Sometimes only the [SiO_4] tetrahedra are considered in deciding whether the structure is a ring, chain, sheet, or framework. At other times the [AlO_4] tetrahedra are also considered; but sometimes they are disregarded whereas other

A

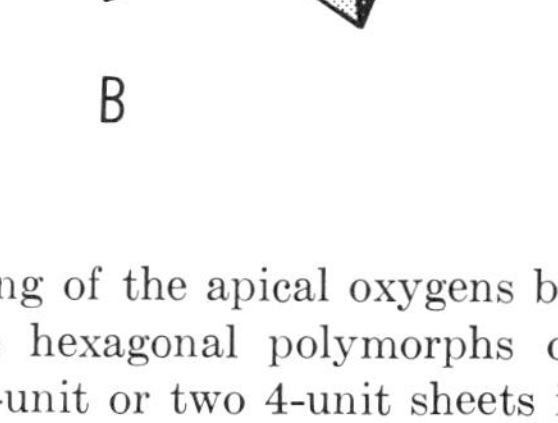

B

FIGURE 9–39

(A) A 2-unit double-sheet structure formed by sharing of the apical oxygens by two 2-unit single sheets. Such are observed in the hexagonal polymorphs of $CaAl_2Si_2O_8$ and of $BaAl_2Si_2O_8$. Similar union of two 3-unit or two 4-unit sheets is theoretically possible but as yet unobserved.
(B) A 6-unit double ring of [SiO_4] tetrahedra as observed in milarite.

TABLE 9–3

Classification of Tetrahedral Structures (after Zoltai)

Types	S^a	Representatives[b]
		I. Groups of Limited Extent
A. Island $[ZO_4]$	1.00	Olivine, $(Mg,Fe)_2\mathbf{SiO_4}$.
	1.00	Garnet, $(Fe^{2+},Ca,Mg,Mn)_3(Al,Fe^{3+},Cr)_2(\mathbf{SiO_4})_3$
	1.00	Sphene, $CaTi\mathbf{SiO_4}(OH)$.
B. Double island $[Z_2O_7]$	1.25	Lawsonite, $CaAl_2(OH)_2\mathbf{Si_2O_7}\cdot H_2O$
	1.25	Tilleyite, $Ca_5(CO_3)_2\mathbf{Si_2O_7}$
C. Single rings $[Z_xO_{3x}]$		
3 unit $[Z_3O_9]$	1.50	Benitoite, $BaTi\mathbf{Si_3O_9}$
4 unit $[Z_4O_{12}]$	1.50	Baoite, $Ba_4Ti_8O_{16}\mathbf{Si_4O_{12}}Cl$
6 unit $[Z_6O_{18}]$	1.50	Tourmaline, $(Na, Ca)(Li, Mg, Al)_3(Al, Fe, Mn)_6(BO_3)_3\mathbf{Si_6O_{18}}(OH, F)_4$
	1.50	Dioptase, $Cu_6\mathbf{Si_6O_{18}}\cdot 6H_2O$
D. Double and multiple rings		None known
E. Mixed		
$[ZO_4]$ and $[Z_2O_7]$	1.11	Idocrase (vesuvianite), $Ca_{10}(Mg, Fe)_2Al_4[\mathbf{Si_2O_7}]_2[\mathbf{SiO_4}]_5(OH, F)_4$
	1.17	Epidote and clinozoisite, $Ca_2(Al, Fe^{3+})Al_2O[\mathbf{Si_2O_7}][\mathbf{SiO_4}](OH)$
$[ZO_4]$ and $[Z_5O_{16}]$	1.33	Zunyite, $Al_{12}(OH, F)_{18}[\mathbf{Si_5O_{16}}][\mathbf{AlO_4}]Cl$
		II. Groups Unlimited Along One Dimension (Chains)
A. Single chains $[ZO_3]_\infty$		
2 unit	1.50	Pyroxenes, $(Mg, Fe)\mathbf{SiO_3}$, etc.
3 unit	1.50	Wollastonite, $Ca\mathbf{SiO_3}$
5 unit	1.50	Rhodonite, $Mn\mathbf{SiO_3}$
7 unit	1.50	Pyroxmangite, $(Mn, Fe)\mathbf{SiO_3}$
B. Double chains		
1 unit $[Z_2O_5]_\infty$	1.75	Sillimanite $Al\mathbf{AlSiO_5}$
2 unit $[Z_4O_{11}]_\infty$	1.62	Amphiboles, $Ca_2(Mg, Fe)_5\mathbf{Si_8O_{22}}(OH)_2$, etc.
3 unit $[Z_6O_{17}]_\infty$	1.58	Xonotlite, $Ca_6\mathbf{Si_6O_{17}}(OH)_2$
3 unit $[Z_6O_{15}]_\infty$	1.75	
4 unit $[Z_8O_{20}]_\infty$	1.75	Narsarsukite, $Na_2TiO\mathbf{Si_4O_{10}}$
		III. Groups Unlimited Along Two Dimensions (Sheets)
A. Single sheets $[Z_{2x}O_{5x}]_\infty$		
2 unit	1.75	Micas, $KAl_2\mathbf{AlSi_3O_{10}}(OH)_2$, etc.
3 unit	1.75	Okenite, $Ca_2\mathbf{Si_4O_{10}}\cdot 4H_2O$
4 unit	1.75	Apophyllite, $KCa_4\mathbf{Si_8O_{20}}F\cdot 8H_2O$
B. Double sheets $[Z_4O_8]$		
2 unit	2.0	α-celsian, $Ba\mathbf{Al_2Si_2O_8}$

TABLE 9-3 (*Continued*)

Types	S^a	Representatives[b]
IV. Groups Unlimited Along Three Dimensions (Frameworks)		
A. No more or less than two tetrahedra share a corner to produce $[Z_xO_{2x}]$ frameworks for which $S = 2.00$		
		Quartz, tridymite, coesite, cristobalite, $\mathbf{SiO_2}$; Scapolite, $(Na, Ca)_4(\mathbf{Al}, \mathbf{Si})_{12}\mathbf{O}_{24}(Cl, CO_3)$; Sodalite, $Na_8(\mathbf{AlSiO_4})_6Cl_2$
		Analcite, $Na\mathbf{AlSi_2O_6}\cdot H_2O$
		Leucite, $K\mathbf{AlSi_2O_6}$; Nepheline, $Na\mathbf{AlSiO_4}$
		Orthoclase, $K\mathbf{AlSi_3O_8}$; Albite, $Na\mathbf{AlSi_3O_8}$
		Anorthite, $Ca\mathbf{Al_2Si_2O_8}$; Beryl, $Al_2\mathbf{Be_3Si_6O_{18}}$; Cordierite, $Mg_2\mathbf{Al_4Si_5O_{18}}$; Chabazite, $CaNa_2\mathbf{Al_2Si_4O_{12}}\cdot 6H_2O$; Natrolite, $Na_2\mathbf{Al_2Si_3O_{10}}\cdot 2H_2O$; Heulandite, $Ca\mathbf{Al_2Si_7O_{18}}\cdot 6H_2O$
B. No more or less than three tetrahedra share a corner to produce $[Z_{3x}O_{4x}]$ frameworks for which $S = 3.00$.		
		Phenacite, $\mathbf{Be_2SiO_4}$; Willemite, $\mathbf{Zn_2SiO_4}$
C. Frameworks for which S lies between 2.0 and 3.0		
	2.75	Hemimorphite, $\mathbf{Zn_4Si_2O_7}(OH)_2\cdot H_2O$

[a] S, the sharing coefficient, is defined by Equation (9-4).

[b] Formulas are in some cases simplified. Boldface type indicates tetrahedrally coordinated cations and the anions tangent to them which together constitute the corner-sharing tetrahedral group. If two or more elements separated by commas are inserted in parentheses, the parentheses indicate a particular site in the structure that may be occupied by any of the parenthesized elements as will be discussed in detail under the topic, *solid solution*, in Chapter 10. In epididymite (Robinson and Fang, 1970), $[BeO_4]$ tetrahedra cross link the $[Si_6O_{15}]$ chains to produce a hybrid chain-framework structure that violates Zoltai's basic assumption (p. 274) and is hence not classifiable.

tetrahedra, like $[BO_4]$, may be accepted. Zoltai suggests that with few exceptions *all the tetrahedra present* should be considered in classifying a silicate even though instead of Si^{4+} they may contain B^{3+}, Be^{2+}, P^{5+}, Al^{3+}, Ge^{4+}, Ga^{3+}, Mg^{2+}, Fe^{3+}, Zn^{2+}, or Ti^{4+}, these being the more important of the cations occasionally observed to be tetrahedrally coordinated in silicates.

The bond between tetrahedrally coordinated cations and their associated anions is frequently the strongest in the structure. Hence the $[ZO_4]$ groups, where Z indicates a tetrahedrally coordinated cation, form the "backbones" of the structure and the nature of their grouping, that is, whether they form frameworks, sheets, or chains, will affect the physical properties of the crystal. Zoltai's classification, which considers $[ZO_4]$ groups rather than just $[SiO_4]$ groups (Table 9-3), more consistently classifies minerals of like properties together because (1) the bond between O^{2-} and non-Si^{4+} tetrahedral occupants is usually strong enough to influence physical properties in the same way as Si^{4+} and (2) it avoids the inconsistencies of classifications that accept the Al^{3+}-filled tetrahedra in classifying feldspars as frameworks but reject them in classifying cordierite as a ring structure. Zoltai thus transfers beryl, $Al_2\mathbf{Be_3Si_6O_{18}}$, and cordierite, $Mg_2\mathbf{Al_4Si_5O_{18}}$, from the ring structures to the framework structures because he considers the $[BeO_4]$ tetrahedra in beryl and the $[AlO_4]$ tetrahedra in cordierite as part of the tetrahedral framework. In these simplified formulas for beryl and cordierite, the ions participating in tetrahedra are in bold type, a convention that will be used henceforward. Beryl and cordierite are thus now classed with the framework mineral quartz, which they megascopically resemble. The proportion of cations to anions in their tetrahedral framework is 1:2 as in quartz.

The Sharing Coefficient. If only $[SiO_4]$ tetrahedra are considered in a silicate structure, only two at most can share a corner. If tetrahedra of other compositions are also considered, however, then sometimes more than two tetrahedra may share a corner. For example, in phenacite ($\mathbf{Be_2SiO_4}$), one $[SiO_4]$ and two $[BeO_4]$ tetrahedra share a corner. In respect to any given silicate structure, Zoltai makes an assumption that agrees with most observations to date: If n is a positive integer that represents the *minimum* number of tetrahedra observed as sharing a common corner in the structure, then $n + 1$ will represent the *maximum* number that will share a common corner. This means that if unshared tetrahedral corners are present in a structure, these corners being regarded by Zoltai as "shared" by only one tetrahedron (hence for them $n = 1$), then any remaining corners in the structure can be shared between at most $(n + 1)$ or two tetrahedra. Similarly, if the minimum number of tetrahedra sharing a common corner is two ($n = 2$), then no corners in the structure will be shared between more than three $(n + 1)$ tetrahedra.

Consider, for example, a length of the 2-unit double chain of tremolite ($Ca_2Mg_5\mathbf{Si_8O_{22}}(OH)_2$) that is just long enough to contain the same number of silicon and oxygen atoms as in tremolite's formula (Fig. 9-40). Points 1 to 12

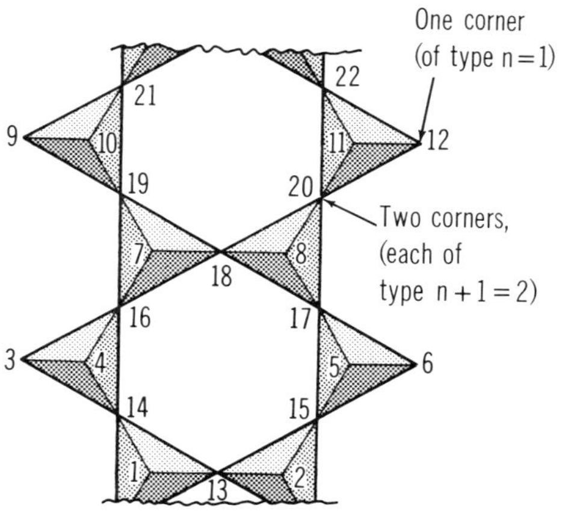

FIGURE 9–40

Portion of the Si_8O_{22} double chain present in amphibole structures. Tetrahedral corners labeled 1 to 12 are of type $n = 1$; that is, they are common to one tetrahedron only and hence are unshared corners. Corners 13 to 22 are of type $(n + 1)$. For them $(n + 1)$ equals 2; that is, these corners are each common to *two* tetrahedra. Corners 13 to 22 thus represent bridging oxygens.

indicate 12 unshared corners that Zoltai regards as "shared" by 1 tetrahedron and thus of type $n = 1$. Points 13 to 22 represent 20 tetrahedral corners, the corners of 2 tetrahedra occurring at each of these points. These 20 corners are of type $n + 1 = 2$, because each is common to 2 tetrahedra. Zoltai uses the symbol T to equal the number of tetrahedral cations represented in a crystal's chemical formula and symbol A to equal the number of anions that bond to these cations. Thus for tremolite T is 8 and A is 22 as confirmed by Figure 9-40 or by its formula $Ca_2Mg_5\mathbf{Si_8O_{22}}(OH)_2$ in which Si represents tetrahedral cations and O the anions bonded to them. Figure 9-40 represents a total of $4T$ or 32 tetrahedral corners of which 12 (37.5 percent) are type $n = 1$ and 20 (62.5 percent) are type $n + 1$ (=2).

For each tetrahedral corner in a structure, we shall assume that a slip of paper is marked n or $n + 1$, according to the type of corner. The result is a population of integers that are either n or $n + 1$. We symbolize the average value of this population as S. Thus for tremolite because there would be 12 slips marked "1" and 20 marked "2," or 32 slips in all,

$$S = \frac{(12 \times 1) + (20 \times 2)}{32} = 1.625 \qquad \text{(Eq. 9-3)}$$

S is called the *sharing coefficient* by Zoltai and always equals the average value of a population of integers that are either n or $n + 1$, that is, integers that differ by 1.0 at most. For a population of integers so confined, the average value can always be written

$$S = n + \left[\frac{x}{4T}\right] \qquad \text{(Eq. 9-4)}$$

where $4T$ symbolizes the total number of integers (or tetrahedral corners) in the population, x is the number of integers (or corners) of type $n + 1$, and thus $x/4T$ is the fraction of the population that is of type $n + 1$. For tremolite, therefore,

$$S = 1 + \left[\frac{20}{32}\right] = 1 + [0.625]$$

The sharing coefficient S thus consists of the integer n plus a decimal fraction, enclosed here in square brackets, which indicates the portion of the total population of corners that are of type $n + 1$. To emphasize this let us consider the set of integers, 1, 1, 1, 1, 1, 1, 1, and 2. Its average is 1.125 or 1 + [0.125] and indicates that n equals 1 and that integers of type $n + 1$ constitute 0.125 or one-eighth of the total population of integers.

In Figure 9-40 points 1 to 12 represent 12 anions or 12 corners of type $n = 1$. Points 13 to 22 represent 10 anions or 20 corners of type $n + 1 = 2$. The number of corners divided by their type, for example, $(12 \div 1)$ or $(20 \div 2)$, equals the number of anions at these corners. Thus A, the number of anions in Figure 9-40, equals 22, that is,

$$A = \frac{12}{1} + \frac{20}{2} = 22$$

Writing this equation in general terms of the total number of tetrahedral corners ($= 4T$), those of type $n + 1$ ($= x$), and those of type n ($= 4T - x$), we obtain

$$A = \frac{4T - x}{n} + \frac{x}{n + 1}$$

Solved for x, this becomes

$$x = 4T(n + 1) - A(n^2 + n) \qquad \text{(Eq. 9-5)}$$

Substituting this value for x into Equation (9-4), there results, after rearrangement of terms, the *general equation* for calculating S, namely

$$S = n + \left[(n + 1)\left(1 - \frac{nA}{4T}\right)\right] \qquad \text{(Eq. 9-6)}$$

As now written this expression for S consists of (a) the integer n, which by definition indicates the minimum number of tetrahedra tangent at a tetrahedral corner, plus (b) the bracketed terms, which yield the decimal fraction of the total tetrahedral corners that are of type $(n + 1)$. To illustrate, for tremolite $n = 1$, $A = 22$, $T = 8$, so that Equation (9-4) becomes

$$S = 1 + [0.625]$$

The integer 1 indicates that tetrahedra of type $n = 1$ exist in the structure, which we already knew. The bracketed decimal fraction indicates that 0.625 or 62.5 percent of the tetrahedral corners are of type $n + 1 = 2$.

If T and A are known for a structure, n can be calculated because

$$\frac{4T}{A} = n + A_{n+1} \qquad \text{(Eq. 9-7)}$$

where A_{n+1} is the fraction of the A anions (but not of tetrahedral corners) of type $n + 1$, that is, anions that are bonded to $n + 1$ tetrahedral cations. Although Equation (9-7) will not be derived, its veracity can be demonstrated for tremolite for which

$$\frac{4T}{A} = \frac{32}{22} = 1 + \frac{5}{11} = 1 + 0.45$$

This indicates that n, as previously defined, equals 1 for tremolite and hence, of course, $n + 1$ equals 2. It also shows that A_{n+1} or A_2 equals 0.45; that is, that 45 percent of the A or O^{2-} anions in tremolite are bonded to *two* tetrahedral cations and hence are *bridging oxygens* (located at points 13 to 22 in Fig. 9-40).

If a crystal's structure is known, its sharing coefficient S can be calculated and used in its classification (as in Table 9-3). In addition, if we can deduce the values of T and A from the chemical formula for an unknown structure, the sharing coefficient offers an important criterion for testing "trial structures," which are educated guesses as to how the atoms are disposed in this unknown structure. Large values for the sharing coefficient S indicate a high degree of sharing of tetrahedral corners in the structure.

Summary and Discussion

Crystals divide into two major types, molecular and nonmolecular. The nonmolecular crystals may be subdivided into at least three types: metallic, covalent, or ionic. Actually, a bond is rarely 100 percent metallic, covalent, or ionic in character and this threefold division was chiefly adopted for convenience of discussion. Most of the classic structural types can be likened to either a hexagonal closest packing or a cubic closest packing of large spheres with smaller spheres filling the octahedral or tetrahedral interstices either completely or fractionally (Table 9-4).

In general the structure adopted by a compound may be influenced by the degree to which its bonding is ionic or covalent. Structures like that of NaCl (Fig. 9-16), CsCl (Fig. 9-15B), TiO_2 (Fig. 9-22B), and CaF_2 (Fig. 9-22A) seem favored by ionic bonding. On the other hand, structures like $CdCl_2$ (Fig. 9-20A), CdI_2 (Fig. 9-21), ZnS (sphalerite, Fig. 9-6A), ZnS (wurtzite, Fig. 9-7A), and NiAs (Fig. 9-17A) appear favored by covalent bonding. In metallic crystals,

TABLE 9–4

Some Classic Structure Types Classified According to Packing

X Ions in CCP Array			
Octahedral Interstices		Tetrahedral Interstices	
100% filled: $\mathbf{A}^o\mathbf{X}^c$ NaCl (halite) (Fig. 9-16)	50% filled: $\mathbf{A}^o\mathbf{X}_2{}^c$ $CdCl_2$ (Fig. 9-20)	50% filled: $\mathbf{A}^t\mathbf{X}^c$ ZnS (sphalerite) (Fig. 9-6)	25% filled: $\mathbf{A}^t\mathbf{X}_2{}^c$ Unknown
X Ions in HCP Array			
Octahedral Interstices		Tetrahedral Interstices	
100% filled: $\mathbf{A}^o\mathbf{X}^h$ NiAs (niccolite) (Fig. 9-17A)	50% filled: $\mathbf{A}^o\mathbf{X}_2{}^h$ CdI_2 (Fig. 9-21)	50% filled: $\mathbf{A}^t\mathbf{X}^h$ ZnS (wurtzite) (Fig. 9-7)	25% filled: $\mathbf{A}^t\mathbf{X}_2{}^h$ Unknown
X Ions in Simple Cubic Array			
Eight-Coordinated (*cb*) Interstices			
	100% filled: $\mathbf{A}^{cb}\mathbf{X}^{sc}$ CsCl (Fig. 9-15)	50% filled: $\mathbf{A}^{cb}\mathbf{X}_2{}^{sc}$ CaF_2 (fluorite) (Fig. 9-22A)	

cubic or hexagonal closest packing of the metal atoms is frequent; body-centered cubic packing (BCC) is also encountered.

Silicate structures were classified by Machatschki (1928) on the basis of how the [SiO_4] tetrahedra share corners in the structure. No structures are known in which these tetrahedra share an edge or a face, the highly charged Si^{4+} ion not being likely to violate Pauling's third rule. Zoltai (1960) reclassified the silicates on the basis of how the [ZO_4] tetrahedra link together, Z not only representing Si^{4+} but other ions such as B^{3+}, Be^{2+}, P^{5+}, Al^{3+}, Ge^{4+}, Ga^{3+}, Mg^{2+}, Fe^{2+}, Fe^{3+}, Zn^{2+}, or Ti^{4+} if coordinated tetrahedrally to four oxygens. Liebau (1962) subclassified the chain and double-chain silicates according to the number of tetrahedra in the repeat unit along the chain.

A radius ratio within the limits of 0.22 to 0.41 is generally assumed to be compatible with tetrahedral coordination; 0.41 to 0.73, with octahedral co-

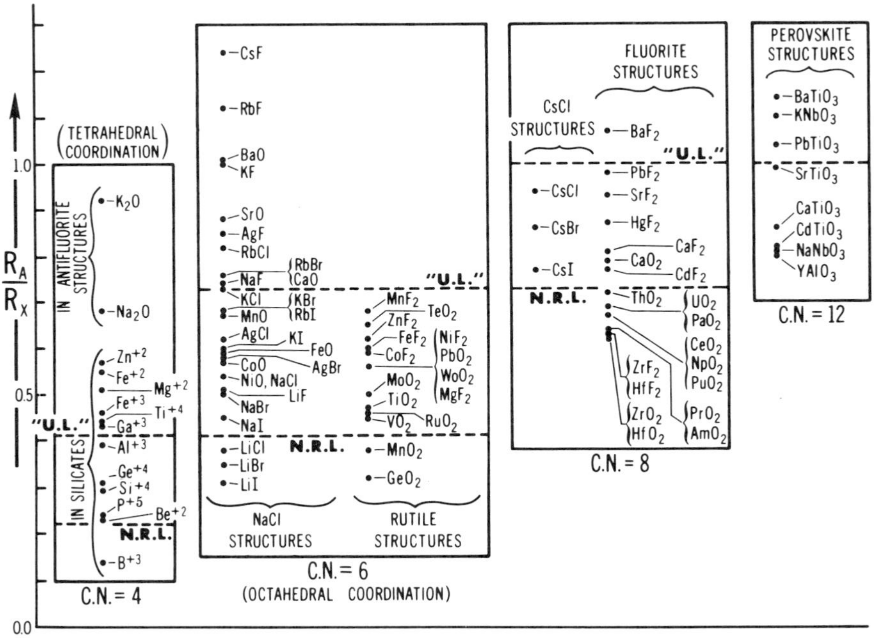

FIGURE 9–41

The radius ratios of cations to anions, R_A/R_X, are here plotted for 4-, 6-, 8-, and 12-fold coordination as observed in various compounds. Ionic radii were taken from Table 8–4 and may differ slightly from the actual radii observed in these compounds. The radius-ratio limits frequently cited for tetrahedral coordination (0.22 to 0.41), for octahedral coordination (0.41 to 0.73), for 8-fold coordination (0.73 to 1), and for 12-fold coordination (*ca.* 1) are seen to be frequently violated; whereas the lower "nonrattle limit" (NRL) is not so frequently violated, the upper limit (UL) is so often violated as to cast doubt on its validity. In short, factors other than radius ratio may also control the type of coordination adopted in ionic compounds.

ordination; 0.73 to 1.00, with 8-fold coordination; and (*ca.*) 1.00, with 12-fold coordination. In each case, however, the upper limit is exceeded so frequently as to be of doubtful validity. For example, K^+ coordinates tetrahedrally to O^{2-} in K_2O even though the radius ratio 0.89 greatly exceeds 0.41. Figure 9-41, which shows the range of radius ratios, possible for some examples of 4-, 6-, 8-, and 12-fold coordination, indicates that radius ratios not only occur above the "upper limits" but also, particularly for 8- and 12-fold coordinations, below the more rigidly observed lower or nonrattle limits (NRL).

It thus appears that a cation may adopt a coordination unrelated to its radius ratio when factors other than geometry become important. For example, Mg^{2+} foregoes its usual, radius-ratio-dictated octahedral coordination to O^{2-} to adopt a tetrahedral coordination in the spinel $MgAl_2O_4$. This defection better satisfies Pauling's second rule for local charge balance, namely, that charges on ions will be neutralized by their first neighbors. Thus in the general spinel structure, $A^tB_2^oO_4$ (Fig. 9-27), one tetrahedral A site and three octahedral B sites surround each O^{2-} ion. If Mg^{2+} occupies the A sites, it directs an electrostatic bond of $+\frac{2}{4}$ or $\frac{1}{2}$ to each of its four O^{2-} neighbors. Similarly, Al^{3+} in an octahedral B site directs an electrostatic bond of $+\frac{3}{6}$ toward each of its six O^{2-} neighbors. Hence the total electrostatic bonds leading to an O^{2-} ion from its one A and three B neighbors is $+2$, precisely enough to balance its -2 charge. On the other hand, if Mg^{2+} had occupied half of the B sites and if Al^{3+} had occupied the remaining B and A sites, then around alternate O^{2-} sites in this spinel structure the first neighbor cations (with the electrostatic bond they direct toward O^{2-} in parentheses) would be alternately

$$\begin{array}{lll} A\ \{Al^{3+} & (+\frac{3}{4}) \\ B \begin{cases} Al^{3+} & (+\frac{3}{6}) \\ Mg^{2+} & (+\frac{2}{6}) \\ Mg^{2+} & (+\frac{2}{6}) \end{cases} \\ \hline \text{Total:} & (1.92) \end{array} \quad \text{and then} \quad \begin{array}{lll} A\ \{Al^{3+} & (+\frac{3}{4}) \\ B \begin{cases} Al^{3+} & (+\frac{3}{6}) \\ Al^{3+} & (+\frac{3}{6}) \\ Mg^{2+} & (+\frac{2}{6}) \end{cases} \\ \hline \text{Total:} & (2.08) \end{array}$$

Failure of the electrostatic bonds leading to an O^{2-} ion to total precisely $+2.0$ indicates that if Mg^{2+} occupied its preferred octahedral sites in spinel, Pauling's second rule would be violated.

In this hypothetical spinel the O^{2-} ion whose electrostatic bonds total 2.08 is said to be *overbonded*; the O^{2-} ion whose total was 1.92 is *underbonded*. The amount of their *overbonding* or *underbonding* is 0.08. In actual silicate structures, particularly those in which highly charged cations such as Si^{4+} and Al^{3+} predominate, amounts of overbonding or underbonding up to 0.3 are observed. Hence Pauling's second rule is not inevitably obeyed. As discussed by Zachariasen (1963), such underbonded O^{2-} ions tend to compensate for the deficiency of positive charges around them by drawing their cation neighbors closer to them, thereby making the positive attractions from these neighbors more effective. Consequently, the bond lengths between underbonded oxygens and their cation neighbors are frequently smaller than usual in proportion to the amount of underbonding. By contrast, bond lengths from overbonded oxygens to their cation neighbors will be longer than usual.

Local charge balance seems to be a factor in the structure of pyrope, $Mg_3Al_2Si_3O_{12}$, in which Si^{4+} coordinates to four O^{2-} ions, Al^{3+} to six, and Mg^{2+} to eight. To each of their O^{2-} neighbors, Si^{4+}, Al^{3+}, and Mg^{2+}, respectively, extend electrostatic bonds of $+\frac{4}{4}$, $+\frac{3}{6}$, and $+\frac{2}{8}$. In pyrope's structure every O^{2-} ion bonds to one Si^{4+}, one Al^{3+}, and two Mg^{2+} ions. Hence the total of the electrostatic bonds leading to an O^{2-} equals $+2$, that is, $+\frac{4}{4}$ plus $+\frac{3}{6}$ plus twice $+\frac{2}{8}$. If Mg^{2+} were to occupy the octahedral sites in pyrope, such local charge balance would not be possible.

Influence of pressure and temperature

For ionic crystals increased pressure reduces the size of the anions more than of the smaller, "less compressible" cations. In effect, therefore, the radius ratio R_A/R_X increases as pressure increases. In the alkali halides the structures adopted at room temperatures and pressures are summarized in Figure 9-42. With increase of pressure, however, the dashed vertical boundary moves to the left because with a pressure-induced increase of radius ratio an 8-fold coordination of the cation (CsCl structure) becomes favored over a 6-fold coordination (NaCl structure). Shannon and Prewitt (1968, p. 273) note that phase transformations resulting from increased pressure involve either (1) a cation adopting a higher coordination number, as for example, when increased pressure causes RbCl, RbBr, or RbI to adopt the CsCl rather than the NaCl structure, or (2) increased

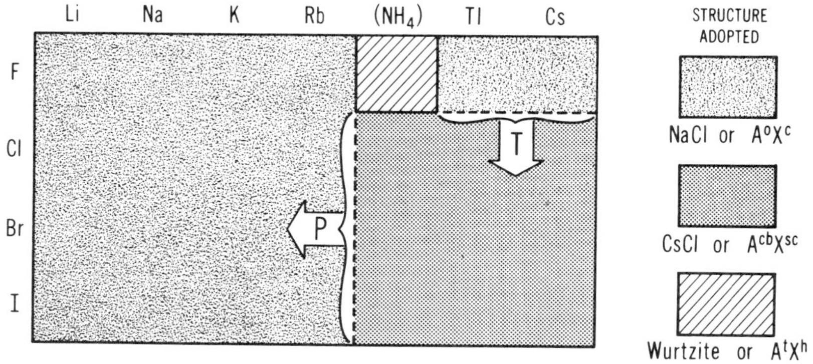

FIGURE 9–42

Structures adopted by alkali halides at normal pressures and temperatures. Arrow P indicates that if pressure is increased sufficiently, the boundary between the NaCl and CsCl structure shifts to the left; that is, the CsCl structure is favored at high pressures and thus its field of stability increases correspondingly. Arrow T indicates that increased temperature will cause the NaCl structure to be favored over the CsCl structure. (This figure is modified from a drawing by Rustum Roy and used by kind permission.)

packing density of the anions without a change in coordination of the cations. Type (2) transformations usually involve small volume changes.

Very high pressures might cause a transition metal ion to adopt a low-spin (smaller radius) state in preference to a high-spin (larger radius) state in conformity to Le Chatelier's principle, which is discussed in Chapter 10. Fyfe (1960), using more definitive arguments, suggests that Fe^{2+} would adopt a low-spin state in olivines deep in the earth's mantle. In Table 8-5 Fe^{2+} in sixfold coordination has radii of 0.61 Å (low spin) and 0.77 Å (high spin), whereas Mg^{2+} has just one spin state and radius (0.72 Å). G. E. Brown and G. V. Gibbs (personal communication) cite the possibility that at low pressures Fe^{2+} (0.77 Å) and Mg^{2+} (0.72 Å) are close enough in size to fill the same niche in the olivine structure, whereas at high pressures they may be sufficiently different in effective size that each fills different sites in the structure (see *ordering*, Chapter 10).

By contrast phase transformations resulting from increased temperature usually involve either (1) a cation adopting a lower coordination number or (2) decreased packing density of the ions or (3) both. In other words increased temperature generally favors more open structures. Thus as increasingly higher temperatures are considered, the dashed horizontal line in Figure 9-42 moves downward; that is, TlCl and CsCl adopt the lower-coordinated NaCl structure at sufficiently high temperatures.

REFERENCES AND RECOMMENDED READING

Bragg, W. L., G. F. Claringbull, and W. H. Taylor (1965) *Crystal Structures of Minerals*, Ithaca: Cornell Univ. Press.

Buerger, M. J. (1954) "The stuffed derivatives of the silica structures," *Am. Mineral.*, **39,** 600–614.

Evans, R. C. (1964) *An Introduction to Crystal Chemistry*, 2d ed. Cambridge: Cambridge Univ. Press.

Fyfe, W. S. (1960) "The possibility of d-electron coupling in olivine at high pressures," *Geochim. Cosmochim. Acta*, **19,** 141–143.

Goldschmidt, V. M. (1937) "The principles of distribution of chemical elements in minerals and rocks," *J. Chem. Soc.*, 655 et seq.

Hauth, W. E., Jr. (1951) "Crystal chemistry in ceramics," *Am. Ceram. Soc. Bull.*, **30,** 5–7, 47–49, 76–77, 137–142, 165–167, 203–205.

Liebau, F. (1961) "Untersuchungen über die Grösse des Si–O–Si-Valenzwinkels," *Acta Cryst.*, **14,** 1103–1109.

——— (1962) "Die systematik der Silikate," *Naturwiss.*, **21,** 481–491.

——— (1968) "Ein Beitrag zur Kristallchemie der Schichtsilikate," *Acta Cryst.*, 690–699.

Lima-de-Faria, J. (1965) "Systematic derivation of inorganic close-packed structures: AX and AX_2 compounds, sequence of equal layers," *Z. Krist.*, **122,** 359–374.

Machatschki, F. (1928) "Zur Frage der Struktur und Konstitution der Feldspate Gleichzeitig vorläufige Mitteilung über die Prinzikien des Baues der Silikate," *Centralblatt Min. Abt. A.*, 97–104.

Osborn, E. F. (1950) "Segregation of elements during the crystallization of a magma," *Jour. Amer. Ceram. Soc.*, **33,** 219–224.

Pauling, Linus (1929) "The principles determining the structure of complex ionic crystals," *Jour. Amer. Chem. Soc.*, **51,** 1010–1026.

Robinson, Paul D., and J. H. Fang (1970) "The crystal structure of epididymite," *Am. Mineral.*, **55,** 1541–1549.

Shannon, R. D., and C. T. Prewitt (1968) "Crystal chemistry of high pressure phases," *Geol. Soc. Amer., abstracts,* 273.

Zachariasen, W. H. (1963) "The crystal structure of monoclinic metaboric acid," *Acta Cryst.*, **16,** 385–389.

Zoltai, Tibor (1960) "Classification of silicates and other minerals, with tetrahedral structures," *Am. Mineral.*, **45,** 960–973.

QUESTIONS AND PROBLEMS

9-1 Determine for an ABX_3 compound with the perovskite structure and O^{2-} ions in the X sites whether Pauling's second rule is better observed if the valencies of the A and B ions are (a) +1 and +5, respectively, (b) +2 and +4, respectively, or (c) +3 and +3.

9-2 What is the percentage of pore space for (a) simple cubic packing? (b) body-centered cubic packing? (c) cubic closest-packing? (d) hexagonal closest-packing? (*Hint*: Make the calculation for a single unit cell and take into consideration that only one-half, one-quarter, or one-eighth of a sphere is owned by a unit cell if this sphere is centered on a face, an edge, or a corner, respectively, of this unit cell.)

9-3 The radius ratios for SrO and BaO, 0.80 and 0.96, respectively, appear to favor a CsCl structure. Why do they adopt the NaCl structure?

9-4 Determine that the fluorite structure and the rutile structure conform to Pauling's second rule and also to the Verwey-Heilmann principle of maximal charge neutralization.

9-5 Demonstrate that the radius ratio representing the lower "no-rattle" limit for tetrahedral coordination is 0.225.

ANSWERS

9-1 Equally well for (a), (b), and (c). Thus in perovskites, each A ion is coordinated by 12 O^{2-} ions, each B ion by 6 O^{2-} ions. The strength of the valency bond they distribute to each oxygen is, respectively, A/12 and B/6, where A and B are their valence charges. But each O^{2-} ion is coordinated by four A and two B ions so that the positive charge from these neighbors is

$$4\left(\frac{A}{12}\right) + 2\left(\frac{B}{6}\right)$$

which simplifies to (A + B)/3. If this equals +2, as it must whenever the valences (A + B) total +6, Pauling's second rule holds precisely.

9-2 (a) 47.64 percent; (b) 31.98 percent; (c) 25.95 percent; (d) 25.95 percent (called "rhombohedral packing" by sedimentary petrologists).

9-3 In the CsCl structure (Fig. 9-15A), the A cations are not well shielded from each other, no anions being directly interposed between them. Sr^{2+} and Ba^{2+} have twice the charge of Cs^+ and the lack of shielding in the CsCl

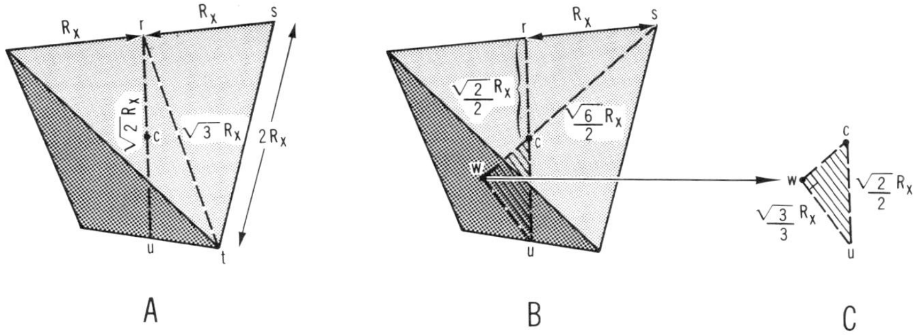

FIGURE 9–43

The dimensions of various lines in a tetrahedron relative to the radius (R_X) of the four, mutually tangent anions centered at its corners.

structure would cause strong cation repulsions. In the NaCl structure, anions interpose directly between cations to reduce cation-cation repulsions.

9-4 In fluorite (Fig. 9-22A), Ca^{2+} is surrounded by eight F^- ions; thus it "directs" an electrostatic bond of $+\frac{2}{8}$ or $\frac{1}{4}$ toward each F^- ion. Four Ca^{2+} cations surround each F^- ion and, in conformity to Pauling's second rule, the sum of their electrostatic bonds, $+\frac{1}{4} \times 4$, precisely equals (and neutralizes) the -1 charge on this F^- ion. Verwey-Heilmann turn the matter around. They note that each F^- ion is surrounded by four Ca ions and thus "directs" an electrostatic bond of $-\frac{1}{4}$ toward each Ca ion. Eight F^- ions surround each Ca^{2+} cation and, in conformity to the Verwey-Heilmann principle, the sum of their electrostatic bonds, $-\frac{1}{4} \times 8$, precisely equals (and neutralizes) the $+2$ charge on this Ca^{2+} ion. The analysis of rutile (Fig. 9-22C) is similar to that (above) for fluorite.

9-5 Four mutually tangent anions of radius R_x define a tetrahedron (Fig. 9-43A) with sides $2R_x$. Applying the Pythagorean theorem to right triangle *trs*, line *tr* equals $\sqrt{3}R_x$. Similarly, from triangle *tur*, line *ur* equals $\sqrt{2}R_x$. A cation tangent to the four anions would be centered at *c*, the midpoint of line *ru*. Thus line *cr* (Fig. 9-43B) equals $(\sqrt{2}/2)R_x$ and, from right triangle *crs*, *cs* must equal $(\sqrt{6}/2)R_x$. Line *cs*, the center-to-center distance between anion and cation, must equal the sum of their radii to avoid "rattle." Thus

$$R_c + R_x = cs = \frac{\sqrt{6}}{2} R_x$$

where R_c is the cation's radius. Dividing through by R_x, there results

$$\frac{R_c}{R_x} + 1 = \frac{\sqrt{6}}{2} \qquad \text{or} \qquad \frac{R_c}{R_x} = \frac{\sqrt{6}}{2} - 1 = 0.2247$$

For a radius ratio R_c/R_x less than 0.2247 "rattling" would occur.

Although not germane to the present problem, tetrahedra are so important in crystal structures that it is worthy of note that right triangle *uwc*, redrawn in Figure 9-43C, permits the length of *cw* to be determined as $(\sqrt{6}/6)R_x$. Distance *sw* thus equals $\frac{2}{3}\sqrt{6}R_x$, or $1.633R_x$.

10 Structural Variations, Composition, and Stability

Thus far crystal structures have been discussed as if ideal, that is, as if devoid of any irregularities so as to possess *perfect short-range* and also *perfect long-range order*. Perfect short-range order exists if the immediate neighbors around any given atom A in the structure are comparable in number, type, and geometric distribution to those around all other A atoms related to the given atom A by translational or rotational symmetry. To understand long-range order, we let V_1, V_2, and V_3 represent a few of the numerous vectors that can translate one identical site (a in Fig. 10-1A) to another (a') in a crystal structure. To the extent that such vectors can be increased n times in length (where n is the integer 1, 2, ... 100, . . . , 1000, . . .) and *yet continue to start and end at identical sites in the structure*, the structure possesses perfect long-range order. For example, V_2 if doubled would end at a'', a site identical to both a and a'.

Structural Imperfections

Mosaic crystals

An actual crystal diffracts X-rays differently than it would if its structure were "perfect," a perfect structure being defined here as possessing perfect long-range and perfect short-range order. To explain these observed diffraction effects mathematically, C. G. Darwin (1914) considered a single crystal as consisting of a mosaic of slightly misoriented blocks with edges on the order of 10^3 to 10^4 Å long. Such a *mosaic structure* (Fig. 10-1B) compared with the perfect structure (Fig. 10-1A) possesses less long-range order but not much less short-range order because, for example, the coordination of X atoms around A would be the same in one block as in the others and would differ only at block boundaries.

A
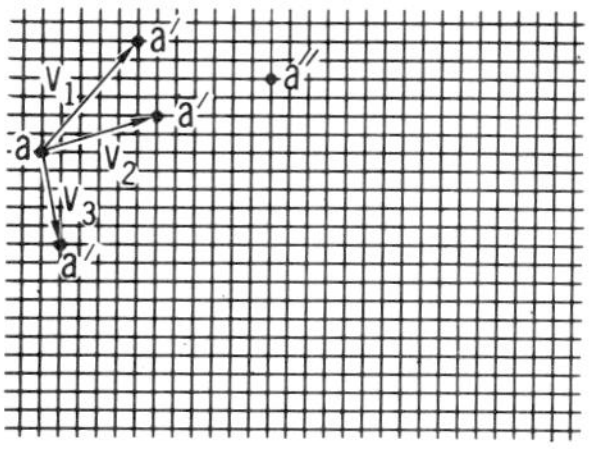

B
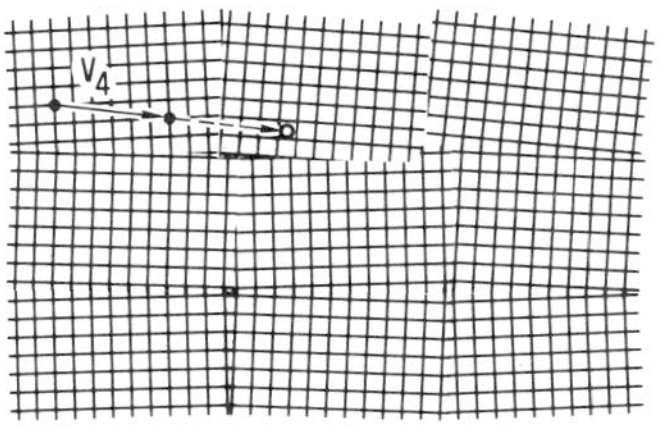

FIGURE 10–1

Two-dimensional representations of the lattice underlying the structure of (A) a perfect crystal and (B) a crystal with mosaic structure, that is, consisting of slightly misoriented blocks, their misorientation being vastly exaggerated. The line intersections mark the lattice points for the structure. The ideally perfect crystal in (A) represents perfect long-range order because any vector that starts and ends at a lattice point, for example, V_1, V_2, or V_3, may be doubled, tripled, or multiplied a thousandfold yet still terminate on a lattice point identical to its starting point. The mosaic structure in (B) lacks perfect long-range order. Thus if vector V_4 is doubled, as indicated by the dashed arrow, it ends in a different block and thus at a point (the hollow circle) which is structurally different from its starting point.

Lineage structures

In 1934 M. J. Buerger introduced an alternative hypothesis that presumes a crystal to consist of a small perfect region or nucleus from which branched regions extend; as they grow outward, these branches twist enough to become slightly misoriented in respect to each other. During growth these branches, which Buerger call *lineages*, grow thicker so as to eliminate the spaces between them, the boundaries between adjacent branches being marked by slight misorientation. Thus the lineage model considers a crystal to consist of small, nearly perfect regions that are slightly misoriented in respect to each other, yet which may be traced back to a common nucleus, in this sense being structurally continuous. On the other hand, the mosaic model regards a crystal to consist of a jumble of slightly misoriented blocks.

Antiphase or out-of-step domains

In Figure 10-2 section (I) is from a hypothetical crystal of composition AB whose structure has perfect long-range and short-range ordering. Referring to the sites occupied by A atoms as α sites and to those occupied by B atoms as β sites, the probability p of encountering an A atom correctly located in an α site is as follows for an AB crystal,

$$p = \frac{\text{number of A atoms in } \alpha \text{ sites}}{\text{total number of } \alpha \text{ sites}} \qquad \text{(Eq. 10-1)}$$

For the perfectly ordered crystal (I), therefore, $p = 1.0$. In other words, the

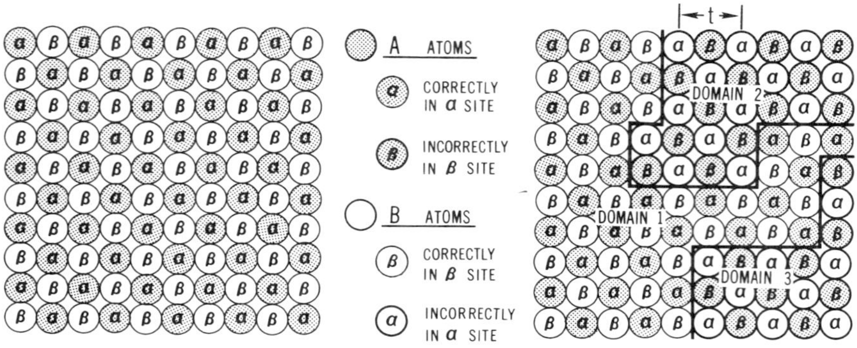

I Perfectly Ordered Structure ($p = 1.0$)

II Disordered Structure ($p = 0.56$; $S = 0.12$)

FIGURE 10–2

(I) Two-dimensional representation of a perfectly ordered structure in which every atom A (shaded) is correctly located in a site labeled α for convenience and every atom B (unshaded) is correctly located in its site labeled β.
(II) A disordered structure for which 22 of those sites labeled α in crystal I are now incorrectly filled by unshaded B atoms. Atoms in incorrect sites are drawn with heavy outlines. This crystal can alternatively be regarded to consist of three domains, each by itself perfectly ordered. Of these, domains 2 and 3, if compared to either domain 1 or the perfectly ordered structure at left, have A atoms where B should be (and vice versa). Domains 2 and 3 are said to be out of phase or antiphase domains in respect to domain 1, "zigging where it zags."

α sites are 100 percent occupied by A atoms. In Figure 10-2 we let (II) represent a section from a disordered AB crystal in which only 28 of the 50 α sites are correctly filled by A atoms and for which, consequently, $p = 0.56$; that is, 56 percent of its α sites are filled by A atoms (shaded) and 56 percent of its β sites by B atoms (unshaded). In general for an AB crystal the probabilities of encountering an A atom correctly located in an α site or incorrectly located in a β site are related as follows:

$$\text{A correctly in } \alpha = \{p\} \qquad \text{B incorrectly in } \alpha = \{1 - p\}$$

$$\text{A incorrectly in } \beta = \{1 - p\} \qquad \text{B correctly in } \beta = \{p\} \qquad \text{(Eq. 10-2)}$$

Complete disorder occurs in an AB compound if p equals 0.50 because this signifies a complete lack of preference of A atoms for α sites and of B atoms for β sites; considering a specific α site (or β site), the chances would then be 50-50 of encountering an A atom (or a B atom) in it.

Measures of Long-Range Order. The value of p, as determined for the entire crystal, is thus a measure of the crystal's long-range order, varying from 1.0 for complete long-range order to 0.5 for complete long-range disorder. As a measure of long-range order in AB compounds, Bragg and Williams (1934) use the quantity s, which they define as

$$s = 2p - 1 \qquad \text{(Eq. 10-3)}$$

For complete order ($p = 1$), s then equals unity; for complete disorder ($p = 0.5$), s equals zero. For a measure of the long-range order in A_mB_n compounds, Lipson (1950) suggests the quantity

$$\frac{p - r}{1 - r} \qquad \text{(Eq. 10-4)}$$

where p is as already defined in Equation (10-1) and r represents that fraction of the total atoms in the structure which are A atoms. Thus for A_mB_n compounds, r equals $m/(m + n)$. For AB compounds, r equals $\frac{1}{2}$ and the quantity cited by Lipson becomes equal to $(2p - 1)$, the Bragg-Williams value for s.

Crystal II in Figure 10-2 can be divided into three areas called *domains*. Within each domain perfect short-range order exists. Thus within each domain each atom has four unlike atoms as closest neighbors, each A being touched by four B atoms and vice versa. At domain boundaries, however, the short-range order decreases because an A atom may have among its nearest neighbors not only B atoms (from its own domain) but A atoms from the neighboring domain.

Domains 2 and 3 are, so to speak, precisely out of step with domain 1, having B atoms (unshaded) in the α sites and A atoms (shaded) in the β sites. Such *out-of-step domains* are called *antiphase domains* by analogy to the fact that when two waves are so out of step that the crests of one coincide with the troughs of the other, they are said to be "out-of-phase."[1] For each domain in crystal II we select the same set of identipoints, for example, the centers of the B (unshaded) atoms. The resultant plane lattices resemble sections of a face-centered cubic lattice with unit translation t. A piece of tracing paper over crystal II will permit the reader to draw these lattices as suggested. When this is done, he will note that if the lattice for domain 1 is extended beyond the domain's borders, its lattice points fall midway between those for domain 2 (or 3). This is characteristic of antiphase domains because of their "out-of-step" nature.

Defect structures

The physical strength observed for a crystal is frequently thousands of times less than if the crystal structure were perfect. Evidently, therefore, a crystal structure contains defects that act as sources of weakness. Because these defects can drastically alter the physical properties of a crystal, evidence for their existence and nature has been eagerly sought. The defects that have received the most attention are of two types: (1) *point defects*, which occur at particular points in the crystal's structure, and (2) *line defects*, which disrupt the structure's regularity along a line.

Point Defects. Various types of point defects have been proposed, some

[1] Wave motion is briefly discussed in Chapter 12.

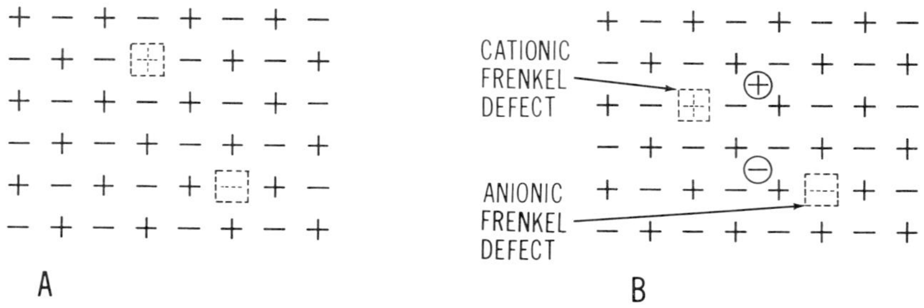

FIGURE 10–3

Schematic, two-dimensional representation of a crystal structure consisting of cations (+) and anions (−). The dashed squares represent vacancies where a cation or anion failed to occupy its proper site. The circles represent a cation or anion incorrectly located in an interstitial site.
(A) A Schottky defect, that is, the paired absence of a cation and anion.
(B) A cationic Frenkel defect (the upper square) represents the absence of a cation from its proper site and its emplacement in an incorrect interstitial site. The lower square and circle represent an anionic Frenkel defect.

bearing the name of the proposer. Thus *Schottky defects* represent the absences of cations and anions from their normal sites in an ionic structure. These defects were proposed in respect to ionic crystals whose cations and anions were charged +1 and −1, respectively. Hence in order to preserve electrical neutrality, total cation vacancies must equal total anion vacancies. In other words a Schottky defect represents a dual imperfection, that is, a cation vacancy coupled with an anion vacancy in another site (Fig. 10-3A). A *Frenkel defect* is also dual in that it represents, in effect, the absence of a cation or anion from its proper site in the structure and its mislocation into the interstices between the structure's component ions (Fig. 10-3B). Such displacement of a cation into an interstitial site is called a *cationic* Frenkel defect. These are more frequently observed than *anionic* Frenkel defects because cations are usually the smallest spheres in an ionic crystal structure and thus can more readily fit into the interstices between the larger anions.

Impurity Defects are point defects resulting from the introduction of a foreign atom or ion either (1) into the interstices between the atoms (or ions) that compose the crystal's structure or (2) in place of one of these atoms or ions. Impurity defects involve a change in chemical composition from that of the "perfect" crystal, whereas Schottky or Frenkel defects are isochemical in nature, involving no change. Actually, however, the amount of impurity may be so small as to defy chemical detection; yet the resultant impurity defects may significantly affect the crystal's properties, for example, its color.

The presence of an electron trapped in a vacant site normally occupied by an anion in the alkali halides has been proposed as a cause of their coloration, this defect being called an *F* center. On the other hand, in covalent crystals an electron may occasionally be absent from the particular electron-pair bond to which it belongs. Such electron-deficient sites are called *holes*, but this term should not be confused with the word "vacancy" which signifies the absence of an atom or ion from its proper site.

Point defects are not necessarily static in a crystal's structure, particularly if the crystal's temperature is elevated. For example, interstitial impurities may diffuse through a crystal by jumping from one interstitial site to the next.

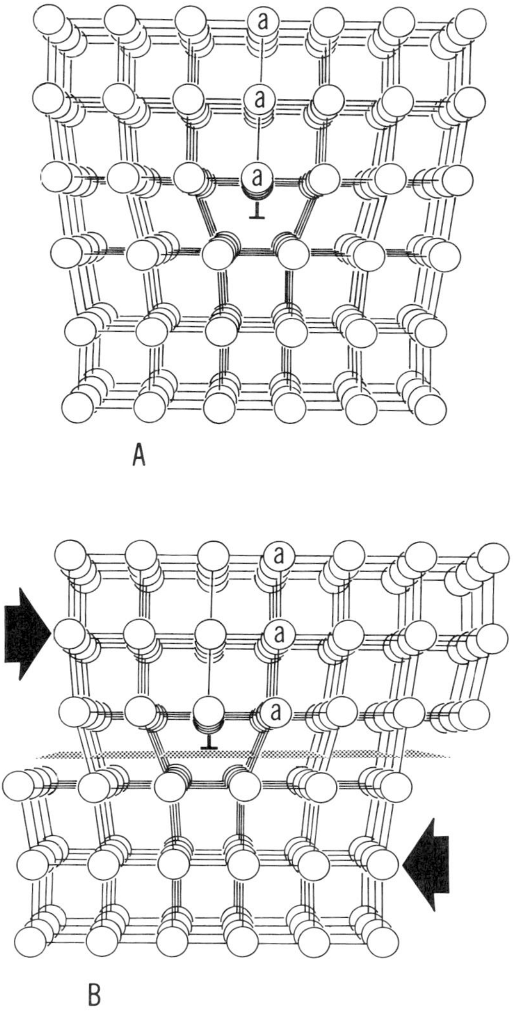

FIGURE 10–4

(A) The atoms labeled *a*, plus those behind them, constitute a plane whose edge terminates at the upside-down T, the common graphic symbol for such edge dislocations.
(B) Migration of the edge dislocation in the crystal in response to stress, the upper part of the crystal slipping (along the shaded *slip plane*) in respect to the lower part. The atoms labeled *a* no longer constitute the edge dislocation.

Similarly, where vacancies exist, as in Schottky or Frenkel defects, a nearby atom A of the type absent from the vacancy may enter the vacancy whereupon A′, a second atom of this type, may enter the site vacated by A. In this way the vacancy has, so to speak, diffused from its original site to that vacated by A′, whereas the atoms have diffused in the opposite direction. In like manner a hole may remedy its charge deficiency by capturing an electron from a nearby electron-pair bond which thus becomes the new site of the hole. Consequently, holes and electrons may be considered to diffuse in opposite directions through the structure of an appreciably covalent crystal.

Line Defects or Dislocations. Line defects represent the concentration of defects along a line, but not necessarily a straight line in the structure. Such linear defects are commonly called *dislocations* because they cause offsets of the lattice where they occur, thereby decreasing the crystal's long-range order. Two types of dislocations have gained attention: (1) *edge dislocations* and (2) *screw*

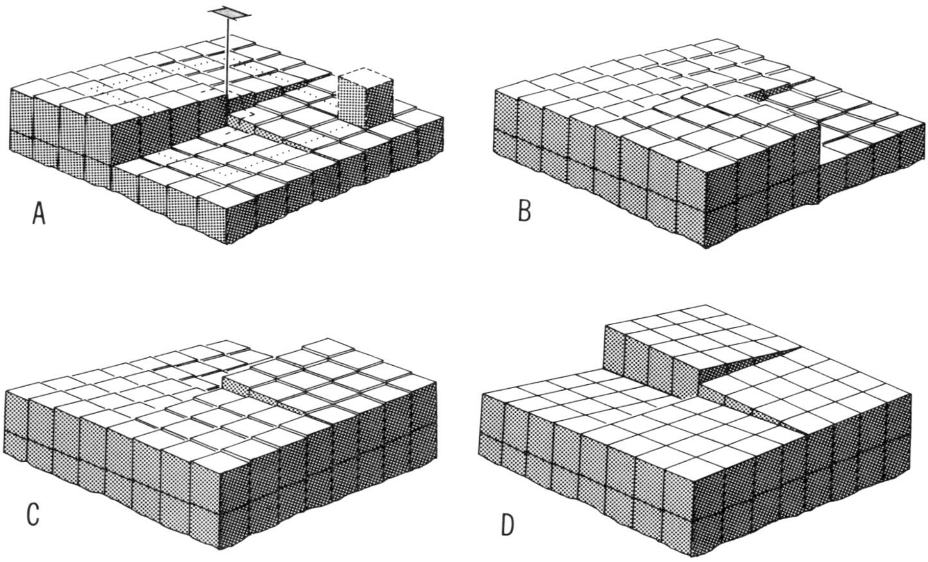

FIGURE 10–5

Location of a screw dislocation (vertical line) on the upper surface of a crystal whose unit cells are shown as cubes.

(A) Dashed lines are drawn atop those cubes whose common edge constitutes the screw dislocation. The crystal grows by adding new material to the edge of the step rather than, for example, starting a new layer in the neighborhood of the dashed unit cell.

(B) (C) (D) Stages in the spiral growth of the crystal's upper surface. For example growth by addition to the front face of the step in (A) will produce (B). (D) is drawn more realistically to show a more gradual transition between atom positions in adjacent cells whereas the neighboring unit cells in (A), (B), and (C) are drawn as if slightly offset, which is less realistic.

dislocations. An edge dislocation results if a plane of atoms or ions in the structure is terminated at a line in the crystal's interior (Fig. 10-4A) rather than continuing through as it would for the perfect crystal. Such edge dislocations permit a crystal, particularly if it is metallic, to deform more readily than if its structure were perfect. For example, if the crystal in Figure 10-4A were subjected to a shearing stress as indicated by the arrows (Fig. 10-4B), its upper part might slide over its lower part as shown. This phenomenon is called *translational slip* and the plane along which it occurs (shaded in Fig. 10-4B) is called the *slip plane*. Such a plane is one of reduced coherence in a crystal because the structural misfit on either side of it is greater than for other planes in the crystal's structure.

Screw Dislocations. Screw dislocations are structural defects caused by the introduction of a screw axis that is not normally present (Fig. 10-5A). The crystal's upper surface now resembles a spiral ramp whose gradient would be greatest immediately around the screw dislocation, that is, along the dashed line atop the unit cells in Figure 10-5A. The gradient becomes increasingly gentle as distance from the screw dislocation increases, for example, along the dotted line, and ultimately decreases to zero so that far from the dislocation the crystal structure again possesses its perfect order. Such spiral ramp-like surfaces are believed to assist crystal growth because the atoms or ions forming the structure are more strongly attracted to the face of the step, particularly where this step is one (or several) unit cells high. For example, there are more bonds attracting them to these sites than to the site of the dashed unit cell in Figure 10-5A, this latter site requiring the nucleation, that is, beginning, of a new layer. Consequently, the crystal may grow more readily by adding to the prominent step in Figure 10-5A and thus producing Figure 10-5B. Similarly, addition to the prominent step in Figure 10-5B produces Figure 10-5C and, in turn, addition to that in Figure 10-5C produces Figure 10-5D, and so on. The crystal can thus grow spirally and thereby avoid the more difficult task of nucleating new layers.

Compositional Variations

The chemical composition of a crystal is summarized by its chemical formula. Thus a crystal consisting of x atoms of iron for every y atoms of oxygen would have the formula Fe_xO_y, the subscripts x and y expressing the relative frequency of iron to oxygen atoms in the structure. Compounds for which the formula subscripts are simple integers are called *daltonide compounds*. To illustrate, FeAsS, $FeCr_2O_4$, and Al_2SiO_5 are daltonide formulas, whereas $FeAs_{1.07}S_{0.93}$ is not. Actually, many crystals, especially of minerals, cannot be assigned a daltonide formula because their structures possess a large number of impurity defects. Thus if crystals of a given mineral are collected from different localities, they may possess compositions that lie between the limits expressed by two or more daltonide formulas (like various shades of gray between black and white.) In such cases the suite of specimens is called a *solid-solutional* series. The nature of such a series, and the significance of the term *solid solution*, will be clarified in discussing the three important types of solid solution: (1) *substitutional solid solution*, (2) *interstitial solid solution*, and (3) *omissional solid solution*.

Substitutional solid solution

The most important and common of the three types is illustrated by the olivine group of minerals. The compositions of different olivine specimens usually lie between the two daltonide formulas Mg_2SiO_4 and Fe_2SiO_4. Evidently, during the growth of an olivine crystal, the structure will accept either Mg^{2+} or Fe^{2+} in the metal atom sites labeled M in Figure 10-6. This is possible because these ions are of equal charge and, more important, of almost identical size. Thus either fits into the M sites. Consequently, most natural crystals of olivine are of compositions expressible by the nondaltonide formula $Mg_{2-x}Fe_xSiO_4$, where x may vary between 0 and 2.0. For example, let us assume that a particular olivine crystal has the formula $Mg_{1.81}Fe_{0.19}SiO_4$; that is, x equals 0.19 in the general

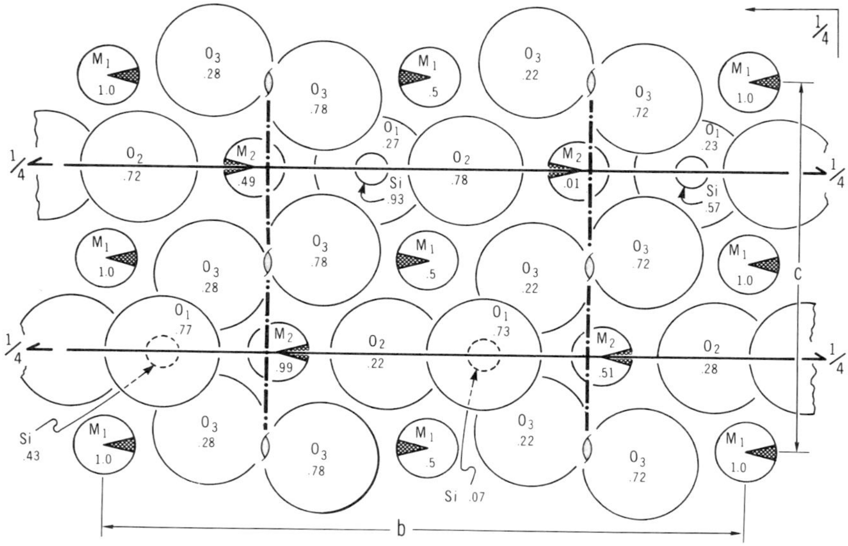

FIGURE 10–6

Structure of olivine, $(Fe, Mg)_2SiO_4$, viewed down the a axis. The numbers represent the percent of the unit of translation along a that each atom lies above a selected plane parallel to (100) and the page. The metal atom sites are labeled M_1 and M_2, the differing subscripts serving to indicate that no symmetry operation in olivine's space group ($Pbnm$) relates M_1 to M_2. In the olivines Fe^{2+} and Mg^{2+} indiscriminately occupy both sites. In the olivine illustrated here 9.5 percent of all the M_1 and M_2 sites in the crystal are occupied by Fe^{2+}, the remainder by Mg^{2+}. To indicate this graphically 9.5 percent of the atom at each M site is shaded to indicate that the probability is 9.5 percent that it will be filled by Fe^{2+}.

formula. In this olivine crystal, therefore, 90.5 percent of all *M* sites[2] are filled by Mg^{2+} and the remaining 9.5 percent by Fe^{2+}. In other words, the chances that a specific *M* site in this crystal will be occupied by Mg^{2+} or Fe^{2+} are 90.5 and 9.5 percent, respectively. To indicate this diagrammatically, 9.5 percent of the area of each ion at an *M* site in Figure 10-6 has been shaded.

Two or more elements that can occupy the same site in a crystal's structure are said to be *diadochic* in that structure. Thus Mg^{2+} and Fe^{2+} are diadochic in the olivine structure. They are said to "proxy" or "substitute" for each other, which actually means that the same site accepts either.

The *end members* for a solid-solution series are, in effect, specimens in which the sites capable of accepting different ions or atoms are occupied by only one type. For the olivine series the end members forsterite (Mg_2SiO_4) and fayalite (Fe_2SiO_4) are recognized; the parenthesized compositions would indicate that the *M* sites are completely filled by Mg^{2+} in one and by Fe^{2+} in the other. For most solid-solution series the end members are more conceptual than actual because even in crystals considered "pure," a small degree of substitution is usually encountered. Thus the great majority of the members of a solid-solution series are intermediate in composition between the end members. These intermediate varieties are usually called *mixed crystals* or, simply, *solid solutions.* Specific compositions are conveniently expressed in terms of the relative percentages of the end members which, if mixed, would chemically correspond to the composition of this mixed crystal. Thus using Fo and Fa as abbreviations for forsterite and fayalite, the formula $Fo_{90.5}Fa_{9.5}$ is preferable to $Mg_{1.81}Fe_{0.19}SiO_4$ for expressing the composition of the olivine crystal previously discussed. An unspecified composition in the series is frequently written as $(Mg, Fe)_2SiO_4$; this indicates that the elements in parentheses proxy for each other, but does not specify their relative proportions in the site.

Solid solutions between molecular crystals are more rare than for ionic crystals. They do occur, however, if a whole molecule is of a shape and size such that it can replace another whole molecule.

Interstitial solid solution

Between the ions or atoms packed in a crystal's structure, interstices exist, for example, the tetrahedral and octahedral sites among closest-packed spheres, within which suitably small ions or atoms may lodge. Thus metal crystals often contain varying amounts of carbon, boron, hydrogen, and nitrogen because these atoms are small enough to lodge in the interstices between the metal atoms. The composition of these metals thus varies according to the extent of this interstitial solid solution. Some crystal structures may possess voids or tubular cavities, several angstroms in diameter, in which extraneous atoms may lodge. For example, the structure of beryl ($Be_3Al_2Si_6O_{18}$), one of whose gem varieties is

[2] These percentages are obtained by dividing 1.81 and 0.19, respectively, by their sum and converting to percentages.

emerald, possesses tubular channels that may contain sodium, potassium, and cesium ions and/or occluded gases. As a consequence its chemical analysis may not agree with that indicated by its idealized formula.

Omissional solid solution

The mineral pyrrhotite consists of hexagonally close-packed sulfur atoms with iron atoms filling the octahedral interstices between the sulfur atoms. As previously discussed, there is one octahedral site for each close-packed sulfur atom. Thus if each octahedral site were filled by iron, the formula for pyrrhotite would be FeS. However, in many pyrrhotite crystals up to 20 percent of the octahedral sites may be vacant; hence the formula for pyrrhotite must be written $Fe_{1-x}S$, where x is between 0 and 0.2. Thus pyrrhotites with various values of x constitute an omissional solid-solutional series. Crystals in which a particular site is incompletely filled are sometimes called *defect structures.* Hence an omissional solid-solution series is called by some a *defect solid-solution series.*

Graphical depiction

A graphic means of depicting all the possible compositional variations in a solid-solutional series is of great help during discussions. If only two end members are involved, a line divided into decimal parts (Fig. 10-7A) permits the composition of a given mixed crystal to be expressed by a point on the line. Thus points 1, 2, and 3 in Figure 10-7A, respectively, denote crystals of compositions $Fa_{9.5}$, Fa_{35}, and Fa_{85} in the olivine series. However, if *three* end members must be recognized, a triangular graph (Fig. 10-7B) is required for the graphic depiction of composition in the series. For example, points 1, 2, 3, 4, and 5 in Figure 10-7B, respectively, represent the following compositions: (1) $Ab_{20}Or_5An_{75}$, (2) $Ab_{30}An_{70}$, (3) $Ab_{60}Or_{30}An_{10}$, (4) $Ab_{25}Or_{55}An_{20}$, (5) $Ab_{20}Or_{80}$. The abbreviations Ab, Or, and An indicate the following end members: albite, $NaAlSi_3O_8$; orthoclase, $KAlSi_3O_8$; and anorthite, $CaAl_2Si_2O_8$. In solid solutions between anorthite and albite (or orthoclase) the substitution of Ca^{2+} for Na^+ (or K^+) requires an accompanying substitution of Al^{3+} for Si^{4+}. This double substitution, [Ca^{2+} and Al^{3+}] for [Na^+ (or K^+) and Si^{4+}], permits the structure to preserve electrical neutrality.

For a solid-solutional series between four end members (that is, four elements can enter a given structural site) points plotted in or on a tetrahedron (Fig. 10-7C) satisfactorily depict compositions. Usually such tetrahedral depictions of composition are used only qualitatively. Thus the extent to which a point plots near a corner of the tetrahedron indicates a composition largely consisting of the end member that this corner represents. Von Philipsborn (1928), however, describes a clever method for plotting compositions quantitatively within the tetrahedron.

Limited solid solution

Let L and S denote two elements that proxy for each other in a given crystal structure, L denoting an atom or ion larger than S. The end members corresponding to 100 percent occupancy by L and 100 percent by S may then differ

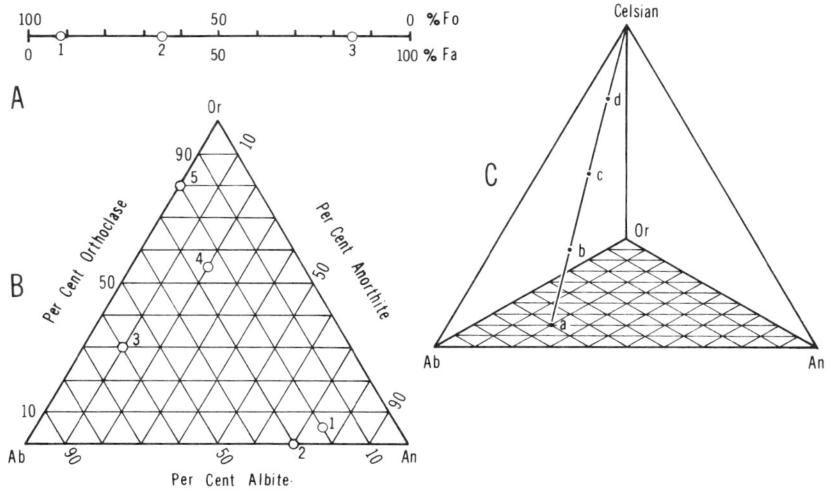

FIGURE 10–7

(A) Graphic depiction of composition for solid solutions between *two* end members, taken here as forsterite, Mg_2SiO_4 (abbreviated Fo), and fayalite, Fe_2SiO_4 (abbreviated Fa). The composition of the olivine whose structure was drawn in Figure 10–6 is indicated by point 1 (open circle).
(B) Depiction of composition for solid solutions between *three* end members as exemplified by the feldspars: albite, $NaAlSi_3O_8$; anorthite, $CaAl_2Si_2O_8$; and orthoclase, $KAlSi_3O_8$ (respective abbreviations, Ab, An, and Or). Standard triangular graph paper can be used.
(C) Depiction for *four* end members; the fourth feldspar end member added here is celsian, $BaAl_2Si_2O_8$. In these solid solutions Ba^{2+} enters the same sites as do Na^+, K^+, and Ca^{2+}. Points *a*, *b*, *c*, and *d* represent compositions (a) $Ab_{60}An_{20}Or_{20}Ce_0$, (b) $Ab_{45}An_{15}Or_{15}Ce_{25}$, (c) $Ab_{30}An_{10}Or_{10}Ce_{50}$, and (d) $Ab_{15}An_5Or_5Ce_{75}$. These four points all represent the same ratio of Ab to An to Or (but differing amounts of Ce) and consequently plot along the straight line representative of this Ab:An:Or ratio.

slightly in structure to the extent of the size difference (or charge difference, if one exists) between L and S. This is particularly true in respect to the locations of atoms or ions that border, and therefore define, the niches that accept L and S. For 100 percent occupancy by L these neighbors may be slightly relocated from their positions in 100 percent S crystals, these positional shifts equaling very small fractions of their radii. Such readjustments enable the niche to be slightly larger in 100 percent L crystals than in 100 percent S crystals. Considering such minute details, the structure of a 95 percent L crystal will more nearly resemble that of a 100 percent L crystal than that of a 100 percent S crystal. Consequently, the 5 percent of S atoms or ions will be a bit too small for the sites they occupy; their presence hence increases potential energy to decrease the crystal's stability. Accordingly, the presence of S atoms beyond a certain amount in a dominantly L crystal may not be tolerated. The same holds true for the presence of L in a dominantly S crystal. Such a phenomenon, in which the only mixed crystals that are stable are those with compositions not much different from the end members, is known as *limited solid solution.* In a *complete solid-solution* series, on the other hand, specimens are stable across the entire compositional range between the two end members. See Figure 10-15A, p. 313.

Effect of temperature

On the basis of similarity in their ionic or atomic radii, it is possible to predict with fair accuracy those elements that may substitute for each other. However, it is not possible to lay down a rigorous rule as to the largest difference in radii beyond which two ions or atoms cannot proxy. Experience shows that different crystal structures have different tolerances and, moreover, physical conditions such as temperature have large effects. Roughly speaking, the more discriminating crystal structures, if grown or maintained at relatively low temperatures, may permit substitution between two elements only if their radii are within 10 percent or less of each other. If a charge difference is involved, this tolerance may be even lower. On the other hand more tolerant structures, if grown at high temperatures, may accept ions in the same niche that differ by roughly 15 percent or more in radii. If bonding is appreciably covalent, two different atoms will be mutually replaceable to the extent that the number and directions of their bonds are similar.

In general, increased temperature of formation enables a structural niche to accept a wider size range of ions. This probably occurs because at elevated temperatures the increased magnitude of the thermal vibrations of the atoms causes the sizes of structural niches to become more flexible.

Chemical Analyses

The chemical analysis of a crystal reports the percentage of its total weight that each of its component elements represents. For arsenopyrite, FeAsS, the analysis might read 34.30 percent iron, 46.01 percent arsenic, and 19.69 percent sulfur. A 100 gram crystal would hence contain 34.30 grams of iron, 46.01 grams of arsenic, and 19.69 grams of sulfur. A crystal's analysis may be a *complete analysis,* which is an attempt to determine the weight percent of every element in the crystal, or a *partial analysis* wherein the attempt is made to determine only a few of the elements present. Ideally, in a complete analysis the determined percentages should total 100 percent as for the arsenopyrite analysis above. Actually, however, analytical errors may cause the total to range between 99.5 and 100.5 percent, even for experienced and competent analysts. Moreover, a total of 100 percent need not necessarily signify an analysis of great accuracy because compensating errors might have occurred, one element being reported erroneously high, another erroneously low.

If Σ, the sum of the percentages for a complete analysis, does not equal 100.0 percent, the "error," that is, the amount that this sum differs from 100.0 percent, can be proportionally distributed by multiplying each reported percentage by $100/\Sigma$. Suppose the analyst had determined arsenopyrite to contain, respectively, 34.13, 45.78, and 19.59 percent (by weight) or iron, arsenic, and sulfur. Thus

$$34.13 + 45.78 + 19.59 = 99.50 \text{ percent} = \Sigma$$

Multiplying both sides of this equation by $100/\Sigma$, that is, by 100.0/99.5, there results

$$34.30 + 46.01 + 19.69 = 100.0 \text{ percent}$$

and the error is seen to be proportionally distributed among the three reported percentages so that they now total 100.0 percent. This method of distributing the error assumes that all values were reported either too high or too low. Such is not always the case; compensating errors may occur.

For minerals containing oxygen, actually the large majority, the chemist usually reports the weight percentages of the oxides of the various elements present. For example, he might report the mineral diopside, $CaMg(SiO_3)_2$, to contain 25.89 percent CaO, 18.62 percent MgO, and 55.49 percent SiO_2. This convention should not be misinterpreted. Minerals that contain oxygen usually consist of oxygen ions in more or less closest packing with various cations occupying the interstices; they are not molecular crystals in which discrete molecules of the various oxides are packed together.

Interconversions between analyses and formulas

If a crystal's complete chemical analysis is known, its chemical formula may be calculated, or vice versa. For example, to determine the formula for arsenopyrite from its complete analysis (column 1 of Table 10-1), we first enter the atomic weights of the elements present into column 2, as done here. As stressed in beginning chemistry, the atomic weight of an element (p. 529) expressed in grams has special significance; it equals the weight of 6.023×10^{23} atoms of the element. Thus 55.85 grams of iron, 74.92 grams of arsenic, or 32.064 grams of sulfur will amount to 6.02×10^{23} atoms for each element. Just as the word "dozen" stands for the number 12, chemists use the word *mole* for the number 6.02×10^{23}. Thus 1 mole of iron equals 6.02×10^{23} atoms or 55.85 grams of iron. Furthermore, if the figures in column 1 are considered as the grams of each element that will be present in 100 grams of the mineral, then each represents the following portion or

TABLE 10–1

Computation of Formula for Arsenopyrite (FeAsS)

	(1) Weight Percent	(2) Atomic Weight	(3) [=(1) ÷ (2)] Atomic Frequency
Fe	34.30	55.85	0.614
As	46.01	74.92	0.614
S	19.69	32.064	0.614

fraction of a mole: (1) 34.30 grams of iron represent 34.30/55.85 or 0.614 mole of iron; (2) 46.01 grams of arsenic represent 46.01/74.92 or 0.614 mole of arsenic; and (3) 19.69 grams of sulfur equal 19.69/32.064 or 0.614 mole of sulfur. The values in column 3 in Table 10-1 thus represent each element's weight percentage (column 1) divided by its atomic weight (column 2). Hence they indicate the number of moles of each element that would be present in 100 grams of the mineral and therefore the three values in column 3 indicate the relative frequency of the atoms of each element in the mineral. Arsenopyrite's formula can hence be roughly written as $Fe_{0.614}As_{0.614}S_{0.614}$ which, if simplified by dividing all subscripts by 0.614, becomes FeAsS.

Determining the formula of an oxygen-containing crystal from its chemical analysis is not unlike the procedure just discussed. Thus to determine the formula for diopside, given its analysis as in column 1 of Table 10-2, we determine the molecular weight for each oxide and enter the value in column 2. Thus the weight of 1 mole of CaO is 56.08, that is, the sum of calcium's atomic weight of 40.08 plus oxygen's atomic weight of 16. The weights of 1 mole of MgO and SiO_2, respectively 40.31 and 60.09 grams, are similarly 24.31 + 16 and 28.09 + (2 × 16). Column 1 divided by column 2 yields column 3 which indicates the number of moles of each oxide in 100 grams of diopside. The values in column 3 thus indicate the relative frequency of the oxides in diopside and permit a tentative formula to be written

$$(CaO)_{0.462} \qquad (MgO)_{0.462} \qquad (SiO_2)_{0.923}$$

This may be simplified by dividing all subscripts by 0.462 to obtain

$$(CaO)\cdot(MgO)\cdot(SiO_2)_2 \qquad \text{or} \qquad CaMgSi_2O_6$$

The formula $CaMgSi_2O_6$ is preferred because it avoids creating the erroneous impression that the crystal is composed of discrete oxide molecules.

If a mineral's chemical formula is known, the percentage of its weight that any of its component elements represent is easily determined. For example, to determine the weight percentage of iron in hematite (Fe_2O_3), we keep in mind that 1 mole of Fe_2O_3 contains 2 moles (= 2 × 55.85 g) of iron plus 3 moles (= 3 × 16.00 g) of oxygen. One mole of Fe_2O_3 thus weighs 159.70 but iron constitutes only 111.70 g of this. Hence the weight percent of iron in Fe_2O_3 is 69.94.

TABLE 10–2

Computation of Formula for Diopside $CaO\cdot MgO\cdot 2SiO_2$ or $CaMg(SiO_3)_2$

	(1) Weight Percent	(2) Molecular Weight of Oxide	(3) [=(1) ÷ (2)] Frequency of Oxides
CaO	25.89	56.08	0.462
MgO	18.62	40.31	0.462
SiO_2	55.49	60.09	0.923

Phase Transformations

Definition of phase

The physical chemist defines a phase to be a homogeneous state of matter that is (conceivably) *physically separable* from any other phase in a system. For example, a sealed container partly filled with a mixture of water and crushed ice contains three phases: (1) a liquid phase (the water), (2) a solid phase (the ice), and (3) a vapor phase that results because H_2O molecules escape from their bonded positions in the ice and water to populate the open space above. The solid and liquid phases could be physically separated from each other by sieving through a fine screen. Similarly, an intimate intergrowth of two or more solid phases could conceivably be crushed fine enough to separate individual grains of the one phase from those of the other.

Stability

Stability is the ability to remain unchanged. Thus a phase that can exist indefinitely within a given range of temperatures and pressures is said to be *stable* over that temperature-pressure range. Ordinarily, if exposed to a temperature or pressure beyond its stability range, a phase will transform into the particular phase (or phases) stable for these new pressure-temperature (P, T) conditions. Thus if ice is subjected to an infinitesimal temperature increase of ΔT degrees above 0°C at atmospheric pressure, it transforms into the liquid phase, that is, it melts. Sometimes if a phase is subjected to a temperature or pressure beyond its stability range, it will transform at an exceedingly slow rate into the phase (or phases) stable at this new temperature-pressure value. Consequently, it can exist for hundreds, thousands, or millions of years at a temperature and pressure beyond its true stability range, all the while *slowly* transforming into the new phase (or phases). In such a case it is said to be *metastable*. Glass is such a metastable material. With age it tends to crystallize or *devitrify*; hence ancient glasses are rarely found in pre-Tertiary rocks, that is, in rocks older than about 70 million years. Meteorites, however, contain glasses of ages up to 4.5 billion years. Unlike natural glasses of terrestrial origins, such glasses are close to $NaAlSi_3O_8$ in composition and thus highly viscous; perhaps they owe their surprisingly greater longevity to this, as well as to the relative absence in outer space of that remarkable catalyst H_2O.

If a solid or liquid were to be confined in a sealed, evacuated container made of a perfectly inert material, then atoms, ions, or molecules would escape from the surface of this solid or liquid to constitute a pressure-exerting gas or vapor filling the remaining space in the container. Eventually the number of escapees breaking their bonds to form this vapor would equal the number of escapees being recaptured from the vapor and again bonded to the main body of the solid or liquid. When such equilibrium is established, the pressure exerted by the vapor represents the substance's *vapor pressure*. Vapor pressure varies from one substance to another, being highest for substances in which bonding is weakest, lowest for substances in which it is strongest. In general, vapor pressure increases with temperature because increased thermal vibrations weaken a substance's chemical bonds, thus making it easier for atoms to escape.

Where two or more polymorphs of a given substance are possible, the one with the lowest vapor pressure for the existent conditions will be stable. For example, under equilibrium conditions at atmospheric pressure, pure sulfur adopts an orthorhombic structure below 95.5°C, a monoclinic structure between 95.5 and 119.25°C, and becomes liquid (melts) above 119.25°C. In Figure 10-8 the curves schematically represent the change in vapor pressure with temperature for orthorhombic sulfur, monoclinic sulfur, and liquid sulfur. Thus at 95.5°C orthorhombic and monoclinic sulfur can stably coexist, their vapor pressures being equal for this temperature. Below 95.5°C, however, monoclinic sulfur would tend to reorganize into the orthorhombic form; between 95.5° and 119.25°C, the reverse would occur.

The phase rule

The word *system* as used by physical chemists signifies a collection of atoms, ions, or molecules that may be considered to be entirely isolated from the rest of the universe. Consequently, they alone can interact to produce the solid, liquid, or gaseous phases present in the system. The *fewest* elements or chemical constituents that, alone or in various proportions, can define the composition of any phase possible for the system represent the system's *components*. Often a system is named by citing its components. For example, the system in which ice,

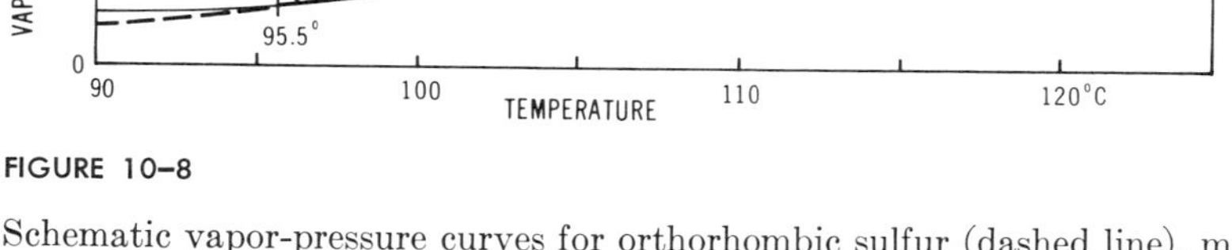

FIGURE 10–8

Schematic vapor-pressure curves for orthorhombic sulfur (dashed line), monoclinic sulfur (solid line), and liquid sulfur (dot-dash line). Vapor pressures are in arbitrary units. For any given temperature, the form that has the lowest vapor pressure at that temperature is stable. Heated slowly, orthorhombic sulfur changes into monoclinic sulfur at 95.5°C. Heated so rapidly that its temperature exceeds 95.5°C before it has time to reorganize into the monoclinic structure, orthorhombic sulfur will melt at 112.8°C. This melt is metastable and, held at 112.8°C, would eventually crystallize as monoclinic sulfur. Monoclinic sulfur on the other hand melts at 119.25°C, the temperature above which liquid sulfur has the lower vapor pressure.

water, and water vapor may occur consists of one component, H_2O, because this by itself describes the composition of each of the three possible phases. Thus unless carried to a temperature so high that H_2O decomposes into H_2 and O_2, (so that one or the other can be independently varied in amount), the system H_2O is a one-component system.

For a system whose phases are in *equilibrium*, that is, no phase is growing at the expense of another, the *phase rule* of J. Willard Gibbs relates p, the number of phases that can coexist in equilibrium, to c, the number of components, and to f, the degrees of freedom, as follows:

$$p + f = c + 2 \qquad \text{(Eq. 10-5)}[3]$$

Actually, the degrees of freedom f represent the number of environmental conditions that can be varied independently of each other without causing the number or types of phases present in the system to change. These conditions include pressure and temperature (the P, T condition), and also chemical composition. Other environmental conditions such as an electrostatic field, magnetic field, or gravitational field apparently have no appreciable effect on the equilibrium between phases in the systems studied to date. If a system were encountered in which one (or more) did have appreciable effect, the constant 2 in Equation (10-5) would need to be increased by one (or more).

One-component system

Figure 10-9, a portion of the phase diagram for the one-component system H_2O, is divided into three areas, the P, T fields for existence of ice, water, and water vapor, respectively. Any point on Figure 10-9 defines (1) a particular combination of P, T values and (2) according to the field in which it falls, the phase (or phases) stable at this P, T combination. If such a point falls at the boundary between two fields for example, point q on line SR, it represents a P, T combination at which *two* phases, in this case ice and water, can coexist indefinitely. Point R, which represents a pressure of 4.58 mm of mercury and a temperature of 0.0100°C, marks the only P, T combination for which *three* phases (ice, water, and vapor) may indefinitely coexist; hence R is called the *triple point.*

For a one-component system, Equation (10-5) reads

$$p + f = 3 \qquad \text{(Eq. 10-6)}$$

This indicates that if three phases are in equilibrium ($p = 3$), then f must equal zero in Equation (10-6). In other words for three phases to coexist indefinitely, P and T must be values that coincide with the nature-dictated triple point R in Figure 10-9. We can change neither P nor T without causing the P, T condition to move away from point R, in which event one or more phases disappear. Thus three phases, the maximum number that can stably coexist in a one-component

[3] The mnemonic device "PFC second class" aids recall of this highly important rule whose derivation is ably discussed by Moore (1962, p. 100).

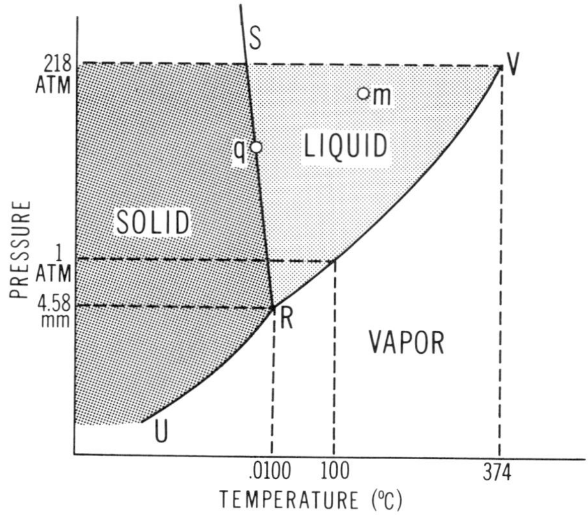

FIGURE 10–9

Schematic diagram for the system H_2O (not drawn to scale). Point R is the nature-dictated triple point, that is, the only P, T point at which all three phases can stably coexist. Point V marks the critical temperature whose significance is discussed on p. 317.

system, do so only if we exercise no degrees of freedom, that is, we vary neither P nor T from the nature-dictated triple point. If only two phases are observed in equilibrium ($p = 2$), Equation (10-6) indicates that f must equal one. In other words *two* phases can coexist indefinitely only at a P, T combination that plots on *boundary lines SR, UR,* or *RV* in Figure 10-9; we can thus exercise the one degree of freedom to choose an arbitrary value for either P or T but the system must, so to speak, be permitted to assume a value for the other such that the resultant P, T point will plot on a boundary line (SR, UR, RV), thereby permitting the two phases to continue their coexistence. If only one phase is observed ($p = 1$), the P, T condition must be represented by a point within one of the three fields in Figure 10-9, for example, point m. In this case both P and T may be varied over a range of values (so that point m moves about in the field) without causing any change in the number or type of phases present in the system; in short, two degrees of freedom exist. If only one phase is observed in a one-component system, Equation (10-6) indicates that $f = 2$, as we have just seen.

Two-component (binary) systems

Two-component systems are those in which two components, alone or in various proportions, can reproduce the composition of any phase possible in the system. These two pure components represent the two extremes of composition for any phase in the system. A system's gaseous phases and, as a rule, its liquid phases may be of compositions grading anywhere between these two extremes. For the solid phases this is not necessarily true. Thus two-component systems will be discussed for which the number of *solid* phases compositionally intermediate between the two extremes is *zero* (as in *eutectic crystallization*), *one* with a specific composition (as in *incongruent melting*, p. 308), or *one* whose composition, like that of a gaseous phase, may range anywhere between those of the two components (as in *crystallization of solid solutions*, p. 311).

No Intermediate Compounds; Eutectic Crystallization. In the system $CaMgSi_2O_6$—$CaAl_2Si_2O_8$, molten liquid phases containing $CaMgSi_2O_6$ and $CaAl_2Si_2O_8$ in any proportion are possible. However, the only solid phases possible are either pure $CaMgSi_2O_6$ (diopside) or pure $CaAl_2Si_2O_8$ (anorthite); no solids of intermediate composition occur.[4] At atmospheric pressure if a crucible containing an intimate mixture of $(100 - X)$ percent diopside plus X percent by weight of anorthite crystals were at a temperature T, the phases present in the crucible would be as follows: (1) for any T, X combination that defines a point in the unshaded field in Figure 10-10A, a single liquid of composition X would alone be present; (2) for any T, X condition falling in the heavily shaded field, a mixture of X weight percent anorthite plus $(100 - X)$ percent diopside crystals would coexist stably; (3) for a T, X condition plotting in the lightly shaded field, diopside crystals plus a melt would coexist; (4) for a T, X condition plotting in the stippled field, anorthite crystals plus a liquid would coexist. The shaded or stippled fields are thus two-phase fields; the unshaded is a single-phase field.

Figure 10-10A is called a *T, X diagram* or a *phase diagram* because it is a plot of temperature T versus composition X that describes the nature, composition, and relative abundance of the phases present for any $CaMgSi_2O_6$—$CaAl_2Si_2O_8$ mixture whose T, X condition is known. Thus point C_1 represents a T, X condition in which a melt containing 75 percent $CaAl_2Si_2O_8$ at 1525°C comprises the entire system. If this system is cooled, the successively new T, X conditions plot along a vertical line of constant composition (dashed in Fig. 10-10A) called an *isopleth.* Thus cooled to 1465°C and below, the system's T, X conditions plot as $C_2 \ldots C_5$. At 1465°C (C_2), anorthite crystals start crystallizing from the liquid. This continues down to 1270°C (C_4) as the system's T, X condition descends into the two-phase field of anorthite plus liquid. For any T, X point such as C_3 (or C_4) in a two-phase field, the composition of the two stable phases is determinable by passing a line of constant temperature called an *isotherm* through this point (short dashes in Fig. 10-10A). This isotherm's two points of exit from the field, for example, S_3 and L_3 (or S_4 and L_4), mark the compositions of the two phases that are stable at the temperature defined by the isotherm. Thus for each isotherm SCL in Figure 10-10A (subscripts omitted here from the symbolism), S and L respectively indicate the composition of the solid and liquid stably coexistent at that temperature, whereas C indicates the system's

[4] The recent discovery that minor amounts of aluminum may substitute for magnesium in diopside is not considered here.

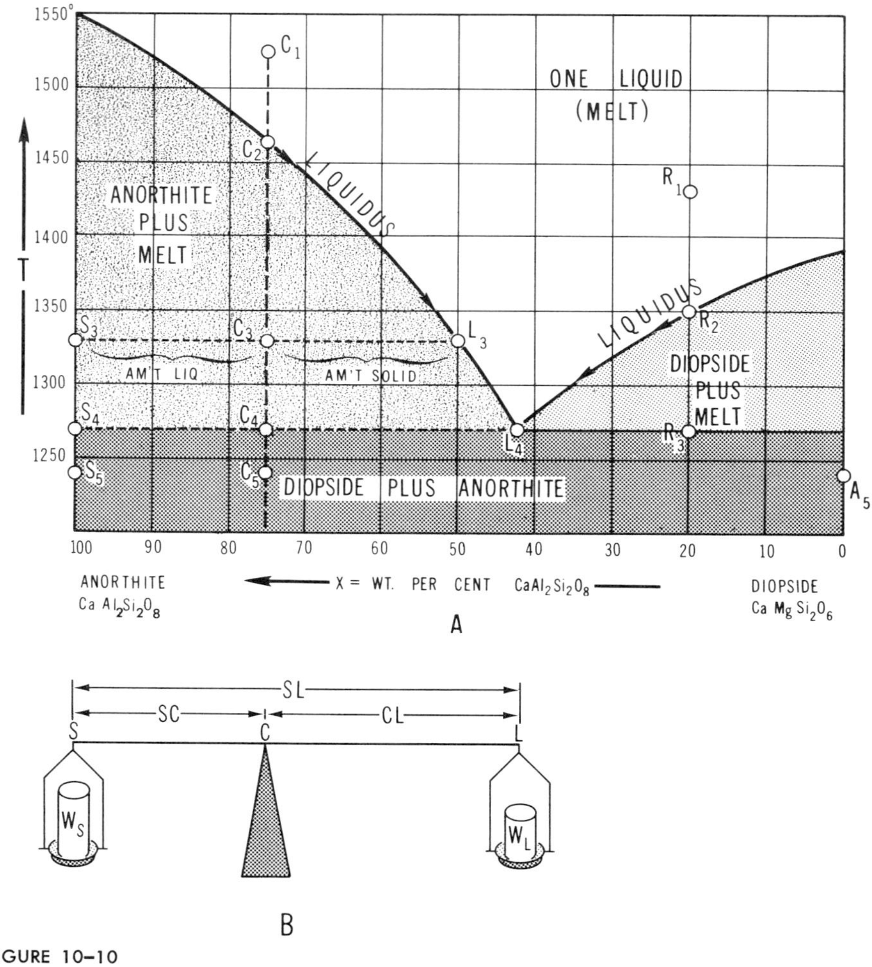

FIGURE 10–10

(A) T, X diagram showing the fields of stability for the phases developed if mixtures of $CaAl_2Si_2O_8$ and $CaMgSi_2O_6$ are heated to various temperatures. For a system containing 75 percent by weight $CaAl_2Si_2O_8$ at 1525°C, its T, X condition plots at C_1, a point in the one-liquid field; the system thus consists of a melt of this composition only. Cooled to 1465°C, the system's T, X condition is now at C_2. Its position on the border of the stippled field indicates that with further cooling, for example, to 1330°C (C_3) anorthite starts to crystallize from the liquid, the liquid's composition following a series of points from C_2 to L_3 along the line labeled *liquidus* as the system's T, X condition changes from C_2 to C_3. For the T, X diagram of a two-component system such as this, the liquidus is a line that represents the boundary between the one-liquid field and the crystal-plus-liquid field. It shows how the composition of the liquid that is stable in contact with the crystals varies with temperature.

A system whose T, X condition plots at R_1 will be completely liquid. Cooled until its T, X condition plots infinitesmally below R_2, crystals of diopside crystallize bulk composition, that is, of the solid and liquid combined. Application of the lever rule (Fig. 10-10B) plus algebraic manipulation indicates that the relative amounts of solid (of composition S) and of liquid (of composition L) are related to line lengths SC, CL, and SL as follows:

$$\text{weight percent solid} = \frac{CL}{SL} \times 100 \text{ percent}$$

$$\text{weight percent liquid} = \frac{SC}{SL} \times 100 \text{ percent}$$

Hence, the system whose T, X point is C_4 consists of pure anorthite crystals

---

out, continuing to do so as the system is cooled until its T, X condition plots at R_3; the composition of the stable liquid simultaneously follows points along the liquidus from R_2 to L_4. After the eutectic liquid composition (L_4) is attained, diopside plus anorthite crystallize simultaneously in the same proportion (42% anorthite–58% diopside) as present in melt L_4. (Diagram after N. L. Bowen, 1928.)

(B) Illustration of the *lever rule*. The weight percent of solid S and liquid L comprising the bulk composition C can be determined from line segments $\overline{SC}$, $\overline{CL}$, and $\overline{SL}$ of the isotherm SCL which compares to isotherm $S_4C_4L_4$ in (A). We can compare isotherm SCL to a lever for which the fulcrum is located under C. If W_S and W_L represent the weight of S and of L, respectively, the lever is in equilibrium (like a precisely balanced beam balance) when

$$W_S \times \overline{SC} = W_L \times \overline{CL}$$

Because the bulk weight W_C of the crucible's contents must equal the weight of solid S plus liquid L, we can write

$$W_L = W_C - W_S$$

and substitute this value for W_L into the lever equation to obtain

$$W_S \times \overline{SC} = (W_C - W_S)\overline{CL} = W_C\overline{CL} - W_S\overline{CL}$$

which becomes

$$W_S(\overline{SC} + \overline{CL}) = W_C\overline{CL}$$

But $(\overline{SC} + \overline{CL})$ equals length $\overline{SL}$ so that we obtain

$$\frac{W_S}{W_C} = \frac{\overline{CL}}{\overline{SL}}$$

Similarly we could show that

$$\frac{W_L}{W_C} = \frac{\overline{SC}}{\overline{SL}}$$

where W_S/W_C and W_L/W_C are the fractions of the bulk weight which are solid and liquid, respectively. To convert to percent, we would multiply W_S/W_C (or $\overline{CL}/\overline{SL}$) by 100. Thus, at the precise moment that the temperature dropped to 1270°C so that isotherm $S_4C_4L_4$ pertained, the weight percent of S in the crucible would equal 100 times $\overline{C_4L_4}/\overline{S_4L_4}$ or 100 times (75–42)/(100–42)—that is, 56.9.

plus a liquid (L_4) containing 42 percent $CaAl_2Si_2O_8$, this liquid comprising $(100 - 75)/(100 - 42) \times 100$ or 43.1 percent by weight of the system.

After the liquid reaches L_4 in composition, further heat loss causes diopside to join anorthite in crystallizing. The two crystallize simultaneously in the same proportion (42 percent anorthite–58 percent diopside) as they are present in melt L_4 whose composition thus remains unchanged during this crystallization. Moreover, this simultaneous crystallization will proceed at a rate proportional to the rate of heat loss from the system so that the heat evolved from crystallization will provide that lost from the system. Consequently, liquid L_4 will remain at 1270°C until it is entirely crystallized into an intergrowth of diopside and anorthite. Such simultaneous crystallization of solids in the same proportion as present in the melt is called *eutectic crystallization*. The liquid from which it occurs, L_4 in this example, is called the *eutectic liquid* and the temperature (1270°) is called the *eutectic temperature*. The composition of the eutectic liquid as well as of the mixture of simultaneously crystallized crystals to which it converts is called the *eutectic mixture*. After crystallization of this eutectic mixture, further subtraction of heat from the system would cause its temperature to drop. At 1240°C the T, X condition of the system would be indicated by C_5, a point in the two-solid field. The amount of anorthite crystals (75 percent) and of diopside (25 percent) for the system at C_5 is identical to their chemical proportions in the original melt at C_1. For example, using the lever rule, weight percent anorthite equals $100 \times$ line length C_5A_5 divided by line length S_5A_5.

In a similar manner, cooling of a system containing 20 percent $CaAl_2Si_2O_8$ at 1430°C (point R_1 in Fig. 10-10A) would cause diopside to start to crystallize at 1350°C and continue down to 1270°C. Such depletion in diopside would cause the residual liquid to follow compositions along the liquidus from R_2 to L_4. On reaching L_4 and 1270°C, anorthite and diopside would crystallize together in eutectic proportions.

Under equilibrium conditions melting histories are the reverse of cooling histories. That for a mixture of 75 percent anorthite and 25 percent diopside crystals is thus reconstructed by considering the nature and relative abundance of the phases present for points C_5 to C_1 (Fig. 10-10A) along the isopleth representing the mixture's bulk composition. Hence this mixture begins to melt at 1270°C with formation of the eutectic liquid L_4, anorthite in excess of the eutectic proportion remaining as unmelted crystals. These melt gradually as the system's temperature increases above 1270°C, the last anorthite crystal melting at 1465°C for the 75 percent anorthite mixture discussed here. In contrast a eutectic mixture containing 42 percent anorthite and 58 percent diopside crystals melts completely at 1270°C. The word *eutectic* stems from the Greek *eutectos* meaning "easily melted" in allusion to the fact that the eutectic mixture melts *completely* at a lower temperature than mixtures of any other proportions.

Control of Eutectic Textures. If heat is withdrawn in just one direction from a eutectic liquid, the less abundant phase may crystallize as parallel platelets or lamellae (Fig. 10-11A) or as parallel needles (Fig. 10-11B) set in a matrix of the more abundant phase. The shape adopted usually minimizes the area of

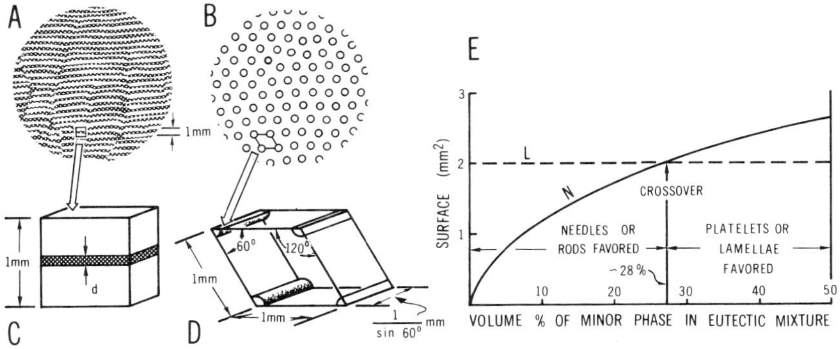

FIGURE 10–11

(A) Cross section normal to eutectically crystallized lamellae of $CuAl_2$ in an aluminum matrix. Center-to-center distance of lamellae is actually about 0.002 mm but is taken here as 1 mm.
(B) Cross section normal to eutectically crystallized needles of Al_3Ni in an aluminum matrix. Center-to-center distance of needles is actually about 0.001 mm but is taken here as 1 mm.
(C) Enlarged, representative cube from the lamellar, eutectic intergrowth in (A).
(D) Polyhedron representative of the eutectic intergrowth of needles in (B), assuming the needles to be in a perfect hexagonal array with a center-to-center distance of 1 mm between them. The edge of the polyhedron parallel to the lengths of the needles is taken as (1/sin 60°) mm so that its volume will be equal to the cube in (C).
(E) Relation between surface area and the volume percent of the minor phase in a crystallized eutectic mixture. Curve N shows the relation if the minor phase crystallized as needles; dashed line L shows the relation if the minor phase crystallized as lamellae. (After Salkind and Lemkey, 1967.)

contact between the two phases and thus their interfacial energy. Salkind and Lemkey (1967, p. 52), following Cooksey and others (1964, p. 745), analyze the problem by comparing two equal volumes, one a representative sample of a eutectic intergrowth in which the minor phase formed lamellae (Fig. 10-11C), the second a sample in which the minor phase formed parallel needles in a hexagonal array (Fig. 10-11D). In each sample the center-to-center distance between lamellae or between needles is assumed to be 1 mm for convenience of discussion. The cube in Figure 10-11C thus has a volume of 1 mm^3 and the interface area between lamella and matrix is 2 mm^2. As d the thickness of the lamella is changed, but the center-to-center distance between lamellae and therefore the cube edge is maintained at 1 mm, the lamella in Figure 10-11C represents differing percentages of the cube's volume whereas its area of contact with the matrix remains unchanged at 2 mm^2 (line L in Fig. 10-11E). In contrast, as the diameter of the needles in Figure 10-11D is increased, not only do they represent an increasing percent of the polyhedron's volume, but their surface of contact increases as shown by curve N in Figure 10-11E. The crossover point for curves N and L occurs where the minor phase represents (ca.) 28 percent by volume of the eutectic mixture. Hence unidirectional cooling of a eutectic liquid containing less than 28 percent by volume of the minor phase will favor the formation of parallel needles of the minor phase; if above 28 percent, parallel platelets or lamellae are likely. Exceptions to this occur, indicating that factors other than minimization of the contact area between phases are also important. For example, alignment of the lamellar phase along those crystallographic planes of the matrix phase for which the two structures fit well together and/or have many atoms in common might permit a lamellar structure to occur far below the 28 percent limit. A crystallized eutectic mixture in which the minor phase occurs as thin parallel needles might have unusual properties suitable for a range of uses from jet engine blades to semiconductor switches.

Incompatibility and Incongruent Melting. If two crystalline materials cannot exist stably together, they are said to be *incompatible*. Such incompatibility usually arises because within the system in which both occur there also occurs an *intermediate compound*, in other words one that is compositionally equivalent to a mixture of the two in a fixed proportion. In the two-component system $KAlSi_2O_6$—SiO_2, for example, the compounds representing the compositional extremes, that is, leucite ($KAlSi_2O_6$) and a crystalline form of SiO_2 such as cristobalite at higher temperatures (but tridymite at lower ones), are incompatible; no T, X field exists in Figure 10-12 in which both are stable together. This is true because in the system there also occurs an intermediate compound, orthoclase or K-feldspar ($KAlSi_3O_8$), whose composition is identical to a mixture containing 21.6 percent SiO_2 and 78.4 percent $KAlSi_2O_6$ by weight and thus plots on the arrow-indicated isopleth at 21.6 percent SiO_2 in Figure 10-12.

To understand better the nature of the incompatibility, we consider the cooling of three crucibles of bulk compositions C_1, C_2, and C_3, respectively. At 1700°C all three contain a liquid melt only. With cooling, leucite begins to crystallize at 1640°C for C_1, at 1530°C for C_2, and at 1360°C for C_3. With further decrease of temperature and, consequently, continued crystallization of leucite from each melt, the melt's composition follows the liquidus from R to I for crucible C_1, from S to I for C_2, and from T to I for C_3. Point I, reached in each case when the crucible is cooled to 1150°C, marks an inflection point on the liquidus known as the *peritectic point*. Point I also marks the composition and temperature of the coolest liquid in the system that can stably coexist with leucite. This liquid, as calories are removed at 1150°C, reacts with the earlier-formed leucite crystals, expending itself to transform them to orthoclase. Crucible C_1's isopleth divides the 1150°C isotherm into the segments labeled Li and Lc in Figure 10-12; applying the lever rule to these, we note that before the transformation occurred crucible C_1 contained 100 (Lc)/(Li + Lc) percent leucite and 100 (Li)/(Li + Lc) percent liquid of composition I. C_1's isopleth divides the 1100°C isotherm into the segments marked Or and Lc′. These indicate that after the transformation, crucible C_1 contains 100 (Or)/(Or + Lc′) percent orthoclase and 100 (Lc′)/(Or + Lc′) percent leucite; this continued existence of leucite after the transformation at 1150°C also indicates that in crucible C_1 the liquid of composition I was entirely expended before all the original leucite could be transformed to orthoclase.

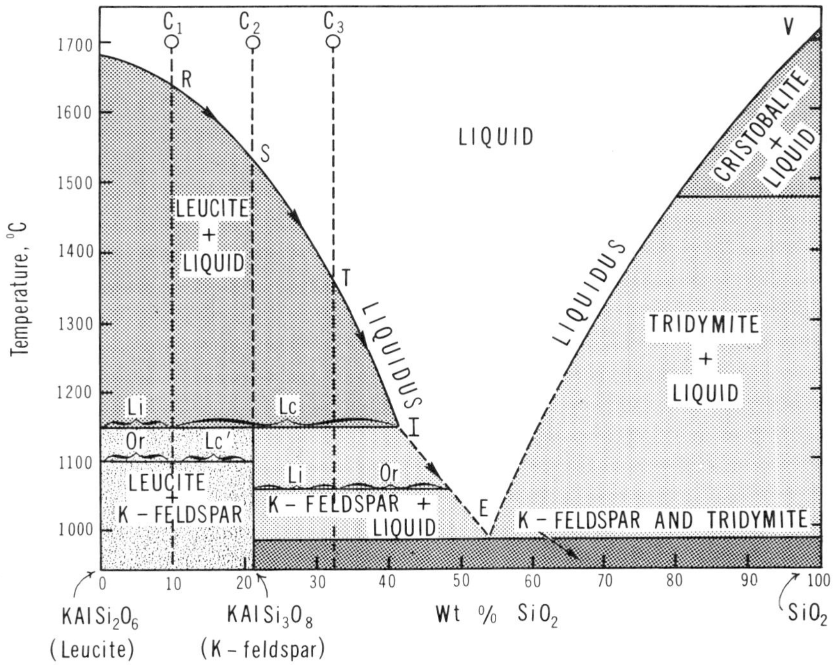

FIGURE 10–12

T, X diagram for the two-component system $KAlSi_2O_6$—SiO_2. Three melts at 1700°C are compared. One's composition (C_1) contains less SiO_2 than $KAlSi_3O_8$, the composition of K-feldspar; the second's composition (C_2) is identical to $KAlSi_3O_8$ in composition; the third (C_3) contains more SiO_2 than $KAlSi_3O_8$. If each melt is cooled so that its *T, X* condition plots lower and lower on its respective isopleth (dashed line), leucite ($KAlSi_2O_6$) starts forming at the temperatures indicated by points *R*, *S*, and *T*, respectively. Continued cooling with concomitant crystallization of leucite causes each liquid's composition to follow a series of points down the liquidus curve until each liquid attains the composition and temperature (1150°) indicated by point *I*. This particular liquid *I* will react with the earlier formed leucite crystals to form K-feldspar ($KAlSi_3O_8$). Depending on whether the original melt was C_1, C_2, or C_3 there is not enough, just enough, or more than enough of this liquid to convert all the earlier formed leucite to K-feldspar. If more than enough (system C_3), the unexpended liquid with further cooling follows a composition along liquidus *IE*, K-feldspar crystallizing out during the cooling; on reaching *E*, SiO_2 and K-feldspar crystallize eutectically.

To the extent that C_1's bulk composition approaches C_2's in value, line segment Lc′ in Figure 10-12 would decrease, vanishing if composition C_2 is attained. Thus as crucible C_2 is cooled below 1150°C, no leucite survives the transformation, there being *just enough* liquid of composition *I* to transform all the earlier formed leucite to orthoclase. Below 1150°C, therefore, crucible C_2 contains only orthoclase. Crucible C_3 has a bulk composition that is more silica rich than orthoclase, its isopleth lying to the right of C_2's isopleth. Consequently, at 1150°C there is more than enough liquid of composition *I* to transform the earlier-formed leucite crystals to orthoclase. With continued cooling to 990°C, orthoclase crystallizes from this excess liquid, the liquid's composition going from *I* to *E*. At *E* a eutectic is reached whereupon orthoclase and tridymite crystallize in the same proportion as they are present in melt *E*. In summary, therefore, melt C_1, the compositional equivalent of a mixture of leucite and orthoclase, consists of these two crystals in the same proportion when completely crystallized. Melt C_2, equivalent to pure orthoclase in composition, when cooled below 1150°C consisted of orthoclase crystals only. Melt C_3, equivalent to a melted mixture of orthoclase and SiO_2, consists of these two crystals in the same proportion when it is completely crystallized. Melts more silica rich than about 54 percent SiO_2 represent simple eutectic crystallization for which a crystalline form of SiO_2 first crystallizes out (until the residual liquid attains the eutectic composition *E* whereupon tridymite and orthoclase crystallize eutectically).

If a crystal of K-feldspar is heated, results are the reverse of the cooling of crucible C_2. Thus orthoclase remains solid until 1150°C is attained; it then melts into a liquid of composition *I* plus leucite crystals. As the temperature is further elevated, the leucite crystals slowly melt, the last disappearing when a tem-

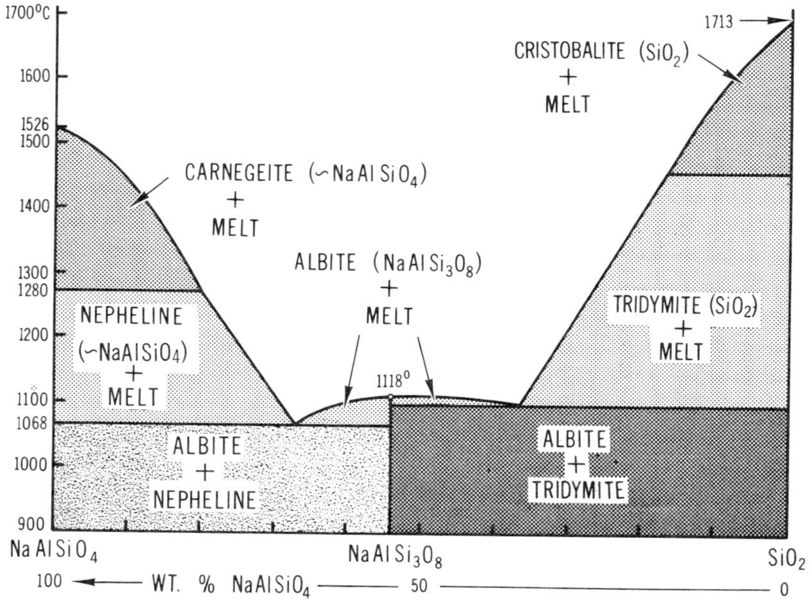

FIGURE 10–13

A phase diagram of the two-component system $NaAlSiO_4$—SiO_2 that has been modified from T. F. W. Barth (1962) by neglecting the compositional variations introduced in the crystalline forms of $NaAlSiO_4$ and of $NaAlSi_3O_8$ by the limited solid solution between them. In essence it represents two eutectic diagrams (similar to Figure 10–10A) one involving eutectic relations between nepheline and albite, the right half involving eutectic relations between albite and tridymite. The liquid field is unshaded.

perature of (ca.) 1530°C is reached. The melting of a crystalline material into a liquid of composition other than itself, as in the case of orthoclase, is known as *incongruent melting*.

The incompatibility of two crystalline materials does not necessarily mean that the compound compositionally intermediate between them melts incongruently. If we ignore the complication of solid solution, the system $NaAlSiO_4$—SiO_2 illustrates this (Fig. 10-13). The observed phases are (1) nepheline ($\sim NaAlSiO_4$), (2) a crystalline form of SiO_2, and (3) a compositionally intermediate compound, albite ($NaAlSi_3O_8$). If this phase diagram is cut along the isopleth at 54 percent $NaAlSiO_4$, that is, at the albite ($NaAlSi_3O_8$) composition, it divides into two simple eutectic diagrams; that to the left involves nepheline and albite, that to the right, albite and tridymite. In neither, however, does a T, X field exist for which tridymite and nepheline can coexist, the two being incompatible even though the intermediate compound albite, when heated to 1118°C, melts congruently to a melt of its same composition.

Crystallization of Solid Solutions. In some systems complete solid solution may occur between the possible crystalline phases. An example is the system Mg_2SiO_4—Fe_2SiO_4 for which the possible crystalline phases, known as the *olivine series*, are not only forsterite (Mg_2SiO_4) and fayalite (Fe_2SiO_4) but solid solutions ($Mg_{2-x}Fe_xSiO_4$) of any composition in between. The phase diagram (Fig. 10-14) possesses liquidus and solidus lines that bound the T, X field (unshaded area) within which olivine and liquid stably coexist. For any T, X condition plotting in this field, for example, point R (1600°C, 42 percent Fe_2SiO_4), the isotherm through this point (line S_RL_R in Fig. 10-14) intersects the solidus and liquidus at two points that denote the compositions of the only solid (point S_R) and liquid (point L_R), respectively, that can stably coexist at this temperature. By the lever rule the percent of liquid is then proportional to the length of line S_RR divided by length S_RL_R.

If a melt whose T, X condition is denoted by point C (1800°C; 42 percent Fe_2SiO_4) is cooled, its T, X condition is denoted by successively lower points on the dashed isopleth CO. At Q, a point barely within the two-phase field, olivine crystals containing 15 percent Fe_2SiO_4 (S_Q) start to form. With further cooling until R is reached, the composition of the olivine that crystallizes out changes from S_Q to S_R; under equilibrium conditions during this cooling, the liquid also reacts with any earlier-formed crystals to change their composition to S_R. As a result the liquid's composition in turn changes from Q to L_R. With further cooling to point M, the system now contains crystals of composition S_M and liquid of composition L_M. From the lever rule, the percent liquid decreases from S_RR/S_RL_R to S_MM/S_ML_M in going from R to M. With continued cooling to N, the percent liquid drops to zero and the stable crystals are the same composition as the original melt (42 percent Fe_2SiO_4). Further cooling to a point such as O merely results in a decrease in temperature of this 42 percent Fe_2SiO_4 solid solution.

If the cooling from C to N were so rapid that the melt lacked sufficient time to react with and transform the earlier-formed crystals to the newly stable compositions, crystals with a core of composition S_Q but with increasingly

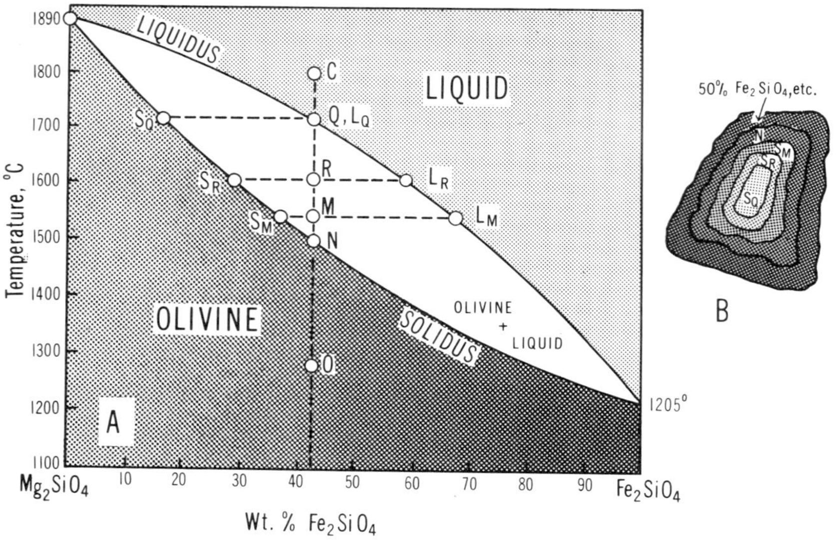

FIGURE 10–14

(A) Typical phase or T, X diagram for a two-component system in which the crystalline forms of the two components are end members of a solid solutional series. Here the two components are Mg_2SiO_4 and Fe_2SiO_4; their crystalline forms are forsterite and fayalite; and solid solutions between them are members of the olivine series. The *liquidus* and *solidus*, which are the two lines that bound the (unshaded) olivine-plus-liquid field, show how the composition of the stable liquid and solid changes as temperature changes. The dashed (horizontal) isotherms connect points on the solidus and liquidus that denote the composition of the solid solution and of the liquid which are stable together at that temperature. Thus a system whose T, X condition plots at R consists of an olivine solid solution (composition S_R) in equilibrium with a melt (composition L_R). (Solidus and liquidus curves are after N. L. Bowen and J. F. Schairer.)
(B) Hypothetical cross section through an olivine crystal formed by such rapid cooling of melt C (in diagram at left) that the early formed core (composition S_Q) lacked time to react with the liquid. Instead it became successively armored by iron-rich compositions, the outer ones eventually becoming even more iron rich than the original liquid denoted by point C.

Fe_2SiO_4-rich rims might result (Fig. 10-14B). In this event the outermost rim would contain more of the lower-melting end member (Fe_2SiO_4) than the original melt C. This feature, a core compositionally closer to the more refractory end member enveloped by rims successively richer in the less refractory end member, is known as *zoning*. *Reverse zoning* in which the core is the less refractory material is more difficult to understand. Zoning thus represents cooling that is too rapid to permit chemical equilibrium to be established in the reaction between the liquid and the earlier-formed crystals. Zoning is sometimes prominent in plagioclase crystals, the cores being more anorthic, the rims more albitic in composition.

Exsolution. Where solid solution is limited between end members, increased temperature usually causes it to be less limited. To illustrate, the compositions of those (Na, K)$AlSi_3O_8$ solid solutions that are stable at 500°C will plot somewhere along either the left or right horizontal bar at 500°C in Figure 10-15A. At 600°C possible compositions are less limited while at 675°C, as indicated by the now continuous bar, unlimited solid solution exists. Similarly, if bars were drawn to indicate the extent of solid solution for each of many temperatures, the bars would coalesce to define the shaded area in Figure 10-15B.

Thus any solid solution whose composition and temperature (T, X point)

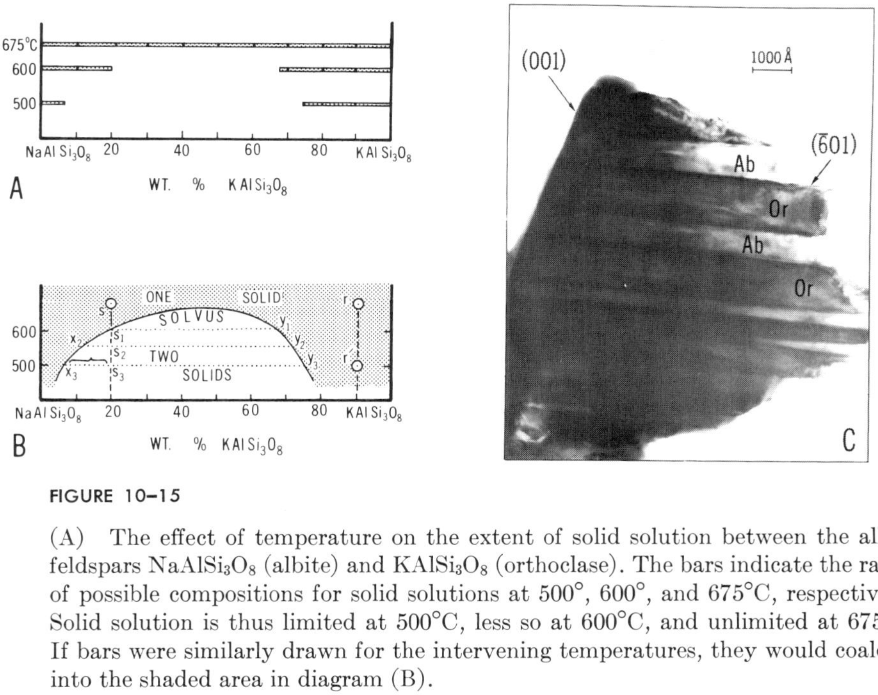

FIGURE 10–15

(A) The effect of temperature on the extent of solid solution between the alkali feldspars $NaAlSi_3O_8$ (albite) and $KAlSi_3O_8$ (orthoclase). The bars indicate the range of possible compositions for solid solutions at 500°, 600°, and 675°C, respectively. Solid solution is thus limited at 500°C, less so at 600°C, and unlimited at 675°C. If bars were similarly drawn for the intervening temperatures, they would coalesce into the shaded area in diagram (B).

(B) A solid solution of temperature and composition as indicated by point s is slowly cooled until the system's T, X condition enters the unshaded area of "forbidden" solid solutional compositions and plots at s_1, s_2, and s_3. This solid solution then exsolves, that is, segregates, into lamellae or patches of differing compositions. Thus at s_2 the original solid solution exsolves into patches or lamellae of compositions x_2, others of compositions y_2. Where they are lamellae, the lamellae may develop parallel to those crystallographic planes along which the two materials, for example, x_2 and y_2, fit the best and/or have many atoms in common. The line that forms the boundary between the two-solid field (unshaded) and the one-solid field (shaded) is called the *solvus*. It shows how the compositions of the two costable solids varies with temperature.

(C) An electron micrograph showing the exsolved lamellae parallel to ($\bar{6}01$) of albite (~$NaAlSi_3O_8$ or Ab) and orthoclase (~$KAlSi_3O_8$ or Or) for a grain lying on the (010) cleavage. (From S. G. Fleet and P. H. Ribbe, *Phil. Mag.*, **8**, 1179, 1963.)

plots in the shaded area in Figure 10-15B is stable. For example, a solid solution at 675°C and containing 20 percent $KAlSi_3O_8$ is seen to be stable, its T, X condition (as indicated by point s) occurring in the one-solid stability field. On cooling to 600°, 550°, and 500°C, this solid solution's T, X conditions, as indicated by s_1, s_2, and s_3, enter the unshaded field within which a solid solution containing 20 percent $KAlSi_3O_8$ is unstable. Consequently, if held, for example, at 550°C for a long enough time, this original solid solution will segregate into patches or lamellae, some of composition x_2, the others of composition y_2 (Fig. 10-15B). Compositions x_2 and y_2 occur at the points of intersection between the dotted 550°C isotherm and the boundary line or *solvus* between the shaded and unshaded areas. This unmixing process, in effect the reverse of solution, has been called *exsolution*. With further cooling, the areas of composition x_2 lose K^+ to areas of composition y_2 but gain Na^+ in return, each area's composition following along the solvus. Thus at 500°C their compositions are points x_3 and y_3, the intersections between the 500° isotherm and the solvus. The proportion of y_3 crystals to total $(x_3 + y_3)$ crystals is x_3s_3 (bracketed line in Fig. 10-15B) divided by x_3y_3, the lever rule again applying. Upon heating, the process is reversed and the exsolved areas approach 20 percent $KAlSi_3O_8$ in composition, becoming identical solid solutions of this composition when 600°C is exceeded. On the other hand, a solid solution containing 90 percent $KAlSi_3O_8$ (point r) remains unchanged if cooled to 500°C (point r').

Like any chemical reaction, the rate of exchange, $Na^+ \leftrightarrows K^+$, decreases markedly as temperature decreases. Consequently, if cooling is rapid, it would be possible for compositions x_3 and y_3 in Figure 10-15B, for example, to persist (metastably) to room temperature. These compositions would, so to speak, be "frozen in" because the exchange rate for $Na^+ \leftrightarrows K^+$ becomes so slow at lower temperatures that the time elapsed while cooling from 500°C is not enough to permit a significant compositional change in x_3 and y_3.

P, T, X Diagrams. The previous two-component systems were discussed under conditions of constant pressure. If the effect of pressure on the system is also to be considered, the fields of stability of the various phases can be shown by P, T, X diagrams. For the system Mg_2SiO_4—SiO_2, the P, T, X diagram (Fig. 10-16) reveals that at 1557°C enstatite ($MgSiO_3$) melts incongruently under 1 atm pressure to yield a liquid of composition I_1 plus the olivine, forsterite (Mg_2SiO_4). An unusual feature of the system at 1 atm is a two-liquid region, which is enlarged in the circular inset. Thus if any liquid of composition and temperature such as indicated by point Y is cooled, it breaks up into two immiscible liquids at the temperature where its isopleth (dashed) enters the two-liquid field. The situation is like that discussed for exsolution. For example, liquid Y cooled to condition Y_1 becomes two immiscible liquids of composition C_1 and C_2, respectively. Further cooled to Y_2, these liquids undergo compositional exchanges so as now to be of composition C_3 and C_4; cooled to Y_3, the liquids are of composition C_5 and C_6. At this point liquid C_6 forms cristobalite (SiO_2) crystals plus liquid C_5; that is, it is as if liquid C_6 were itself exsolving into a mixture of cristobalite and liquid C_5. While this occurs, three phases are present in the system: (1) liquid C_5; (2) cristobalite; and (3) such liquid C_6 as has not yet "exsolved."

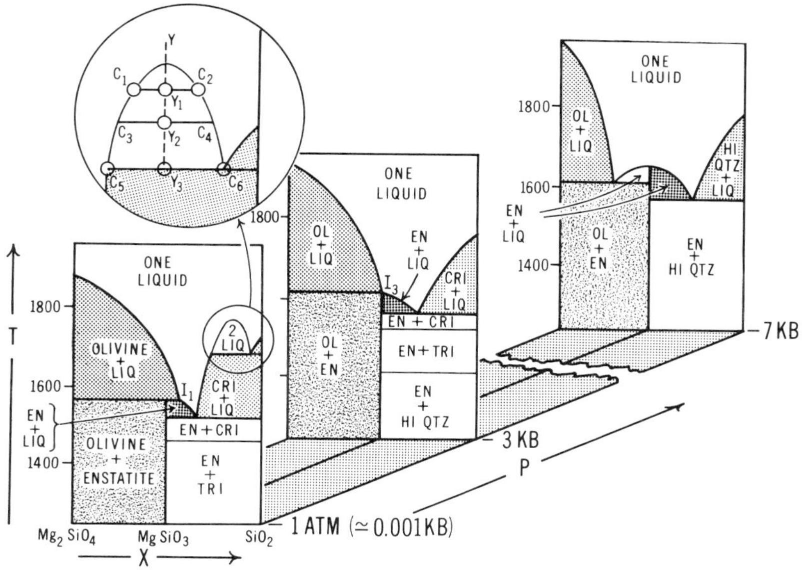

FIGURE 10–16

A T, X diagram at three different pressures of the two component system Mg_2SiO_4—SiO_2 (not to scale). At 1 atm pressure $MgSiO_3$ (enstatite) melts incongruently to form forsterite (Mg_2SiO_4) plus a liquid of composition I_1. As pressure increases, $MgSiO_3$ melts to a liquid increasingly close to $MgSiO_3$ in composition. Thus at 3 kbar pressure it melts congruently to a liquid I_3 of the same composition as itself. The relations at 7 kbar are similar to those in Figure 10–13. At 1 atm, systems whose T, X condition plots as a point in the two-liquid field (enlarged in the circular inset) will segregate into two immiscible liquids. Thus liquid Y, if cooled to Y_1, segregates into two liquids of composition C_1 and C_2, respectively. Cristobalite and tridymite, the high temperature polymorphs of SiO_2, are abbreviated CRI and TRI.

At elevated pressures, enstatite no longer melts incongruently. Thus at 3 kbar (Fig. 10-16), the peritectic point I_3 falls on the vertical isopleth corresponding to $MgSiO_3$. Hence enstatite melts to form liquid I_3 whose composition $MgSiO_3$ is identical to enstatite's. At 7 kbar the situation is like that discussed for the system $NaAlSiO_4$—SiO_2 (Fig. 10-13). In every case, however, olivine (Mg_2SiO_4) is incompatible with any crystalline form of SiO_2. Kushiro, Yoder, and Nishikawa (1968) show that if water is present, enstatite continues to melt incongruently to at least 30 kbar water pressure.

Figure 10-17 is a hypothetical P, T, X diagram for a two-component system in which no solid solution exists between crystals of component A and those of component B, these solid phases being symbolized as S_A and S_B. The upper T, X plane is an *isobaric section*, that is, a plane representing equal pressure, that reveals normal eutectic crystallization. Parallel (but lower) isobaric sections would also illustrate eutectic crystallization, the only change being that the eutectic point shifts (for this example) toward composition B as pressure is lowered. For very low pressures, however, these horizontal T, X planes intersect the (stippled) field for the gas or vapor phase, which may then appear under

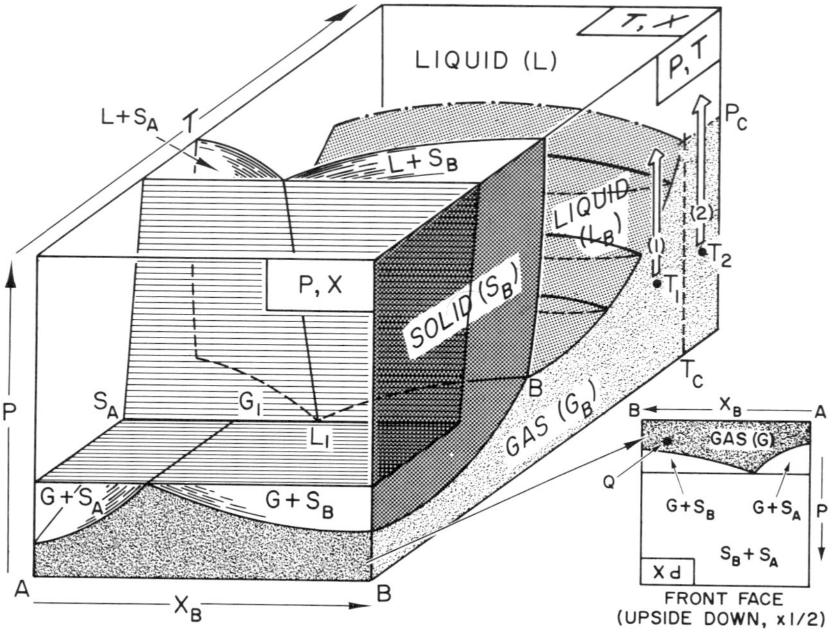

FIGURE 10–17

A hypothetical P, T, X diagram for a two-component system. S_A and S_B represent solids whose composition corresponds to the pure components, A and B, respectively. The heavy line at the intersection between the two ruled planes is called a *tie line*. In phase diagrams a tie line is defined as the straight line that joins the composition points, in this case, S_A, G_1, L_1, and S_B, for all the phases stable at the particular temperature and/or pressure that the tie line represents. This particular tie line represents the only P, T condition in the system at which *four* phases are stable. To the lower right the front face of the P, T, X diagram is redrawn upside down (and on a smaller scale) to emphasize its similarity to Figure 10–10 and thus that it represents eutectic crystallization from a gas.

suitable T, X conditions. In systems involving silicate crystals, and for most oxides in the absence of water or other volatile materials, a pressure of 1 atm or more is usually well above the vapor field so that at pressures above 1 atm gases will not be present under equilibrium conditions. The front face, labeled P, X in Figure 10-17, is an isothermal section. Turned upside down (lower right), it is better seen to represent eutectic crystallization of solids S_A and S_B from a gas (G) of composition intermediate between components A and B. Thus if pressure were sufficiently increased for the P, X condition denoted by Q, first S_B (and ultimately a eutectic mixture of S_B and S_A) would crystallize from the gas.

The *isocompositional plane* labeled P, T in Figure 10-17 represents a one-component system where B is the component. The plane parallel to it, but through composition A, similarly represents the one-component system for component A. Isocompositional planes between these two do not represent one-component systems, their compositions corresponding to a mixture of the two components A and B. The one-component system corresponding to B reveals the usual triple point. The boundary between the vapor and liquid phase terminates abruptly at a point (the "+" sign in Fig. 10-17), called the *critical point,* whose P, T coordinates represent the *critical temperature* Tc and the *critical pressure* Pc. For any gas at a temperature below Tc, for example, at T_1 in Figure 10-17, a steady increase of pressure (arrow 1) causes it to enter the liquid field whereupon it condenses from gas to liquid. However, for a gas above Tc, for example, at T_2, a steady increase of pressure (arrow 2) causes its properties to change, *gradually and imperceptibly,* from those of a gas to those of a liquid; it no longer abruptly condenses into a liquid.

The volume above and to the front of the two ruled planes in Figure 10-17 is the field for stable coexistence of the two solids, S_A and S_B. The line of intersection between these two ruled planes represents the P, T value for which four phases (S_A, S_B, liquid L_1, and gas G_1) can stably coexist in any mixture of A plus B. Only the upper surface (shaded gray) is drawn of the lens-shaped volume representing the field of stability for gas and liquid, the dashed, concave-upward lines being on the lower surface of the lens. The rear edge of this field is dot dashed; points along it represent critical temperatures beyond which the physical properties of gas and liquid grade into each other.

Le Chatelier's principle

Let a and b represent two phases that can transform from (or into) phases c and d in a multicomponent system. At T_1 and P_1, the temperature and pressure at which all four can coexist in equilibrium,

$$a + b \leftrightarrows c + d \qquad \text{(Eq. 10-7)}$$

that is, the amount of a and b transforming into c and d precisely equals the reverse transformation of c and d into a and b. We assume that the transformation of a and b into c and d is endothermic, that is, uses up a certain number of heat calories symbolized as ΔH. We also assume c plus d to be less dense than a plus b. If so, the transformation a and b into c and d will result in an increase in volume (ΔV) because in c and d the matter is less tightly packed. To include this additional information, we rewrite Equation (10-7) as

$$a + b + \Delta H \leftrightarrows c + d + \Delta V \qquad \text{(Eq. 10-8)}$$

Le Chatelier's principle states that *for a system in equilibrium,* as are phases a and b with c and d at temperature T_1 and pressure P_1, if the conditions are changed, the reaction favored is that which tends to restore the original conditions. For example, any increase of temperature of our hypothetical system, that is, addition of heat ΔH to the system, would drive the reaction (Eq. 10-8) to the right because transformation of a and b into c and d uses up heat and tends to lower the temperature. In contrast an increase of pressure would drive the reaction to the left because as *c* and *d* transform into the denser phases *a* and *b*, the total volume occupied by the system decreases and pressure is thus reduced.

Comparing a crystal, a liquid, and a gas of identical compositions, the crystal usually represents the densest packing of atoms and the gas, the least. Consequently, Le Chatelier's principle indicates that increase of pressure will favor the formation of the solid at the expense of the liquid and the liquid at the expense of the vapor. The system H_2O is unusual because the solid form (ice) is less dense than the liquid form (water); consequently, increased pressure causes water to form at the expense of ice.

Types of Polymorphic Transformations

Attention will be confined here to some of the mechanisms in which one polymorph transforms into a second.

Displacive transformations

Also called *high-low inversions* or *α-β inversions*, displacive transformations occur with great rapidity, the structure "snapping" over into that of a second polymorph immediately upon entering the second's *P*, *T* field of stability. Before and after such transformations, near neighbors remain near neighbors, in other words, *no bonds between atoms are broken* during a displacive transformation. However, bond directions may change as near neighbors shift slightly, that is, are *displaced*, relative to each other. An excellent example of a displacive transformation is the inversion at 573°C between high quartz (class 622) and low quartz (class 32). The stable form of SiO_2 below 573°C is low quartz (Fig. 10-18A). If heated to 573°C, low quartz transforms displacively (and instantaneously) into the high-quartz structure (Fig. 10-18B), this being the modification of SiO_2 that is stable between 573°C and 870°C at 1 atm pressure. Similarly, upon cooling below 573°C, high quartz instantaneously transforms into low quartz. In the course of such low-high or high-low inversions, a quartz crystal retains its original right handedness or left handedness.

Some of the principal characteristics of displacive or high-low polymorphism as cited byR. B. Sosman (1965, p. 75) are: (1) a high-low inversion is completed, promptly, often within a few seconds after attaining the inversion temperature (or inversion pressure); (2) a high-low inversion is reversible, proceeding in the opposite direction if the temperature (or pressure) is returned to its pre-inversion value; (3) little difference in energy content exists between the high and low polymorphs; (4) the presence of any other phases in external contact with the crystal will neither alter or catalyze its high-low inversion; and (5) a temperature hysteresis exists, that is, for falling temperatures the inversion occurs at a slightly lower temperature than for the reverse situation of rising temperatures, the average of these two temperatures usually being quoted as the inversion tem-

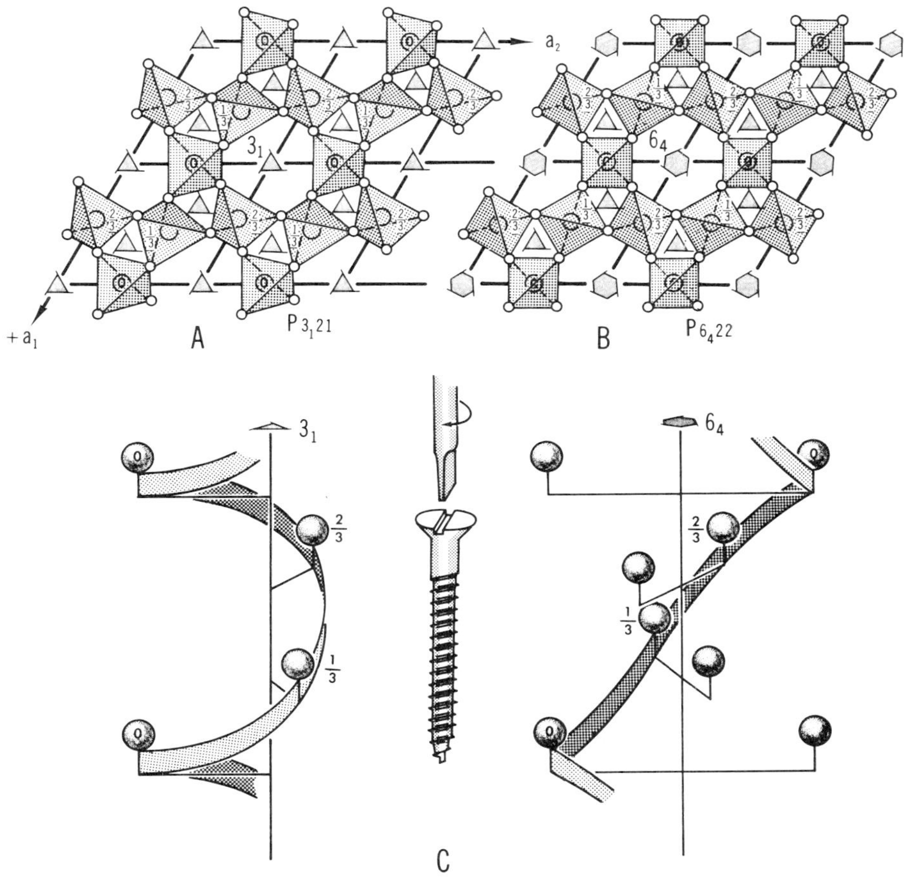

FIGURE 10–18

(A) The structure of right-handed low quartz (space group: $P3_121$) projected onto a (0001) plane. For illustrative purposes the silicon atoms are drawn as larger circles than the oxygens. Heights of these silicon atoms above the plane of the paper are cited as fractions of c_0 (5.40 Å).

(B) If heated above 573°C, the structure in (A) transforms displacively into right-handed high quartz $P6_422$, whose structure is as shown. The atoms shift position only slightly in "snapping" from (A) to (B) or vice versa. In (B) the 2-fold axes that occur normal to the paper (at each silicon site) have been omitted.

(C) The right- or left-handed nature of a screw axis is determinable by imagining a tape to join the *closest* atoms (or sites) related by this screw axis, then comparing this tape to the thread of a right-handed screw (which, by definition, draws itself into the wood if turned clockwise). So compared, 3_1 is right handed but 6_4 is left handed. Interestingly the transformation of right-handed low quartz ($P3_122$) to right-handed high quartz ($P6_422$) changes the set of screw axes at the origin, that is, at the intersections of the bold lines in (A) or (B), from 3_1 (right handed) to 6_4 (left handed). Evidently the set of 3_1 axes not at the origin, which survives from (A) to (B), determines the "handedness" of the quartz.

perature. A similar pressure hysteresis may also exist but remains unproved. Buerger (1951, p. 105) observes that in high-low transformations the high-temperature form, as compared to the low, will be the more open structure, its atoms being packed less compactly and less tightly bonded to each other; it will also have the larger heat capacity and the higher symmetry, the low form usually being a derivative structure (p. 256) of the high.

Reconstructive transformations

Also known as *diffusive transformations* or *sluggish inversions,* reconstructive transformations involve the transformation of one polymorph into another by means of two steps: (1) the breakage of bonds and a general disruption of the old structure and (2) the reorganization and diffusion of the atoms into their positions in the newly stable structure. Step (1) requires the introduction of considerable energy to break the bonds and thus constitutes a high-energy barrier that opposes the progress of any reconstructive transformation, causing it to be slow and sluggish. In fact some reconstructive transformations are so slow that the old polymorph may persist unstably even when held in the P, T field of the new. The energy barrier inherent in step (1) permits its metastable existence beyond its own stability field. Reconstructive transformations are particularly deterred by rapid cooling. Brought suddenly to a temperature at which the low-temperature polymorph is stable, but so low that breakage of bonds (step 1) is difficult and rate of diffusion (step 2) is very slow, a high-temperature polymorph will transform to the low-temperature structure at a negligible rate. Thus it survives metastably at this low temperature and is said to be "quenched in." It is less definitely known whether a sudden pressure drop can "quench in" a high-pressure structure.

High quartz, if heated above 867°C, the upper limit of its thermal stability, reconstructively transforms into tridymite or cristobalite. This transformation is so sluggish that its rate becomes appreciable only if a "mineralizer" or "flux", usually Li_2O, Na_2O, K_2O, MgO, or CaO, is added. Direct conversion of quartz into tridymite by heat alone has not been conclusively demonstrated. (Some investigators do not regard tridymite as a polymorph of SiO_2, arguing that traces of the fluxes are present in *and necessary to* the tridymite structure, tridymite thus being compositionally distinct from SiO_2 and hence not a polymorph of it.)

In contrast to displacive transformations, reconstructive transformations have the following principal characteristics: (1) the transformation occurs sluggishly, thus tridymite and cristobalite may persist on the earth for millions of years at temperatures at which low quartz is stable; (2) considerable difference in the energy content between the two polymorphs may exist; (3) mineralizers may be necessary to permit the transformation to occur at an observable rate; (4) the temperature (or pressure) of transformation may be considerably undershot or overshot so that a polymorph may persist, sometimes indefinitely, within a P, T field for which it is unstable. On the other hand, in displacive transformations, that is, α-β inversions, the high-temperature form may never be quenched in.

Order-disorder transformations

Order-disorder transformations are divisible into two subtypes: (1) rotational transformations and (2) interchange transformations. Rotational transformations may occur in crystal structures containing NO_3^-, CO_3^{2-}, or CN^- groups. For example, the structure of $NaNO_3$ at 25°C (Fig. 10-19A, B) trans-

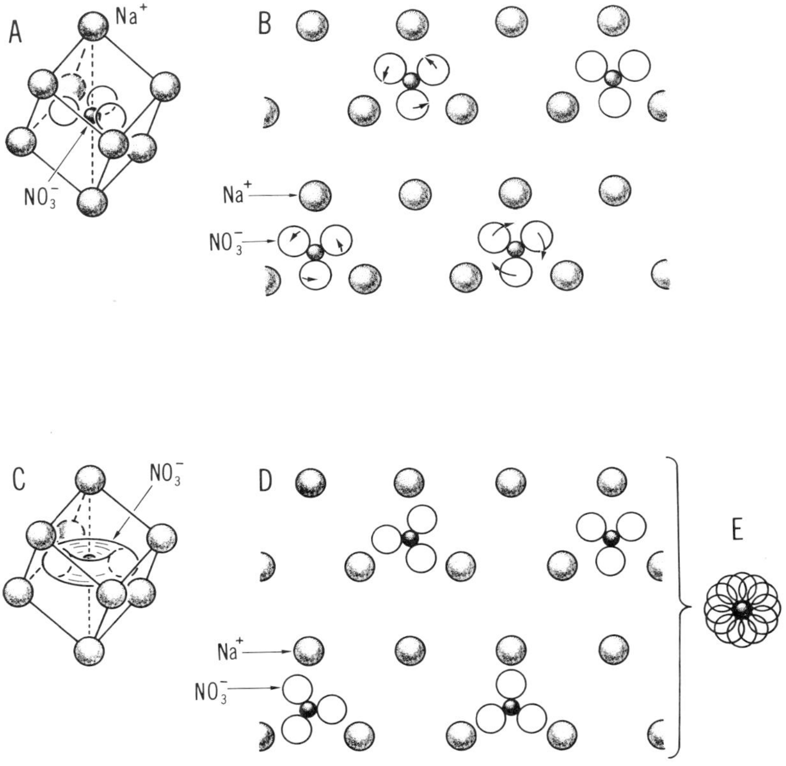

FIGURE 10–19

(A) Rhombohedral unit cell for the $NaNO_3$ structure at 25°C.
(B) A thin layer of the $NaNO_3$ structure at 25°C as viewed down the $\bar{3}$ axis of the rhombohedron. The curved arrows indicate how the oxygens start to rotate out of position as the structure is heated from 180° to 275°C. The gradual transition of the structure during heating through this range is subject to two interpretations (C or D).
(C) The oxygens are considered to rotate continuously as temperatures approaching 275°C are approached, their positions averaged over a given time defining the doughnut shape shown.
(D) Each NO_3^- is considered to have rotated about the $\bar{3}$ axis in different directions (and finite amounts). Thus if the four NO_3^- groups in (D) are superimposed, with their orientations preserved, then (E) results.
(E) The resultant spatial average of the positions of the NO_3^- groups in the structure is indistinguishable from (C).

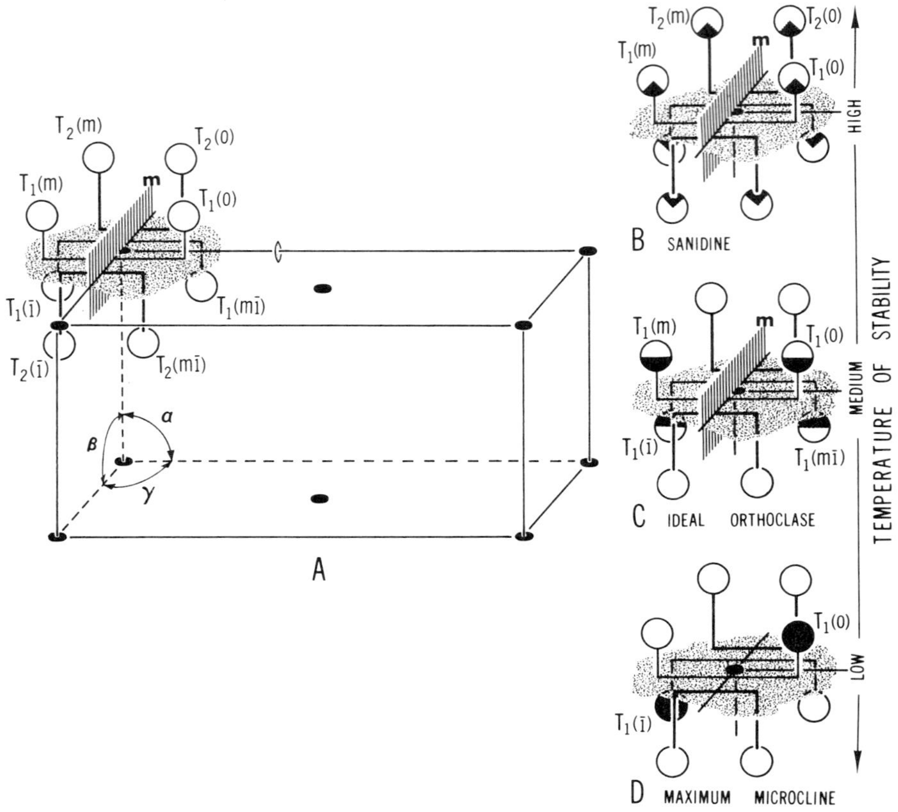

FIGURE 10–20

(A) Unit cell for sanidine ($KAlSi_3O_8$; $C\,2/m$) showing the sites, commonly called T_1 and T_2, of atoms in tetrahedral coordination to oxygen in the structure. Translationally identical sets of sites exist about each of the other lattice points but, for simplicity, only those about one lattice point are drawn. No symmetry operation present in sanidine can relate an atom in a T_1 site to one in a T_2 site; however, the four atoms in T_1 sites *are* symmetrically related to each other. For example, taking the atom at $T_1(0)$ as the original starting point, $T_1(m)$ is related to it by the plane of symmetry m, and $T_1(\bar{1})$ is related to it by the center of symmetry located at the lattice point. Similarly, $T_1(m\bar{1})$ is related to $T_1(m)$ by this center, or to $T_1(0)$ by the 2 fold. The four T_2 sites are similarly related to each other by this symmetry.
(B) In the structure of sanidine the T_1 and T_2 sites are indiscriminately filled by aluminum and silicon, the probability being 25 percent that each site will contain aluminum, 75 percent that it will contain silicon as indicated by the shading.
(C) The structure of ideal orthoclase ($KAlSi_3O_8$; $C\,2/m$) differs from that of sanidine in that the aluminum is concentrated in the T_1 sites and depleted from T_2, the shading indicating a 50 percent chance that any particular T_1 site is filled by aluminum.
(D) In the structure of maximum microcline ($KAlSi_3O_8$; "C" $\bar{1}$) aluminum is largely confined to the $T_1(0)$ and $T_1(\bar{1})$ sites. Consequently, the mirror plane and

forms continuously over the temperature range 180 to 275°C, the oxygens seeming to rotate about the 3-fold axis of the NO_3^- ion (Fig. 10-19C). Crystallographic evidence alone cannot disclose whether the oxygens in the high-temperature structure are truly in continuous rotation (as on a nonstop merry-go-round) or whether, compared to the low-temperature structure (Fig. 10-19B), a positional disorder of the oxygens has been introduced (Fig. 10-19D). Because X-ray techniques would yield the position of the oxygen as averaged over the entire crystal structure in Figure 10-19D, this "space average" (Fig. 10-19E) yields the same results as would a "time average" of their positions if the oxygens were in continuous rotation (Fig. 10-19C).

Interchange transformations are illustrated by the polymorphs of $KAlSi_3O_8$. All have unit cells as in Figure 10-20A. Two sets of tetrahedrally coordinated sites, usually labeled T_1 and T_2, occur in the structure, only those about one lattice point being shown in Figure 10-20A. No symmetry operation ever relates a T_1 site to a T_2 site, hence their differentiation by subscripts. In sanidine, the monoclinic high-temperature polymorph, aluminum and silicon randomly occupy both sets of sites (Fig. 10-20B). Consequently, as is indicated by shading 25 percent of each site, the probability of finding aluminum in each site is 25 percent as compared to 75 percent for silicon. In the monoclinic medium-temperature polymorph, orthoclase, aluminum preferentially occupies the T_1 sites. In ideal orthoclase, which is a concept rather than a reality, aluminum would occupy 50 percent of the T_1 sites but none of the T_2 sites (Fig. 10-20C). No ideal orthoclases have been found as yet, the T_2 sites never being entirely free of aluminum. Thus aluminum occupancy in a typical orthoclase might be 35 percent for T_1 sites and 15 percent for T_2. In microcline, the low-temperature polymorph of $KAlSi_3O_8$, aluminum is concentrated in the $T_1(0)$ and $T_1(\bar{1})$ sites. In the extreme case, "maximum" microcline, aluminum is *entirely* confined to the $T_1(0)$ and $T_1(\bar{1})$ sites (Fig. 10-20D). This represents a maximum degree of ordering in which aluminum occupies only the $T_1(0)$ and $T_1(\bar{1})$ sites throughout the crystal's untold millions of unit cells. By contrast maximum disorder exists in sanidine because aluminum randomly occupies the T_1 and T_2 sites and hence decreases both short- and long-range order.

As illustrated for maximum microcline (Fig. 10-20D) the ordering of aluminum into the $T_1(0)$ and $T_1(\bar{1})$ sites destroys both the symmetry plane and

---

2-fold axes are degraded so that only the center of symmetry at the lattice point persists. Not only does microcline's point symmetry thus degrade to $\bar{1}$ but because Al^{3+} is larger than Si^{4+}, the angles α and γ differ from 90°. However, even if Al^{3+} and Si^{4+} were identical in size, so that microcline's unit cell were to maintain monoclinic geometry, that is, have angles α and γ equal 90°, the distribution of the silicon and aluminum atoms would make it triclinic (since it would possess only $\bar{1}$ symmetry).

Although microcline is triclinic, and convention dictates that a triclinic unit cell should be chosen to be primitive, microcline's unit cell is chosen so that its edges and general appearance are like those for sanidine's unit cell. This produces the anomalous situation of a C-face-centered triclinic unit cell, that is, a space group $C\bar{1}$.

2-fold axis inherent to sanidine and orthoclase leaving only a center of symmetry (at the lattice point), a possibility first recognized by T. F. W. Barth (1934). The concentration of the larger-sized aluminum ions into the $T_1(0)$ and $T_1(\bar{1})$ sites not only changes the point-group symmetry to $\bar{1}$ but slightly changes the shape of microcline's unit cell. Thus the interaxial angles α and γ deviate from 90° to the extent that aluminum is ordered into the $T_1(0)$ and $T_1(\bar{1})$ sites and the smaller silicon ions into the remaining sites. Between different microclines the values of α and γ may vary to the extent that these microclines represent different degrees of ordering of aluminum. For maximum microcline α and γ would deviate most from their monoclinic values of 90°C. For this reason maximum microcline would be said to exhibit maximum "triclinicity." If microcline is heated and held long enough at 1100°C, aluminum from the shaded sites in Figure 10-20D interchanges with silicon from the unshaded sites until the disordered sanidine structure (Fig. 10-20B) is attained.

Laves (1950, p. 548) has shown that orthoclase crystals consist of intergrowths between two types of submicroscopic domains: (1) domains consisting of unit cells with aluminum ordered into the $T_1(0)$ and $T_1(\bar{1})$ sites as in Figure 10-20D and (2) domains consisting of unit cells with silicon in these $T_1(0)$ and $T_1(\bar{1})$ sites and with the aluminum concentrated in the remaining two (unshaded) T_1 sites in Figure 10-20D. Figure 10-20D would thus apply to these latter unit cells if the shading of the T_1 sites were reversed. For a crystal consisting of approximately equal volumes of these two types of intimately intergrown domains, the average aluminum content for each of the T_1 sites would then be as represented in Figure 10-20C.

Interchange transformations thus involve a breakage of bonds and an exchange of positions as for aluminum and silicon in the microcline-to-sanidine transformation. As a consequence they can be as sluggish as reconstructive transformations. With falling temperatures they may not occur at all. Thus the transformation sanidine to microcline, theoretically achievable by cooling sanidine to a sufficiently low temperature, has not as yet been accomplished in the laboratory. The high-temperature phase, sanidine, is the disordered, more symmetric structure whereas the low-temperature phase, microcline, is the ordered, less symmetric structure in which aluminum is ordered in the $T_1(0)$ and $T_1(\bar{1})$ sites rather than distributed equally over all the T_1 and T_2 sites.

Crystal Intergrowths

Ideally, any stable crystal structure represents a packing of the constituent atoms which minimizes the crystal's potential energy. In actual crystals, however, atoms are not always located at their ideal sites; indeed sometimes the site may even be occupied by a "wrong" atom. Such deviations may occur in the first, second, third, and higher coordination spheres of neighbors that surround a particular atom. In this event this particular atom's contribution to the crystal's overall potential energy may increase markedly if the "error" occurred in its

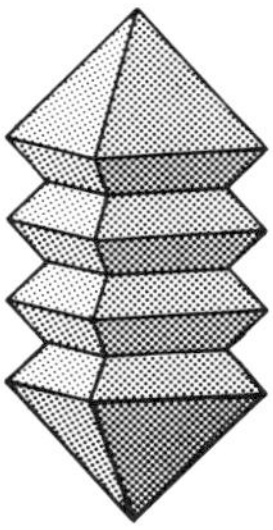

FIGURE 10–21

A parallel growth of alum, $KAl(SO_4)_2 \cdot 12H_2O$. Although it appears to represent four crystals, the same internal crystal structure, unchanged in orientation, extends throughout the specimen so that it is actually a single crystal.

first coordination sphere, less markedly for "errors" in the second, and almost negligibly for errors in higher coordination spheres.

For atoms comprising the surface of a crystal, their first coordination sphere is either incomplete or profoundly modified. As a consequence their contribution to the crystal's potential energy is unduly large and a crystal's surface is the site of unfulfilled chemical bonds and thus high activity; it is "where the action is." For large crystals the proportion of surface to volume is less than for small crystals. Consequently, large crystals possess less surface energy in proportion to their bulk and are hence inherently more stable than small crystals of the same material, growing at their expense whenever conditions permit.

An aggregate of identical crystals with their crystallographic axes and faces mutually parallel is called a *parallel growth* (Fig. 10-21). Actually it represents a single crystal because the same internal structure, unchanged in orientation, extends throughout the specimen. Alternatively, two or more identical crystals may start growth at different points but grow together into chance contacts. In these the axes and crystal structure of the one individual will then be at random angles to those of the other. Consequently, atoms at the contact surface will coordinate to some atoms from the one individual, some from the other; hence their first coordination spheres will deviate from the ideal. A third type of intergrowth between identical crystals occurs if the individuals on either side of the contact are related by a symmetry operation that is not normally present along that direction in either individual. Such crystals are said to be *twinned.* Indeed, if related by a plane of symmetry, their relation is analogous to Siamese twins in humans. Twinning is so important in crystallography that it merits a fuller description.

Twins

Definition. Two crystals of a given crystalline material are said to be *twins* if they occur intergrown so that all crystallographic directions of the first (for example, axes a_1, b_1, and c_1) are related to the corresponding directions of the second (for example, axes a_2, b_2, and c_2) by the operation of either (1) a mirror

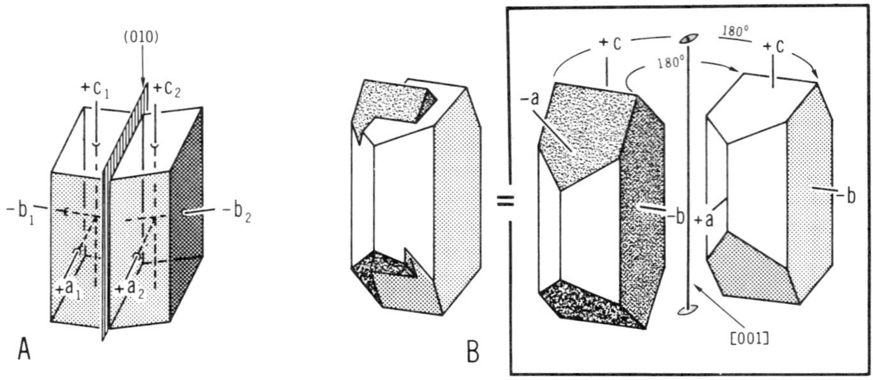

FIGURE 10–22

(A) Reflection twin of albite, $NaAlSi_3O_8$ (class: $\bar{1}$). The vertically ruled mirror plane {010}, which reflects the axes and crystallographic directions of the one crystal into those of the other, is added for illustrative purposes.
(B) Rotation twin of orthoclase, $KAlSi_3O_8$ (class: $2/m$). At right the two intergrown individuals are drawn separately, but with their orientation preserved, to show that a 180° rotation about a direction parallel to the twin axis [001] relates one individual to the other.

plane of symmetry (ruled in Fig. 10-22A), (2) a 2-fold axis of symmetry (Fig. 10-22B) or less frequently a 3-, 4-, or 6-fold axis, or (3) a center of symmetry $\bar{1}$. Such a geometric relationship between two individuals must be observed in several instances so as to eliminate the possibility of a chance contact between two independent crystals which grew at close quarters.

Classification by Symmetry. According to which of the above three types of symmetry relate the two individuals, the twins are called (1) *reflection twins,* (2) *rotation twins,* or (3) *centrosymmetric twins,* respectively. The plane of symmetry relating reflection twins is called the *twin plane*; the axis of symmetry relating rotation twins is called the *twin axis*; and the center of symmetry relating centrosymmetric twins is called the *twin center*.

In *reflection twins* the twin plane is always parallel to a possible lattice plane (hkl). The twin plane's crystallographic direction is thus defined by stating the indices for this plane (in relation to either of the twinned individuals). For *rotation twins* the twin axis is a 2-fold axis in such an overwhelming number of instances that it is sometimes called an *axis of hemitropy* (half a turn). The direction of a twin axis in a crystal may generally be specified by designating the indices $[uvw]$ of the zone to which it is parallel or if it parallels no zones, by stating the indices (hkl) of the face to which it is perpendicular, that is, defining it as the $\perp (hkl)$. Occasionally neither method suffices and a twin axis can only be described as that direction within the plane (hkl) that is perpendicular to $[uvw]$, a lattice direction also within this plane (see Fig. 10-23). In this latter event the twin-axis direction is written $\perp [uvw]/(hkl)$.

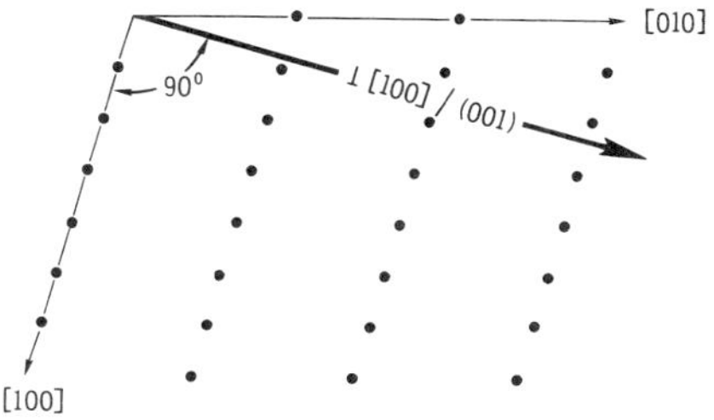

FIGURE 10–23

The plane of the page represents the (001) plane from a triclinic lattice. The direction indicated by the bold arrow, if extended outward from the origin, intersects no additional lattice points and hence cannot be described by the usual symbol [*uvw*]. However, it is perpendicular to [100] and can therefore be described as "the perpendicular to [100] in the (001) plane" and symbolized as ⊥[100]/(001).

Classification by Spatial Relationship. Twins are also classified and named according to the way they are intergrown. If the two twinned individuals are in contact along a well-defined plane (Fig. 10-24A) so that a cut along this plane would entirely separate the two, they are called *contact twins* and the plane of contact is called the *composition plane.* On the other hand, two twinned individuals may either be in contact along an irregular surface or be so intergrown (Fig. 10-24B, C) that no compositional plane exists along which a cut could be made to separate cleanly the two twinned individuals, each to its own side of the plane. Such twins, in which a composition plane is lacking, are called *penetration twins* or *interpenetrant twins.*

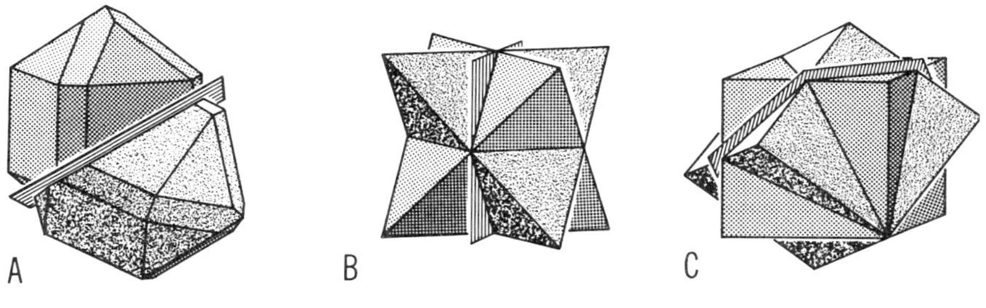

FIGURE 10–24

(A) Contact twin of cassiterite SnO_2 with ruled plane $(01\bar{1})$ as both the twin plane and composition plane.
(B) Interpenetration twin of tetrahedrite $Cu_{12}Sb_4S_{13}$ with (010) as the twin plane. Neither individual, the one being stippled to distinguish it, falls entirely to one side of (010); thus (010) is not a composition plane.
(C) Interpenetration twin of fluorite CaF_2. The two cubes, one stippled, may be regarded as symmetrically related by (a) the (ruled) mirror plane $(11\bar{1})$, here added, or (b) by 180° rotation about $[11\bar{1}]$, the line normal to $(11\bar{1})$. Thus they may be described as either reflection twins or rotational twins.

Repetitive Twinning. In some materials the twin-producing symmetry operation may occur repeatedly so that numerous individuals and composition planes occur even within one relatively small grain. If the several composition planes are mutually parallel, the twinning is called *polysynthetic twinning* (Fig. 10-25A). If the several composition planes are at an angle θ to each other such that the ratio $360°/\theta$ is very close in value to 3, 4, or 6, *cyclic twins* result (Fig. 10-25B). In cyclic twins the several composition planes may intersect in a line that may resemble a 3-, 4-, or 6-fold axis of symmetry, whereas, in reality, such a symmetry operation is not present in single crystals of the same material. An example is the pseudo 6-fold axis in Figure 10-25B which results because the two composition (and twin) planes, (110) and ($1\bar{1}0$), have an interfacial angle near to 60°.

Repetitive twinning is not limited to twins of the contact type. In cyclic twins, for example, the composition planes may degrade into extremely irregular surfaces so that the twins are more properly described as penetration twins.

Types of Rotation-Contact Twins. Rotation twins of the contact types may be subdivided into three kinds: (a) normal twins, (b) parallel twins, or (c) complex twins. They differ in respect to the geometrical attitude of their twin axis to the composition plane. In *normal twins* the twin axis is normal to the composition plane (Fig. 10-26A). In both *parallel twins* and *complex twins* the twin axis is parallel to the composition plane. In respect to a particular direction [*uvw*] within the composition plane, however, the twin axis is parallel to it for parallel twins (Fig. 10-26B) but perpendicular to it for complex twins (Fig. 10-26C). In respect to these three types of rotation twins the direction of the twin axis may be symbolized as (a) $\perp(hkl)$, (b) $[uvw]$, and (c) $\perp[uvw]/(hkl)$, where (hkl) defines the composition plane and $[uvw]$ a specific direction within

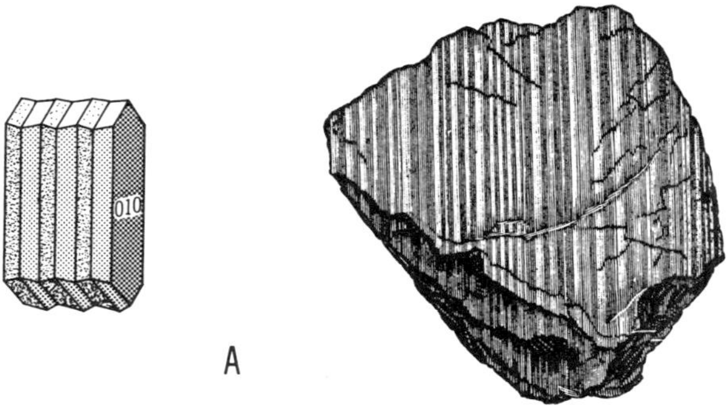

FIGURE 10–25

(A) The twinning illustrated in Figure 10–22A has occurred here repeatedly to produce polysynthetic twinning on {010} which serves as both the twin and composition plane between the stippled and nonstippled individuals. This twinning occurs so commonly in albite it is called "albite twinning." The small re-entrant angles appear as fine striae in hand specimens, for example, that at the right. (After W. E. Ford, *Dana's Textbook of Mineralogy*, Wiley, 1932, p. 545.)

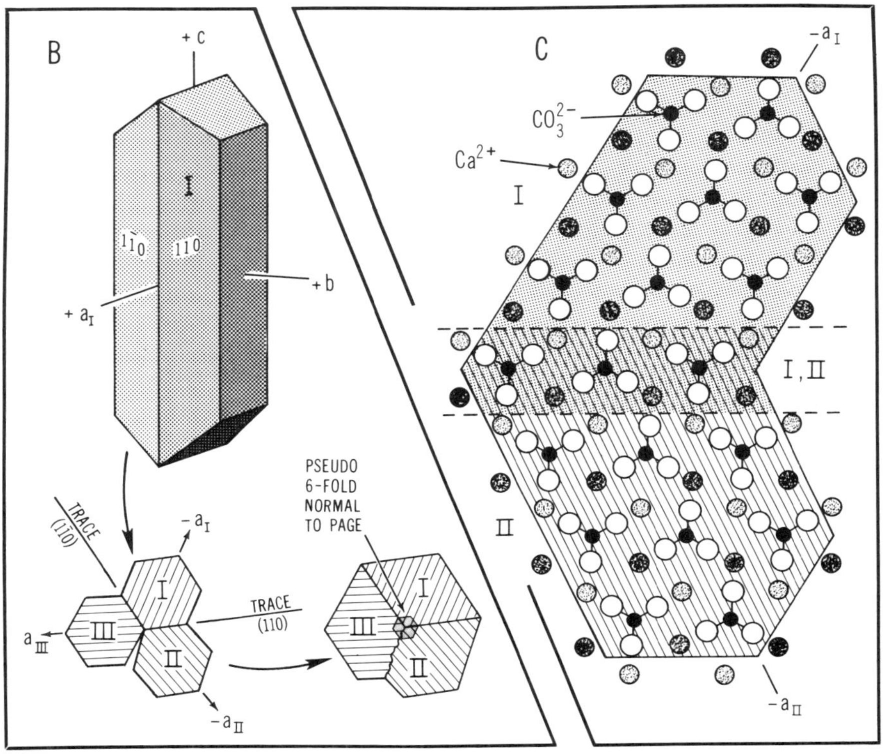

FIGURE 10-25 Continued

(B) Cyclic twinning in aragonite, $CaCO_3$. Crystal I represents an untwinned individual. *Below, left*: crystal I is viewed from above as it would appear if twinned on (110) to crystal II and twinned on (1$\bar{1}$0) to crystal III. *Below, right*: the three cyclically twinned crystals are shown after continued growth has filled in the re-entrant angles to produce a pseudo 6-fold axis where shown.
(C) Structural explanation of aragonite twinning. (After Bragg.) The structures, projected onto (001), are shown for crystal I (on shaded background) and for crystal II (on ruled background). The slice of structure between the dashed lines, and parallel to face (110) for crystal I, is common to both structure I and structure II. Consequently, all atoms within this slice or twin boundary have the same first coordination sphere of neighbors as in the untwinned structure. Detailed examination reveals that only their higher coordination spheres differ from those in the stable untwinned structure. Hence aragonite twins possess little more energy than single crystals, being almost as stable and therefore occurring with great frequency.

it. The symbol used to describe a complex twin such as $\perp[uvw]/(hkl)$ may be read "the direction perpendicular to $[uvw]$ within the plane (hkl)."

Symmetry-Imposed Restrictions. A twin plane or axis can never coincide in direction with any symmetry operation (in the untwinned crystal structure) for which this twinning operation is either identical or a subgroup. For example,

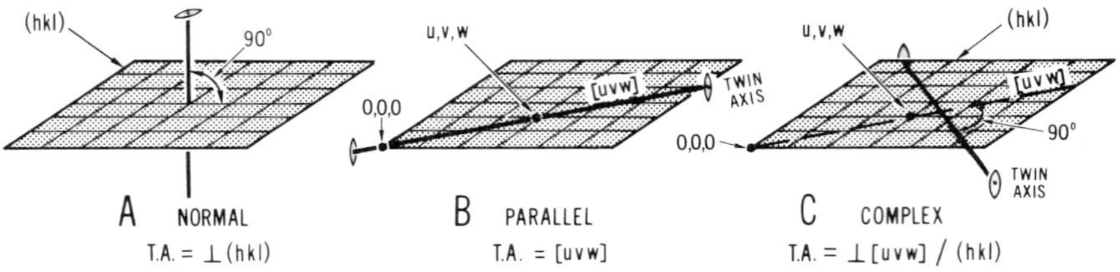

FIGURE 10–26

Relation of the twin axis (TA) to the composition plane (*hkl*), and to lattice directions within this plane, is shown for the three types of rotation-contact twins. The twinned crystals, disposed on opposite sides of the composition plane, are not drawn. The twin axis, which if operative on one causes its crystallographic axis to coincide in all respects with its twin, is assumed to be a 2 fold, as is usually true. If it is a 2-fold axis for the three cases shown here, then the crystal on one side of composition plane (*hkl*) would, if rotated about this twin axis, become identical in orientation to the crystal on the opposite side of (*hkl*).
(A) A *normal twin* in which the twin axis is perpendicular to the composition plane (*hkl*).
(B) A parallel twin in which the twin axis coincides with a direction [*uvw*] in the composition plane.
(C) A complex twin for which the twin axis lies in the composition plane (*hkl*) but coincides with no rational [*uvw*] direction within it, coinciding only with the perpendicular to a direction [*uvw*] in this plane.

a twin plane can never occur parallel to a plane of symmetry in the crystal as demonstrated structurally in Figure 10-27A and macroscopically in Figure 10-27B. For similar reasons a twin axis of the 2-fold type (*hemitropic*) can never occur along a 2-, 4-, or 6-fold axis of symmetry in the structure because any 180° rotation about these latter axes leaves the crystal structure unchanged instead of producing a second individual of differing orientation. By contrast, a 3 or $\bar{3}$ axis of symmetry in the structure can serve as a hemitropic twin axis because a 2-fold axis is not a subgroup of it. Thus the 3-fold axis labeled *b* in Figure 10-27C can be regarded as the twin axis about which a 180° rotation relates crystals III and IV. In summary, a twin operation must always introduce a symmetry operation not normally present along that direction in the crystal. It is for this reason that an intimately twinned crystal may appear to have higher symmetry than the untwinned individual.

FIGURE 10–27

(A) In the NaCl structure epitomized by just the single row L_1 from along its *a* axis, endowing a random (*hkl*) plane with the mirror-plane operation of a twin plane produces a differently oriented crystal, that is, a twin, whose orientation is epitomized by row L_2. By contrast plane (010), which is parallel to a plane of symmetry in the NaCl structure, is hence already endowed with mirror-plane symmetry. Thus its mirror operation on the crystal epitomized by row L_3 merely yields the crystal orientation epitomized by row L_4. But L_4 is merely the extension of L_3,

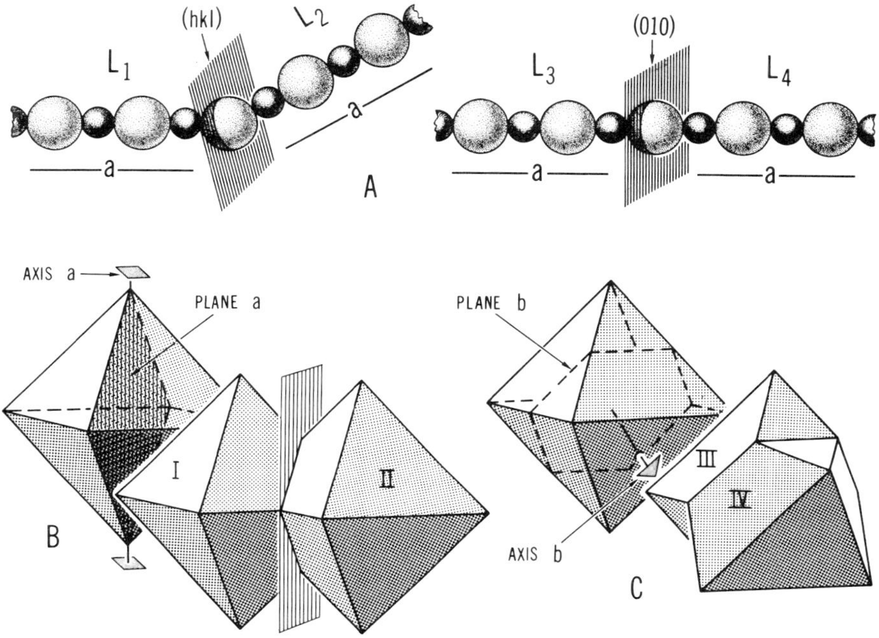

that is, the two together constitute the same row from a single untwinned crystal, and hence (010) cannot be a twin plane.

(B) To demonstrate macroscopically that a symmetry plane can never be a twin plane, we assume octahedra I and II to be related by a twin plane (ruled) whose direction coincides with a symmetry plane *a* in the untwinned crystal behind I. In the untwinned crystal the crystal structure obviously remains unchanged in orientation as one passes from the crystal's left half, across plane *a*, and into its right half. So too, therefore, must the crystal structure remain unchanged in orientation in passing from crystal I, across the ruled mirror plane, and into II. Thus crystals I and II are identical in structure *and orientation* and merely represent a parallel growth but not a twin.

(C) Plane *b*, because it is not a symmetry plane of the untwinned octahedron behind III, *can* serve as a twin plane. Endowing it with mirror-plane symmetry permits it to relate octahedron III to IV, where IV clearly has a different crystallographic orientation from III. Alternatively, III can be considered to be related to IV by a 180° rotation about the 3-fold axis *b*, $[11\bar{1}]$, as a twin axis. However, if *b* were an even-fold axis of symmetry for the untwinned crystal, a 180° rotation about it would leave the structure unchanged and *b* could not serve as a 2-fold twin axis. The twin represented by octahedra III and IV is so common in spinel $MgAl_2O_4$, it is known as *spinel twinning*.

The alternative choice of twin laws, that is, twin plane *b* or twin axis *b*, inevitably arises for centrosymmetric crystals. For these, to endow a plane with mirror-plane symmetry is the same as endowing the axis normal to this plane with 2-fold symmetry. This is true because as demonstrated earlier (Fig. 1–12), if two of the following three symmetry elements are present, a center, a plane, and a 2-fold normal to the plane, then the third is necessarily present as well.

In point-group symmetry if a center of symmetry is located on a plane of symmetry, a 2-fold axis must exist normal to this plane. Conversely, if a center of symmetry is located on a 2-fold axis of symmetry, a symmetry plane must exist perpendicular to this 2-fold axis. Consequently, twinned crystals that individually possess a center of symmetry will, if related by a twin plane, also be related by a 2-fold twin axis normal to this plane. Conversely, if related by a twin axis of 2-fold symmetry, they will also be related by a twin plane disposed normal to the twin axis. Thus the spinel twins shown in Figure 10-27C may be considered either as reflection twins with the $(11\bar{1})$ *plane as the twin plane* (plane *b*) or as rotation twins with the $(11\bar{1})$ *direction as the twin axis* (axis *b*).

Given a face (*hkl*) with indices identical to a direction [*hkl*] and excluding special instances in which one or more indices equals zero, the nonisometric systems differ from the isometric system in that the plane (*hkl*) is not perpendicular to the direction [*hkl*], a point discussed earlier (p. 139). In nonisometric crystals, therefore, the direction normal to plane (*hkl*) is not necessarily a rational direction describable by a zone symbol such as [*hkl*]; conversely, a plane perpendicular to the rational direction [*hkl*] is not necessarily a rational plane described by rational indices such as (*hkl*). In those nonisometric crystals that possess a center of symmetry, therefore (a) reflectional twins, related by a twin plane (*hkl*), could be less desirably described as rotational twins with the direction $\perp$ (*hkl*), a nonrational direction, as the hemitropic twin axis; and (b) rotational twins, related by a twin axis [*hkl*], could be less desirably described as reflectional twins with the nonrational plane normal to [*hkl*] serving as the twin plane.

Formation of twins

Buerger (1945) classifies twins according to their method of formation into (1) *growth twins,* in which twinning develops as an accident of growth; (2) *transformation twins,* which form during a high-to-low phase transition; and (3) *mechanical or gliding twins,* which form in response to mechanical stress.

Growth Twins. Growth twins represent *primary twinning* because they develop chiefly from "accidental" emplacement of atoms or groups of atoms into incorrect sites *during* crystal growth. For example, within the structure of phlogopite micas there occurs a layer of essentially close-packed oxygens and hydroxyls to which Mg^{2+} ions coordinate (Fig. 10-28A). During growth a second such close-packed layer of oxygens and hydroxyls is added, these atoms centering above the sites labeled 4, 8, and 12, in allusion to their 4, 8, or 12 o'clock direction from the nearest hydroxyl in the lower layer (Fig. 10-28B). This upper layer may form with its hydroxyls at either 12 o'clock sites (Fig. 10-28C), at 4 o'clock sites (Fig. 10-28D), or at 8 o'clock sites (Fig. 10-28E); in each case oxygens (not shown) locate over the remaining numbered sites to complete the close-packed upper layer as well as the coordination octahedra of oxygens and hydroxyls around Mg^{2+}.

In an untwinned crystal, once the hydroxyls overlay the 12 o'clock sites, for example, they do so throughout the entire crystal. Sometimes, however, the

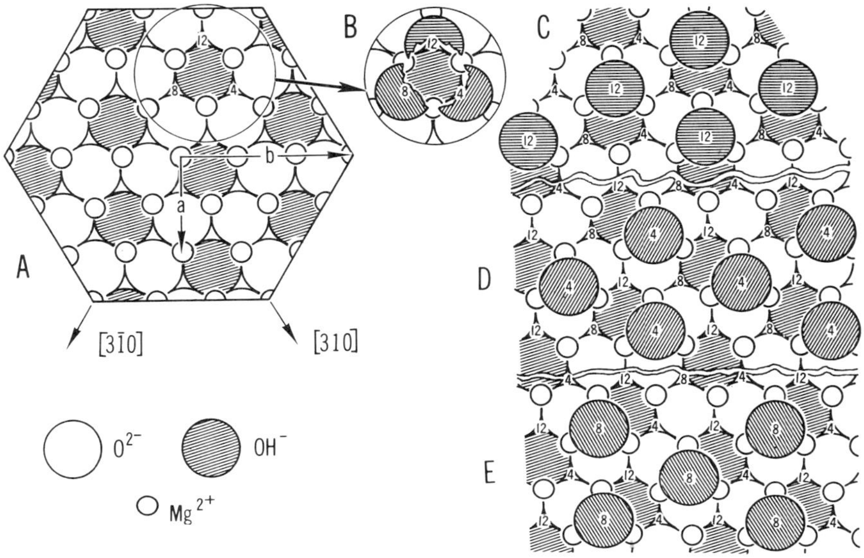

FIGURE 10–28

(A) The closest-packed layer consisting of two-thirds oxygens and one-third hydroxyls which occurs parallel to (001) in phlogopite, $KMg_3AlSi_3O_{10}(OH)_2$. The smaller Mg^{2+} ions nestle into the indentations where shown. A second closest-packed layer of oxygens and hydroxyls, identical to this first, nestles into the indentations marked 4, 8, and 12, these indentations being so labeled because, as shown in (B), they are located at 4, 8, and 12 o'clock in respect to the hydroxyl of the lower layer. The hydroxyls of this second closest-packed array can fit into the 12 o'clock indentations, the 4 o'clock, or the 8 o'clock indentations (C, D, or E); the oxygens that fit in the remaining numbered sites are omitted from (C), (D), and (E). If section (C) is rotated 120° clockwise about an axis normal to the page, it becomes identical to (D); if rotated 120° counterclockwise, it becomes identical to (E).

hydroxyls may locate above 12 o'clock sites in one area of a growing mica flake but, because of an accident of growth, above 4 o'clock sites in another region. In this event the two areas will bear the same twinned relationship to each other as exists between Figures 10-28C and 10-28D. Thus Figure 10-28C can be brought to the same orientation as Figure 10-28D by either (a) a 120° clockwise rotation about an axis normal to the close-packed layer or (b) a 180° rotation about [310], a direction illustrated in Figure 10-28A. Either axis could be chosen as the twin axis but [310] is preferred because it preserves the notion that twin axes are inevitably 2-fold axes. It is of interest that [310] is the common twin axis for micas, an indication that "mistakes" in hydroxyl emplacement occur fairly frequently.

Processes that modify a crystal after its formation are sometimes called *secondary* processes. In this sense transformation twinning and mechanical twinning both represent *secondary twinning*.

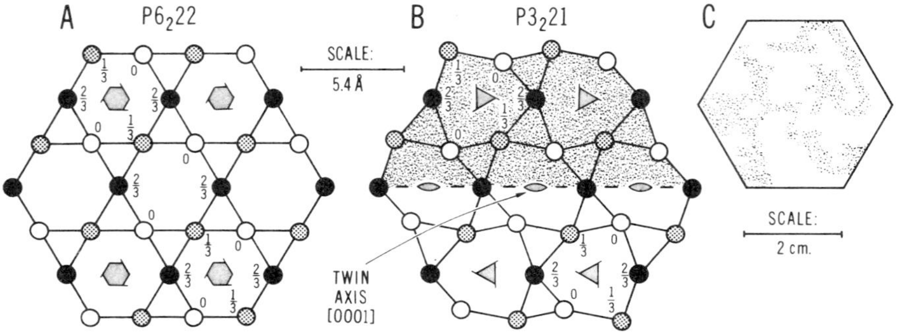

FIGURE 10–29

(A) The location of the silicon atoms in the structure of left-handed high quartz.
(B) Upon cooling below 573°C the high-quartz structure at left transforms to the left-handed low-quartz structure in either the orientation above the dashed line or, with equal ease, to the orientation below the dashed line. These two orientations bear a twinned relationship to each other because a 180° rotation about [0001] brings one orientation into coincidence with the other. Such twinning in quartz is known as *Dauphine twinning.*
(C) A section normal to the *c* axis of a quartz crystal, the stippled and unstippled areas being related by Dauphine twinning.

Transformation Twinning. Transformation twinning often results if an earlier formed crystal, as its temperature is lowered, transforms to a structure of lower symmetry.[5] For example, if cooled below 573°C, high quartz (Fig. 10-29A) may transform into low quartz of an orientation like that on the stippled background in Figure 10-29B or with equal ease like that on the unstippled background. Thus a quartz crystal may transform in some regions into the one orientation (stippled, Fig. 10-29C) but in other regions into the equally likely alternative orientation (unstippled). As Figure 10-29B shows, these two orientations are symmetrically related to each other by a 180° rotation about [0001], which is thus the twin axis, this twinning in quartz being called *Dauphine twinning.* If a quartz crystal containing Dauphine twinning is subjected to pressure, the twin boundaries may shift spasmodically. If the crystal is heated above 573°C, Dauphine twinning will disappear and under controlled cooling may be prevented from reappearing.

In transformation twinning those elements of point-group symmetry that are lost during the high-to-low transformation are likely to be the twin laws. For example, in Dauphine twinning the transformation from high quartz (point group, 622) to low quartz (point group, 32) degrades the [0001] direction from a 6-fold to a 3-fold axis by suppression of the 2-fold operation inherently present as a subgroup of a 6-fold axis. As already discussed Dauphine twins are related to

[5] In solid-state transformations it is an almost universal rule that the forms which are stable at high temperatures are more symmetrical than those which are stable at lower temperatures.

each other by a 2-fold operation about [0001], this being the very symmetry lost when [0001] transformed from a 6-fold to a 3-fold axis. This tendency of transformation twins to be related by the symmetry lost during the high-to-low transformation could cause an intimate, submicroscopic intergrowth of transformation twins to appear to have a symmetry equal to that of the high-temperature form.

Mechanical Twinning. Mechanical twinning can be produced in calcite by careful pressure from a razor blade (Fig. 10-30). In fact lamellar twinning with $\{01\bar{1}2\}$ occurring repeatedly as both twin plane and contact plane is common in calcite, particularly if it has been subjected to mechanical deformation (as in marble).

For further reading R. W. Cahn's comprehensive review of twinning (1954, p. 363) is recommended.

Oriented growth between unlike crystals

In solid-state reactions in which a substance remains solid while undergoing a chemical or structural change, Bernal and Mackay (1965, p. 331) distinguish three types of reactions in which the crystallographic orientation of the reaction products may be inherited or controlled by that of a pre-existing crystal reactant. These are (1) polymorphic transformations, which involve no compositional changes; (2) solid-state reactions, which involve compositional changes and where reagents or products may enter or leave the system; and (3) epitaxial growths as defined next.

Epitaxy. Epitaxy signifies the oriented overgrowth of a crystal A upon the surface of a second phase B, this second phase upon which growth occurs being sometimes called the *substrate.* Crystals A and B are not necessarily related chemically (for example, the epitaxy of vitamin B_{12} on quartz), but their lattice dimensions along their common plane must be fairly close. Thus crystal A can maintain either (a) a continuous contact with B along this plane, A's structure in the layers closest to B being sometimes distorted to improve the fit or (b) if the lattice dimensions of A cannot be altered so as to permit a better fit and a continuous contact, then only small crystals of A will develop on substrate B, larger crystals of A then growing from these and developing a less intimate contact where they abut against B. Interestingly, the geometry of the substrate may sometimes cause an epitaxial overgrowth to adopt an abnormal polymorphic form, one not stable for the existent pressure, temperature, or compositional conditions.

Syntaxy. As originally defined by Ungemach, syntaxy signifies an oriented intergrowth of two polymorphs of a given substance. For example, coquimbite, $Fe_2(SO_4)_3 \cdot 9H_2O$, occurs as lamellar intergrowths with paracoquimbite, $Fe_2(SO_4)_3 \cdot 9H_2O$, $\{0001\}$ being their plane of contact and the a and c axes of one being parallel to those of the other. Gabrielle and J. D. H. Donnay recommend that this definition be extended to include oriented intergrowths between chemically unlike crystals.

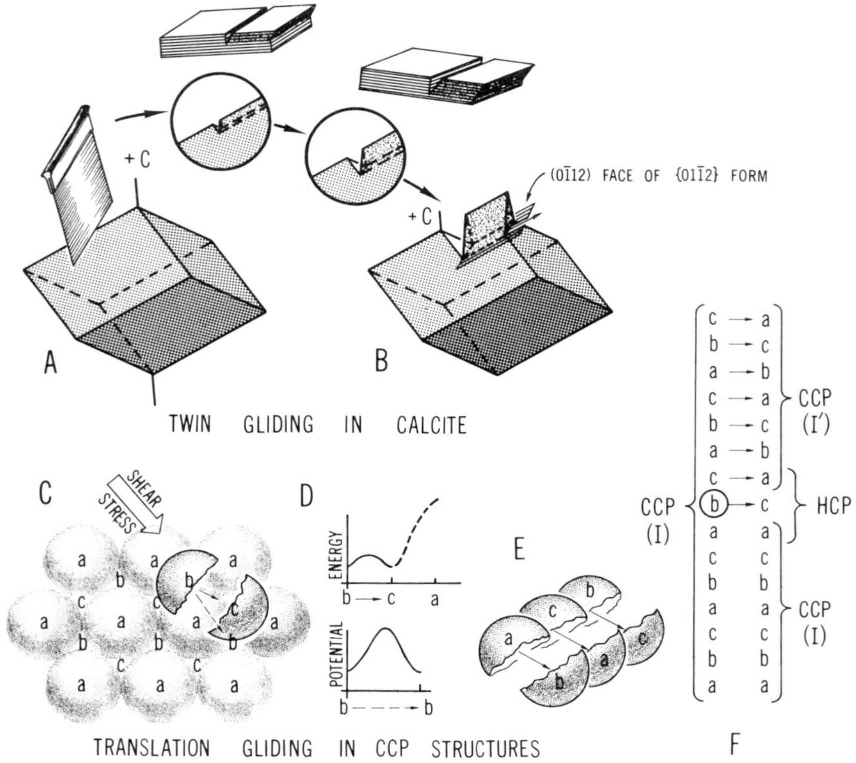

FIGURE 10–30

(A)(B) Careful pressure with a razor blade on a rhombohedron of calcite will produce mechanical twinning with a plane of the form {01$\bar{1}$2} as the twin plane, this plane being ruled in (B). The circular inserts represent intermediate stages. This twinning can only result from the ability of layers parallel to {01$\bar{1}$2} in the calcite structure to slide past each other, like the cards in the decks immediately above the circular inserts, yet not break along these planes. This process is known as *slip* or *gliding* and those planes along which it occurs are called *slip planes* by metallurgists but *glide planes* by mineralogists. The term *glide plane,* so used, differs in meaning from the identical term when used in space-group symmetry. Slip or gliding occurs not only along certain planes, but also along specific directions, called *glide directions* or *glide lines,* within the plane. In (B), for example, slip occurs along the arrow in the (0$\bar{1}$12) plane.

(C) The reason for confinement of gliding to specific directions within a glide plane is illustrated here by a cubic closest-packed metal structure. In it the atoms *a* represent a single closest-packed layer parallel to {111}. Atom *b* represents the closest-packed layer immediately above the *a* atoms; the remaining *b* atoms (not drawn) would also nestle into the indentations labeled *b*. A third closest-packed layer, called *c,* would overlie the *b* layer, nestling in the *b* layer's indentations so that each *c* atom is centered above the letter *c*. If the crystal is subjected to a shearing stress along the hollow arrow, each *b* atom could slip, as shown by the solid arrow *bc,* into the indentation labeled *c* in the *a* layer. Less likely is slippage from *b* to *b* sites (dashed arrow *bb*).

Topotaxy. The term topotaxy is used to indicate that the product crystals of a reaction have inherited their orientations from the reagent crystals. Structural features of the reagent crystals, frequently the packing of the bulkiest atoms or ions, survive relatively unchanged in the product crystals and control their crystallographic orientation. For example, the orthorhombic mineral groutite MnO(OH) if heated at 300°C in air, transforms topotactically into the tetragonal mineral pyrolusite MnO_2. The *a*, *b*, and *c* axes of groutite become, respectively, the *a*, *a*, and *c* axes of pyrolusite. In topotactic reactions involving oxides or silicates the packing of the oxygens frequently survives unchanged to control the orientation of the product crystals with only the smaller cations diffusing through the oxide framework to assume new positions or participate in the chemical interchanges. A number of topotactic transformations of mineralogic significance are listed by Glasser, Glasser, and Taylor (1963, p. 201). Occasionally the geometrical and symmetry conditions of topotactic reactions are sufficiently powerful to permit formation of phases not normally stable.

The distinction of epitaxy from topotaxy is clear-cut if the substrate remains inert as the epitaxial overgrowths crystallize on it. However, if the overgrowth crystals form by decomposition or reaction of a substrate that imparts the initial orientation to these overgrowths but subsequently disappears (completely or in part), it may be difficult to decide whether the reaction was epitaxial or topotactic. According to Bernal and Mackay (1965, p. 336), "... it should be designated as an *epitaxial reaction* if the initial and final products do not resemble each other in structure and only fit each other in two dimensions (although fully

---

(D) The change in potential energy is chematically shown as *b* atoms slip from *b*-to-*c* positions (upper curve) or from *b*-to-*b* sites (lower curve). This potential energy is very roughly proportional to the degree to which the *b* atoms must rise above the plane of the page in slipping from one site to the next. Thus potential energy reaches a lesser maximum for the *b*-to-*c* slip than for the *b*-to-*b* slip. This *b*-to-*c* slip, which represents gliding along ⟨110⟩ in the {111} plane, is common in cubic closest-packing metal structures. The *b*-to-*c* slip would not occur for a length greater than arrow *bc* in (C); if it did, the *b* atom would ride up over the *a* atom and potential energy would increase as shown by the dashed portion of the upper curve in (D). Thus the *b*-to-*c* slip occurs along numerous parallel planes in the structure, each thin lamina (frequently about 10^{-3} to 10^{-5} cm thick) sliding beyond the lamina below it by an amount and direction comparable to vector *bc* in (C).

(E) The gliding of a plane of close-packed *b* atoms into *c* sites carries the atoms of overlying *c* layers into *a* sites, and the overlying *a* layers into *b* sites as is here shown for just one atom of each close-packed layer.

(F) Schematic depiction of a cubic close-packed structure, each letter representing a close-packed layer of metal atoms. If the atoms in the (circled) *b* layer glide into *c* positions, the overlying layers are translated into the positions indicated by the arrows. Even after this gliding, the translated upper portion of this structure (I') retains the same orientation as the original (I). The gliding has created an ultrathin layer of hexagonal close-packed structure in the region of the glide plane. Gliding in which the translated part of the crystal attains a twinned position relative to the untranslated portion is called *twin gliding* (see (B)). If the translated part maintains a parallelism of its axes with those of the untranslated portion (as in (F)), the phenomenon is called *translation gliding.*

oriented). [Whereas], if the reagent substrate and the overgrowth crystals show a three-dimensional accord, then the reaction can be called *topotactic* or *surface topotactic.*"

Polytypism

Polytypism is a special variety of polymorphism in which the polymorphs differ only in the manner of stacking of identical layers of structure. Consequently, unit cell lengths will not differ between polytypes for directions parallel to these layers whereas for directions normal (or approximately normal) to these layers they will be multiples or submultiples of each other. The first coordination sphere about each atom is not significantly changed from one polytype to the other; hence the various polytypes of a substance show no measurable differences in density even though they may crystallize in different systems. The polytypism of SiC, ZnS, and of the micas and clay minerals has been studied in detail. An illustrated explanation of polytypism is deferred to the discussion of layer silicates in a companion volume (in preparation). The subject of polytypism is discussed in great detail by Verma and Krishna (1966).

Crystal Morphology

Crystals of a given substance gathered from many different locales may exhibit certain forms or *assemblages of forms* that are important by virtue of either greater frequency of occurrence or possession of areally larger faces. Thus garnet crystals are usually bounded by the dodecahedron {110} and/or the trapezohedron {211} with the hexoctahedron {321} being present occasionally; faces of the cube and octahedron are rare and, if present, usually small in area. Many hypotheses have been advanced to explain the presence (or absence) of the various forms on crystals of a specific substance. Early among these was the "law" of Bravais which inferred that areal extent would be greatest for faces of high reticular density, that is, for faces parallel to those lattice planes that contain more lattice points per unit area than others. Such planes thus represent the closest packing of the component atoms per area of plane. Donnay and Harker (1937) later demonstrated that screw axes and glide planes necessitated a modification and extension of the law of Bravais.

Periodic bond chains

More recently Hartman and Perdok (1955) emphasized the importance of *periodic bond chains* (PBC) in controlling external form. They define these as unbroken chains of strong crystallization bonds[6] that uninterruptedly link one

[6] Strong crystallization bonds are those which (1) formed *during* the crystallization process and (2) released high amounts of energy, thereby indicating a relatively strong attraction between the two atoms forming the bond (as compared to the bonds between other atoms in the structure).

atom to its neighbor in the structure. The atoms so linked resemble a string of beads and repeat their sequence along this string with a period equal to the spacing of the identipoints along the string (Fig. 10-31A). The line joining two identipoints on the same bond chain is the direction of its *PBC* vector. Hartman and Perdok state that, "If a bond chain contains bonds of different type, its

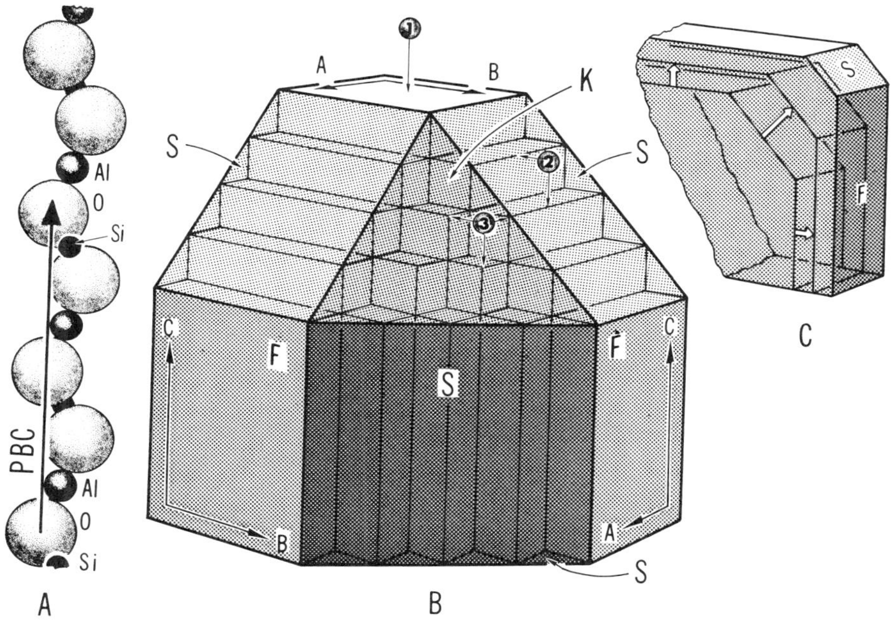

FIGURE 10–31

(A) The chain of strongly bonded atoms that form a *periodic bond chain* (PBC) along the *a* axis in pyrope, $Mg_3Al_2(SiO_4)_3$. The repeat distance as well as the direction of the periodic bond-chain vector for this chain is as indicated by the arrow.
(B) Hypothetical crystal with three mutually perpendicular periodic bond-chain vectors *A*, *B*, and *C*. (After Hartman and Perdok, 1955.) Faces parallel to at least two periodic bond chains are called *F* or *flat* faces; those parallel to only one are called *S* or *stepped* faces; those parallel to none are called *K* or *kinked* faces. Atom (1) is attracted to an *F* face by only one PBC bond (short downward arrow). Atom (2) is attracted into a position on the *S* face by two PBC bonds. Atom (3) is attracted to the *K* face by *three* bonds, being thus more strongly attracted into position than either atom (1) or (2).
(C) Two earlier stages in growth are shown inside a crystal. In growing from stage 1 to stage 2, the unshaded upper *F* face plus shaded faces *F* and *S* grew at rates proportional to the hollow arrows drawn on the large front face. Because atoms are attracted to it more strongly, the *S* face has a greater displacement velocity, that is, growth rate perpendicular to itself, than either of the (shaded or unshaded) *F* faces. This greater growth rate causes *S* to grow itself out of existence, its area steadily decreasing from the interior growth stage to the outermost growth stage. Conversely, the slower growing *F* faces are becoming the most important in areal extent.

influence on the shape of the crystal is determined by the weakest bond present in the chain"; and, "Important zone directions must be parallel to bond chains containing strong bonds only." This latter statement thus implies that if there is a prominent zone on a crystal, then the corresponding zone axis is parallel to the *PBC* vector for a chain consisting of relatively strong bonds all along its length. Hartman and Perdok divide crystal faces into three classes, depending on their attitude to the *PBC* vectors corresponding to the stronger chains in the structure. Thus class *F* or *flat faces* are parallel to two or more coplanar *PBC* vectors (Fig. 10-31B), class *S* or *stepped faces* are parallel to only one *PBC* vector, and class *K* or *kinked faces* are parallel to no important *PBC* vector.

Figure 10-31B contrasts three atoms (or building units) that are being attracted into their correct sites on the surface of an *F* face (atom 1), an *S* face (atom 2), and a *K* face (atom 3). Atoms 1, 2, and 3 are attracted by bonds (indicated by arrows) from the ends of one, two, or three periodic bond chains, respectively. Thus atom 3's attraction to the *K* face exceeds atom 2's attraction to the *S* face by one bond and atom 1's attraction to the *F* face by two bonds. Consequently, the displacement velocity for a face, that is, its rate of growth in the direction normal to itself, is greatest for *K* faces, least for *F* faces. The areal extent of a face is inversely proportional to its displacement velocity as is illustrated by the various stages of growth for faces *S* and *F* in Figure 10-31C, *S* being the face that has the higher displacement velocity. Consequently, as a crystal grows, its *F* faces usually become the larger dominant forms; its *K* faces will often be unobserved or very rare forms, and all other faces will likely be *S* faces.

Habit

The shape imparted to a crystal by the relative development of its various forms is referred to as *habit*. Small needlelike crystals have an *acicular* habit (Fig. 10-32A); if the needles are so fine as to be flexible, the crystals are *fibrous* in habit. Other possible habits (Fig. 10-32) are *columnar* (or *prismatic*), *equant*, *bladed*, and *tabular* (or *lamellar*). Needlelike crystals usually indicate a structurally dominant periodic bond chain parallel to the needle axis. Tabular crystals

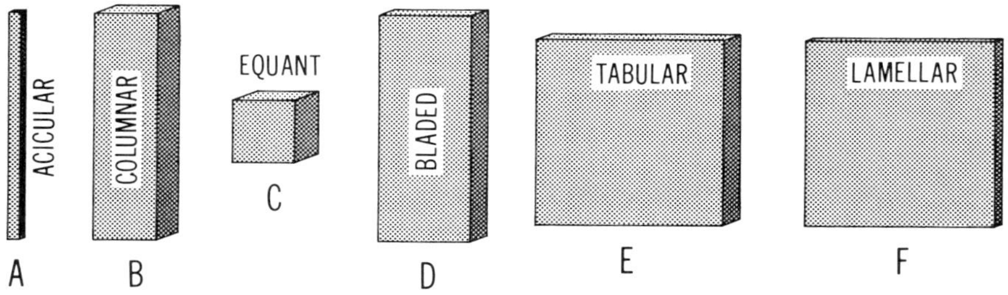

FIGURE 10–32

The various types of crystal habit. (A) Acicular. (B) Columnar or prismatic. (C) Equant. (D) Bladed. (D) Tabular. (F) Lamellar.

indicate strong periodic bond chains parallel to two or more directions within the crystal flake but none normal to it. Prismatic habit indicates one strong periodic bond chain along the crystal's length and two lesser periodic bond chains at a high angle to the crystal's length. Habit is not exclusively controlled by internal structure, however, environmental conditions during growth also influencing a crystal's habit. Thus foreign ions present in the crystallizing solution may significantly affect crystal habit. For example, $Pb(NO_3)_2$ crystallizes as octahedra from a pure water solution but as cubes if the solution is saturated with methylene blue. Possibly these foreign ions or atoms temporarily adhere to the growth surface and become weaker links in certain periodic bond chains in the structure causing the less important periodic bond chains along other directions to become important by default.

REFERENCES AND RECOMMENDED READING

Barth, T. F. W. (1934) "Polymorphic phenomena and crystal structure," *Am. Jour. Sci.*, **27,** 273–286.

Bernal, J. D., and A. L. Mackay (1965) "Topotaxy," *Tschermaks Min. Petr. Mitt.*, **10,** 331–340.

Bragg, W. L., and E. J. Williams (1934) "The effect of thermal agitation on atomic arrangement in alloys," *Proc. Royal Soc. (London)*, **145,** 669–730.

Buerger, M. J. (1945) "The genesis of twin crystals," *Am. Mineral.*, **30,** 469–482.

——— (1951) *Phase Transformation in Solids,* (R. Smoluchowski, ed.) New York: Wiley.

——— (1961) "Polymorphism and phase transformations," *Fortschr. Mineral.*, **39,** 9–24.

Cahn, R. W. (1954) "Twinned crystals," *Adv. in Phys.*, **3,** 363–445.

Cooksey, D. J. S. *et al.* (1964) "The freezing of some continuous binary eutectic mixtures," *Phil. Mag.*, **10,** 745–769.

Darwin, C. G. (1914) "The theory of X-ray reflection," *Phil. Mag.*, **27,** 315, 675.

Donnay, J. D. H., and D. Harker (1937) "A new law of crystal morphology extending the law of Bravais," *Am. Mineral.*, **22,** 446–467.

Glasser, L. S. D., F. P. Glasser, and H. F. W. Taylor, (1963) "The role of oriented transformations in mineralogy," Special Paper 1, Min. Soc. Am., 200–203.

Hartman, P., and W. G. Perdok, (1955) "On the relations between structure and morphology of crystals," *Acta Cryst.*, **8,** 49–52, 521 *et seq.*

Kushiro, I., H. S. Yoder, and M. Nishikawa, (1968) "Effect of water on the melting of enstatite," *Bull. Geol. Soc. Am.*, **79,** 1685–1692.

Lipson, H. (1950) "Order-disorder changes in alloys," *Progress in Metal Physics, II,* B. Chalmers (ed.). London: Butterworth, 1–52.

Moore, Walter J. (1962) *Physical Chemistry,* 3d. ed. Englewood Cliffs, N. J.: Prentice-Hall, 844.

von Philipsborn, H. (1928) *Neues Jahrb. f. Min., Beil. Bd.*, 1001.

Salkind, M., and F. Lemkey (1967) "Metals with grown-in whiskers," *International Sci. and Tech.*, 51–56.

Sosman, R. B. (1965) *The Phases of Silica.* New Brunswick, N. J.: Rutgers Univ. Press, 388.

Verma, A. R., and P. Krishna (1966) *Polymorphism and Polytypism in Crystals.* New York: Wiley.

QUESTIONS AND PROBLEMS

10-1 What is the weight percent of (a) Na in NaCl? (b) Al_2O_3 in kyanite, Al_2SiO_5? (c) U_3O_8 in becquerelite, $UO_3 \cdot 2H_2O$? (d) MgO in an olivine of composition $Fo_{40}Fa_{60}$?

10-2 What is the chemical formula for a crystal whose analysis is reported as 16.47 percent (by weight) of CO_2 and 83.51 percent PbO?

10-3 Give the formula for an olivine crystal whose complete chemical analysis is reported as MgO, 31.58 percent; FeO, 31.65 percent; SiO_2, 36.77 percent (by weight).

10-4 Why can twinning with {010} as the twin plane, that is, albite twinning, occur in albite, $NaAlSi_3O_8$, but not in sanidine or orthoclase, $KAlSi_3O_8$?

10-5 Given some metal atoms whose radii equal 1 Å: (a) Assuming a cubic closest packing of these atoms, what is the value of a_0? (b) Assuming a hexagonal closest packing, what are the values of a_0 and c_0? (c) For hexagonal closest-packed structures, what is the axial ratio, c/a?

10-6 Given a system whose T, X condition is indicated by point C_3 in Figure 10-10: (a) What are the compositions and relative amounts of the phases present? (b) Same question as (a) if T, X condition is C_5?

10-7 Assuming liquid L_4 in Figure 10-10 is at 1 atm pressure, the temperature of liquid L_4 will remain constant at 1270°C, as heat is withdrawn from the system, until L_4 has entirely crystallized. Why?

ANSWERS

10-1 (a) 39.34 percent; (b) 62.92 percent; (c) 87.16 percent; (d) 18.06 percent. The answer for (c) was obtained by dividing the molecular weight of U_3O_8 by triple the molecular weight of $UO_3 \cdot 2H_2O$. By custom the percentage of uranium in minerals is reported in terms of U_3O_8.

10-2 $PbCO_3$.

10-3 $Fo_{64}Fa_{36}$ or $(Mg_{1.28}Fe_{0.72})SiO_4$.

10-4 In the structure of the monoclinic crystals, sanidine and orthoclase (see Fig. 10-20), {010} is a plane of symmetry and hence cannot be a twin plane.

10-5 (a) 2.83 Å; (b) a_0 = 2 Å; c_0 = 3.27 Å; (c) c/a = 1.633.

10-6 (a) 50 percent solid plus 50 percent liquid. The solid is pure anorthite, the liquid is of composition L_3; (b) anorthite crystals, 75 percent; diopside crystals, 25 percent.

10-7 While the eutectic crystallization is in progress, diopside, anorthite, and liquid L_4 coexist for this two-component system. Substituting $p = 3$ and $c = 2$ into the phase rule (Eq. 10-5), the degree of freedom is seen to equal 1, this single degree of freedom being used up when we selected the pressure to be 1 atm. Thus the eutectic temperature at a given pressure for a two-component system must remain at its nature-dictated value (1270°C in this example) until the number of phases drops back to 2, that is, until the liquid has entirely crystallized.

11 Physical Properties

A crystal's physical properties are determined by its structure, its chemical composition, and the nature and abundance of any structural defects. Desirable physical properties may give a substance economic value. Moreover, a few relatively easy tests of a material's physical properties sometimes permit its identification without resort to more cumbersome tests.

A crystal's physical properties are of two types: (1) *nondirectional* and (2) *directional.* Nondirectional properties lack a sense of direction. Examples include a crystal's temperature, its density, and such sensate properties as taste and odor. In contrast, directional properties are measurable along *specific directions* in a crystal. If their value changes with direction of measurement, the crystal is *anisotropic* in respect to this directional property; if not, the crystal is *isotropic* in respect to it. A crystal that is isotropic for some physical properties may be anisotropic for others.

Nondirectional Properties

Density and specific gravity

Density, by definition, equals mass per unit volume, that is,

$$\rho = \frac{m}{V} \qquad \text{(Eq. 11-1)}$$

where ρ, m, and V, respectively, represent the density, mass, and volume of a given fragment of a mineral. Thus if a piece of quartz has a mass of 79.5 grams and a volume of 30 cm^3, its density is 2.65 g/cm^3. As its temperature is increased, a mineral usually expands in volume whereas its mass remains constant; hence as a general rule density decreases as temperature increases. On the other hand density increases slightly as pressure increases. In precise work the temperature at which the density was measured, usually 20°C, is recorded. For example, the symbol ρ_{20° indicates the density of a substance at 20°C, that is, the mass of 1 cm^3 of it at 20°C.

The *specific gravity* of a substance, usually symbolized as G, represents its density divided by the density of water at 4°C, the temperature at which water possesses its maximum density of 0.999973 g/cm^3. Thus

$$G_{4^\circ}^{20^\circ} = \frac{\rho\,[\text{substance at } 20^\circ\text{C}]}{0.999973 \text{ gram/cm}^3} \qquad \text{(Eq. 11-2)}$$

Because 0.999973 is so close to unity, specific gravity $G_{4^\circ}^{20^\circ}$ differs from density ρ_{20° in only the fourth or fifth decimal place. Thus for work of ordinary accuracy no distinction need be made; however, specific gravity is a pure number, being the ratio of two densities, whereas a density value must be expressed as mass per unit of volume. Specific gravity may be alternatively defined as a substance's weight divided by the weight of an equal volume of water. Thus the numerator on the right side of Equation (11-2) could be the weight of 1 cm^3 of the substance at 20°C, and the denominator could be the weight of 1 cm^3 of water at 4°C.

Methods of Measurement.[1] The specific gravity or density of a crystal may aid in its identification and is hence frequently measured. Methods of measurement depend on whether the crystal is available as a suitably large fragment containing neither pores nor inclusions of foreign materials or whether it is only available as small fragments or as a powder. A large single fragment may be weighed (1) in air and then (2) while immersed in water, these weights being recorded as m_1 and m_2, respectively. Their difference, that is, $(m_1 - m_2)$, represents the water's buoyant effect on the fragment during the water-immersion weighing. Thus as per Archimedes' principle, $(m_1 - m_2)$ equals the weight of the water displaced by, and therefore equal in volume to, the sample. The crystal's specific gravity is thus

$$G = \frac{m_1}{m_1 - m_2} \qquad \text{(Eq. 11-3)}$$

that is, the material's weight m_1 divided by the weight of an equal volume of water $(m_1 - m_2)$.

The *Jolly Balance* can be used to determine the specific gravity of crystals by the method given above. In the original model (Fig. 11-1A) the degree of distention of a spiral spring is recorded first as the crystal fragment rests on

[1] Hidnert and Peffer (1950, Circular C487 of U.S. National Bureau of Standards) describe 11 different methods for determining density and indicate the expectable accuracy of each.

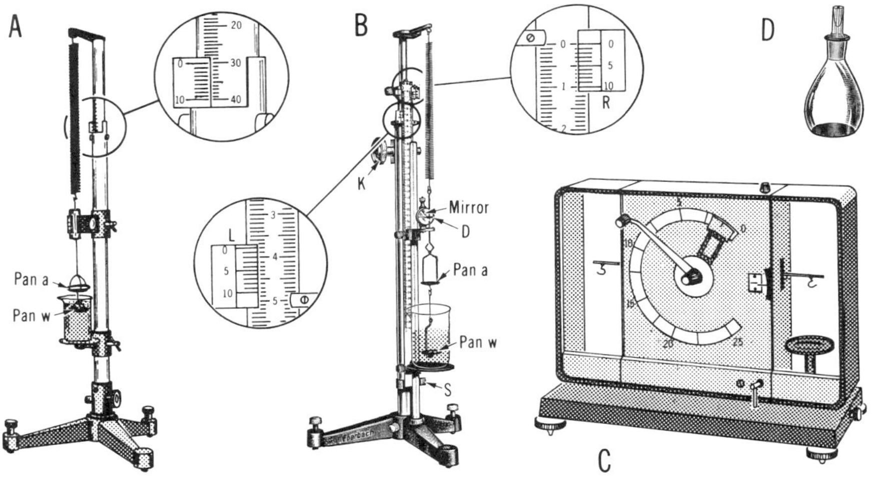

FIGURE 11–1

(A) Jolly balance. (B) Jolly balance as modified by E. H. Kraus. (C) Berman balance. (D) Pycnometer.

pan *a*, and again when it rests, fully immersed, on pan *w*. The two scale readings correspond to m_1 and m_2 of the preceding paragraph and inserted into Equation (11-3) permit calculation of G. The *Kraus-Jolly balance*, as improved by E. H. Kraus, has two sliding indicators, L and R in Figure 11-1B. With the crystal fragment on pan *a* and with set screw S loosened, knob K is adjusted until the edge view of disk D returns to its original zero point (observed when neither pan *a* or pan *w* was loaded). Indicator L now reads a value proportional to m_1 and tightening set screw S will lock L at this reading. The fragment is then transferred to pan *w* and knob K is again adjusted until the edge view of disk D returns to the zero point. Along the right side of the scale, indicator R now reads a value directly corresponding to $(m_1 - m_2)$. Thus from Equation (11-3), G equals the reading for indicator L divided by that for R.[2]

The Jolly balance permits accuracy to within 1 percent but this drops off rapidly to the extent that the fragment is so small as to yield a reading on the L scale of less than 20, which is half of the full-scale value.

The Berman Balance. H. S. Berman (1939, p. 434) adapted a torsion balance (Fig. 11-1C) to measure specific gravity to within 0.1 percent for crystals that weigh as little as 15 mg. Specific gravity is computed from the weight of the crystal in air and while immersed in toluene, a liquid whose density is well known. (Detailed instructions are provided by the distributor, the Bethlehem Instrument Company of Bethlehem, Pennsylvania.)

The Pycnometer. For substances available only in grains too small for the

[2] Complete details for the operation of the Kraus-Jolly balance are supplied by its manufacturer, the Eberbach Corp., P.O. Box 1024, Ann Arbor, Michigan, 48106.

Berman balance, the density of an aggregate or powder of these grains can be measured using a glass vessel with a ground-glass stopper known as a *pycnometer* (Fig. 11-1D). Procedural steps include (1) obtaining the sample weight[3] in air, (2) weighing the pycnometer while it is filled to capacity with water, and (3) weighing the pycnometer while it contains the grain aggregate plus enough water to fill it again to capacity. At step (3) the pycnometer contains less water than in step (2), a volume having been displaced that is equal to the aggregate of volumes for all the grains comprising the sample.[4] The weights measured in steps (1), (2), and (3), symbolized m_1, m_2, and m_3, represent the following:

$$\begin{aligned} m_1 &= \text{sample} \\ m_2 &= \text{pycnometer} + \text{its full capacity of water} \\ m_3 &= \text{sample} + \text{pycnometer} + \text{undisplaced water} \\ (m_1 + m_2) - m_3 &= \text{weight of water displaced by sample} \end{aligned}$$

Thus Equation (11-3) becomes for the sample

$$G = \frac{m_1}{(m_1 + m_2) - m_3} \qquad \text{(Eq. 11-4)}$$

For a sample whose aggregate weight is as small as 10 mg, a special micro-pycnometer is suggested by May and Marinenko (1966, p. 931).

The *suspension method* permits the density of a substance to be measured from just a few small grains. These are placed in a small test tube containing, for example, methylene iodide, an organic liquid whose specific gravity is 3.32 at 20°C. If the grains float in this liquid, their specific gravity is less than 3.32; in this case a second miscible liquid, for example, *N*,*N*-dimethylformamide, whose specific gravity is 0.95, may be gradually stirred in. When the specific gravity of the liquid mixture exactly matches that of the grains, the grains will remain suspended at whatever depth in the liquid they occur, neither sinking or rising. For near matches the upward or downward motion of the grains may be so slow that it is best detected with a microscope. When a match is obtained, the density of the liquid solution may be measured by a Westphal balance or by using a pycnometer. If the pycnometer's weight (1) alone, (2) filled to capacity with distilled water, and (3) filled to capacity with the matching solution are known, that is, m_1, m_2, and m_3, then, for the mineral and matching solution,

$$G = \frac{m_3 - m_1}{m_2 - m_1} \qquad \text{(Eq. 11-5)}$$

Meyrowitz, Cuttitta, and Levin (1960, p. 1278) discuss the health precautions to be observed in handling dimethylformamide.

[3] Determine the difference between the weight of the pycnometer when empty and when containing the sample.

[4] This is true if no air bubbles remain entrapped between grains. Such air bubbles are best eliminated by adding a wetting agent to the water used for steps (2) and (3) or by boiling the water prior to completion of step (3).

Calculated Densities. Densities that are measured in the manner just discussed are symbolized ρ_{obs}, the subscript signifying *observed*. They contrast with ρ_{calc}, the density calculated for a crystal if M_g, the mass of its unit cell in grams, is divided by V_{cm^3}, this cell's volume in cubic centimeters.

Calculated Cell Volume. If unit cell dimensions in angstroms (1 Å equals 10^{-8} cm) are substituted into one of the following formulas,

$$\begin{aligned}
&\text{Isometric:} && V = a_0^3 \\
&\text{Tetragonal:} && V = a_0^2c_0 \\
&\text{Hexagonal:} && V = a_0^2c_0 \sin 60° \\
&\text{Orthorhombic:} && V = a_0b_0c_0 \\
&\text{Monoclinic:} && V = a_0b_0c_0 \sin\beta \\
&\text{Triclinic:} && V = a_0b_0c_0 \\
& && \times \sqrt{1 - \cos^2\alpha - \cos^2\beta - \cos^2\gamma + 2\cos\alpha\cos\beta\cos\gamma}
\end{aligned} \qquad \text{(Eqs. 11-6a)}$$

cell volumes in cubic angstroms (Å³) result. These convert to cubic centimeters if multiplied by 10^{-24} cm³/Å³. Thus

$$V_{cm^3} = V_{(\text{Eq. 11-6a})} \times 10^{-24} \frac{\text{cm}^3}{\text{Å}^3} \qquad \text{(Eqs. 11-6b)}$$

As an example the volume for the orthorhombic mineral forsterite (dimensions 4.76 × 10.20 × 5.98 Å) is 290.34 Å³ or 290.34 × 10^{-24} cm³.

Calculated Cell Mass. A unit cell's mass represents the sum of the masses of the various atoms it contains. Consequently, it can be computed if the mineral's chemical formula as well as Z, the number of formula units within the unit cell, are known. Z is simply an integer, usually a small number such as 1, 2, 3, or 4; if it is used to multiply each subscript in the mineral's chemical formula, it produces numbers that indicate the numbers of the various atoms within the accepted unit cell. For example, in forsterite (Mg_2SiO_4) with unit cell as already cited, Z equals 4. Thus its unit cell contains 8 magnesiums, 4 silicons, and 16 oxygens. As discussed in Chapter 10, the weight of an individual atom is merely its atomic weight in grams divided by 6.023×10^{23}. Thus the weight of the 8 magnesiums, 4 silicons, and 16 oxygens can be computed; their total is then M_g, the weight in grams for the unit cell. Expressed compactly, this resolves to

$$M_g = \frac{MZ}{6.023 \times 10^{23}} \qquad \text{(Eq. 11-7)}$$

where M is a molecular weight in grams corresponding to the mineral's correct chemical formula.

The value of M to be substituted into Equation (11-7) can be determined for a solid solution between end members $A, B, C, \ldots,$ since

$$M = x_A M_A + x_B M_B + x_C M_C + \cdots \qquad \text{(Eq. 11-8)}$$

where $M_A, M_B, M_C, \ldots,$ are molecular weights corresponding to each end member's formula and $x_A, x_B, x_C, \ldots,$ represent the fraction of a mole of end members A, B, C, . . ., present in 1 mole, that is, M grams, of the solid solution. Thus 1 mole, M, of an olivine solid solution of composition $Fo_{30}Fa_{70}$ consists of 0.3 mole forsterite and 0.7 mole fayalite, that is,

$$M = 0.3M_{Fo} + 0.7M_{Fa}$$

As this equation illustrates, the sum of the mole fractions of all the end members in 1 mole of solid solution must equal unity. In other words

$$x_A + x_B + x_C + \cdots = 1 \qquad \text{(Eq. 11-9)}$$

Calculation of Density. Density can now be calculated since

$$\rho_{calc} = \frac{M_g}{V_{cm^3}} \qquad \text{(Eq. 11-10)}$$

Substituting into Equation (11-10) the values of M_g and V_{cm^3} as indicated by Equations (11-7) and (11-6b), respectively, there results

$$\rho_{calc} = \frac{MZ}{V}\,1.6602 \text{ g cm}^{-3} \qquad \text{(Eq. 11-11)}$$

where V represents the unit cell volume in cubic angstroms and M the molecular weight corresponding to the crystal's formula. The value 1.6602, actually 10^{24} divided by 6.023×10^{23}, may be regarded as a conversion factor. Inserting the values for forsterite in Equation (11-11) (that is, $M = 140.71$ grams, $Z = 4$, and $V = 290.34$ Å^3), ρ_{calc} is 3.218 g cm^{-3}.

Density and composition in solutions

Density and composition in solutional series are usually nonlinearly related (Fig. 11-2). In fact the oft-assumed straight-line or linear relationship represents only a special case. This can be demonstrated for a series of solutions between, for simplicity, just two end members, A and B. Thus Equations (11-8) and (11-9) simplify to

$$M = x_A M_A + x_B M_B \qquad \text{(Eq. 11-8')}$$

$$x_A = 1 - x_B \qquad \text{(Eq. 11-9')}$$

Combining Equations (11-8′) and (11-9′), there results

$$M = (M_B - M_A)x_B + M_A \qquad \text{(Eq. 11-12)}$$

a linear equation relating the solution's molecular weight M to its composition x. This equation is of the type $y = kx + c$, where slope k compares to $(M_B - M_A)$ and the constant c compares to M_A. We now let v represent the volume of 1 mole of a solution containing end members A, B, . . . (molar volumes, v_A, v_B, . . .). Assuming that v, like M, is linearly related to composition, then

$$v = x_A v_A + x_B v_B + x_C v_C + \cdots \qquad \text{(Eq. 11-13)}$$

For a solution between just two end members, Equation (11-9′) holds and

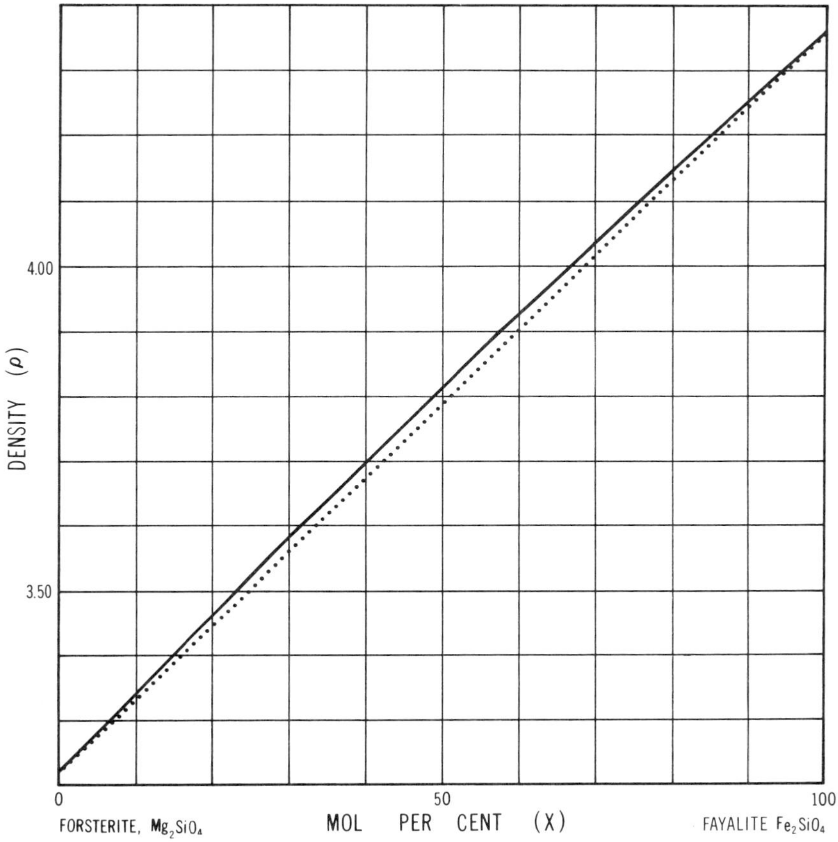

FIGURE 11–2

Relationship between density and composition for the solid solution series between forsterite (Mg_2SiO_4) and fayalite (Fe_2SiO_4). The slightly curved solid line is the actual relationship. The straight (dotted) line is the relationship often, but usually erroneously, assumed to be true for a binary solid-solution series.

Equation (11-13) thus becomes

$$v = (v_B - v_A)x_B + v_A \qquad \text{(Eq. 11-14)}$$

This is another linear equation of type $y = kx + c$.

The density ρ of 1 mole of mixed crystal or solution equals its molecular weight M divided by its molar volume v. Thus Equation (11-12) divided by Equation (11-14) yields

$$\rho = \frac{M}{v} = \frac{(M_B - M_A)x_B + M_A}{(v_B - v_A)x_B + v_A} \qquad \text{(Eq. 11-15)}$$

This is the general relationship between density and composition (in mole fractions) for a two-component solutional series whose molar volume is linearly related to composition as in Equation (11-14). Equation (11-15) can be compared to a function y_3 that results from division of a linear function y_1 by a second linear function y_2. For example, if we plot on graph paper the two linear functions

$$y_1 = x + 2 \qquad \text{and} \qquad y_2 = 3x - 1$$

plus their quotient y_3, that is,

$$y_3 = \frac{y_1}{y_2} = \frac{x + 2}{3x - 1}$$

then y_3 proves to be nonlinear. Specifically, Equation (11-15) will plot as (1) a convex upward curve (Fig. 11-3A), (2) a straight line (Fig. 11-3B), or (3) a concave upward curve (Fig. 11-3C), if the value of δ is (1) positive, (2) zero, or (3) negative, respectively, where we define δ as

$$\delta = \left(\frac{M_A + M_B}{v_A + v_B}\right) - \left(\frac{\rho_A + \rho_B}{2}\right)$$

and where ρ_A and ρ_B are the densities of the pure end members. Actually, δ represents the difference at midcomposition ($x_B = 0.5$) between the value of ρ

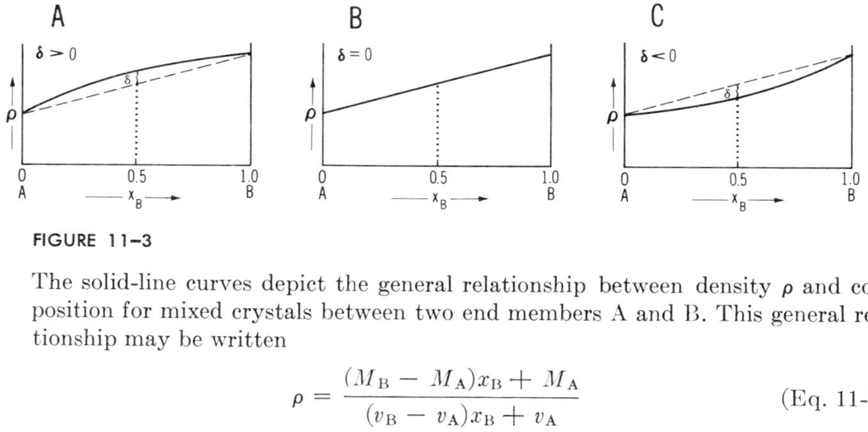

FIGURE 11–3

The solid-line curves depict the general relationship between density ρ and composition for mixed crystals between two end members A and B. This general relationship may be written

$$\rho = \frac{(M_B - M_A)x_B + M_A}{(v_B - v_A)x_B + v_A} \qquad \text{(Eq. 11-15)}$$

where M_B and M_A represent the molecular weights and v_B and v_A represent the volume of a mole of pure A and pure B, respectively. Also ρ represents the density of any mixed crystal between A and B and x_B is the mole fraction of pure B present in one mole of this mixed crystal. Letting ρ_A and ρ_B represent the densities of the pure end members A and B, then δ the deviation of Equation (11-15) from linearity (dashed line) at the compositional midpoint ($x_B = 0.5$) can be computed since

$$\delta = \left(\frac{M_A + M_B}{v_A + v_B}\right) - \left(\frac{\rho_A + \rho_B}{2}\right)$$

Solid-line curves A, B, and C represent the three possible shapes that Equation (11-15) can take when plotted. The shape depends upon whether, for the solid solutional series, $\delta > 0$ (convex upward curve), $\delta = 0$ (true linearity), or $\delta < 0$ (concave upward curve).

as determined by Equation (11-15) and that determined by the commonly assumed straight-line relationship (dashed in Fig. 11-3A, C).

Equation (11-15) is valid only to the extent that Equation (11-14) holds, that is, to the extent that volume is linearly related to composition. If the volume-composition relationship is nonlinear, there is even less likelihood of the density-composition relation being linear. Interestingly, the linear volume-composition relationship expressed by Equation (11-14) is a necessary property to all solutions that are *ideal* in the thermodynamic sense. Hence demonstration that Equation (11-15) does not hold for a solutional series is proof of the thermodynamic nonideality of these solutions. However, demonstration that Equation (11-15) *does hold* is not necessarily proof of ideality because it may hold fortuitously for some nonideal solutions as well.

The composition of an ideal two-component solid solution can be calculated provided its density ρ as well as the densities ρ_A and ρ_B and molecular weights M_A and M_B of its two end members are known. Thus

$$x_B = \frac{1}{1 + [(M_B\rho_A)/(M_A\rho_B)][(\rho_B - \rho)/(\rho - \rho_A)]} \qquad \text{(Eq. 11-16)}$$

Equation (11-16) results from manipulation of Equation (11-15) after substitution of M_A/ρ_A for v_A and of M_B/ρ_B for v_B.

For a thermodynamically ideal solution containing i end members, the density of a mixed crystal is

$$\rho = \frac{x_A M_A + x_B M_B + \cdots + x_i M_i}{x_A v_A + x_B v_B + \cdots + x_i v_i} \qquad \text{(Eq. 11-17)}$$

Practical Uses. A substance's measured density often provides an important clue to its identity. Mursky and Thompson (1958, p. 273) provide a list of over 1400 minerals arranged in order of increasing specific gravity. In the laboratory, mixtures of materials of differing densities may be separated through use of liquids of above-average density such as bromoform ($\rho = 2.8 - 2.9$ g/cm^3), acetylene tetrabromide ($\rho = 2.96$), or diiodomethane (3.32). A sand, for example, can be stirred into one of these heavy liquids whereupon the bulk of its grains such as quartz ($\rho = 2.65$), calcite ($\rho = 2.72$), and dolomite, $CaMg(CO_3)_2$ ($\rho = 2.85$) will float, whereas minerals with higher densities will sink. Gradual addition of a miscible, low-density liquid to this heavy liquid permits minerals within a certain density range to be separated out, these sinking when the liquid's density becomes less than theirs.

The packing index

The packing index (PI) was defined by Fairbairn (1943, p. 1305) as

$$\text{PI} = \frac{V_I}{V_C} 10 \qquad \text{(Eq. 11-18)}$$

where V_C is the volume of the unit cell for the crystal and V_I represents the total of the volumes of the individual ions within this unit cell, these volumes being calculated from their ionic radii (Table 8-5). For most ionic crystals the packing index lies between 3.5 and 7.4, that is, 35 to 74 percent of the cell volume is occupied by ions, the remainder representing "pore" space. A crystal's density depends on (1) its packing index; (2) the density of its component atoms or ions, namely, their weight divided by their volume; and (3) the degree to which defects, flaws, or cracks are present. In low-to-high-pressure transformations, the high-pressure polymorph is more dense and has the higher packing index.

Sensate Properties

Sensate properties are more chemical than physical in nature but are included here because they sometimes aid in gross identification of crystals.

Taste

The tongue may interact temporarily with specific flavor factors in substances to produce a momentary burst of energy; this travels along a nerve to the brain where it is interpreted as "taste." The taste of a water-soluble crystal is a blend of the four primary taste qualities, sour, bitter, salty, and sweet. The salty taste predominates in NaCl (table salt or halite), whereas in KCl (sylvite) a slight bitter taste accompanies the salty taste and permits sylvite to be distinguished from halite. The following tastes or taste blends, with examples in parentheses, often serve as descriptive terms: acid or sour (HCl, vinegar); alkaline or soapy (sodium bicarbonate); astringent or puckering (alum); bitter (epsom salts); cooling ($NaNO_3$, KNO_3); salty (NaCl); metallic (iron sulfate, well water).

Odor

Even at room temperature the thermal vibrations may allow surface molecules to break their van der Waals bonds with neighboring molecules and escape from a molecular crystal. These molecules may thus reach the nose to be interpreted as an "odor." Most minerals and inorganic crystals are sufficiently strongly bonded (ionically, covalently, or metallically) so as to lack any odor unless heated to temperatures at which decomposition or burning occurs. Terms used to describe odors are argillaceous (moist clay), bituminous (asphalt), fetid (rotten eggs), garlic (heated arsenic compounds), horseradish (heated selenium compounds), acrid or sulfurous (SO_2).

Directional Properties

Laue groups

Directional properties are of two types: (1) *centrosymmetric* and (2) *noncentrosymmetric.* A centrosymmetric property by its very nature prevents any distinction between a direction $[uvw]$ and its opposite $[\bar{u}\bar{v}\bar{w}]$, the thermal

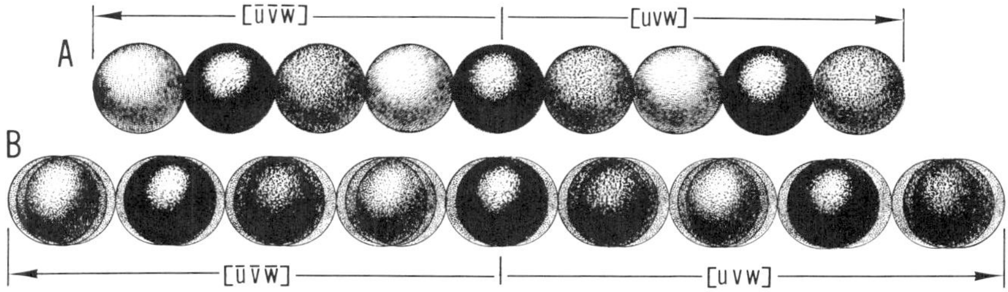

C

THE 11 LAUE GROUPS

System	Centrosymmetric class, noncentrosymmetric classes	Centrosymmetric class, noncentrosymmetric classes
TRICLINIC	$\bar{1}$, 1	
MONOCLINIC / ORTHORHOMBIC	2/m, 2, m	2/m 2/m 2/m, 222, mm2
TETRAGONAL	4/m, 4, $\bar{4}$	4/m 2/m 2/m, 422, 4mm, $\bar{4}$2m
HEXAGONAL	6/m, 6, $\bar{6}$	6/m 2/m 2/m, 622, 6mm, $\bar{6}$m2
HEXAGONAL (TRIGONAL SUB-SYSTEM)	$\bar{3}$, 3	$\bar{3}$ 2/m, 32, 3m
ISOMETRIC OR CUBIC	2/m $\bar{3}$, 23	4/m $\bar{3}$ 2/m, 432, $\bar{4}$3m

FIGURE 11–4

(A)(B) Thermal expansion as an example of a centrosymmetric property. A single row of atoms is shown extending along directions [*uvw*] and [*ūv̄w̄*] within a crystal structure which, seen in entirety, would lack a center of symmetry.

(A) Crystal at absolute zero, vibrational motion being minimal.

(B) Crystal at an elevated temperature which imposes increased thermal vibrations on the atoms. For illustrative purposes these vibrations are exaggerated and confined to directions [*uvw*] and [*ūv̄w̄*]. Note the impossibility of attributing the thermal expansion any more to [*uvw*] than to [*ūv̄w̄*]. It is more likely that the crystal expanded outward from its precise geometric center by the same amount along [*uvw*] as along [*ūv̄w̄*]; thus thermal expansion is seen to cause the crystal to respond as if it possessed a center of symmetry. In brief, thermal expansion is a *centrosymmetric property*.

(C) The 11 Laue groups or Friedel classes. Within a given group or box, if a center of symmetry were added to the noncentrosymmetric crystal classes (small type), they become equal to the centrosymmetric class (large type). Measurement of a centrosymmetric property on a noncentrosymmetric crystal causes the crystal to respond as if it possessed a center of symmetry. Thus all classes in the same Laue group will respond alike to a centrosymmetric property.

expansion of a crystal being an example of a centrosymmetric property (Fig. 11-4A, B). In effect a centrosymmetric property imparts a center of symmetry to a crystal, that is, the crystal seems to have one, even if one is absent. Thus on the basis of thermal expansion measurements alone, the crystal in Figure 11-4A would respond as if of class $\bar{1}$, although actually of class 1, or as if of class $2/m$, although actually of class 2 or m, and so on. Consequently, a centrosymmetric property only permits a crystal to be assigned to one of the 11 *Laue groups* or *Friedel classes* (Fig. 11-4C). Each Laue group contains a centrosymmetric crystal class (large type) plus those classes that become equivalent to it upon addition of a center of symmetry.

Neumann's principle

If we draw vectors outward from a common origin in a crystal and if we make the lengths of these vectors proportional to the magnitude of a direction-dependent property as measured along each vector, the tips of these vectors will then define a three-dimensional surface whose point-group symmetry will, according to Neumann's principle, include the symmetry elements present in the crystal's point group. In other words, the symmetry observed by any physical property of a crystal must equal, or be a supergroup of, the crystal's point-group symmetry.

Cohesive Properties

Fracture

When struck or crushed, a crystal may break along an irregular or curved surface, called a *fracture surface*. Such surfaces are called (1) *conchoidal*, if smoothly curved as on broken plate glass; (2) *uneven* or *irregular*, if rough textured; (3) *hackly*, if they possess jagged, sharp-edged irregularities as for copper and some other metals, and (4) *splintery*, if splinters protrude from them like from broken wood. *These surfaces bear no well-defined angular relationship to the crystallographic axes.*

Cleavage

Within most crystalline structures certain planar directions exist that cut across fewer and often weaker bonds than any others. Consequently, the crystal breaks more easily along these planes, which are called *cleavage planes* or, simply, *cleavages*. In contrast to fracture surfaces, *cleavage planes are invariably parallel to a possible crystal face, usually one with simple Miller indices.* Moreover if a crystal can have cleavage parallel to one face of a possible crystal form, it can then possess cleavages of identical quality parallel to the remaining faces of the form. Such groups of symmetry-related cleavage planes constitute a *cleavage form*. Thus a {001} cleavage signifies *cubic cleavage* for an isometric crystal but a *basal cleavage*, that is, *basal pinacoidal cleavage*, for nonisometric crystals. A crystal's cleavage form is not necessarily the same as the form it adopts upon growth. Thus fluorite, CaF_2, usually grows as cubes {100} but always cleaves parallel to the faces of the octahedron {111} and is consequently said to have octahedral cleavage.

A crystal's cleavage is described by citing (1) the symbol of each cleavage form present; (2) the ease of production of cleavage planes for this form; and (3) the perfection of the resultant surfaces, that is, the degree to which they are so smooth as to be brilliantly shiny. Item (1) thus stipulates the direction of the cleavage, whereas items (2) and (3) stipulate its quality. Such quality may be described as *perfect* if the crystal breaks only along planes of the cleavage form to produce extensive, brilliantly reflecting surfaces, or *good*, if cleavage along faces of the form are frequently interrupted by other fractures, thus reducing their areal extent and brilliance. As interruptions become more frequent, areal extent and brilliance are correspondingly reduced and cleavage quality degrades from *good* to *distinct* to *indistinct*. Such degradation of cleavage quality is symptomatic of a lessening of structural control. As Fairbairn notes (1939, p. 358), "... the terms *cleavage* and *fracture* have no precise meaning. Common usage defines cleavage as a rupture which is closely controlled by the structure, and fracture as a rupture in which the structural control is weak."

If the broken surface of a crystal consists of steplike combinations of two or more cleavage directions, the several directions may represent one or more cleavage forms. Cleavage directions belonging to the same form are necessarily of equal quality and, provided the direction of stress did not favor one direction, of equal smoothness and areal development (Fig. 11-5A). If they represent more than one cleavage form, that is, the different cleavage directions are not related to each other by any symmetry operation, they will likely be of different qualities. In this latter case the cleavage of highest quality will be represented by the largest and smoothest surfaces. In Figure 11-5B, for example, the cleavages would be qualitatively rated {001} > {010} > {100}.

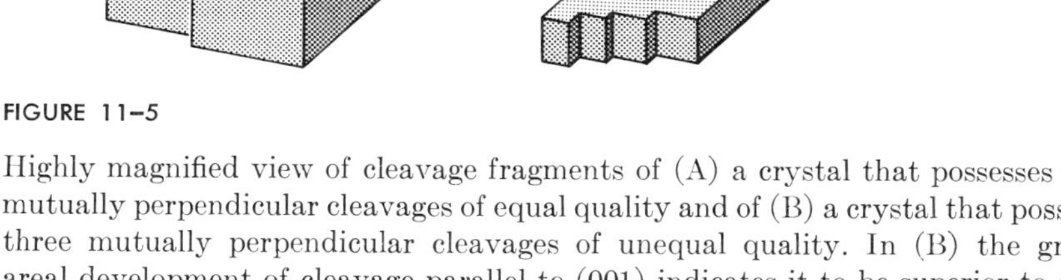

FIGURE 11–5

Highly magnified view of cleavage fragments of (A) a crystal that possesses three mutually perpendicular cleavages of equal quality and of (B) a crystal that possesses three mutually perpendicular cleavages of unequal quality. In (B) the greater areal development of cleavage parallel to (001) indicates it to be superior to (010) which is in turn superior to (100).

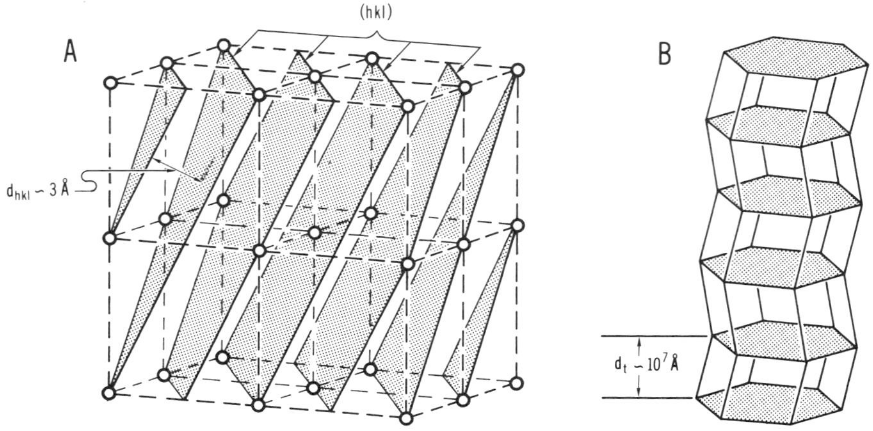

FIGURE 11–6

(A) Illustration that a cleavage parallel to an (*hkl*) plane in a crystal may occur along *any* of the innumerable, translationally equivalent (*hkl*) planes that pass through each of the multitude of identipoints in the structure.
(B) A polysynthetically twinned crystal whose composition planes (shaded) are zones of weakness and are spaced about 0.1 cm apart. The parting can only occur along the specific (*hkl*) planes that are shaded, not along the millions of parallel (*hkl*) planes intervening between them.

If a plane of weakness (*hkl*) passes through a given point in a crystal's structure, then parallel (and identical) weakness planes must also pass through all points in the structure that are translationally equivalent to this first (Fig. 11-6A). Hence cleavage can occur along any of a large family of parallel weakness planes. The perpendicular distance between neighboring planes in this family of (*hkl*) planes, known as the *interplanar spacing* d_{hkl}, is usually smaller than the cell edges. Thus the thinnest possible cleavage flake would theoretically be of thickness d_{hkl}. Even if attainable, however, this would be too thin to see.

Parting

Occasionally certain specimens of a crystalline substance may fracture along a specific plane (*hkl*), whereas most do not. Such fractures, known as *partings*, often occur in specimens that are polysynthetically twinned, the composition planes constituting planes of weakness along which fractures may occur. Thus the minimum spacing between two parallel parting planes cannot be less than d_t, the thickness of an individual of the twin (Fig. 11-6B). If a specimen with parting is crushed into grains with diameters less than d_t, some of these grains would not be bounded, even on one side, by a parting plane. Crushed finer and finer, the proportion of flat-sided grains would decrease steadily for a crystal with parting, but not for one with cleavage.

Tenacity

Tenacity describes the manner with which a crystal reacts to striking, crushing, bending, or cutting. Crystals are called *brittle* if they break or powder easily when struck or crushed, but *malleable* if they can be hammered into thin sheets, never breaking or powdering. Thus a grain of fool's gold (pyrite, FeS_2) becomes a powder when struck or crushed between two hammers, whereas one of true gold becomes flatter and thinner. A material that can be cut smoothly with a knife is *sectile*; one that can be drawn out into a wire is called *ductile*. An *elastic* material springs back into its original position if released after bending whereas *flexible* materials, once bent, remain bent even after release of pressure.

Hardness

Hardness (H) of a solid may mean (1) the ability of one material to scratch or abrade another, this being called *Mohs hardness*; (2) the resistance to indentation under a steadily applied pressure, this being *indentation hardness*; or (3) the resistance to indentation upon impact, this being called *scleroscopic* or *rebound hardness*. Each is a measure of a crystal structure's resistance to mechanical breakdown, more strongly bonded crystals having greater hardness. W. F. Brace (1964, p. 257) notes that in metals hardness is a measure of their ductility or plasticity under compression; in glasses it correlates with tensile strength.

In 1824 Mohs chose the following ten minerals to serve as standards of scratch hardness.

Mohs Scale of Hardness

1 Talc	2 Gypsum	3 Calcite	4 Fluorite	5 Apatite
6 Orthoclase	7 Quartz	8 Topaz	9 Corundum	10 Diamond

Each mineral can scratch (and is, therefore, harder than) a lower-numbered one but will not scratch a higher-numbered one. Solids of equal hardness can scratch each other. A Mohs hardness of $5\frac{1}{2}$ would indicate a solid of hardness between apatite and orthoclase, although not necessarily midway between; this solid could scratch apatite, but not orthoclase, yet apatite could not scratch it. Mineralogists use hardness to aid in rapid identification of minerals, often employing their fingernail ($H \cong 2\frac{1}{2}$) and pocketknife ($H = 5$ or $5\frac{1}{2}$) as convenient standards. A greasy feeling marks solids of hardness 1.

Indentation hardness is more quantitative than Mohs hardness but requires greater sample preparation. A plunger weighted with a standard load, for example, 100 g, is lowered to gentle contact with the polished surface of a solid. After 10 seconds, during which this surface bears the full weight, the plunger is withdrawn. The depth to which the plunger's tip, which consists of diamond or other hard material, indents this surface provides a quantitative value for hardness. This indentation hardness is called (1) Brinell hardness, (2) Vickers hardness, or (3) Knoop hardness according to whether the shape of the indenter's tip is (1)

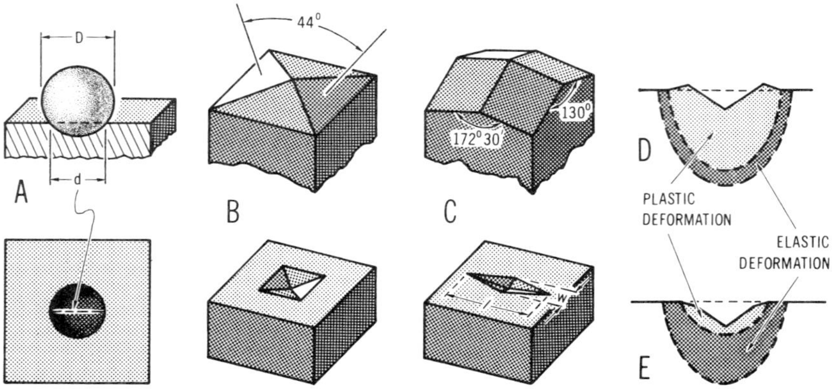

FIGURE 11–7

Shapes of indenter (upper) and the indentation they make (lower) for (A) Brinell, (B) Vickers, and (C) Knoop microhardness testers. In (A) a spherical indenter of diameter D enters the material tested to form a circular indentation of diameter d. In (C) the indentation's length l is about seven times the width w, this width being exaggerated here.
(D)(E) Cross sections through the indentation made by a conical indenter in (D) a soft material and in (E) a hard material. These indentations are drawn as if the indenter were still in place and would thus be measures of the *unrecovered hardness*. After withdrawal of the indenter, the indentation sizes decrease (because the elastic deformation is recovered) and are thus measures of the *recovered hardness*. Size change is particularly marked for the hard material (E) because, while the indenter was in place, a larger proportion of its deformation was elastic (recoverable) deformation. For a soft material, the high amount of plastic deformation causes a ridge to form around the indentation. For the indentation shown in (C), the recovered and unrecovered measurements of l usually differ little, whereas those of w may differ significantly.

spherical (Fig. 11-7A), (2) a shallow tetragonal pyramid (Fig. 11-7B), or (3) an orthorhombic pyramid capable of producing an elongated indentation (Fig. 11-7C). Rockwell hardness tests employ either a steel ball indenter or a cone-shaped diamond. An indentation changes in size and shape upon removal of the indenter and the subsequent elastic rebound of the material. For soft materials in which the nonrecoverable plastic deformation greatly exceeds the recoverable elastic deformation (Fig. 11-7D), this change is small; in hard materials (Fig. 11-7E), it is large. Hard materials particularly, therefore, possess one indentation size and hardness, the *unrecovered hardness,* while the indenter is in place, but a lesser indentation size and hardness, the *recovered hardness,* after its removal. The theory and measurement of indentation hardness are discussed in more detail by Tweeddale (1964, pp. 142–171).

Mohs' choice of hardness standards was exceedingly canny. He very nearly attained his objective of having his ten basic minerals represent relatively equal hardness intervals, particularly if the log of the indentation hardness is accepted as its quantitative measure (Fig. 11-8). Mohs recognized that the gap between corundum and diamond was unduly large, but no materials of intermediate hardness were at hand.

The ability to deform by translation gliding or glide twinning (Fig. 10-30) reduces hardness but increases malleability and ductility. Thus in metals translation gliding occurs most easily along planes of closest packing (Fig. 10-30D). Four such planes of closest packing, parallel to the faces of the {111} octahedron, occur in cubic closest-packing structures. Only one, {0001}, occurs in hexagonal closest-packing structures and none in body-centered cubic structures. Hence unless other factors intervene, metals with cubic closest-packing structures, for example, copper, gold, silver, nickel, γ-iron, and α-cobalt, tend to be more soft, malleable, and ductile than those with hexagonal closest-packing or body-centered cubic structures such as chromium, vanadium, tungsten, molybdenum, α- and β-iron, or β-cobalt. The indentation hardness of metals is thus a measure of plasticity or ductility under compression. Interestingly, hardness values for binary solid solutions are often slightly higher than for either end member.

Either a scratch or the long diagonal of a Knoop identation can be oriented

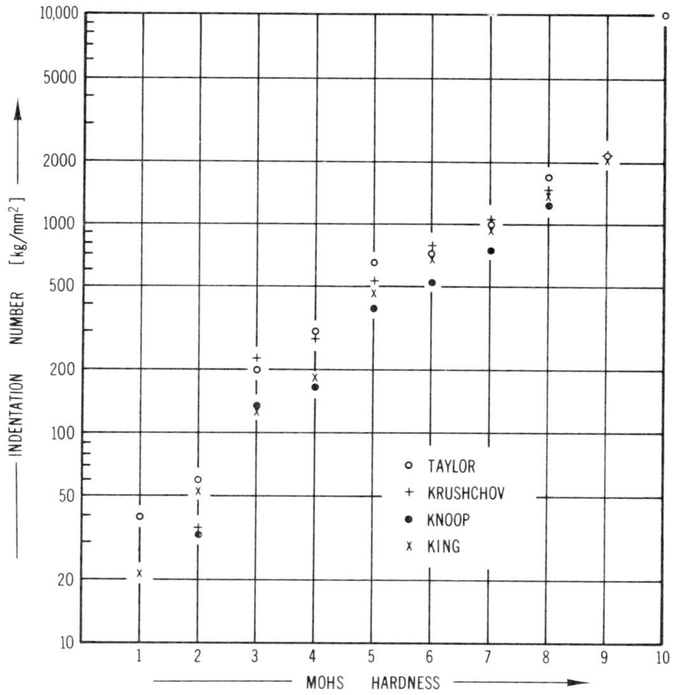

FIGURE 11-8

Plot of Mohs' hardness against logarithm of the average indentation hardness. (After Plendl and Gielisse, 1963, from data by Knoop, King, Taylor, and Khrushchov.)

along a given direction $[uvw]$ on a crystal face; a spherical, conical, or Vickers indenter cannot. Consequently, scratch hardness or Knoop hardness can be measured not only on different planes but for different directions in these planes. For example, H. Winchell (1945, p. 583) observed the following Knoop hardnesses along the cited directions in kyanite Al_2SiO_5,

[100] on (001) : 1205	[100] on (010) : 1665	
[010] on (001) : 184		[010] on (100) : 462
	[001] on (010) : 1260	[001] on (100) : 1120

Knoop hardnesses measured along different directions on the (001) plane of a synthetic and of a natural mica (Fig. 11-9A) disclose maximum hardness along [010], minimum along [100]. The results display symmetry that conforms to the

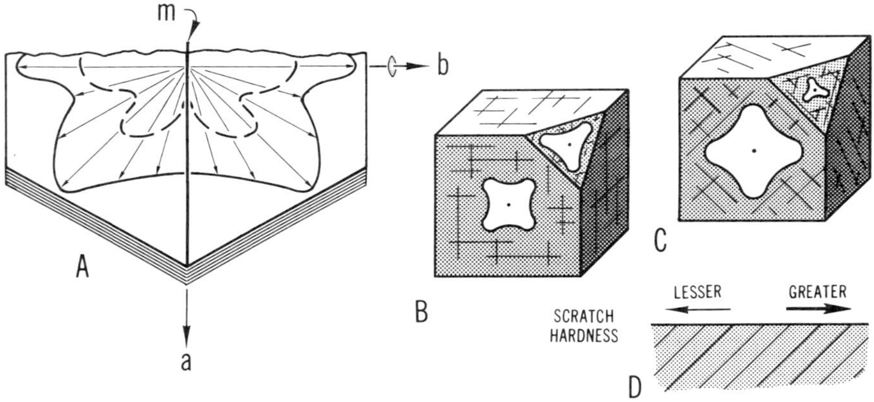

FIGURE 11–9

(A) Drawing of a flake of mica with arrows drawn on its upper surface proportional in length to the measured Knoop indentation hardnesses (with 85 units subtracted) observed by Bloss, Shekarchi, and Shell for synthetic fluorphlogopite, $KMg_3[AlSi_3O_{10}]F_2$. If similar arrows were drawn for natural muscovite $KAl_2[AlSi_3O_{10}](OH)_2$, the arrow tips would define the dashed curve. The hardness measurements disclose the plane of symmetry m in each case. Fluorphlogopite's greater hardness compared to natural muscovite is attributed to the presence of F^- rather than $(OH)^-$ in its structure.

(B)(C) Anisotropy of scratch hardness on the (100) and (111) faces of a NaCl crystal (B) and of a fluorite (CaF_2) crystal (C). A vector drawn from the central dot to the edge of the unshaded figure will be proportional to the scratch hardness measured in this direction. On the (111) faces such a vector $[uvw]$ differs in length from the opposite direction $[\bar{u}\bar{v}\bar{w}]$. Thus scratch hardness, unlike indentation hardness, is a noncentrosymmetric property. (After Correns.)

(D) Noncentrosymmetry of scratch hardness occurs if the two opposite directions (arrows) are at unequal angles to any cleavage, traces of which are shown as diagonal lines in this hypothetical cross section of a crystal. The direction of lesser scratch hardness is at an angle to the cleavage such that plucking action is favored during scratching.

crystal class ($2/m$) for these micas and also to a pseudo-hexagonal symmetry. This latter reflects the quasi-hexagonal symmetry of the [$AlSi_3O_{10}$] sheets (Fig. 9-37A) which lie parallel to (001) in the mica structure. The anisotropy of scratch hardness on the (100) and (111) faces is shown for sodium chloride (Fig. 11-9B) and for fluorite (Fig. 11-9C). As shown in Figure 11-9D scratch hardness along a direction [*uvw*] may be less than along the opposite direction [$\bar{u}\bar{v}\bar{w}$] if [*uvw*] is more favorably oriented relative to an existent cleavage direction.

Westbrook and Jorgensen (1968) observed Vickers indentation microhardness to increase, as a rule, if measured on surfaces that were stripped of adsorbed atmospheric water by heating for 2 hours in dry argon at 300°C, then quenched and indented in fresh anhydrous toluene at room temperature. Such dried surfaces could be returned to their previous "wet" condition (and microhardness) by several hours of exposure to air of 40 percent relative humidity, by immersion in water for a few minutes, or by a few seconds exposure to steam from boiling water. Using *W* and *D* to distinguish "wet" from "dry" Vickers hardnesses, some of Westbrook and Jorgensen's results in kilograms per square millimeters are as follows:

1. Quartz on the *r* $\{10\bar{1}1\}$ face, *W*935, *D*1240; on *z*$\{01\bar{1}1\}$, *W*1020, *D*1245; on $\{10\bar{1}0\}$, *W*870, *D*1140.
2. Corundum on {0001} *W*2630, *D*3070; on $\{10\bar{1}0\}$ *W*2350, *D*2600; on $\{11\bar{2}0\}$ *W*2320, *D*2680.
3. Silicon carbide on {0001} *W*2850, *D*3270; on $\{000\bar{1}\}$ *W*2780, *D*3250, on $\{10\bar{1}0\}$ *W*1860, *D*2730; on $\{11\bar{2}0\}$, *W*2240, *D*2740.
4. Calcite on $\{10\bar{1}1\}$ cleavage, *W*126, *D*135; on $\{10\bar{1}0\}$ *W*131, *D*145; on $\{21\bar{3}1\}$ *W*146, *D*147.

Tensors and Tensor Properties

Physical properties represent the response of a crystal to a physical stimulus. This stimulus may be a physical quantity that is fully described by (1) its magnitude alone, whereupon it is a *scalar quantity*; by (2) its magnitude *and* direction, whereupon it is called a *vector quantity* and can be represented by a vector; or by (3) a more complex means called a *tensor*. The nature of tensors will be clarified in the following discussion on electrical conductivity in crystals.

Electrical conductivity

We let **E** in Figure 11-10A be a vector whose direction and length indicate the direction and magnitude of a force or electric field causing electricity to flow from point *O* in a crystal.[5] We also let x_1, x_2, and x_3 be three mutually perpendicular axes drawn outward from *O*. Then $\mathbf{E}_1$, $\mathbf{E}_2$, and $\mathbf{E}_3$ are components of **E** along these axes and may replace **E** in our thinking. The amount of electricity

[5] This discussion follows that of Nye (1957, p. 3) which is recommended for further reading.

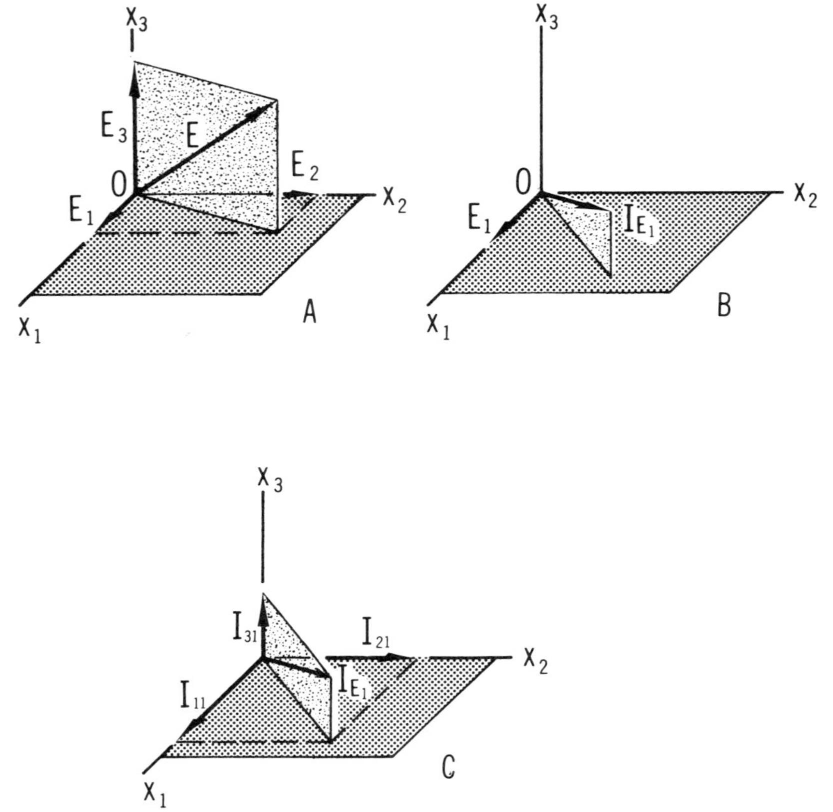

FIGURE 11–10

(A) Resolution of the vector representing the direction and strength of an electic field **E** into its components $\mathbf{E}_1$, $\mathbf{E}_2$, and $\mathbf{E}_3$ on the three mutually perpendicular reference axes, x_1, x_2, and x_3.
(B) The component $\mathbf{E}_1$, if alone applied along x_1 in an anisotropic crystal, would cause a current $\mathbf{I}_{E1}$ that, for the general case, is at a small angle to x_1. For illustrative purposes the angle between vectors $\mathbf{E}_1$ and $\mathbf{I}_{E1}$ is greatly exaggerated.
(C) Resolution of vector $\mathbf{I}_{E1}$ onto axes x_1, x_2, and x_3. The first number in the subscript indicates the axis onto which $\mathbf{I}_{E1}$ was resolved. The second number indicates the axis along which the potential ($\mathbf{E}_1$) was applied to produce $\mathbf{I}_{E1}$.

flowing outward from O will depend not only on **E**, the driving force, but also upon σ, which measures the crystal's ability to conduct electricity. Crystals are *isotropic* if σ has the same value regardless of direction of flow, but *anisotropic* if σ has different values for different directions of flow.

A sufficiently strong electric field $\mathbf{E}_1$ applied along x_1 in an anisotropic crystal causes a current $\mathbf{I}_{E1}$ which usually differs from x_1 in direction (Fig. 11-10B). Thus $\mathbf{I}_{E1}$ has components $\mathbf{I}_{11}$, $\mathbf{I}_{21}$, and $\mathbf{I}_{31}$ along x_1, x_2, and x_3 (Fig. 11-10C). Let $\mathbf{I}_{rs}$ be the general symbol for such components, where r identifies the axis along which the *response* component occurred and s is the axis along which the *stimulus*, or component thereof, was applied to produce this response. The alphabetical sequence of subscripts, *response* axis first, *stimulus* axis second, is a convenient mnemonic aid that will be maintained throughout this chapter. For example, $\mathbf{I}_{23}$ indicates the component along x_2 of the response ($\mathbf{I}_{E3}$) from a stimulus ($\mathbf{E}_3$) applied along axis x_3. Returning to components $\mathbf{I}_{11}$, $\mathbf{I}_{21}$, and $\mathbf{I}_{31}$, their magnitudes are proportional to (1) the stimulus $\mathbf{E}_1$ which produced them and to (2) the crystal's conductivity for this component. Thus we now *define* the crystal's conductivities for currents produced along x_1, x_2, and x_3 by a stimulus $\mathbf{E}_1$ applied along x_1, that is, conductivities σ_{11} along x_1, σ_{21} along x_2, and σ_{31} along x_3, by the following equations:

$$\mathbf{I}_{11} = \sigma_{11}\mathbf{E}_1$$
$$\mathbf{I}_{21} = \sigma_{21}\mathbf{E}_1 \qquad \text{(Eqs. 11-19a)}$$
$$\mathbf{I}_{31} = \sigma_{31}\mathbf{E}_1$$

Similarly, a field $\mathbf{E}_2$ applied along x_2 would produce $\mathbf{I}_{E2}$, a current noncoincident in direction with x_2, whose components and conductivities along x_1, x_2, and x_3 would be

$$\mathbf{I}_{12} = \sigma_{12}\mathbf{E}_2$$
$$\mathbf{I}_{22} = \sigma_{22}\mathbf{E}_2 \qquad \text{(Eqs. 11-19b)}$$
$$\mathbf{I}_{32} = \sigma_{32}\mathbf{E}_2$$

Application of a field $\mathbf{E}_3$ along x_3 would produce a current $\mathbf{I}_{E3}$ whose components would be

$$\mathbf{I}_{13} = \sigma_{13}\mathbf{E}_3$$
$$\mathbf{I}_{23} = \sigma_{23}\mathbf{E}_3 \qquad \text{(Eqs. 11-19c)}$$
$$\mathbf{I}_{33} = \sigma_{33}\mathbf{E}_3$$

A field $\mathbf{E}$ applied along a direction other than x_1, x_2, and x_3 has components $\mathbf{E}_1$, $\mathbf{E}_2$, and $\mathbf{E}_3$ along these axes, respectively (Fig. 11-10A). The current $\mathbf{I}$ caused by $\mathbf{E}$ has components $\mathbf{I}_1$, $\mathbf{I}_2$, and $\mathbf{I}_3$ along x_1, x_2, and x_3, respectively. Component $\mathbf{I}_1$ can be regarded as the sum of the currents produced along x_1 by $\mathbf{E}_1$, $\mathbf{E}_2$, and $\mathbf{E}_3$. Thus

$$\left.\begin{aligned} \mathbf{I}_1 &= \mathbf{I}_{11} + \mathbf{I}_{12} + \mathbf{I}_{13} = \sigma_{11}\mathbf{E}_1 + \sigma_{12}\mathbf{E}_2 + \sigma_{13}\mathbf{E}_3 \\ \mathbf{I}_2 &= \mathbf{I}_{21} + \mathbf{I}_{22} + \mathbf{I}_{23} = \sigma_{21}\mathbf{E}_1 + \sigma_{22}\mathbf{E}_2 + \sigma_{23}\mathbf{E}_3 \\ \mathbf{I}_3 &= \mathbf{I}_{31} + \mathbf{I}_{32} + \mathbf{I}_{33} = \sigma_{31}\mathbf{E}_1 + \sigma_{32}\mathbf{E}_2 + \sigma_{33}\mathbf{E}_3 \end{aligned}\right\} \qquad \text{(Eqs. 11-20)}$$

Once the nine conductivities σ_{11}, σ_{12}, . . . , σ_{33} are known for an anisotropic crystal, we can determine the magnitude and direction of the current $\mathbf{I}$ (components $\mathbf{I}_1$, $\mathbf{I}_2$, and $\mathbf{I}_3$ on the reference axes) which results from any applied electric field $\mathbf{E}$

whose angles (and therefore its components, $\mathbf{E}_1$, $\mathbf{E}_2$, and $\mathbf{E}_3$) to axes x_1, x_2, and x_3 are known. The nine conductivities that determine the direction and magnitude of **I**, the crystal's response to an applied field **E**, are the components of an array,

$$\begin{bmatrix} \sigma_{11} & \sigma_{12} & \sigma_{13} \\ \sigma_{21} & \sigma_{22} & \sigma_{23} \\ \sigma_{31} & \sigma_{32} & \sigma_{33} \end{bmatrix} \qquad \mathrm{T}(1)$$

which is known as a *tensor*. Because this tensor contains 3^2 components, it is called a *second-rank tensor*.[6] The number of subscripts needed to specify a component also indicates a tensor's rank. The nine constants in tensor T(1) fully describe how the crystal relates a stimulus, as represented by vector **E** (components $\mathbf{E}_1$, $\mathbf{E}_2$, and $\mathbf{E}_3$), to its response, the current **I** (components $\mathbf{I}_1$, $\mathbf{I}_2$, and $\mathbf{I}_3$ on the reference axes).

Those familiar with matrices will observe that Equations (11-20) may be represented in matrix form as

$$\underset{\text{Vector }\mathbf{I}}{\overset{\text{Response}}{\begin{bmatrix} \mathbf{I}_1 \\ \mathbf{I}_2 \\ \mathbf{I}_3 \end{bmatrix}}} = \underset{\mathrm{T}(1)}{\overset{\text{Tensor}}{\begin{bmatrix} \sigma_{11} & \sigma_{12} & \sigma_{13} \\ \sigma_{21} & \sigma_{22} & \sigma_{23} \\ \sigma_{31} & \sigma_{32} & \sigma_{33} \end{bmatrix}}} \underset{\text{Vector }\mathbf{E}}{\overset{\text{Stimulus}}{\begin{bmatrix} \mathbf{E}_1 \\ \mathbf{E}_2 \\ \mathbf{E}_3 \end{bmatrix}}} \qquad \text{(Eq. 11-20')}$$

where response vector **I** and stimulus vector **E** are depicted, in terms of their components on axes x_1, x_2, and x_3, as column matrices. The two matrices representing T(1) and **E** in Equation (11-20′), if multiplied in the standard manner for matrices (p. 412), yield the elements to the right of the equality signs in Equation (11-20). The matrices in Equation (11-20′) are tensors whose components represent physical quantities and thus have physical units; for example, $\mathbf{I}_1$, $\mathbf{I}_2$, or $\mathbf{I}_3$ are in amperes; $\mathbf{E}_1$, $\mathbf{E}_2$, and $\mathbf{E}_3$ are in volts.

The column matrices representing response vector **I** and stimulus vector **E** in Equation (11-20′) are thus tensors. In fact they are *first-rank tensors* because each contains 3^1 components. All physical properties that may be represented by vectors are thus conveniently represented mathematically as first-rank tensors. In like manner a scalar physical property such as temperature, which has magnitude but not direction, can be represented by a single number or scalar (to which units must be ascribed). This scalar may be regarded as a *zero-rank tensor* because it contains 3^0, that is, one, component.

Certain components of a tensor become equal or become zero if the reference axes x_1, x_2, and x_3 are selected to coincide with a symmetry axis or lie in a symmetry plane. Axes so selected are then called *principal axes*. To understand this

[6] Third-rank and fourth-rank tensors (in three-dimensional space), respectively, contain 3^3 and 3^4 components.

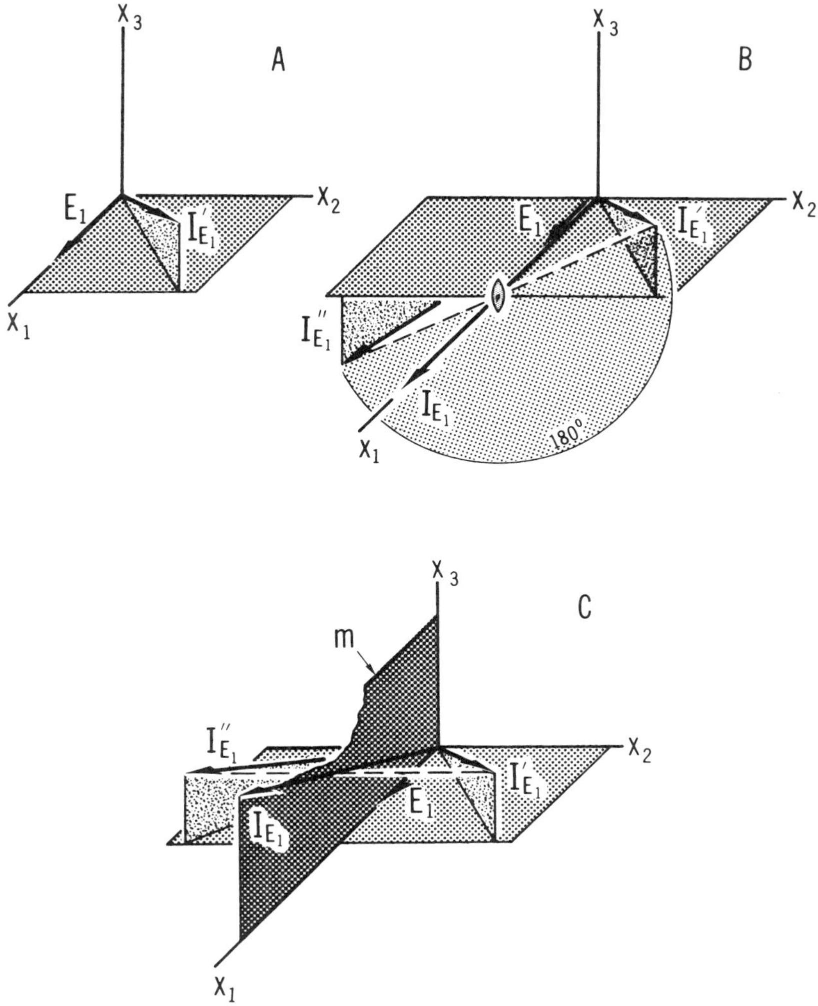

FIGURE 11–11

(A) The case where x_1 is neither a symmetry axis nor lies in a symmetry plane. A field $\mathbf{E}_1$ applied along x_1 produces a current $\mathbf{I}_{E1}'$ which does not coincide with x_1, this divergence again being exaggerated.
(B) The case where axis x_1 coincides with a 2-fold axis. If field $\mathbf{E}_1$ again produces a current at $\mathbf{I}_{E1}'$, the 2-fold symmetry of x_1 would also require an identical current at $\mathbf{I}_{E1}''$. The resultant of vectors $\mathbf{I}_{E1}'$ and $\mathbf{I}_{E1}''$ would be $\mathbf{I}_{E1}$; thus if x_1 is an axis of symmetry, a potential $\mathbf{E}_1$ applied along it produces a current $\mathbf{I}_{E1}$ also along it.
(C) The case where x_1 lies in a plane of symmetry. If application of $\mathbf{E}_1$ along x_1 produced a current $\mathbf{I}_{E1}'$, the heavily shaded mirror plane would also require it to produce the equivalent current $\mathbf{I}_{E1}''$. Their resultant $\mathbf{I}_{E1}$ thus lies in the symmetry plane but does not necessarily coincide with x_1.

we compare the directions of the current $\mathbf{I}_{E1}$ resulting from applying a field $\mathbf{E}_1$ along x_1 for three cases: (1) x_1 is not a principal axis (Fig. 11-11A) and thus $\mathbf{I}_{E1}$ and x_1 differ in direction; (2) x_1 coincides with a symmetry axis, for example, the 2-fold axis in Figure 11-11B; or (3) x_1 lies in a symmetry plane (Fig. 11-11C). For case (2) we assume $\mathbf{E}_1$ caused a current $\mathbf{I}_{E1}'$ whose direction differs from x_1; because x_1 is now a symmetry axis, it requires stimulus $\mathbf{E}_1$ to produce currents symmetrically equivalent to $\mathbf{I}_{E1}'$; for example, if x_1 is a 2-fold axis (Fig. 11-11B), a second current $\mathbf{I}_{E1}''$ related to $\mathbf{I}_{E1}'$ by a 180° rotation about x_1 would also be required. The resultant of all such symmetrically equivalent currents, for example, $\mathbf{I}_{E1}$ in Figure 11-11B, inevitably coincides with the symmetry axis x_1, regardless of whether x_1 is a 2-, 3-, 4-, 6-fold axis. For case (3) if $\mathbf{E}_1$ produced a current not in the symmetry plane, for example, if $\mathbf{I}_{E1}'$ in Figure 11-11C resulted, the symmetry plane would require $\mathbf{E}_1$ to produce an equivalent current such as $\mathbf{I}_{E1}''$; the resultant of $\mathbf{I}_{E1}'$ and $\mathbf{I}_{E1}''$, namely, $\mathbf{I}_{E1}$, must thus lie in the symmetry plane but need not coincide with x_1.

In summary: (1) *a stimulus* **E** *applied along a symmetry axis causes a response* **I** *also along this axis*; (2) *a stimulus* **E** *applied along a direction in a symmetry plane causes a response* **I** *along a direction in this plane.*

The restrictions imposed on conductivities $\sigma_{11}, \sigma_{21}, \ldots, \sigma_{33}$, if axes x_1, x_2, or x_3 either coincide with symmetry axes or lie in symmetry planes, can be examined for monoclinic crystals. Electrical conductivity, like all properties describable by a second-rank tensor, is centrosymmetric; that is, current is conducted along direction $[uvw]$ and its opposite $[\bar{u}\bar{v}\bar{w}]$ with equal facility. Hence in respect to conductivity all monoclinic crystals belong to Laue class $2/m$. The reference axes x_1, x_2, and x_3 can become principal axes only if x_3 is made to coincide with the 2-fold axis of this Laue group, whereupon x_1 and x_2 lie in the mirror plane (Fig. 11-12A). Applying field $\mathbf{E}_3$ along x_3 produces a current $\mathbf{I}_{E3}$ *along* x_3; this current has components $\mathbf{I}_{13} = 0$ (on x_1), $\mathbf{I}_{23} = 0$ (on x_2), and $\mathbf{I}_{33} = \mathbf{I}_{E3}$ (on x_3). By definition of conductivity (Eq. 11-19a),

$$\mathbf{I}_{13} = \sigma_{13}\mathbf{E}_3 \qquad \mathbf{I}_{23} = \sigma_{23}\mathbf{E}_3 \qquad \mathbf{I}_{33} = \sigma_{33}\mathbf{E}_3$$

Because $\mathbf{I}_{13}$ and $\mathbf{I}_{23}$ both equal zero, whereas $\mathbf{E}_3$ has a finite value, the preceding equations indicate that σ_{13} and σ_{23} each equals zero. A field $\mathbf{E}_2$ applied along x_2 (Fig. 11-12B) produces $\mathbf{I}_{E2}$, a current confined to the mirror plane. Thus $\mathbf{I}_{E2}$ is perpendicular to x_3; consequently its component on x_3, that is, $\mathbf{I}_{32}$, and its conductivity σ_{32} both equal zero. Similarly, applying $\mathbf{E}_1$ along x_1 (Fig. 11-12C) produces $\mathbf{I}_{E1}$ which is also in the mirror plane and thus perpendicular to x_3 (so that $\mathbf{I}_{31}$ and hence σ_{31} must equal zero). Thus if x_3 is chosen to coincide with the 2- or $\bar{2}$-fold axis of a monoclinic crystal in the first setting, the tensor relating **I** to **E** becomes

$$\begin{bmatrix} \sigma_{11} & \sigma_{12} & 0 \\ \sigma_{21} & \sigma_{22} & 0 \\ 0 & 0 & \sigma_{33} \end{bmatrix} \qquad \mathrm{T}(2)$$

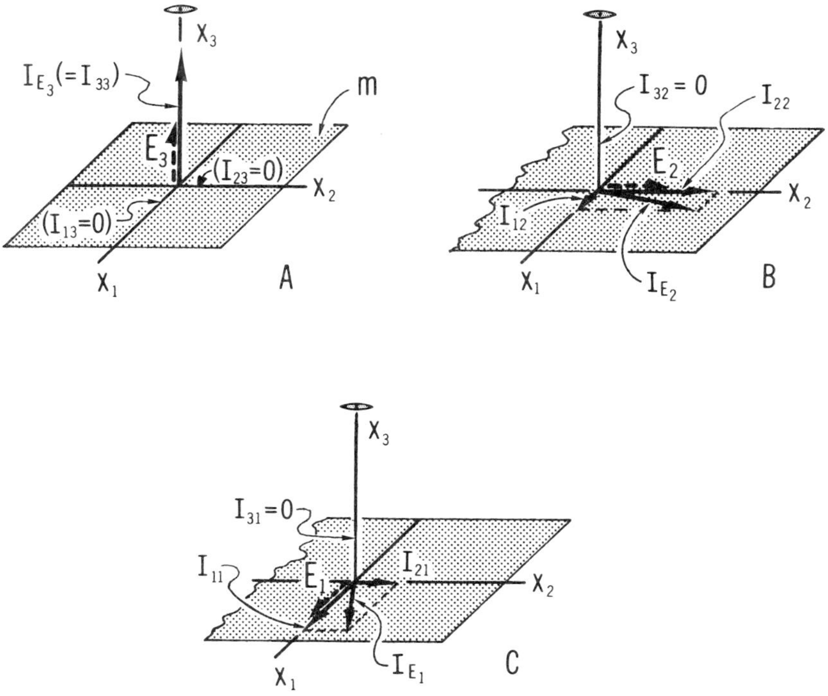

FIGURE 11–12

All three drawings represent Laue class $2/m$ where axis x_3 has been chosen to coincide with the single 2-fold that is present and consequently axes x_1 and x_2 lie in the mirror plane.

(A) The current $\mathbf{I}_{E3}$, produced by applying field $\mathbf{E}_3$ (dashed vector) along the 2-fold axis, is shown resolved into its components $\mathbf{I}_{13}$, $\mathbf{I}_{23}$, and $\mathbf{I}_{33}$ on axes x_1, x_2, and x_3, respectively.

(B) The current $\mathbf{I}_{E2}$ produced by applying a field $\mathbf{E}_2$ (dashed vector) along x_2 lies in the mirror plane and is similarly resolved into components along the reference axes, x_1, x_2, and x_3.

(C) The current $\mathbf{I}_{E1}$ produced by applying $\mathbf{E}_1$ (dashed vector) along x_1 is similarly resolved into components.

For a centrosymmetric property, all orthorhombic crystals belong to Laue group $2/m\,2/m\,2/m$ and if axes x_1, x_2, and x_3 are chosen to coincide with the 2 folds, their conductivity tensor simplifies to

$$\begin{bmatrix} \sigma_{11} & 0 & 0 \\ 0 & \sigma_{22} & 0 \\ 0 & 0 & \sigma_{33} \end{bmatrix} \qquad \mathrm{T}(3)$$

as the reader should verify. For Laue groups $4/m$, $6/m$, and $\bar{3}$, if x_3 coincides with the symmetry axis, the conductivity tensor for each becomes

$$\begin{bmatrix} \sigma_{11} & \sigma_{12} & 0 \\ -\sigma_{12} & \sigma_{11} & 0 \\ 0 & 0 & \sigma_{33} \end{bmatrix} \qquad \text{T(4)}$$

This resembles T(2) but implies that σ_{22} equals σ_{11} and that σ_{21} equals $-\sigma_{12}$. To verify T(4) for Laue group $4/m$, we consider that fields of equal strength $\mathbf{E}_1$ and $\mathbf{E}_2$ are applied along x_1 and x_2 to produce currents $\mathbf{I}_{E1}$ and $\mathbf{I}_{E2}$, respectively (Fig. 11-13A). The 4-fold axis x_3 symmetrically relates x_1 to x_2 and because $\mathbf{E}_1$ equals $\mathbf{E}_2$, $\mathbf{I}_{E1}$ must equal $\mathbf{I}_{E2}$. Let us resolve $\mathbf{I}_{E1}$ (Fig. 11-13B) and $\mathbf{I}_{E2}$ (Fig. 11-13C) into their components on axes x_1, x_2, and x_3. In Figure 11-13C component $\mathbf{I}_{12}$ is along the negative x_1 axis and is hence given a negative sign. A 90° counterclockwise rotation about x_3 would cause Figure 11-13B to be identical to Figure 11-13C. Thus $\mathbf{I}_{11}$ is symmetrically equivalent to $\mathbf{I}_{22}$ and $\mathbf{I}_{21}$ to $-\mathbf{I}_{12}$. In other words σ_{22} equals σ_{11} and σ_{21} equals $-\sigma_{12}$. The proof of T(4) for Laue groups $\bar{3}$ or $6/m$ is given by Wooster (1949, p. 11).

For Laue groups $4/m\ 2/m\ 2/m$, $6/m\ 2/m\ 2/m$, and $\bar{3}\ 2/m$, if x_3 is chosen as the unique 4, 6, or $\bar{3}$ axis and if x_1, x_2 are 2-fold or $\bar{2}$-fold axes, the conductivity tensor becomes

$$\begin{bmatrix} \sigma_{11} & 0 & 0 \\ 0 & \sigma_{11} & 0 \\ 0 & 0 & \sigma_{33} \end{bmatrix} \qquad \text{T(5)}$$

For the isometric Laue groups $2/m\ \bar{3}$ and $4/m\ \bar{3}\ 2/m$, it becomes

$$\begin{bmatrix} \sigma_{11} & 0 & 0 \\ 0 & \sigma_{11} & 0 \\ 0 & 0 & \sigma_{11} \end{bmatrix} \qquad \text{T(6)}$$

For triclinic crystal classes no symmetry above $\bar{1}$ exists. Hence tensor T(1) pertains.

Let us consider a tensor T(1) whose values refer to the reference axes x_1, x_2, and x_3. A tensor that will similarly relate stimulus $\mathbf{E}$ to response $\mathbf{I}$ for the crystal for a second set of mutually perpendicular axes x_1', x_2', and x_3' can then be calculated if the angles between each new and old axis are known. Thus in Figure 11-14A the arcs indicating the angles between new axis x_1' and each of the old axes are marked with symbols (a_{11}, a_{12}, and a_{13}) that indicate the cosine of each angle. The convention used for the subscripts is that the first number denotes the new axis, the second denotes the old axis (note alphabetical order: new, old).

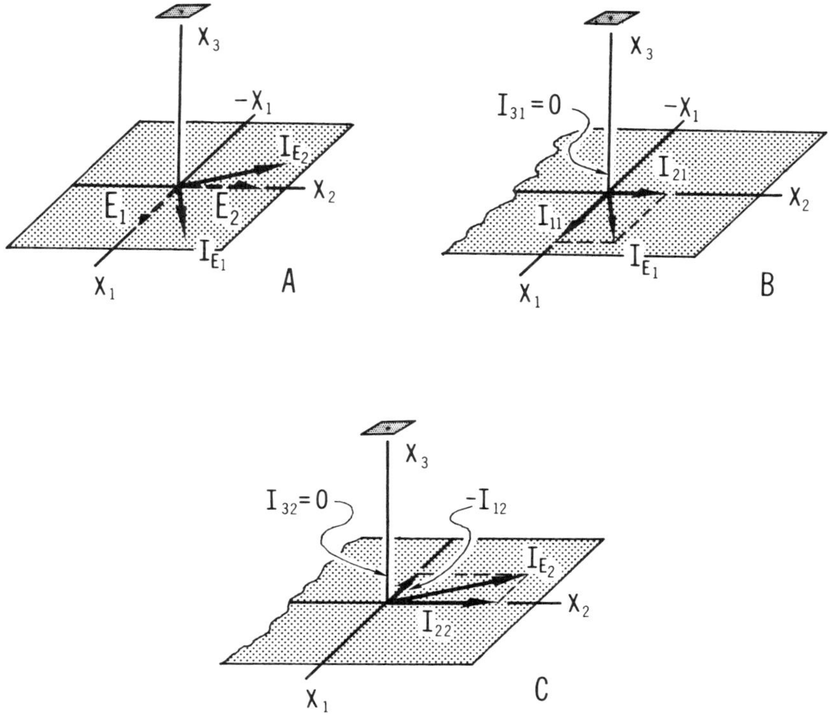

FIGURE 11–13

Illustration of Laue class $4/m$ with x_3 selected to coincide with the 4-fold axis whereupon axes x_1 and x_2 lie in the symmetry plane (shaded).
(A) Fields of equal strength represented by dashed vectors $\mathbf{E}_1$ and $\mathbf{E}_2$ are applied along axes x_1 and x_2 to produce currents $\mathbf{I}_{E1}$ and $\mathbf{I}_{E2}$, respectively. These lie in the symmetry plane, the angle between $\mathbf{E}_1$ and $\mathbf{I}_{E1}$ equaling that between $\mathbf{E}_2$ and $\mathbf{I}_{E2}$. Since $\mathbf{E}_1$ equaled $\mathbf{E}_2$ and the 4-fold axis requires x_1 to be equivalent to x_2, $\mathbf{I}_{E1}$ and $\mathbf{I}_{E2}$ must also be symmetrically equivalent (and at 90° to each other).
(B) Resolution of $\mathbf{I}_{E1}$ into its components.
(C) Resolution of $\mathbf{I}_{E2}$ into its components. Note that a 90° anticlockwise rotation of drawing (B) about its 4-fold axis would cause it to be identical to (C). In other words, the 4-fold symmetry relates $\mathbf{I}_{11}$ to $\mathbf{I}_{22}$ and $\mathbf{I}_{21}$ to $-\mathbf{I}_{12}$.

Thus a_{32} represents the cosine of the angle between x_3' and x_2 (Fig. 11-14B), and so on. The new tensor whose components relate to new axes x_1', x_2' and x_3', that is,

$$\begin{bmatrix} \sigma_{11}' & \sigma_{12}' & \sigma_{13}' \\ \sigma_{21}' & \sigma_{22}' & \sigma_{23}' \\ \sigma_{31}' & \sigma_{32}' & \sigma_{33}' \end{bmatrix} \qquad \text{T(7)}$$

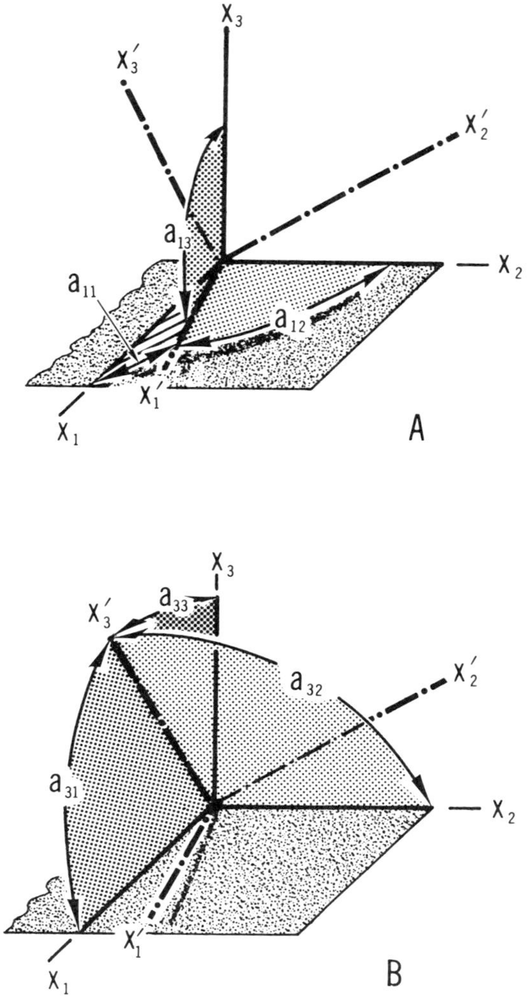

FIGURE 11–14

Relation of a new set of mutually perpendicular axes x_1', x_2', and x_3' to the old set x_1, x_2, and x_3.

(A) The angles between new axis x_1' and old axes x_1, x_2, and x_3 are marked by symbols a_{11}, a_{12}, and a_{13} which will also be used to represent the cosines of these angles.

(B) The angles between new axis x_3' and old axes x_1, x_2, and x_3 are similarly marked. In like manner, the angles between new axis x_2' and the old axes could be similarly shown and labeled.

can now be calculated from the components of the old tensor since

$$\begin{aligned}\sigma_{ij}' = {} & a_{i1}a_{j1}\sigma_{11} + a_{i1}a_{j2}\sigma_{12} + a_{i1}a_{j3}\sigma_{13} \\ & + a_{i2}a_{j1}\sigma_{21} + a_{i2}a_{j2}\sigma_{22} + a_{i2}a_{j3}\sigma_{23} \\ & + a_{i3}a_{j1}\sigma_{31} + a_{i3}a_{j2}\sigma_{32} + a_{i3}a_{j3}\sigma_{33}\end{aligned}$$

For example, to evaluate σ_{23}' we substitute 2 for i and 3 for j in the above equation.

For a stimulus tensor of rank s, a response tensor of rank r, and relating them a crystal-property tensor of rank c, their ranks are related as follows:

$$s + r = c \qquad \text{(Eq. 11-21)}$$

For example, stimulus **E** and response **I**, both vectorial properties representable by tensors of rank 1, were related by T(1), T(2), . . ., or T(6), a crystal-property tensor of rank 2. However, some stimuli and responses are more complex than **E** or **I** and must themselves be represented by a second-rank tensor. For such a stimulus and response, $(s + r)$, the rank of the crystal-property tensor that relates them, will exceed 2 in value. Space does not permit discussion of these more complex crystal-property tensors which, if of rank 3 or 4, contain 3^3 or 3^4 components.

Semiconductors. Crystals consisting of covalently bonded atoms with all bonds saturated are usually insulators because the overlapped orbitals of neighboring atoms are filled to capacity and accept no transient electrons. For most substances at normal temperatures, the electrons in this valence band of overlapped orbitals lack sufficient energy to occupy the next higher energy band, which thus remains vacant (Fig. 11-15A). *Intrinsic semiconductors* are materials whose valency band is filled but for which the energy level of the next higher band is sufficiently small (Fig. 11-15B) so that thermal excitation can cause some valence electrons to "jump" into this higher band, called the conduction band, and thence migrate through the crystal. The electron deficiency left behind in a valence band is called a *hole.* As electrons travel in one direction, the "holes" travel in the opposite direction.

In contrast to intrinsic semiconductors, *impurity semiconductors* owe their ability to conduct electricity to the presence of foreign atoms or ions in substitutional or interstitial solid solution. In the so-called n-type of semiconductor these foreign atoms possess electrons that (1) do not participate in the bonding and (2) are at energy levels not far below those in the conduction band of the host's atoms (Fig. 11-15C). Such foreign atoms are called *donors* because elevated temperature and/or an applied field **E** can raise the energy of their nonbond-forming electrons until they enter, that is, are donated to, the conductivity band of the host's atoms and become free to move about. In p-type semiconductors the impurity atoms are called *acceptors* because they have a vacant level just above the top of the valence band (Fig. 11-15D). An electron from the valence band of a neighboring atom may enter this vacancy level leaving behind a "hole." Electrons from neighboring atoms then enter this hole, the hole thus traveling in one direction, the electrons in the opposite.

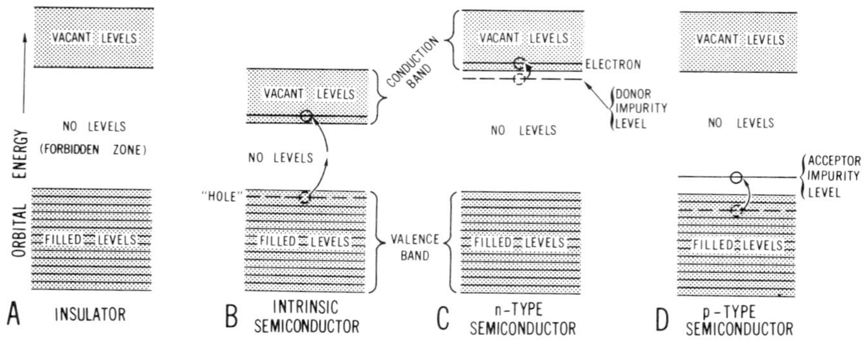

FIGURE 11–15

Schematic diagrams to illustrate the energy gaps between electrons in the valence bands (lower shaded areas) and those less tightly held in conduction bands (upper shaded areas).

(A) For insulators the gap is so large, often about 6 or 7 eV, that an applied potential **E** cannot cause electrons to jump from valence to conduction bands and thence to neighboring atoms.

(B) For intrinsic semiconductors the gap is inherently small (about 1.10 eV for crystalline silicon and 0.72 eV for germanium) so that, particularly as temperature is increased, a large number of electrons can be caused to jump in response to an applied field **E**. The jump of an electron from a filled level to the conduction band leaves a "hole" like those discussed in Chapter 10 (p. 288).

(C) Both *n*- and *p*-type semiconductors contain a small number of foreign atoms in solid solution. In *n*-type semiconductors these foreign atoms possess electrons that do not take part in the chemical bond and are hence free to move into the vacant levels of the conduction bands of neighboring atoms. The foreign atoms thus act as donors of electrons.

(D) In *p*-type semiconductors the foreign atoms lack an electron from their valence band, the energy level of this electron being somewhat higher than the filled levels in the valence bands of their neighboring atoms. Thus electrons jump from valence levels in these neighboring atoms into the foreign atom's, that is, acceptor's, valence band.

The deliberate addition of foreign atoms or ions to a semiconductor's strucuure is called *doping*. Conductivity of semiconductors depends not only on the impurity concentration but also on the temperature. Unlike metals, higher temperatures increase the conductivity of semiconductors. Thus in *n*-type semiconductors the thermal energy causes more nonbond-forming electrons of the donor atoms to enter the conductivity band. Semiconducting devices called *thermistors* are used in electrical control devices because their conductivity increases strongly with temperature.

Thermal properties

Heat Conduction. The mathematical form of the theory of electrical conduction in crystals is the same as for heat conduction. To apply the preceding discussion to thermal conductivity, we interpret vector **E** as representing the direction and temperature difference between two infinitesimally close points in a crystal and vector **I** as representing the crystal's response in terms of an amount of heat that flows across a unit cross section in a unit of time. This done, the constants σ_{11}, σ_{21}, and so on, now represent thermal rather than electrical conductivities in tensors T(1) to T(6).

Thermal Expansion. Upon increase in temperature (ΔT), most solids undergo a change in dimensions, and often shape, called *thermal expansion* or *thermal dilatation.* Upon cooling ($-\Delta T$), the effect is reversed. The decimal fraction by which the length of a line along a direction [*uvw*] changes in response to a 1°C increase in temperature is represented by $\alpha_{[uvw]}$, the crystal's *thermal coefficient of expansion* along [*uvw*]. For example, a 1°C rise in temperature of a calcite crystal at 40°C would cause a 1 cm line along its *c* axis to become 1.000025 cm, whereas a 1 cm line along the *a* axis would become (1.000000 − 0.000006) cm long. Thus for calcite at 40°C, $\alpha_c = 25 \times 10^{-6}$ and $\alpha_a = -6 \times 10^{-6}$ per 1°C, the negative value indicating a decrease in length for a 1° rise in temperature. At temperatures different from 40°C, α_c and α_a would have different values for a 1° rise in temperature; in short, a crystal's thermal coefficient of expansion may change with temperature (Fig. 11-16A). Such curves, wherein either α or volume is plotted against temperature, sometimes display an abrupt inflection at the temperature for which the substance transforms from one polymorph to another (Fig. 11-16B).

Figure 11-17A represents an infinitesimally small cube cut from a non-isometric crystal. We assume the cube's edges, denoted as axes x_1, x_2, and x_3, coincide with no symmetry directions. Subjected to a 1°C increase in tempera-

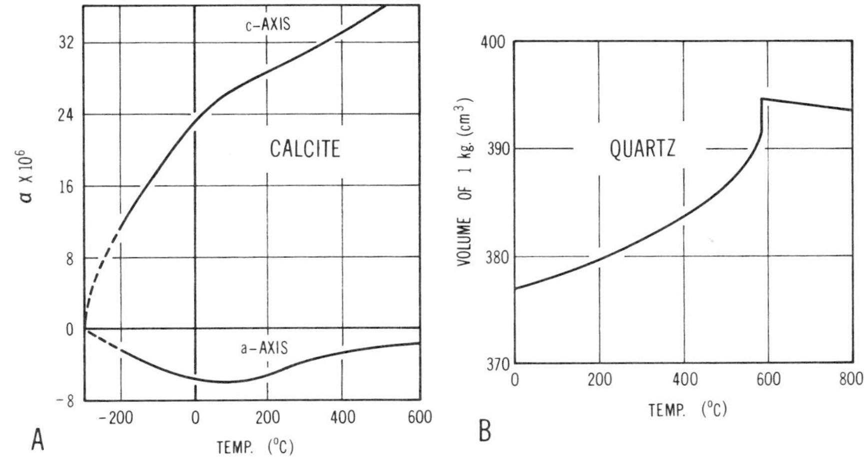

FIGURE 11–16

(A) Values of the thermal coefficient of expansion α, as measured along the *c* and *a* axis, at different temperatures for calcite.

(B) Change with temperature of the volume occupied by 1 kg of quartz. Note sharp inflection at the inversion temperature of 573°C. (After Austin, 1952.)

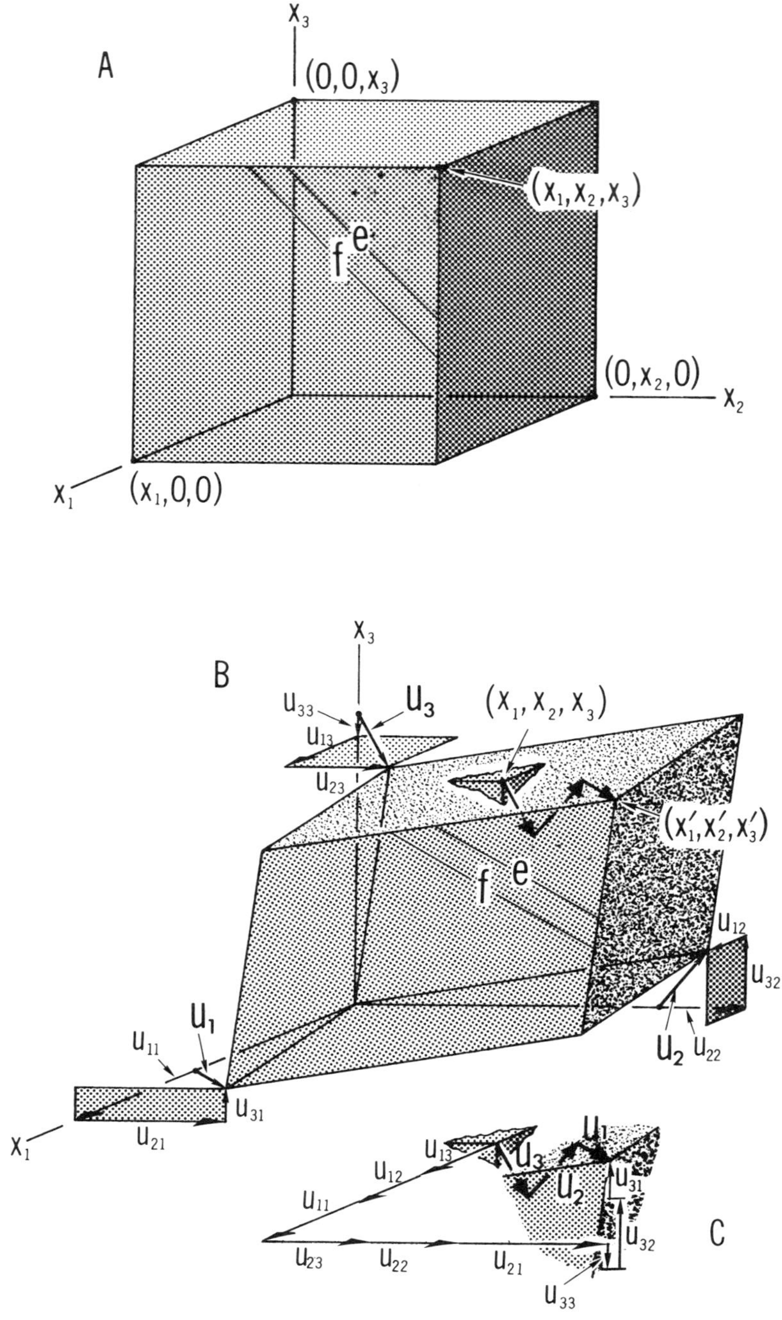

FIGURE 11–17

(A) An infinitesimally small cube *cut, not grown,* from a *nonisometric* crystal. Lines e and f on its front surface are merely for reference. The coordinates of four of its corners are as shown.

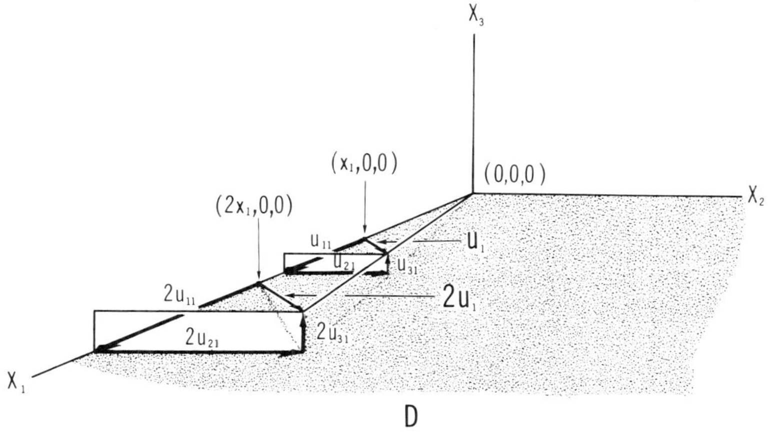

FIGURE 11–17 continued

(B) Deformation, highly exaggerated, of the cube in response to a 1°C rise in temperature. As a result of the deformation the polyhedron is no longer a cube and its edges no longer coincide with axes x_1, x_2, and x_3. Displacement of these cube corners are indicated as vectors $\mathbf{u}_1$, $\mathbf{u}_2$, and $\mathbf{u}_3$. The corner with original coordinates x_1, x_2, x_3 is displaced by the resultant of all three vectors $\mathbf{u}_1$, $\mathbf{u}_2$, and $\mathbf{u}_3$.
(C) Isolated drawing of this corner to show how its displacement can be considered to result from the components of $\mathbf{u}_1$, $\mathbf{u}_2$, and $\mathbf{u}_3$ along the x_1, x_2, and x_3 axes.
(D) The line between points (0, 0, 0) and (x_1, 0, 0) represents the undeformed edge of the original cube in (A). Vector $\mathbf{u}_1$ represents the thermal displacement of the cube corner at (x_1, 0, 0) for a 1°C rise in temperature as previously shown in (B). If this edge were doubled in size but still the cube remained infinitesimally small, the undeformed edge would extend from point (0, 0, 0) to ($2x_1$, 0, 0) and its thermal displacement for a 1°C rise in temperature is correspondingly doubled to become $2\mathbf{u}_1$. Correspondingly, the components of vector $2\mathbf{u}_1$ along axes x_1, x_2, and x_3 are doubled to become $2\mathbf{u}_{11}$, $2\mathbf{u}_{21}$, and $2\mathbf{u}_{31}$. Thus the thermal displacement of point (x_1, 0, 0) relative to point (0, 0, 0) is proportional to its distance from (0, 0, 0) as long as we consider distances and displacements that are infinitesimally small.

ture, the cube deforms *homogeneously*, which means that straight parallel lines like *e* and *f* remain so after deformation (Fig. 11-17B) and a sphere within the cube would become an ellipsoid. Let us consider the cube corner at coordinates (O, O, O) to be fixed in space during this thermal deformation whereas, relative to (O, O, O) the corners at (x_1, O, O), (O, x_2, O), and (O, O, x_3) are displaced as shown by vectors $\mathbf{u}_1$, $\mathbf{u}_2$, and $\mathbf{u}_3$ (Fig. 11-17B). These displacements have components along axes x_1, x_2 and x_3 which are $\mathbf{u}_{11}$, $\mathbf{u}_{21}$ and $\mathbf{u}_{31}$ (for $\mathbf{u}_1$), $\mathbf{u}_{12}$, $\mathbf{u}_{22}$, and $\mathbf{u}_{32}$ (for $\mathbf{u}_2$), and $\mathbf{u}_{13}$, $\mathbf{u}_{23}$, and $\mathbf{u}_{33}$ (for $\mathbf{u}_3$). The corner at (x_1, x_2, x_3) in deforming to (x_1', x_2', x_3') undergoes a displacement equal to the resultant of vectors $\mathbf{u}_1$, $\mathbf{u}_2$, and $\mathbf{u}_3$ (Fig. 11-17C). This resultant, a vector which extends from point (x_1, x_2, x_3) to point (x_1', x_2', x_3'), is not drawn in Figure 11-17C; however, it has (1) a component along axis x_1 equal to the vector sum ($\mathbf{u}_{11} + \mathbf{u}_{12} + \mathbf{u}_{13}$) in Figure 11-17C, (2) a component along x_2 equal to ($\mathbf{u}_{21} + \mathbf{u}_{22} + \mathbf{u}_{23}$), and (3) a component along x_3 equal to ($\mathbf{u}_{31} + \mathbf{u}_{32} + \mathbf{u}_{33}$). In Figure 11-17C, $\mathbf{u}_{33}$ is a negative value because it is directed along the $-x_3$ axis.

Suppose the cube in Figure 11-17A were doubled in size (but still remained infinitesimally small). In this case thermal displacements $\mathbf{u}_1$, $\mathbf{u}_2$, and $\mathbf{u}_3$—as well as the components of each along axes x_1, x_2, and x_3—would also be doubled. This is illustrated in Figure 11-17D for $\mathbf{u}_1$ and its components $\mathbf{u}_{11}$, $\mathbf{u}_{21}$, and $\mathbf{u}_{31}$ which would be doubled in magnitude if the cube edge were doubled from x_1 to $2x_1$. In other words, the thermal displacement of a point such as $(x_1, 0, 0)$ relative to fixed point (0, 0, 0) is proportional to its distance (x_1) from (0, 0, 0). To obtain parameters which are independent of such distance, we define *strain* to be (1) the *displacement* of a point relative to a neighboring point in the crystal *divided by* (2) its *original distance* from this neighbor, that is, from (0, 0, 0) in our example. Thus we now define thermal strains $\mathbf{a}_1$, $\mathbf{a}_2$, and $\mathbf{a}_3$ as

$$\mathbf{a}_1 = \frac{\mathbf{u}_1}{x_1} \qquad \mathbf{a}_2 = \frac{\mathbf{u}_2}{x_2} \qquad \mathbf{a}_3 = \frac{\mathbf{u}_3}{x_3} \qquad \text{(Eqs. 11-22)}$$

where x_1, x_2, and x_3 here represent the distances from (0, 0, 0) of points $(x_1, 0, 0)$, $(0, x_2, 0)$, and $(0, 0, x_3)$ in Figure 11-17A. In Figure 11-17D, for example, $\mathbf{a}_1$ equals $\mathbf{u}_1/x_1$ for point $(x_1, 0, 0)$ and $2\mathbf{u}_1/2x_1$ for point $(2x_1, 0, 0)$; thus $\mathbf{a}_1$ has the same value in both cases. Similarly, the components of $\mathbf{u}_1$ on the x_1, x_2, and x_3 axes, if divided by distance x_1, will become the components of $\mathbf{a}_1$ on these axes, that is,

$$\mathbf{a}_{11} = \frac{\mathbf{u}_{11}}{x_1} \qquad \mathbf{a}_{21} = \frac{\mathbf{u}_{21}}{x_1} \qquad \mathbf{a}_{31} = \frac{\mathbf{u}_{31}}{x_1} \qquad \text{(Eqs. 11-23)}$$

Consequently, the components of thermal strains $\mathbf{a}_1$, $\mathbf{a}_2$, and $\mathbf{a}_3$ along axes x_1, x_2, and x_3 are as summarized in Table 11-1. These components define the tensor denoting the deformation caused by a 1°C change in temperature, namely

$$\begin{bmatrix} a_{11} & a_{12} & a_{13} \\ a_{21} & a_{22} & a_{23} \\ a_{31} & a_{32} & a_{33} \end{bmatrix} \qquad \text{T(8)}$$

Figure 11-18A shows the $\mathbf{a}_{ij}$ components corresponding to the $\mathbf{u}_{ij}$ components of $\mathbf{u}_1$, $\mathbf{u}_2$, $\mathbf{u}_3$ in Figure 11-17B. Because $\mathbf{a}_{33}$ (and $\mathbf{u}_{33}$) is opposite to the positive direction along axis x_3, its value is negative and it indicates shortening. Similarly, $\mathbf{a}_{12}$ and $\mathbf{a}_{21}$ are each positive, being directed along the $+x_1$ and $+x_2$ axes, respectively. Had all a_{ij} values except a_{12} and a_{21} equaled zero and if $-a_{21}$ equaled a_{12}, this would merely signify a clockwise rotation (Fig. 11-18B) of the cube around axis x_3. A cross section of the cube normal to axis x_3 (Fig. 11-18C) thus remains square because points in this plane did not undergo strain, that is, did not change position *relative to each other*. They do undergo strain, however, if $a_{21} = a_{12}$ (Fig. 11-18D, E). In like manner, $a_{23} = -a_{32} \neq 0$ would indicate a clockwise rotation about axis x_1 with a corresponding lack of strain in the plane normal to x_1. Similarly, $a_{31} = -a_{13}$ represents (1) clockwise rotation about axis x_2 and (2) a lack of strain in the plane normal to x_2.

A general deformation will consist of three components: (1) translation, (2) rigid body rotations, and (3) strain. Now any second-rank tensor may be

TABLE 11–1

Components of $\mathbf{a}_1$, $\mathbf{a}_2$, and $\mathbf{a}_3$

	Of $\mathbf{a}_1$	Of $\mathbf{a}_2$	Of $\mathbf{a}_3$
On x_1	$\mathrm{a}_{11} = \frac{\mathrm{u}_{11}}{x_1}$	$\mathrm{a}_{12} = \frac{\mathrm{u}_{12}}{x_2}$	$\mathrm{a}_{13} = \frac{\mathrm{u}_{13}}{x_3}$
On x_2	$\mathrm{a}_{21} = \frac{\mathrm{u}_{21}}{x_1}$	$\mathrm{a}_{22} = \frac{\mathrm{u}_{22}}{x_2}$	$\mathrm{a}_{23} = \frac{\mathrm{u}_{23}}{x_3}$
On x_3	$\mathrm{a}_{31} = \frac{\mathrm{u}_{31}}{x_1}$	$\mathrm{a}_{32} = \frac{\mathrm{u}_{32}}{x_2}$	$\mathrm{a}_{33} = \frac{\mathrm{u}_{33}}{x_3}$

expressed as the sum of a *symmetrical tensor* for which $a_{ij} = a_{ji}$, that is, $a_{12} = a_{21}$, $a_{13} = a_{31}$, and so on, and an *antisymmetrical* or *skew tensor* for which $a_{ij} = -a_{ji}$, that is, $a_{12} = -a_{21}$. For example, deformation tensor T(8), which summarizes the thermal deformation produced by a 1°C rise in temperature in a crystal, may be represented as the sum of an antisymmetric tensor or rotational tensor, which summarizes all the rigid body rotation involved, namely,

$$\begin{bmatrix} 0 & \frac{a_{12} - a_{21}}{2} & -\left(\frac{a_{31} - a_{13}}{2}\right) \\ -\left(\frac{a_{12} - a_{21}}{2}\right) & 0 & \frac{a_{23} - a_{32}}{2} \\ \frac{a_{31} - a_{13}}{2} & -\left(\frac{a_{23} - a_{32}}{2}\right) & 0 \end{bmatrix} \qquad \mathrm{T}(9)$$

plus a symmetric tensor, which summarizes the strain, that is, change in shape of the body, namely,

$$\begin{bmatrix} a_{11} & \frac{a_{12} + a_{21}}{2} & \frac{a_{31} + a_{13}}{2} \\ \frac{a_{12} + a_{21}}{2} & a_{22} & \frac{a_{23} + a_{32}}{2} \\ \frac{a_{31} + a_{13}}{2} & \frac{a_{23} + a_{32}}{2} & a_{33} \end{bmatrix} \qquad \mathrm{T}(10)$$

The off-diagonal symbols in strain tensor T(10) are cumbersome so we substitute simpler symbols for these quantities. For example, we represent

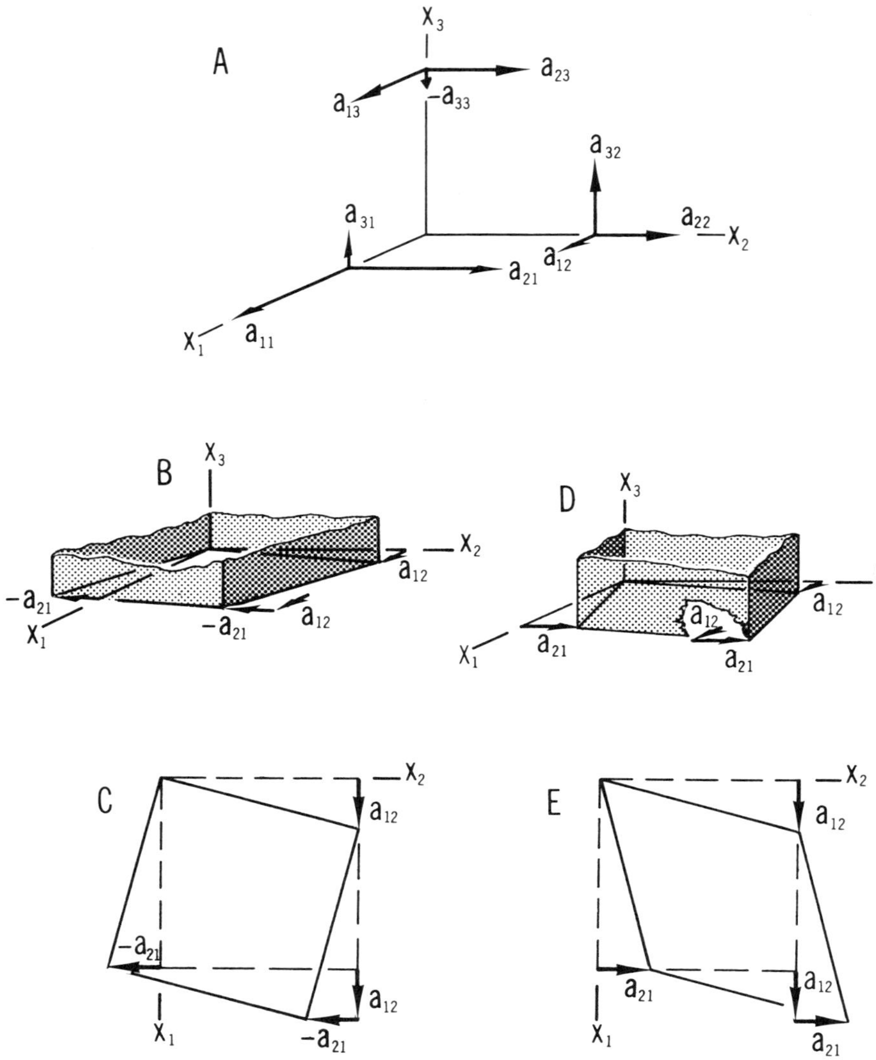

FIGURE 11-18

(A) The thermal strains $\mathbf{a}_1$, $\mathbf{a}_2$, and $\mathbf{a}_3$, which respectively correspond to thermal displacements $\mathbf{u}_1$, $\mathbf{u}_2$, and $\mathbf{u}_3$ in Figure 11-17, are here represented by their components along x_1, x_2, and x_3 (as defined in Table 11-1). Thus $\mathbf{a}_1$ is represented by $\mathbf{a}_{11}$, $\mathbf{a}_{21}$, and $\mathbf{a}_{31}$; $\mathbf{a}_2$ by $\mathbf{a}_{12}$, $\mathbf{a}_{22}$, and $\mathbf{a}_{32}$; $\mathbf{a}_3$ by $\mathbf{a}_{13}$, $\mathbf{a}_{23}$, and $\mathbf{a}_{33}$.
For another cube cut from a crystal, let us assume that (1) all foregoing $\mathbf{a}_{ij}$ components equal zero except that (2) $\mathbf{a}_{12} = -\mathbf{a}_{21} \neq 0$, the negative sign indicating that $-\mathbf{a}_{21}$ is reversed in direction from $\mathbf{a}_{21}$ in (A). If conditions (1) and (2) were fulfilled, the cube's response to the thermal strains would be to rotate about the x_3 axis

$(a_{12} + a_{21})/2$ by either α_{12} or α_{21}, depending on its position in the tensor, to obtain the *thermal expansion tensor*

$$\begin{bmatrix} \alpha_{11} & \alpha_{12} & \alpha_{13} \\ \alpha_{21} & \alpha_{22} & \alpha_{23} \\ \alpha_{31} & \alpha_{32} & \alpha_{33} \end{bmatrix} \quad \text{or} \quad \begin{bmatrix} \alpha_{11} & \alpha_{12} & \alpha_{13} \\ & \alpha_{22} & \alpha_{23} \\ & & \alpha_{33} \end{bmatrix} \qquad \mathrm{T}(11)$$

whose elements α_{ij} are the *coefficients of thermal expansion*. The abbreviated form at the right emphasizes this tensor's inevitable symmetry by omitting α_{21} $(=\alpha_{12})$, α_{31} $(=\alpha_{13})$, and α_{32} $(=\alpha_{23})$.

If one or more of the axes x_1, x_2, and x_3 coincide with (or are related by) a symmetry axis or plane, a simplification occurs in the thermal expansion tensor referenced to them. Thus for a monoclinic crystal we let x_2 coincide with a 2-fold axis. In such case $\mathbf{u}_2$ in Figure 11-17B would coincide with x_2, the situation being analogous to that whereby $\mathbf{I}_{E1}$ coincided with x_1 in Figure 11-11B, and hence $\mathbf{u}_{12}$ and $\mathbf{u}_{32}$ must now equal zero. This causes a_{12} and a_{32} to equal zero. Because thermal expansion is a centrosymmetric property (Fig. 11-4A), this monoclinic crystal must react as if it belonged to class $2/m$, that is, as if it also possessed a plane of symmetry normal to x_2. The x_1 and x_3 axes in Figure 11-17B would lie within this mirror plane and, as a result, $\mathbf{u}_1$ and $\mathbf{u}_3$ would be constrained to lie within this mirror plane. In other words their components parallel to axis x_2, namely $\mathbf{u}_{21}$ and $\mathbf{u}_{23}$, would equal zero. Hence $\mathbf{a}_{21}$ and $\mathbf{a}_{23}$ must equal zero. Substituting zeros for a_{12}, a_{21}, a_{32}, and a_{23} in T(10), we obtain the form of T(11), which is the thermal expansion tensor for a monoclinic crystal in the second setting, namely

$$\begin{bmatrix} \alpha_{11} & 0 & \alpha_{13} \\ & \alpha_{22} & 0 \\ & & \alpha_{33} \end{bmatrix} \qquad \mathrm{T}(12)$$

For an orthorhombic crystal in which x_1, x_2, and x_3 coincide with the a, b, and c axes, respectively, these being 2-fold axes, the thermal expansion tensor is

$$\begin{bmatrix} \alpha_{11} & 0 & 0 \\ & \alpha_{22} & 0 \\ & & \alpha_{33} \end{bmatrix} \qquad \mathrm{T}(13)$$

---

without change of shape as illustrated (B) by the cube's lower portion or (C) by a view down the x_3 axis. Such rotation without change of shape is called *rigid body rotation*. All cubes are assumed to be infinitesimally small.

For still another cube, assume (1) all thermal strain components $\mathbf{a}_{ij} = 0$ except that (2) $\mathbf{a}_{12} = \mathbf{a}_{21} \neq 0$. The cube then undergoes a change of shape as shown (D) by its lower portion or (E) by viewing it down the x_3 axis.

In tetragonal, hexagonal, or trigonal crystals, the x_3 axis customarily coincides with the c axis; its symmetry then requires α_{22} to equal α_{11}. Thus the tensor becomes

$$\begin{bmatrix} \alpha_{11} & 0 & 0 \\ & \alpha_{11} & 0 \\ & & \alpha_{33} \end{bmatrix} \qquad \mathrm{T}(14)$$

In isometric crystals in which axes x_1, x_2, and x_3 correspond to crystal axes a_1, a_2, and a_3, a 3-fold axis equates α_{11}, α_{22}, and α_{33}. Thus the thermal expansion tensor becomes

$$\begin{bmatrix} \alpha_{11} & 0 & 0 \\ & \alpha_{11} & 0 \\ & & \alpha_{11} \end{bmatrix} \qquad \mathrm{T}(15)$$

Tensors become diagonalized, that is, all off-diagonal terms become zero, as in tensors T(13) to T(15), if their reference axes x_1, x_2, and x_3 coincide with *principal axes*. The principal axes for thermal expansion represent three mutually perpendicular directions such that if the cube's edges and hence axes x_1, x_2, and x_3 in Figure 11-17 were to coincide with these directions, then $\mathbf{u}_1$ would lie along x_1, $\mathbf{u}_2$ along x_2, and $\mathbf{u}_3$ along x_3. Thus all u_{ij} terms, where $i \neq j$, would become zero. For crystals of orthorhombic symmetry or higher, tensors T(13) to T(15) indicate that the x_1, x_2, and x_3 axes chosen represent principal axes. If the thermal expansion tensor for monoclinic or triclinic crystals is referenced to principal axes, tensors T(12) and T(11) diagonalize to one like T(13). However, location of the principal axes is tedious in triclinic crystals, trial and error methods being required because no symmetry directions exist to pinpoint their location. Their location is less difficult in monoclinic crystals; one always coincides with the 2-fold or b axis (second setting assumed), the remaining two lying in the (010) plane.

The variation in thermal expansion for different directions in a crystal may be represented graphically if the crystal's principal expansion coefficients, α_{11}, α_{22}, α_{33}, are known for its diagonalized tensor. Consider any point (coordinates x_1, x_2, x_3) on a sphere of unit radius drawn about the origin in the crystal. Thus

$$x_1^2 + x_2^2 + x_3^2 = 1 \qquad \text{(Eq. 11-24)}$$

which is the equation for a sphere of radius 1. The thermal expansion resultant from a 1°C increase in temperature deforms the sphere by causing points on it to move to new coordinates (X_1, X_2, X_3), where

$$\begin{aligned} X_1 &= (1 + \alpha_{11})x_1 \\ X_2 &= (1 + \alpha_{22})x_2 \\ X_3 &= (1 + \alpha_{33})x_3 \end{aligned} \qquad \text{(Eq. 11-25)}$$

Solving Equations (11-25) for x_1, x_2, and x_3, then substituting these values in Equation (11-24), there results

$$\frac{X_1^2}{(1+\alpha_{11})^2} + \frac{X_2^2}{(1+\alpha_{22})^2} + \frac{X_3^2}{(1+\alpha_{33})^2} = 1 \qquad \text{(Eq. 11-26)}$$

This is the equation of an ellipsoid (Fig. 11-19A) with semiaxes of lengths $(1 + \alpha_{11})$, $(1 + \alpha_{22})$, and $(1 + \alpha_{33})$ and pertains to all orthorhombic, monoclinic, and triclinic crystals for which x_1, x_2, and x_3 are principal axes. For tetragonal, hexagonal, or trigonal crystals, because α_{11} equals α_{22}, this ellipsoid will be either oblate ($\alpha_{11} > \alpha_{33}$ as in Fig. 11-19B) or prolate ($\alpha_{33} > \alpha_{11}$ as in Fig. 11-19C) and will possess, normal to the *c* or x_3 axis, a circular cross section of radius $(1 + \alpha_{11})$. For isometric crystals ($\alpha_{11} = \alpha_{22} = \alpha_{33}$) the ellipsoid becomes a sphere (Fig. 11-19D) of radius $(1 + \alpha_{11})$. A sphere inscribed in an isometric crystal thus remains a sphere, but of slightly different radius, at the new temperature.

The length of a line drawn outward from the origin until it intercepts one of the surfaces in Figure 11-19 (line OP, for example), will equal $(1 + \alpha)$, the factor by which this line changed in length as temperature increased 1°C. For isometric crystals (Fig. 11-19D), this factor $(1 + \alpha)$ equals $(1 + \alpha_{11})$ regardless of the direction of OP. For tetragonal, hexagonal, or trigonal crystals, if ρ is the angle between OP and principal axis x_3, then

$$(1+\alpha) = \frac{(1+\alpha_{11})(1+\alpha_{33})}{\sqrt{(1+\alpha_{11})^2\cos^2\rho + (1+\alpha_{33})^2\sin^2\rho}} \qquad \text{(Eq. 11-27)}$$

For triclinic, monoclinic, and orthorhombic crystals, if ϕ is the angle that the x_2 principal axis makes to the plane containing line OP and axis x_3 (Fig. 11-20A), and if OP is again at an angle ρ to x_3, then

$$(1+\alpha) = \frac{1}{\sqrt{\dfrac{\sin^2\rho\,\sin^2\phi}{(1+\alpha_{11})^2} + \dfrac{\sin^2\rho\,\cos^2\phi}{(1+\alpha_{22})^2} + \dfrac{\cos^2\rho}{(1+\alpha_{33})^2}}} \qquad \text{(Eq. 11-28)}$$

If a crystal is oriented in stereographic projection so that x_3 plots at $\rho = 0°$ and x_2 at $\phi = 0°$, $\rho = 90°$, then the spherical coordinates, ϕ and ρ, for any line plotted in this stereogram are the angles cited in Equation (11-28).

Let us assume that a 1°C rise in temperature causes a point P (x_1, x_2), which lies in a crystal section parallel to axes x_1 and x_2 (Fig. 11-20B), to move to P' $(x_1 + \alpha_{11}x_1,\ x_2 + \alpha_{22}x_2)$, line OP thus moving to OP'. Line increment PP' has components $\alpha_{11}x_1$ and $\alpha_{22}x_2$ and a slope $a_{22}x_2/a_{11}x_1$. If $\alpha_{22} \neq \alpha_{11}$, this slope differs from x_2/x_1, the slope of OP; hence OP and OP' will differ in direction (Fig. 11-20B). However, if $\alpha_{22} = \alpha_{11}$, this slope equals x_2/x_1 and hence OP and OP' have the same direction (Fig. 11-20C). In most crystals α is a value such as 3×10^{-5} per 1°C. Thus length PP' in Figure 11-20B, C was exaggerated about 10^5 times relative to OP. At true scale PP' would be a tiny dot and hence OP and OP', even in Figure 11-20B, would be essentially the same direction. Of course,

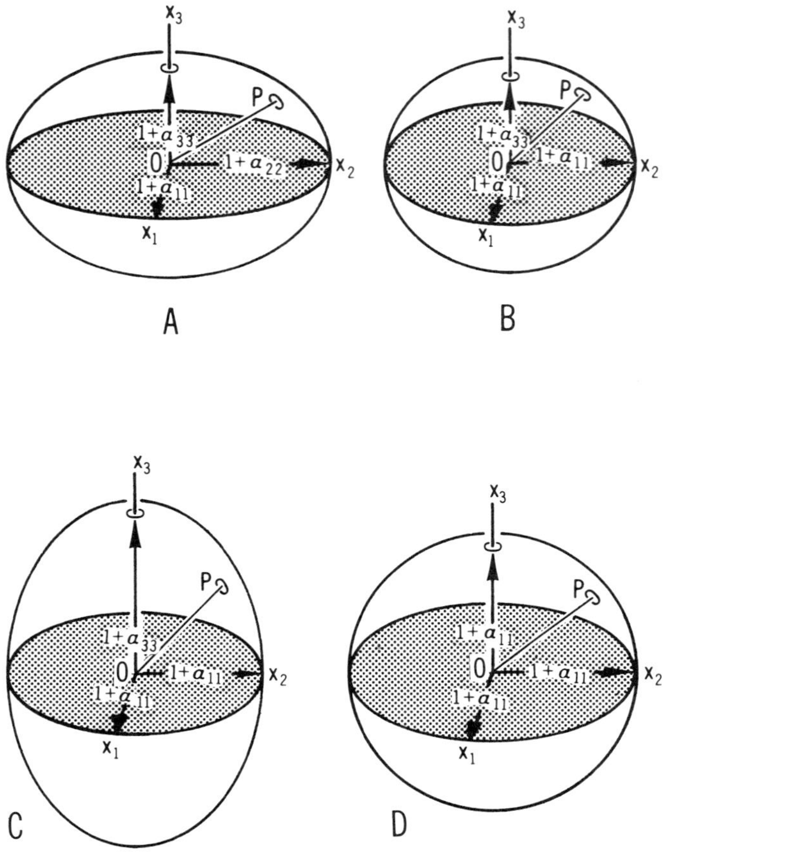

FIGURE 11–19

Quadric surfaces formed by drawing lines proportional in length to $(1 + \alpha)$ for different directions in a crystal, then connecting the ends of these lines, all of which have been erased except OP, with a smooth surface. For triclinic, monoclinic, and orthorhombic crystals the resultant surface is a triaxial ellipsoid (A) such that a line along the three principal axes x_1, x_2, and x_3 would have lengths proportional to $(1 + \alpha_{11})$, $(1 + \alpha_{22})$, and $(1 + \alpha_{33})$, respectively; these principal axes inevitably coincide with symmetry axes, if such are present. For tetragonal and hexagonal crystals a line along x_1 or x_2 would have the same length $(1 + \alpha_{11})$, whereas one along x_3, which always coincides with the crystallographic c axis, would have length $(1 + \alpha_{33})$. Tetragonal crystals, or hexagonal crystals, divide into two types: (1) those for which $(1 + \alpha_{33}) < (1 + \alpha_{11})$ as in (B) or (2) those for which $(1 + \alpha_{33}) > (1 + \alpha_{11})$ as in (C). In (B) and (C) the shaded plane is a circle of radius $(1 + \alpha_{11})$. Surface (D) which applies to isometric crystals or to amorphous substances is a sphere of radius $(1 + \alpha_{11})$.

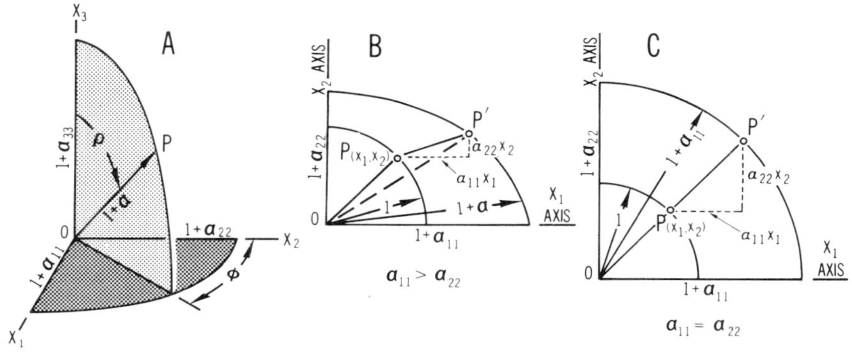

FIGURE 11–20

(A) Definition of the spherical coordinates ϕ and ρ of a line OP when x_3 is located at $\rho = 0°$ and x_2 at $\phi = 0°$.
(B) The movement, highly exaggerated, of a point P to P' in a crystal in response to a 1° temperature change. Accepting the origin O as a fixed reference point, line OP changes in direction to become OP'. The change in direction of OP' in respect to OP vanishes to the extent that α_{11} and α_{22} are small in value or that $\alpha_{11} = \alpha_{22}$ so that the slope of line OP, that is, x_2/x_1, equals that of line PP', that is, $\alpha_{22}x_2/\alpha_{11}x_1$.
(C) If P is located in a circular section of the quadric in Figure 11-19B, C, or in an isometric crystal, then OP and OP' inevitably coincide in direction because the slope of PP' invariably equals x_2/x_1 (because $\alpha_{22} = \alpha_{11}$).

for a 1000°C increase in temperature, PP' would be about 1000 times the value for a 1° increase and the directional change of OP to OP' would be more obvious. Extending our discussion to three dimensions, lines, which include face normals, do not change directions in a heated isometric crystal (because $\alpha_{11} = \alpha_{22} = \alpha_{33}$), or if these lines are normal to the c axis in tetragonal, hexagonal, or trigonal crystals. Thus interfacial angles between (hkl) faces remain unchanged on heating isometric crystals, whereas only those between $(hk0)$ faces are unchanged for tetragonal or hexagonal crystals. For orthorhombic, monoclinic, or triclinic crystals, only directions or face normals coincident with axes x_1, x_2, and x_3 are unchanged on heating. For small temperature increases the directional changes may be small. For large increases they may be appreciable, but for such large increases the values α_{11}, α_{22} and α_{33} may themselves change in value (Fig. 11-16A) because they can only be regarded as constants at temperatures not far removed from those at which they were originally determined.

In crystal structures the atoms, although bound to certain sites, vibrate about these sites with increasing amplitude as temperature is increased. Such vibration smears the electron density of the atom over a larger volume, in effect making the atom or ion larger so that interatomic distances increase. The shape of this volume will resemble one of the shapes in Figure 11-19, depending on the site's symmetry. Thermal expansion, not accompanied by any great increase of interatomic distances, may result from changes in bond angles as temperature is increased, particularly if the angle can vary without violating the structure's symmetry. Silicon carbide is an example. For diamond to remain isometric on heating, the tetrahedral bond angles must remain unchanged and thermal expansion then results from lengthening of the covalent C—C bonds. In calcite, heating causes expansion along the *c* axis but contraction along directions normal to it. Presumably, the disk-like CO_3 groups, which are initially parallel to (0001), rotate out of this plane with increasing temperature thereby expanding the structure normal to (0001) but contracting it along directions within (0001).

As temperature increases, strongly bonded atoms separate less than weakly bonded atoms. Thus coefficients of thermal expansion are less along strong periodic bond-chain vectors in a structure, but greater at right angles to them. For example, chain silicates exhibit lesser values of α parallel to the chains than at 90° to them. In sheet silicates two or more strong periodic bond-chain vectors occur parallel to the sheets; accordingly, directions parallel to the sheets exhibit the smaller values of α, those normal to them the largest. Where anisotropy of thermal expansion may be ignored, the volume coefficient of thermal expansion α_v is used and is defined as

$$\alpha_v = \alpha_{11} + \alpha_{22} + \alpha_{33}$$

For a tetragonal crystal it would equal $(2\alpha_{11} + \alpha_{33})$ for an isometric $(3\alpha_{11})$. B. J. Skinner (in Clark, 1966) tabulates thermal-expansion data for a large number of naturally occurring substances.

The force exerted by thermal expansion can be very great. Expressing this force in terms of the pressure necessary to prevent expansion, Austin (1952, p. 244) notes that if MgO is heated from room temperature to 100°C, a pressure of 70,000 lb/in.^{2} is required to prevent any increase in volume. The forces resultant from such large thermal stresses may be important during the cooling of an igneous rock or of any metal or ceramic bodies of appreciable size.

Magnetic Properties

Magnetism

Charged particles create magnetic fields when in motion. For example, an electric current **I** flowing through a wire loop of area A will create, normal to the loop's area, two magnetic poles (N and S in Fig. 11-21A). If freely suspended, the loop will rotate so that pole N will point toward the earth's north magnetic pole whereupon pole S points south. A current-bearing loop thus becomes a *magnetic dipole,* so called because it possesses *two* magnetic poles of opposite tendencies. The intrinsic intensity of a dipole's magnetic field, which is usually indicated by a vector **m** directed along its N-seeking pole, is called its *magnetic moment.* The magnitude of **m** equals IA, where I and A indicate the loop's current and area. If the direction of the current in the loop is reversed, its N and S poles are also reversed (Fig. 11-21B). Both loops together (Fig. 11-21C) would thus nullify each other; that is, their combined magnetic moment would be zero.

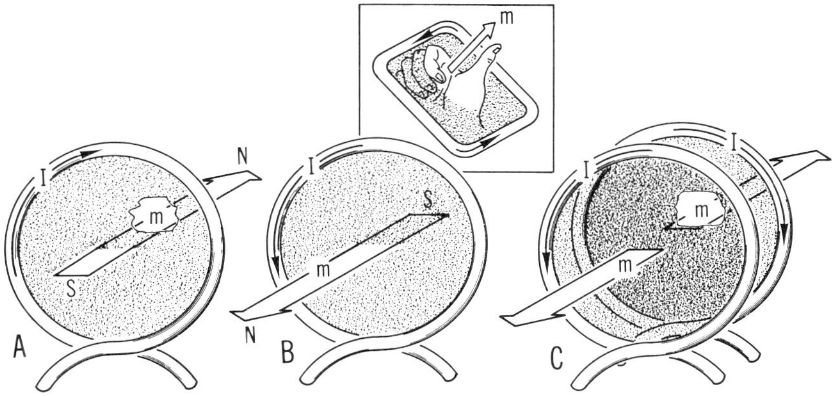

FIGURE 11–21

(A) Flow of an electric current I through a wire loop, converting the loop to a magnet. The large hollow arrow normal to the loop's stippled area marks the direction of the loop's N-seeking magnetic pole. The arrow's nonbarbed end denotes the loop's S-seeking magnetic pole. Thus flow of electricity through a loop converts it to a *magnetic dipole*. The strength of the magnetic field for this loop is known as its *dipole moment* **m** and is here graphically indicated by the direction and length of the hollow arrow (vector). The length of vector **m**, that is, the strength of the loop's magnetic field, is proportional to the product of the current **I** times the area A (stippled) of the loop.
(B) Reversing the current reverses the direction of its N- and S-seeking poles. The dipole moment is thus reversed. To determine the direction of a dipole moment for a loop, clench the right hand (see square inset) with fingers pointing in direction of the current flow which is, by convention, opposite to the actual flow of electrons. The extended right thumb then points in the positive direction (arrowhead) for the dipole moment **m**.
(C) For two closely adjacent and identical loops with equal but opposite current flows, the equal but opposite dipole moments cancel each other. In other words, their two magnetic fields mutually nullify each other. This is analogous to two oppositely spinning electrons in the same orbital in an atom.

Individual atoms or ions contain extremely small current loops of three types: (1) the relatively slow spin and rotation of the protons in the nucleus; (2) the orbital motion of electrons about this nucleus; and (3) the spin of these electrons. Of these, (2) and (3) far exceed (1) in contributing to an atom's magnetic moment. Two electrons paired within the same orbital necessarily orbit and spin in opposition to each other, the *Pauli exclusion principle* requiring this. Thus their magnetic moments nullify each other (somewhat like the loops in Fig. 11-21C). Hence atoms or ions possess an effective magnetic moment, usually symbolized μ_{eff}, in proportion to the number of unpaired electrons in their outer orbitals.

A natural unit for expressing an atom's magnetic moment is the *Bohr magneton* ($=0.92732 \times 10^{-23}$ ampere meters^2), which is approximately equal to

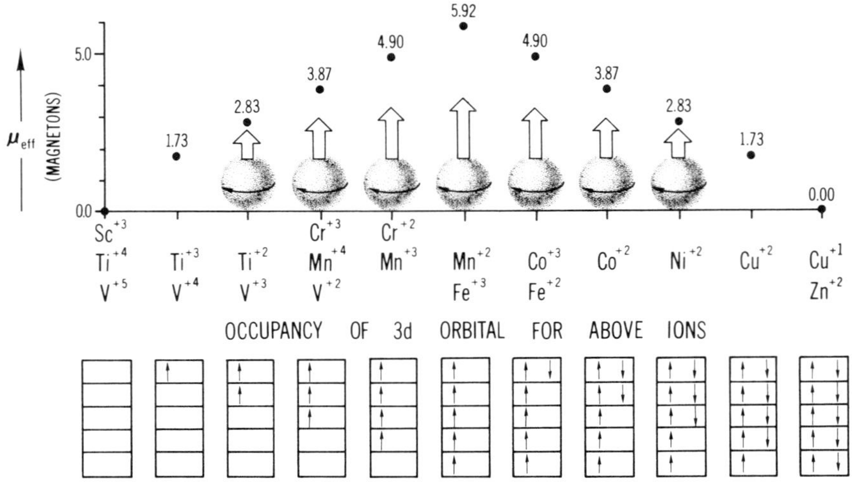

FIGURE 11–22

Calculated effective magnetic moments μ_{eff} for ions of the transition elements of the iron family. (After A. S. Povarennykh.) Each of the five $3d$ suborbitals is schematically shown by a small rectangle. Filling of the suborbital by two oppositely spinning electrons is shown by insertion of two oppositely directed vectors (magnetic moments) in the rectangle.

the magnetic moment resultant from the spin of a single unpaired electron. For example, the effective magnetic moment μ_{eff}, which is a measure of ability to produce (or be affected by) a magnetic field, is proportional to the number of unpaired electrons in the $3d$ orbital for the ions of the transitional elements of the iron family (Fig. 11-22). Thus Fe^{3+} and Mn^{2+}, each with five unpaired electrons, are among the most magnetic ions known. Some common ions which, if present in a crystal, particularly increase its tendency to be affected by a magnetic field are Fe^{3+}, Mn^{2+}, Fe^{2+}, Mn^{3+}, Mn^{4+}, Cr^{3+}, Co^{2+}, Ni^{2+}, Cu^{2+}, efficacy decreasing from Fe^{3+} to Cu^{2+}. Less common contributors to magnetism are U^{4+}, Dy^{3+}, Er^{3+}, Gd^{3+}, Nd^{3+}, and Ce^{2+}.

A unit of positive charge experiences a sideward force **F** as it travels with velocity **v** through a magnetic field of intensity and direction **B**. This force **F** equals $\mathbf{v} \times \mathbf{B}$.[7] For example, we let the arrows labeled **v** in Figure 11-24A represent the velocity of flow of a unit of positive current[8] in each of four wires located

[7] The cross product of **v** and **B**, two vectors at an angle θ to each other, is symbolized as $\mathbf{v} \times \mathbf{B}$ (read "v cross B"). *By definition,* $\mathbf{v} \times \mathbf{B}$ represents a new vector (labeled **F** in Fig. 11-23A) which is normal to the plane containing **v** and **B** and whose magnitude equals $vB \sin \theta$, where v and B are the magnitudes (lengths) of vectors **v** and **B**. The sense of positive direction of **F** derives from the *right-hand convention* for a vector cross product. If $\mathbf{v} \times \mathbf{B}$ is desired, (1) point the right thumb and forefinger along **v** and **B**, respectively (Fig. 11-23B), then (2) the middle finger can be bent only toward the positive direction of **F**. Note that $\mathbf{B} \times \mathbf{v}$ would yield a vector opposite to **F**, namely, $-\mathbf{F}$.

[8] Convention dictates electricity to flow from positive to negative. Actually, electrons flow opposite to arrows **v** in Figure 11-24A.

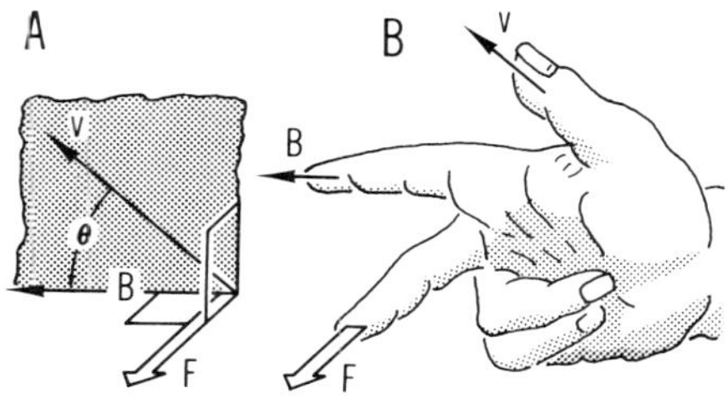

FIGURE 11-23

(A) Significance of the vector cross product of vectors **v** and **B**. Thus, by definition, **v** × **B** defines a second vector **F** which is perpendicular to **v** and **B** and whose magnitude (length) equals $vB \sin \theta$, that is, the product of the magnitudes of **v** times **B** times the sine of the angle between them.
(B) The right-hand convention which determines the positive end of **F**. For **v** × **B**, align the right thumb along **v**, the forefinger along **B**; then the middle finger can only bend in the direction of positive **F**. Note that **B** × **v** would be equal but opposite to **F**.

within a magnetic field with intensity and direction as indicated by **B**. Forces $\mathbf{F}_1$, $\mathbf{F}_2$, $\mathbf{F}_3$, and $\mathbf{F}_4$ thus arise. If these segments were joined in a single loop (Fig. 11-24B), the current flow converts the loop to a magnetic dipole whose magnetic moment **m**, which is normal to the stippled plane, is directed upward. By convention the positive direction for **m** is determined by encircling the loop with the right hand so that the fingers point in the direction of the positive current (Fig. 11-24C); the right thumb then points toward positive **m**. This convention relates direction of a positive current in a loop to the direction of the magnetic moment produced. Substitute the *left hand* if direction of electron flow is given; for example, the curved arrows in Figure 11-22 represent directions of electron flow, and the hollow arrows represent the positive directions for magnetic moment **m**.

If the loop in Figure 11-24B were suspended from a string somehow attached to the loop's center and if θ_1 equals 90°, forces $\mathbf{F}_1$ and $\mathbf{F}_3$, but not $\mathbf{F}_2$ and $\mathbf{F}_4$, would rotate the loop about the dashed arrow as an axis. The direction of this axis and the torque or twisting force about it are indicated by (dashed) vector $\boldsymbol{\tau}$, where $\boldsymbol{\tau} = \mathbf{m} \times \mathbf{B}$. The length of vector $\boldsymbol{\tau}$ in Figure 11-24B equals the magnitude of the twisting force, that is, $mB \sin \phi$, where ϕ is the angle between **m** and **B**. The direction of vector $\boldsymbol{\tau}$ defines the axis about which twisting occurs; its positive (barbed) end is located by applying the right-hand rule (Fig. 11-23B). The loop or dipole in Figure 11-24B would thus rotate until its magnetic moment **m** coincided with **B**, the direction of the magnetic field. At this point angle ϕ (and thus the magnitude of the torque, $mB \sin \phi$) equals zero.

Diamagnetism

Repulsion by a magnet is called *diamagnetism*. It results from the presence of paired, oppositely spinning electrons in the same orbital. For these (e_1 and e_2 in Fig. 11-25A) their equal but opposite magnetic moments $\mathbf{m}_1$ and $\mathbf{m}_2$ nullify each other. However, if an atom containing two such paired electrons is approached by a magnetic field of intensity and direction as summarized by **B** in Figure 11-25B, then Lenz's law applies. In essence the law states that movement of a magnetic field toward a closed circuit or loop induces in the loop a current whose associated magnetic field will oppose (repel) the approaching one. Thus regarding each orbital in Figure 11-25B as a loop, the approaching magnetic field slows e_1 but speeds e_2, thereby decreasing moment $\mathbf{m}_1$ but increasing $\mathbf{m}_2$ so that their resultant $(\mathbf{m}_2 - \mathbf{m}_1)$ opposes **B**. The effect is feebly present in all materials.

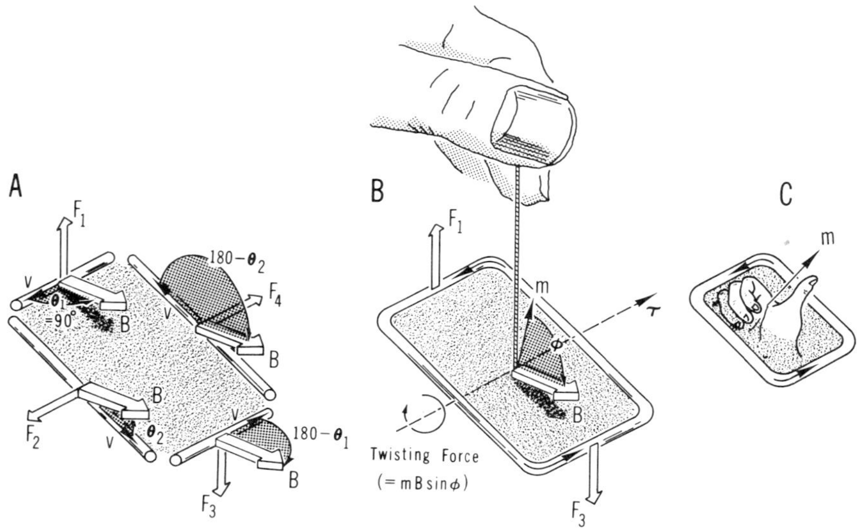

FIGURE 11–24

(A) If a unit of positive charge flows with velocity **v** at an angle θ to a magnetic field (vector **B**), a force **F** is exerted comparable to $\mathbf{v} \times \mathbf{B}$. The direction of **F** is illustrated for a positive current along each of four mutually perpendicular segments of wire in the same magnetic field. We now consider angle θ_1 to equal 90°. In such case the shaded planes denoting angles θ_2 and 180-θ_2 will be perpendicular to the line which joins the midpoints of the two long wires. In consequence vectors $\mathbf{F}_2$ and $\mathbf{F}_4$ will lie along this line and, being oppositely directed, nullify each other.
(B) If the four wires are joined to form a rectangular loop, forces $\mathbf{F}_1$ and $\mathbf{F}_3$, which do not nullify each other, cause the loop to rotate about the dashed axis. This dashed axis represents the line between the midpoints of the two long wires in (A). The direction of rotation about this dashed axis will cause the loop's dipole moment **m**, which is normal to the loop's stippled area, to decrease its angle ϕ relative to vector **B**. The magnitude of the twisting force $\boldsymbol{\tau}$ equals $mB \sin \phi$ (because $\boldsymbol{\tau}$ equals $\mathbf{m} \times \mathbf{B}$).
(C) Determination of the positive direction of the loop's dipole moment **m**. Clench the right fist so that the fingers point in the direction of flow of the positive current (this is opposite from the electron flow in the loop). The unclenched thumb will then point along the positive direction for **m**.

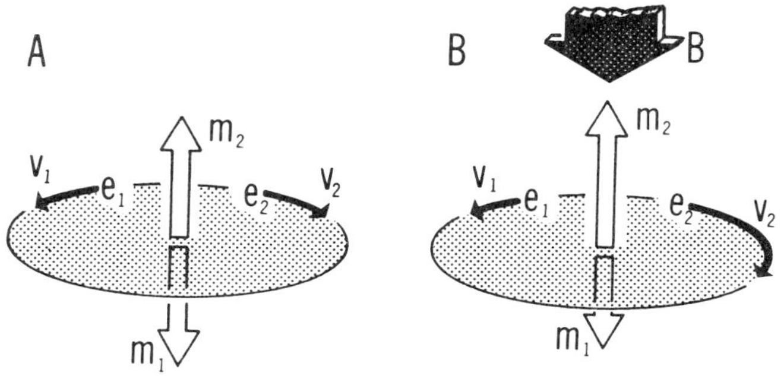

FIGURE 11–25

(A) Schematic drawing showing dipole moments caused by e_1 and e_2, two oppositely spinning (and here oppositely orbiting) electrons paired in the same orbital of one atom. Because these are electron-flow directions, which are opposite from electric-current directions, we use a left-hand rule to determine the direction of dipole moments $\mathbf{m}_1$ and $\mathbf{m}_2$.
(B) If a magnetic field with intensity and direction as indicated by **B** is brought toward the atom, then as demanded by Lenz's law, the current whose magnetic moment opposes **B** is intensified. In other words, electron e_2 speeds up and e_1 slows down so that the resultant of dipole moments $\mathbf{m}_1$ and $\mathbf{m}_2$ is no longer zero as it was in (A). Such repulsion of an atom by a magnetic field is called *diamagnetism.*

However, in the paramagnetic and ferromagnetic materials to be discussed next the presence of diamagnetism is completely masked by the more powerful effects of magnetic attraction.

Paramagnetism

If we compare a rod consisting of atoms whose magnetic moments are randomly oriented (Fig. 11-26A) to one whose atoms are diamagnetic and lack an effective magnetic moment (Fig. 11-26B), we find that neither rod possesses a net magnetic field. Assume each is delicately suspended in a magnetic field whose strength is greater near the magnet's poles, regions of increased field strength being indicated in Figures 11-26B,C by a closer spacing of the lines of force. The diamagnetic rod is feebly repelled toward the dashed position (Fig. 11-26B), whereas the other rod is weakly attracted (Fig. 11-26C) toward the poles. This weak attraction results from this rod's *paramagnetism,* that is, the ability of its randomly oriented magnetic moments to develop a slight degree of parallelism while in a magnetic field. This preferred orientation, though slight, causes the rod itself to develop a magnetic moment $\mathbf{m}_p$ and thus be attracted by the magnet's poles. The strength of this paramagnetically induced magnetic moment $\mathbf{m}_p$ could theoretically be determined from the minimum force necessary to prevent the rod in Figure 11-26C from moving toward the dashed position. A rod twice this volume would require twice this force, that is, $\mathbf{m}_p$ would be doubled

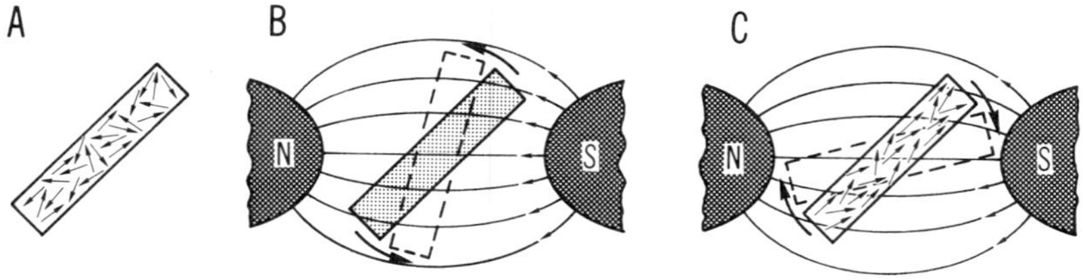

FIGURE 11–26

(A) A rod consisting of atoms, each with an effective magnetic moment, but with no significant orientation of these magnetic moments (small arrows) from one atom to the next.
(B) The effect of a magnetic field on a rod consisting of atoms wherein all electrons are paired and oppositely spinning so that each atom lacks an effective magnetic moment. Because of the diamagnetism of its individual atoms as discussed in Figure 11-25B, the rod would tend to rotate toward the dashed position; that is, it is repelled by the magnetic poles of a magnet toward regions where the magnetic lines of force are less dense.
(C) The rod in (A), while in a magnetic field, develops a degree of alignment of the effective magnetic moments of *its* component atoms and hence is attracted toward the magnetic poles, this phenomenon being called *paramagnetism*.

in value. To eliminate this volume effect, a solid's paramagnetically induced moment $\mathbf{m}_p$ is usually divided by its volume V. This ratio $\mathbf{m}_p/V$ is called the solid's *magnetization* **M**. Magnetization **M** is thus a measure of the number (and degree of mutual parallelism) of the magnetic moments aligned in a unit volume of a solid while in a magnetic field. Magnetization **M** also represents the strength of the magnetic field induced in the solid because of this alignment. Pierre Curie in 1885 determined empirically that

$$\mathbf{M} = C\frac{\mathbf{B}}{T} \qquad \text{(Eq. 11-29)}$$

where C is a constant, T is the absolute temperature (°K) of the solid, and **B** is the intensity of the magnetic field in which it was placed. The inverse relation between **M** and T indicates that thermal vibrations tend to misalign magnetic moments.

Let us consider a long coil of wire or solenoid (Fig. 11-27A) whose current can be varied to change the strength of **H**, the magnetic field at its center. A rod-shaped crystal inserted into this solenoid (Fig. 11-27B) develops a magnetization **M**. We let vectors **H** and **B** represent the magnetic field strength at the solenoid's center *before* and *after* insertion of the crystal. Then

$$\mathbf{B} = \mu_0\mathbf{H} + \mathbf{M} \qquad \text{(Eq. 11-30)}$$

where μ_0 is the primary magnetic constant or permeability of a vacuum and equals $4\pi 10^{-7}$. For paramagnetic materials the induced field **M** is directly proportional to the applied field **H** (for low values of **H**); in other words,

$$\mathbf{M} = \mu_0 k \mathbf{H} \qquad \text{(Eq. 11-31)}$$

where k is a constant called the *magnetic susceptibility* or, more specifically, the volume susceptibility. For paramagnetic substances k is a positive number usually about 10^{-5} in value, but larger for materials that are more easily magnetized. For diamagnetic materials k is negative with values of about 10^{-6}; insertion of a diamagnetic rod in the solenoid thus weakens the magnetic field within the volume occupied by the rod. In contrast, insertion of a paramagnetic rod would strengthen this field.

Combining Equations (11-30) and (11-31), there results

$$\mathbf{B} = (1 + k)\mu_0 \mathbf{H} \qquad \text{(Eq. 11-32)}$$

Customarily the quantity $(1 + k)\mu_0$ is given the symbol μ and called the *permeability* of the material. The quantity $(1 + k)$ is the factor by which the material's permeability to lines of magnetic force exceeds (or is less than) μ_0, the permeability defined for a vacuum. Thus for paramagnetic or diamagnetic materials

$$\mathbf{B} = \mu \mathbf{H} \qquad \text{(Eq. 11-33)}$$

and **B** and **H** are linearly related.

In anisotropic crystals the magnetic susceptibility k varies with direction. A few crystals may be paramagnetic along one direction but diamagnetic along another; that is, a rod inserted in the solenoid would give a positive value for k if cut along the one crystallographic direction but a negative value if cut along the other. In anisotropic crystals the applied field, vector **H**, and the crystal's response, its magnetization as represented by vector **M**, are generally not parallel and are related by the susceptibility tensor k_{ij} as follows (Nye, 1957, p. 54):

$$\mathbf{M}_i = \mu_0 k_{ij} \mathbf{H}_j \qquad \text{(Eq. 11-34)}$$

This is a short-hand way of summarizing the tensor relationships. Expanded by

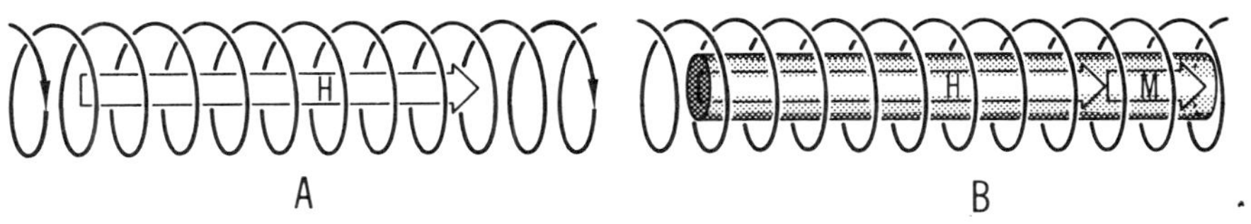

FIGURE 11–27

(A) A current flow in a coil of wire or solenoid develops a magnetic field whose moment or direction equals vector **H**. Note that the solenoid can be regarded as a stacked sequence of loops. For illustrative purposes the solenoid was not tightly wound.
(B) Insertion of a rod of paramagnetic material into the solenoid. Because of partial alighment of the magnetic moments of its component atoms, the rod develops a magnetic field of its own **M**. Thus the magnetic field **B** existent at the solenoid's center equals the vector sum of **H** and **M** (as stated in Eq. 11-30).

letting i and j take values from 1 to 3, Equation 11-34 becomes

$$\mathbf{M}_1 = \mu_0(k_{11}\mathbf{H}_1 + k_{12}\mathbf{H}_2 + k_{13}\mathbf{H}_3)$$
$$\mathbf{M}_2 = \mu_0(k_{21}\mathbf{H}_1 + k_{22}\mathbf{H}_2 + k_{23}\mathbf{H}_3) \qquad \text{(Eqs. 11-34}')$$
$$\mathbf{M}_3 = \mu_0(k_{31}\mathbf{H}_1 + k_{23}\mathbf{H}_2 + k_{33}\mathbf{H}_3)$$

Thus $\mathbf{M}_1$, $\mathbf{M}_2$, and $\mathbf{M}_3$ are the components along three mutually perpendicular axes (x_1, x_2, and x_3 in the crystal) of the vector **M**, which, in turn, represents the magnetization resultant from application of a magnetic field, vector **H**, whose components along x_1, x_2, and x_3 are $\mathbf{H}_1$, $\mathbf{H}_2$, and $\mathbf{H}_3$. The susceptibility k_{12} represents the crystal's magnetic susceptibility along axis x_1 in response to an applied field $\mathbf{H}_2$ along x_2, and so on.

Ferromagnetism

If certain materials are placed in an applied magnetic field **H**, they develop magnetic fields that may exceed by a billionfold those of paramagnetic materials placed in this same field. Because iron is prominent among them, they are known as *ferromagnetics.* As the applied field **H** brings the dipole moments of neighboring atoms into mutual parallelism in a ferromagnetic, an effect known as *exchange coupling* "locks" them together to preserve this parallelism even after the field **H** is removed. Consequently, ferromagnetics can become permanent magnets. The preference of neighbors to have parallel spins and thus magnetic moments such as ↑ ↑ or →→ results not from magnetic interactions between neighbors but rather from a much stronger effect whose cause lies in the quantum mechanics of the participating atoms. If we heat a ferromagnetic material into which permanent magnetism has been induced, the increasing temperatures may uncouple increasing numbers of the exchange couples. When a temperature known as the *Curie temperature* T_c is attained, the last of the exchange coupling disappears. Above T_c, therefore, all ferromagnetic materials respond paramagnetically. Crystals of some selected elements and alloys have Curie temperatures as follows: Fe (770°C); Co (1127°C); Ni (358°C); Gd (16°C); Dy (−168°C); MnBi (357°C); MnSb (314°C); Thus crystals of dysprosium are paramagnetic at room temperatures and only become ferromagnetic below −168°C (105°K).

Magnetization **M** can be more easily induced in a ferromagnetic crystal if the field **H** is applied along certain crystallographic directions (Fig. 11-28). In body-centered cubic iron, for example, the iron atoms "prefer" their magnetic moments to be directed along ⟨100⟩, that is, along [100], [$\bar{1}$00], [010], [0$\bar{1}$0], [001], or [00$\bar{1}$]. Next preferred is ⟨110⟩; least is ⟨111⟩. A single, unmagnetized iron crystal may consist of microscopic regions called *domains.* One magnetic domain in the crystal might consist of billions of atoms, all with magnetic moments pointing along [100]. In another domain they might all point along [010]; in still another, along [$\bar{1}$00]; in another, along [001]; and so on. Less frequent would be domains with alignments along ⟨110⟩. The crystal thus displays no magnetism because the different magnetic domains cancel each other. *Bloch walls* are transitional zones, often hundreds of unit translations wide, between

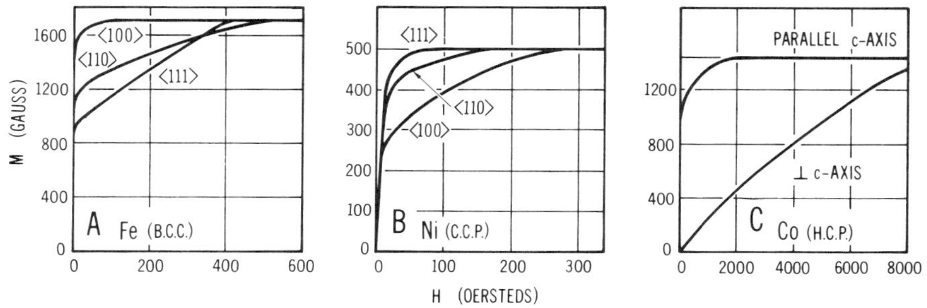

FIGURE 11–28

Development of magnetization **M** in single crystals if the magnetic field **H** is aligned along different directions in the crystal.
(A) For body-centered cubic iron, **M** develops most readily if **H** is applied along one of the ⟨100⟩ directions, whereas ⟨110⟩ and ⟨111⟩ are relatively hard directions.
(B) For cubic close-packed nickel crystals, the ⟨111⟩ direction is the easy direction of magnetization, whereas ⟨100⟩ is hard.
(C) In hexagonally close-packed crystals of cobalt, the easy direction is along the *c* axis, those normal to *c* being hard directions. (From Kittel after Honda and Kaya.)

domains. Going through a Bloch wall from a [100] to a [010] domain, for example, the angle between [100] and the magnetic moment for the first atom encountered might be 0.3°; for the second, 0.6°; for the third, 0.9°; . . . and 89.7° for the 299th atom whose moment is almost parallel to [010]. Actually, fractions of a degree much smaller than 0.3° might be involved.

Placed in a strong magnetic field, those domains of a crystal that have magnetic moments most nearly parallel to the applied field will grow by accretion at their edges at the expense of neighboring domains. The crystal thus becomes magnetized. Presumably the magnetic moments of Bloch wall atoms bordering the growing domain are slightly rotated to become parallel to those in the domain, thereby becoming a part of the domain. This slight rotation proceeds so as to rotate, by this same amount, Bloch-wall atoms successively farther away from the growing domain. Ultimately the Bloch-wall boundary with a neighboring domain is affected; indeed, an atom *in*, but at the edge of, this neighboring domain is similarly rotated to become part of the Bloch wall. The Bloch wall thus moves outward from the growing domain and toward the diminishing domain. Such migration of domain boundaries is frequently observable microscopically.

Hysteresis Loops. For paramagnetic crystals, **M** was both small and directly proportional to **H**. Hence Equation (11-30) became Equation (11-33), a linear relationship between **B** and **H**. Thus the plot of **B** against **H** for a paramagnetic crystal, the so-called *magnetization curve,* is a straight line with slope μ, particularly for low values of **H** (Fig. 11-29A). In contrast the magnetization curve for a ferromagnetic crystal subjected to a magnetic field of increasing intensity **H** is nonlinear (Fig. 11-29B). The circular insets represent graphic

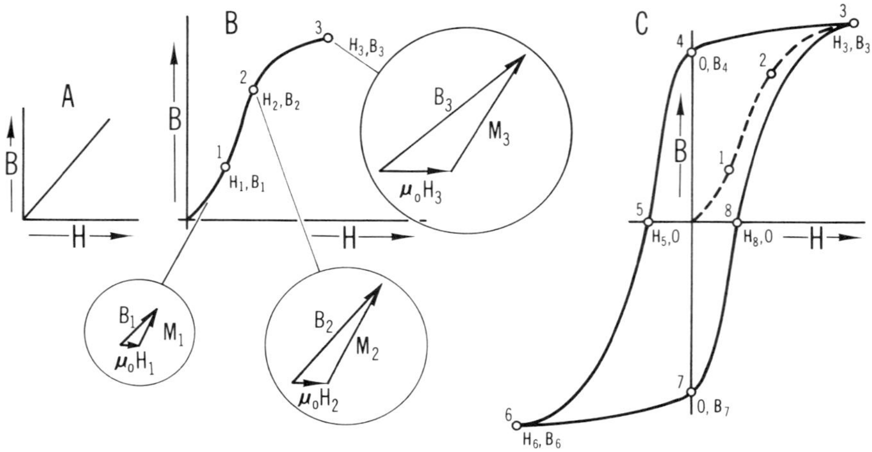

FIGURE 11-29

(A) Relation between the applied magnetic field **H** and the field **B**, which is the resultant of both **H** and **M**, **M** being the magnetization developed in a paramagnetic crystal subjected to field **H**. At low values of **H**, the relationship between **B** and **H** is linear for paramagnetic materials as shown here.
(B) The comparable magnetization curve for a ferromagnetic material is characteristically nonlinear. Initially, **B** increases very markedly as **H** is increased only slightly, this occurring because the material's magnetization **M** at first increases markedly in response to small increases in **H**. However, as more and more magnetic domains become aligned, a point of diminishing returns is reached so that further increase of **H** can produce little increase of **M** (and thus of **B**). This can be seen by comparing the circular insets that illustrate the relationships between vectors $\mu_0\mathbf{H}$, **M**, and **B** at points 1, 2, and 3. Ultimately, the curve to the right of point 3 becomes a horizontal line because all domains have become aligned so that further increase of **H** is not accompanied by any increase in **M** (nor thus in **B**).
(C) Hysteresis loop for a ferromagnetic material showing how, as **H** is decreased from $\mathbf{H}_3$ to zero, the value of **B** does not similarly return to zero but to $\mathbf{B}_4$. This nonreversibility of the dashed magnetization curve 1, 2, 3 is called *hysteresis*. To return **B** to zero it is necessary to reverse the field **H** until $\mathbf{H}_5$ is attained. $\mathbf{B}_4$ is called the *remanence*; $\mathbf{H}_5$ the reverse field necessary to undo it entirely is called the *coercive force*. Continued increase of **H** in this reverse direction traces out a curve to point 6. Now reversal of **H** causes the curve 6, 7, 8, 3 to result to complete the loop.

depictions of Equation (11-30) at three points on the curve. Magnetization vector $\mathbf{M}_2$ greatly exceeds $\mathbf{M}_1$ as more and more domains align. At point 3 most domains are now aligned; hence further increases in **H** produce little increase in the magnetization **M**. After attaining point 3, if the current in the electromagnet that supplies field **H** is gradually decreased to zero, the curve between points 3 and 4 in Figure 11-29C results. Even though **H** at point 4 equals zero, the crystal retains a residual magnetism $\mathbf{B}_4$, called the *remanence*, because some domains remain partially aligned. The crystal, now a permanent magnet, will remain so until its thermal vibrations gradually destroy the domain alignments. Heating to the Curie temperature will do this almost immediately. Another way to destroy the alignment is to reverse the current (and thus field **H**) for the electromagnet, the curve going from point 4 to 5. At point 5 all magnetism is lost. Continued increase of the current and thus of **H** in this direction produces the curve between points 5 and 6. Now, returning the current and field **H** to the original direction, the curve containing points 6, 7, 8, and 3 results.

As **H** was lowered the curve containing points 1, 2, and 3 in Figure 11-29B did not retrace itself but proceeded from 3 to 4 (Fig. 11-29C). Such irreversibility is called *hysteresis* and the solid curves in Figure 11-29C constitute a *hysteresis loop*, a characteristic feature of all ferromagnetic materials.

Antiferromagnetism

In antiferromagnetic materials, exchange coupling causes neighbors to adopt antiparallel spins, that is, ↑ ↓ or →←. Hence the macroscopic magnetic moment for the crystal equals zero. Above a temperature known either as the Néel temperature T_N or, less frequently, as the antiferromagnetic Curie temperature, antiparallel exchange coupling disappears and the material becomes paramagnetic. In most cases ferromagnetic and antiferromagnetic substances consist of elements whose atomic numbers lie between 23 and 29, vanadium, chromium, manganese, iron, cobalt, nickel, and copper, or to alloys or compounds of these. In these atoms the $3d$ and $4s$ electrons are close in energy; this perhaps permits exchange coupling to occur. Two of the many types of antiferromagnetic ordering observed in the face-centered cubic distribution of metal ions in NaCl-type structures are shown in Figure 11-30.

Ferrimagnetism

Ferrimagnetism is in effect a compromise between antiferromagnetism and ferromagnetism in which the antiparallel magnetic moments do not completely compensate because those in one direction are either more numerous or stronger

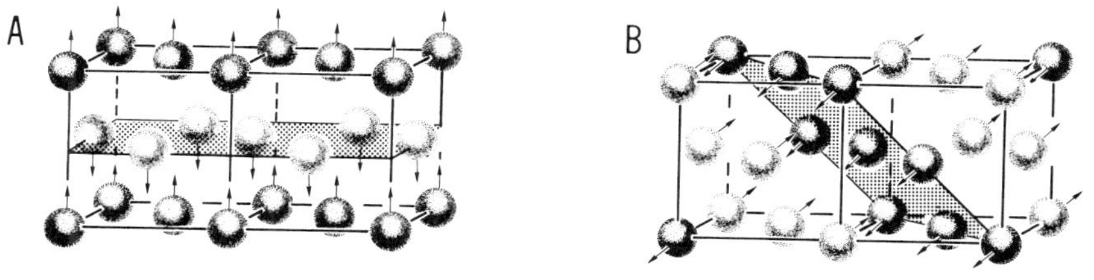

FIGURE 11–30

Two of the numerous types of antiferromagnetic ordering observed in face-centered cubic structures.
(A) Alternate (001) planes of [001] and [$00\bar{1}$] moments.
(B) Alternate ($\bar{1}11$) planes of [$1\bar{1}\bar{1}$] and [$\bar{1}11$] moments. Atoms with opposed magnetic moments have been differently shaded for illustrative purposes.
Another type of antiferromagnetic ordering (not shown) consists of alternate ($\bar{1}11$) planes of [001] and [$00\bar{1}$] moments. It can be visualized by imagining the atoms in (B) to have magnetic moments parallel to those for the comparably shaded atoms in (A).

than those opposed. Hence the crystal possesses a magnetic moment and, consequently, ferromagnetic properties. Ferrimagnetism is prominent in the ferrite compounds. These possess inverse spinel structures (Fig. 9-27) in which all the tetrahedral and half the normally occupied octahedral sites are filled by Fe^{3+}, the remaining octahedral sites being filled by divalent metal ions Me^{2+}. Using superscripts to indicate tetrahedral and octahedral coordinations for cations as well as cubic closest packing of the oxygens, the ferrite formula reads

$$[Fe^{3+}\uparrow]^t[Fe^{3+}\downarrow Me^{2+}\uparrow]^o O_4^c$$

The arrows indicate that the magnetic moments of Fe^{3+} in tetrahedral sites oppose those for Fe^{3+} in octahedral sites, mutual nullification thereby resulting. The possible Me^{2+} occupants have magnetic moments approximately proportional to their number of unpaired electrons (parenthesized) as follows: Mn^{2+} (5), Fe^{2+} (4), Co^{2+} (3), Ni^{2+} (2), Cu^{2+} (1), Zn^{2+} (0), Cd^{2+} (0), Mg^{2+} (0). If Me^{2+} represents Mn^{2+}, as in manganese ferrite, ferromagnetism is very high. On the other hand, zinc, cadmium, or magnesium ferrites lack ferromagnetic properties.

Effect of shape and size

Elongate particles tend to develop magnetic moments parallel to their lengths even if these lengths are not parallel to the external field **H**. Also, below a certain diameter, (ca.) 10^{-3} cm for the iron ferrite Fe_3O_4 better known as magnetite, grains of ferromagnetic or ferrimagnetic crystals become too small for Bloch walls to exist and hence are *monodomains*. Magnetization of aggregates of such small crystals requires the applied field **H** to be made intense enough to reverse or rotate the magnetic moments in whole domains at a time, the gradual atom-by-atom shifts afforded by the presence of Bloch walls being no longer possible. On the other hand once entire monodomains are so rotated, fewer lose this orientation in response to thermal vibrations after the external field **H** is removed. In other words finely crystalline materials, once magnetized, become permanent magnets.

Magnetic symmetry

Ferromagnetic, ferrimagnetic, and antiferromagnetic crystals possess ordered arrays of spin magnetic moments. The directions of these moments and the nature of their ordering in the crystal structure can be determined by neutron diffraction studies. Such knowledge requires extension of the classic concepts of symmetry operations to include *complementary operations*. To understand these, we assume two identical atoms, x and y in Figure 11-31A, to be related by a mirror plane m in a crystal structure. Later neutron diffraction studies might reveal the electron spins and resultant magnetic moments for these two atoms to be related as for atoms x' and y' in Figure 11-31B or for atoms x'' and y''. In this case the plane separating the atoms remains a normal symmetry plane m. Relating s to s' and p to p' by its operation, plane m reverses electron-spin direction sp in atom x' to $s'p'$ in atom y'. If the left-hand rule is applied, such

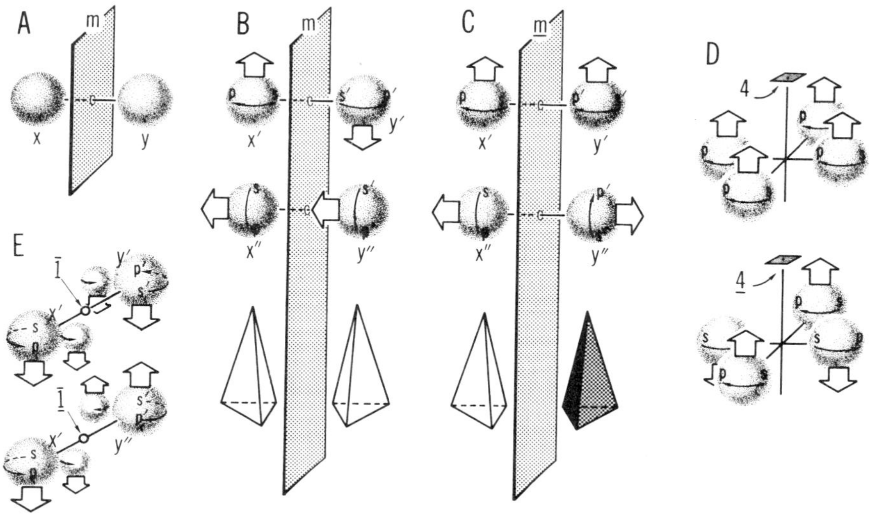

FIGURE 11–31

(A) Two identical atoms x and y related by a plane of symmetry in a crystal's structure.
(B) Neutron-diffraction studies might reveal their effective magnetic moments (hollow arrows) and the uncompensated spins sp which produced these moments to be related, among other possibilities, as for x' and y' or as for x'' and y''. Here the spins observe the standard mirror-plane operation; that is, s reflects to s' and p to p'. However, the magnetic moments, whose directions are determinable by the left-hand rule, bear a visually anomalous relation to the mirror plane.
(C) Alternatively, the spins and moments of atoms x and y might be related as between x' and y' or as between x'' and y'' as shown here. Now what seemed to be a symmetry plane in (A) is revealed to be a *complementary mirror plane*. This new operation, which is sometimes needed to describe the symmetry between atoms if their magnetic spins are also considered, combines reflection across the mirror plane with a spin reversal. In (C), for example, the spins $s'p'$ are reversed from what they would be if m were an ordinary mirror plane. Such complementary magnetic symmetry operations are usually underlined and represent "black-white" symmetry. To illustrate we consider the two geometric figures in (B) to be hollow and made of cardboard that is white on the form's outside but black on its inside. As shown in (C) the complementary mirror plane m not only reflects one figure into the other but turns it inside-out to reverse its color.
(D) Four atoms with spins and moments that obey a true 4-fold axis (upper); those below obey a $\underline{4}$ axis. In the case of the 4-fold axis, both the spins and the moments visually conform to the 4-fold symmetry; for the $\underline{4}$ fold they do not.
(E) At the lower left shows some atoms that are related by a $\bar{1}$ axis as well as some related by a $\underline{1}$ axis. Note the seemingly anomalous distribution of the magnetic moments in respect to the $\bar{1}$ axis. This is always true for ordinary operations that involve a change of hand (or for the complementary operations of those that do not, for example, $\underline{4}$.)

reversal results in reversal of the magnetic moment for atom y'. Note the anomaly that although the spin directions visually conform to the symmetry plane, the magnetic moments make it appear as if plane m were not a true symmetry plane.

But what if the spins of atoms x and y in Figure 11-31A proved to be related as for atoms x' and y' (or x'' and y'') in Figure 11-31C? We then need to define a new symmetry operation to relate x' to y' (or x'' to y''), namely, reflection across the plane plus *reversal of the spin*. Except for this reversal of spin, atom y' in Figure 11-31C would be identical to atom y' in Figure 11-31B. Reflection plus reversal is defined as a *complementary mirror plane* operation and symbolized as $\underline{m}$. Complementary symmetry operations are thus a normal symmetry operation coupled with reversal of spin. The symbols for such complementary operations are underscored to distinguish them from the corresponding classic symmetry operations. To illustrate further, the magnetic moments and spins are compared for four atoms related by a 4-fold axis and a $\underline{4}$-fold axis in Figure 11-31D. We note that each individual operation of the $\underline{4}$-fold axis includes reversal of spin. In Figure 11-31E two atoms related by a $\bar{1}$ axis and a $\underline{\bar{1}}$ axis are compared. Such complementary operations are also called operations of antisymmetry or black-white symmetry. A. V. Shubnikov, the pioneer in this field, illustrates the concepts by means of three-dimensional shapes that can be regarded as made from paper that is black on one side and white on the other. Figure 11-31B shows two such shapes, folded with the white on their outsides and related by mirror plane m. In Figure 11-31C complementary mirror plane $\underline{m}$ reflects one shape into the other *and* produces a reversal, which is schematically depicted by turning it inside out so that it becomes black on the outside and white inside.

The inconsistent distribution of the magnetic moment vectors in respect to classical symmetry plane m (Fig. 11-31B) has already been discussed. Such inconsistencies in distribution of the moment vectors occur for all classic symmetry operations that involve a change of hand, for example, a $\bar{1}$ axis or m plane (Fig. 11-31B, E). However, for complementary change-of-hand operations, for example $\underline{\bar{1}}$ or $\underline{m}$ in Figure 11-31, the magnetic moment vectors appear to obey the classic symmetry operation whereas the spins do not.

Substitution of complementary symmetry operations for some of their classical counterparts in the 32 point groups permits the development of 58 additional point groups. Considering the spin directions of atoms, a crystal's symmetry must then conform to one of the 90 magnetic point groups (Table 11-2). Similar substitution of complementary operations into the 230 space groups expands them into the 1651 Shubnikov groups.

If unmagnetized, a body-centered cubic crystal of iron will consist of intermixed domains whose magnetic moments variously point along (1) any of the six ⟨100⟩ directions, that is, [100], [$\bar{1}$00], [001], and so on, (2) any of the twelve ⟨110⟩ directions, or (3) any of the eight ⟨111⟩ directions. Such randomness of magnetic moment results in no overall moments averaged across the structure; consequently, the crystal (Fig. 11-32A) has point-group symmetry $4/m\,\bar{3}\,2/m$. Strongly magnetized so that domains with moments parallel to [001] grow at the

TABLE 11–2

The 90 Magnetic Point Groups[a]

	Triclinic
(1)	1
(2)	$\bar{1}$
(3)	$\underline{\bar{1}}$
	Monoclinic
(4)	m
(5)	$\underline{m}$
(6)	2
(7)	$\underline{2}$
(8)	$2/m$
(9)	$\underline{2}/m$[b]
(10)	$2/\underline{m}$[b]
(11)	$\underline{2}/\underline{m}$
	Orthorhombic
(12)	$mm2$
(13)	$m\underline{m}\underline{2}$
(14)	$\underline{m}\underline{m}2$
(15)	222
(16)	$\underline{2}\underline{2}2$
(17)	$2/m\,2/m\,2/m$
(18)	$\underline{2}/m\,\underline{2}/m\,2/\underline{m}$[b]
(19)	$2/\underline{m}\,2/\underline{m}\,2/\underline{m}$[b]
(20)	$\underline{2}/\underline{m}\,\underline{2}/\underline{m}\,2/m$
	Tetragonal
(21)	4
(22)	$\underline{4}$
(23)	$\bar{4}$
(24)	$\underline{\bar{4}}$
(25)	$4/m$
(26)	$\underline{4}/\underline{m}$[b]
(27)	$4/\underline{m}$[b]
(28)	$\underline{4}/m$
(29)	$4mm$
(30)	$4\underline{m}\underline{m}$
(31)	$\underline{4}m\underline{m}$
(32)	$\bar{4}2m$
(33)	$\underline{\bar{4}}2\underline{m}$
(34)	$\underline{\bar{4}}\underline{2}m$
(35)	$\bar{4}\underline{2}\underline{m}$
(36)	422
(37)	$4\underline{2}\underline{2}$
(38)	$\underline{4}2\underline{2}$
(39)	$4/m\,2/m\,2/m$
(40)	$\underline{4}/\underline{m}\,\underline{2}/m\,2/\underline{m}$[b]
(41)	$4/\underline{m}\,\underline{2}/m\,\underline{2}/m$[b]
(42)	$4/m\,\underline{2}/\underline{m}\,\underline{2}/\underline{m}$
(43)	$4/\underline{m}\,2/\underline{m}\,2/\underline{m}$[b]
(44)	$\underline{4}/m\,2/m\,\underline{2}/\underline{m}$
	Trigonal
(45)	3
(46)	$\bar{3}$
(47)	$\underline{\bar{3}}$[b]
(48)	$3m$
(49)	$3\underline{m}$
(50)	32
(51)	$3\underline{2}$
(52)	$\bar{3}\,2/m$
(53)	$\bar{3}\,\underline{2}/\underline{m}$
(54)	$\underline{\bar{3}}\,\underline{2}/m$[b]
(55)	$\underline{\bar{3}}\,2/\underline{m}$[b]
	Hexagonal
(56)	$\bar{6}$
(57)	$\underline{\bar{6}}$
(58)	6
(59)	$\underline{6}$
(60)	$6/m$
(61)	$6/\underline{m}$[b]
(62)	$\underline{6}/m$[b]
(63)	$\underline{6}/\underline{m}$
(64)	$\bar{6}m2$
(65)	$\bar{6}\underline{m}\underline{2}$
(66)	$\underline{\bar{6}}m\underline{2}$
(67)	$\underline{\bar{6}}\underline{m}2$
(68)	$6mm$
(69)	$6\underline{m}\underline{m}$
(70)	$\underline{6}\underline{m}m$
(71)	622
(72)	$6\underline{2}\underline{2}$
(73)	$\underline{6}2\underline{2}$
(74)	$6/m\,2/m\,2/m$
(75)	$6/\underline{m}\,\underline{2}/m\,\underline{2}/m$[b]
(76)	$6/m\,\underline{2}/\underline{m}\,\underline{2}/\underline{m}$
(77)	$6/\underline{m}\,2/\underline{m}\,2/\underline{m}$[b]
(78)	$\underline{6}/m\,\underline{2}/m\,2/\underline{m}$[b]
(79)	$\underline{\bar{6}}/\underline{m}\,2/m\,\underline{2}/\underline{m}$
	Cubic
(80)	23
(81)	$2/m\,\bar{3}$
(82)	$2/\underline{m}\,\underline{\bar{3}}$[b]
(83)	$\bar{4}3m$
(84)	$\underline{\bar{4}}3\underline{m}$
(85)	432
(86)	$\underline{4}3\underline{2}$
(87)	$4/m\,\bar{3}\,2/m$
(88)	$4/\underline{m}\,\bar{3}\,2/\underline{m}$[b]
(89)	$\underline{4}/m\,\bar{3}\,2/\underline{m}$
(90)	$\underline{4}/\underline{m}\,\bar{3}\,2/m$[b]

[a] Each operation on an atom of those symmetry elements underlined is accompanied by a reversal of the spin of the atom.

[b] The center of symmetry present is $\underline{\bar{1}}$ rather than $\bar{1}$.

expense and to the exclusion of all others, the crystal adopts the symmetry $4/m\,\underline{2}/\underline{m}\,\underline{2}/\underline{m}$ (Fig. 11-32B). Magnetized along a $\langle 110 \rangle$ direction it adopts $\underline{2}/\underline{m}\,\underline{2}/\underline{m}\,2/m$; if along a $\langle 111 \rangle$ direction, $\underline{\bar{3}}\,2/m$.

In ferrimagnetic and antiferromagnetic crystals, reversals of magnetic moments occur; consequently, the edges of the magnetic unit cell will be multiples of the chemical unit cell. To illustrate by analogy we consider a row of white marbles, each touching its neighbors on either side. The repeat distance along this row then equals d, the center-to-center distance of the marbles. We now assume that the magnetic moment of every second marble points down instead

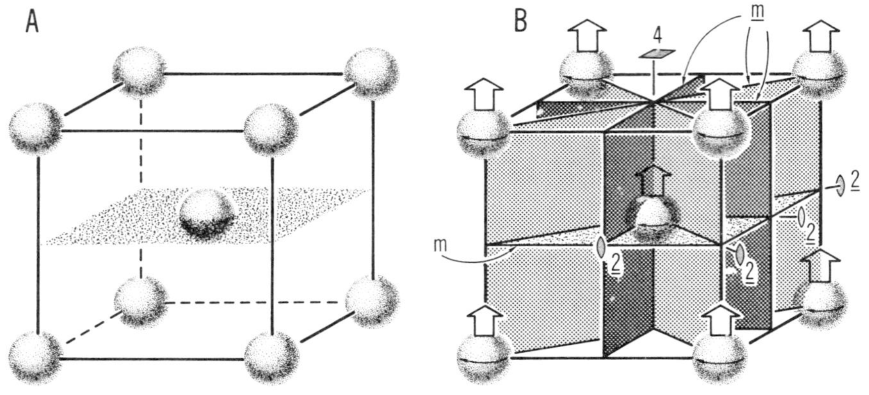

FIGURE 11–32

(A) One unit cell of body-centered cubic iron. Throughout the structure the magnetic moments of the iron atoms may be variously oriented parallel to ⟨100⟩, ⟨110⟩, or ⟨111⟩. No particular direction is preferred; thus the structure is magnetically disordered and its resultant magnetic moment is zero.
(B) Subjected to a strong magnetic field **H** along [001], the domains with moments along [001] would grow at the expense of the others. This magnetic ordering causes the magnetic crystal class to be $4/m\ \underline{2}/\underline{m}\ \underline{2}/\underline{m}$.

of up, or replace every second marble with a black one. The magnetic repeat distance along the row then equals $2d$.

Electric Behavior of Dielectric Crystals

Crystals that conduct electricity so poorly as to be called *insulators* or *dielectrics* consist of a set of negative charges, the electrons, interspersed among a set of positive charges, the atomic nuclei. Both sets occupy relatively fixed sites in space. Although dispersed in space, such sets of negative or positive charges act externally almost as if concentrated at a point, called here the *centroid.* Crystals, atoms, or ions whose positive and negative centroids coincide are electrically nonpolar. Those whose centroids do not coincide are electrically polarized so as to constitute *electric dipoles,* the negative centroid marking their negative pole, the positive centroid their positive pole. The strength and direction for such a dipole are summarized by a vector **p**, which is called the *electric dipole moment.* By definition **p** points from the negative to the positive centroid. Accepting the approximation that a set of like charges can be treated as if the entire charge q were located at the centroid, we define the magnitude of **p** to equal qd, where d is the distance between the positive and negative centroids and q is the charge for each set.

Usually a single atom (or an ion consisting of a solitary charged atom) will consist of a cloud of electrons so distributed about the nucleus that the atom's positive and negative centroids coincide. The atom or ion is thus nonpolar (Fig. 11-33A). Subjected to an electric field **E**, this atom's nucleus is attracted upward along **E** in Figure 11-33B, whereas its electron cloud is distorted downward. Such separation of the positive and negative centroids for an atom or ion is called *electronic polarization*, each atom developing a dipole moment **p** whose value equals

$$\mathbf{p} = \alpha \mathbf{E} \qquad \text{(Eq. 11-35)}$$

where the constant α is a property of the atom called the *atomic polarizability* and **E** is the strength of the applied field. The more easily polarized atoms have higher values of α. Purcell (1965, p. 311) cites values of α in units of 10^{-24} cm^3 for selected elements as follows: hydrogen (0.66), helium (0.21), lithium (12), beryllium (9.3), carbon (1.5), neon (0.4), sodium (27), argon (1.6), potassium (34). Atomic polarizabilities are seen to be low for noble gases, for example,

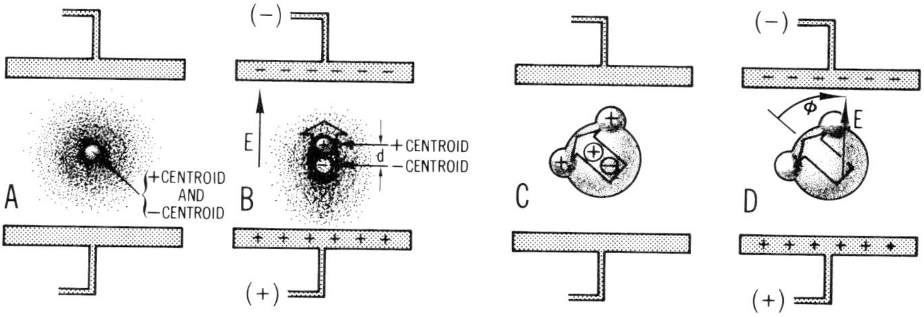

FIGURE 11–33

(A) Schematic cross section through an atom for which the positive centroid, centered in its nucleus, coincides with the negative centroid or "center of gravity" for the electron cloud (stippled) surrounding this nucleus. Because the two centroids coincide, the atom is not an electric dipole.
(B) This same atom placed in a strong electric field **E** undergoes *electronic polarization,* its cloud of negative charge being attracted downward and its positive nucleus upward to separate the positive and negative centroids by a distance *d*. The atom develops a positive pole, that is, an excess of positive charge relative to negative charge, at its upper end and a corresponding negative pole at is lower end. By convention, the electric dipole moment for the atom (large hollow arrow) always points from the negative to the positive centroid. It is proportional in magnitude to the distance between centroids and to the charge effectively centered at each.
(C) The H_2O molecule shown here is a natural dipole. Its positive centroid (encircled) is displaced from the oxygen atom's center because of the two H^+ protons (overemphasized bumps). The negative centroid would be close to the center of the oxygen atom. The dipole moment for H_2O is indicated by the hollow arrow which again follows the convention by pointing from the negative (and toward the positive) centroid.
(D) Placed in an electric field **E**, a polar molecule such as H_2O tends to rotate so as to reduce the value of ϕ, the angle between the molecule's dipole moment, and **E**. The twisting force τ equals $\mathbf{p} \times \mathbf{E}$, where **p** is the dipole moment of the molecule.

helium, neon, and argon, but large for alkali atoms such as lithium, sodium, and potassium. Evidently, loosely bound valence electrons promote easy polarizability. Also, as the sequence helium, neon, argon, or the sequence lithium, sodium, potassium each indicates, polarizability increases as the number of electron shells increases. Zhdanov (1965, p. 252) cites values for the polarizabilities of individual atoms in covalent compounds and of ions in ionic compounds.

Nonpolar molecules also become polarized in an electric field. Their induced dipole moment is calculable from Equation (11-35) if one substitutes α_M, the *molecular polarizability*, for α. Interestingly, for a molecule consisting of three different atoms, 1, 2, and 3, α_M does not equal the sum of their atomic polarizabilities, α_1, α_2, and α_3; the bonds between atoms 1, 2, and 3 evidently alter their individual polarizabilities. Polar molecules such as H_2O (Fig. 11-33C), if subjected to an electric field **E**, tend to rotate as shown (Fig. 11-33D). The calculation of torque $\boldsymbol{\tau}$ for such an electric dipole

$$\boldsymbol{\tau} = \mathbf{p} \times \mathbf{E} \qquad \text{(Eq. 11-36)}$$

is analogous to that for a magnetic dipole. For example, Figure 11-24B applies if the labels **p** and **E** are substituted for **m** and **B**, respectively.

Ferroelectricity

Subjected to an electric field **E**, a molecular crystal may develop polarity because of (1) electronic polarization of individual atoms and (2) rotation of any polarized molecules, or parts thereof, which it contains. For example, these molecules, like the individual H_2O molecule in Figure 11-33C, rotate (to the extent the structure permits) until their dipole moments align as nearly parallel to **E** as possible. This phenomenon, called *orientational polarization*, usually causes a larger separation of the crystal's positive and negative centroids than does the ever-present electronic polarization of the individual atoms.

Ionic crystals placed in an electric field **E** may develop polarity because of electronic polarization of individual ions and, more significantly, because of *ionic polarization*. This latter frequently occurs if a cation is undersized in respect to the structural site it occupies. To illustrate, in Figure 11-34A (circular inset) the cations may occupy any of four positions (12, 3, 6, or 9 o'clock) relative to the cavity center *c*. These four positions represent minima or "potential wells" in the curves which relate the cation's changing contribution ΔE to the crystal's potential energy as the cation shifts from one position to another within the cavity (Fig. 11-34B). In the absence of an electric field, domains in which 12 o'clock sites are occupied will be counterbalanced by domains of 3, 6, or 9 o'clock types. Hence although each domain possesses polarity, the crystal as a whole does not. Neighboring domains would be separated by thin walls comparable to but usually less than one hundredth as thick as the Bloch walls between magnetic domains. In the domain wall between a 12 and a 3 o'clock domain (Fig. 11-34A), the cations would occupy positions intermediate between the 12 and 3 o'clock sites. This particular wall would be called a "90° wall" because it separates two

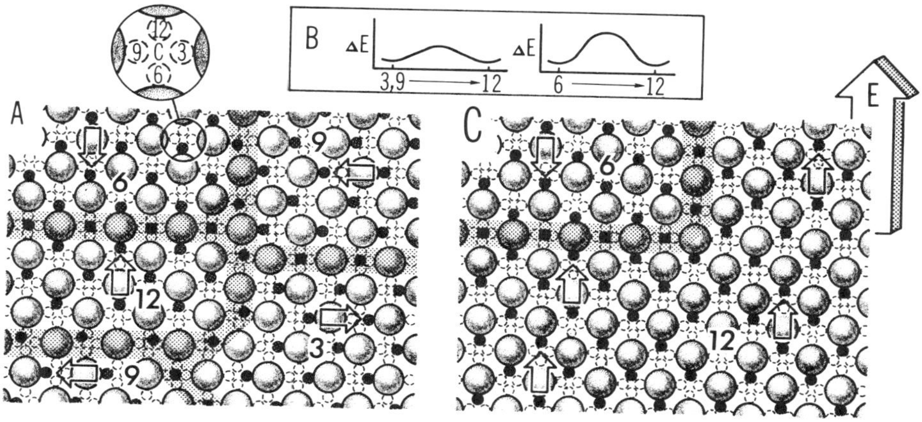

FIGURE 11–34

(A) Layer from a hypothetical crystal structure in which the cations, being undersized for the cavity, do not locate at its center *c* but rather in one of four positions (see circular enlargement) at 3, 6, 9, or 12 o'clock. If the cations were all located at the centers *c*, the positive centroid for them would coincide with the negative centroid resulting from the anions. However, the crystal is divided here into domains where 3, 6, 9, or 12 o'clock occupancies predominate, the boundaries between domains being shaded. In the 6 o'clock domain (upper left) the collection of cations are all shifted downward from the centers *c*; in other words, the positive centroid for this domain is below the negative centroid for the anions in this domain. This domain thus constitutes an electric dipole whose electric-dipole moment is shown as the hollow arrow pointing from its negative to positive centroids. The crystal as a whole is not an electric dipole, however, because its constituent domains have oppositely directed dipole moments which cancel each other.
The shaded transitional regions between domains, which compare to Bloch walls but are much narrower, contain cations in sites hybrid between the 3, 6, 9, or 12 o'clock sites.
(B) Assuming the 3, 6, 9, or 12 o'clock sites are those of least potential energy for the crystal, the change in energy (ΔE) is schematically shown for the shift, by the most direct route, of a cation from 3 (or 9) to 12 o'clock and from 6 to 12 o'clock. Greatest increase occurs when the cation is midway between 6 and 12 (at *c* in the circular inset at the left) and thus not tangent to any anions.
(C) If the crystal in (A) is subjected to a strong enough electric field (arrow **E**), positive charges will be constrained to move as arrow **E** points (and negative charges opposite to it). Consequently, cations in the 3 and 9 o'clock domains tend to shift to 12 o'clock sites so that the 12 o'clock domain grows at their expense. However, the 6 o'clock domains would likely survive until larger values of **E** were applied, because to move from a 6 to a 12 o'clock site requires overcoming a larger energy barrier [refer to (B) again] than to move from a 3 or 9 o'clock site to 12 o'clock.

domains whose moments are at 90°. The domain wall between a 12 and a 6 o'clock domain would thus be a 180° wall (shaded in Fig. 11-34C).

In a sufficiently strong electric field **E**, those domains in Figure 11-34A with dipole moments closest to **E** in direction grow at the expense of less favorably oriented domains (Fig. 11-34C). Such movement of cations for a limited distance along the general direction of **E** relative to the anions in the crystal's structure is called *ionic polarization* and imparts an electric dipole moment **P** to the crystal. We let **D** represent the electric field from the combined effect of the applied field **E** and the induced field **P**, then

$$\mathbf{D} = \epsilon_0\mathbf{E} + \mathbf{P} \qquad \text{(Eq. 11-37)}$$

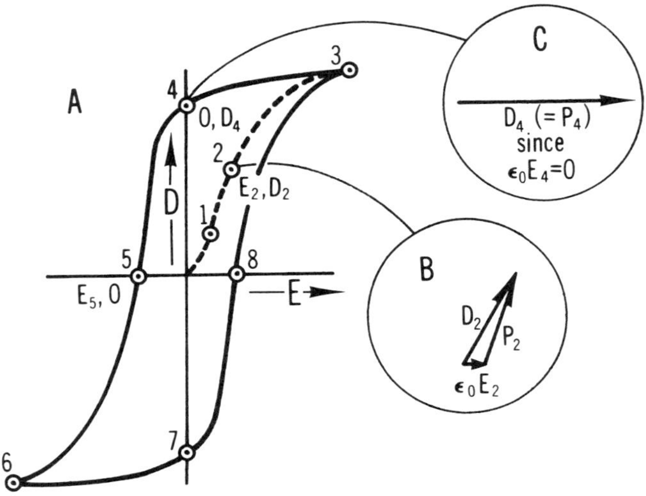

FIGURE 11–35

(A) A graph in which the magnitude of **E**, the electric field applied to a crystal, is plotted against the magnitude of **D** (defined below). Thus, subjected to an electric field **E**, the domains with dipole moments parallel or near parallel to **E** grow at the expense of others. Their contribution to **P**, the dipole moment per unit volume of crystal, thus increases. We let vector **D** represent the electric field resulting from the combined effect of applied field **E** and of **P**, the field caused by domain alignment. Consequently, **D** is the vector sum of **P** and of $\epsilon_0\mathbf{E}$ (as in circular inset *B*), ϵ_0 being a constant. Dashed line 1, 2, 3 represents the crystal's *polarization curve*. Increasing the applied field to $\mathbf{E}_3$ (the abscissa of point 3) causes all domains to become perfectly aligned relative to $\mathbf{E}_3$ so that **P** attains its maximum possible value (and its maximum parallelism to $\mathbf{E}_3$). Increase of **E** above $\mathbf{E}_3$ cannot increase either the magnitude of **P** or its parallelism to **E**; thus $\mathbf{D}_3$ is the maximum attainable value for **D** no matter how much **E** is increased above $\mathbf{E}_3$. Upon reducing **E** to zero, the crystal retains a remanent polarization (point 4 and circular inset *C*) because *all* domains do not become disaligned until a reversed field direction equal to $\mathbf{E}_5$ is applied. This hysteresis loop is comparable to that for ferromagnetic crystals (Fig. 11-29C) and, by analogy, crystals capable of retaining a remanent polarization have been called *ferroelectrics* even though iron is rarely a significant component.

where ϵ_0, the *permittivity constant*, equals 8.85×10^{-12} coulombs^2 per newton-(meter)^2. Equation (11-37) is analogous to Equation (11-30) with **D**, **E**, **P**, respectively, substituted for **B**, **H**, and **M**. Crystals for which a plot of **D** against different values of **E** produces a hysteresis loop (Fig. 11-35A) are called *ferroelectric crystals* or simply *ferroelectrics* because of the analogy to the behavior of ferromagnetic materials (Fig. 11-29C). Thus even after field **E** is reduced to zero (point 4 in Fig. 11-35A), ferroelectric materials retain an electric dipole moment because the favored domains persist, and thus $\mathbf{D}_4$ is entirely due to $\mathbf{P}_4$ (Fig. 11-35C), until the material's Curie temperature is attained or until an oppositely directed field ($\mathbf{E}_5$ in Fig. 11-35A) disorders them. Despite the name, there is no association of the phenomenon of ferroelectricity with iron.

Ferroelectric crystals that can be polarized along only one axis include rochelle salt, $NaK(C_4H_4O_6) \cdot 4H_2O$, and related tartrates; potassium dihydrogen

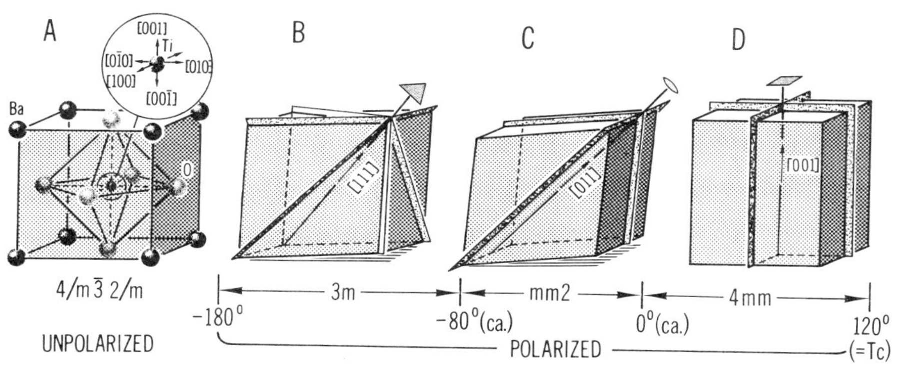

FIGURE 11–36

(A) The structure of barium titanate which is a perovskite ($CaTiO_3$) structure. The small Ti^{4+} ion (circled) is too small for octahedral coordination to the six oxygens (lightly shaded) that are in closest packing with the heavily shaded bariums. Ti^{4+} may thus be off-centered within its coordination octahedron. Direction of off-centering may be along one of the six ⟨001⟩ directions (circular enlargement) or along one of the eight ⟨111⟩ or twelve ⟨110⟩ directions. If throughout the structure there is no preferred direction of off-centering, the crystal is unpolarized because the resultant of the dipole moments of all its component domains is zero.
(B) However, a crystal between −180 and −80°C is most readily polarized by applying a field **E** along [111], the Ti^{4+} in the crystal's component domains then becoming off-centered along [111] to change the crystal class from $4/m\,\bar{3}\,2/m$ for the unpolarized state to $3m$ in this polarized state.
(C) Between −80 and 0°C, the crystal will polarize most readily along ⟨011⟩. Here application of **E** along [011] causes the domains to adopt this direction of Ti^{4+} off-centering, the crystal now being orthorhombic ($mm2$).
(D) From 0 to 120°C, easy polarization is along ⟨100⟩. Thus a crystal can be made to be tetragonal by polarizing it so that all Ti^{4+} ions are offset along [001]. Heated above 120°C, its Curie temperature (T_c), the crystal returns to its unpolarized state.

phosphate, KDP or KH_2PO_4, and related isostructural compounds; colemanite, $CaB_3O_4(OH)_3 \cdot H_2O$; and the triglycine sulfate (TGS) group. Those polarizable along several axes that are symmetrically equivalent in the nonpolar state include $BaTiO_3$ and other perovskite-type structures. In these perovskite ($CaTiO_3$) structures, the anions together with Ca^{2+}, or similarly large cation, form a cubic closest-packing array (Fig. 9-25A). Small ions like Ti^{4+} or Nb^{5+} cannot occupy the tetrahedral interstices in this array without touching the large cation. Instead, therefore, they occupy the octahedral interstices for which they are undersized. Consequently, rather than being centered in this cavity, they are off-centered toward one end of the cavity or the other. In barium titanate, off-centering occurs along any of the eight $\langle 111 \rangle$ directions at temperatures below $-80°C$, along any of the twelve $\langle 110 \rangle$ directions between -80 and $0°C$, or along any of the six $\langle 100 \rangle$ directions (Fig. 11-36A) from 0°C to the Curie temperature of 120°C. An unpolarized single crystal hence may consist of eight differently off-centered domains at $-80°C$, of twelve at $-80°C$ to 0°C, or of six at 0° to 120°C. In such unpolarized single crystals the "space-average" of the titanium position for the entire crystal would lead one to believe that Ti^{4+} is centered in the cavity and that it is larger than it actually is. Such nonpolarized crystals would seem to belong to class $4/m\,\bar{3}\,2/m$ and have a cubic unit cell. Application of a field **E** along a specific direction—for example, along [111] at $< -80°C$, along [110] at -80 to 0°C, or along [100] at 0° to 120°C—could, under suitable conditions, cause one domain to grow to the eventual replacement of domains less favorably oriented to **E**. A monodomain crystal would result whose crystal class and unit cell would be as shown in Figures 11-36B, C, and D; the arrows indicate the direction of off centering for Ti^{4+} throughout the monodomain as well as the direction of elongation of the unit cell compared to the cubic unit cell of the previously nonpolarized crystal.

If a crystal's positive and negative centroids move apart in response to an applied electric field **E**, its dimensions necessarily change. Such dimensional change in response to an applied electric field **E** is called *electrostriction*. Similarly, a dimensional change in response to an applied magnetic field H is called *magnetostriction*.

A monodomain's crystal class is a subgroup of the crystal class to which the nonpolarized, multidomained crystal seems to belong. Moreover, the multidomains are twinned in respect to each other by some of the symmetry operations that drop out in going to the monodomain's crystal class.

Antiferroelectrics

In the perovskite-structured $PbZrO_3$, the Zr^{4+} ions in the octahedral sites are off centered by about 0.2 Å in an antiparallel manner, for example, along the directions that were $[1\bar{1}0]$ and $[\bar{1}10]$ in ideal perovskite (Fig. 11-37A). Such crystals, for which each single monodomain consists of equal numbers of antiparallel dipoles, are called *antiferroelectrics* by analogy with antiferromagnetics. This off centering produces orthorhombic symmetry (space group, $Pba2$; $a_0 = 5.87$, $b_0 = 11.74$, $c_0 = 8.20$ Å) with unit cell as boldly outlined.

Polar directions and classes

Some basic definitions of symmetry must first be made or reviewed to facilitate discussion. By definition, a *polar direction* in a crystal is one whose end [uvw] is not symmetrically related to its oppositely directed end [$\bar{u}\bar{v}\bar{w}$]. No polar directions can thus occur in the 11 crystal classes that contain a center of symmetry. In the 21 noncentrosymmetric classes, however, polar directions occur (Table 11-3). If a vector is aligned along one such polar direction [uvw] in a given noncentrosymmetric class, the class's symmetry then operates on this vector to require a "form" of vectors along $\langle uvw \rangle$ (Fig. 11-37B). For 11 of these noncentrosymmetric classes, including class 222 (Fig. 11-37B), **p** the resultant or vector sum of this form of vectors is zero. For the remaining 10 noncentrosymmetric classes (that is, 1, 2, 3, 4, 6, *m*, *mm*2, 3*m*, 4*mm*, and 6*mm*) **p** is not equal to zero and these 10 classes are consequently called *polar classes*. For example, **p** is not equal to zero in class 2 (Fig. 11-37C), class 4 (Fig. 11-37D), and so on. The resultant **p** inevitably coincides with the solitary 2-, 3-, 4-, or

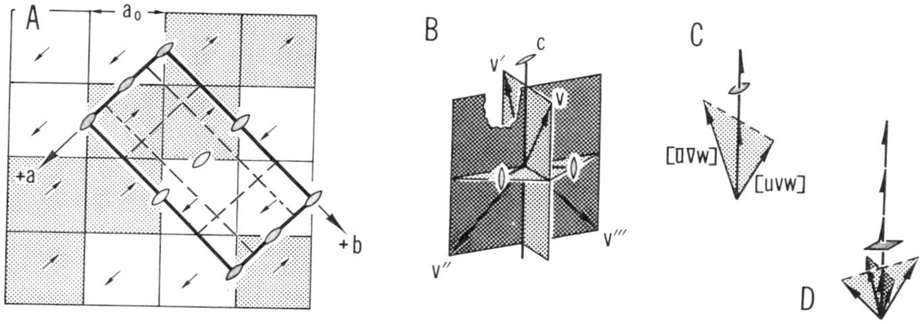

FIGURE 11-37

(A) View down the *c* axis of the antiferroelectric structure proposed for $PbZrO_3$ by Sawaguchi et al. Sixteen unit cells of the underlying (cubic) perovskite structure are shown. In the eight shaded cells the Zr^{4+} ions are off-centered along what was [$\bar{1}10$] in the original perovskite cube but is now [$\bar{1}00$] for the orthorhombic cell (*Pba*2) boldly outlined. The eight unshaded cells represent off-centering in the opposite direction.
(B) Given vector **V** along [uvw] in class 222, the symmetry requires the presence of the remaining three vectors as well. The resultant of **V** and **V′** points upward along the *c* axis. The resultant of **V″** and **V‴** is equal but opposite along the *c* axis. The resultant of vector form ⟨**uvw**⟩ is thus zero. A crystal class for which the resultant of the vector form ⟨**uvw**⟩ does not equal zero is called a *polar class*. Ten of the twenty-one noncentric classes are polar classes.
(C) Two vectors, [**uvw**] and [**ūv̄w**] related by a 2-fold axis. Their resultant, the single-barbed vector, necessarily coincides with the 2-fold axis.
(D) The form of four vectors related by a 4-fold axis also has a resultant which coincides with the 4-fold axis. In general the resultant of a vector form ⟨**uvw**⟩ related by a single X-fold axis where X exceeds 1 in value will coincide with this axis. For a $\bar{X}$ axis, the resultant will equal zero except when $\bar{X}$ equals $\bar{2}$.

TABLE 11–3

Polar Directions in the Twenty-One Noncentrosymmetric Classes

Class[a]	Polar Directions	Class	Polar Directions
Triclinic		Trigonal	
1	All	**3**	All
Monoclinic			
2	All not ⊥2	32	All not ⊥2
m	All not ⊥*m*	**3m**	All not ⊥*m*
Orthorhombic		Hexagonal	
mm2	All not ⊥2 or ⊥*m*	**6**	All not ⊥6
222	All not ⊥2	$\bar{6}$	All but $\bar{6}$
Tetragonal			
4	All not ⊥4	622	All not ⊥6 or ⊥2
$\bar{4}$	All not ‖ or ⊥$\bar{4}$	**6mm**	All not ⊥6
422	All not ⊥4 or ⊥2	$\bar{6}m2$	All not ⊥2
		Isometric	
4mm	All not ⊥4	23	All not ⊥2
$\bar{4}2m$	All not ⊥$\bar{4}$ or ⊥2	432	All not ⊥4 or ⊥2
		$\bar{4}3m$	All not ⊥$\bar{4}$

[a] The ten polar classes among these are in boldface type. The symbols ‖ and ⊥ should be read "parallel to" and "perpendicular to," respectively.

6-fold axis present in a polar class. These axes are thus called *unique polar axes* because they alone, of all the polar axes in the crystal, must coincide with the resultant **p**. For class *m* the resultant **p** must lie in the mirror plane.

Pyroelectricity

Crystals that belong to one of the 10 polar classes may develop an electric dipole moment along the resultant **p** in response to a uniform temperature change, the phenomenon being known as *pyroelectricity*. This dipole moment or polarization direction, that is, the line joining the separated positive and negative centroids, will inevitably coincide with the *unique polar axis* in classes 2, 3, 4, 6, *mm*2, 3*m*, 4*mm*, and 6*mm*. For class *m* the developed dipole moment must lie in the mirror plane; for class 1 it may develop along any direction. A potential difference will exist between two faces at opposite ends of, but perpendicular to, the polarization direction. This difference is detectable by Kundt's method (Fig. 5-30B) only immediately after the temperature change because, with time, it is soon neutralized by electrical leakage and capture of stray charges at these surfaces.

The similar development of a magnetic moment in response to a uniform temperature change is called *pyromagnetism*.

A uniform temperature change whereby no temperature gradients develop in a crystal has no assignable direction. If this change causes a dipole moment to develop along one direction in a crystal, the crystal's symmetry (Neumann's principle) will require it to be developed to an equal extent along all directions symmetrically equivalent to this first. Only for the 10 polar classes will the resultant of this "form" of moments have a value other than zero. Hence the restriction of pyroelectricity and pyromagnetism to these 10 classes and more particularly to directions along the unique polar axis (Fig. 11-37B, C) or, for class m, in the mirror plane.

Piezoelectricity

Piezoelectricity (pronounced "pie-eezo-electricity") is defined as the electrical polarity **P** that develops along a crystal's polar axis in proportion to a mechanical strain along this axis. The direction of **P**, the electric dipole moment per unit volume, reverses if the stress is changed from tension to compression. For crystals possessing a polar axis, therefore, a mechanical stress, representable by a 2-rank tensor rather like the strain tensor illustrated by Figures 11-17 and 11-18, causes the crystal to develop a polarity **P**, which is represented by a 1-rank tensor. As Equation (11-21) indicates, this 2-rank stimulus tensor must be related to the 1-rank response tensor by a 3-rank crystal-property tensor. This tensor's 27 components d_{ijk} are the *piezoelectric moduli*. The symmetry of the crystal class may reduce many of these moduli to zero. In fact for class 432 and those that possess a center of symmetry, all 27 moduli reduce to zero and piezoelectricity is no longer possible.[9] Nye (1957, p. 113) discusses (a) the restrictions imposed by symmetry on the values of the piezoelectric moduli for each of the 21 noncentrosymmetric classes and (b) the mathematics of 3-rank tensors as applied to piezoelectricity.

If a crystal's temperature is changed nonuniformly so that thermal gradients and stresses arise in the crystal, these thermally produced stresses cause piezoelectric effects known as *false pyroelectricity*. Such may occur in the 20 noncentric classes exclusive of 432. *True pyroelectricity* is caused by change of temperature alone and is restricted to the 10 *polar* noncentrosymmetric classes.

If an alternating current is applied along a polar axis, the crystal will expand along this axis as the current flows one way, but contract as it reverses. This *converse piezoelectricity* has great practical value. For example, properly cut plates from quartz crystals develop polarity along a given axis if compressed along a direction at 90° to this axis. By inserting this plate into an electrical condenser, the alternations in its degree of polarity as caused by the incidence on it of sound waves or vibrations from a phonograph needle can be converted into varying currents leading outward from the condenser. These currents, electronically amplified, may then be reconverted into high-amplified sounds as in phonographs or in the field of ultrasonics.

[9] In fact such crystals cannot possess rank 3 tensor properties of any kind.

Physical Properties, Bonding, and Structure

The physical properties of a crystal depend largely upon (1) the strength and type of bonds present and (2) the nature of the spatial distribution of these bonds as embodied in the crystal structure. In general, ionic and covalent bonds are usually strong and hence produce hard crystals with high melting points and low coefficients of thermal expansion. On the other hand metallic bonding may be strong between some metal atoms, weak between others. Thus some metals are fairly hard, others are soft; some have high melting points, others have low. Molecular crystals, whose molecules are held one to the other by the relatively weak van der Waals bonds, are usually soft, have low melting points and large coefficients of thermal expansion.

Between two oppositely charged ions in a structure, the bond strength increases as (1) their center-to-center distance or bond length decreases (Table 11-4) and as (2) their charge increases (Table 11-5). Such increase in bond strength causes increases in the crystal's Mohs hardness (M) and melting point (MP) plus a decrease in α_v, the volume coefficient of thermal expansion. Thus if F^- or $(OH)^-$ substitute in place of O^{2-} in a crystal structure, weaker bonds and softer crystals result.

Hardness and melting point are closely related properties of a crystal,

TABLE 11–4

Physical Properties of Oxides with NaCl-Type Structures

	——Increasing Bond Strength——→ BaO	SrO	CaO	MgO
Bond length	2.76 Å	2.56 Å	2.40 Å	2.12 Å
H (Mohs)	2.0	3.5	4.0	5.5
M.P. (°C)	1933°C	2430°	2570°	2815°
$\alpha_v \cdot 10^5$ deg^{-1}			2.8[a]	2.7[a]

[a] Average volume expansion over 20 to 100°C range.

TABLE 11–5

Effect of Increase of Ionic Charge on Hardness and Melting Point of Two NaCl-Type Structures

	NaF[a]	CaO
Bond Length[a]	2.35 Å	2.40 Å
H	3.2	4.0
M.P.	992°	2570°

[a] The cation-to-anion distance of 2.35 Å is practically identical to that in CaO (2.40 Å). Thus the bond between Ca^{2+} and O^{2-} in CaO is stronger than that between Na^+ and F^- in NaF largely because of the greater charges on the Ca^{2+} and O^{2-} ions (as compared to Na^+ and F^-).

measuring the resistance of the crystal's bonding to disruption under mechanical stress or thermal vibrations, respectively. However, melting is a solid $\leftrightarrows$ liquid equilibrium and hence depends not only on bond strengths in the solid but also on their values relative to those within the liquid that "competes" with this solid for existence.

Anisotropy in the structure of a crystal, which results if atoms or ions are more closely packed or more strongly (or differently) bonded along some directions than others, produces an attendant anisotropy of such directional physical properties as electrical or thermal conductivity, cleavage, thermal coefficients of expansion, directional hardness tests, and so on. For example, these properties vary with direction more significantly for chain or layer silicates than for the less anisotropic framework silicates. In all cases, however, Neumann's principle is obeyed.

Appendix: Matrix Algebra

Today's "computer revolution" has made matrix algebra an important research tool. Those concepts of matrix algebra that pertain most frequently in crystallography will now be reviewed briefly. The inexpensive circular by Harding (1967) is recommended for further reading.

Definitions

A *matrix* is a rectangular array of numbers of the form

$$\begin{array}{l} \text{Row 1} \rightarrow \\ \\ \text{Row 2} \rightarrow \\ \\ \text{Row 3} \rightarrow \end{array} \begin{bmatrix} a_{11} & a_{12} & a_{13} & a_{14} \\ \\ a_{21} & a_{22} & a_{23} & a_{24} \\ \\ a_{31} & a_{32} & a_{33} & a_{34} \end{bmatrix}$$
$$\begin{array}{cccc} \uparrow & \uparrow & \uparrow & \uparrow \\ \text{Col.} & \text{Col.} & \text{Col.} & \text{Col.} \\ 1 & 2 & 3 & 4 \end{array}$$

This particular matrix, denoted as A_{ij}, consists of three rows and four columns. Each individual number a_{ij} of a matrix is called an *element*. For such an element, for example, a_{23}, the *first* subscript denotes the *row*, *the second the column* to which this element belongs. The mnemonic phrase, RC cola, helps one remember this sequence. Similarly, the matrix A_{ij} is one with i rows and j columns. For the example, i thus equals 3 and j equals 4. Matrices are of various forms. The three following, that is,

$$\begin{bmatrix} 1 & 4 \\ \\ 5 & 9 \end{bmatrix} \qquad \begin{bmatrix} 3 \\ \\ 2 \end{bmatrix} \qquad \text{and} \qquad \begin{bmatrix} 6 & 8 \end{bmatrix}$$

are, respectively, a square matrix ($i = j$), a column matrix ($j = 1$), and a row matrix ($i = 1$).

Multiplication

A given matrix A_{im} can be multiplied by a second B_{mj} to yield a third matrix P_{ij} only if m the number of columns in the first equals the number of rows in the second. Rules for multiplication are best explained by example.

$$\underset{A_{im}}{\begin{bmatrix} a_{11} & a_{12} & a_{13} \\ \mathbf{a}_{21} & \mathbf{a}_{22} & \mathbf{a}_{23} \end{bmatrix}} \cdot \underset{B_{mj}}{\begin{bmatrix} \mathbf{b}_{11} & b_{12} \\ \mathbf{b}_{21} & b_{22} \\ \mathbf{b}_{31} & b_{32} \end{bmatrix}} = \overbrace{\begin{bmatrix} \underbrace{(a_{11}\mathbf{b}_{11} + a_{12}\mathbf{b}_{21} + a_{13}\mathbf{b}_{31})}_{p_{11}} & \underbrace{(a_{11}b_{12} + a_{12}b_{22} + a_{13}b_{32})}_{p_{12}} \\ \underbrace{(\mathbf{a}_{21}\mathbf{b}_{11} + \mathbf{a}_{22}\mathbf{b}_{21} + \mathbf{a}_{23}\mathbf{b}_{31})}_{p_{21}} & \underbrace{(\mathbf{a}_{21}b_{12} + \mathbf{a}_{22}b_{22} + \mathbf{a}_{23}b_{32})}_{p_{22}} \end{bmatrix}}^{P_{ij}}$$

For illustrative purposes only, the second row of A_{im} and the first column of B_{mj} are in boldface type. Thus we can better see that each individual element p_{ij} of the product matrix P_{ij} equals the sum of those products that result from an element-by-element multiplication of the ith row of A_{im} by the jth column of B_{mj}. Thus p_{21} is such a summation for the second row (in boldface) of A_{im} so multiplied by the first column (also in boldface) of B_{mj}. Similarly, p_{12} involves the first row of A_{im} times the second column of B_{mj}. The first matrix, A_{im} in this case, is called the *premultiplier,* whereas the second, B_{mj}, is called the *postmultiplier.* Thus A_{im} was *postmultiplied* by B_{mj} or, conversely, B_{mj} was *premultiplied* by A_{im}.

A numerical example will confirm understanding. Thus

$$\begin{bmatrix} 3 & 4 & 1 \\ 5 & 2 & 0 \end{bmatrix} \cdot \begin{bmatrix} 1 & 0 \\ 2 & 4 \\ 3 & 5 \end{bmatrix} = \begin{bmatrix} 14 & 21 \\ 9 & 8 \end{bmatrix}$$

But the reverse multiplication

$$\begin{bmatrix} 1 & 0 \\ 2 & 4 \\ 3 & 5 \end{bmatrix} \cdot \begin{bmatrix} 3 & 4 & 1 \\ 5 & 2 & 0 \end{bmatrix} = \begin{bmatrix} 3 & 4 & 1 \\ 26 & 16 & 2 \\ 34 & 22 & 3 \end{bmatrix}$$

yields different results because in contrast to numbers (which are commutative in respect to multiplication; for example, 7×8 equals 8×7) *matrices* are *not* usually *commutative* for multiplication.

A *unit matrix,* usually symbolized I, is a square matrix for which the main diagonal's elements all equal unity whereas off-diagonal elements all equal zero.

The three-by-three ($i = j = 3$) unit matrix is thus

$$\begin{bmatrix} 1 & 0 & 0 \\ 0 & 1 & 0 \\ 0 & 0 & 1 \end{bmatrix} = I$$

A given matrix A, with subscripts omitted here for convenience, multiplied by unit matrix I yields a product matrix identical to the given matrix. Thus $A \cdot I = A$. For example,

$$[uvw] \cdot \begin{bmatrix} 1 & 0 & 0 \\ 0 & 1 & 0 \\ 0 & 0 & 1 \end{bmatrix} = [uvw]$$

For this reason the unit matrix is called the *identity matrix* and symbolized as I.

Inverse matrices

For any square matrix A there may exist a second matrix, called its *inverse* and denoted as A^{-1}, such that the product of the two, whether $A \cdot A^{-1}$ or $A^{-1} \cdot A$, yields the identity matrix I. Thus

$$\underset{\text{Matrix } A^{-1}}{\begin{bmatrix} x_{11} & x_{12} & x_{13} \\ x_{21} & x_{22} & x_{23} \\ x_{31} & x_{32} & x_{33} \end{bmatrix}} \cdot \underset{\text{Matrix } A}{\begin{bmatrix} a_{11} & a_{12} & a_{13} \\ a_{21} & a_{22} & a_{23} \\ a_{31} & a_{32} & a_{33} \end{bmatrix}} = \begin{bmatrix} 1 & 0 & 0 \\ 0 & 1 & 0 \\ 0 & 0 & 1 \end{bmatrix}$$

Performing the above multiplication, we obtain three sets of three linear equations, each with three unknowns. Solving these for the values of x_{11}, x_{12}, and so on, there results

$$x_{11} = \frac{a_{22}a_{33} - a_{32}a_{23}}{\Delta} \qquad x_{12} = \frac{a_{32}a_{13} - a_{12}a_{33}}{\Delta} \qquad x_{13} = \frac{a_{12}a_{23} - a_{22}a_{13}}{\Delta}$$

$$x_{21} = \frac{a_{23}a_{31} - a_{21}a_{33}}{\Delta} \qquad x_{22} = \frac{a_{11}a_{33} - a_{31}a_{13}}{\Delta} \qquad x_{23} = \frac{a_{21}a_{13} - a_{11}a_{23}}{\Delta}$$

$$x_{31} = \frac{a_{21}a_{32} - a_{31}a_{22}}{\Delta} \qquad x_{32} = \frac{a_{12}a_{31} - a_{32}a_{11}}{\Delta} \qquad x_{33} = \frac{a_{11}a_{22} - a_{12}a_{21}}{\Delta}$$

where the determinant Δ equals $a_{11}a_{22}a_{33} + a_{12}a_{23}a_{31} + a_{13}a_{21}a_{32} - a_{31}a_{22}a_{13} - a_{21}a_{12}a_{33} - a_{11}a_{32}a_{23}$. If Δ equals zero for matrix A, then A^{-1} does not exist. The multiplication of two inverse matrices *is* commutative, that is, $A \cdot A^{-1} =$

$A^{-1} \cdot A = I$. The reader should confirm that

$$\begin{bmatrix} 3 & 2 & 6 \\ 1 & 1 & 2 \\ 2 & 2 & 5 \end{bmatrix} \quad \text{and} \quad \begin{bmatrix} 1 & 2 & -2 \\ -1 & 3 & 0 \\ 0 & -2 & 1 \end{bmatrix}$$

are inverses of each other (1) by multiplying them together and (2) by computing the elements of one (x_{ij}) from those of the other (a_{ij}) using the nine equations given above.

Addition

Two matrices can be added only if one has the same number of rows (and of columns) as the other. Thus

$$\underset{\text{Matrix } A}{\begin{bmatrix} 1 & 0 & 2 \\ -1 & 4 & 3 \end{bmatrix}} + \underset{B}{\begin{bmatrix} 5 & 6 & 7 \\ 8 & 9 & -2 \end{bmatrix}} = \underset{\text{Matrix } S}{\begin{matrix} (1+5)\,(0+6)\,(2+7) \\ (-1+8)\,(4+9)\,(3-2) \end{matrix}} = \underset{S}{\begin{bmatrix} 6 & 6 & 9 \\ 7 & 13 & 1 \end{bmatrix}}$$

Matrix addition is simply an element-wise addition as illustrated above. We note that *matrices are commutative in respect to addition*. Thus $A + B = B + A = S$.

Transpose of a matrix

For a given matrix A its transpose, denoted A^t, is related to A by an interchange of columns with rows, thus

$$\underset{A}{\begin{bmatrix} a_{11} & a_{12} & a_{13} \\ a_{21} & a_{22} & a_{23} \\ a_{31} & a_{32} & a_{33} \end{bmatrix}} \qquad \underset{A^t}{\begin{bmatrix} a_{11} & a_{21} & a_{31} \\ a_{12} & a_{22} & a_{32} \\ a_{13} & a_{23} & a_{33} \end{bmatrix}}$$

REFERENCES AND RECOMMENDED READING

Austin, J. B. (1952) "Thermal expansion of nonmetallic crystals," *J. Am. Ceramic Soc.*, **35,** 243–253.

Battey, M. H., and S. I. Tomkeieff, eds. (1964) *Aspects of Theoretical Mineralogy in the U.S.S.R.* New York: Macmillan.

Berman, Harry S. (1939) "A torsion microbalance for the determination of specific gravities of minerals," *Am. Mineral.*, **24,** 434–440.

Bhagavantan, S. (1966) *Crystal Symmetry and Physical Properties.* New York: Academic Press.

Brace, W. F. (1964) "Indentation hardness of minerals and rocks," *N. Jb. Mineral. Mh.*, 257–269.

Clark, S. P., Jr. (1966) *Handbook of Physical Constants.* Memoir 97, Geol. Soc. Amer.

Fairbairn, H. W. (1939) "Correlation of quartz deformation with its crystal structure," *Am. Mineral.*, **24,** 351–368.

——— (1943) "Packing in ionic minerals," *Bull. Geol. Soc. Am.*, **54,** 1305–1374.

Harding, Katherine (1967) *Basic Properties of Matrix Algebra,* U.S. Bureau of Mines, Information Circular 8332, U. S. Government Printing Office, Washington, D. C. 20402.

Koerber, George G. (1962) *Properties of Solids.* Englewood Cliffs, N. J.: Prentice-Hall.

May, Irving, and John Marinenko (1966) "A micropycnometer . . . ," *Am. Mineral,* **51,** 931.

Megaw, Helen D. (1957) *Ferroelectricity in Crystals.* London: Methuen.

Meyrowitz, R., F. Cuttitta, and B. Levin (1960) "N,N-Dimethylformamide . . . ," *Am. Mineral.*, **45,** 1278.

Mursky, G. A., and R. M. Thompson (1958) "A specific gravity index for minerals," *Can. Mineral.*, **6,** part 2, 273–287.

Nye, J. F. (1957) *Physical Properties of Crystals.* London: Oxford Univ. Press.

Plendl, J. N., and P. J. Gielisse (1963) "Atomistic expression of hardness" *Z. Krist.*, 404–421.

Purcell, E. M. (1963) *Electricity and Magnetism.* New York: McGraw-Hill (Vol. II in the Berkeley Physics Course).

Tabor, D. (1954) "Hardness of solids," *Endeavour,* **13,** 27.

Tweeddale, J. G. (1964) *The Mechanical Properties of Metals,* Chapter 11. New York: American Elsevier.

Westbrook, J. H., and P. J. Jorgensen (1968) *Effects of Water Desorption on Indentation Hardness Anisotropy in Minerals,* Report No. 68-C-041. Schenectady, New York: General Electric Research and Development Center.

Winchell, H. (1945) "The Knoop microhardness tester as a mineralogical tool," *Am. Mineral.*, **30,** 583–595.

Wooster, W. A. (1949) *A Text-Book on Crystal Physics.* Cambridge: Cambridge Univ. Press.

Zhdanov, G. S. (1965) *Crystal Physics* (Trans. A. F. Brown). New York: Academic Press.

QUESTIONS AND PROBLEMS

11-1 At 20°C α-iron (BCC) is stable and its cell edge a_0 equals 2.866 Å. From 909 to 1403°C, γ-iron (FCC) is stable and at 950° its cell edge a_0 equals 3.656 Å. Measuring a_0 at other temperatures within the 909 to 1403°C range and extrapolating the curve obtained to 20°C, γ iron at 20°C would have a cell edge a_0 of 3.571 Å (ca.). Above 1403°C δ iron (BCC) is stable and at 1425°C has a_0 of 2.94 Å. Compute the densities of (a) α-iron at 20°C; (b) γ-iron at 20°; (c) γ-iron at 950°; and (d) δ iron at 1425°C.

11-2 Compute the effective radius, that is, one-half the center-to-center distance, for the iron atoms in (a) to (d) above.

11-3 Determine the packing index for (a) body-centered cubic iron; (b) face-centered cubic iron; (c) any hexagonal closest-packed element.

11-4 Calculate the density of an olivine crystal Fo_4Fa_{96} with cell edges a_0 4.81, b_0 10.56, c_0 6.09 Å.

ANSWERS

11-1 (a) 7.877 g cm^{-3}; (b) 8.144; (c) 7.589; (d) 7.297.
11-2 (a) 1.241 Å; (b) 1.263 Å; (c) 1.293 Å; (d) 1.273 Å.
11-3 (a) 6.8; (b) 7.4; (c) 7.4.
11-4 ρ_{calc} 4.321 g cm^{-3}.

12 Crystals and Light

Those physical properties relating to a crystal's response to light constitute its *optical properties.* Textbooks devoted to optical crystallography [for example, Bloss (1961)] describe in detail the properties associated with light transmission as well as the use of the polarizing microscope in their measurement. In view of this prior coverage, the present chapter only briefly describes the nature of light and the physical significance of the optical constants n, ϵ, ω, α, β, γ, and $2V$. Methods of their measurement will not be discussed.

Nature of Light

Gamma rays, X-rays, ultraviolet rays, visible light, infrared, and radio waves are all portions of the electromagnetic spectrum (Fig. 12-1). Each of these wave types travels in a vacuum at the common velocity c of 3×10^{10} cm/sec, the speed of light; each has a slightly different wavelength λ and frequency ν from its nearest neighbor. Visible light represents a relatively limited band of wavelengths within the electromagnetic spectrum, ranging from 3900 to 7700 Å. By photochemical processes not yet understood, when light of a particular wavelength within this range is incident upon a normal human retina, it produces a signal that the brain interprets as a particular color. The expanded scale of Figure 12-1 denotes the wavelength limits set by Hardy and Perrin (1932, p. 16) for the seven distinctive "colors of the rainbow" recognized by Sir Isaac Newton. Light of a particularly limited range of wavelengths is called *monochromatic*; and the more limited the range, the more highly monochromatic it is. Thus a light beam whose wavelengths range from 5800 to 5900 Å would not be as monochromatic as light from a sodium vapor lamp that consists chiefly of wavelengths

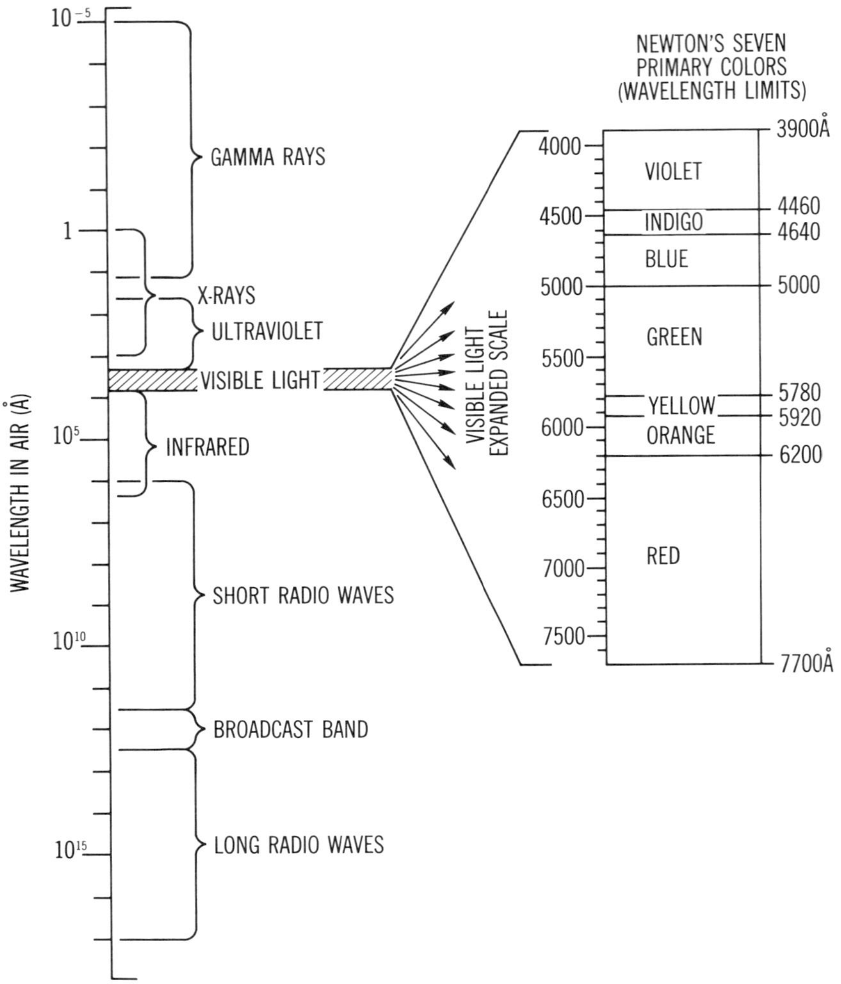

FIGURE 12–1

The relation of visible light to the electromagnetic spectrum is shown on the left. The expanded scale on the right illustrates the range in wavelengths assigned by Hardy and Perrin (1932) to Newton's seven principal colors.

5890 and 5896 Å. If light of the entire visible range, 3900 to 7700 Å, simultaneously enters the eye, it is interpreted by the brain as "white light."

Polarized light

Light energy may be considered to travel by a transverse wave motion in which the vibrations are usually perpendicular to the direction of travel of the energy. A plucked guitar string, for example, vibrates approximately perpendicular to its length, whereas the energy travels up and down the string's length.

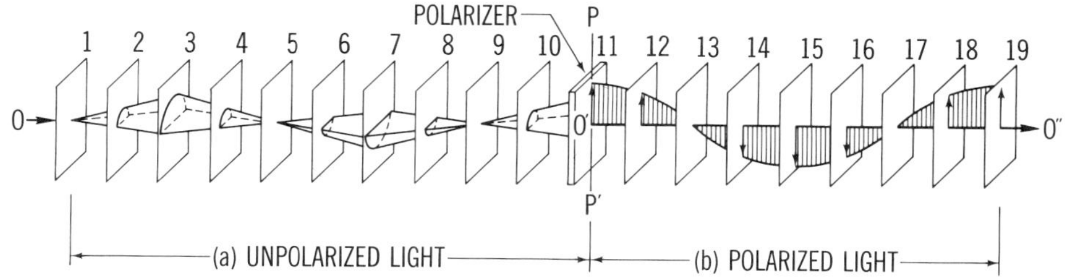

FIGURE 12–2

Unpolarized light traveling through imaginary planes 1 and 10 is polarized by a polarizer located at 11. After passing through the polarizer, the light vibrates exclusively parallel to PP', the privileged direction of the polarizer.

As schematically shown on the left half of Figure 12-2, ordinary (unpolarized) sodium light is considered to vibrate in numerous directions, all of which are at right angles to its ray path. As the ray of light energy travels from plane 1 to plane 11, its vibration directions may be imagined to trace out a three-dimensional figure resembling a series of canoes, end to end, with alternate canoes upside down. As shown, the ray's vibration directions within any plane perpendicular to its path may be represented by a semicircle of radius equal to the vibration of the wave within that plane. Planes 1, 5, and 9 are exceptions because they are located at points on the ray path for which the vibration is nil (at the particular instant in time which Fig. 12-2 represents). Light whose vibrations are restricted to a single direction in space, for example, the ray between O' and O'' in Figure 12-2, is said to be *plane polarized.* The *plane of vibration* of the light may be defined as the plane parallel to both the ray's path and vibration direction; in Figure 12-2, therefore, it is the ruled plane parallel to the lines $O'O''$ and PP'.

Wavelength λ may be defined as the distance between two neighboring points experiencing vibrations of the same amount and in the same direction, such points being said to be "in phase." In Figure 12-2 wavelength equals the distance between construction planes 1 and 9 (or 2 and 10) for the unpolarized light or between 11 and 19 for the polarized light. The frequency ν of a wave motion represents the number of wavelengths that would pass by a fixed point during a 1-sec interval. The standard relation between frequency ν, wavelength λ, and velocity c for wave motion is

$$c = \nu\lambda \qquad \text{(Eq. 12-1)}$$

Transmission of Light

Refractive index

The index of refraction n of a particular material may be defined as

$$n = \frac{c}{c_m} \qquad \text{(Eq. 12-2)}$$

where c and c_m symbolize the velocity of light in a vacuum and in the material, respectively. For light c_m is always less than c; consequently, a material's refractive index always exceeds 1.0 in value. Air, through which light travels almost as fast as in a vacuum, has a refractive index of 1.0003. Because it represents the ratio of two velocities, refractive index n is a dimensionless (pure) number.

Isotropic and anisotropic media

Those materials through which monochromatic light travels with the same speed, *regardless of its direction of vibration*, are called *isotropic* media. In addition to glasses and to crystals of the isometric system, a vacuum, all gases, and most liquids are isotropic with respect to light. Other materials, mainly the crystals of any nonisometric system, are *anisotropic* with respect to light; through them a light ray may travel with considerably different speeds for different directions of vibration within the crystal.

The optical indicatrix

The optical indicatrix illustrates how the refractive index of a transparent material varies according to the vibration direction of the light wave in the material (monochromatic light assumed). Let us consider an infinite number of vectors radiating outward in all directions from a common point within the crystal. Each vector is drawn proportional in length to the crystal's refractive index for light vibrating parallel to that vector direction. *The indicatrix is the surface connecting the tips of these vectors.* The vectors themselves are generally

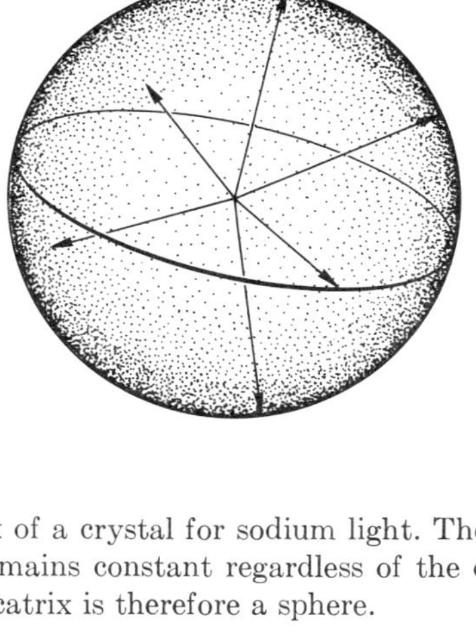

FIGURE 12–3

Isotropic indicatrix of a crystal for sodium light. The crystal's refractive index for sodium light n_D remains constant regardless of the direction in which the light is vibrating; the indicatrix is therefore a sphere.

omitted when this connecting surface is drawn but in Figure 12-3 a few are shown for illustration. Figure 12-3 represents the optical indicatrix for an isotropic material; in such, the value of the refractive index is independent of the vibration direction and the indicatrix is hence a sphere of radius n.

Light of one wavelength will pass through a material at a velocity different from light of another wavelength. In other words the refractive index of a material is different for different wavelengths. For precise work, monochromatic light corresponding to the C, D, or F Fraunhofer wavelengths (C = 6563 Å, D = 5893 Å, and F = 4861 Å) is used as the illuminant. For example, sodium chloride possesses the indices $n_C = 1.541$, $n_D = 1.544$, and $n_F = 1.553$, the subscripts indicating the Fraunhofer wavelength for which the refractive index value was measured. Usually refractive indices are measured and cited for D light in precise work or in white light for less precise. If white light is used, no subscript is added. Thus $n = 1.545$ represents a refractive index measured with white light.

For anisotropic crystals the refractive index, even for monochromatic light, varies with vibration direction in the crystal. Consequently, the optical indicatrix is not a sphere but an ellipsoid, which by definition is a surface or solid so shaped that all *central sections,* that is, cross sections through its center, are either ellipses or, for one or two special orientations, *circular sections.* Two basically different types of ellipsoids exist, those with *one* circular section or those with *two*. In optical crystallography the direction perpendicular to such a circular section is called an *optic axis.* Thus an indicatrix is called *uniaxial* or *biaxial* according to whether it is represented by an ellipsoid with one circular section or two. Of these indicatrices, the simpler uniaxial type will be discussed first.

The uniaxial indicatrix

For monochromatic light vibrating parallel to the c axis, crystals of the hexagonal and tetragonal systems exhibit a unique index of refraction customarily symbolized as ϵ. In Figure 12-4A, therefore, a vector proportional in length to the value of ϵ has been drawn parallel to the c axis to indicate this. On the other hand, for all vibration directions at 90° to the c axis, the crystal's refractive indices equal a common value symbolized as ω.[1] By constructing vectors proportional in length to ω along these vibration directions, a circle of radius ω is defined (Fig. 12-4A); this circular section is, of course, always perpendicular to the c axis. For light vibrating at a random angle θ to the c axis (Fig. 12-4A), the crystal exhibits a refractive index somewhere between ω and ϵ in value; these intermediate indices are symbolized as ϵ'. The value of ϵ' can be computed from the formula

$$\epsilon' = \frac{\omega\epsilon}{\sqrt{\omega^2 \cos^2\theta + \epsilon^2 \sin^2\theta}} \qquad \text{(Eq. 12-3)}$$

[1] Other symbols sometimes used for ϵ and ω are ϵ: n_ϵ, N_ϵ, E, n_E, N_E; and ω: n_ω, N_ω, O, n_O, N_O.

Plane 1 in Figure 12-4B indicates by the vector lengths how, as θ (for a vibration direction) varies from θ_1 to θ_3, the corresponding crystal index varies from ϵ_1' to ϵ_3'. The tips of all such vectors within plane 1 fall along an ellipse (for which Eq. (12-3) is the equation in polar coordinates). Similarly, within any other plane containing the crystal's *c* axis (for example, plane 2) the vectors describing the ϵ' values fall along an ellipse identical with that for plane 1. Thus the ellipse of plane 1 can be rotated about the *c* axis as a hinge to coincide with any of these identical ellipses. As a result an ellipsoid of rotation (Fig. 12-4C) is formed. The length of any radius of this ellipsoid indicates the crystal's refractive index for light vibrating parallel to this radius. Figure 12-4C is called the *indicatrix* or, more precisely, the *uniaxial indicatrix* (because there exists only one axis through the indicatrix center that is perpendicular to a circular section).

Optic Sign. The indices ω and ϵ represent the maximum and minimum, or minimum and maximum, refractive indices measurable in a tetragonal or

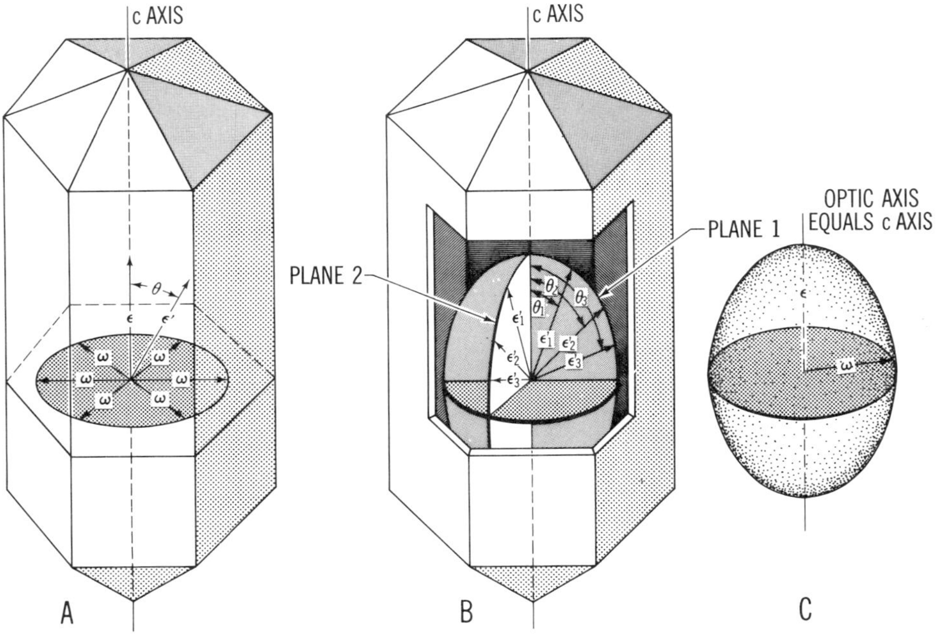

FIGURE 12–4

(A) Angular relationships to the *c* axis of the vibrations corresponding to indices ϵ, ω, and of one particular value of ϵ'. All ω vibrations lie in the plane perpendicular to the *c* axis.
(B) Vector lengths in planes 1 and 2 indicate the variation in index ϵ' of the crystal for light vibrating parallel to them. Their tips outline identical ellipses in all planes through the *c* axis (for example, planes 1 and 2).
(C) The uniaxial indicatrix for the crystal.

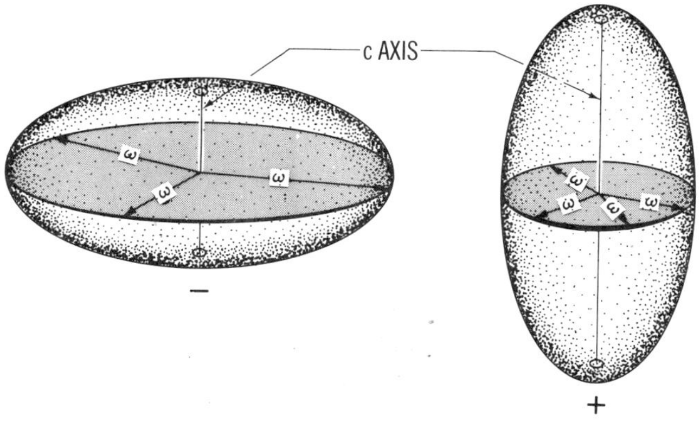

FIGURE 12–5

Comparison of the negative and positive uniaxial indicatrices. The circular section for each is shaded dark gray. In each case the crystal's refractive index for light vibrating parallel to the *c* axis (= optic axis) is ϵ. For negative crystals $\epsilon < \omega$ so that $(\epsilon - \omega)$ is negative in value. For positive crystals $\epsilon > \omega$ so that $(\epsilon - \omega)$ is positive in value.

hexagonal crystal. Collectively they may be referred to as the crystal's *principal indices.* Uniaxial materials fall naturally into two categories: (1) *uniaxial positive,* in which by definition the value $(\epsilon - \omega)$ is positive in sign, and (2) *uniaxial negative* crystals, in which $(\epsilon - \omega)$ is negative in sign (that is, $\epsilon < \omega$). If the indicatrix for each of these two types is drawn in its customary orientation, that is, with the *c* axis vertical (Fig. 12-5), the negative indicatrix (an oblate spheroid) may be mnemonically linked to a "−" sign and the positive indicatrix (a prolate spheroid) to the vertical stroke of a "+" sign. Regardless of sign, there is only one circular section in a uniaxial indicatrix; its radius is always equal to principal index ω, and it lies in a plane normal to the *c* axis.

The biaxial indicatrix

A biaxial crystal belongs to the orthorhombic, monoclinic, or triclinic system and possesses *three* significant indices of refraction, commonly symbolized as α, β, and γ. Of these, α and γ represent, respectively, the smallest and largest refractive indices exhibited by the crystal; β is intermediate in value between them.[2]

As in uniaxial crystals, the index of refraction for monochromatic light

[2] Alternative symbols are as follows:

Smallest index	α	n_α	n_p	n_x	nX	X	n_a	N_x	n_1
Intermediate index	β	n_β	n_m	n_y	nY	Y	n_b	N_y	n_2
Largest index	γ	n_γ	n_g	n_z	nZ	Z	n_c	N_z	n_3

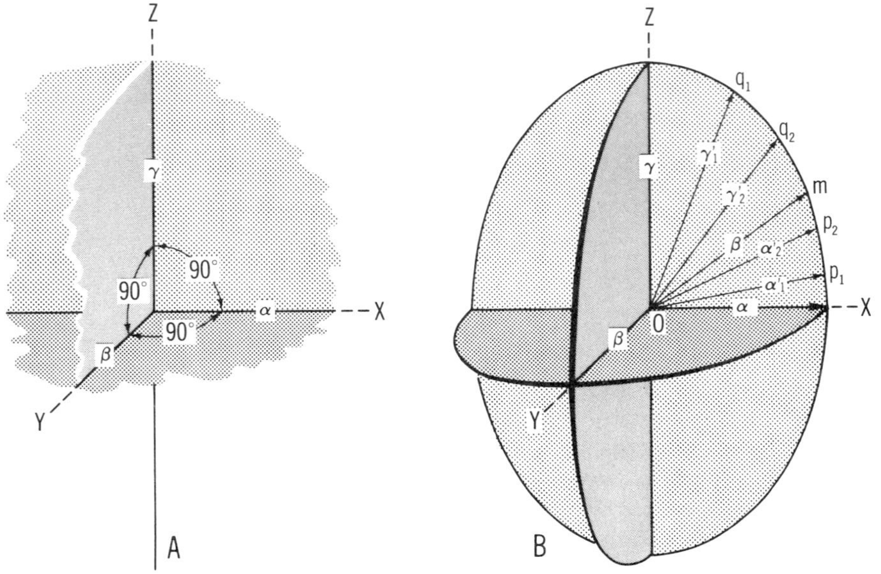

FIGURE 12–6

(A) The three, mutually perpendicular, principal vibration axes X, Y, and Z, and the common symbols for the indices of refraction of a biaxial mineral for light vibrating parallel to them.
(B) Elliptical distribution of the index of refraction (as shown by the vector lengths) for light vibrating parallel to Op_1, Op_2, Om, Oq_2, Oq_1 within the ZX plane.

varies with its vibration direction in biaxial crystals. Within such crystals, two unique mutually perpendicular directions exist for which the crystal exhibits its greatest and least refractive indices, γ and α, respectively. For a third direction, at right angles to these two, the crystal exhibits the intermediate refractive index β. These three directions, commonly symbolized as X, Y, and Z, are called the three *principal vibration axes*. The correspondence between these directions and the refractive indices that the crystals exhibit for light vibrating parallel to them is always as illustrated in Figure 12–6A.

For vibration directions not coinciding with X, Y, or Z the corresponding refractive indices lie between α and β or between β and γ in value. The former index is symbolized as α', the latter as γ'. Consequently, the following relationship always holds true,

$$\alpha < \alpha' < \beta < \gamma' < \gamma$$

We now consider a few of the random vibrations within the XZ plane in Figure 12-6B. The varying lengths of these radii, drawn to represent the crystal's refractive index for that vibration, define an ellipse whose semiminor axis is α and whose semimajor axis is γ. One of these radii, for example, Om must exactly equal β. Similarly, the lengths of random vibrations in the lightly shaded YZ plane also define an ellipse (semiminor axis, β; semimajor axis, γ), and those lying in the (horizontal) XY plane define a third ellipse (semiminor axis, α; semimajor axis, β). Figure 12-6B thus illustrates the basic framework of the biaxial indicatrix.

Countless planes hinging on the Z axis (in addition to the XZ and YZ planes of Fig. 12-6B) could be drawn; in Figure 12-7A unshaded planes 1, 2, and 3 are examples. As is demonstrated in detail for plane 2, the crystal's refractive indices vary with vibration direction in each such plane. The vectors representing the crystal's indices for light vibrating parallel to them vary in length like radii of an ellipse. The semiaxes of this ellipse for plane 1 are γ and α_1'; for plane 2 they are γ and α_2'; and for plane 3 they are γ and α_3'. In each case the semiminor axis of the ellipse, that is, α_1', α_2', or α_3', is also a radius of the horizontal $\alpha\beta$ ellipse. Since $\alpha' < \beta < \gamma$, in each of the ellipses hinging on Z there exist radii

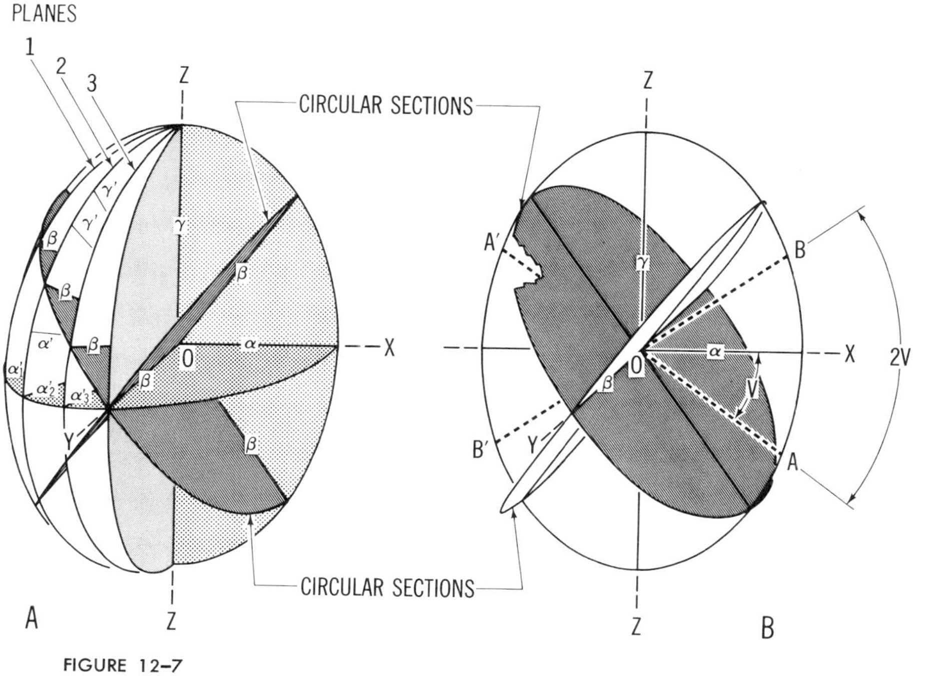

FIGURE 12–7

(A) Variation of refractive index in a biaxial crystal as shown by ellipses (whose radii are proportional in length to the crystal's refractive index for light vibrating parallel to them). Planes 1, 2, and 3 are typical of the numerous ellipses that could be drawn to hinge on the Z axis. All such ellipses contain one radius equal to β in length, these radii lying in the same planes (ruled). These planes, the circular sections, intersect in principal axis Y.
(B) The biaxial indicatrix; that is, simply an imaginary three-dimensional ellipsoid whose radii are proportional to the crystal's refractive indices for light vibrating parallel to them. The normals to the circular sections, namely AA' and BB', are the two optic axes. The circular section perpendicular to AA' has been left unshaded.

equal to β in length. All of these β radii lie in the same plane and define two circles (of radius β), each closely ruled in Figure 12-7A.

The shape of the biaxial indicatrix has now emerged (Fig. 12-7B). It is a three-dimensional ellipsoid, all central sections of which are ellipses except for two. These two, one unshaded and one closely ruled in Figure 12-7B, are the *circular sections*, the radius of each being β. The two normals to these circular sections, AA' and BB', are called the (primary) *optic axes* (a term often abbreviated as O.A.). Since there are two such optic axes (each comparable with the unique optic axis of the uniaxial indicatrix), the origin of the name "biaxial indicatrix" becomes obvious.[3] The two optic axes always lie within the XZ ($\alpha\gamma$) plane, which is therefore called the *optic plane*. Since the principal vibration direction Y is always normal to the optic plane, it is called the *optic normal* (abbreviated O. N.). The acute angle between the two optic axes is called the *optic axial angle* or simply $2V$. The obtuse axial angle, that is, angle BOA', equals $180° - 2V$.

The principal vibration axis that bisects the (acute) optic axial angle (for example, X in Fig. 12-7B) is called the *acute bisectrix* (abbreviated A. B. or symbolized as Bx_a). The principal vibration axis that bisects the obtuse axial angle is called the *obtuse bisectrix* (abbreviated O. B. or symbolized as Bx_o). In Figure 12-7A, $X(\alpha)$ is shown to be the acute bisectrix whereas $Z(\gamma)$ is the obtuse bisectrix. For many minerals this will be reversed; that is, Z will be the acute bisectrix and X will be the obtuse bisectrix. In biaxial minerals the optic sign depends on this relationship. Biaxial (+) crystals are defined to be those for which $Z(\gamma)$ is the acute bisectrix; biaxial (−) crystals are those for which $X(\alpha)$ is the acute bisectrix. For (+) minerals, β is closer in value to α than to γ. For (−) minerals, β is closer to γ than to α. The biaxial indicatrix of a mineral is usually drawn with Z vertical.

The angle between the two optic axes, if denoted as $2V$, is by definition the acute angle between them. An increasingly popular convention, however, uses the symbols $2V_x$ or $2V_z$ to denote the angle between the optic axes according to whether it is measured across the X direction or across the Z bisectrix (Fig. 12-8). The angles so defined may then exceed 90°. The sum of $2V_x$ plus $2V_z$ for a given crystal, for example, either crystal in Figure 12-8, always equals 180°. The statement that a biaxial crystal is optically negative in sign with $2V = 37°$ may be succinctly restated in this new symbolism as $2V_x = 37°$ or, alternatively, as $2V_z = 143°$. Either of these latter statements implies that X bisects the acute angle between the optic axes and that Z bisects the obtuse angle between them, and that the mineral is optically negative.

For anisotropic minerals the difference between their two most divergent refractive indices, that is, $(\epsilon - \omega)$ or $(\omega - \epsilon)$ for uniaxial crystals and $(\gamma - \alpha)$ for biaxial crystals, is known as their *birefringence*, the word "refringence" being a synonym for refraction.

[3] In geometrical terminology the biaxial indicatrix is a triaxial ellipsoid in reference not to the optic axes but to its three, mutually perpendicular, principal axes, X, Y, and Z.

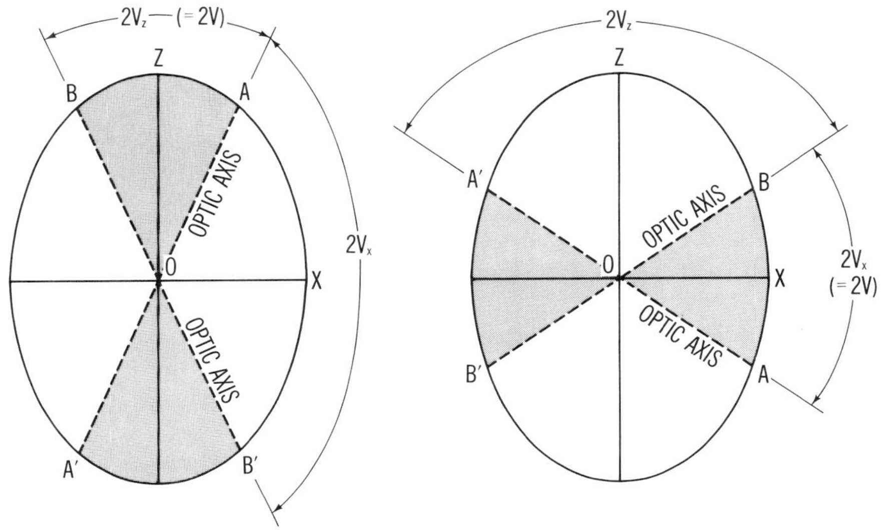

FIGURE 12–8

Relationship between $2V_x$ and $2V_z$. By definition, $2V$ equals the acute optic angle, that is, $2V_x$ or $2V_z$, whichever is smaller. The dashed lines represent the optic axes.

Transmitted colors and absorption

The intensity of a light beam decreases with passage through a material, some of the light energy being converted to heat while in transit. This effect, called *light absorption*, is more pronounced in some media than in others. Many substances exhibit a "general absorption" of all the wavelengths of visible light (Fig. 12-9, glass C). Sunlit stained-glass windows are examples of the color effects produced by selective absorption. The blue glass preferentially absorbs the red wavelengths of the incident sunlight but transmits the blue, relatively unabsorbed, to the room interior. Red glass, on the other hand, preferentially transmits red light and absorbs the blue wavelengths (see Fig. 12-9, glass B).

Selective absorption is responsible for the transmission colors of certain minerals and crystals; the color transmitted is often a diagnostic property. In crystals the color is said to be *idiochromatic* if it is produced through selective absorption by the crystal itself, and *allochromatic* if it is the result of selective absorption (or scattering) of light by minute impurities dispersed through the crystal.

The degree of absorption of light by a crystal is strongly dependent on its thickness. The relationship is expressed in Lambert's law,

$$\frac{I}{I_0} = e^{-kt} \qquad \text{(Eq. 12-4)}$$

where I_0 and I, respectively, signify the intensity of a beam of monochromatic

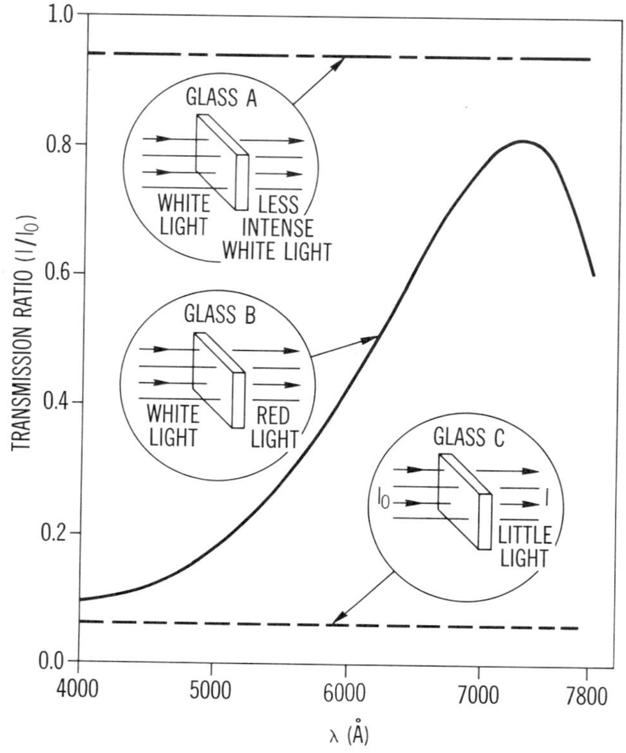

FIGURE 12–9

Light transmission curves for plates of transparent colorless glass (A), red glass (B), and dark gray, almost opaque glass (C). I_0 and I, respectively, signify the intensity of the light before entry and after passage through the glass.

light before and after passage through a thickness t. The symbol e is the base of the natural logarithms and k is the *absorption coefficient* or *absorbance index* of the material in which the beam is traveling. Equation (12-4) may also be expressed as

$$\ln \frac{I}{I_0} = -kt \qquad \text{(Eq. 12-4')}$$

where "ln" signifies a natural logarithm.

The value of the absorption coefficient may vary for light of different wavelengths. For example, if we were to plot for glass B in Figure 12-9 the value of k (that is, $\ln I/I_0 \div -t$) on the ordinate instead of I/I_0, the resultant curve would slope downward to the right. In other words, for glass B the absorption coefficient decreases steadily toward the longer (red) wavelengths of light. The advantage of plotting the absorption coefficient k instead of the transmission ratio is that k is independent of thickness.

The thickness of the crystal may even affect the eye's interpretation of the transmitted color. For example, if glass B were 10,000 times thicker than the plate whose transmission curve is shown in Figure 12-9, its transmission curve would be a nearly horizontal line even below the (dashed) curve for glass *C*; in this thickness, glass *B* would transmit little light for any wavelength. This illustrates why some crystals appear black and opaque for certain thicknesses, but at their thin edges may transmit a particular color. On the other hand, for a plate of glass *B* which is only 10^{-4} times as thick as the plate whose transmission curve is shown in Figure 12-9, the transmission curve would be similar to that for colorless glass *A*. Thus some minerals that are faintly colored in large masses may appear to be colorless in small grains or thicknesses, whereas minerals that are black in large masses may appear lighter or even white when ground into a powder.

Optical Dispersion and Orientation[4]. The angular divergence and resolution into spectral colors of white light as it obliquely crosses an interface between two media (Fig. 12-10A) constitutes *optical dispersion*. The divergence between light rays of different wavelengths (λ_1, λ_2, λ_3, and so on) increases to the extent that the intersecting *dispersion curves* for the two media, that is, curves that relate their change in refractive index n with wavelength λ, differ in steepness of slope (Fig. 12-10B). If the relation of n to λ is plotted on Hartmann dispersion net paper, particularly for media with refractive indices below 1.65, the dispersion curves usually become straight lines (Fig. 12-10C).

The dispersive ability of an isotropic material may be expressed as its *coefficient of dispersion* $(n_F - n_C)$ or preferably by its *dispersive power* $(n_D - 1)/(n_F - n_C)$. Sometimes, unfortunately, *dispersive power* is taken as the reciprocal of the foregoing value, that is, as $(n_F - n_C)/(n_D - 1)$. In general, dispersion curves have steeper slopes (1) for liquids as compared to solids, (2) for materials with high rather than low refractive indices, and (3) for substances containing transition elements such as iron and titanium.

For uniaxial crystals the dispersion curves for ϵ and ω are seldom parallel. In the rare instance of metatorbernite, $Cu(UO_2)_2P_2O_8$, the ϵ and ω curves intersect (Fig. 12-11A) so that metatorbernite is isotropic for $\lambda = 5120$ Å, uniaxial (−) for $\lambda < 5120$ Å, and uniaxial (+) for $\lambda > 5120$ Å. For biaxial crystals a dispersion curve exists for α, for β, and for γ (Fig. 12-11B). As a result $2V$ differs for different wavelengths of light because its value depends on the values of α, β, and γ as follows,

$$\cos V_Z = \frac{\alpha}{\beta}\sqrt{\frac{(\gamma + \beta)(\gamma - \beta)}{(\gamma + \alpha)(\gamma - \alpha)}} \qquad \text{(Eq. 12-5)}$$

In consequence of the dispersion of α, β, and γ, the optic axes occupy different positions within the crystal for different wavelengths of light. This *dispersion of the optic axes* may be so pronounced for some crystals it can be observed with a polarizing microscope. Biaxial crystals showing dispersion of the optic axes are divisible into two classes: (1) those for which $2V$ for red light noticeably exceeds

[4] This discussion is abbreviated from a more complete treatment (Bloss, 1961, 1971).

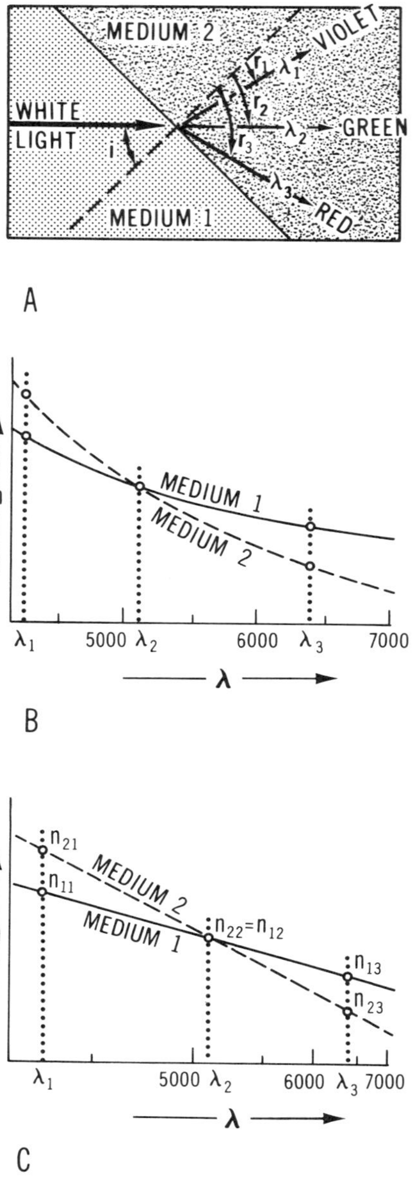

FIGURE 12–10

(A) Dispersion of a ray of white light into colored rays after it strikes the interface between media 1 and 2 at an *angle of incidence* i (measured from the dashed perpendicular to this interface). The refracted rays are at differing *angles of refraction* (r_1, r_2, r_3) relative to the dashed perpendicular. Any ray of monochromatic light, as it crosses an interface between two isotropic media, must obey Snell's law which

$2V$ for the shorter violet or blue wavelengths, this being expressed as $2V_r > 2V_v$ or simply $r > v$ and (2) those for which $r < v$. Extreme separation of the optic axes for red and violet or blue light may be indicated, for example, as $r > v$ strong, or as $r \gg v$.

Optical orientation denotes the orientation of a crystal's principal vibration axes relative to its crystallographic axes. For isotropic media the optical indicatrix is a sphere and thus lacks any unique axes to relate to the crystallographic axes. For hexagonal, trigonal, or tetragonal crystals, the unique principal vibration axis, that is, the vibration direction for which the crystal exhibits the index ϵ, always coincides with the c axis. In orthorhombic crystals the X, Y, and Z axes coincide (for all wavelengths of light) with the crystallographic axes, according to one of the six following permutations:

	(1)	(2)	(3)	(4)	(5)	(6)
$a =$	X	X	Y	Y	Z	Z
$b =$	Y	Z	X	Z	X	Y
$c =$	Z	Y	Z	X	Y	X

For monoclinic crystals a principal axis X, Y, or Z coincides (for all wave-

---

reads

$$n_i \sin i = n_r \sin r$$

In this equation i and r are the ray's angle of incidence and angle of refraction, whereas n_i and n_r, respectively, symbolize the index of refraction for the medium containing the incident ray and for the medium containing the refracted ray (for this particular wavelength of monochromatic light). The variation with wavelength of $\sin r/\sin i$ $(= n_i/n_r$, as per Snell's law) indicates that a medium has different refractive indices for light of different wavelengths.
(B) The change in refractive index n relative to wavelength λ, that is, dispersion curves, for media 1 and 2 are plotted on ordinary graph paper and (C) on Hartmann dispersion net paper. As shown, dispersion curves plotted on the Hartmann paper frequently become straight lines and are fixed if two points on them are known. For light of wavelengths λ_1, λ_2, and λ_3 whose angles of refraction were r_1, r_2, and r_3 in (A), let n_{11}, n_{12}, n_{13}, respectively, represent the refractive indices of medium 1 and n_{21}, n_{22}, n_{23} those for medium 2. If i represents the angle of incidence of the non-dispersed ray, then according to Snell's law,

$$\frac{n_{11}}{n_{21}} = \frac{\sin r_1}{\sin i}; \qquad \frac{n_{12}}{n_{22}} = \frac{\sin r_2}{\sin i}; \qquad \frac{n_{13}}{n_{23}} = \frac{\sin r_3}{\sin i}$$

(C) If n_{11}, n_{12}, n_{13}, n_{21}, n_{22}, and n_{23} have the relative values plotted here, the above equations indicate that $r_1 < i$ (because $n_{11} < n_{21}$), that $r_2 = i$ (because $n_{12} = n_{22}$), and that $r_3 > i$ (because $n_{13} > n_{23}$). Consequently, $r_1 < r_2 < r_3$ and thus the ray of white light shown in (A) is dispersed into a spectrum of colored rays of which only three rays (representing wavelengths λ_1, λ_2, and λ_3) are drawn in (A). Colored rays representing wavelengths intermediate between λ_1 and λ_2 (or between λ_2 and λ_3) would occur between the rays in (A) but have not been drawn. Figures after Bloss (1971), courtesy *Encyclopedia of Earth Sciences*, R. W. Fairbridge, ed. Princeton, N. J.: Van Nostrand-Reinhold.

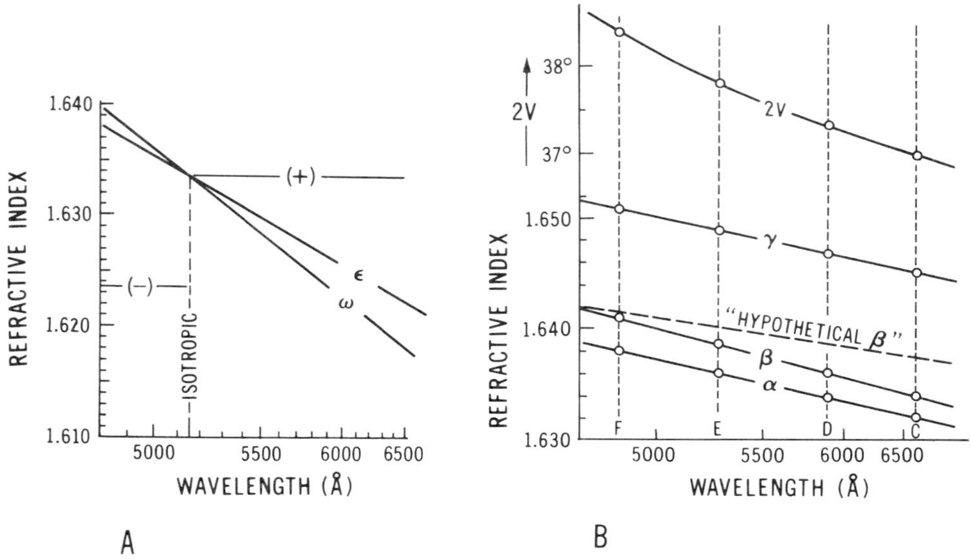

FIGURE 12–11

(A) Dispersion curves for ϵ and ω in metatorbernite show it to be isotropic for wavelength 5120 Å, uniaxial (+) for wavelengths $>$5120 Å, and uniaxial (−) for wavelengths $<$5120 Å.
(B) Dispersion curves for α, β, and γ and resultant change in $2V$ with wavelength for the orthorhombic mineral barite $BaSO_4$. If β had changed in value as indicated by the dashed line labeled "hypothetical β," the slope of the $2V$ curve would be reversed so that then $2V_C > 2V_F$ or $r > v$.

lengths of light) with the single 2- or $\bar{2}$-fold axis selected to be the b axis (second setting) or c axis (first setting). The remaining two principal axes then lie in the plane perpendicular to this 2- or $\bar{2}$-fold axis; they occupy slightly different positions within this plane for different wavelengths of light. To specify the optic orientation of a monoclinic crystal, we state (1) the optical identity of the symmetry-defined axis, for example, $b = X$ or $b = Z$ and (2) the angle between one of the remaining crystallographic axes and principal vibration axes. To illustrate, for gypsum ($CaSO_4 \cdot 2H_2O$), $b = Y$ and $Z \wedge c = 52°32'$. The angle 52°32′ is usually given a plus sign to indicate that Z lies within obtuse angle β (Fig. 12-12), whereas −52°32′ would locate it in the acute angle (180° − β). Some authorities, unfortunately, adopt a reverse convention. The angle 52°32′ might vary slightly according to the wavelength of light being used. In other words there is a dispersion of the principal vibration axis Z in gypsum and thus of X, which is perpendicular to Z. *Any principal vibration axis that does not coincide with a 2- or $\bar{2}$-fold axis in a biaxial crystal may exhibit dispersion.* Often this dispersion is so slight as to escape detection.

In triclinic crystals the X, Y, and Z axes may each exhibit dispersion and none coincides, except by chance, with a crystallographic axis.

Optical activity, *rotatory power*, or *rotatory polarization* denotes the ability of a crystal to rotate the plane of polarization even for light beams traveling normal to a circular section of the crystal's indicatrix. After emergence from the crystal, as the light travels toward the observer the beam's plane of polarization may have been rotated either to the left (counterclockwise) or to the right (clockwise) relative to its plane upon entry. If the former, the crystal is *levorotatory*; if the latter, it is *dextrorotatory*. In anisotropic crystals such rotation of the plane of polarization is most readily observed for beams traveling along the optic axes (whereas any direction suffices for isometric crystals). Optical activity appears restricted to those crystals that lack a center of symmetry. Thus it has been observed in classes 1, 2, 222, 4, 422, 3, 32, 6, 622, 23, 432, $\bar{4}$, and $\bar{4}2m$ (but not as yet in m or $mm2$ although they too lack a center). Observation of optical activity for a crystal presumably excludes the presence of a center of symmetry but nonobservation does not (because the effect may have been so weak as to escape detection, especially in small crystals). Interestingly, solutions containing

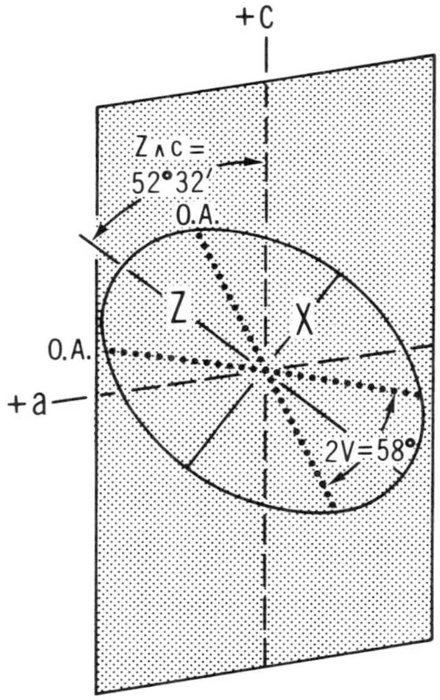

FIGURE 12–12

Section parallel to (010) of the monoclinic crystal gypsum to show the angles between its principal vibration axes X, Y, and Z and crystallographic axes a, b, and c. The obtuse angle β is the angle between the $+a$ and $+c$ axes. The acute angle $180 - \beta$ lies between $+c$ and $-a$. Axis Y $(= b)$ is normal to the page. Directions X, Y, and Z and the two optic axes (dotted) belong to the crystal's indicatrix for sodium light. The crystal's indicatrix for light of another wavelength would have its Y axis in common with this indicatrix but its X and Z axes slightly rotated relative to those for sodium light. After Bloss (1971), courtesy *Encyclopedia of Earth Sciences*, R. W. Fairbridge, ed. Princeton, N. J.: Van Nostrand-Reinhold.

molecules that lack a center of symmetry, inversion axis, or plane of symmetry will sometimes exhibit optical activity. Thus aqueous solutions of the sugar dextrose are dextrorotatory, whereas those of levulose are levorotatory.

The rotatory power of crystals is usually expressed in degrees of rotation per millimeter of travel. This value will vary with direction of travel in a crystal but in anisotropic crystals is usually reported for light traveling along an optic axis because optical activity along these directions is experimentally easier to measure. Rotatory power also varies with temperature and wavelength. Usually it is greater for the shorter wavelengths. Thus for sodium light traveling along an optic axis in quartz, α_D the rotatory power at 20°C is 21°43′/mm of travel; for blue light α would have a larger value than 21°43′, whereas for red light it would be less.

Reflection of Light

Body color

Body color represents the color of a sizable crystal or fragment. If viewed in daylight or the equivalent, it may aid in the crystal's identification, particularly for *idiochromatic* (self-colored) crystals. This is, however, less true for *allochromatic* crystals, which by definition contain a finely dispersed extraneous colorant that imparts their body color. Because the identity and color of the dispersed material may vary, different specimens of the same allochromatic crystal, particularly if they formed under different environmental conditions, may exhibit widely different body colors. For example, quartz (SiO_2), which is clear and colorless if pure, may contain small flakes of hematite, small needles of rutile or tourmaline, or even bubblelike cavities containing gas or liquid and thus appear pinkish, bluish, or opaque white, respectively.

Even idiochromatic crystals may exhibit variations in color. For example, if they are members of a solid solution series between two differently colored end members, their color will depend on their composition within the series. Thus olivine specimens range from green if near forsterite in composition to brown or black if near fayalite. Moreover, exposure to ultraviolet rays or X-rays may change the color of a specimen. Clear colorless quartz becomes black (smoky) quartz if exposed to X-rays. In the Andes Mountains and in desert areas, ultraviolet rays from the sun soon impart a violet color to previously colorless glasses.

In pure idiochromatic crystals undamaged by radiation, the ions or groups of ions responsible for their color are called *chromophores*. In crystalline structures, elements 22 to 29, that is, the transitional metals titanium, vanadium, chromium, manganese, iron, cobalt, nickel, or copper, are especially effective chromophores, perhaps because of (1) the loosely held electrons in their outer shells and (2) their ability to adopt several different valence states. The presence of an element in two valence states in a crystal may produce particularly intense coloration. For example, crystals that contain only Fe^{2+} or only Fe^{3+} may possess relatively pale colors, whereas if both Fe^{2+} and Fe^{3+} are present, deep green, blue, or black colors result. The mineral vivianite, $Fe_3(PO_4)_2 \cdot 8H_2O$, contains only ferrous iron and is thus practically colorless when first mined; in air, however, probably because some of the ferrous iron oxidizes to ferric iron, the color becomes deep blue or green.

Despite the foregoing complexities, color may be an important clue to a crystal's identity. Observations on freshly fractured surfaces avoid the possibility that a surface tarnish is masking the true color. Such tarnishes occasionally develop rapidly as a crystal's surface is oxidized in air, particularly in moist air. Thus the surface of a specimen of bornite (Cu_5FeS_4) is reddish copper in color when freshly fractured but rapidly tarnishes to an iridescent bluish purple.

Streak

A pencil point drawn across paper leaves a black line consisting of powdery graphite. Similarly, a mineral drawn across a plate of white unglazed porcelain leaves a line that is called the mineral's *streak*. The color of a mineral's streak is more reliable than body color as an aid in identification. However, minerals harder than the streak plate ($\mathbf{H} = 7$) leave behind no powdery residue. If no streak plate is at hand, the specimen may be struck and bruised with a geological hammer or scratched with a knife, thereby producing a powder whose color will be that of the mineral's streak.

Luster

The nature and degree of light reflectance from a crystal's surface determine its *luster*, that is, its general appearance in reflected light. Luster is divisible on the basis of transparency into two basic types: (1) *metallic luster*, which is associated with opacity to light and therefore with crystals whose streaks are either black, strongly colored, or themselves metallic and (2) *nonmetallic luster*, which is possessed by crystals relatively transparent to light whose streaks are therefore white or weakly colored. Mineral examples of metallic luster are pyrite, gold, silver, and copper. Several varieties of nonmetallic luster are recognized, namely the following:

Adamantine. A brilliant luster as for diamond

Vitreous. A bright luster as for glass

Greasy. A luster as if thinly coated with oil or grease

Resinous. A luster as of resin

Pearly. An iridescent whitish luster as for pearls (observed on surfaces that are parallel to cleavage directions between which air separations exist)

Silky. A luster as for silk (observed for aggregates of parallel fibers or needles of a mineral)

The percentage of light reflected has a major influence on luster and depends on several factors including the smoothness of the surface examined. On a crystal, smoothness and thus luster may vary between faces belonging to different forms or even along different directions in the same face. Consequently, luster may be demonstrably anisotropic; for example, apophyllite's luster is pearly on {001} but vitreous for other faces. Specimens that consist of millions of randomly oriented, submicroscopic crystals possess surfaces that are submicroscopically rough and that therefore exhibit an "earthy luster", that is, a dullness that is actually just the opposite of luster.

If equally smooth surfaces of different materials are compared, R (the percentage of light reflected after perpendicularly striking each smooth surface) is related to the material's refractive index n and to its absorption coefficient k by the Bohr equation

$$R = \frac{(n - 1)^2 + n^2k^2}{(n + 1)^2 + n^2k^2} \qquad \text{(Eq. 12-6)}$$

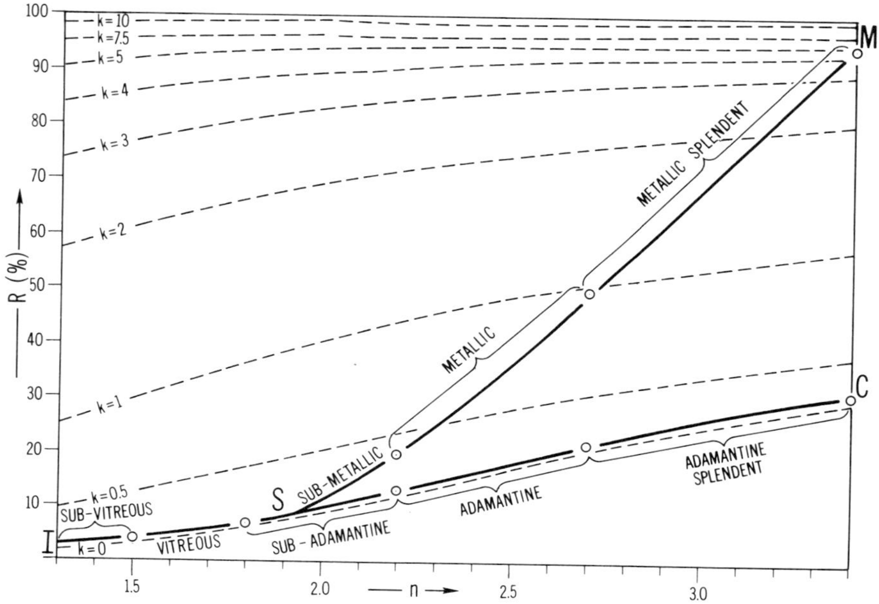

FIGURE 12–13

The dashed lines represent the Bohr relationship (Eq. 12-6) between refractive index n and R, where R is the percentage of light reflected from normal incidence on a surface for materials whose absorption coefficients k are 0, 0.5, 1.0, ... Solid line ISC shows the relationship to n (and to R) of the five classes of luster for transparent crystals in which k is near zero in value. Line SM, which might better be drawn as a wide band, shows the relationship to R (but only approximately to n and k) of the grades of luster for colored materials in which the value of k becomes appreciable. The limits for the five grades of luster in transparent or semi-transparent solids are those set by Povarennykh (1964), namely: (1) *subvitreous* ($n = 1.3 - 1.5$); (2) *vitreous* ($n = 1.5 - 1.8$); (3) *subadamantine* ($n = 1.8 - 2.2$); (4) *adamantine* ($n = 2.2 - 2.7$); and (5) *adamantine splendent* ($n = 2.7 - 3.4$). For slightly translucent to opaque materials, his limits are (3) *submetallic* ($R = 8$ to 20 percent), (4) *metallic* ($R = 20$ to 50 percent), and (5) *metallic splendent* ($R = 50$ to 95 percent).

The dashed curves in Figure 12-13 are based on this equation and illustrate the relation between R and n for selected values of k. For exceedingly transparent minerals, that is, those for which k practically equals zero, Equation (12-6) becomes essentially

$$R = \frac{(n-1)^2}{(n+1)^2} \qquad \text{(Eq. 12-6')}$$

Povarennykh (in Battey and Tomkeieff, 1964, p. 488) divides transparent minerals into five grades of luster on the basis of their reflectivity R as shown along line ISC in Figure 12-13. This five-fold division is essentially based on refractive index upon which, for transparent crystals, the value of R depends almost solely. As a rule covalent bonding results in a higher value for refractive index than does ionic bonding. Hence minerals in which covalent bonding predominates would usually possess lusters corresponding to reflectivities along line SC in Figure 12-13, whereas for ionically bonded minerals subvitreous or vitreous lusters would predominate. Opaque minerals possess metallic bonding or bonding that grades between metallic and covalent and these are subdivided by Povarennykh into three classes of luster (along line SM in Fig. 12-13). Minerals with pure metallic bonding have lusters near M; those in which such bonding becomes increasingly covalent possess lusters nearer to S on line SM in Figure 12-13. For his scale of luster, Povarennykh cites index minerals as follows: *subvitreous,* fluorite; *vitreous,* topaz; *subadamantine,* zircon; *adamantine,* diamond; *adamantine splendent,* cinnabar; *submetallic,* ilmenite; *metallic,* molybdenite; *metallic splendent,* gold.

In a very general way the relationship between a crystal's refractive index n, light absorption coefficient k, its reflectivity R, and the type of bonding dominantly present is

	n	k	R
Metallic bonding	High	High	High
Covalent bonding	High	Low	Moderate
Ionic bonding	Low	Low	Low

Structural Controls of Optical Properties

A crystal's optical properties depend on its structure, composition, and bonding. Indeed they provide clues as to the nature of an unknown structure. W. L. Bragg's (1924) explanation of the relation between a crystal's refractive indices and its structure requires first a closer look at the nature of light. Maxwell's equations indicated that a light wave represented a time-varying electric field **E** inextricably associated with a similarly varying magnetic field **B** at 90° to it (Fig. 12-14A). The "vibration directions" alluded to earlier in the chapter in fact represented this time-varying field **E**. For simplicity the associated magnetic field **B** was omitted from both discussions and drawings, a practice that will

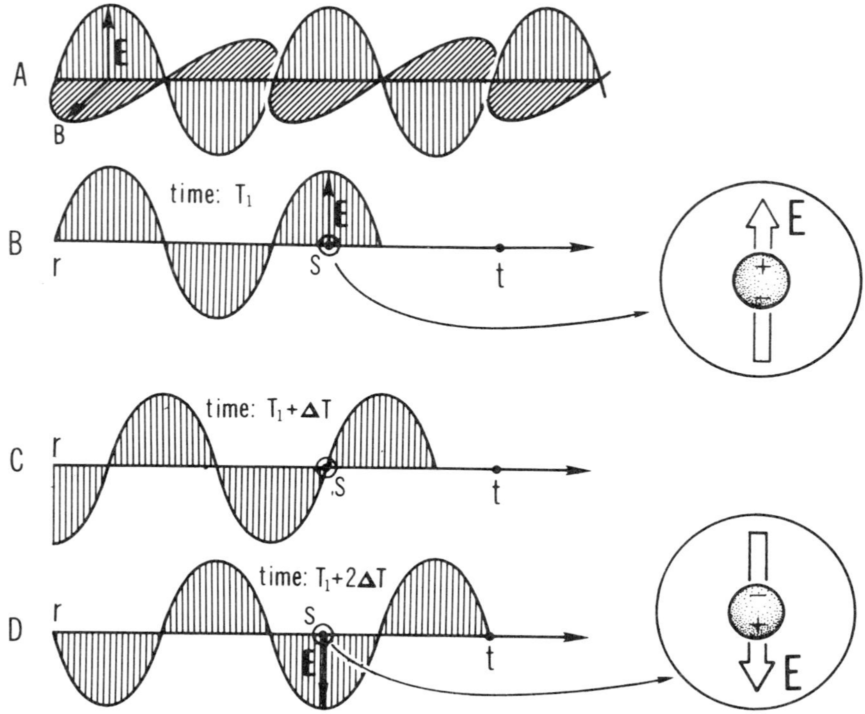

FIGURE 12–14

(A) A polarized light wave is drawn according to the Maxwell concept. Maxwell considered a light wave to be a time-varying electric field **E** inextricably associated with a similarly varying magnetic field **B** at 90° to it. For simplicity only the field **E** is depicted in drawings (B) to (D).
(B) A polarized wave (in air $\lambda = 6000$ Å) passes from r toward t in a crystal ($n = 1.50$). If rt coincided with a row of tangent oxygen ions, one wavelength of light would span, that is, extend along (ca.) 3300 oxygen ions. Considering a specific oxygen ion (at s) during the point in time when the maximum amplitude of **E** reaches s, this ion (circular inset) is *electronically polarized* as shown. By convention, vector **E** points opposite to the direction in which it constrains electrons to move; thus the cloud of electrons would be shifted slightly downward so that the atom's negative centroid is now located below its positive centroid (see Fig. 11-33A,B on p. 401).
(C) As a node passes the ion at s, the negative centroid returns to its rest position and the ion is momentarily not polarized.
(D) One-half wavelength or one-half of the wave's period after (B) the ion at s is polarized in the opposite direction.

continue. Thus we let Figures 12-14B, C, and D represent the sinusoidally varying electric field **E** inherent to a plane-polarized wave of monochromatic light ($\lambda = 6000$ Å) as it travels from r to t in a crystal. This field **E** is shown at three instants in time: T_1 (Fig. 12-14B); $T_1 + \Delta T$ (Fig. 12-14C); and $T_1 + 2\Delta T$ (Fig. 12-14D) where ΔT equals 0.5×10^{-15} sec. A given atom at point s in the crystal is *electronically polarized* by field **E**, the "center of gravity" of the electron cloud about its nucleus shifting first one way (Fig. 12-14B), then oppositely (Fig. 12-14D) as the field **E** reverses itself every 1.0×10^{-15} sec (for $\lambda = 6000$ Å).

Atomic polarizability

Such separation between the centroids of positive and negative charge for an atom or ion results largely from shifts of the electron cloud, the positive nucleus being too massive to follow the almost instantaneous reversals of **E**. This *electronic polarization* of an atom or ion thus depends not only on field strength **E**, but also on the atom or ion's susceptibility to such polarization, that is, its *atomic polarizability* α, which is usually cited in units of volume ($Å^3$ or 10^{-24} cm^3). Because the electron cloud surrounding the atomic nucleus is chiefly affected, high values of α are observed for atoms with loosely held orbital electrons, namely, (a) those with high atomic numbers and therefore numerous shells of orbital electrons or (b) anions as opposed to cations, because anions have orbital electrons in excess of the positive charge on their nucleus and thus hold their orbital electrons more loosely than do cations. For example, the anion I^- has a high value of α because it conforms to characteristics (a) and (b).

Refractive index relative to polarizability

The refractive index n of a material increases to the extent that it contains atoms or ions of high polarizabilities. For an inert gas, Hartshorne and Stuart (1960, p. 147), following Bragg, show that

$$\frac{n^2 - 1}{n^2 + 2} = \frac{4}{3}\pi N'\alpha \qquad \text{(Eq. 12-7)}$$

where N' is the number of atoms per unit volume, here assumed to be 1 cm^3, and α is the polarizability of these individual gas atoms expressed in these same volume units. Inserting 1, 2, or 3 for refractive index n in Equation (12-7), the value of $\frac{4}{3}\pi N'\alpha$ increases from 0.0 to 0.17 to 0.43. Conversely an increase in the value of $\frac{4}{3}\pi N'\alpha$ (that is, of $N'\alpha$) would require an increase in refractive index. For an inert gas, therefore, two factors may increase refractive index: (1) an increase of N', that is, in the density of packing of the atoms in a unit volume, or (2) an increase of polarizability α for these atoms.

Equation (12-7) has been applied to crystals but with limited success in predicting refractive indices from polarizabilities, primarily because an atom or ion's polarizability α may change considerably when it is in close proximity to other atoms. However, the main trends are observed. Thus *the denser polymorph of a material always has a higher refractive index than the less dense one*. Moreover, silicate and oxide crystals in which the O^{2-} ions are closest packed will usually have refractive indices above 1.7, whereas for more open framework structures like quartz or feldspar the indices are below 1.6. Factor (2) in the preceding paragraph is observable in solid solutions wherein the more polarizable Fe^{2+} ion substitutes for Mg^{2+}, refractive index then increasing. For metallic atoms and in crystals in which metallic bonding predominates, the electron clouds around the atomic nuclei are so loosely held, and polarizabilities therefore so high, that electrons can actually be removed by the force of an electric vector (electrical conductivity). Thus metallic luster, electrical conductivity, and high refractive index are associable properties.

Anisotropy of polarizability

The polarizability α' for an atom or ion in a crystal structure will differ from α, its polarizability if its neighbors are randomly disposed about it as in a gas. Moreover α', and thus refractive index n, varies in value according to the direction of **E** in respect to the structure. Illustrating the principle with a simplified example, we note that each individual atom in a line of atoms is more susceptible to polarization if vector **E** is at 0° to this line (Fig. 12-15A) than if at 90° (Fig. 12-15B). For **E** parallel to the line, the negative pole of one atom faces the positive pole of its neighbor; thus the resultant attraction between them reinforces the polarization of each. For **E** at 90° to the line, the favorable negative-to-positive arrangement no longer exists, the positive poles of neighboring atoms now being closer together and unshielded from each other so as to detract from each other's polarity. For **E** at 90° to the line of atoms, their individual polarizabilities may thus be significantly reduced.

Following Hartshorne and Stuart (1960, p. 148), α' is related to α as follows:

$$\alpha' = \frac{\alpha}{1 - [(3\cos^2\phi - 1)/d^3]\alpha} \qquad \text{(Eq. 12-8)}$$

where d is the center-to-center distance between the atoms and ϕ is the angle of electric vector **E** to the line joining their centers. They then set $\alpha = 1 \times 10^{-24}$ cm^3 and successively insert values of 2, 3, 4, and 5 Å for d into Equation (12-8)

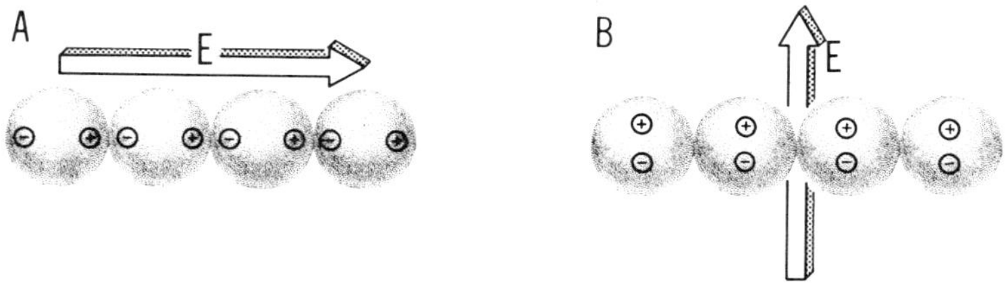

FIGURE 12–15

The polarization of the individual cations constituting a row is greatest if the electric field **E** is directed along the row (A), least if **E** is perpendicular to the row (B), this being graphically indicated by the greater separation of the encircled (+) and (−) centroids in (A) as compared to (B). This separation is greatly exaggerated in both (A) and (B). In (A) note that the positive centroid of one anion is brought closer to the negative centroid of its neighbor, a situation that makes the polarization easy by reinforcing it. In (B) this situation does not exist and polarization of the individual ions is hence more difficult to achieve.

to compute the following values of α':

	Computed α'			
	$d = 2$ Å	$d = 3$ Å	$d = 4$ Å	$d = 5$ Å
$\phi = 0°$	1.33	1.08	1.03	1.02
$\phi = 90°$	0.89	0.96	0.98	0.99
Average	1.11	1.02	1.00	1.00

Note that (1) the average value of α' decreases as interatomic distance d increases and that (2) α' differs most, between $\phi = 0°$ and $\phi = 90°$, for the smaller values of d.

In crystal structures consisting of sheets of closely packed atoms the average distance d between atoms in the sheets is less than that between sheets. Hence these atoms have a higher polarizability α' (and the crystal a larger refractive index) if vibration **E** is parallel to the sheets (Fig. 12-16A) rather than perpendicular to them. Thus crystals with sheet structures tend to be optically negative, their lesser refractive index occurring for vibrations perpendicular to the sheets. In contrast, structures containing chains of closely packed atoms have their largest index for vibrations parallel to the chain and tend to be positive in optic sign (Fig. 12-16B). If no planes or directions in a crystal are more closely packed than any others, the crystal will be isotropic or, if anisotropic, will possess low birefringence.

In a crystal structure two oxygens that represent the shared edge between two coordination polyhedra frequently have a smaller than usual center-to-center distance d. Thus they will have large polarizabilities and help produce higher refractive indices. Gibbs and Smith (1965, p. 2029) attribute the high densities and refractive indices of garnets in general, and of pyrope, $Mg_3Al_2(SiO_4)_3$, in particular, to the large percentage of such shared edges in the structure. In stishovite, a high-pressure polymorph of SiO_2, shared edges occur parallel to the c axis and account for the much larger index ($\epsilon = 1.845 \pm 0.005$) parallel to c as opposed to that perpendicular to c ($\omega = 1.800 \pm 0.005$).

Density and refractive index

Increased density, like refractive index, results from (1) closer packing and/or (2) substitution of atoms or ions with higher atomic numbers (and, in all likelihood, polarizabilities) for those with lesser. The sympathetic increase of refractive index with density has been described by the following equations (among others):

The Gladstone-Dale equation,

$$\frac{n-1}{\rho} = K_G \qquad \text{(Eq. 12-9)}$$

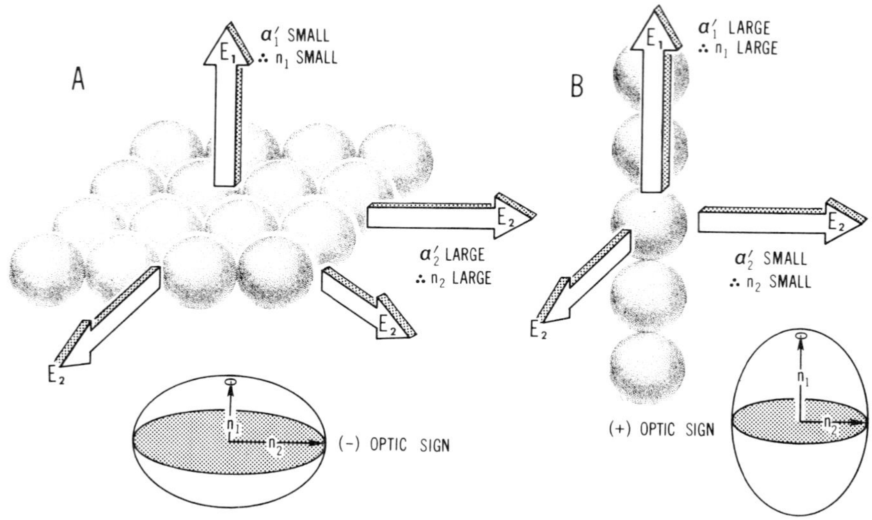

FIGURE 12-16

(A) If the electric field or vibration direction of light is perpendicular to a sheet of closely packed ions as, for example, $\mathbf{E}_1$, then the polarization α_1' of the individual ions is small because the situation resembles Figure 12-15B. Thus the crystal's refractive index n_1 for light vibrating parallel to $\mathbf{E}_1$ is also small. For light vibrating in a direction parallel to the sheet, for example, along any of the $\mathbf{E}_2$ directions shown, polarization α_2' of individual ions is large (see Fig. 12-15A) and consequently the refractive index of n_2 for this vibration direction would be large. Crystal structures consisting of closely packed sheets of atoms, stacked one above the other, possess their smallest refractive index for light vibrating (more or less) normal to the sheets. Such crystals are usually negative in optic sign.

(B) In structures containing chains or rows of densely packed atoms all parallel to a given direction, the atoms exhibit their greatest polarizability α_1' (and thus the structure possesses its greatest refractive index n_1) when the vibration of the light $\mathbf{E}_1$ is parallel to these rows. The atoms exhibit their least polarizability α_2' (and the structure possesses its least refractive index n_2) for light vibrating along directions which (like $\mathbf{E}_2$) are perpendicular to these rows. Usually such crystals are optically positive.

The Lorentz-Lorenz equation,

$$\frac{(n^2 - 1)}{(n^2 + 2)} \frac{1}{\rho} = K_{\mathrm{L}} \qquad \text{(Eq. 12-10)}$$

The Lichtenecker equation,

$$\frac{\log n}{\rho} = K_{\mathrm{Li}} \qquad \text{(Eq. 12-11)}$$

The Drude or Newton-Drude equation,

$$\frac{n^2 - 1}{\rho} = K_D \qquad \text{(Eq. 12-12)}$$

The Allen (1956) equation

$$\frac{n^3}{\rho} = K_A \qquad \text{(Eq. 12-13)}$$

where the constants K are known as the *specific refractivity* (K_G), *specific refraction* (K_L), the *Drude constant* (K_D), and the *specific refractive capacity* (K_A), respectively. Here the constant K_{Li} is called the *Lichtenecker constant* but generally has no accepted name. Of these equations only the Lorentz-Lorenz and the Drude equations have a theoretical basis in the physics of dielectrics. For anisotropic media the arithmetic mean refractive index, $(\epsilon + 2\omega)/3$ or $(\alpha + \beta + \gamma)/3$, is commonly used for n except in the Allen equation where $\epsilon\omega^2$ or $\alpha\beta\gamma$ are used in place of n^3.

Defining the *mean atomic weight* of a compound as its molecular weight M divided by the number of atoms p in its chemical formula (thus M/p for SiO_2 is

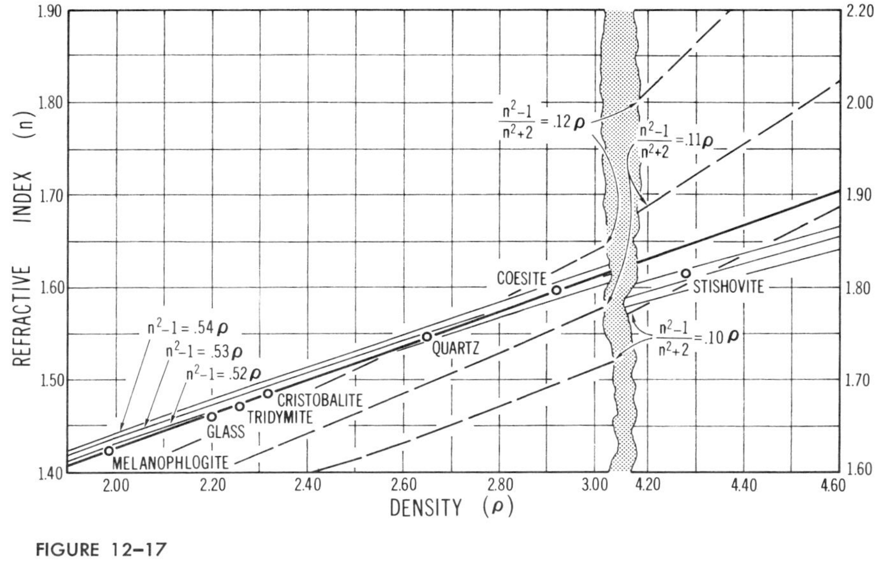

FIGURE 12–17

Relationship between density and refractive index for the polymorphs of SiO_2. The heavy solid line represents the linear Gladstone-Dale relation to which all SiO_2 polymorphs except stishovite conform. The three lighter solid lines represent the Newton-Drude equation (Eq. 12-12) if K_D equals 0.54, 0.53, or 0.52; the middle line of these, $K_D = 0.53$, fits the data fairly well for the polymorphs more dense than 2.6. The three dashed lines represent the Lorentz-Lorenz equation (Eq. 12-10) where K_L equals 0.10, 0.11, or 0.12.

20.03), Birch (1961) notes that many compounds in the earth's mantle have M/p values near 21. Anderson and Schreiber (1965) show that for minerals whose anion is oxygen and whose mean atomic weight is near 21, density is essentially related to index of refraction by the linear Gladstone-Dale equation, particularly at low densities. This is illustrated for the polymorphs of SiO_2 by the bold line in Figure 12-17. Skinner and Fahey (1963) believe that the deviation of stishovite from this linear relationship may result from the coordination change of Si^{4+} from 4 to 6 in stishovite. Actually the Allen equation best fits the data at all densities. The Lorentz-Lorenz equation (dashed lines in Fig. 12-17) and the Drude equation (fine lines) deviate at high densities in opposite directions from the linear relationship (bold line).

Crystal field theory

Crystal field theory relates the energy levels of electrons in the five outer *d* orbitals of a transition metal ion to the geometric distribution of negative charge and thus of neighbors (*ligands*) around this ion in a crystal structure. For a free ion devoid of neighbors the five *d* orbitals are *degenerate*, that is, of equal energies (Fig. 12-18A). If the ion is surrounded spherically by a negative charge, the *d*

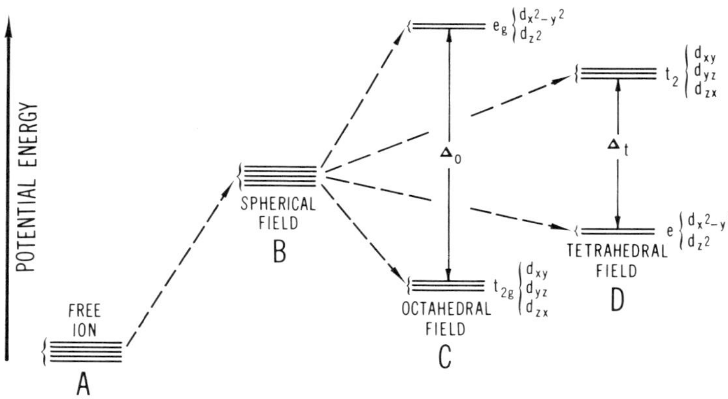

FIGURE 12–18

Schematic picture of the energy levels of the five *d* orbitals for a transition metal ion according to whether the ion is (A) a free ion that lacks neighbors like one in the gaseous state, (B) surrounded by a negative charge uniformly disposed as if on a spherical shell around it, (C) surrounded by six neighbors (*ligands*) that are located at the corners of a perfect octahedron centered around the metal ion, or (D) surrounded by four ligands at the corners of a perfect tetrahedron centered around the metal ion. From the equal energy or *degenerate states* shown in (A) and (B), the *d* orbitals split into two sets of unequal energy, e_g and t_{2g} in (C) or e and t_2 in (D), for reasons discussed in the text. If the octahedron or tetrahedron is located in a crystal structure, all energy levels in (C) and (D) are lowered so that even the higher energy levels, namely e_g in (C) and t_2 in (D), are below the levels in (A) because of electrostatic interactions with other ions.

orbitals remain degenerate but their energy levels will exceed those for the free ion because the spherical negative field electrostatically repels electrons in these orbitals and adds to their potential energy (Fig. 12-18B). If this same strength of negative field were supplied by six anions *octahedrally coordinated* about the metal ion, the resultant nonuniform field would raise the energy levels of the d_{z^2} and $d_{x^2-y^2}$ orbitals (called the e_g set), but lower those of d_{xy}, d_{yz}, and d_{zx} (set t_{2g}) relative to their levels for a spherical field (Fig. 12-18C). This split of the d orbitals into two energetically nonequivalent sets e_g and t_{2g} results because (1) set e_g (d_{z^2} and $d_{x^2-y^2}$) orbitals have lobes that are directed *toward* the anions (Fig. 12-19A) and hence encounter greater concentrations of negative charge than for the spherical field, whereas (2) the lobes for d_{xy}, d_{yz}, and d_{zx} (set t_{2g}) extend *between* these anions (Fig. 12-19B) to encounter lesser concentrations of negative charge than for the spherical field. Relative to their level for the spherical field, therefore, e_g is above it whereas t_{2g} is below it (Fig. 12-18C); the energy difference between e_g and t_{2g} is customarily symbolized Δ_o, where the subscript o indicates octahedral coordination.

For a tetrahedral distribution of anions around the metal ion, the two sets of d orbitals are conventionally symbolized as e and t_2, the g subscripts now being omitted. Picturing this tetrahedral distribution by locating the anions at alternate corners of a cube and the metal ion at the cube's center, we note that now the e orbitals extend between the anions so as to be the low-energy orbitals (Fig. 12-19C) whereas the t_2 orbitals extend more nearly toward the anions (Fig. 12-19D) to be the higher-energy set. The energy difference or split between the t_2 and e orbitals imposed by tetrahedral coordination is symbolized as Δ_t. The e_g orbitals approach anions more closely (Fig. 12-19A) than do t_2 (Fig. 12-19D). Consequently, e_g is at a higher level than t_2 in Figure 12-18C, D and thus Δ_o exceeds Δ_t.

For tetrahedra or octahedra in which the anions depart from their idealized regularity, some anions are closer to the metal ion than others. Consequently, energy differences may arise between individual orbitals in a set, for example, splitting of a t_{2g} orbital may occur.

Letting Δ represent either Δ_o or Δ_t, its magnitude also depends on the anions involved. Small highly charged anions produce strong fields around the metal ion and strongly repel any electrons in the orbitals directed toward these anions, raising their energy levels so as to cause Δ to be large. In contrast, large anions with low charges represent lesser concentrations of negative charge (and thus weaker fields) which permit Δ to be smaller. As discussed earlier (p. 217), a d^4, d^5, d^6, or d^7 metal ion adopts a low-spin state if Δ is large but a high-spin state if Δ is small. Thus, directing a remark momentarily to the subject of magnetism, a crystal structure that sets up a weak field around a d^4 to d^7 transition metal ion will cause this ion to adopt its high-spin, more magnetic state.

The wavelength of light absorbed by a transition metal ion depends on the energy difference Δ between its t_{2g} and e_g (or t_2 and e) orbitals. Wavelengths for which a photon equals Δ in value will be preferentially absorbed by the ion, an electron then being elevated from a t_{2g} to an e_g (or an e to a t_2) orbital through

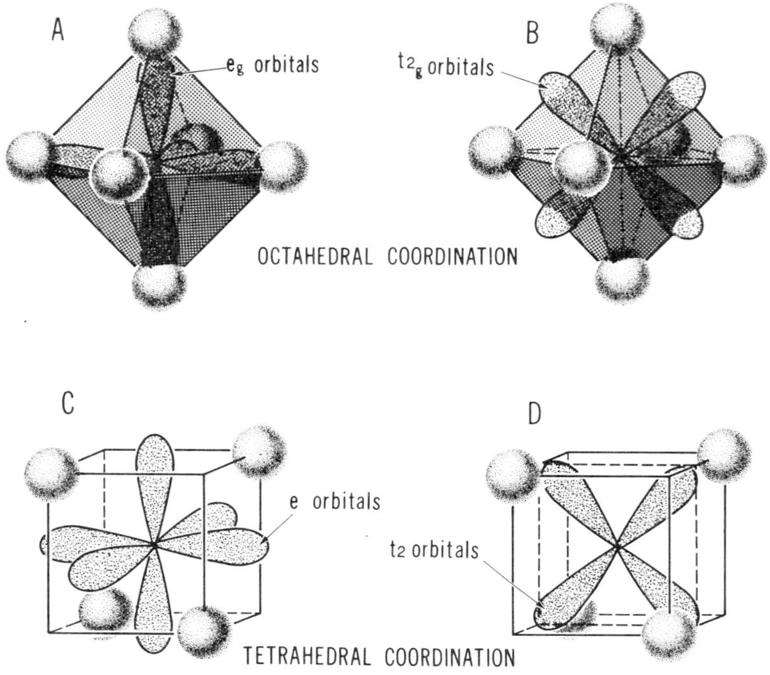

FIG. 12–19

Some of the orbitals emanating from a transition metal ion (but not the cation itself) are drawn relative to the anions surrounding this cation. The drawing has been simplified by omitting the annular segment of d_{z^2} (see Fig. 8-4) in drawing the e_g (or e) orbitals (namely $d_{x^2-y^2}$ and d_{z^2}) and by only drawing the d_{yz} orbital in representing the t_{2g} (or t_2) orbitals (d_{xy}, d_{yz}, d_{zx}).

For an octahedral coordination of anions around the metal cation, (A) shows that the e_g orbitals will be directed *toward* the anions and thus into concentrated regions of negative charge, whereas (B) shows that the t_{2g} orbitals, using d_{yz} as an example, are directed *between* the anions and thus into regions of weak negative charge. Hence for octahedral coordination the e_g orbitals are of higher energy levels than t_{2g}.

For a tetrahedral coordination of anions around the cation, (C) shows that the e orbitals are more distant from the anions and thus of lower energy levels than are the t_2 orbitals as illustrated for d_{yz} in (D). For tetrahedral coordination, therefore, t_2 is a higher energy state than e.

absorption of this light energy. For example, the intense absorption of infrared wavelengths near 1 μ (= 10,000 Å) observed in many iron compounds is attributed by Farrell and Newnham (1967) to transitions of electrons from t_{2g} to e_g levels for Fe^{2+} in an undistorted octahedral environment. Hence studies of absorption spectra like that for cordierite (Fig. 12-20A, B, C), particularly for the visible ($0.4 - 0.77\ \mu$), near infrared ($0.77 - 3.0\ \mu$), and ultraviolet regions ($0.2 - 0.4\ \mu$), yield valuable information concerning the coordination of ligands about a transition metal ion in a crystal structure. To cite an extreme example,

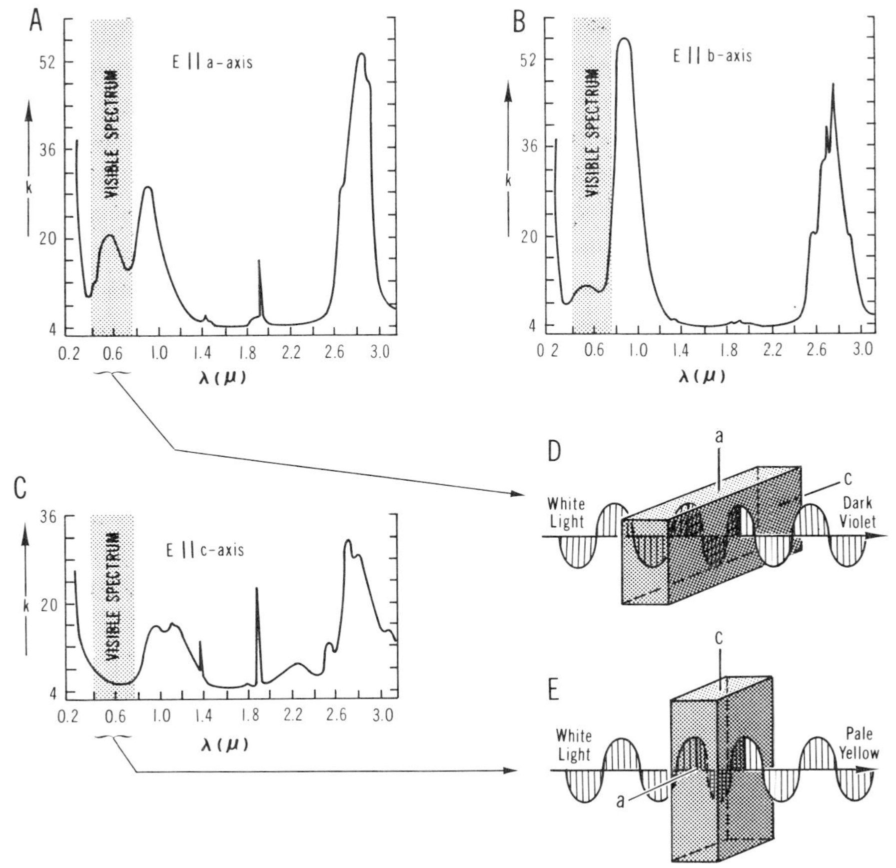

FIGURE 12–20

The room-temperature absorption spectra of cordierite for radiation polarized parallel to (A) the *a* axis, (B) the *b* axis, and (C) the *c* axis. (After Farrell and Newnham, 1967.)

(D) A beam of polarized white light that vibrates parallel to the *a* axis while passing through a (010) plate of cordierite emerges as a dark violet beam. This occurs because, as illustrated in (A), the green wavelengths near the center of the visible spectrum are strongly absorbed, whereas the violet and red ends of the spectrum are less absorbed and are thus transmitted more readily.

(E) The (010) plate has been rotated 90° so that the entering polarized white light vibrates parallel to the *c* axis while passing through the crystal. The emerging light beam is pale yellow because the absorption curve in (C) now applies and indicates considerably less absorption for all visible wavelengths and least absorption of all for the yellow wavelengths. This (010) plate, placed in a beam of plane-polarized white light and rotated on its *b* axis from position (D) to (E), changes its transmission color from dark violet to pale yellow, a phenomenon called *pleochroism*.

Co^{2+} usually produces a deep blue color in compounds in which it is tetrahedrally coordinated to four oxygen ions but a pink color if octahedrally coordinated to six oxygen ions.

The relatively broad absorption peaks in absorption spectra (Fig. 12-20A, B, and C) each represent absorption over a correspondingly broad range of wavelengths. Thus photons corresponding to a broad range in energies are being absorbed. Consequently Δ, whether Δ_o or Δ_t, must be changing value over this same range to account for this absorption. Such variation in Δ is ascribed to thermal agitation in the crystal structure. For example, thermal agitation would cause the anions in Figures 12-19A and B to be at times closer and at times farther from the cation. When they are closer, Δ, the difference in energy (splitting) between e_g and t_{2g} is increased and shorter wavelengths of light are absorbed. When the anions are farther from the cation, Δ is reduced in value and the longer wavelengths are absorbed. The maximum in a particular absorption peak thus represents the splitting energy that corresponds approximately to the position of the anions averaged over space and time. If an absorption spectrum is made of a crystal at a reduced temperature, the absorption curves become sharper because thermal agitation is reduced.

Pleochroism

A crystal structure's ability to absorb a particular wavelength of radiation passing through it depends on the radiation's direction of vibration within the structure. For example, infrared wavelength 1.9 μ in Figure 12-20A, B, and C is absorbed most for light vibrating parallel to c, less if along a, and least if along b. In like manner individual wavelengths of the visible spectrum are also differentially absorbed for different directions of vibration in cordierite. Thus if plane-polarized white light entered a thick (010) plate of cordierite so as to vibrate parallel to the a axis, the central wavelengths of the visible spectrum would be more strongly absorbed than the red and violet ends (Fig. 12-20A). The (non-absorbed) light that emerges would hence be a deep shade of violet (Fig. 12-20D). However, if this plane-polarized white light vibrated parallel to the c axis while in the crystal, the violet end of the visible spectrum would be preferentially absorbed (Fig. 12-20C) so that its complementary color, yellow (Fig. 12-20E) would be transmitted. If the crystal plate in Figure 12-20D were rotated about the ray path as an axis until it reached its position in Figure 12-20E, the color of light transmitted by it would change from violet to yellow. Such change in color as an anisotropic crystal is rotated in a beam of plane-polarized white light is known as *pleochroism*. The relationship between pleochroism and crystal structure offers a fertile field for investigation.

Optical Properties of Solid Solutions

If the variation of optical properties with composition is known for a given solid solutional series, measurement of these properties for a mixed crystal of the series may determine, fairly closely, the composition of this crystal. The variation of optical properties with composition for the solid solutional series between albite ($NaAlSi_3O_8$) and anorthite ($CaAl_2Si_2O_8$) is instructive (Fig. 12-21). We observe that the refractive indices α, β, and γ change nonuniformly, the position of β between α and γ changing with composition so that the angle $2V$ changes as well. For compositions An_0 to An_{16} and for An_{38} to An_{70}, the value of β is closer to α than to γ, and the crystals are hence optically (+) in sign. The remaining compositions are (−) in sign.

Schiller or iridescence

Certain intermediate varieties of plagioclase, notably labradorite (An_{50}—An_{70}) exhibit schiller (iridescence). Laves, Nissen, and Bollman (1965) observed such specimens to consist of alternating lamellae of slightly different refractive index and thickness. In the colored light reflected by these specimens they observed the wavelength of maximum intensity to be related by the Bragg

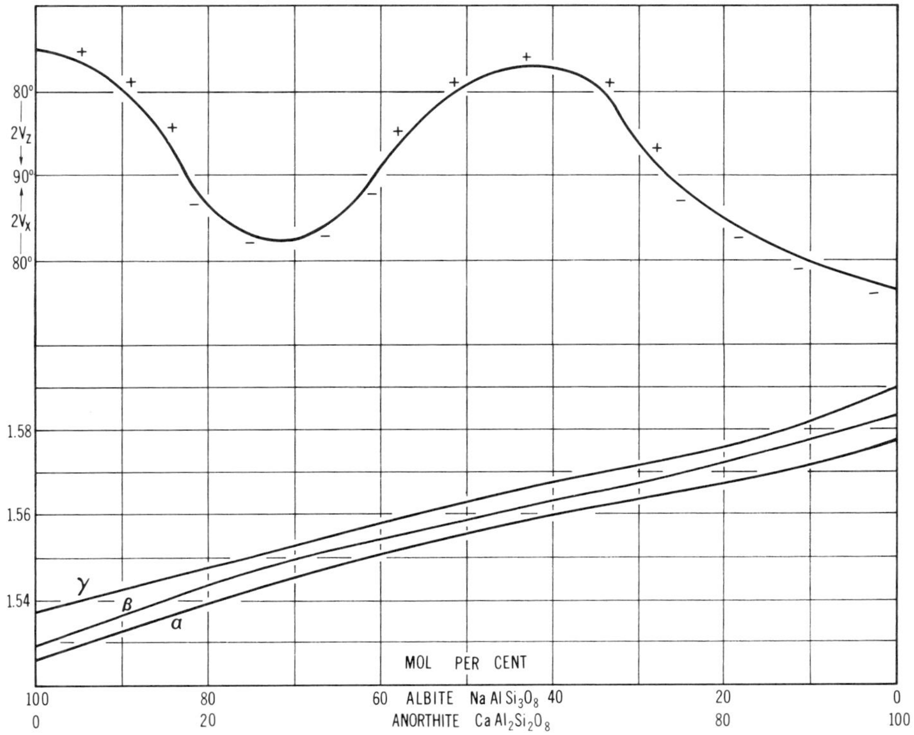

FIGURE 12–21

Variation of the refractive indices α, β, γ and of the optical angle $2V$ for the plagioclase series, that is, for solid solutions between albite ($NaAlSi_3O_8$) and anorthite ($CaAl_2Si_2O_8$). Note that when β is closer to α than to γ in value, the crystals are (+) in sign as indicated by the + signs on the curve for $2V$.

law (Eq. 13-10) to the average thickness of these lamellae. Bolton, Bursill, McLaren, and Turner (1966) confirmed in detail that schiller colors in labradorite result from optical interference between light rays reflected from lamellae surfaces successively deeper in the specimen. The schiller observed by Fleet and Ribbe (1963, 1965) in moonstone (alternating lamellae of orthoclase and albite), and in peristerite (alternating lamellae of two plagioclases of composition An_0 and An_{25}) may be similarly interpreted. The whitish sheen of the moonstone schiller is attributed to a greater variation in lamellae thickness as compared to that for labradorite whose lamellae were observed to be more uniformly of the order of 1000 Å thick by Bolton et al.

Light Emission

Incandescence

To the extent that a body's temperature exceeds absolute zero, its component atoms or ions undergo thermal vibrations and, in consequence, the body radiates energy. Below 550°C, this radiation is chiefly as heat. Above 550°C, radiation of light increasingly accompanies that of heat (Fig. 12-22). Thus at 550°C an object glows dull red. With continued increase in temperature it glows yellow and, eventually, becomes "white hot."

Luminescence

Luminescence signifies a phenomenon other than incandescence whereby a material absorbs one form of energy, then re-emits it as visible light. As may be best observed in total darkness, some crystals become faintly luminous if struck, scratched, rubbed, or crushed (*triboluminescence*) or if heated to temperatures above 50°C (*thermoluminescence*). In the latter event, luminescence may occur only within a restricted temperature range, light usually ceasing to be emitted above around 475°C, a temperature below that of incandescence. Some crystals emit light if exposed to (1) a beam of electrons or cathode rays

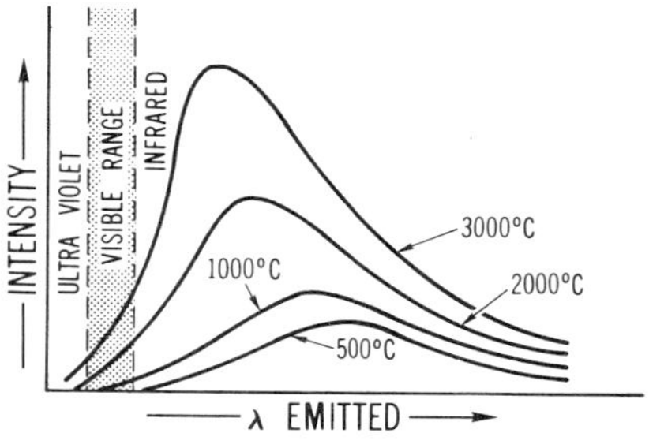

FIGURE 12–22

Spectral distribution and relative intensities of the wavelengths emitted because of incandescence at 500, 1000, 2000, and 3000°C.

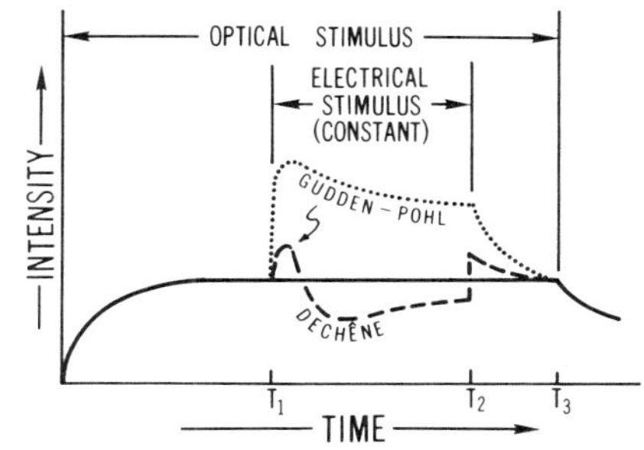

FIGURE 12-23

The intensity of a particular wavelength being emitted because of the electroluminescence of a material is plotted against time. The optical and the electrical stimulus were applied during the times indicated by the double-ended arrows. If no electrical stimulus were applied, the solid line curve shows how intensity of the emitted light is related to the application of the optical stimulus. With an electrical stimulus simultaneously imposed, a part of the solid line curve would be replaced by either the dotted line (if the phosphor is well insulated from the electrodes supplying the electrical stimulus) or the dashed line (if the phosphor is permitted to transmit appreciable current).

(*cathodoluminescence*), (2) X-rays or gamma rays (*roentgenoluminescence*), (3) products of radioactive decay, primarily the alpha particles (*radioluminescence*), or (4) ultraviolet or visible light (*photoluminescence*). Photoluminescence which ceases within 10^{-8} sec after the stimulating rays are turned off is called *fluorescence*; if it endures beyond 10^{-8} sec (sometimes for hours), it is called *phosphorescence*. Materials that exhibit photoluminescence are called *phosphors*; however, some authors restrict the term to materials showing phosphorescence. The gradual decrease in intensity of photoluminescence after cessation of the stimulus is called *photoluminescent decay*. Rapid acceleration of the rate of this decay, whether achieved by heating or irradiating the specimen, is called *quenching*.

In *electroluminescence* a high-intensity electric field **E**, either constant or alternating, is applied to the phosphor while it is being stimulated optically. The intensity of the light being emitted as the result of this double stimulus is summarized in Figure 12-23 for application of a constant field **E**. Two effects may occur: (1) the *Gudden-Pohl effect*, an intensification of the light being emitted by the phosphor because of the electrical stimulus or (2) the *Déchêne effect*, which represents a partial reduction or quenching of the light being emitted. The Gudden-Pohl effect is observable when the phosphor is well insulated from the electrodes supplying the field **E**, whereas the Déchêne effect seems enhanced when the phosphor passes an appreciable current.

During photoluminescence, the ultraviolet stimulus represents a stream of bulletlike energy packets called *photons*, the energy in each equaling $h\nu$ (or $hc \div \lambda$), where h is *Planck's constant* (6.6256×10^{-34} kg-m²/sec) and ν, c, and λ are the frequency, velocity, and wavelength in a vacuum of the stimulant. Struck by these photons, a solid's component atoms or ions may absorb the energy packet $h\nu$ by (1) increasing their rate of thermal vibration and thus the body's temperature and heat energy or (2) elevating an electron from a valence band into an orbital associated with the higher energy conduction band (see Fig. 11-15B). Process (2) can occur only if the conduction band exceeds the energy of the valence band by precisely $h\nu$. When (2) occurs, the atom or ion is in an "excited" state. The atom may return to its unexcited or "ground" state if an electron from the conduction band descends into the valence band to restore full occupancy there. The descent may be by a single downward step whereupon the atom emits a photon identical in energy and frequency to the original stimulus, this being called *resonant fluorescence*. Much more frequently, however, the electron descends by more than one step, momentarily lodging at energy levels intermediate between the crystal's valence and conduction bands, these intermediate levels being furnished by impurity atoms in the host like those described for *n*- or *p*-type semiconductors (Fig. 11-15C, D). Such impurities are then called "activators" because without them, or even with too many of them, the crystal will not fluoresce. If an electron falls through an intermediate step, light is emitted whose energy ($h\nu_i$ or $hc \div \lambda_i$) is less than for a photon of the stimulus ($hc \div \lambda_s$). In other words the emitted wavelength λ_i exceeds that λ_s of the stimulating radiation (*Stokes' rule*). It is never less.

The ability to fluoresce and that of semiconduction are frequently allied. Often the activators are atoms or ions with nonpaired spins (paramagnetic atoms or ions) which, to a limited extent, substitute in the host structure for atoms or ions with paired spins (diamagnetic atoms or ions). Photoluminescence is not observed in metallic crystals because the energy gap between their valence band and conduction band is too small to conform to the energy of a photon of visible light.

Fluorescence is routinely observed using either a *short-wave* (wavelengths around 2537 Å predominate) or a *long-wave* ultraviolet lamp (3200 to 4000 Å predominate). The energy in the photons of light emitted by one or the other of these lamps may equal the energy difference between the valence and conduction bands in ionic or covalent-ionic crystals. If activators are present in these crystals, they can then fluoresce. A given specimen may fluoresce different colors, or not at all, according to which lamp is used. Scheelite ($CaWO_4$) fluoresces a brilliant blue under short-wave ultraviolet but not at all under the long wave. The color fluoresced may also depend on the kind and quantity of trace elements (activators) that occupy structural sites in the host crystal. The bright green fluorescence of willemite (Zn_2SiO_4) from Franklin, New Jersey, results from Mn^{2+} in the Zn^{2+} sites. Note that Mn^{2+} has five unpaired 3*d* electrons, Zn^{2+} has none (Fig. 11-22).

Of apparently identical specimens from different, or even the same, locales, some may fluoresce, others will not. Hence fluorescent colors, being so variable and dependent upon minute impurities, seldom serve as a means of identification. Under short-wave ultraviolet, the following fluorescent colors, or *whitish* shades thereof, occur among minerals: *blue* (anorthoclase, aragonite, celestite, cerussite, scheelite); *green* (gypsum, opal, willemite, and various secondary uranium minerals); *yellow* (apatite, barite, pectolite, wernerite); *orange* (sphalerite, zircon); *red* (calcite, halite, spodumene, spinel). Minerals that apparently have no dominant fluorescent color, or various colors in addition to those listed above, are adularia, aragonite, calcite, chalcedony, dolomite, fluorite, gypsum, magnesite, opal, and sphalerite.

REFERENCES AND RECOMMENDED READING

Allen, R. D. (1956) "A new equation relating index of refraction and specific gravity," *Am. Mineral.*, **41**, 245–267.

Anderson, Orson L., and E. Schreiber (1965) "The relation between refractive index and density of minerals related to the earth's mantle," *J. Geophys. Research.*, **70**, 1463–1471.

Birch, Francis (1961) "The velocity of compressional waves in rocks to 10 kilobars," *J. Geophys. Research*, **66**, 2199–2224.

Bloss, F. D. (1961) *An Introduction to the Methods of Optical Crystallography*. New York: Holt, Rinehart and Winston.

——— (1971) "Dispersion—optical" and "Optical orientation" in *Encyclopedia of Earth Sciences* (R. W. Fairbridge, ed.). Princeton, N. J.: Van Nostrand-Reinhold.

Bolton, H. C., L. A. Bursill, A. C. McLaren, and R. G. Turner (1966) "On the origin of the colour of labradorite," *Phys. Stat. Sol.*, **18**, 221–230.

Bragg, W. L. (1924) "The refractive indices of calcite and aragonite," *Proc. Royal Soc.*, **105**, 370–386.

Farrell, E. F., and R. E. Newnham (1967) "Electronic and vibrational absorption spectra in cordierite," *Am. Mineral.*, **52**, 380–388.

Fleet, S. G., and P. H. Ribbe (1963) "An electron-microscope investigation of a moonstone," *Phil. Mag.*, **8**, No. 91, 1179–1187.

——— (1965) "An electron-microscope study of peristerite plagioclases," *Mineral. Mag.*, **35**, p. 165–176.

Gibbs, G. V., and J. V. Smith (1965) "Refinement of the crystal structure of synthetic pyrope," *Am. Mineral.* **50**, 2023–2039.

Hardy, A. C., and F. H. Perrin (1932) *The Principles of Optics*. New York: McGraw-Hill.

Hartshorne, N. H., and A. Stuart (1960) *Crystals and the Polarising Microscope*. London: Edw. Arnold.

Laves, F., H.-U. Nissen, and W. Bollmann (1965) "On schiller and submicroscopical lamellae of labradorite . . .," *Naturwiss.*, **14**, 427–428.

Povarennykh, A. S. in M. H. Battey and S. I. Tomkeieff (1964) *Aspects of Theoretical Mineralogy in the U.S.S.R.* New York: Macmillan.

Skinner, B. J., and J. J. Fahey (1963) "Observations on the inversion of stishovite to silica glass," *J. Geophys. Research*, **68**, 5595–5604.

13 Introductory X-ray Crystallography

The Nature of X-Rays

X-rays may be defined as electromagnetic radiations whose wavelengths lie in the 0.1 to 10 Å range, this range partly overlapping those assigned to gamma and ultraviolet rays by some authorities (Fig. 12-1, p. 418). X-rays are generated whenever a stream of high-energy electrons strikes a material. In an X-ray tube (Fig. 13-1A) a stream of electrons from the cathode is focused to strike a metal target on the positively charged anode. If the electrons in this stream possess sufficient energy, the target emits X-rays.

At its point of impact on the metal target, an electron of the stream possesses the energy eV, where e is its charge and V is the voltage difference maintained electronically between cathode and anode. If this energy is sufficiently large, the electron penetrates deep into the atoms of the target, dislodging some tightly held orbital electrons from their innermost K shells. Thus excited, the atom almost immediately restores its K shell occupancy by pirating an electron from one of its own outer shells (L, M, N). Usually this is from the neighboring L shell, or the next neighboring M shell (Fig. 13-1B) and the pirated electron is then said to undergo an L-K transition or an M-K transition. As this electron "drops" from a high-energy outer shell into a lower-energy inner shell, X-rays are emitted whose photons, now called *quanta,* possess energy equal to ΔE, the difference in energy between the outer and inner shells involved in the transition. As Figure 13-1B shows, an electron usually "drops" to the K shell from either an M shell level or from two slightly different energy levels of the L shell, L_{III} and L_{II}. The frequency ν of the radiation emitted when an electron drops from an outer to an inner shell must fulfill the *Einstein frequency condition,* namely,

$$\nu = \frac{\Delta E}{h} \qquad \text{(Eq. 13-1)}$$

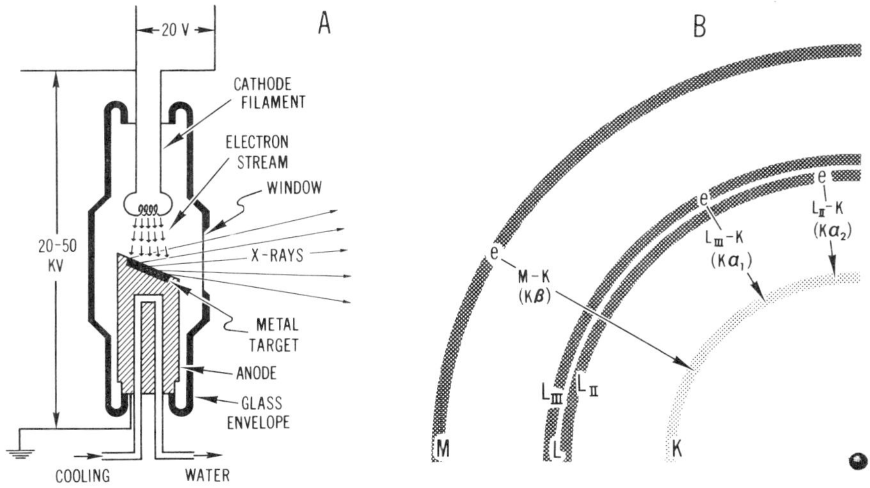

FIGURE 13-1

(A) Generalized cross section through an X-ray tube. High-energy electrons from the cathode strike the metal target on the anode, exciting the metal atoms by "knocking" electrons out of their *K* shells.
(B) Schematic energy-level diagram illustrating how the excited *K* shell captures electrons from outer shells, the *M–K* transition producing $K\beta$ X-rays, the *L–K* transitions producing $K\alpha_1$ and $K\alpha_2$ X-rays.

where h is Planck's constant and ΔE is as already stated. By utilizing Equation (12-1), Equation (13-1) can be rewritten as

$$\lambda = \frac{ch}{\Delta E} \qquad \text{(Eq. 13-2)}$$

which indicates that as ΔE becomes larger, the wavelengths of the emitted radiation become shorter.

After excitation the atoms in the target chiefly undergo (1) *M-K*, (2) L_{III}-*K*, or (3) L_{II}-*K* transitions and, correspondingly, emit X-rays called (1) $K\beta$, (2) $K\alpha_1$, or (3) $K\alpha_2$ radiation. A plot of their relative intensities for a typical X-ray tube (Fig. 13-2) indicates peaks corresponding to the wavelengths for $K\beta$, $K\alpha_1$, and $K\alpha_2$. The precise wavelengths of $K\beta$, $K\alpha_1$, and $K\alpha_2$ radiation are characteristic of the metal used for the target (Table 13-1) and are hence called *characteristic wavelengths*. In all cases the $K\beta$ peak possesses an intensity of about one fourth that for $K\alpha_1$ and one half that for $K\alpha_2$. These relative intensities indicate that *M-K* transitions are less frequent than *L-K* transitions. However, a single *M-K* transition releases more energy ΔE than an *L-K* transition. Thus it follows from Equation (13-2) that $K\beta$ radiation is shorter in wavelength than $K\alpha_1$ or $K\alpha_2$.

If the voltage differential between the cathode and anode of an X-ray tube is less than a particular value, called the *excitation potential*, an electron striking

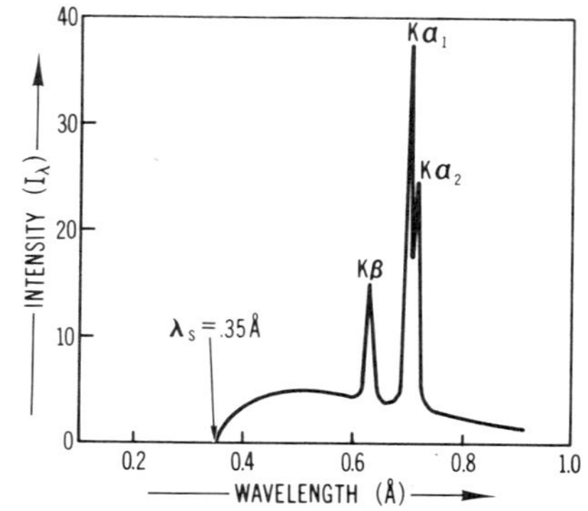

FIGURE 13–2

Relative intensities I_λ for the various wavelengths of X-rays emitted by an X-ray tube with a molybdenum target operated at 35 kV.

the target lacks sufficient energy to penetrate the outer electron shells of the target atoms so as to dislodge orbital electrons from their inner K shells. Thus an X-ray tube held below the excitation potential emits no $K\alpha_1$, $K\alpha_2$, or $K\beta$ radiation. The excitation potential of a tube varies with its target. Targets consisting of atoms with high atomic numbers, that is, atoms whose highly charged positive nuclei tightly hold the electrons in the inner K shells, require a higher excitation potential than those consisting of atoms with low atomic numbers. Excitation potentials for the common target materials are molybdenum, atomic number **42** (20 kV); copper, atomic number **29** (8.86 kV); nickel, **28** (8.29 kV); cobalt, **27** (7.71 kV); iron, **26** (7.10 kV); and chromium, **24** (5.98 kV).

In addition to $K\beta$, $K\alpha_1$, and $K\alpha_2$ wavelengths, an X-ray tube emits "white radiation" which, like white light, embraces a continuous range of wavelengths. In Figure 13-2 white radiation accounts for the intensities at wavelengths other than $K\beta$, $K\alpha_1$, and $K\alpha_2$. Such white radiation results if the electrons streaming from the cathode are slowed or abruptly halted as they hit the target. Just as the kinetic energy of a bullet is transformed into mechanical and heat energy when it hits a wall, if an electron is stopped the energy is transformed into X-radiation

TABLE 13–1

Characteristic Wavelengths for the Common Target Metals

λ	Target Metal				
	Mo	Cu	Co	Fe	Cr
$K\beta$	0.63225	1.39217	1.62073	1.75653	2.08479
$K\alpha_1$	0.70926	1.54051	1.78892	1.93597	2.28962
$K\alpha_2$	0.71354	1.54433	1.79279	1.93991	2.29351
$K\alpha$	0.7107	1.5418	1.7902	1.9373	2.2909

(plus heat). The higher the energy that the electron possessed, the shorter will be the wavelengths of the X-rays produced by its abrupt halt. The shortest (and most energetic) wavelength present in the white radiation (λ_s in Fig. 13-2) can be calculated from the formula

$$\lambda_s = \frac{12.4}{V} \qquad \text{(Eq. 13-3)}$$

where λ_s is in angstroms and V is in kilovolts (kV). As discussed earlier an electron's energy at impact is proportional to eV; thus V in Equation (13-3) is proportional to the energy of an electron as it strikes the target. The derivation of Equation (13-3) is discussed by Henry, Lipson, and Wooster (1960, p. 25).

If "white" X-rays pass through a thickness of metal foil, the degree of absorption of its component wavelengths by this foil will increase with increasing wavelength to a maximum at λ_A (Fig. 13-3A), the so-called *absorption edge*. For wavelengths slightly longer than λ_A, absorption decreases abruptly, then once more rises as wavelengths increasingly longer than λ_A are considered. This feature permits suitably chosen metal foils to filter out the $K\beta$ wavelength from an X-ray beam. For example, if the metal foil consists of an element whose atomic number is one or two less than that of the metal constituting the tube's target, the absorption edge of this foil (Fig. 13-3B) will fall between the target's $K\beta$ and $K\alpha$ peaks. Hence the foil will strongly absorb the $K\beta$ wavelength but not $K\alpha_1$ and $K\alpha_2$, particularly if of optimum thickness. Optimum filters and thickness (Azároff

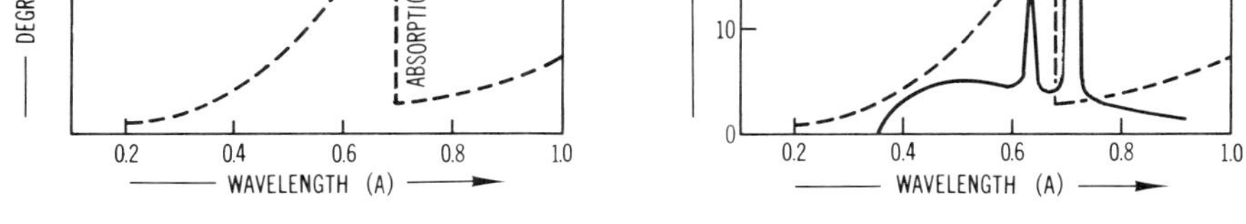

FIGURE 13–3

(A) Absorption curve for zirconium foil for X-rays of wavelengths between 0.2 and 1.0 Å. The abrupt decrease in ability to absorb at $\lambda = 0.687$ Å marks zirconium's *absorption edge*.

(B) Curve showing the intensities of X-rays emitted by a molybdenum tube operated at 35 kV on which is superimposed the absorption curve for zirconium foil. The absorption edge for zirconium (atomic number, 40) falls between the $K\beta$ and $K\alpha$ peaks for molybdenum (atomic number, 42). This permits zirconium foil to filter out the Mo$K\beta$ radiation from the X-rays emitted by a molybdenum tube.

and Buerger, 1958, p. 34) are as follows: Molybdenum target (0.108 mm thick zirconium foil); copper target (0.021 mm thick nickel foil); cobalt target (0.018 mm iron foil); iron target (0.016 mm manganese foil); chromium target (0.016 mm vanadium foil).

In Figure 13-3B, the solid line resembles the outline of a whale from which three spouts are rising (the smaller $K\beta$ peak plus the overlapped $K\alpha_1$ and $K\alpha_2$ peaks). The area under the $K\beta$ peak (but above the "whale," which is actually the background of white radiation) represents the integrated $K\beta$ intensity. Similarly, the total area under the two overlapped $K\alpha$ peaks (but above the "whale") represents the integrated $K\alpha$ intensity. The filters and thicknesses previously cited as optimum will reduce the (integrated) intensity ratio $K\beta/K\alpha$ to 1/600, whereas for unfiltered radiation this ratio lies between 90/600 and 180/600, depending on the target metal. The filtered $K\alpha$ intensity will be less than 50 percent of its unfiltered intensity.

Unfiltered radiation thus contains $K\beta$ plus the $K\alpha_1$ and $K\alpha_2$ wavelengths, whereas suitably filtered radiation contains chiefly $K\alpha_1$ and $K\alpha_2$. These latter two peaks are so close in wavelength that one cannot be filtered out by ordinary means without severely attenuating the other. Radiation consisting of $K\alpha_1$ and $K\alpha_2$ wavelengths only is called "$K\alpha$ radiation." The wavelength of $K\alpha$ radiation is assumed to be the intensity weighted average of the $K\alpha_1$ and $K\alpha_2$ peaks. Thus since the $K\alpha_1$ peak is twice the intensity of the $K\alpha_2$ peak, the $K\alpha$ wavelength is calculated as

$$K\alpha = \frac{2K\alpha_1 + K\alpha_2}{3} \qquad \text{(Eq. 13-4)}$$

For a copper target, for example,

$$\text{Cu}K\alpha = \frac{2(1.54051) + 1.54433}{3} = 1.54178\ \text{Å}$$

The $K\alpha$ wavelengths for the various targets are summarized in Table 13-1.

The Nature of Diffraction

An electron in the path of an X-ray beam will oscillate in response to the periodically changing electric field of the X-rays. This oscillation causes the electron to emit (in all directions) X-rays of the same wavelength and frequency as those in the primary X-ray beam (Fig. 13-4). The electron thus *scatters* a slight amount of the directed energy of the primary beam, dissipating this energy in all directions. For an atom in an X-ray beam each of its orbital electrons scatters the primary beam. Thus the (X-ray) *scattering power* of an atom or ion increases with its number of orbital electrons. In general practice an atom may be considered to act like a point source of X-rays, that is, as if its orbital electrons were concentrated at a central point in the atom.

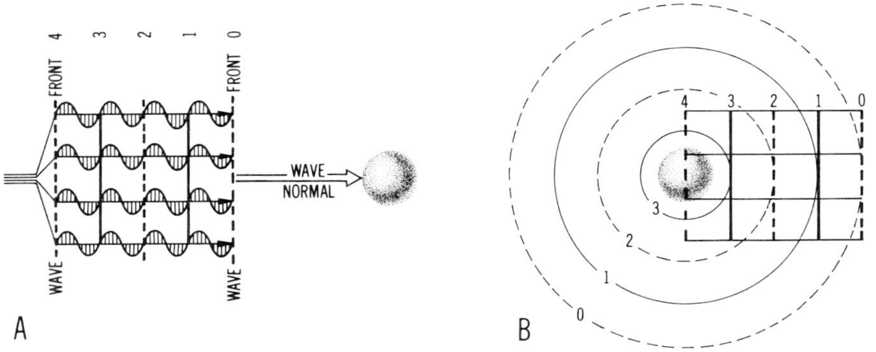

FIGURE 13–4

(A) Four almost infinitely close rays in an X-ray beam approaching an atom are drawn as if widely separated so as to show their wave motion. Surfaces that join points on these rays which are in phase are called *wave fronts*. For example, the solid and dashed lines represent five wave fronts, labeled 0, 1, 2, 3, 4, which successively differ in phase by 1λ. The perpendicular to these wave fronts, called their *wave normal*, is indicated by the hollow arrow.
(B) Primary wave fronts 0, 1, 2, and 3 in (A) have passed through the atom and have caused its orbital electrons to scatter a minute fraction of this primary energy, that is, radiate in all directions exceedingly weak *secondary* rays (none drawn). Joining points on these secondary rays that are in phase, the weak secondary wave fronts so defined would be seen to be spheres concentric with the atom. (Phase shift of $\pi/2$ between the primary wave front and its resultant secondary wave front has been omitted as unnecessary to this discussion.)

The Laue equations

Assume that a primary beam of X-rays of wavelength λ (rays, p_1, p_2, p_3 in Fig. 13-5A) impinges at an angle $\bar{\mu}$ (Fig. 13-5B) upon a row of periodically spaced atoms for which t is the center-to-center or repeat distance. Of the rays scattered in all directions by each atom, we consider a family of parallel rays such as c_1, c_2, and c_3; these rays, where joined by the bold perpendiculars or wave fronts, differ in phase by $n\lambda$ (n being an integer) and hence reinforce each other. For such adjacent, parallel, scattered rays to differ in phase by $n\lambda$, their angle $\bar{\nu}$ to the atom row must fulfill the following relationship to λ, $\bar{\mu}$, and t,

$$n\lambda = t(\cos \bar{\nu} - \cos \bar{\mu}) \qquad \text{(Eq. 13-5)}$$

This equation, the general Laue equation, can be derived by considering primary rays p_2 and p_3 (Fig. 13-5B), which are precisely in phase at the points connected by the dashed perpendicular. We note that ray path p_3c_3 exceeds ray path p_2c_2 by $(r - s)$. Thus if

$$(r - s) = n\lambda \qquad \text{(Eq. 13-6)}$$

then rays c_1 and c_2, where connected by the dashed perpendicular, will differ in phase by $n\lambda$ and hence reinforce each other. However, $r = (t \cos \bar{\nu})$ and $s =$

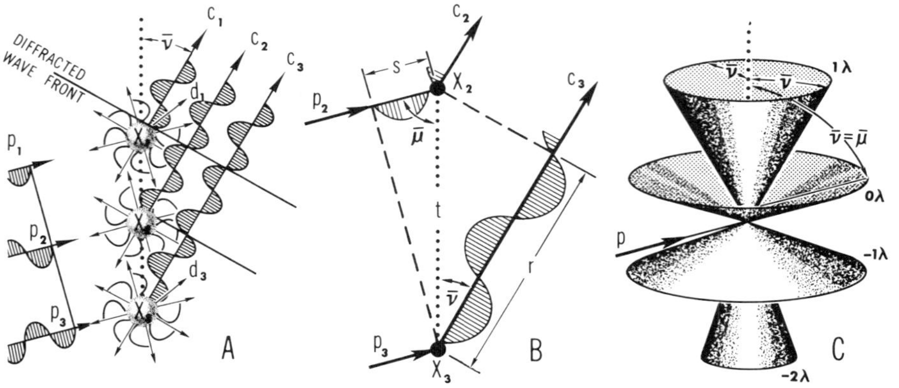

FIGURE 13–5

(A) A primary beam of X-rays obliquely striking a row of atoms. Of the X-rays spherically scattered by each atom, there exists a family of parallel rays, c_1, c_2, c_3, that is in phase. This can be recognized because a perpendicular drawn from one ray to its neighbor joins points that are in the same state of vibration; these perpendicular lines thus constitute wave fronts for the diffracted rays. Neighboring rays c_1, c_2, and c_3 thus reinforce each other to produce a diffracted wave of appreciable intensity. Next consider ray d_1 of those scattered by the uppermost atom x_1, as well as the ray parallel to d_1 of the rays scattered by the lower atoms, x_2 and x_3. If these three parallel rays were extended and their wave motion schematically illustrated as for rays c_1, c_2, and c_3, then a perpendicular dropped from a point on d_1 to its parallel neighbor would not connect points in phase; consequently, destructive interference would cause these two rays to annihilate each other to a large extent.

(B) The distance t between atoms x_2 and x_3 in (A) enlarged to permit a geometric analysis of the relation between (1) $\bar{\mu}$, the angle between the incident beam and the atom row; (2) t, the spacing between identipoints along this atom row; (3) $\bar{\nu}$, the angle between the atom row and the diffracted X-ray beam; and (4) λ, the wavelength of the X-rays involved. Thus incident rays p_2 and p_3 are precisely in phase where joined by the dashed perpendicular, the amplitudes of vibrations being suppressed for illustrative purposes. Similarly, c_2 and c_3 must be precisely in phase where joined by the dashed line perpendicular to them if they are to constitute mutually reinforcing rays in a diffracted beam. This can only be true if $(r - s)$, the extra distance traveled by ray c_3 relative to c_2, equals an integral number of wavelengths, that is, $n\lambda$, for the X-rays used. Note that $r = t \cos \bar{\nu}$ and that $s = t \cos \bar{\mu}$. Thus if a primary beam strikes an atom row at an angle $\bar{\mu}$, diffracted beams can develop at an angle $\bar{\nu}$ to the row only if $n\lambda = (r - s) = t\,(\cos \bar{\nu} - \cos \bar{\mu})$.

(C) The row of atoms, drawn so small that each atom appears as a dot of the dotted line, is struck at an angle $\bar{\mu}$ by the primary beam p. Cones of diffracted rays are hence emitted by the row; $\bar{\nu}$, the angle that any particular ray in the cone makes with the atom row, must fulfill the equation $n\lambda = t\,(\cos \bar{\nu} - \cos \bar{\mu})$. A cross section and enlargement of the cone labeled 1λ could be represented by (B), the ray closer to the cone's interior, c_2 in this case, being 1λ ahead of c_3 where

(t cos $\bar{\mu}$). Substituting these values into Equation (13-6), there results Equation (13-5). Consequently, for an atom row struck by a beam of X-rays at angle $\bar{\mu}$, diffraction cones result (Fig. 13-5C) whose ray paths are at angles $\bar{\nu}$ to the atom row such that Equation (13-5) is fulfilled.

Diffraction of X-rays by a crystal structure may be explained by considering the diffraction cones for three intersecting lattice rows, all three not in the same plane. Let us assume that a beam of primary X-rays strikes a crystal at the angles $\bar{\mu}_1$, $\bar{\mu}_2$, and $\bar{\mu}_3$ to nonparallel lattice rows 1, 2, and 3, respectively. This crystal can produce a secondary diffracted beam (whose angles are $\bar{\nu}_1$, $\bar{\nu}_2$, and $\bar{\nu}_3$ to these lattice rows) only if the Laue equations

$$t_1(\cos \bar{\nu}_1 - \cos \bar{\mu}_1) = m\lambda$$

$$t_2(\cos \bar{\nu}_2 - \cos \bar{\mu}_2) = p\lambda \qquad \text{(Eqs. 13-5}')$$

$$t_3(\cos \bar{\nu}_3 - \cos \bar{\mu}_3) = q\lambda$$

are satisfied, where t_1, t_2, and t_3 are the repeat distances along rows 1, 2, and 3, respectively, and m, p, and q are integers. If lattice rows 1, 2, and 3 represent the a, b, and c axes, then in Equations (13-5′) distances t_1, t_2, and t_3, respectively, equal a_o, b_o, and c_o and for m, p, and q one can write Miller indices h, k, and l. Considered graphically, Equations 13-5′ indicate that a set of cones like those in Figure 13-5C is centered around each of the three lattice rows, 1, 2, and 3. Wherever a cone centered around row 1, a cone centered around row 2, and a cone centered around row 3 intersect in a common line, a diffracted X-ray beam travels outward along this common line.

The Bragg equation

In practical work an equation developed by W. L. Bragg proves more useful than the Laue equations. Bragg observed that beams of X-rays diffracted from a crystal could be likened to reflections from the crystal's lattice planes. For example, let us consider an NaCl crystal, cut and polished parallel to (111), to be rotated on an axis normal to the page in Figure 13-6A. The incident light beam I is reflected at all possible values of θ_I, which is the *glancing angle* between I and plane (111). From the *law of reflection* the angle between reflected ray R and (111), θ_R, invariably equals θ_I. If we replace beam I with a beam of Cu$K\alpha$ X-rays and again rotate the crystal as before, a diffracted beam of X-rays is emitted only when θ_I equals 13.70°, 28.27°, 45.27°, or 71.30° (Figs. 13-6B) but *not at values in between*. In each instance of diffraction, θ_R, the angle between the diffracted beam and (111) equals θ_I, the glancing angle; this similarity to the

joined by the dashed line; thus n would equal 1 in the above equation. The extensions of rays p_2 and p_3 beyond atoms X_2 and X_3 in (B) are among the rays constituting the $O\lambda$ cone in (C). The angle $\bar{\nu}$ for rays in the $O\lambda$ cone precisely equals $\bar{\mu}$ since $n\lambda = O\lambda = t$ ($\cos \bar{\nu} - \cos \bar{\mu}$). On the other hand, setting n equal to -1 in the equation $n\lambda = (r - s) = t$ ($\cos \bar{\nu} - \cos \bar{\mu}$) requires length s to exceed r by 1λ, that is, $\bar{\nu}$ must exceed $\bar{\mu}$ in value.

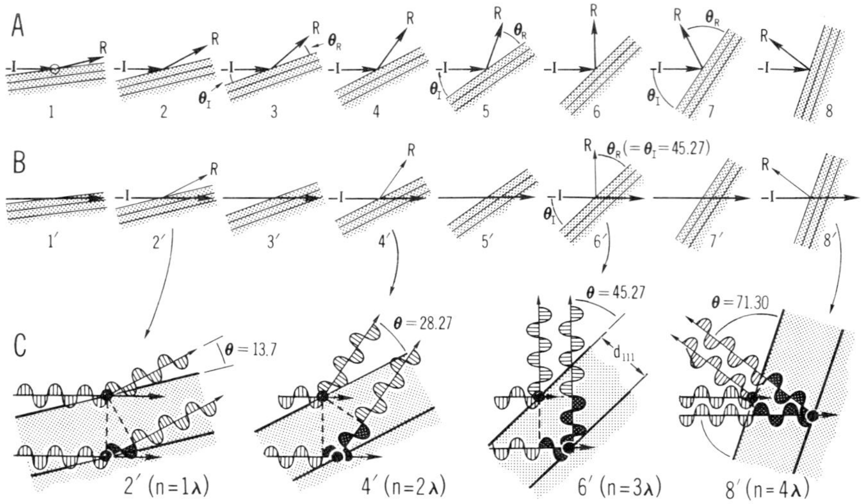

FIGURE 13–6

(A) Reflection of an incident light beam I by the polished (111) face of an NaCl crystal, as the crystal is rotated about an axis normal to the page so as to change the value of the glancing angle θ_I from left to right. In all cases a beam R is reflected whose angle to (111), that is, θ_R, equals θ_I for that position.
(B) Similar rotation of this NaCl crystal so that its (111) face makes different glancing angles θ_I relative to I, a beam of incident Cu$K\alpha$ X-rays. A diffracted beam R, bearing the same geometric relation to I as if it had been reflected, arises only when θ_I equals 13.7°, 28.27°, 45.27°, or 71.30°. At in-between positions no diffracted beam occurs. In actuality the beam at 45.27° would be extremely weak but for reasons not pertinent to this discussion.
(C) For the four positions that produced diffracted X-ray beams in (B) a section of the crystal is drawn enlarged. The two bold lines from which the X-rays appear to be "reflected" represent cross sections through neighboring lattice planes of the (111) stack; the distance between them d_{111} equals 3.255 Å. The extra distance traveled by the ray reflected from the lower 111 plane is shown to equal 1λ, 2λ, 3λ, or 4λ, the extra wavelengths of travel being shaded more heavily; the maximum value for this extra distance traveled is $2d_{111}$ and would occur for $\theta = 90°$.

law of reflection permits these diffracted beams to be regarded as "reflections" of the incident beam.

A necessary condition for such "reflection" of X-rays by a stack of parallel lattice planes is that neighboring rays in each "reflected" beam differ in phase by $n\lambda$. Thus the diffracted beams at 13.70°, 28.27°, 45.27°, and 71.30° are known as the first-, second-, third-, and fourth-order (111) reflections, respectively, because adjacent rays in these beams (see Fig. 13-6C) differ by 1λ, 2λ, 3λ, and 4λ. By convention the nth order of reflection from an (hkl) face is called the $nh.nk.nl$ reflection, for example, the first-, second-, third-, and fourth-order reflections from (111) are called the 111, 222, 333, and 444 reflections, respectively. The absence of parentheses and the frequent presence of a common factor n distinguish these *reflection indices* from Miller indices. Thus 264 commonly represents the second-order (132) reflection.

Rays 1 and 2 in Figure 13-7A represent parallel X-rays in a beam incident at the glancing angle θ to a stack of parallel lattice planes. Atom M "reflects" ray 1 to 1′ and, on the next lower lattice plane, atom N "reflects" ray 2 to 2′. A diffracted beam, of which reflected rays 1′ and 2′ are representative, arises only if adjacent rays in this beam, for example, 1′ and 2′, are in phase. For rays 1′ and 2′ this is true only if the extra distance for path 2-2′ exceeds that for path

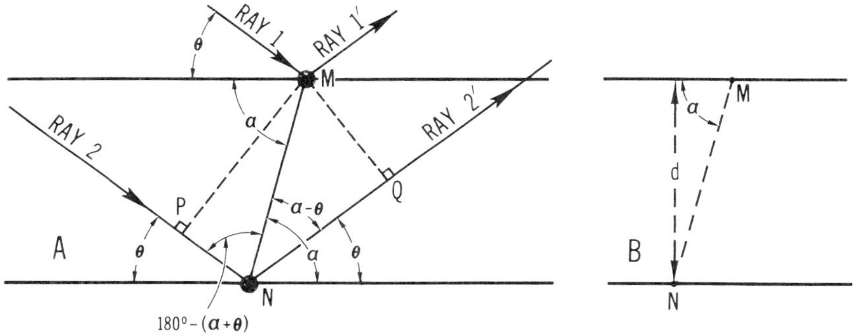

FIGURE 13–7

(A) Rays 1 and 2 of an incident beam of X-rays strike two successive lattice planes (heavy horizontal lines) at the angle θ, thus causing atoms such as M and N to scatter a minute portion of the X-rays in all directions. Of these scattered rays, only 1′ and 2′ are drawn, because we assume that at points M and Q they differ in phase by $n\lambda$ and thus reinforce each other to constitute a diffracted X-ray beam. Stated another way, we assume that $(PN + NQ)$, the extra distance traveled by ray 2′, equals $n\lambda$, that is,

$$n\lambda = PN + NQ \qquad \text{(Eq. 13-7)}$$

Let θ represent both the glancing angle and the "reflecting" angle and let α represent the angle of MN to the lattice planes; the remaining angles are then as marked. Observe that $PN = MN \cos 180° - (\alpha + \theta) = MN \cos (\alpha + \theta)$ and that $NQ = MN \cos (\alpha - \theta)$. Thus Equation (13-7) becomes

$$n\lambda = MN [\cos (\alpha - \theta) - \cos (\alpha + \theta)]$$

The standard double-angle formulas in trigonometry indicate the difference between the terms in square brackets to equal $[2 \sin \alpha \sin \theta]$, hence

$$n\lambda = MN [2 \sin \alpha \sin \theta] \qquad \text{(Eq. 13-9)}$$

(B) Distance MN in (A) is seen to equal $d/\sin \alpha$ where d is the perpendicular distance between the two lattice planes. Substituting this value of MN into Equation (13-9), it becomes the Bragg equation,

$$n\lambda = 2d \sin \theta$$

1-1′ by precisely $n\lambda$; that is,

$$n\lambda = PN + NQ \quad \text{(Eq. 13-7)}$$

In Figure 13-7A, we note that

$$PN + NQ = MN \cos [180° - (\alpha + \theta)] + MN \cos (\alpha - \theta)$$

Because angles $[180° - (\alpha + \theta)]$ and $(\alpha + \theta)$ are supplementary, their cosines differ in sign only, thus

$$PN + NQ = MN[-\cos (\alpha + \theta) + \cos (\alpha - \theta)] \quad \text{(Eq. 13-8)}$$

However, from classic trigonometric formulas

$$\cos (\alpha + \theta) = \cos \alpha \cos \theta - \sin \alpha \sin \theta$$

$$\cos (\alpha - \theta) = \cos \alpha \cos \theta + \sin \alpha \sin \theta$$

Hence Equation (13-8) becomes

$$n\lambda = MN [2 \sin \alpha \sin \theta] \quad \text{(Eq. 13-9)}$$

But as shown in Figure 13-7B the interplanar spacing d between neighboring lattice planes in the stack equals $MN \sin \alpha$. Thus Equation (13-9) becomes

$$n\lambda = 2d \sin \theta \quad \text{(Eq. 13-10)}$$

This, the well-known Bragg equation, defines the conditions for "reflection" of X-rays by a set of parallel lattice planes of interplanar spacing d.

Returning to the example of Cu$K\alpha$ rays ($\lambda = 1.5418$ Å) incident on (111) for NaCl ($d_{111} = 3.2555$ Å), substitution of these values for λ and d in Equation (13-10) permits us to determine that the first-order reflection ($n = 1$) from (111) will occur when $\theta = 13.70°$, the second order at $\theta = 28.27°$, the third at $\theta = 45.27°$, and the fourth at $\theta = 71.30°$. No fifth-order reflection from (111) could occur because solution of the Bragg equation, namely

$$5(1.5418) = 2(3.2555) \sin \theta \quad \text{(Eq. 13-10′)}$$

leads to a value for $\sin \theta$ in excess of 1.0, which is impossible. The maximum order of reflection from a crystal plane (hkl) is limited to the largest integer below $(2d_{hkl}/\lambda)$ in value. For example, for Cu$K\alpha$ reflections from the (111) plane of NaCl, this ratio is $2(3.2555)/1.5418$ or 4.2. Thus a fourth-order reflection is the maximum that is possible.

If a plane (hkl) passes through a lattice point, the lattice translations require identical planes to pass through each remaining point of the lattice. For a primitive lattice the resultant set of parallel (hkl) planes cuts the a, b, and c axes at intervals a_o/h, b_o/k, and c_o/l, respectively, as demonstrated for the specific examples (110) and (230) in Figure 13-8A, B. For convenience a reflection index $nh.nk.nl$ may similarly be regarded to represent a set of planes that cuts the axes at intervals a_o/nh, b_o/nk, and c_o/nl, respectively. For example, reflection index 220 could be regarded to result from reflection by the set of planes

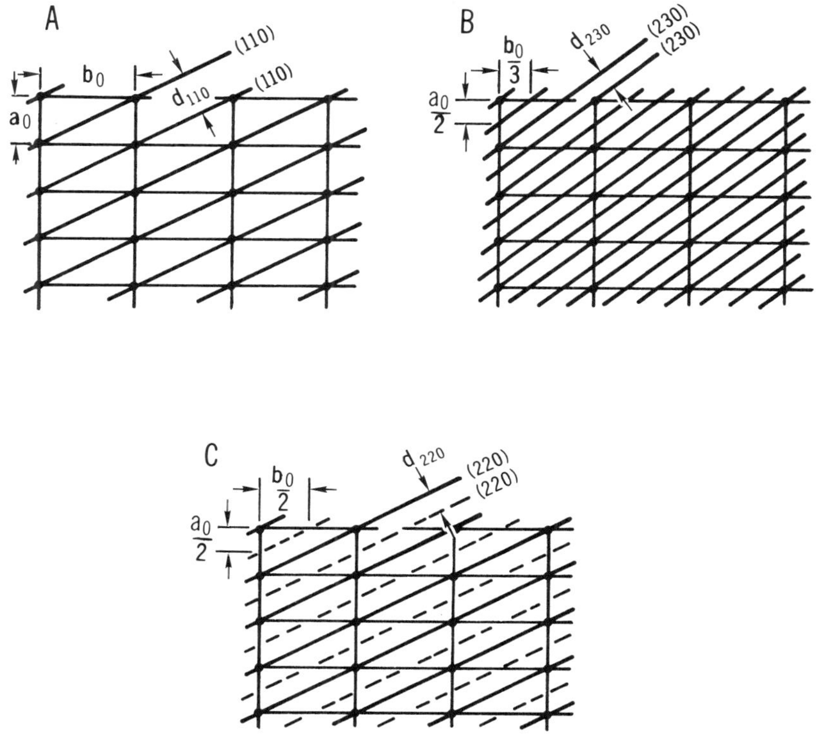

FIGURE 13–8

A primitive orthorhombic lattice viewed down the $+c$ axis with cell edges a_0 and b_0 being as marked.

(A) If a plane parallel to (110) passes through a lattice point, the translation symmetry inherent to the lattice requires similar planes to pass through all other lattice points as shown. The perpendicular distance between neighboring planes in this set of parallel planes is called the *interplanar spacing* and symbolized as d_{110}.

(B) Operation of the lattice's translation symmetry upon a (230) plane similarly generates the set of planes with interplanar spacing d_{230}. It is important to observe that if all the planes translationally equivalent to an (hkl) plane are drawn for a lattice, they divide the a_0 cell edge into h parts, the b_0 edge into k parts and, although not illustrated here, the c_0 edge into l parts.

(C) Assume the solid and dashed lines to constitute a single, translationally equivalent set of planes parallel to the (110) planes in (A). (In truth no translational operation of this primitive lattice can relate a dashed plane to a solid-line plane, but for purposes of discussion ignore this). Because this set of planes intercepts the a axis at $a_0/2\,(= a_0/h)$ intervals and the b axis at $(b_0/2)\,(= b_0/k)$, we conclude its reflection index to be 220. We observe that d_{220} equals $d_{110}/2$, where d_{110} is as shown in (A). In general, as discussed in the text, $d_{nh \cdot nk \cdot nl} = d_{hkl}/n$.

in Figure 13-8C. Actually, every second plane of this set is not a lattice plane because it passes through no lattice points, whereas by definition *a lattice plane must pass through at least three noncollinear lattice points.* Disregarding this and accepting this 220 set as a set of lattice planes with an interplanar spacing d_{220} (Fig. 13-8C), we note that

$$d_{220} = \frac{d_{(110)}}{2}$$

We could similarly demonstrate that

$$d_{330} = \frac{d_{(110)}}{3}; \qquad d_{440} = \frac{d_{(110)}}{4}; \qquad \text{and so on}$$

In general if a reflection index $nh.nk.nl$ is obtained by multiplying a Miller index (hkl) by n, then

$$d_{nh.nk.nl} = \frac{d_{(hkl)}}{n} \qquad \text{(Eq. 13-11)}$$

Substituting $d_{nh.nk.nl}$ for d/n in Equation (13-10), the Bragg equation becomes

$$\lambda = 2d_{nh.nk.nl} \sin \theta \qquad \text{(Eq. 13-12)}$$

In Equation (13-12) $d_{nh.nk.nl}$ the interplanar spacing between "lattice planes" for a reflection index such as 111, 222, 333, and so on, can be written as d_{hkl} because the absence of parentheses around hkl marks it as a Bragg reflection index with a possible common factor n. The Bragg equation now becomes

$$\lambda = 2d_{hkl} \sin \theta \qquad \text{(Eq. 13-12')}$$

with n eliminated (because the nth-order reflection from a set of (hkl) lattice planes can be regarded as the first-order reflection from a set of hkl "lattice planes" where hkl possess a common factor n).

Table 13-2 summarizes the formulas for calculating an interplanar spacing d_{hkl} if the reflection indices hkl and the unit cell's edges and angles are known. The formula for orthorhombic crystals is derived in this chapter's appendix.

P. P. Ewald devised a useful graphical solution for the Bragg equation. To illustrate it we let IC represent a beam of Cu$K\alpha$ radiation striking the (111) plane of NaCl at C in Figure 13-9. Around C we construct an imaginary sphere, called the *sphere of reflection,* with a radius proportional to $1/\lambda$. To plot this reciprocal of a distance, a graphic scale like that in Figure 13-9 must be employed. For example, if λ equaled 1, 2, 3, or 4 Å, then its reciprocal $1/\lambda$ would be plotted as 1, 0.5, 0.33, or 0.25 reciprocal angstroms ($Å^{-1}$), respectively, which are the units marked off on the graphic scale in Figure 13-9. If λ equals 1.5418 Å, as for Cu$K\alpha$ radiation, then $1/\lambda$ equals 0.6486 $Å^{-1}$ units on this scale, this length being the radius of the sphere of reflection in Figure 13-9. At O, where direct beam IC leaves the sphere of reflection, we draw a line normal to the (111) plane. On this

TABLE 13-2

Interplanar Spacing d_{hkl} as Related to Reflection Indices and Unit Cell Dimensions

Crystal System	$\dfrac{1}{d_{hkl}^2}$
Isometric	$\dfrac{1}{a_o^2}(h^2 + k^2 + l^2)$
Tetragonal	$\dfrac{h^2 + k^2}{a_o^2} + \dfrac{l^2}{c_o^2}$
Orthorhombic	$\dfrac{h^2}{a_o^2} + \dfrac{k^2}{b_o^2} + \dfrac{l^2}{c_o^2}$
Hexagonal:	
Hexagonal indices	$\dfrac{4}{3a_o^2}(h^2 + hk + k^2) + \dfrac{l^2}{c_o^2}$
Rhombohedral indices	$\dfrac{1}{a_r^2}\dfrac{(h^2 + k^2 + l^2)\sin^2\alpha + 2(hk + kl + lh)(\cos^2\alpha - \cos\alpha)}{1 - 2\cos^3\alpha + 3\cos^2\alpha}$
Monoclinic	$\dfrac{\dfrac{h^2}{a_o^2} + \dfrac{k^2}{b_o^2} - \dfrac{2hk\cos\gamma}{a_o b_o}}{\sin^2\gamma} + \dfrac{l^2}{c_o^2}$ (first setting)
	$\dfrac{\dfrac{h^2}{a_o^2} + \dfrac{l^2}{c_o^2} - \dfrac{2hl\cos\beta}{a_o c_o}}{\sin^2\beta} + \dfrac{k^2}{b_o^2}$ (second setting)
Triclinic	$\left[\dfrac{h^2}{a_o^2}\sin^2\alpha + \dfrac{k^2}{b_o^2}\sin^2\beta + \dfrac{l^2}{c_o^2}\sin^2\gamma + \dfrac{2hk}{a_o b_o}(\cos\alpha\cos\beta - \cos\gamma) + \dfrac{2kl}{b_o c_o}(\cos\beta\cos\gamma - \cos\alpha) + \dfrac{2lh}{c_o a_o}(\cos\gamma\cos\alpha - \cos\beta)\right] \div [1 - \cos^2\alpha - \cos^2\beta - \cos^2\gamma + 2\cos\alpha\cos\beta\cos\gamma]$

line the point 111 is plotted at a distance from O corresponding to $1/d_{111}$, employing the same scale as used for the sphere's radius. Since d_{111} is 3.2555 Å for NaCl, $1/d_{111}$ plots at 0.307 Å^{-1} units from O. Point 111 so plotted represents the reciprocal projection of face (111), being of the class of projections based on the normal to a face (see Fig. 4-1). Similarly, it follows from Equation (13-11) that reflection indices 222, 333, 444, and 555, respectively, would plot at $2/d_{111}$, $3/d_{111}$, $4/d_{111}$, and $5/d_{111}$ outward from O as shown. Thus $2/d_{111} = 1/d_{222}$ in Figure 13-9A.

We now consider that as the crystal rotates about an axis normal to the page at C in Figure 13-9A, the line OP rotates synchronously so as to be at all times perpendicular to (111). Figure 13-9A shows the position just as point

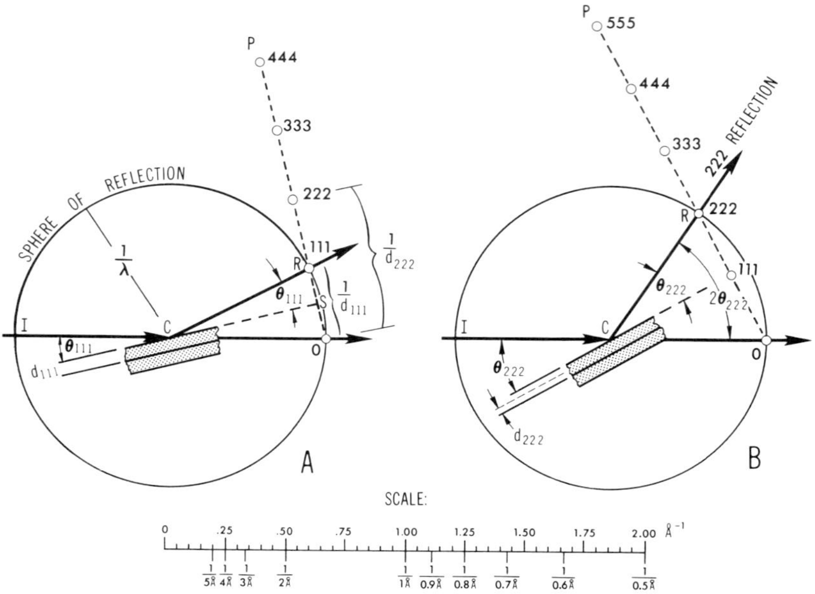

FIGURE 13–9

(A) The NaCl crystal with polished (111) face in Figure 13-6 is centered at C within an imaginary sphere called the *Ewald sphere* or the *sphere of reflection.* Lengths in angstroms are plotted according to the length of their reciprocals; thus lengths of 4, 1, or 0.5 Å would be plotted as lengths corresponding to 0.25, 1, or 2, respectively, on the graphic scale. The sphere's radius is drawn proportional to $1/\lambda$, where λ equals 1.5418 Å for the Cu$K\alpha$ rays of incident beam ICO, and hence $1/\lambda$ approximately equals 0.65 Å^{-1} units on the scale. At O, where beam ICO leaves the sphere, we draw OP perpendicular to dashed line CS; thus OP is perpendicular to the (111) plane. Along OP we locate point 111 at R, a distance from O corresponding to $1/d_{111}$; this distance is thus 0.31 Å^{-1} units on the scale (because d_{111} equals 3.26 Å. Point 111 is called the *reciprocal lattice projection* (abbreviated *relp*) of the (111) lattice planes. Interplanar spacings d_{222}, d_{333}, and d_{444} are one-half, one-third, and one-fourth of d_{111}, respectively. Hence the 222, 333, and 444 relps plot 2, 3, or 4 times as far as 111 from O.
(B) The crystal rotated about an axis normal to the page at C whereupon OP rotated in like manner about O to maintain its perpendicularity to (111). When relp 222 touches the sphere, a diffracted beam, called the 222 "reflection" travels from C toward relp 222. This occurs because, whenever an *hkl* relp touches the Ewald sphere, the Bragg condition (Eq. 13-12) is satisfied as discussed in the text.

111 intersects the sphere of reflection. At this instant a diffracted X-ray beam CR travels from C to point 111 and is at an angle 13.70° (θ_{111}) to the (111) plane. With continued rotation of the crystal at C until point 222 intersects the sphere (Fig. 13-9B), the diffracted beam from C to point 222 is at an angle 28.27° to (111) as was previously determined by inserting values into Equation 13-10.

With continued rotation of the crystal at C and a corresponding rotation of line OP about O, points 333 and 444 successively intersect the sphere to define a diffraction angle of 45.27° for 333 and of 71.30° for 444.

This method of solving the Bragg equation can be analyzed geometrically. In isosceles triangle COR in Figure 13-9A, $CR = CO = 1/\lambda$. Line CS, the edge view of plane (111) extended, is by construction perpendicular to OR and divides triangle COR into two congruent right triangles COS and CSR whose smallest leg (SR or SO) equals one half of length $1/d_{111}$, that is, $1/2d_{111}$. From triangle CSR

$$\sin \theta_{111} = \frac{SR}{CR} = \frac{\lambda}{2d_{111}}$$

which is thus Bragg's law as rephrased in Equation (13-12).

No matter how the crystal is rotated in Figure 13-9B, 555 can never intersect the sphere of reflection. In other words, as discussed earlier the fifth-order reflection from (111) is not possible. If we changed from Cu$K\alpha$ to the shorter wavelength Mo$K\alpha$ ($\lambda = 0.7107$), the sphere of reflection would have its radius $1/\lambda$ more than doubled and hence 555 could intersect this sphere. Thus 555 and other high-order (111) reflections could be observed with Mo$K\alpha$ radiation.

The reciprocal lattice

To construct a reciprocal lattice, we select one point of the crystal's space lattice to be the origin O. Outward from this origin extend normals (dotted in Fig. 13-10) to each set of lattice planes. In Figure 13-10 this is done only for the (100), (110), and (010) sets of planes. As was done for the (111) planes of NaCl in Figure 13-9, along each such normal we plot points at distances outward from O corresponding to $1/d_{hkl}$ for each set of planes. The hollow dots labeled 100, 110, and 010 were so plotted and, by definition, represent the reciprocal projections of planes (100), (110), and (010), respectively. The hollow dots at 020, 030, and so on, may be regarded as reciprocal projections of hypothetical lattice planes (020) and (030). As discussed for Figure 13-8C no such set of lattice planes may exist, but it is a useful fiction because these 020 and 030 reflections serve to account for the second- and third-order reflections from the (010) face. If all possible lattice planes and corresponding higher-order reflections are reciprocally plotted as just described, the result is a lattice of points known as the *reciprocal lattice* to distinguish it from the direct lattices with which we have dealt exclusively until now. The individual points constituting the reciprocal lattice are called *reciprocal lattice points* or *projections* (or, for short, *relps*).

Ewald's reciprocal lattice concept greatly clarifies the results of X-ray crystallography. Thus for any crystal being struck by an X-ray beam, we can imagine (1) that it is centered within a sphere of reflection of radius $1/\lambda$ and (2) that its reciprocal lattice, *correctly oriented with respect to the crystal,* has its origin at the point where the incident X-ray beam exits from the sphere. If a *relp* touches the sphere of reflection, a diffracted X-ray beam will proceed from

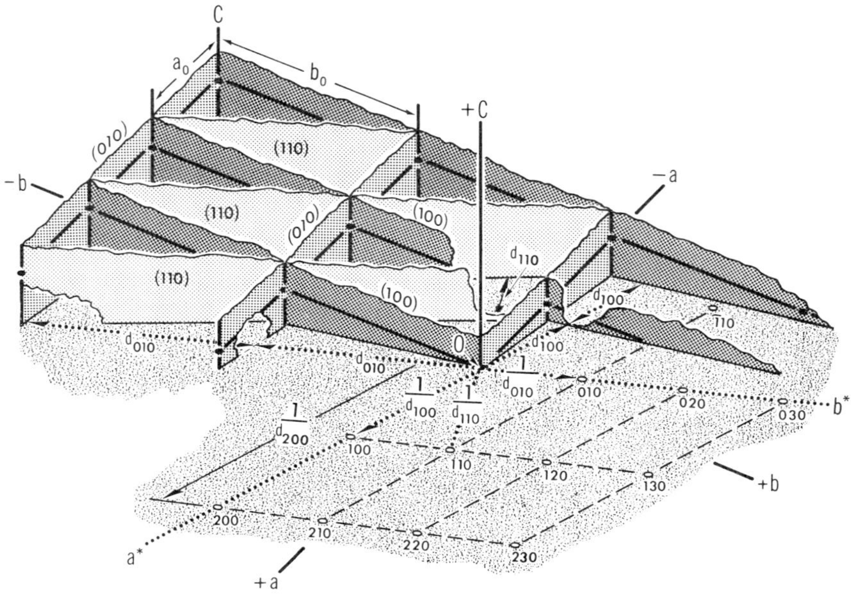

FIGURE 13–10

A few points (black dots) of the direct lattice of a triclinic crystal shown to the left rear. Drawn through these points are a set of (100) planes, of (010) planes, and of (110) planes with interplanar spacings d_{100}, d_{010}, and d_{110}, respectively. From point O, which is taken as the origin of direct axes a, b, and c, perpendiculars to planes (100) and (010) are drawn as dotted lines and labeled a^* and b^*, respectively. Relp 100 is now located along a^* at a distance from the origin equal to $1/d_{100}$. For example, if $d_{100} = 2\text{Å}$, relp 100 would be located at 0.50 Å^{-1} units from point O (see scale, Fig. 13-9). Relp 010 would similarly locate along b^* at a distance from O equal to $1/d_{010}$. Along the dotted line normal to the (110) planes, we similarly plot relp 110 at a distance from O equal to $1/d_{110}$. Relps 020, 030, 120, 130, . . . were plotted in like manner even though the sets of parallel planes corresponding to these relps are not shown. Thus relp 020 represents a set of parallel planes that include not only the set of (010) planes drawn here but also planes midway between them. Any given relp thus completely describes a set of parallel planes in the direct lattice by disclosing (1) their spatial orientation [because these planes will always be perpendicular to the line formed by joining their relp to the origin O] and (2) their interplanar spacing because the relp is at a distance $1/d_{hkl}$ from O. In this drawing the relps were drawn only for sets of planes parallel to the c axis, that is, for ($hk0$), ($h00$), and ($0k0$) planes. The normals to these planes are all perpendicular to the c axis; hence the stippled plane containing relps 100, 110, 020, 130, and so on, is perpendicular to the c axis, as are reciprocal axes a^* and b^*, which lie in this stippled plane.

Direct lattice points could have been plotted at distances of c_0, or multiples thereof, directly above or below those already shown as black dots, thus depicting the crystal's direct lattice in three dimensions. Sets of parallel (hkl) planes, each at the crystal to this *relp* during the time of such tangency. This occurs because for the set of lattice planes that the *relp* represents, such tangency indicates that the modified Bragg equation (Eq. 13-12) is satisfied (Fig. 13-9). A few examples will demonstrate the utility of the reciprocal lattice concept in practical X-ray crystallography.

The Powder Method

The *powder* or *Debye-Scherrer method* is a powerful, easily performed method of identifying crystalline compounds. A few milligrams of the unknown are crushed in an agate mortar (Fig. 13-11A), then sieved (Fig. 13-11B) through a 200-mesh sieve (openings, 0.074 mm). The material that passed through is transferred to a glass slide and mixed with a drop of a cementing agent such as gum tragacanth, collodion, or acetate-base cement (Fig. 13-11C). The resultant mixture is allowed to set until still plastic but no longer tacky. At this point it is scooped onto a single-edged razor blade, then rolled between two clean ground-glass slides,[1] to form a rod approximately 10 mm long and 0.2 to 0.3 mm in diameter (Fig. 13-11D). The rod should be allowed to harden at least 15 minutes before handling. Substances that are affected by the solvents in the cementing agent, or which deliquesce, dehydrate, or otherwise alter in contact with air, cannot be made into rods. Instead they must be packed into fine capillary tubes of lead-free boron glass that are sealed at one end (Azároff and Buerger, 1958, p. 38-39). After this tube is filled, its remaining end can be sealed. For very small samples a sliver cut from a gelatin capsule may be moistened until tacky so that powder grains adhere to it to produce a very delicate rod. Alternatively, a thin boron-glass fiber lightly coated with vaseline can be rolled in the powder to produce a serviceable rod.

The powder rod (or filled capillary tube) is affixed with modeling clay to a special brass pin, the axes of the rod and pin being maintained parallel (Fig. 13-11E). Using bent-nose pliers or tweezers, this rod-bearing brass pin is inserted into the powder camera at *C* (Fig. 13-12). The collimator and the beam catcher, both removed prior to this insertion, are then replaced. A small motor, which is located behind the camera and thus not shown in Figure 13-12, will cause this brass pin (and powder rod) to rotate on its axis. This axis must coincide precisely with the axis of the squat metal cylinder that constitutes the powder camera.

---

different angles to the *c* axis, could then be drawn and plotted reciprocally at distances from *O* corresponding to $1/d_{hkl}$ along their normals. If this were done for numerous sets of *hkl* planes such as (111), (121), (131), (112), (122), and so on, their relps would be regularly spaced to produce a second three-dimensional lattice known as the *reciprocal lattice*.

[1] A single or double thickness of Scotch tape stuck onto opposite ends of the lower slide produces a more cylindrical rod whose diameter equals the thickness of the tape.

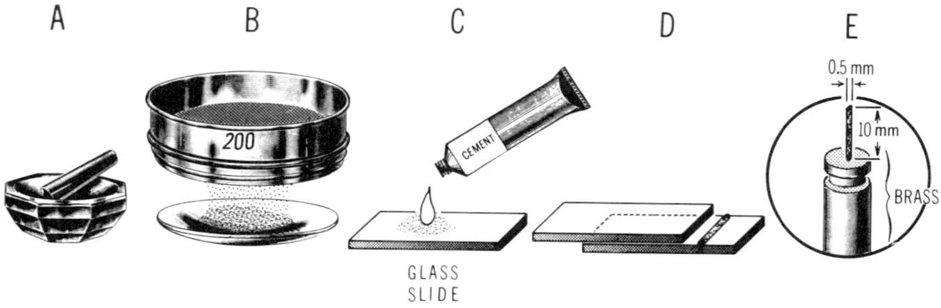

FIGURE 13–11

Steps in preparing a powder rod of a crystal.
(A) Grind crystal to a fine powder in an agate mortar.
(B) Sieve to obtain grains finer than 200 mesh.
(C) Thoroughly mix these fine grains with a small amount of a cement such as gum tragacanth, collodion, or acetate base cement.
(D) After the mixture has "set up" so that it is yet plastic but no longer sticky or tacky, roll it between two clean ground-glass slides to form a rod approximately 10 mm long and 0.2 to 0.3 mm in diameter.
(E) After allowing rod to harden for at least 15 minutes, insert into a special brass pin that can in turn be inserted into the powder camera at *C* in Figure 13-12.

After verifying and, if necessary, adjusting this,[2] the powder camera is taken into a photographic darkroom where the collimator and beam catcher are again removed and then replaced after insertion of a strip of photographic film (see Fig. 13-12). This film strip is punched to accommodate collimator and beam catcher, then snugly fit against the camera walls. After the camera is so loaded, its lid is inserted to make it light tight.

The loaded camera is correctly positioned in respect to a source of, for example, Cu$K\alpha$ radiation, so that the collimator permits only a thin beam of X-rays to strike the rod. The major portion of this beam's energy passes through the rod to enter the beam catcher and to be absorbed there without exposing the film. But a minor portion undergoes Bragg diffraction by the myriads of randomly oriented crystals in the rod. Hence O, the point where the primary beam leaves the sphere of reflection, represents the origin for as many differently oriented reciprocal lattices as there are differently oriented crystals being struck by the X-ray beam. Considering the same *hkl* relp in each of these randomly oriented reciprocal lattices, they define the surface of a sphere of radius $1/d_{hkl}$ (Fig. 13-13A), this sphere being called a *relp sphere*. For example, selecting the 321 relp

[2] This is done by replacing the beam catcher on one of the collimators with the small lens that is provided. Viewed through this lens and with the camera open and brightly illuminated, the rod should (ideally) appear stationary during a 360° rotation of the sample holder. If not, then each time it almost leaves the top of the field of view, the rotating of the sample holder is stopped and the centering screw (Fig. 13-12) is turned clockwise to center the rod. Before resuming the rotation of the sample holder, this screw is turned counterclockwise to return it to its original position.

from each randomly oriented lattice, they fall on a relp sphere of radius $1/d_{321}$. Those relps that lie on a relp sphere's circle of intersection (dotted in Fig. 13-13A) with the sphere of reflection are relps from crystals correctly oriented to produce a Bragg reflection. Thus diffracted X-rays proceed from the point where the primary X-ray beam strikes the powder rod to all points on the dotted circle. The result is a cone of secondary X-rays (Fig. 13-13B) whose axis is the primary beam.

A single crystal of NaCl correctly oriented at different attitudes to the primary beam can yield the following relatively intense Bragg reflections: 111, 200, 220, 311, 222, 400, 331, 420, 422, 511, 333, 440, 531, 600, 620, 533, 622, and 444. For reasons to be discussed later, other *hkl* reflections, for example, 210, 211, 110, and so on, are absent for NaCl. Thus for a powder of randomly oriented NaCl crystals struck by Cu$K\alpha$ (λ = 1.5418 Å) radiation, we need consider only

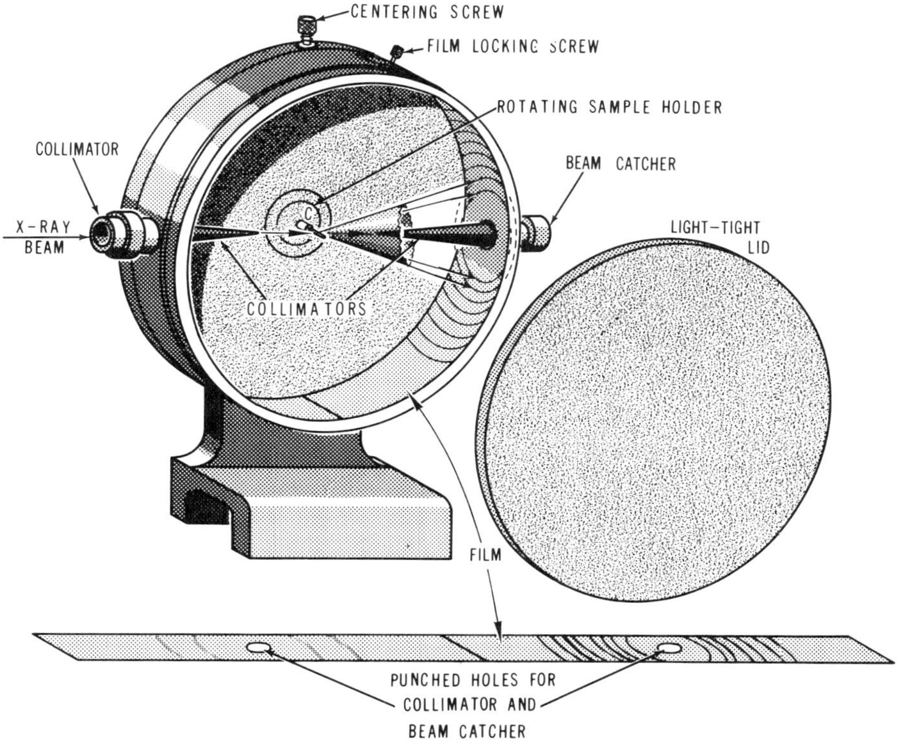

FIGURE 13–12

Schematic view of a powder camera with its light-tight lid removed. Its collimator is placed against the exit port of the X-ray tube and blocks out all X-rays except a fine beam that impinges on the powder rod. This powder rod is rotated about its axis by a small motor which is not shown. For reasons to be discussed in Figure 13-13, cones of diffracted X-rays emanate from the powder rod where it is struck by the direct beam, these cones exposing the film in a series of arcs which, after the film is developed, might appear as shown here.

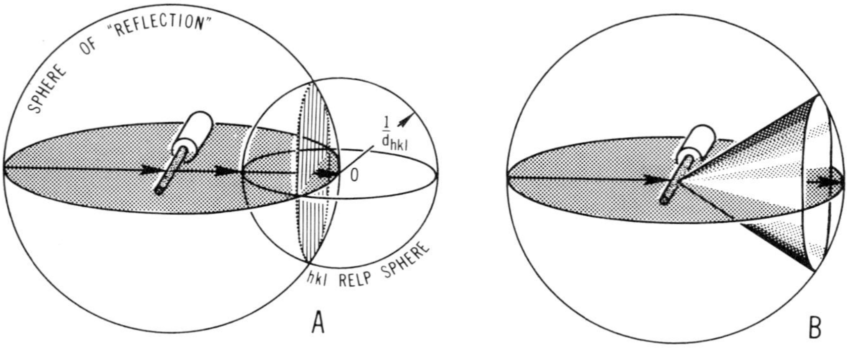

FIGURE 13–13

(A) Where a powder rod is struck by the direct beam, we locate the center of the imaginary sphere of reflection (radius, $1/\lambda$). The origin of the reciprocal lattice for each of the variously oriented grains in the powder rod is at O, the point where the direct beam emerges from this sphere of reflection. If the same *hkl* relp were plotted, for example, relp 213, for each of these variously oriented grains, a relp sphere would result of radius $1/d_{hkl}$, that is, of radius $1/d_{213}$ for our specific example. This relp sphere intersects the sphere of reflection in the dotted circle (vertically ruled). (B) From the powder rod, diffracted rays travel toward those relps located on the dotted circle to form the cone shown. Numerous other relp spheres, each with a different radius, for example, $1/d_{100}$, $1/d_{111}$, $1/d_{234}$, could be drawn centered at O and would similarly intersect the sphere of reflection in circles to which cones of diffracted rays would proceed from the powder rod.

the circles of intersection between the sphere of reflection (radius, 1/1.5418 Å) and relp spheres 111, 200, 220, 311, 222, 400, and so on. Figure 13-14A shows the cones of diffracted X-rays leading to these circles of intersection. One of the crystals which produce the arrow-indicated ray in the 220 cone (Fig. 13-14A) is drawn enlarged in Figure 13-14B. A crystal thus oriented produces a second-order reflection from its (110) planes. This diffracted beam is (1) at the Bragg angle θ_{220} to the (110) planes and (2) at an angle $2\theta_{220}$ to the direct beam. The orientations of those other grains in the powder rod whose diffracted rays also belong to the 220 cone would be reproduced by rotating Figure 13-14B (up to 360°) around the line labeled "direct beam." Thus two diffracted rays on opposite sides of the same cone will be at an angle 4θ, in this case $4\theta_{220}$, to each other.

Let us assume a powder camera of 360 mm circumference. The film strip within it, which has been exposed where struck by each cone of diffracted rays, is developed, dried, and flattened against the illuminated ground-glass surface of a film-measuring device (Fig. 13-14C). This device's cursor can be slid back and forth over a fixed scale to locate the center points for each pair of arcs from the same cone, points x_2 and x_2' constituting such points for cone 2. This is done for all the diffraction cones and the results are recorded on a previously prepared form (Fig. 13-15). Returning to Figure 13-14A, a powder rod struck by an X-ray beam produces: (a) cones of *forward reflections* whose rays are at a 2θ angle of 0° to 90° to the direct beam and (b) cones of *back reflections* whose backward-

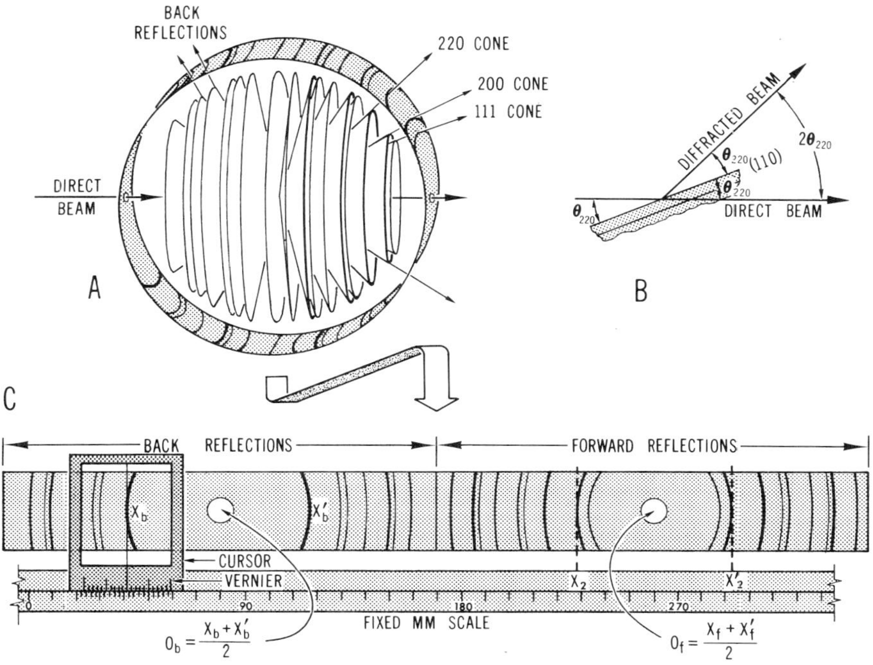

FIGURE 13–14

(A) The numerous cones of diffracted rays that emanate from a powder rod (hidden by the cones) expose the film strip in a powder camera to produce a series of arcs. The powder camera itself is not shown. The powder rod would be rotating on its axis, which is also the axis of the squat cylinder formed by the film strip. (B) Cross section (enlarged 100,000,000X) of a grain whose diffracted beam lies along the arrow on the 220 cone in (A). As this cross section is rotated 360° around the direct beam as an axis, it successively represents the cross section of each grain in the rod that produces a diffracted ray lying in the 220 cone. If we compare this cross section with itself after a 180° rotation around the direct beam, we see that 4θ represents the cone angle, that is, the angle between the two most divergent rays in the same diffraction cone.
(C) Enlarged drawing of the film in (A) after its processing and mounting onto a device for measuring the distance between arcs on a powder film. For example, the cursor now indicates the midpoint of arc x_b to be above 40.6 on the fixed millimeter scale (whose finer divisions have been largely omitted). The cursor could next be slid to coincide with the midpoint of x_b'. The distance in millimeters between x_b and x_b', corrected for film shrinkage, equals 4θ for the cone which produced these arcs, *if a powder camera was used whose circumference equals 360 mm.* Similarly, we could determine 4θ for arcs resultant from forward reflections, for example, the distance between x_2 and x_2'. Note that O_b, the arithmetic average of two arcs from the same back-reflected diffraction cone $[\frac{1}{2}(x_b + x_b')]$ marks the point where the direct beam enters the film in (A). Similarly O_f, the arithmetic average of two arcs from the same forward-reflected cone, is the coordinate of the point where the direct beam leaves the powder camera (through a hole punched in the film). The intensities of the back reflections are greatly exaggerated.

LINE NO.	I	x	x'	x+x'	ARC LENGTH		$\frac{C(x'-x)}{2} = 2\theta$	d	$\frac{1}{d^2}$	hkl
					MEASURED $x'-x$	CORRECTED $C(x'-x)=4\theta$				

FIGURE 13–15

Sample of form (after Azároff and Buerger, 1958) for recording measurements on a powder film such as discussed for Figure 13-14. It is convenient to number the arcs or cones sequentially 1, 2, 3, . . . , starting with the smallest 2θ angle.

directed rays are at a 2θ angle of 90° to 180° to the direct beam. The forward reflections are usually more intense and thus produce darker arcs on the film.

Let x_f and x_f' represent the measured coordinates of the center points of two arcs from a forward reflected cone. Their average $(x_f + x_f')/2$ represents the coordinate of O_f, the point (at $2\theta = 0°$) where the direct beam leaves the sphere of reflection. Similarly, if x_b and x_b' represent the measured coordinates of center points for the two arcs from a back-reflected cone, then $(x_b + x_b')/2$ represents the coordinate of O_b, the point (at $2\theta = 180°$) where the direct beam entered the sphere of reflection. The column $(x + x')$ in Figure 13-15 also checks the accuracy of measured values x and x' because $(x + x')$ should be constant in value; moreover $(x + x')/2$ locates O_f or O_b. Their difference $(O_b - O_f)$ should equal 180 mm because (1) a 180° angle exists between the entrance and exit points of the direct beam and (2) the camera's circumference was 360 mm (1 mm per degree of arc). However, film shrinkage during development and drying may reduce $(O_b - O_f)$ to a distance less than 180 mm. Hence distances measured along the film, for example, $(x' - x)$, the distance in millimeters between the center points of two arcs from the same cone, must be corrected for film shrinkage through multiplication by *correction factor* C, where C equals 180 mm$/(O_b - O_f)$. This corrected value, $(x' - x)C$ in millimeters, numerically equals 4θ, which is the angle (see Fig. 13-14C) for the cone of rays that struck the film at x' and x, respectively. Thus θ is known and Equation (13-12) can be solved to determine d, the interplanar spacing for the set of planes that produced the reflection. Usually the d value corresponding to the measured 2θ value is looked up in one of the existing tables (Fig. 13-19, p. 485).

Identifying crystalline unknowns

From the powder photograph of an unknown, 2θ and therefore d may be determined for the three most intense (blackest) lines. These are frequently selected just by visual inspection. However, line intensities may be quantitatively measured with an optical densitometer or by comparison with lines of known intensities on a previously prepared film standard. If such quantitative measurements are made, the ratio of each line's intensity to that of the most intense line on this powder photograph, that is, I/I_{max}, may be determined.

Once d is determined for the three most intense lines on a powder photograph, a card identifying the unknown can usually be located in the *Powder Diffraction File* published by the American Society for Testing and Materials. Suppose, for example, d is determined as 3.34 Å for the most intense line and as 10.0 and 4.99 Å for the two next most intense lines. The cards in the file are subdivided into *Hanawalt groups* according to the d spacing of the most intense line (Table 13-3). Our unknown thus falls into Group No. 50, but No. 51 remains a possibility because 3.34 Å is borderline. Within each Hanawalt group, cards are filed from front to back in order of decreasing d spacing of the *second* most intense line. Sorting through Group 50, which, of course, changes from year to year as new cards are added or old data corrected, we might locate a card sequence like that

TABLE 13–3

Number Code and d Values of Hanawalt Groups

Number	d	Number	d	Number	d
1	Under 0.80	30	2.30–2.34	59	4.00–4.09
2	0.80–0.89	31	2.35–2.39	60	4.10–4.19
3	0.90–0.99	32	2.40–2.44	61	4.20–4.29
4	1.00–1.04	33	2.45–2.49	62	4.30–4.39
5	1.05–1.09	34	2.50–2.54	63	4.40–4.49
6	1.10–1.14	35	2.55–2.59	64	4.50–4.59
7	1.15–1.19	36	2.60–2.64	65	4.60–4.69
8	1.20–1.24	37	2.65–2.69	66	4.70–4.79
9	1.25–1.29	38	2.70–2.74	67	4.80–4.89
10	1.30–1.34	39	2.75–2.79	68	4.90–4.99
11	1.35–1.39	40	2.80–2.84	69	5.00–5.24
12	1.40–1.44	41	2.85–2.89	70	5.25–5.49
13	1.45–1.49	42	2.90–2.94	71	5.50–5.74
14	1.50–1.54	43	2.95–2.99	72	5.75–5.99
15	1.55–1.59	44	3.00–3.04	73	6.00–6.49
16	1.60–1.64	45	3.05–3.09	74	6.50–6.99
17	1.65–1.69	46	3.10–3.14	75	7.00–7.49
18	1.70–1.74	47	3.15–3.19	76	7.50–7.99
19	1.75–1.79	48	3.20–3.24	77	8.00–8.49
20	1.80–1.84	49	3.25–3.29	78	8.50–8.99
21	1.85–1.89	50	3.30–3.34	79	9.00–9.49
22	1.90–1.94	51	3.35–3.39	80	9.50–9.99
23	1.95–1.99	52	3.40–3.44	81	10.0–10.9
24	2.00–2.04	53	3.45–3.49	82	11.0–11.9
25	2.05–2.09	54	3.50–3.59	83	12.0–13.9
26	2.10–2.14	55	3.60–3.69	84	14.0–15.9
27	2.15–2.19	56	3.70–3.79	85	16.0–17.9
28	2.20–2.24	57	3.80–3.89	86	18.0–19.9
29	2.25–2.29	58	3.90–3.99	87	20.0 or over

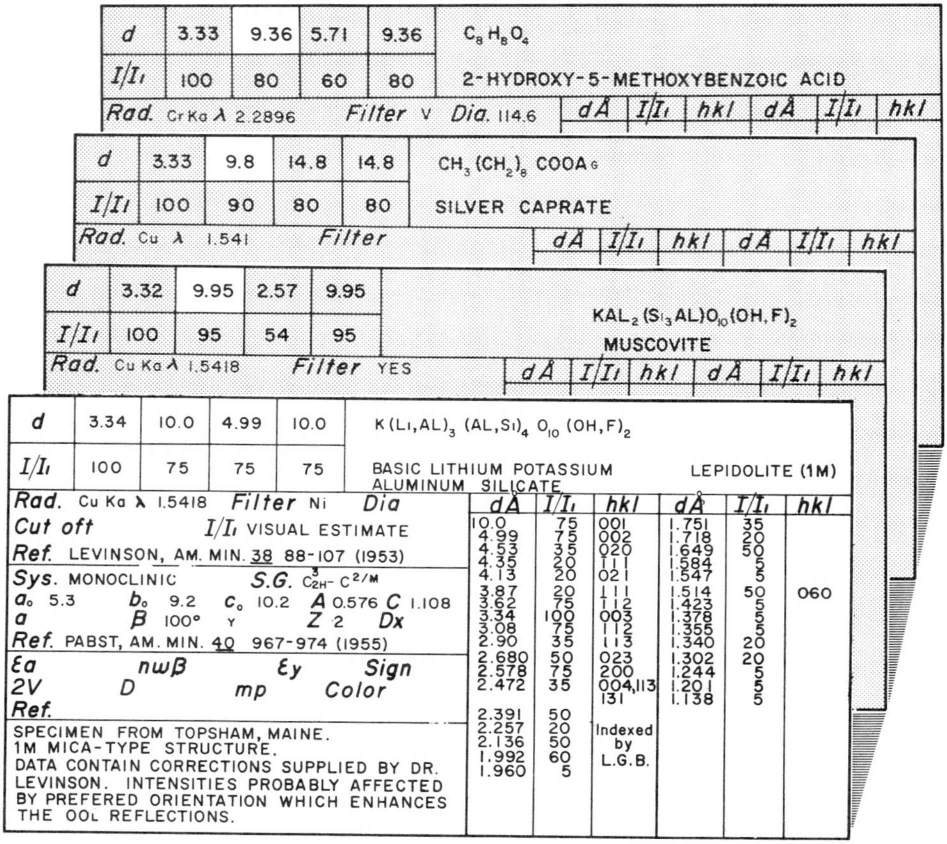

d	3.33	9.36	5.71	9.36	$C_8H_8O_4$
I/I_1	100	80	60	80	2-HYDROXY-5-METHOXYBENZOIC ACID

Rad. Cr Kα λ 2.2896 Filter V Dia. 114.6 | dÅ | I/I_1 | hkl | dÅ | I/I_1 | hkl

d	3.33	9.8	14.8	14.8	$CH_3(CH_2)_8COOAg$
I/I_1	100	90	80	80	SILVER CAPRATE

Rad. Cu λ 1.541 Filter | dÅ | I/I_1 | hkl | dÅ | I/I_1 | hkl

d	3.32	9.95	2.57	9.95	$KAL_2(Si_3AL)O_{10}(OH,F)_2$
I/I_1	100	95	54	95	MUSCOVITE

Rad. Cu Kα λ 1.5418 Filter YES | dÅ | I/I_1 | hkl | dÅ | I/I_1 | hkl

d	3.34	10.0	4.99	10.0	$K(Li,AL)_3(AL,Si)_4O_{10}(OH,F)_2$
I/I_1	100	75	75	75	BASIC LITHIUM POTASSIUM ALUMINUM SILICATE — LEPIDOLITE (1M)

Rad. Cu Kα λ 1.5418 Filter Ni Dia
Cut off I/I_1 VISUAL ESTIMATE
Ref. LEVINSON, AM. MIN. 38 88-107 (1953)
Sys. MONOCLINIC S.G. C^3_{2H}- C2/M
a_0 5.3 b_0 9.2 c_0 10.2 A 0.576 C 1.108
α β 100° γ Z 2 Dx
Ref. PABST, AM. MIN. 40 967-974 (1955)
εα nωβ εγ Sign
2V D mp Color
Ref.
SPECIMEN FROM TOPSHAM, MAINE.
1M MICA-TYPE STRUCTURE.
DATA CONTAIN CORRECTIONS SUPPLIED BY DR. LEVINSON. INTENSITIES PROBABLY AFFECTED BY PREFERED ORIENTATION WHICH ENHANCES THE OOL REFLECTIONS.

dÅ	I/I_1	hkl	dÅ	I/I_1	hkl
10.0	75	001	1.751	35	
4.99	75	002	1.718	20	
4.53	35	020	1.649	50	
4.35	20	$\bar{1}11$	1.584	5	
4.13	20	021	1.547	5	
3.87	20	111	1.514	50	060
3.62	75	$\bar{1}12$	1.423	5	
3.34	100	003	1.378	5	
3.08	75	112	1.355	5	
2.90	35	113	1.340	20	
2.680	50	023	1.302	20	
2.578	75	200	1.244	5	
2.472	35	004,113 131	1.201	5	
			1.138	5	
2.391	50	Indexed by L.G.B.			
2.257	20				
2.136	50				
1.992	60				
1.960	5				

FIGURE 13–16

Cards selected from the *Powder Diffraction File* published by the American Society for Testing and Materials, that is, ASTM cards. These represent crystalline substances whose three most intense lines or arcs have d spacings close to 3.34, 10.0, and 4.99 Å, the d spacings of the three most intense lines determined for a powdered unknown. The unknown thus appears to be lepidolite.
For each material the first three of the four d values at the top of the card represent its three most intense powder lines. The fourth, for example, 9.95 for muscovite, is its largest spacing. The ASTM cards are filed into Hanawalt groups (Table 13-3) according to the d value of their most intense line. These four cards are from Hanawalt group 50 (d = 3.30 to 3.34). Within a given Hanawalt group the cards are filed according to decreasing spacing of their second most intense line (left unshaded in the three cards to the rear to emphasize this filing order).

in Figure 13-16. The unknown thus appears to be lepidolite. To confirm the identification the d spacings and relative intensities of the remaining lines on the unknown's powder photograph can now be compared with those reported on the card by A. A. Levinson. Of course, it is always possible that the unknown is an entirely new substance for which there is no card in the file. Another possibility is that the unknown represents a mixture of two or more compounds. In this case the procedure suggested by Nuffield (1966, p. 134) may be followed.

The intensities of the various lines on a powder photograph are affected by preferred orientation of the crystals. In the case of the mica lepidolite it is difficult to prevent the cleavage flakes from orienting parallel to the axis of the powder rod which in turn causes certain areas of the 001, 002, and 003 relp spheres to be more densely populated as shown for 001 in Figure 13-17A. This preferred orientation thus increases the intensities of the $00l$ reflections as recorded on the powder film. Consequently, the relative intensities of the various reflections may fluctuate from one rod mount to another of the same material.

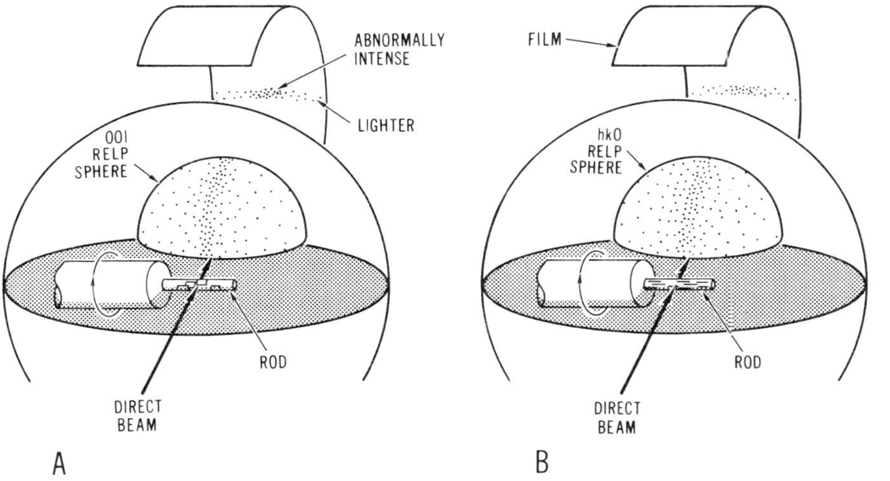

FIGURE 13–17

Effects of preferred orientation of grains in powder rods.
(A) The (001) flakes of a clay or mica mineral tend to orient parallel to the axis of the powder rod as shown for a few grains whose size is greatly exaggerated relative to the rod. Consequently, the 001, 002, 003, and 004 relps spheres, of which only the 001 sphere is drawn, are each no longer evenly populated but instead contain concentrations of relps (black dots) where shown. The powder film, shown only in part, records the arcs in the regions where they are more intense. If complete randomness of orientation were obtained the intensity of the 001, 002, 003, . . ., $00l$ lines recorded on the film would be less because the relps would be evenly distributed over the entire relp sphere rather than concentrated where shown.
(B) Needle-shaped grains adopt a preferred orientation so as to be more frequently oriented parallel to the rod's axis. If the needles are developed along the crystal's c axis as for {110} cleavage fragments of chain silicates, the $hk0$ relps for each grain will lie in planes perpendicular to the needles and thus, assuming preferred orientation as described, more $hk0$ relps will lie in planes perpendicular to the powder rod than if random orientation existed. Selecting any given $hk0$ relp sphere, the $hk0$ lines show enhanced intensities to the extent that crystals tend to orient preferentially with their c axis parallel to the rod's axis. (In like manner, for a crystal with two directions of cleavage that cause it to pulverize into needles parallel to the a axis, preferred orientation would then enhance the $0kl$ intensities.)

TABLE 13–4

Sets of d_1^2/d^2 Quotients for Indexing Powder Diffraction Patterns of Isometric Crystals

	True Reflection Index Corresponding to d_1						
hkl for *d*	100 (1)	110 (2)	111 (3)	200 (4)	210 (5)	211 (6)	220 (7)
100	1						
110^I	2	1					
111^F	3	1.5	1				
$200^{I,F}$	4	2	1.33	1			
210	5	2.5	1.67	1.25	1		
211^I	6	3	2	1.5	1.2	1	
$220^{I,F}$	8	4	2.67	2	1.6	1.33	1
300, 221	9	4.5	3	2.25	1.8	1.5	1.12
310^I	10	5	3.33	2.5	2	1.67	1.25
311^F	11	5.5	3.67	2.75	2.2	1.83	1.37
$222^{I,F}$	12	6	4	3	2.4	2	1.5
320	13	6.5	4.33	3.25	2.6	2.17	1.62
321^I	14	7	4.67	3.5	2.8	2.33	1.75
$400^{I,F}$	16	8	5.33	4	3.2	2.67	2
410, 322	17	8.5	5.67	4.25	3.4	2.83	2.12
411, 330^I	18	9	6	4.5	3.6	3	2.25
331^F	19	9.5	6.33	4.75	3.8	3.17	2.37
$420^{I,F}$	20	10	6.67	5	4	3.33	2.5
421	21	10.5	7	5.25	4.2	3.5	2.62
332^I	22	11	7.33	5.5	4.4	3.67	2.75
$422^{I,F}$	24	12	8	6	4.8	4	3
500, 430	25	12.5	8.33	6.25	5	4.17	3.12
510, 431^I	26	13	8.67	6.5	5.2	4.33	3.25
511, 333^F	27	13.5	9	6.75	5.4	4.5	3.37
520, 432	29	14.5	9.67	7.25	5.8	4.83	3.62
521^I	30	15	10	7.5	6	5	3.75
$440^{I,F}$	32	16	10.67	8	6.4	5.33	4
522, 441	33	16.5	11	8.25	6.6	5.5	4.12
530, 433^I	34	17	11.33	8.5	6.8	5.67	4.25
531^F	35	17.5	11.67	8.75	7	5.83	4.37
600, $442^{I,F}$	36	18	12	9	7.2	6	4.5
610	37	18.5	12.33	9.25	7.4	6.17	4.62
611, 532^I	38	19	12.67	9.5	7.6	6.33	4.75
$620^{I,F}$	40	20	13.33	10	8	6.67	5
443, 621, 540	41	20.5	13.67	10.25	8.2	6.83	5.12
541^I	42	21	14	10.5	8.4	7	5.25
533^F	43	21.5	14.33	10.75	8.6	7.17	5.37
$622^{I,F}$	44	22	14.67	11	8.8	7.33	5.5
630, 542	45	22.5	15	11.25	9	7.5	5.62
631^I	46	23	15.33	11.50	9.2	7.67	5.75

TABLE 13–4 (*Continued*)

hkl for *d*	100 (1)	110 (2)	111 (3)	200 (4)	210 (5)	211 (6)	220 (7)
	True Reflection Index Corresponding to d_1						
$444^{I,F}$	48	24	16	12	9.6	8	6
700, 632	49	24.5	16.33	12.25	9.8	8.17	6.12
710, 550, 543^I	50	25	16.67	12.50	10	8.33	6.25
711, 551^F	51	25.5	17	12.75	10.2	8.5	6.37
$640^{I,F}$	52	26	17.33	13	10.4	8.67	6.5
720, 641	53	26.5	17.67	13.25	10.6	8.83	6.62
721, 633, 552^I	54	27	18	13.50	10.8	9	6.75
$642^{I,F}$	56	28	18.67	14	11.2	9.33	7
722, 544	57	28.5	19	14.25	11.4	9.5	7.12
730^I	58	29	19.33	14.5	11.6	9.67	7.25
731, 553^F	59	29.5	19.67	14.75	11.8	9.83	7.37
650, 643	61	30.5	20.33	15.25	12.2	10.17	7.62
732, 651^I	62	31	20.67	15.5	12.4	10.33	7.75
$800^{I,F}$	64	32	21.33	16	12.8	10.67	8
810, 740, 652	65	32.5	21.67	16.25	13	10.83	8.12
811, 741, 554^I	66	33	22	16.50	13.2	11	8.25
733	67	33.5	22.33	16.75	13.4	11.17	8.37
820, $644^{I,F}$	68	34	22.67	17	13.6	11.33	8.5
821, 742	69	34.5	23	17.25	13.8	11.5	8.62
653^I	70	35	23.33	17.5	14	11.67	8.75
822, $660^{I,F}$	72	36	24	18	14.4	12	9
830, 661	73	36.5	24.33	18.25	14.6	12.17	9.12
831, 750, 743^I	74	37	24.67	18.5	14.8	12.33	9.25
751, 555^F	75	37.5	25	18.75	15	12.5	9.37
$662^{I,F}$	76	38	25.33	19	15.2	12.67	9.5
832, 654	77	38.5	25.67	19.25	15.4	12.83	9.62
752^I	78	39	26	19.5	15.6	13	9.75
$840^{I,F}$	80	40	26.67	20	16	13.33	10
900, 841, 744, 663	81	40.5	27	20.25	16.2	13.5	10.12
910, 833^I	82	41	27.33	20.5	16.4	13.67	10.25
911, 753^F	83	41.5	27.67	20.75	16.6	13.83	10.37
$842^{I,F}$	84	42	28	21	16.8	14	10.5
920, 760	85	42.5	28.33	21.25	17	14.17	10.62
921, 761, 655^I	86	43	28.67	21.50	17.2	14.33	10.75
$664^{I,F}$	88	44	29.33	22	17.6	14.67	11

^{I,F} Indices on the lines superscripted *I* and/or *F* are confined to centered lattices *I* and/or *F*. However, all indices, superscripted or not, are possible for a primitive (*P*) lattice.

Indexing powder photographs

The lepidolite card in Figure 13-16 has reflection indices *hkl* assigned to its various lines. The lines are thus said to be "indexed." Techniques of indexing the lines of powder photographs are discussed by Azároff and Buerger (1958, p. 60) and by Nuffield (1966, p. 118), among others. Powder photographs of isometric crystals are especially easy to index. To do so, the film's measured d values are listed from the largest d_1 to the smallest, then each is squared to obtain

$$d_1^2, d_2^2, d_3^2, \ldots, d_n^2$$

Next each of these values is divided into d_1^2 to obtain

$$\frac{d_1^2}{d_1^2}, \frac{d_1^2}{d_2^2}, \frac{d_1^2}{d_3^2}, \ldots, \frac{d_1^2}{d_n^2}$$

The resultant set of quotients will compare to those in one of the columns, 1, 2, . . ., or 7 in Table 13-4. If it includes integers only, it compares to column 1. If its values also include $\frac{1}{2}$, $\frac{1}{3}$, $\frac{1}{4}$, $\frac{1}{5}$, $\frac{1}{6}$, or $\frac{1}{8}$ intervals between integers, it will compare to columns 2, 3, 4, 5, 6, or 7, respectively. Once the correct column for comparison is identified, the *hkl* indices associated in Table 13-4 with the values in this column can be assigned to the lines whose d_1^2/d^2 quotients are identical to these values.

To illustrate the method, we consider the 2θ values measured from the lines of a Debye-Scherrer photograph of synthetic spinel ($MgAl_2O_4$) taken with Cu$K\alpha$ radiation (Table 13-5). Such radiation consists of Cu$K\alpha_1$ (1.54051 Å) and Cu$K\alpha_2$ (1.54433 Å). Hence two spheres of reflection exist: (1) a $K\alpha_1$ sphere of radius 1/1.54051 and (2) a slightly smaller $K\alpha_2$ sphere of radius 1/1.54433. The smallest relp sphere for a powdered material will have the radius $1/d_1$ if d_1 is the largest interplanar spacing observable. This $h_1k_1l_1$ relp sphere will be centered on O_1 in respect to the $K\alpha_1$ sphere (Fig. 13-18A) but on O_2 for the $K\alpha_2$ sphere, O_1 and O_2 being the points where the direct beam exits from each sphere. For this $h_1k_1l_1$ relp sphere the $K\alpha_1$ and $K\alpha_2$ cones of diffraction are practically identical, their difference being highly exaggerated in Figure 13-18A for illustrative purposes. Thus the $K\alpha_1$ and $K\alpha_2$ cones expose the film as two overlapping arcs. A much larger relp sphere, for example, the $h_nk_nl_n$ sphere in Figure 13-18B, where d_n is the smallest interplanar spacing observed, will intersect the $K\alpha_1$ and $K\alpha_2$ sphere of reflection so as to produce $K\alpha_1$ and $K\alpha_2$ cones that are sufficiently distinct to be recorded as two distinct arcs on the film. In Table 13-5 such resolution (non-overlap) of the $K\alpha_1$ and $K\alpha_2$ reflections occurs for all relp spheres for which 2θ exceeds 90°. Thus the lines at 99.37° and at 99.72°, respectively, represent the $K\alpha_1$ and $K\alpha_2$ reflections from the same set of *hkl* planes.

Each measured 2θ value in Table 13-5 was looked up in the Fang-Bloss tables to obtain the d value corresponding to it (Fig. 13-19). For overlapped $K\alpha_1$ and $K\alpha_2$ lines for the same *hkl* plane, wavelength was taken as Cu$K\alpha$ (1.5418 Å) in looking up the plane's d value. For resolved $K\alpha_1$ and $K\alpha_2$ lines, Cu$K\alpha_1$ and

TABLE 13–5

Observed 2θ and Calculated d_1^2/d^2 Values for a Powder Film of Synthetic Spinel, $MgAl_2O_4$

$2\theta_{obs}$	d	d^2	d_1^2/d^2	hkl	N^a	a_{calc}^b
19.02α	4.6658	21.7697	1	111	3	8.0814
31.30α	2.8577	8.1664	2.666	220	8	8.0828
36.894α	2.4365	5.9365	3.667	311	11	8.0810
44.84α	2.0212	4.0852	5.329	400	16	8.0848
55.702α	1.6501	2.7232	7.994	422	24	8.0838
59.427α	1.5553	2.4190	8.999	511/333	27	8.0816
65.287α	1.4291	2.0423	10.659	440	32	8.0842
68.736α	1.3656	1.8649	11.673	531	35	8.0790
74.19α	1.2781	1.6335	13.327	620	40	8.0834
77.388α	1.2331	1.5205	14.317	533	43	8.0860
82.691α	1.1670	1.3619	15.984	444	48	8.0852
85.78α	1.1327	1.2830	16.968	711/551	51	8.0891
94.171α_1	1.0517	1.1063	19.678	731/553	59	8.0783
94.451α_2	1.0520					8.0806
99.37α_1	1.0102	1.0203	21.337	800	64	8.0816
99.72α_2	1.0101					8.0808
107.92α_1	0.95257	0.90729	23.994	822/660	72	8.0828
108.327α_2	0.95248					8.0821
111.26α_1	0.93318	0.87094	24.995	751/555	75	8.0816
111.656α_2	0.93330					8.0826
116.944α_1	0.90365	0.81640	26.665	840	80	8.0825
117.45α_2	0.90345					8.0807
120.50α_1	0.88719	0.78582	27.667	911/753	83	8.0827
121.074α_2	0.88686					8.0797
130.744α_1	0.84733	0.71797	30.321	931	91	8.0836
131.366α_2	0.84734					8.0831
138.028α_1	0.82497	0.68056	31.987	844	96	8.0830
138.785α_2	0.82495					8.0828

[a] $N = h^2 + k^2 + l^2$
[b] $a = d\sqrt{N}$

Cu$K\alpha_2$ wavelengths, respectively, were used in determining the d values. The two d values thus obtained were then averaged to obtain d for this set of planes. On occasion a reflection may be so weak that only its $K\alpha_1$ line will be sufficiently intense to be measurable. Once the d values are determined, then d^2 and d_1^2/d^2 can be calculated as done in Table 13-5. The resultant set of d_1^2/d^2 quotients contains integers plus values at one-third intervals between integers. Hence this set is comparable to column 3 in Table 13-4 and the hkl indices corresponding to each of these quotients can thereby be assigned, these being entered in Table 13-5 in the hkl column.

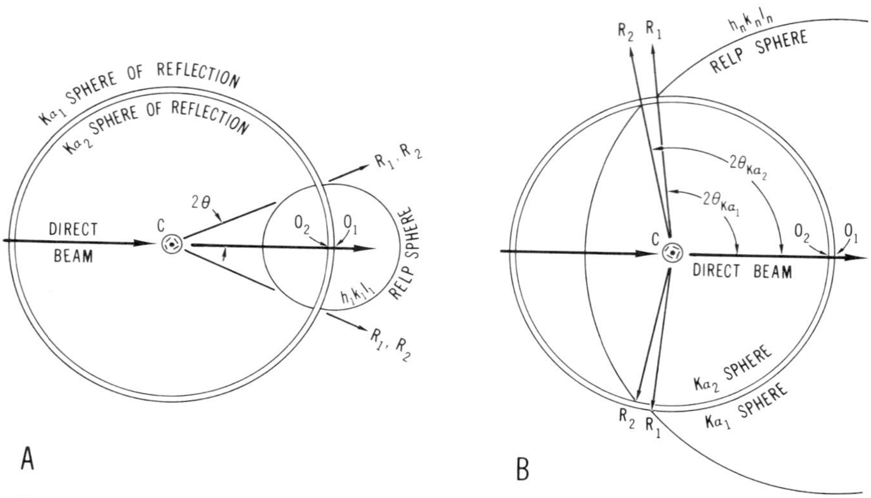

FIGURE 13–18

End views of a powder rod at C around which are constructed spheres of reflection for the $K\alpha_1$ and $K\alpha_2$ radiations normally present in an X-ray beam, the size differences between these spheres being exaggerated. The direct beam leaves the $K\alpha_1$ sphere at O_1 and the $K\alpha_2$ sphere at O_2. Hence for a crystal at C, the origin of its reciprocal lattice will be at O_1 for the $K\alpha_1$ radiation but at O_2 for $K\alpha_2$.

(A) The smallest relp sphere for the powder rod, for example, relp sphere $h_1k_1l_1$ of radius $1/d_1$, where d_1 is the largest interplanar spacing within this crystal, is drawn centered at O_1 to locate this relp sphere's intersection with the $K\alpha_1$ sphere and again at O_2 to locate its intersection with $K\alpha_2$ sphere. Diffracted beams CR_1 and CR_2 are seen to differ little in direction. Thus the cones of diffraction for $K\alpha_1$ and $K\alpha_2$ radiations are practically identical at low values of 2θ and the $h_1k_1l_1$ arcs for $K\alpha_1$ and for $K\alpha_2$ radiation overlap to appear as one line on the powder film.

(B) A much larger relp sphere $h_nk_nl_n$ for this same powder rod. If centered at O_1, it intersects the $K\alpha_1$ sphere at R_1; if centered at O_2, it intersects the $K\alpha_2$ sphere at R_2. Diffracted beams CR_1 and CR_2, which respectively represent diffraction of $K\alpha_1$ and $K\alpha_2$ wavelengths by the same set of $h_nk_nl_n$ planes, now differ significantly. *Thus the $K\alpha_1$ line and the $K\alpha_2$ lines, as recorded for a given reflection hkl on a powder film, become increasingly separated as 2θ increases in value.* The reader is invited to test this thesis further by doing a similar construction using a relp sphere that is even larger than the $h_nk_nl_n$ relp sphere shown here.

In these drawings the powder rod and direct beam were assumed to be so thin that their intersection was a point C. To the extent that this is not true, point C would be dispersed over the volume defined by the intersection between rod and direct beam. The spheres of reflection for $K\alpha_1$ and $K\alpha_2$ could then no longer be centered at a single point C, but at many different points. The resultant "fuzziness" in the $K\alpha_1$ and $K\alpha_2$ sphere surfaces would hence broaden the widths of the lines recorded on the powder film. A thinner rod thus permits greater resolution of the lines on powder films.

94.01–94.50°

Interplanar spacing (d) corresponding to 2Θ for various wavelengths

2Θ	$\sin^2\Theta$	Cu				Fe				Mo				Cr		
		$K\alpha_1$	$K\alpha_2$	$K\alpha$	$K\beta$	$K\alpha_1$	$K\alpha_2$	$K\alpha$	$K\beta$	$K\alpha_1$	$K\alpha_2$	$K\alpha$	$K\beta$	$K\alpha_1$	$K\alpha_2$	$K\alpha$
94.01	.53497	1.0531	1.0557	1.0540	.95170	1.3234	1.3261	1.3243	1.201	.48486	.48778	.48583	.4322	1.5652	1.5679	1.5661
94.02	.53505	1.0530	1.0556	1.0539	.95162	1.3233	1.3260	1.3242	1.201	.48482	.48774	.48579	.4322	1.5651	1.5677	1.5660
94.03	.53514	1.0529	1.0555	1.0538	.95154	1.3232	1.3259	1.3241	1.201	.48478	.48770	.48575	.4321	1.5649	1.5676	1.5658
94.04	.53523	1.0528	1.0555	1.0537	.95147	1.3231	1.3258	1.3240	1.200	.48474	.48766	.48571	.4321	1.5648	1.5675	1.5657
94.05	.53531	1.0528	1.0554	1.0536	.95139	1.3230	1.3257	1.3239	1.200	.48470	.48762	.48568	.4321	1.5647	1.5674	1.5656
94.06	.53540	1.0527	1.0553	1.0535	.95131	1.3229	1.3256	1.3238	1.200	.48466	.48758	.48564	.4320	1.5646	1.5672	1.5655
94.07	.53549	1.0526	1.0552	1.0535	.95123	1.3228	1.3255	1.3237	1.200	.48462	.48754	.48560	.4320	1.5644	1.5671	1.5653
94.08	.53557	1.0525	1.0551	1.0534	.95116	1.3227	1.3254	1.3236	1.200	.48458	.48750	.48556	.4320	1.5643	1.5670	1.5652
94.09	.53566	1.0524	1.0550	1.0533	.95108	1.3226	1.3253	1.3235	1.200	.48454	.48746	.48552	.4319	1.5642	1.5668	1.5651
94.10	.53575	1.0523	1.0549	1.0532	.95100	1.3225	1.3252	1.3234	1.200	.48450	.48742	.48548	.4319	1.5641	1.5667	1.5649
94.11	.53584	1.0522	1.0549	1.0531	.95093	1.3224	1.3251	1.3233	1.200	.48446	.48739	.48544	.4319	1.5639	1.5666	1.5648
94.12	.53592	1.0522	1.0548	1.0530	.95085	1.3223	1.3250	1.3232	1.200	.48442	.48735	.48540	.4318	1.5638	1.5665	1.5647
94.13	.53601	1.0521	1.0547	1.0529	.95077	1.3222	1.3248	1.3230	1.200	.48438	.48731	.48536	.4318	1.5637	1.5663	1.5646
94.14	.53610	1.0520	1.0546	1.0529	.95069	1.3220	1.3247	1.3229	1.200	.48434	.48727	.48532	.4318	1.5635	1.5662	1.5644
94.15	.53618	1.0519	1.0545	1.0528	.95062	1.3219	1.3246	1.3228	1.199	.48430	.48723	.48528	.4317	1.5634	1.5661	1.5643
94.16	.53627	1.0518	1.0544	1.0527	.95054	1.3218	1.3245	1.3227	1.199	.48427	.48719	.48524	.4317	1.5633	1.5660	1.5642
94.17	.53636	1.0517	1.0543	1.0526	.95046	1.3217	1.3244	1.3226	1.199	.48423	.48715	.48520	.4316	1.5632	1.5658	1.5641
94.18	.53645	1.0517	1.0543	1.0525	.95038	1.3216	1.3243	1.3225	1.199	.48419	.48711	.48516	.4316	1.5630	1.5657	1.5639
94.19	.53653	1.0516	1.0542	1.0524	.95031	1.3215	1.3242	1.3224	1.199	.48415	.48707	.48512	.4316	1.5629	1.5656	1.5638
94.20	.53662	1.0515	1.0541	1.0523	.95023	1.3214	1.3241	1.3223	1.199	.48411	.48703	.48508	.4315	1.5628	1.5654	1.5637
94.21	.53671	1.0514	1.0540	1.0523	.95015	1.3213	1.3240	1.3222	1.199	.48407	.48699	.48504	.4315	1.5627	1.5653	1.5635
94.22	.53679	1.0513	1.0539	1.0522	.95008	1.3212	1.3239	1.3221	1.199	.48403	.48695	.48501	.4315	1.5625	1.5652	1.5634
94.23	.53688	1.0512	1.0538	1.0521	.95000	1.3211	1.3238	1.3220	1.199	.48399	.48691	.48497	.4314	1.5624	1.5651	1.5633
94.24	.53697	1.0511	1.0537	1.0520	.94992	1.3210	1.3237	1.3219	1.199	.48395	.48687	.48493	.4314	1.5623	1.5649	1.5632
94.25	.53705	1.0511	1.0537	1.0519	.94985	1.3209	1.3236	1.3218	1.198	.48391	.48683	.48489	.4314	1.5622	1.5648	1.5630
94.26	.53714	1.0510	1.0536	1.0518	.94977	1.3208	1.3234	1.3217	1.198	.48387	.48679	.48485	.4313	1.5620	1.5647	1.5629
94.27	.53723	1.0509	1.0535	1.0518	.94969	1.3207	1.3233	1.3215	1.198	.48383	.48675	.48481	.4313	1.5619	1.5646	1.5628
94.28	.53732	1.0508	1.0534	1.0517	.94961	1.3205	1.3232	1.3214	1.198	.48379	.48671	.48477	.4313	1.5618	1.5644	1.5627
94.29	.53740	1.0507	1.0533	1.0516	.94954	1.3204	1.3231	1.3213	1.198	.48376	.48667	.48473	.4312	1.5616	1.5643	1.5625
94.30	.53749	1.0506	1.0532	1.0515	.94946	1.3203	1.3230	1.3212	1.198	.48372	.48663	.48469	.4312	1.5615	1.5642	1.5624
94.31	.53758	1.0505	1.0531	1.0514	.94938	1.3202	1.3229	1.3211	1.198	.48368	.48660	.48465	.4312	1.5614	1.5640	1.5623
94.32	.53766	1.0505	1.0531	1.0513	.94931	1.3201	1.3228	1.3210	1.198	.48364	.48656	.48461	.4311	1.5613	1.5639	1.5622
94.33	.53775	1.0504	1.0530	1.0512	.94923	1.3200	1.3227	1.3209	1.198	.48360	.48652	.48457	.4311	1.5611	1.5638	1.5620
94.34	.53784	1.0503	1.0529	1.0512	.94915	1.3199	1.3226	1.3208	1.198	.48356	.48648	.48453	.4311	1.5610	1.5637	1.5619
94.35	.53792	1.0502	1.0528	1.0511	.94908	1.3198	1.3225	1.3207	1.197	.48352	.48644	.48450	.4310	1.5609	1.5635	1.5618
94.36	.53801	1.0501	1.0527	1.0510	.94900	1.3197	1.3224	1.3206	1.197	.48348	.48640	.48446	.4310	1.5608	1.5634	1.5617
94.37	.53810	1.0500	1.0526	1.0509	.94892	1.3196	1.3223	1.3205	1.197	.48344	.48636	.48442	.4310	1.5606	1.5633	1.5615
94.38	.53819	1.0499	1.0526	1.0508	.94885	1.3195	1.3222	1.3204	1.197	.48340	.48632	.48438	.4309	1.5605	1.5632	1.5614
94.39	.53827	1.0499	1.0525	1.0507	.94877	1.3194	1.3221	1.3203	1.197	.48336	.48628	.48434	.4309	1.5604	1.5630	1.5613
94.40	.53836	1.0498	1.0524	1.0506	.94869	1.3193	1.3220	1.3202	1.197	.48332	.48624	.48430	.4308	1.5603	1.5629	1.5611
94.41	.53845	1.0497	1.0523	1.0506	.94862	1.3192	1.3218	1.3201	1.197	.48329	.48620	.48426	.4308	1.5601	1.5628	1.5610
94.42	.53853	1.0496	1.0522	1.0505	.94854	1.3191	1.3217	1.3199	1.197	.48325	.48616	.48422	.4308	1.5600	1.5627	1.5609
94.43	.53862	1.0495	1.0521	1.0504	.94846	1.3189	1.3216	1.3198	1.197	.48321	.48612	.48418	.4307	1.5599	1.5625	1.5608
94.44	.53871	1.0494	1.0520	1.0503	.94839	1.3188	1.3215	1.3197	1.197	.48317	.48608	.48414	.4307	1.5598	1.5624	1.5606
94.45	.53879	1.0494	1.0520	1.0502	.94831	1.3187	1.3214	1.3196	1.197	.48313	.48605	.48410	.4307	1.5596	1.5623	1.5605
94.46	.53888	1.0493	1.0519	1.0501	.94823	1.3186	1.3213	1.3195	1.196	.48309	.48601	.48406	.4306	1.5595	1.5622	1.5604
94.47	.53897	1.0492	1.0518	1.0501	.94816	1.3185	1.3212	1.3194	1.196	.48305	.48597	.48403	.4306	1.5594	1.5620	1.5603
94.48	.53906	1.0491	1.0517	1.0500	.94808	1.3184	1.3211	1.3193	1.196	.48301	.48593	.48399	.4306	1.5593	1.5619	1.5601
94.49	.53914	1.0490	1.0516	1.0499	.94800	1.3183	1.3210	1.3192	1.196	.48297	.48589	.48395	.4305	1.5591	1.5618	1.5600
94.50	.53923	1.0489	1.0515	1.0498	.94793	1.3182	1.3209	1.3191	1.196	.48293	.48585	.48391	.4305	1.5590	1.5616	1.5599

FIGURE 13–19

Partial page (much reduced in size) from a standard table for rapid solution of the Bragg equation to convert 2θ to d values. For example, we assume that the 731 reflection for a synthetic spinel is determined from a powder film to be at 2θ of 94.171° for Cu$K\alpha_1$ and at 94.451° for Cu$K\alpha_2$. The table indicates d values of 1.0517 Å (α_1) and 1.0520 Å (α_2), experimental error accounting for their difference. The $K\beta$ column for chromium and the $L\alpha_1$ column for tungsten are omitted here but appear in the original reference. (From *X-ray Diffraction Tables* by J. H. Fang and F. D. Bloss, courtesy of Southern Illinois University Press, Carbondale, Illinois.)

Determination of lattice parameters

After the lines of a measured powder photograph are indexed relative to a set of crystallographic axes, the lattice parameters, that is, the repeat distances along these axes as well as the interaxial angles, can then be determined. Determination of the parameters of a tetragonal or hexagonal lattice (a_o and c_o), of an orthorhombic (a_o, b_o, c_o), of a monoclinic (a_o, b_o, c_o, β), and of a triclinic lattice (a_o, b_o, c_o, α, β, γ), requires measurement of at least two, three, four, or six hkl reflections, respectively. In practice a large number of reflections are measured and indexed so that statistical methods can be used to minimize the errors, both random and systematic, in the calculated parameters. Nelson and Riley (1945, p. 160) list the systematic errors in determining d_{hkl} as (1) *the absorption error*,

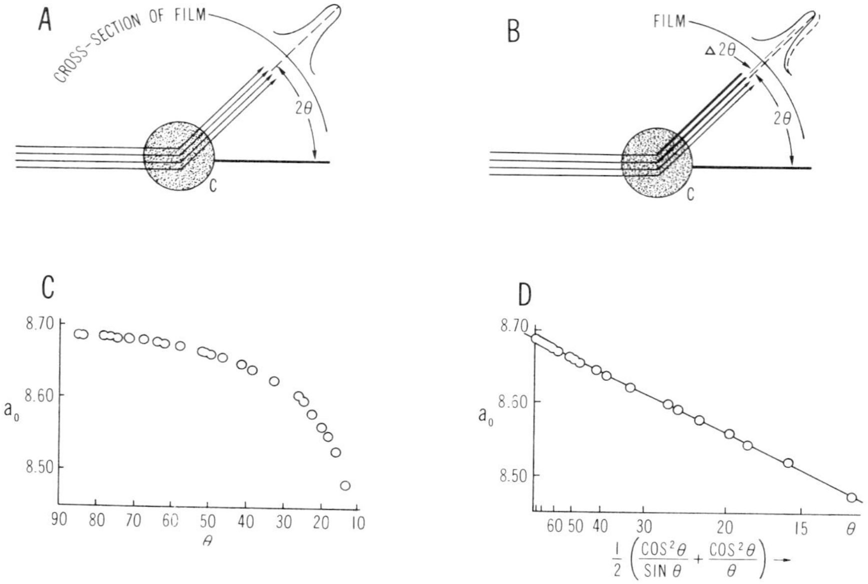

FIGURE 13–20

(A) Cross section normal to the rotation axis of a powder rod that is diffracting X-rays that are assumed to be not absorbed at all by passage through the crystals. If a densitometer were used to measure the density of the diffraction line recorded on film, the range of intensities from one edge to the other might be as shown by the curve that is symmetrically disposed about the dashed line marking the true value of 2θ.

(B) The situation in (A) reproduced except that now we assume the crystals to absorb X-rays strongly. The intensity curve (solid line) for the line recorded on the film is now skewed toward 2θ values larger than true 2θ. This occurs because the upper rays represent lesser distances of travel through the rod and thus lesser absorption as compared with the lower rays. Thus to the extent that a powder rod is thick and its crystals absorb X-rays, the 2θ values recorded by the lines on the film tend to be larger (by $\Delta 2\theta$) than the true values; hence the d values obtained from them are less than the true values. This absorption-caused error in 2θ decreases to zero for diffraction lines to the extent that 2θ approaches 180°.

(C) The a_0 values calculated from Equation 13-13 for twenty-three indexed d_{hkl} spacings for an isometric crystal are plotted against θ_{hkl} (after Nelson and Riley, 1945). The resultant curve indicates that a_0 will equal 8.68 Å at $\theta = 90°$. This value for a_0 is relatively free of absorption error because such errors vanish if θ equals 90°.

(D) Same data as in (C) except that instead of plotting a_0 against θ, a_0 is plotted against

$$\frac{1}{2}\left(\frac{\cos^2\theta}{\sin\theta} + \frac{\cos^2\theta}{\theta}\right)$$

which is known as the *extrapolation function*. For $\theta > 30°$ the curve now becomes a straight line and thus can be more easily extrapolated to $\theta = 90°$ than the curve in (C) to obtain the most correct a_0 value.

which is caused by the powder rod's increased absorption of the diffracted beam as θ approaches 0°, (2) *the eccentricity error,* which occurs if the powder rod's rotation axis is not exactly at the camera's center, and (3) *inaccurate knowledge of camera constants* such as its radius. These three errors vanish at $\theta = 90°$. Even for lines at small values of θ, errors (2) and (3) are minimal if the rod is carefully centered and the camera is precisely made.

The Nelson-Riley Method. J. B. Nelson and D. P. Riley minimized the effect of absorption error (Fig. 13-20A, B) in calculating lattice parameters. To illustrate their method, they made powder photographs of γ-Cu_9Al_4 (isometric) using rods of differing thickness. From the equation on page 467,

$$a_o = d_{hkl}\sqrt{h^2 + k^2 + l^2} \qquad \text{(Eq. 13-13)}$$

they calculated a_o values for each powder line (using its d_{hkl} spacing and *hkl* index). Plotting this a_o value against θ for each line, they obtained a curve (Fig. 13-20C). However, if they plotted these same a_o values against what they named the *extrapolation function,* namely,

$$\frac{1}{2}\left(\frac{\cos^2\theta}{\sin\theta} + \frac{\cos^2\theta}{\theta}\right)$$

the relation would be linear for $\theta > 30°$ (Fig. 13-20D) and the a_o value of this line at $\theta = 90°$ would be relatively free of systematic absorption error. If a plot of the a_o values against the extrapolation function is nonlinear for $\theta > 30°$, the systematic error in measuring d_{hkl} is likely to be *eccentricity error* rather than *absorption error.* In such cases a plot of a_o against $\cos^2\theta$ will be linear for $\theta > 60°$. To eliminate the need of calculating the extrapolation function, Nelson and Riley present its values at 0.1° intervals of θ (Table 13-6).

The X-ray diffractometer

Increasingly supplementing or displacing the powder camera, the X-ray diffractometer is used for measuring the intensities and 2θ values of the various *hkl* reflections from a crystalline powder. Sample preparation for a "smear mount" includes spreading the powder evenly over a 15 × 15 mm area on a 27 × 46 mm petrographic-type glass slide. Next a drop of acetone (in which a little acetate-base cement may be dissolved if a mount of greater permanence is desired) is added to the powder and stirred in with a dissecting needle so as to preserve uniformity of powder distribution. After a suitable drying time, overruns beyond the 15 × 15 mm area are scraped clean with a razor blade. The glass slide[3] (or other type of sample holder) is clamped at the center G of a goniometer (Fig. 13-21A). Relative to the direct beam, the motor-driven gears of the goniometer rotate the slide through an angle θ and simultaneously cause an X-ray detector to scan through an angle 2θ.

[3] An identifying number marked on this slide facilitates its storage for any future re-examination.

TABLE 13–6

Table of $\frac{1}{2}\left(\frac{\cos^2\theta}{\sin\theta} + \frac{\cos^2\theta}{\theta}\right)$

θ	0.0	0.1	0.2	0.3	0.4	0.5	0.6	0.7	0.8	0.9
10	5.572	5.513	5.456	5.400	5.345	5.291	5.237	5.185	5.134	5.084
11	5.034	4.986	4.939	4.892	4.846	4.800	4.756	4.712	4.669	4.627
12	4.585	4.544	4.504	4.464	4.425	4.386	4.348	4.311	4.274	4.238
13	4.202	4.167	4.133	4.098	4.065	4.032	3.999	3.967	3.935	3.903
14	3.872	3.842	3.812	3.782	3.753	3.724	3.695	3.667	3.639	3.612
15	3.584	3.558	3.531	3.505	3.479	3.454	3.429	3.404	3.379	3.355
16	3.331	3.307	3.284	3.260	3.237	3.215	3.192	3.170	3.148	3.127
17	3.105	3.084	3.063	3.042	3.022	3.001	2.981	2.962	2.942	2.922
18	2.903	2.844	2.865	2.847	2.828	2.810	2.792	2.774	2.756	2.738
19	2.921	2.704	2.687	2.670	2.653	2.636	2.620	2.604	2.588	2.572
20	2.556	2.540	2.525	2.509	2.494	2.479	2.464	2.449	2.434	2.420
21	2.405	2.391	2.376	2.362	2.348	2.335	2.321	2.307	2.294	2.280
22	2.267	2.254	2.241	2.228	2.215	2.202	2.189	2.177	2.164	2.152
23	2.140	2.128	2.116	2.104	2.092	2.080	2.068	2.056	2.045	2.034
24	2.022	2.011	2.000	1.989	1.978	1.967	1.956	1.945	1.934	1.924
25	1.913	1.903	1.892	1.882	1.872	1.861	1.851	1.841	1.831	1.821
26	1.812	1.802	1.792	1.782	1.773	1.763	1.754	1.745	1.735	1.726
27	1.717	1.708	1.699	1.690	1.681	1.672	1.663	1.654	1.645	1.637
28	1.628	1.619	1.611	1.602	1.594	1.586	1.577	1.569	1.561	1.553
29	1.545	1.537	1.529	1.521	1.513	1.505	1.497	1.489	1.482	1.474
30	1.466	1.459	1.451	1.444	1.436	1.429	1.421	1.414	1.407	1.400
31	1.392	1.385	1.378	1.371	1.364	1.357	1.350	1.343	1.336	1.329
32	1.323	1.316	1.309	1.302	1.296	1.289	1.282	1.276	1.269	1.263
33	1.256	1.250	1.244	1.237	1.231	1.225	1.218	1.212	1.206	1.200
34	1.194	1.188	1.182	1.176	1.170	1.164	1.158	1.152	1.146	1.140
35	1.134	1.128	1.123	1.117	1.111	1.106	1.100	1.094	1.098	1.083
36	1.078	1.072	1.067	1.061	1.056	1.050	1.045	1.040	1.034	1.029
37	1.024	1.019	1.013	1.008	1.003	0.998	0.993	0.988	0.982	0.977
38	0.972	0.967	0.962	0.958	0.953	0.948	0.943	0.938	0.933	0.928
39	0.924	0.919	0.914	0.909	0.905	0.900	0.895	0.891	0.886	0.881
40	0.877	0.872	0.868	0.863	0.859	0.854	0.850	0.845	0.841	0.837
41	0.832	0.828	0.823	0.819	0.815	0.810	0.806	0.802	0.798	0.794
42	0.789	0.785	0.781	0.777	0.773	0.769	0.765	0.761	0.757	0.753
43	0.749	0.745	0.741	0.737	0.733	0.729	0.725	0.721	0.717	0.713
44	0.709	0.706	0.702	0.698	0.694	0.690	0.687	0.683	0.679	0.676
45	0.672	0.668	0.665	0.661	0.657	0.654	0.650	0.647	0.643	0.640
46	0.636	0.632	0.629	0.625	0.622	0.619	0.615	0.612	0.608	0.605
47	0.602	0.598	0.595	0.591	0.588	0.585	0.582	0.578	0.575	0.572
48	0.569	0.565	0.562	0.559	0.556	0.553	0.549	0.546	0.543	0.540
49	0.537	0.534	0.531	0.528	0.525	0.522	0.518	0.515	0.512	0.509

TABLE 13–6 (*Continued*)

θ	0.0	0.1	0.2	0.3	0.4	0.5	0.6	0.7	0.8	0.9
50	0.506	0.504	0.501	0.498	0.495	0.492	0.489	0.486	0.483	0.480
51	0.477	0.474	0.472	0.469	0.466	0.463	0.460	0.458	0.455	0.452
52	0.449	0.447	0.444	0.441	0.439	0.436	0.433	0.430	0.428	0.425
53	0.423	0.420	0.417	0.415	0.412	0.410	0.407	0.404	0.402	0.399
54	0.397	0.394	0.392	0.389	0.387	0.384	0.382	0.379	0.377	0.375
55	0.372	0.370	0.367	0.365	0.363	0.360	0.358	0.356	0.353	0.351
56	0.349	0.346	0.344	0.342	0.339	0.337	0.335	0.333	0.330	0.328
57	0.326	0.324	0.322	0.319	0.317	0.315	0.313	0.311	0.309	0.306
58	0.304	0.302	0.300	0.298	0.296	0.294	0.292	0.290	0.288	0.286
59	0.284	0.282	0.280	0.278	0.276	0.274	0.272	0.270	0.268	0.266
60	0.264	0.262	0.260	0.258	0.256	0.254	0.252	0.250	0.249	0.247
61	0.245	0.243	0.241	0.239	0.237	0.236	0.234	0.232	0.230	0.229
62	0.227	0.225	0.223	0.221	0.220	0.218	0.216	0.215	0.213	0.211
63	0.209	0.208	0.206	0.204	0.203	0.201	0.199	0.198	0.196	0.195
64	0.193	0.191	0.190	0.188	0.187	0.185	0.184	0.182	0.180	0.179
65	0.177	0.176	0.174	0.173	0.171	0.170	0.168	0.167	0.165	0.164
66	0.162	0.161	0.160	0.158	0.157	0.155	0.154	0.152	0.151	0.150
67	0.148	0.147	0.146	0.144	0.143	0.141	0.140	0.139	0.138	0.136
68	0.135	0.134	0.132	0.131	0.130	0.128	0.127	0.126	0.125	0.123
69	0.122	0.121	0.120	0.119	0.117	0.116	0.115	0.114	0.112	0.111
70	0.110	0.109	0.108	0.107	0.106	0.104	0.103	0.102	0.101	0.100
71	0.099	0.098	0.097	0.096	0.095	0.094	0.092	0.091	0.090	0.089
72	0.088	0.087	0.086	0.085	0.084	0.083	0.082	0.081	0.080	0.079
73	0.078	0.077	0.076	0.075	0.075	0.074	0.073	0.072	0.071	0.070
74	0.069	0.068	0.067	0.066	0.065	0.065	0.064	0.063	0.062	0.061
75	0.060	0.059	0.059	0.058	0.057	0.056	0.055	0.055	0.054	0.053
76	0.052	0.052	0.051	0.050	0.049	0.048	0.048	0.047	0.046	0.045
77	0.045	0.044	0.043	0.043	0.042	0.041	0.041	0.040	0.039	0.039
78	0.038	0.037	0.037	0.036	0.035	0.035	0.034	0.034	0.033	0.032
79	0.032	0.031	0.031	0.030	0.029	0.029	0.028	0.028	0.027	0.027
80	0.026	0.026	0.025	0.025	0.024	0.023	0.023	0.023	0.022	0.022
81	0.021	0.021	0.020	0.020	0.019	0.019	0.018	0.018	0.017	0.017
82	0.017	0.016	0.016	0.015	0.015	0.015	0.014	0.014	0.013	0.013
83	0.013	0.012	0.012	0.012	0.011	0.011	0.010	0.010	0.010	0.010
84	0.009	0.009	0.009	0.008	0.008	0.008	0.007	0.007	0.007	0.007
85	0.006	0.006	0.006	0.006	0.005	0.005	0.005	0.005	0.005	0.004
86	0.004	0.004	0.004	0.003	0.003	0.003	0.003	0.003	0.003	0.002
87	0.002	0.002	0.002	0.002	0.002	0.002	0.001	0.001	0.001	0.001
88	0.001	0.001	0.001	0.001	0.001	0.001	0.001	0.000	0.000	0.000

[a] From J. B. Nelson and D. P. Riley (1945, p. 177).

The random orientation of grains in the powder sample produces relp spheres with radii of $1/d_{hkl}$. Short segments of these relp spheres are drawn where they intersect the sphere of reflection in Figure 13-21B, diffracted rays $a \ldots o$ proceeding from the sample to these intersections. Consequently, whenever the detector has moved into the 2θ position for one of these diffracted rays (Fig.

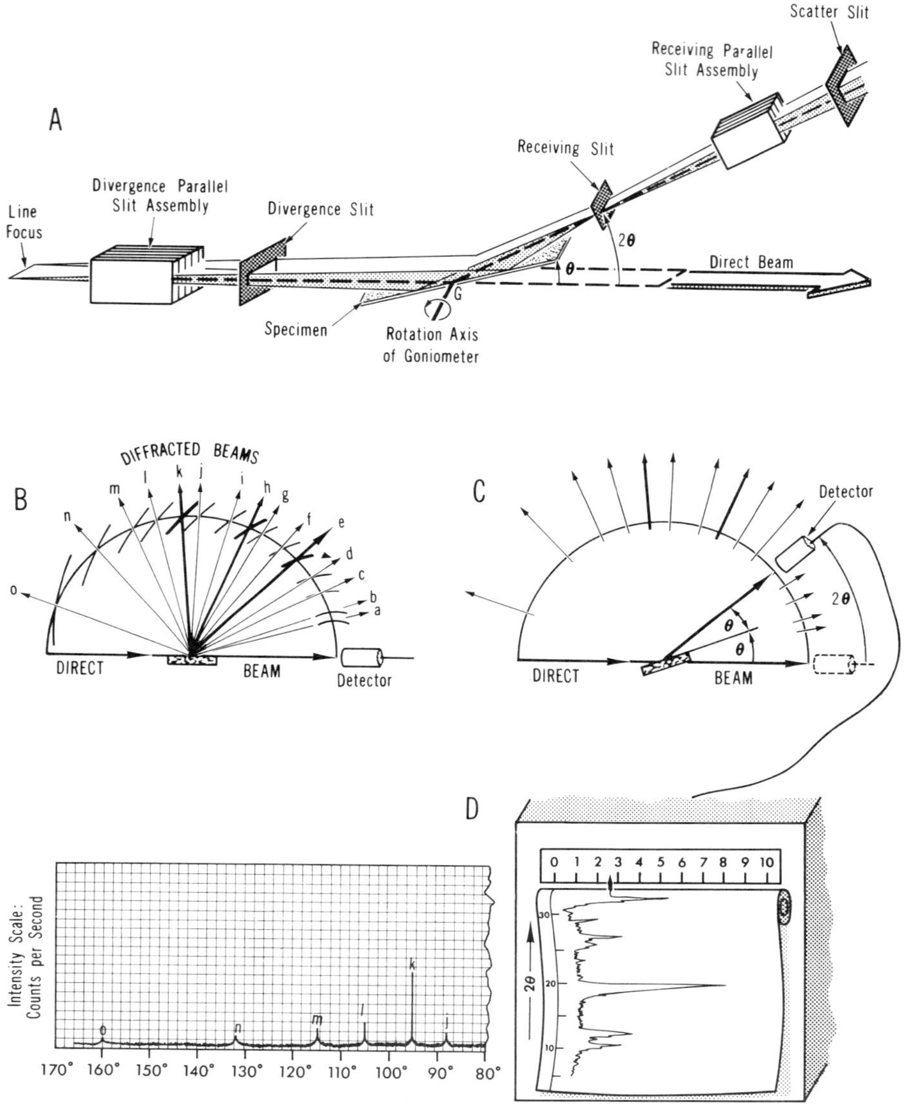

FIGURE 13–21

(A) Schematic drawing of the principal parts of an x-ray diffractometer. X-rays from the tube's line focus are collimated by passage through a parallel slit assembly (Soller slit) and a divergence slit before striking the powder (stippled) on a glass slide or sample holder. While the slide (length exaggerated) rotates about axis G so as to be at an angle θ to the direct beam, a detector (stripped of its housing to reveal its receiving, Soller, and scatter slits) scans through the angle 2θ.
(B) Segments of relp spheres $a, b, \ldots m, n, o$ for the crystalline powder intersect the sphere of reflection (semicircle) to produce diffracted rays $a, b, \ldots m, n, o$.
(C) While the sample rotates through an angle θ about the axis G, which is now

13-21C), this ray enters the detector and, according to its intensity, deflects the pen of a strip-chart recorder (Fig. 13-21D). The chart paper rolls past this pen at chart speed such that, for example, 1 in. of paper passes the pen while the detector scans a 2θ angle of 1° (or 2°, 4°, or 8° as desired). The various *hkl* Bragg reflections are thus recorded as a series of peaks that correspond to the lines of the powder photograph (Fig. 13-22A, B). By reducing the scanning speed but increasing chart speed, the entire diffraction record for a crystal, or a small selected segment of it (Fig. 13-22C), may be spread out so as to separate lines that appeared to be a single line on the powder photograph.

Intensity and Location of Peaks. The intensity of an *hkl* peak is approximated by its height above background (Fig. 13-22D) but its area above background is the more precise measure. The 2θ value for a peak is taken as its midpoint measured at one-half peak height by some but at two-thirds height by others (Fig. 13-22D).

As a general rule adjustments that increase the intensity of the diffracted rays entering the detector decrease the resolution of closely adjacent peaks. However, intensity can be increased with only a slight sacrifice in resolution by (1) replacing the receiving slit in Figure 13-21A with a wider slit or (2) using a thicker layer of powder. The latter may require packing a sample holder with powder rather than thinly coating a glass slide.

Comparison of the diffractometer and powder camera methods

A comparison of the diffractometer and the powder camera methods reveals the diffractometer to possess greater resolving power, to eliminate film processing, to allow quick estimation of *hkl* intensities through peak heights, and to permit a diffraction record to be investigated in great detail either in its entirety or over a small range in 2θ values. By contrast the powder camera requires less sample, is much less expensive, allows weak reflections to be readily detected, and affords a more compact diffraction record that facilitates storage of data and convenient comparison of several diffraction records. The systematic errors in 2θ resulting from absorption are easily obviated for a powder film by applying the Nelson-Riley extrapolation function. Use of a very small diameter rod increases the resolving power of a powder camera but then necessitates a much longer exposure time and concurrent increase of background radiation on the film.

---

normal to the page but unlabeled, the detector synchronously rotates through the angle 2θ relative to the direct beam. Thus diffracted rays $a, b, \ldots m, n, o$ successively enter the detector. Proportional to the intensity of each such diffracted ray, a signal from the detector actuates the recording pen of a strip chart recorder like that in (D), where the pen is recording a deflection of about 2.7 on the scale.

(D) While the detector traverses through an angle 2θ, the chart rolls downward past this pen at a fixed rate so that each diffracted ray records as a peak. The peak's height is roughly proportional to the ray's intensity, whereas the peak's location on the chart, which is typically of graph paper like that to the left, discloses the ray's 2θ value.

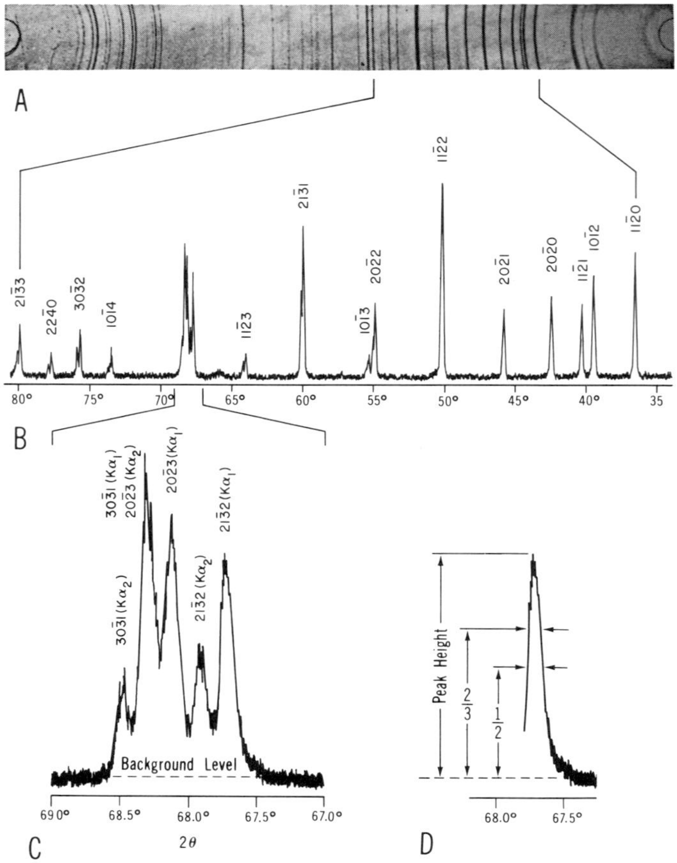

FIGURE 13–22

(A) A powder photograph of quartz using Cu$K\alpha$ radiation.
(B) Diffractometer chart of quartz from 2θ values of 34° to 80°.
(C) Increasingly detailed diffractometer chart of quartz from 2θ values of 67° to 69°. The $K\alpha_1$ and $K\alpha_2$ peaks for reflection index $21\bar{3}2$ are completely resolved, the $K\alpha_1$ peak having about twice the intensity of the $K\alpha_2$ peak. This is the normal case and results from a similar intensity ratio of $K\alpha_1$ to $K\alpha_2$ wavelengths (see Fig. 13-2) in the primary X-ray beam. The tallest peak represents almost complete overlap of the $30\bar{3}1$ ($K\alpha_1$) and $20\bar{2}3$ ($K\alpha_2$) peaks. If not overlapped, the $30\bar{3}1$ ($K\alpha_1$) peak would be twice as intense as $30\bar{3}1$ ($K\alpha_2$) and the $20\bar{2}3$ ($K\alpha_2$) peak would have one-half the intensity of $20\bar{2}3$ ($K\alpha_1$).
(D) The $21\bar{3}2$ $K\alpha_1$ peak drawn separately. The intensity of the $21\bar{3}2$ $K\alpha_1$ ray would be proportional to the area of this peak above the dashed background line. The precise 2θ values for this resolved $K\alpha_1$ line would be obtained by finding the midpoint between the two horizontal arrows located at one half the peak height above the background line. For lesser resolutions in which the $K\alpha_1$ and $K\alpha_2$ peaks overlap, the 2θ value for the $K\alpha_1$ line is frequently taken as the midpoint at two-thirds peak height, that is, the midpoint of the line between the two arrows at two-thirds peak height. (Figures modified from *X-ray Diffraction Methods* by E. W. Nuffield, 1966, Wiley, by kind permission).

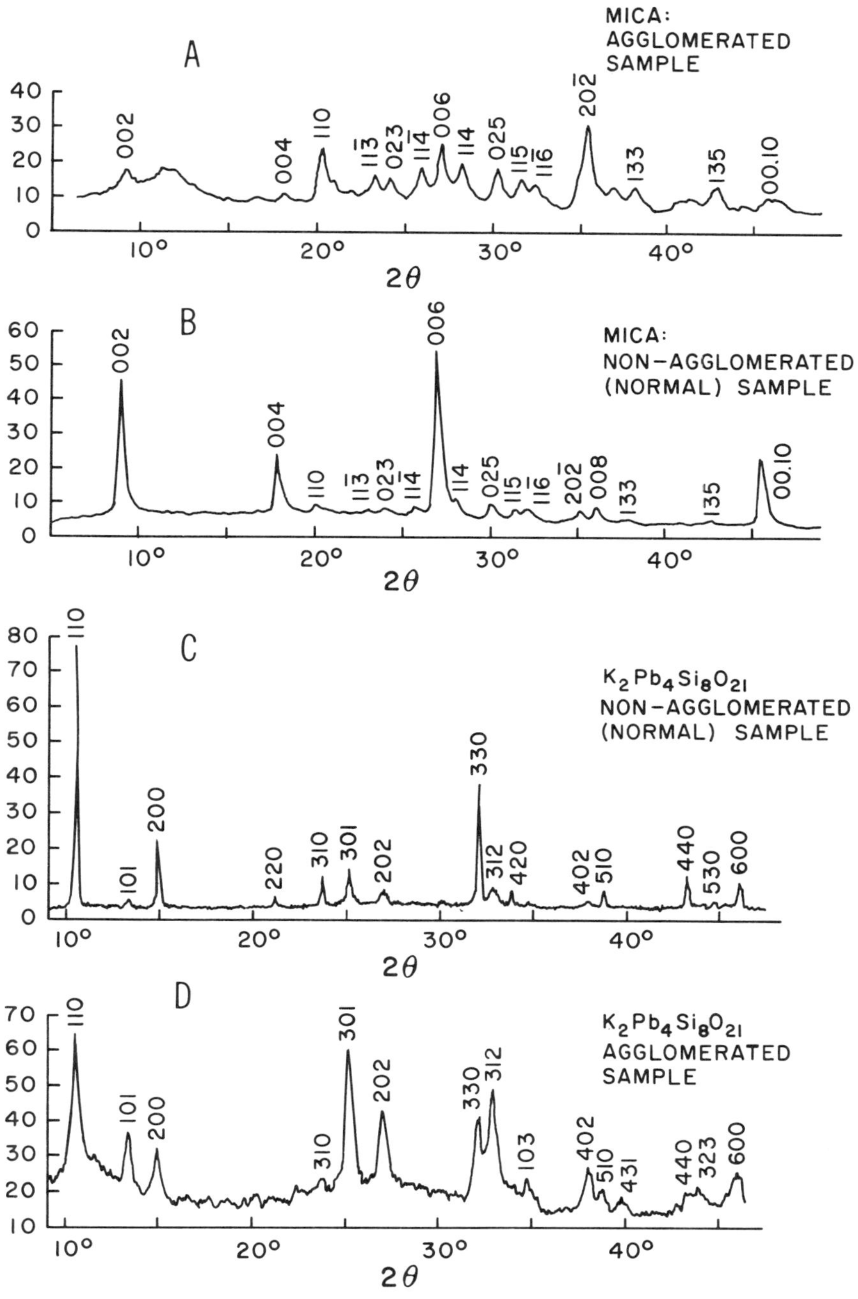

FIGURE 13–23

Comparison of the $00l$ intensities relative to hkl intensities for a diffractogram of muscovite for which the powder was agglomerated to reduce preferred orienta-

(*Figure legend continued on next page.*)

Minimizing Preferred Orientation. Most crystals break (cleave) more readily along certain planes than others. Consequently, their grain fragments tend to orient with these cleavage planes parallel to the axis of the powder rod (Fig. 13-17) or parallel to the plane of the diffractometer mount. This is particularly true if the cleavage is pronounced. Bloss, Frenzel, and Robinson (1967) greatly reduced such preferred orientation by directing one or two bursts of an aersol plastic spray[4] onto the powder from a distance of 8 in. The individual droplets of plastic clumped the crystal grains into aggregates of randomly oriented grains, which could then be seived out from the unagglomerated grains. Diffractograms of a mica powder thus agglomerated compared to the same mica when untreated (Fig. 13-23A, B) indicate that the high intensities of $00l$ peaks for untreated mica samples result (in part) from preferred orientation of the (001) cleavage flakes. Similarly, for a crystal with a {110} cleavage that produced needlelike fragments elongated parallel to [001], the intensities of $hk0$ peaks were reduced by agglomerating the grains with the plastic spray (Fig. 13-23C, D).

Diffraction by Single Crystals

To interpret X-ray diffraction by a single crystal, we must relate its direct lattice (parameters: a_o, b_o, c_o, α, β, γ in Fig. 13-24A) to its reciprocal lattice (parameters: a^*, b^*, c^*, α^*, β^*, γ^* in Fig. 13-24B). A parallelism in nomenclature exists. Thus (100) is a plane of direct lattice points that is parallel to direct axes b and c and, similarly, $(100)^*$ is a plane of relps that is parallel to reciprocal axes b^* and c^*. By definition of the reciprocal lattice, its axes a^*, b^*, and c^* are perpendicular to direct lattice planes (100), (010), and (001), respectively. Hence as Figures 13-24C, D, and E illustrate, a^* is perpendicular to direct axes c and b, b^* to a and c, and c^* to a and b. This also means that direct axis a is perpendicular to b^* and c^* and thus to $(100)^*$; b to a^* and c^* and thus to $(010)^*$; c to a^* and b^* and thus to $(001)^*$. To summarize, *a direct axis and a reciprocal axis, if they have different letter symbols, will be perpendicular.* Thus a and b^* are always mutually perpendicular, but not a and a^*.

As just discussed, direct axis a whose alternate symbol is [100] is perpendicular to $(100)^*$; axis b or [010] is perpendicular to $(010)^*$; axis c or [001] is perpendicular to $(001)^*$. In general a direction indexed as $[uvw]$ in the direct lattice will be perpendicular to the similarly indexed reciprocal lattice plane $(uvw)^*$. Actually $(uvw)^*$ represents a set of parallel planes in the reciprocal lattice, their interplanar spacing being $d_{(uvw)^*}$. For example, in Figure 13-24B

tion (A) and for a normal mount (B) where preferred orientation of the (001) cleavage flakes greatly enhances the intensities of $00l$ peaks.

(C) (D) Comparison of the $hk0$ intensities for diffractograms of synthetic $K_2Pb_4Si_8O_{21}$, a tetragonal material that yields needle-shaped cleavage fragments because of a pronounced {110} cleavage. Note the significant reduction of $hk0$ peaks in the agglomerated sample.

[4] Krylon and Blair Spray-clear, as constituted then, were both satisfactory.

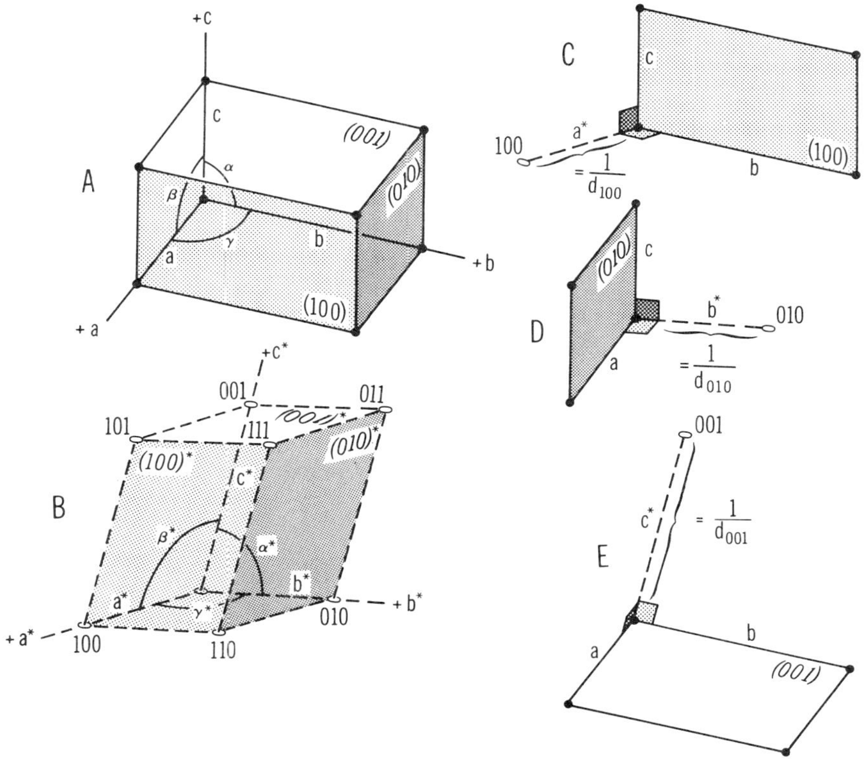

FIGURE 13–24

(A) Direct lattice and unit cell for a triclinic crystal.
(B) The reciprocal lattice and unit cell for this same crystal. Planes $(hkl)^*$ in the reciprocal lattice pass through reciprocal lattice points and bear relations to the reciprocal axes a^*, b^*, and c^* that are analogous to those which direct planes (hkl) bear to direct lattice points and axes. For example $(001)^*$ is parallel to the a^* and b^* axes. The lower plane of this $(001)^*$ set contains relps 100, 110, and 010;
(C) (D) (E) A reciprocal axis (dashed) is always perpendicular to two direct axes. As shown in (C), a^* represents the normal to (100) and thus it is perpendicular to the direct b and c axes. Relp 100 was located along a^* at a distance proportional to $1/d_{100}$.

the family $(100)^*$ is represented by the rear face, which contains relps 001, 011, and 010, among others, and by the front face, whose *hkl* relps include 101, 111, 100, 110, and so on. The perpendicular distance between this front and rear face is $d_{(100)^*}$. If outward from a given point, we drew normals to all such $(uvw)^*$ planes of a reciprocal lattice and if we then spaced points along these normals at intervals corresponding to $1/d_{(uvw)^*}$, a line of direct lattice points (and thus the direct lattice itself) would be regenerated. In other words, although it will not be proved, the reciprocal of a reciprocal lattice is its parent direct lattice.

In constructing the original reciprocal lattice, the spacings between points along the normals to (100), (010), and (001), that is, along reciprocal axes a^*, b^*, and c^*, are

$$a^* = \frac{1}{d_{(100)}} \qquad b^* = \frac{1}{d_{(010)}} \qquad c^* = \frac{1}{d_{(001)}} \qquad \text{(Eqs. 13-14)}$$

Similarly, because the direct lattice is a reciprocal of the reciprocal lattice, the spacing between direct lattice points (identipoints) along the direct axes are

$$a_o = \frac{1}{d_{(100)^*}} \qquad b_o = \frac{1}{d_{(010)^*}} \qquad c_o = \frac{1}{d_{(001)^*}} \qquad \text{(Eqs. 13-15)}$$

In general,

$$t_{[uvw]} = \frac{1}{d_{(uvw)^*}} \qquad \text{(Eq. 13-16)}$$

where $t_{[uvw]}$ is the spacing between identipoints along the direction $[uvw]$ in the direct lattice and $d_{(uvw)^*}$ is the interplanar spacing between the set of reciprocal lattice planes normal to $[uvw]$.

Figure 13-25A represents several (hkl) planes (unshaded) each parallel to $[uvw]$ and thus satisfying the zone condition (p. 60) that

$$hu + kv + lw = 0$$

Each of these (hkl) planes represents a set of parallel lattice planes at spacing intervals d_{hkl} in the crystal lattice; their relps all lie in a plane that is perpendicular to $[uvw]$ and that passes through the origin of the reciprocal lattice.

The relps for (hkl) planes not parallel to $[uvw]$, although not proved here, must also lie in planes perpendicular to $[uvw]$ but at different levels above (or below) the origin (Fig. 13-25B). In respect to a given direction $[uvw]$, such levels are called 0, 1, 2, $\bar{1}$ levels, and so on, according to whether for all hkl relps in the level, the sum $(hu + kv + lw)$ equals 0, 1, 2, $\bar{1}$, and so on (Fig. 13-25C). In other words, if

$$hu + kv + lw = n$$

then n is the level in which a relp hkl lies in respect to a direction $[uvw]$. For example, relp 211 lies in the fourth level in respect to zone axis [111] but in the second level in respect to [100].

If a crystal is placed with a zone axis $[uvw]$ parallel to the axis of a cylindrical camera (Fig. 13-26), then rotated on $[uvw]$ while a beam of X-rays strikes normal to $[uvw]$, the various relp levels normal to $[uvw]$ intersect the sphere of reflection as shown. During the crystal's rotation, its reciprocal lattice rotates accordingly and whenever a relp such as l_1 (or l_n) touches the sphere of reflection, a diffracted X-ray beam travels from C to l_1 (or l_n) to expose the film at L_1 (or L_n). Consequently, after the film is developed and processed, lines of diffraction spots known as *layer lines* mark the exposures caused by the various relp levels relative to the rotation axis $[uvw]$.

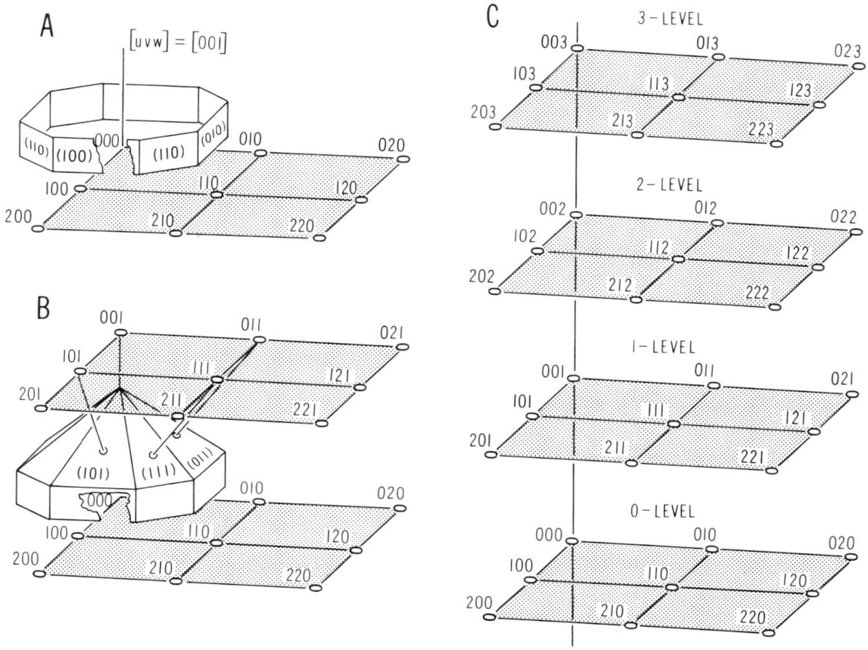

FIGURE 13–25

(A) As illustrated for zone axis [001], any faces parallel to a particular zone axis [*uvw*] will have their normals and therefore their relps in a plane (shaded) which is perpendicular to this zone axis and passes through the origin 0, 0, 0. For any *hkl* relp in this plane,

$$hu + kv + lw = 0$$

and this plane of relps is accordingly called the 0 level in respect to the zone axis perpendicular to it.

(B) The 001, 101, 201, 011, 111, 021 relps are also plotted for the crystal along the normals to the faces with comparable Miller indices of which only (101), (111), and (011) appear on the crystal. Although not proved here, all these relps lie in a second plane that is parallel to the 0-level. For any *hkl* relp in this plane

$$hu + kv + lw = 1$$

and this plane of relps is called the 1 level relative to the zone axis perpendicular to it.

(C) The reciprocal lattice for the crystal at the left is drawn to show the 0, 1, 2, and 3 levels. Its $\bar{1}$ level would be a similar plane below the 0 level.

The direct-lattice repeat distance along rotation axis [*uvw*], that is, t_{uvw}, is determinable from such a rotation photograph because $d_{(uvw)^*}$ can be determined from it and Equation (13-16) applied. From Figure 13-26 we note that whenever any relp in the 1 level touches the sphere of reflection, a diffracted beam occurs at an angle θ_1 to the 0-level plane. Hence if R is the camera radius and y_1 is the film distance between the 0- and 1-layer lines, $\tan \theta_1 = y_1/R$. From triangle

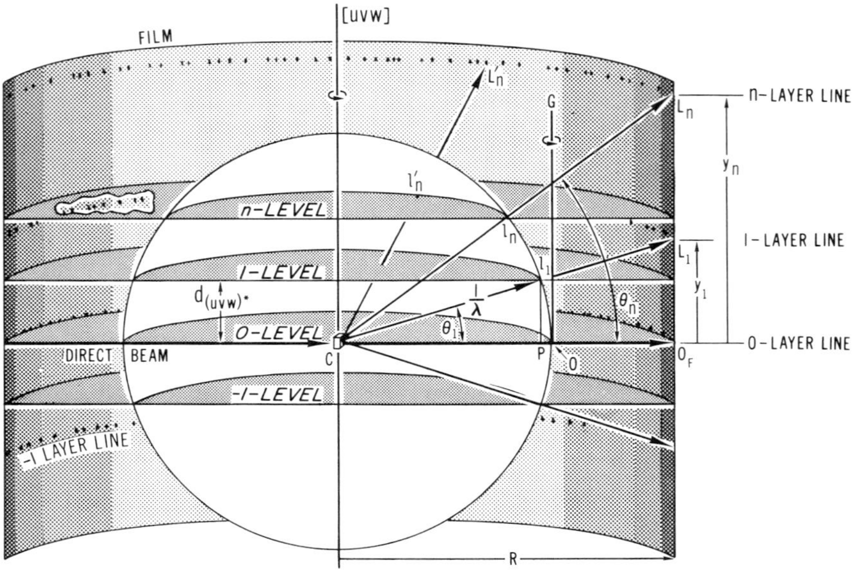

FIGURE 13–26

A crystal at C is mounted so that its zone axis $[uvw]$ coincides with the rotational axis of a rotation camera. The 0, 1, . . . , and n levels of its reciprocal lattice relative to $[uvw]$ are hence normal to the camera's rotation axis. As the crystal at C is rotated about $[uvw]$ as an axis, the horizontal reciprocal lattice levels rotate about OG as an axis, so that the reciprocal lattice at all times maintains a correct orientation relative to the crystal. Each level 0, 1, . . . n would be a net (p. 144) containing relps, but for simplicity only relps l_n', l_n and l_1 are located. Whenever the rotation of the crystal and attendant rotation about OG of the reciprocal lattice levels cause a relp to touch the sphere of reflection, for example, at points l_1, l_n, or l_n', a diffracted ray proceeds from C to this relp; for example, a ray proceeds from C to l_1, to l_n, or to l_n' to be recorded on the film as a diffraction spot like those at L_1, L_n, or L_n'. Thus the diffraction spots for the relps in a given level will record on the film as a particular layer line. Observe that the tangent of θ_n, the angle which a diffracted ray recording at the n layer makes relative to the 0 level, equals O_FL_n, the height of the n layer above the 0 layer on the film, divided by the camera radius R.

COl_1 in Figure 13-26,

$$\sin\theta_1 = \frac{d_{(uvw)^*}}{1/\lambda} = \lambda d_{(uvw)^*} \qquad \text{(Eq. 13-17)}$$

Combining Equations (13-16) and (13-17)

$$t_{uvw} = \frac{1}{d_{(uvw)^*}} = \frac{\lambda}{\sin\theta_1} \qquad \text{(Eq. 13-18)}$$

where θ_1 equals $\tan^{-1} y_1/R$. In actual practice the distance between the n- and $\bar{n}$-layer lines is usually measured on the film, then divided by two to get y_n

whereupon (Fig. 13-26) $\tan \theta_n = y_n/R$. As could be demonstrated by dropping a perpendicular from l_n to line CO in Figure 13-26,

$$\sin \theta_n = \frac{nd_{(uvw)^*}}{Cl_n} = \frac{nd_{(uvw)^*}}{1/\lambda} \qquad \text{(Eq. 13-17')}$$

then

$$t_{uvw} = \frac{1}{d_{(uvw)^*}} = \frac{n\lambda}{\sin \theta_n} \qquad \text{(Eq. 13-18')}$$

If axis $[uvw]$ is actually [100], [010], or [001], that is, the crystal's a, b, or c axis, the lattice spacings a_o, b_o, or c_o can then be measured.

Angular relationships

The angular relationships between direct and reciprocal lattice axes are readily seen in stereographic projection. Let points a, b, and c in Figure 13-27A represent the cyclographic projections of a set of triclinic axes, the great circles

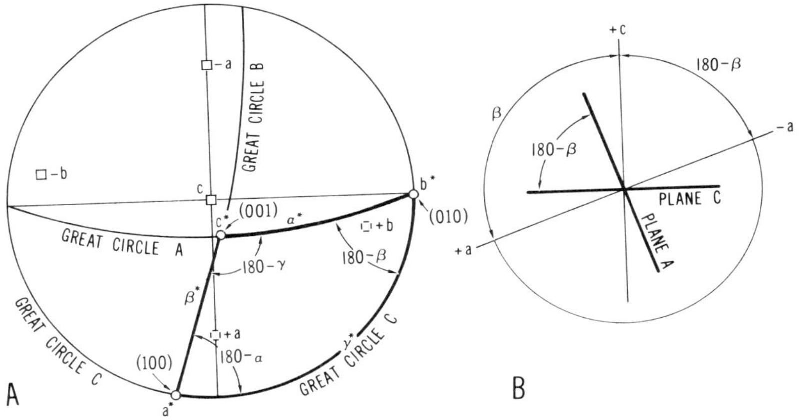

FIGURE 13–27

(A) Triclinic axes, a, b, and c are projected cyclographically as points and stereographically as great circles A, B, and C. Points of intersection between great circles A, B, and C locate the a^* axis or (100) face pole, b^* or (010), and c^* or (001). The spherical triangle outlined with bold lines has sides equal to reciprocal lattice angles α^*, β^*, and γ^*. Angle α^* thus equals interfacial angle (010) $\wedge$ (001); β^* equals (100) $\wedge$ (001); γ^* equals (100) $\wedge$ (010). Angles 180-α, 180-β, and 180-γ, of the spherical triangle represent the angles measured between planes that are normal to the direct axes.

(B) The significance of angle 180-β as demonstrated by drawing direct axes a and c within the page (so that axis b^* as well as planes A and C, which are respectively perpendicular to axes a and c, are perpendicular to the page). The angle between planes A and C can thus be proved to equal 180-β, as marked between the edge views of these planes (bold lines) or as marked in the stereogram to the left between the bold great circles which represent either the cyclographic projections of planes A and C or the stereographic projections of axes a and c.

being their stereographic projections. The points of intersection between these great circles represent the face poles (100), (010), and (001) or, alternatively, the cyclographic projections of reciprocal axes a^*, b^*, and c^*, respectively. The heavy segments of the great circles define a spherical triangle in Figure 13-27A whose three sides are angles α^*, β^*, and γ^* and whose three angles are $(180 - \alpha)$, $(180 - \beta)$, $(180 - \gamma)$. Great circles A, B, and C in Figure 13-27A also represent the cyclographic projection of relp planes $(100)^*$, $(010)^*$, and $(001)^*$, respectively. Their interplanar angles equal the angles of the spherical triangle, namely,

$$(180 - \alpha) = B \wedge C = (010)^* \wedge (001)^*$$

$$(180 - \beta) = A \wedge C = (100)^* \wedge (001)^*$$

$$(180 - \gamma) = A \wedge B = (100)^* \wedge (010)^*$$

If three sides, α^*, β^*, γ^*, or three angles, $(180 - \alpha)$, $(180 - \beta)$, $(180 - \gamma)$, of the spherical triangle in Figure 13-27A are known, the remaining three can be calculated. Thus if the three sides are known, the law of cosines yields the three angles as follows:

$$\cos\,(180 - \alpha) = \frac{\cos\alpha^* - \cos\beta^* \cos\gamma^*}{\sin\beta^* \sin\gamma^*} \qquad \text{(Eq. 13-19a)}$$

$$\cos\,(180 - \beta) = \frac{\cos\beta^* - \cos\alpha^* \cos\gamma^*}{\sin\alpha^* \sin\gamma^*} \qquad \text{(Eq. 13-19b)}$$

$$\cos\,(180 - \gamma) = \frac{\cos\gamma^* - \cos\alpha^* \cos\beta^*}{\sin\alpha^* \sin\beta^*} \qquad \text{(Eq. 13-19c)}$$

Similarly, if the three angles are known, the sides can be calculated because

$$\cos\alpha^* = \frac{\cos\,(180 - \alpha) + \cos\,(180 - \beta)\cos\,(180 - \gamma)}{\sin\beta \sin\gamma} \qquad \text{(Eq. 13-20a)}$$

$$\cos\beta^* = \frac{\cos\,(180 - \beta) + \cos\,(180 - \alpha)\cos\,(180 - \gamma)}{\sin\alpha \sin\gamma} \qquad \text{(Eq. 13-20b)}$$

$$\cos\gamma^* = \frac{\cos\,(180 - \gamma) + \cos\,(180 - \alpha)\cos\,(180 - \beta)}{\sin\alpha \sin\beta} \qquad \text{(Eq. 13-20c)}$$

If two sides and the included angle are known, the third side can be determined from Equation (13-19). If two angles and the side between are known, the third angle can be determined from Equation (13-20). In Table 13-7, Equations (13-19) and (13-20) were simplified by substituting $-\cos\alpha$ for $\cos 180 - \alpha$, $-\cos\beta$ for $\cos 180 - \beta$, and $-\cos\gamma$ for $\cos 180 - \gamma$. Applying the sine formula to the spherical triangle in Figure 13-27A there results

$$\frac{\sin\alpha^*}{\sin\,(180 - \alpha)} = \frac{\sin\beta^*}{\sin\,(180 - \beta)} = \frac{\sin\gamma^*}{\sin\,(180 - \gamma)} \qquad \text{(Eqs. 13-21)}$$

which reduces to the form in Table 13-7 because $\sin\,(180 - \alpha)$ equals $\sin\alpha$, and so on.

TABLE 13–7

Dimensional Relationships between the Direct and Reciprocal Triclinic Unit Cell

Interaxial Angles

$$\cos \alpha = \frac{\cos \beta^* \cos \gamma^* - \cos \alpha^*}{\sin \beta^* \sin \gamma^*} \qquad \cos \alpha^* = \frac{\cos \beta \cos \gamma - \cos \alpha}{\sin \beta \sin \gamma}$$

$$\cos \beta = \frac{\cos \alpha^* \cos \gamma^* - \cos \beta^*}{\sin \alpha^* \sin \gamma^*} \qquad \cos \beta^* = \frac{\cos \alpha \cos \gamma - \cos \beta}{\sin \alpha \sin \gamma}$$

$$\cos \gamma = \frac{\cos \alpha^* \cos \beta^* - \cos \gamma^*}{\sin \alpha^* \sin \beta^*} \qquad \cos \gamma^* = \frac{\cos \alpha \cos \beta - \cos \gamma}{\sin \alpha \sin \beta}$$

$$\frac{\sin \alpha}{\sin \alpha^*} = \frac{\sin \beta}{\sin \beta^*} = \frac{\sin \gamma}{\sin \gamma^*}$$

Cell Edges

$$a_o = \frac{b^*c^* \sin \alpha^*}{V^*} = \frac{1}{d_{(100)}^*} \qquad a^* = \frac{b_oc_o \sin \alpha}{V} = \frac{1}{d_{(100)}}$$

$$b_o = \frac{a^*c^* \sin \beta^*}{V^*} = \frac{1}{d_{(010)}^*} \qquad b^* = \frac{a_oc_o \sin \beta}{V} = \frac{1}{d_{(010)}}$$

$$c_o = \frac{a^*b^* \sin \gamma^*}{V^*} = \frac{1}{d_{(001)}^*} \qquad c^* = \frac{a_ob_o \sin \gamma}{V} = \frac{1}{d_{(001)}}$$

Cell Volumes

$$\begin{aligned} V &= a_ob_oc_o\sqrt{1 - \cos^2 \alpha - \cos^2 \beta - \cos^2 \gamma + 2 \cos \alpha \cos \beta \cos \gamma} \\ &= a_ob_oc_o \sin \alpha \sin \beta \sin \gamma^* \\ &= a_ob_oc_o \sin \alpha \sin \beta^* \sin \gamma \\ &= a_ob_oc_o \sin \alpha^* \sin \beta \sin \gamma \end{aligned}$$

$$\begin{aligned} V^* &= a^*b^*c^*\sqrt{1 - \cos^2 \alpha^* - \cos^2 \beta^* - \cos^2 \gamma^* + 2 \cos \alpha^* \cos \beta^* \cos \gamma^*} \\ &= a^*b^*c^* \sin \alpha^* \sin \beta^* \sin \gamma \\ &= a^*b^*c^* \sin \alpha^* \sin \beta \sin \gamma^* \\ &= a^*b^*c^* \sin \alpha \sin \beta^* \sin \gamma^* \end{aligned}$$

Systematic Absences

After all the Bragg reflections from a single crystal are indexed, usually from photographs taken with a Buerger precession camera or a Weissenberg camera,[5] it may be found, for example, that no hkl reflections occur for which $(h + k)$ is odd. Such systematic absences reveal (1) the crystal's space lattice

[5] See M. J. Buerger (1942, 1964), Nuffield (1966), and Azároff (1968) for descriptions of these techniques.

to be P, I, F, A, B, C, or R, in respect to the selected crystallographic axes, as well as (2) the presence or absence of screw axes or glide planes in the crystal's structure. In short, the systematic absences will disclose the crystal's space group or narrow it down to a few possibilities that can be further reduced if the crystal can be shown (p. 136) to lack a center of symmetry.

Systematic absences from lattice centering

If a plane passes through a given lattice point, the lattice's translational symmetry requires identical (and hence parallel) planes to pass through each remaining lattice point. If the resultant set of translationally equivalent parallel planes divide the edges of a particular unit cell into h, k, and l parts, then hkl will be the index of the Bragg reflection from this set of planes. To illustrate, a set of translationally equivalent planes parallel to Miller plane (210) occupies each point of a primitive lattice (Fig. 13-28A) and divides the cell edges into 2, 1, and 0 parts, respectively; consequently, their Bragg reflection indexes as 210. For a C face-centered lattice, however, the number of translationally equivalent planes in the set is doubled (Fig. 13-28B) and their interplanar spacing is halved, and the planes hence divide the cell edges into 4, 2, and 0 parts. Their Bragg reflection thus indexes as 420. For the lattice in Figure 13-28B, no 210 reflection can occur because no set of parallel planes can occupy all points of a C face-centered lattice and yet divide the cell edges into 2, 1, and 0 parts. The 420 reflection represents a first-order reflection from the set of planes in Figure 13-28B; its second- and third-order reflections would be 840 and 12.6.0.

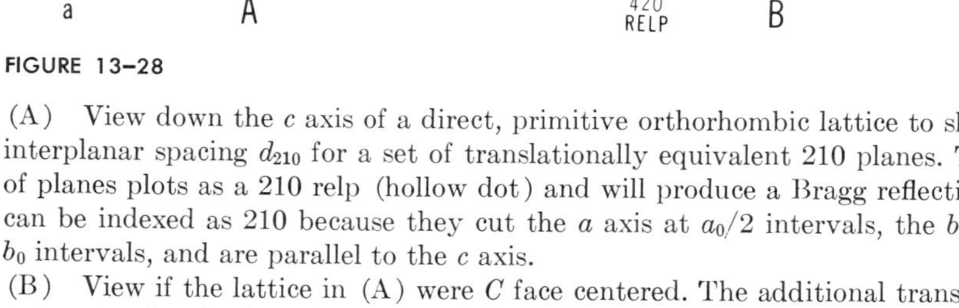

FIGURE 13–28

(A) View down the c axis of a direct, primitive orthorhombic lattice to show the interplanar spacing d_{210} for a set of translationally equivalent 210 planes. This set of planes plots as a 210 relp (hollow dot) and will produce a Bragg reflection that can be indexed as 210 because they cut the a axis at $a_0/2$ intervals, the b axis at b_0 intervals, and are parallel to the c axis.

(B) View if the lattice in (A) were C face centered. The additional translational symmetry thereby introduced requires members of the set of planes to pass through the C face centers as well as corners, there being twice as many planes in the set as compared to those in (A). This set of planes thus plots as a 420 relp which is twice as far from the origin as the 210 relp in (A). The reflection index would be 420 for this set of planes because it intersects the a axis at intervals equal to $a_0/4$, the b axis at $b_0/2$, and is parallel to the c axis.

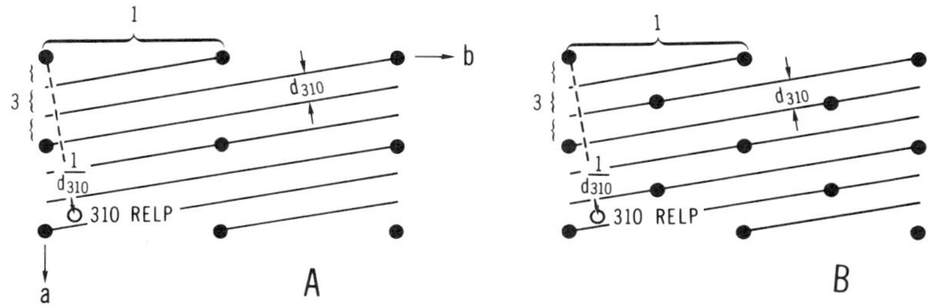

FIGURE 13–29

Views down the c axis of a direct lattice that is primitive in (A) and C face centered in (B). In contrast to the 210 planes in Figure 13-25 the 310 set shown here contains the same number of planes for a C face-centered lattice as for a P lattice. This is because the 310 set in (A) already passes through the point where C face-centering is added. Hence this additional translation required no additional planes in the set. If two of a face's indices add up to be an even number, the translationally equivalent set will pass through the center of that unit cell face which is parallel to the two crystallographic axes to which these two indices refer. Considering the 310 set, we note that $3 + 1$ yields an even number and hence the 310 set already passes through the center of the (001) face of the primitive cell. Hence C face centering does not add to the number of planes in the 310 set.

In contrast a set of translationally equivalent lattice planes parallel to (310) in a primitive lattice (Fig. 13-29A) is not increased in number by adding C face centering (Fig. 13-29B). This is because those in Figure 13-29A already pass through the point where the C face-centered point is added. Consequently, this set of planes cuts the a, b, and c cell edges into 3, 1, and 0 parts for either lattice, which is to say that a 310 Bragg reflection is possible for either.

We shall now formulate a general concept to explain the absences of certain hkl reflections as caused by lattice centering. As illustrated for 430 in Figure 13-30, any set of translationally equivalent lattice planes associable with the reflection index hkl will divide (1) the a, b, and c cell edges into h, k, and l parts, respectively; (2) the diagonals for cell faces (100), (010) and (001) into $(k + l)$, $(h + l)$, and $(h + k)$ parts, respectively; and (3) the body diagonal into $(h + k + l)$ parts. Furthermore, a plane of this hkl set must pass (1) through the midpoint of cell edge a, b, or c if respectively h, k, or l is even, (2) through the midpoint of the face diagonal and thus through the face center of (100) if $(k + l)$ is even, of (010) if $(h + l)$ is even, and of (001) if $(h + k)$ is even, and (3) through the midpoint of the body diagonal and thus through the body-centered point for the unit cell if $(h + k + l)$ is even. The converses also hold true. For example, if a plane of a translationally equivalent set passes through the face center for (100), then $(k + l)$ must be even for the reflection index hkl corresponding to this set. Consequently, if a lattice point exists at the center of the (100) face (as for an A-centered lattice) a plane of the set hkl must pass through the center of the (100) face and consequently $(k + l)$ for their reflection index

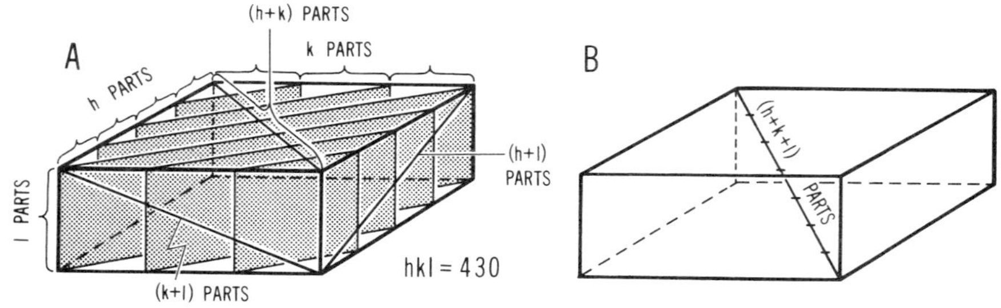

FIGURE 13–30

(A) A set of translationally equivalent planes *hkl* in a primitive direct lattice is seen to divide the *a*, *b*, and *c* cell edges into, *h*, *k*, and *l* parts, respectively. It divides a diagonal on the cell's (100) face into $(k + l)$ parts, on its (010) face into $(h + l)$ parts, and on its (001) face into $(h + k)$ parts. If *h*, *k*, or *l* are even numbers, as is *h* in this example, the set of *hkl* faces must pass through the midpoint of the cell edge to which the even digit refers. Similarly, if $(k + l)$, $(h + l)$, or $(h + k)$ totals an even number, planes of this *hkl* set will pass through the midpoint of the (100), (010), or (001) cell face, respectively. Thus for the 430 example $h + l$ is even and a plane of the set passes through the center of the (010) face of the direct cell.
(B) An *hkl* set of planes (not shown) divides the direct cell's body diagonal into $(h + k + l)$ parts. If this total is an even number, then a plane of the *hkl* set must pass through the midpoint of the body diagonal and thus through the unit cell's body center.

must be even. Similarly, for a *C*-face-centered lattice, the lattice point centered on the (001) face would require $(h + k)$ to be even for all *hkl* reflections and so on.

The conditions imposed on reflection indices by the various types of space lattices are summarized in Table 13-8.

TABLE 13–8

Lattice-Imposed Conditions for Bragg Reflections

Lattice Type	Conditions for Possible Reflection
P, R	None
A	$k + l = 2n$
B	$h + l = 2n$
C	$h + k = 2n$
F	$\left\{\begin{matrix} k + l = 2n \\ h + l = 2n \\ h + k = 2n \end{matrix}\right\} = h, k, l$ all odd or all even
I	$h + k + l = 2n$
R (if indexed on hexagonal axes)	$\left\{\begin{matrix} -h + k + l = 3n \\ \text{or} \\ h - k + l = 3n \end{matrix}\right.$

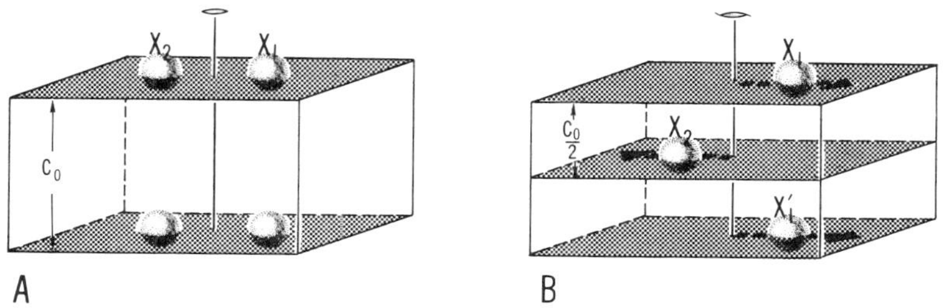

FIGURE 13–31

(A) In an orthorhombic unit cell a 2-fold axis parallel to the c axis is shown to relate an atom X_1 to a second atom X_2, both atoms then lying in the plane normal to this 2-fold axis. This plane and the atoms in it are related to those directly below by the c_0 translation inherent to the lattice symmetry.
(B) The 2_1 axis substituted for the 2-fold axis would relate any atom in the upper plane, for example, X_1, to a second atom X_2 in a plane $c_0/2$ below the first. This plane, which passes through the unit cell's center, would be identical to the upper plane except that the direction along which a man 1 Å tall would walk to see an identical sequence of events or atoms (the dashed arrows) would be reversed in the central plane. The 2_1 axis similarly relates atom X_2 (and its plane) to atom X'_1 (and its plane). Thus if a 2_1 axis lies along a crystallographic axis, the c axis in this case, the set of translationally equivalent planes normal to this axis must cut this axis at intervals corresponding to half the unit cell's edge, $c_0/2$ in this case. For this second example, therefore, any $00l$ reflection index must have l as an even number, the constraint that planes of the set cut the c axis at $c_o/2$ intervals requiring this.

Absences due to screw axes

Any set of lattice planes normal to a screw axis will have an interplanar spacing equal to the translational component of this screw axis. For example, a set of lattice planes perpendicular to the c axis in a primitive lattice will be spaced at intervals c_0 if the c axis is a 2-fold (Fig. 13-31A) but at intervals $c_0/2$ (the screw translation) if the c axis is a 2_1 (Fig. 13-31B), 4_2, or 6_3 axis. In Figure 13-31B the shaded planes divide cell edge c into *two* parts and hence correspond to reflection index 002. (Their second- and third-order reflections would index as 004 and 006). If the c axis in Figure 13-31B were a 3_1, 3_2, 6_2, or 6_4 axis, the shaded planes would be spaced at intervals $c_0/3$ and divide cell edge c into *three* parts so that their first order reflection index would be 00*3. If the c axis in Figure 13-31B were a 4_1 or 4_3 screw axis the resultant (shaded) set of lattice planes would be spaced at intervals $c_0/4$ and divide cell edge c into *four* parts so as to have 004 as a first order reflection index. If the c axis in Figure 13-31B were a 6_1 or 6_5 axis, the set of lattice planes normal to the c axis would be spaced at $c_0/6$ intervals and correspond to a first order reflection index of 00*6. Their second order reflection would be 00*12.

The systematic absences caused by the presence of screw axes in various orientations are summarized in Table 13-9 (p. 508).

Absences due to glide planes

As previously discussed, a set of parallel planes corresponding to the reflection index hkl cuts through the midpoints of only those cell edges in respect to which its index is an even number. For example, planes 340 in Figure 13-32A intercept the midpoints of edge b but not of a because $k = 2n$, whereas h is odd. Similarly, a set of parallel planes cuts through the center of a unit cell's face (marked "+" in Fig. 13-32B) only if the sum of its indices in respect to the two cell edges that define this face is an even number. This is true for planes 330 intercepting face (001) in Figure 13-32B [because $(h + k) = 2n$].

An a, b, or c axial glide plane introduces a translation of $a/2$, $b/2$, or $c/2$ in addition to the lattice's unit translations. The a glide plane in Figure 13-33A translates and reflects oblique plane p_1 into p_2, p_2 into p_3, and so on. Two sets of parallel planes result, (a) set $p_1, p_3, \ldots$, and (b) set $p_2, p_4, \ldots$, each set intercepting the a axis at intervals corresponding to cell edge a. However, if plane p_1 is perpendicular to the a glide (Fig. 13-33B), then planes $p_1, p_2, p_3, p_4, \ldots$,

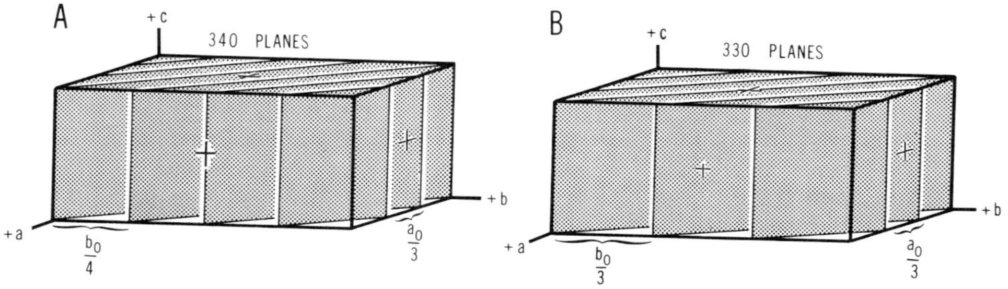

FIGURE 13-32

The traces on the (100), (010), and (001) faces of a unit cell shown for a set of planes that can produce (A) a 340 reflection and (B) a 330 reflection. These demonstrate the following mutual constraints that exist between a reflection index hkl and the set of planes which can produce it: (1) If a digit in the reflection index is even, the set must pass through the midpoint of the cell edge to which the even digit refers; thus a plane of set 340 cuts through the midpoint of the b cell edge. (2) If the total of two digits of the reflection index is even, a plane of the set must pass through the center of that unit cell face which is parallel to the two axes to which the digits refer; thus if $(k + l)$ is even, as for set 340, the set occupies the (100) face center; if $(h + l)$ is even, the (010) face center is occupied; and if $(h + k)$ is even the (001) face center is occupied as for set 330 but not set 340. (3) If the sum of all three digits is even, a member of the set must pass through the cell's body center as discussed in Figure 13-30B.
Conversely, if a set of parallel planes passes not only through a chosen unit cell's corners but also through either (1) the midpoints of one of its edges, (2) the center of a face, or (3) its body center, then hkl, the reflection index for this set of planes relative to the unit cell chosen, must contain either (1) an even digit in the position pertaining to this centered edge, (2) two digits that refer to the two edges parallel to the centered cell face and whose total will be even, or (3) have a total, $h + k + l$, which equals an even number.

constitute a single set of parallel planes spaced at half-cell-edge intervals along the a axis. Hence any set of parallel planes perpendicular to an axial glide plane must cut the midpoints of the cell edge along which the glide component occurs; consequently, its reflection index relative to this cell edge must be even, not odd. Given a glide plane oriented parallel to (010) as in Figure 13-33C, then for non-triclinic crystals $h0l$ will represent a set of planes perpendicular to (010) and thus to the glide plane. If this glide is an a glide as in Figure 13-33C, then h must be even, not odd. If it is a c glide, then l must be even. Axial glides hence cause systematic absences as summarized in Table 13-10.

A diagonal glide parallel to (100) has two components of translation, $b/2 + c/2$; these translate a point (or plane) at the origin to one that is face centered on (100). The members of a set of lattice planes perpendicular to (100), which in the orthorhombic or more symmetric systems correspond to the reflection index $0kl$, are hence constrained to pass through the face center of the (100) face as well as through lattice points. To fulfill this constraint $(k + l)$ must be even. Similar constraints imposed on $h0l$ indices by a diagonal glide parallel to (010) or on $hk0$ reflection indices by a diagonal glide parallel to (001) are sum-

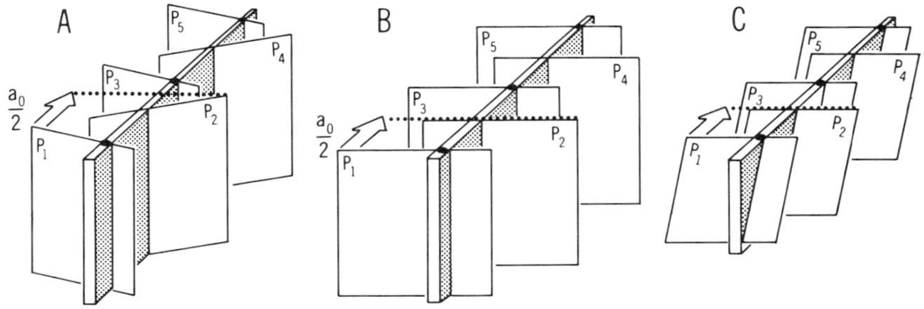

FIGURE 13–33

A shaded plane representing a glide plane in an orthorhombic crystal. In all cases this glide plane is oriented parallel to (010) with a glide component of $a_0/2$ as indicated by the hollow arrow. The black dots on this plane's upper edge represent lattice points spaced at intervals equal to a_0. The operation of this glide plane, that is, a translation of $a0/2$ (hollow arrow) followed by reflection (dotted line), relates plane p_1 (and any point in it) to plane p_2, plane p_2, to p_3, and so on. In (A) plane p_1 has been drawn parallel to $(h\bar{k}0)$ and is thus oblique to this (010) glide plane. Because of this obliquity, the glide plane's repeated operation on p_1 produces *two* sets of parallel planes, set p_1, p_3, p_5, . . . , and set p_2, p_4, ... Each set intercepts the glide direction (a axis) at a_0 intervals. Hence no constraints are imposed on the value of h in reflection indices $h\bar{k}0$ (or $hk0$).
Plane p_1 is drawn parallel to (100) in (B) and parallel to $(h0l)$ in (C) and is thus in each case perpendicular to the (010) glide plane. Consequently, the glide plane's repeated operation on p_1 produces a *single* set of parallel planes, p_1, p_2, p_3, . . . , which intercept the a axis at intervals equal to $a_0/2$. As a result h is constrained to even for reflection indices $h00$ or for $h0l$.

marized in Table 13-10. For a diagonal glide oriented parallel to (110) in a tetragonal or isometric crystal, the translation components $(a/2 + b/2 + c/2)$ translate a point (or plane) at the origin to the cell's body center. Members of a set of planes perpendicular to (110), and thus with reflection index *hhl*, must thus pass through the body center as well as the origin; hence the sum $(h + h + l)$ must be even, a point already discussed relative to Figure 13-30B.

Diamond glides (d) involve translation components equaling one quarter of the cell edges. The constraints imposed by these glides on the reflection indices of the sets of planes normal to them can be deduced in the same way as for the diagonal glides. These constraints are summarized in Table 13-10.

The modified diffraction symbol; space-group determination

Buerger (1942, 1964), Nuffield (1966), Azároff (1968), and others have described in detail the Weissenberg and precession techniques for recording X-ray diffraction by a single crystal. From such films a large number of reflections of types *hkl, hhl, hk*0, *h*0*l*, 0*kl*, *h*00, 0*k*0, and 00*l* may be recorded and indexed so that the restrictions on the observed values of h, k, and l may be determined. From these restrictions we can deduce the lattice type (Table 13-8) and, within limits, the presence and nature of any screw axes (Table 13-9) and/or glide planes (Table 13-10) that are present in the crystal's structure. Such knowledge permits the crystal's space group to be identified or to be narrowed to a limited number of space groups that produce identical restrictions on h, k, and l. Space groups that impose identical restrictions on h, k, and l and thus cannot be distinguished by X-ray techniques alone are grouped together under the same

TABLE 13–9

Conditions for Reflection Imposed by Screw Axes

Screw Axis			Conditions for
Type	Orientation	Translation[a]	Possible Reflection
	Nonhexagonal Crystals		
2_1, 4_2	[100]	$a/2$	For *h*00: $h = 2n$
2_1, 4_2	[010]	$b/2$	For 0*k*0: $k = 2n$
2_1, 4_2	[001]	$c/2$	For 00*l*: $l = 2n$
2_1	[110]	$a/2 + b/2$	For *hh*0: $h = 2n$
4_1, 4_3	[100]	$a/4$	For *h*00: $h = 4n$
4_1, 4_3	[010]	$b/4$	For 0*k*0: $k = 4n$
4_1, 4_3	[001]	$c/4$	For 00*l*: $l = 4n$
	Hexagonal Crystals		
6_3	[0001]	$c/2$	For 000*l*: $l = 2n$
3_1, 3_2, 6_2, 6_4	[0001]	$c/3$	For 000*l*: $l = 3n$
6_1, 6_5	[0001]	$c/6$	For 000*l*: $l = 6n$

[a] The "o" subscripts are omitted from a_o, b_o, and c_o in this column, this being a common practice for which the reader should be prepared.

diffraction symbol. A crystal's diffraction symbol as defined by Buerger (1942) consists of its Laue-class symbol plus its space-group symbol (but with dots replacing those symmetry elements that cannot be unequivocally identified from the observed restrictions on h, k, and l). In this text the Laue-class symbol is omitted from the diffraction symbol. An example will clarify both this modified diffraction symbol and its utility.

Let us assume that for a tetragonal crystal *all* the observed hkl and hhl reflections obey the following limiting conditions:

$$hkl: \quad h + k + l = 2n$$

$$hhl: \quad 2h + l = 4n$$

All other conditions limiting reflection may be attributed to these two main restrictions. For example, because for hkl reflections $h + k + l = 2n$, then for $hk0$, $(h + k = 2n)$; for $h0l$, $(h + l = 2n)$; for $0kl$, $(k + l = 2n)$; for $h00$, $(h = 2n)$. By convention the parentheses around these restrictions indicate them

TABLE 13–10

Conditions Imposed on Reflection Indices by Glide Planes

Glide Plane			Permissible Reflection Indices[b]
Orientation	Symbol	Translation[a]	
		Axial Glides	
{100}	b	$b/2$	For $0kl$, $k = 2n$
{100}	c	$c/2$	For $0kl$, $l = 2n$
{010}	a	$a/2$	For $h0l$, $h = 2n$
{010}	c	$c/2$	For $h0l$, $l = 2n$
{001}	a	$a/2$	For $hk0$, $h = 2n$
{001}	b	$b/2$	For $hk0$, $k = 2n$
$\{1\bar{1}0\}$	c	$c/2$	For hhl, $l = 2n$
$\{1\bar{1}00\}$	c	$c/2$	For $h\overline{h2h}l$, $l = 2n$
$\{11\bar{2}0\}$	c	$c/2$	For $h\bar{h}0l$, $l = 2n$
		Diagonal Glides	
{100}	n	$b/2 + c/2$	For $0kl$, $k + l = 2n$
{010}	n	$a/2 + c/2$	For $h0l$, $h + l = 2n$
{001}	n	$a/2 + b/2$	For $hk0$, $h + k = 2n$
$\{1\bar{1}0\}$	n	$a/2 + b/2 + c/2$	For hhl, $(2h+)\, l = 2n$
		Diamond Glides	
{100}	d	$b/4 + c/4$	For $0kl$, $k + l = 4n$
{010}	d	$a/4 + c/4$	For $h0l$, $h + l = 4n$
{001}	d	$a/4 + b/4$	For $hk0$, $h + k = 4n$
$\{1\bar{1}0\}$	d	$a/4 + b/4 + c/4$	For hhl, $2h + l = 4n$

[a] "o" subscripts omitted.

[b] The reflectional indices cited in this column represent sets of lattice planes that (a) are perpendicular to the glide plane and thus (b) are doubled or quadrupled in number by the glide-plane translation as discussed in Figure 13-33.

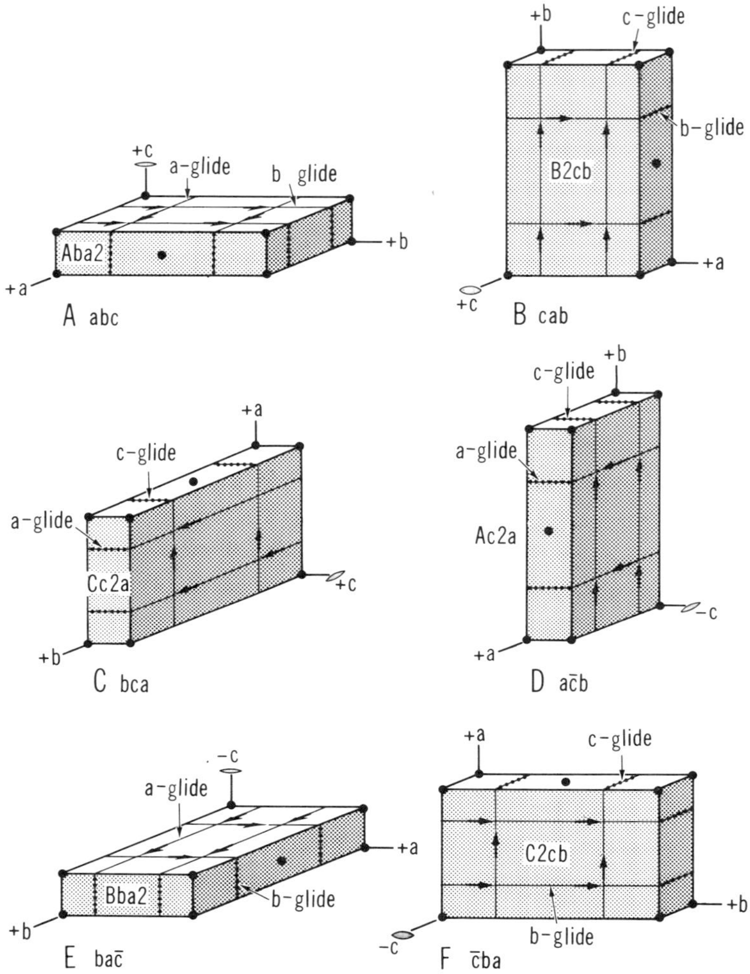

FIGURE 13–34

A unit cell representing orthorhombic space group number 41 (but with not all symmetry shown) is shown with its conventionally chosen **a**, **b**, and **c** axes in six different orientations. Let us suppose that an investigator assumes that for each of these six cases (1) the axis directed toward the reader is $+a$, (2) the axis directed toward the reader's right is $+b$, and (3) the upward directed axis is $+c$. Under each of these six orientations we list the conventional symbols for the axis in the order that the investigator called them "his" a, b, and c axes. For example, the **cab** under (B) indicates that the investigator chose (1) conventional **c** as "his" a axis, (2) conventional **a** as "his" b axis, and (3) conventional **b** as "his" c axis. In (A) he correctly chose **a** as his a axis, **b** as b, and **c** as c so that space group 41 has its standard symbol $Aba2$. In contrast, if he oriented the space lattice as in (B), then his labeling of the axes indicates lattice centering on (010), that is, a B face-centered lattice. Similarly, his a axis is a 2-fold; his (010) plane, which intersects to be the result of a more general restriction, in this case $h + k + l = 2n$. Similarly, the limiting condition on observed hhl reflections, $2h + l = 4n$ requires for $00l$ reflections that $(l = 4n)$. As discussed on page 162, a tetragonal space-group symbol consists of four parts. These identify (1) its lattice type as well as the nature of any symmetry axis parallel to and/or any symmetry plane perpendicular to (2) [001], (3) [010], and (4) [110]. Using Table 13-11, which combines the earlier tables so as to indicate the alternative interpretations for certain constraints, the restriction $h + k + l = 2n$ for observed hkl reflections unequivocally establishes the space lattice as I. Similarly, the restriction $2h + l = 4n$ for hhl reflections indicates that a d glide exists along $(1\bar{1}0)$, which is the plane normal to all (hhl) planes for tetragonal crystals. Thus only parts (1) and (4) of the space-group symbols, but not (2) and (3), are known unequivocally. We denote this by the diffraction symbol $I..d$, where dots replace the nonestablished parts in the *short form* of the space-group symbol. This diffraction symbol, located under the tetragonal section of Table 13-12, reveals the space group to be either $I4_1md$ or $I\bar{4}2d$. Each of these space groups requires that for hkl reflections, $h + k + l = 2n$ and for hhl reflections, $2h + l = 4n$. Table 13-12 thus divides the 230 space groups into those with common diffraction symbols, that is, into those that impose identical limitations on h, k, and l for all observed Bragg reflections.

As a further example we shall assume that an orthorhombic crystal gives only the following reflections:

hkl:	$h + l = 2n$	$hk0$:	$k = 2n\ (h = 2n)$
$0kl$:	$(l = 2n)$	$h0l$:	$h = 2n$; $(l = 2n)$ or
			$l = 2n\ (h = 2n)$

The condition for hkl reflections that $h + l = 2n$ unequivocally determines the lattice as type B. Successively we now consider the nature of the restrictions on $0kl$, $h0l$, and $hk0$ to determine whether there are any glide planes parallel to (100), (010), or (001) because the space-group symbol for an orthorhombic crystal lists (1) its lattice type, as well as (2), (3), (4) the symmetry axis parallel (or any symmetry plane perpendicular) to axes [100], [010], and [001]. The restriction of $(l = 2n)$ for $0kl$ reflections is parenthesized because it follows from the lattice restriction that $h + l = 2n$. Consequently, on consulting Table 13-11 we cannot determine whether (100) is a c glide or not. The first two parts of our modified diffraction symbol are thus $B.$ (at this point). Considering the $h0l$ limitations, we note the existence of two alternate possibilities between which we cannot distinguish. Because $h + l = 2n$, it follows that if h is even, then l

---

the unshaded upper face as a dotted line, is a c-glide; and his (001) plane, which intersects the darkest face as a dotted line, is a b-glide. Thus, relative to his choice of axes, which we accept if we accept the unit cell in (B) to be in the standard orientation ($+a$ toward reader, $+b$ to his right, and $+c$ vertically upward), the space group symbol would be $B2cb$ as shown on the front face of the unit cell. Similar remarks could be made for the remaining orientations (C) through (F).

TABLE 13–11

Interpretation of the Limitations on *h*, *k*, and *l* Obeyed by Various Reflections

	Limitations on h, k, and l	Alternatives Causes of Observed Limitations: Lattice Type	Glide Planes	Screw Axes
hkl:	none	P		
	$h+k+l=2n$	I		
	$k+l=2n$	A } F		
	$h+l=2n$	B } F		
	$h+k=2n$	C } F		
$0kl$:	$k+l=2n$	I, A	$(100)=n$	
	$k=2n$	C } F	$(100)=b$	
	$l=2n$	B	$(100)=c$	
	$k+l=4n$		$(100)=d$	
$h0l$:	$h+l=2n$	I, B	$(010)=n$	
	$h=2n$	C } F	$(010)=a$	
	$l=2n$	A	$(010)=c$	
	$h+l=4n$		$(010)=d$	
$hk0$:	$h+k=2n$	I, C	$(001)=n$	
	$h=2n$	B } F	$(001)=a$	
	$k=2n$	A	$(001)=b$	
	$h+k=4n$		$(001)=d$	
$h00$:	$h=2n$	I, B, C, F	(010) or $(001)=n$	$[100]=2_1, 4_2$
	$h=4n$		(010) or $(001)=d$	$[100]=4_1, 4_3$
$0k0$:	$k=2n$	I, A, C, F	(100) or $(001)=n$	$[010]=2_1, 4_2$
	$k=4n$		(100) or $(001)=d$	$[010]=4_1, 4_3$
$00l$:	$l=2n$	I, A, B, F	(100), (010), or $(1\bar{1}0)=n$	$[001]=2_1, 4_2$
	$l=4n$		(100), (010), or $(1\bar{1}0)=d$	$[001]=4_1, 4_3$
	For cubic and tetragonal crystals the following may also apply[a]			
hhl:	$l=2n$	I	$(1\bar{1}0)=c$	
	$2h+l=4n$		$(1\bar{1}0)=d$	
	$h+l=2n$		$(1\bar{1}0)=n$	
$hh0$:	$h=2n$		$(1\bar{1}0)=d$	$[110]=2_1$
	Hexagonal crystals if indexed on P lattice (Fig. 6-8B)			
$hk{*}l$:	none	hexagonal P		
	$-h+k+l=3n$	R		
	or			
	$h-k+l=3n$	R		
$hh{*}l$:	$l=2n$		$(1\bar{1}{*}0)=c$	
$h\bar{h}{*}l$:	$l=2n$		$(11{*}0)=c$	
$00{*}l$:	$l=2n$			$[00{*}1]=6_3$
	$l=3n$			$[00{*}1]=3_1, 3_2, 6_2, 6_4$
	$l=6n$			$[00{*}1]=6_1, 6_5$
	Hexagonal crystals if indexed on R lattice (Fig. 6-8D)			
pqr:	none	R		
	$p+q+r=3n$	hexagonal P		

[a] Only in the cubic and tetragonal systems will $(1\bar{1}0)$ be perpendicular to lattice planes of type hkl and $hh0$ and thus be able if it is a glide plane to produce restrictions on h and l, a point discussed for Figure 13-33B. Similarly, $[110]$ is perpendicular to $hh0$ planes, and thus can be a screw axis, only in the tetragonal or cubic system.

TABLE 13–12

Diffraction Symbols for the 230 Space Groups[a]

Triclinic (Laue Class $\bar{1}$)

Diffraction Symbol	Crystal Class: 1	Crystal Class: $\bar{1}$
$P.$	$P1$ (1)	$P\bar{1}$ (2)

Monoclinic (Laue Class $2/m$): First Setting

Diffraction symbol	Crystal Class: 2	Crystal Class: m	Crystal Class: $2/m$
$P...$	$P2$ (3)	Pm (6)	$P2/m$ (10)
$P..b$		Pb (7)	$P2/b$ (13)
$P..2_1$	$P2_1$ (4)		$P2_1/m$ (11)
$P..2_1/b$			$P2_1/b$ (14)
$B...$	$B2$ (5)	Bm (8)	$B2/m$ (12)
$B..b$		Bb (9)	$B2/b$ (15)

Second Setting

Diffraction symbol	Crystal Class: 2	Crystal Class: m	Crystal Class: $2/m$
$P...$	$P2$ (3)	Pm (6)	$P2/m$ (10)
$P.c.$		Pc (7)	$P2/c$ (13)
$P.2_1.$	$P2_1$ (4)		$P2_1/m$ (11)
$P.2_1/c.$			$P2_1/c$ (14)
$C...$	$C2$ (5)	Cm (8)	$C2/m$ (12)
$C.c.$		Cc (9)	$C2/c$ (15)

[a] In these tables the parenthesized numbers following (or below) the space-group symbols represent the space group's standard number. The first part of all diffraction symbols, namely, the crystal's Laue symbol is omitted in order to save space. The standard orthorhombic space group symbols are given in bold face type.

TABLE 13-12 (*Continued*)

Orthorhombic[a] (Primitive Lattices) Laue Class: *mmm*

Space Group Number	Orientation					
	abc	cab	bca	a$\bar{c}$b	ba$\bar{c}$	$\bar{c}$ba
	P . . .	*P* . . .	*P* . . .	*P* . . .	*P* . . .	*P* . . .
25	**Pmm2**	*P2mm*	*Pm2m*	*Pm2m*	*Pmm2*	*P2mm*
16	**P222**	*P222*	*P222*	*P222*	*P222*	*P222*
47	**Pmmm**	*Pmmm*	*Pmmm*	*Pmmm*	*Pmmm*	*Pmmm*
	$P\,.\,.\,2_1$	$P2_1\,.\,.$	$P\,.\,2_1\,.$	$P\,.\,2_1\,.$	$P\,.\,.\,2_1$	$P2_1\,.\,.$
17	$\mathbf{P222_1}$	$P2_122$	$P22_12$	$P22_12$	$P222_1$	$P2_122$
	$P2_12_1\,.$	$P\,.\,2_12_1$	$P2_1\,.\,2_1$	$P2_1\,.\,2_1$	$P2_12_1\,.$	$P\,.\,2_12_1$
18	$\mathbf{P2_12_12}$	$P22_12_1$	$P2_122_1$	$P2_122_1$	$P2_12_12$	$P22_12_1$
19	$\mathbf{P2_12_12_1}$	$P2_12_12_1$	$P2_12_12_1$	$P2_12_12_1$	$P2_12_12_1$	$P2_12_12_1$
	P . c .	*P . . a*	*Pb . .*	*P . . b*	*Pc . .*	*P . a .*
26	$\mathbf{Pmc2_1}$	$P2_1ma$	*Pb2m*	$Pm2_1b$	$Pcm2_1$	$P2_1am$
28	*P2cm*	*Pm2a*	*Pbm2*	*P2mb*	*Pc2m*	**Pma2**
51	*Pmcm*	**Pmma**	*Pbmm*	*Pmmb*	*Pcmm*	*Pmam*
	P . n .	*P . . n*	*Pn . .*	*P . . n*	*Pn . .*	*P . n .*
31	$\mathbf{Pmn2_1}$	$P2_1mn$	$Pn2_1m$	$Pm2_1n$	$Pnm2_1$	$P2_1nm$
59	*Pmnm*	**Pmmn**	*Pnmm*	*Pmmn*	*Pnmm*	*Pmnm*
	Pcc.	*P . aa*	*Pb . b*	*Pb . b*	*Pcc .*	*P . aa*
27	**Pcc2**	*P2aa*	*Pb2b*	*Pb2b*	*Pcc2*	*P2aa*
49	**Pccm**	*Pmaa*	*Pbmb*	*Pbmb*	*Pccm*	*Pmaa*
	Pca .	*P . ab*	*Pc . b*	*Pb . a*	*Pbc .*	*P . ca*
29	$\mathbf{Pca2_1}$	$P2_1ab$	$Pc2_1b$	$Pb2_1a$	$Pbc2_1$	$P2_1ca$
57	*Pcam*	*Pmab*	*Pcmb*	*Pbma*	**Pbcm**	*Pmca*

	Pba.	*P.cb*	*Pc.a*	*Pc.a*	*Pba.*	*P.cb*
32	**Pba2**	*P2cb*	*Pc2a*	*Pc2a*	*Pba2*	*P2cb*
55	**Pbam**	*Pmcb*	*Pcma*	*Pcma*	*Pbam*	*Pmcb*
	Pnc.	*P.na*	*Pb.n*	*Pn.b*	*Pcn.*	*P.an*
30	**Pnc2**	*P2na*	*Pb2n*	*Pn2b*	*Pcn2*	*P2an*
53	*Pncm*	**Pmna**	*Pbmn*	*Pnmb*	*Pcnm*	*Pman*
	Pna.	*P.nb*	*Pc.n*	*Pn.a*	*Pbn.*	*P.cn*
33	**Pna2₁**	$P2_1nb$	$Pc2_1n$	$Pn2_1a$	$Pbn2_1$	$P2_1cn$
62	*Pnam*	*Pmnb*	*Pcmn*	**Pnma**	*Pbnm*	*Pmcn*
	Pnn.	*P.nn*	*Pn.n*	*Pn.n*	*Pnn.*	*P.nn*
34	**Pnn2**	*P2nn*	*Pn2n*	*Pn2n*	*Pnn2*	*P2nn*
58	**Pnnm**	*Pmnn*	*Pnmn*	*Pnmn*	*Pnnm*	*Pmnn*
54	**Pcca**	*Pbaa*	*Pbcb*	*Pbab*	*Pccb*	*Pcaa*
61	**Pbca**	*Pbca*	*Pbca*	*Pcab*	*Pcab*	*Pcab*
56	**Pccn**	*Pnaa*	*Pbnb*	*Pbnb*	*Pccn*	*Pnaa*
50	**Pban**	*Pncb*	*Pcna*	*Pcna*	*Pban*	*Pncb*
60	**Pbcn**	*Pnca*	*Pbna*	*Pcnb*	*Pcan*	*Pnab*
52	**Pnna**	*Pbmn*	*Pncn*	*Pnan*	*Pnnb*	*Pcnn*
48	**Pnnn**	*Pnnn*	*Pnnn*	*Pnnn*	*Pnnn*	*Pnnn*

[a] For orthorhombic space groups, the modified diffraction symbol is given for the six possible orientations (p. 523) of the space groups to which it applies, these space groups being listed immediately below the diffraction symbol. A line of space separates those space groups which have different diffraction symbols. Where a diffraction symbol belongs uniquely to one space group (and the space group symbol is identical to the modified diffraction symbol), a line of space will occur above and below a *single line* of symbols. This is true for space group 19 (on p. 514), for space groups 54, 61, 56, 50, 60, 52, 48 (on p. 515), and for 73 and 70 (on p. 517).

TABLE 13–12 (Continued)

Orthorhombic (Centered Lattices) Laue Class: *mmm*

Space Group Number	Orientation abc	cab	bca	$a\bar{c}b$	$ba\bar{c}$	$\bar{c}ba$
	$C\ldots$	$A\ldots$	$B\ldots$	$B\ldots$	$C\ldots$	$A\ldots$
21	**C222**	$A222$	$B222$	$B222$	$C222$	$A222$
35	**Cmm2**	$A2mm$	$Bm2m$	$Bm2m$	$Cmm2$	$A2mm$
38	$Cm2m$	**Amm2**	$B2mm$	$Bmm2$	$C2mm$	$Am2m$
65	**Cmmm**	$Ammm$	$Bmmm$	$Bmmm$	$Cmmm$	$Ammm$
	$C..2_1$	$A2_1..$	$B.2_1.$	$B.2_1.$	$C..2_1$	$A2_1..$
20	**C222$_1$**	$A2_122$	$B22_12$	$B22_12$	$C222_1$	$A2_122$
	$C.c.$	$A..a$	$Bb..$	$B..b$	$Cc..$	$A.a.$
36	**Cmc2$_1$**	$A2_1ma$	$Bb2_1m$	$Bm2_1b$	$Ccm2_1$	$A2_1am$
40	$C2cm$	$Am2a$	$Bbm2$	$B2mb$	$Cc2m$	**Ama2**
63	**Cmcm**	$Amma$	$Bbmm$	$Bmmb$	$Ccmm$	$Amam$
	$A\binom{b}{c}..$	$B.\binom{c}{a}.$	$C..\binom{a}{b}$	$A\binom{c}{b}..$	$B.\binom{a}{c}.$	$C..\binom{b}{a}$
39	**Abm2**	$B2cm$	$Cm2a$	$Ac2m$	$Bma2$	$C2mb$
67	$Abmm$	$Bmcm$	**Cmma**	$Acmm$	$Bmam$	$Cmmb$
	$A\binom{b}{c}a.$	$B.\binom{c}{a}b$	$Cc.\binom{a}{b}$	$A\binom{c}{b}.a$	$Bb\binom{a}{c}.$	$C.c\binom{b}{a}$
41	**Aba2**	$B2cb$	$Cc2a$	$Ac2a$	$Bba2$	$C2cb$
64	$Acam$	$Bmab$	$Ccmb$	$Abma$	$Bbcm$	**Cmca**
	$Ccc.$	$A.aa$	$Bb.b$	$Bb.b$	$Ccc.$	$A.aa$
37	**Ccc2**	$A2aa$	$Bb2b$	$Bb2b$	$Ccc2$	$A2aa$
66	**Cccm**	$Amaa$	$Bbmb$	$Bbmb$	$Cccm$	$Amaa$
68	**Ccca**	$Abaa$	$Bbcb$	$Bbab$	$Cccb$	$Acaa$

	$I...$	$I...$	$I...$	$I...$	$I...$	$I...$
23	**I222**	$I222$	$I222$	$I222$	$I222$	$I222$
24	**$\mathbf{I2_12_12_1}$**	$I2_12_12_1$	$I2_12_12_1$	$I2_12_12_1$	$I2_12_12_1$	$I2_12_12_1$
44	**Imm2**	$Imm2$	$Im2m$	$I2mm$	$I2mm$	$Im2m$
71	**Immm**	$Immm$	$Immm$	$Immm$	$Immm$	$Immm$
	$I.\binom{a}{c}.$	$I..\binom{b}{a}$	$I\binom{c}{b}..$	$I..\binom{a}{b}$	$I\binom{b}{c}..$	$I.\binom{c}{a}.$
46	**Ima2**	$I2mb$	$Ic2m$	$Im2a$	$Ibm2$	$I2cm$
74	$Imam$	$Immb$	$Icmm$	**Imma**	$Ibmm$	$Imcm$
	$I\binom{b}{c}\binom{a}{c}.$	$I.\binom{c}{a}\binom{b}{a}$	$I\binom{c}{b}.\binom{a}{b}$	$I\binom{c}{b}.\binom{a}{b}$	$I\binom{b}{c}\binom{a}{c}.$	$I.\binom{c}{a}\binom{b}{a}$
45	**Iba2**	$I2cb$	$Ic2a$	$Ic2a$	$Iba2$	$I2cb$
72	**Ibam**	$Imcb$	$Icma$	$Icma$	$Ibam$	$Imcb$
73	**Ibca**	$Ibca$	$Ibca$	$Icab$	$Icab$	$Icab$
	$F...$	$F...$	$F...$	$F...$	$F...$	$F...$
22	**F222**	$F222$	$F222$	$F222$	$F222$	$F222$
42	**Fmm2**	$F2mm$	$Fm2m$	$Fm2m$	$Fmm2$	$F2mm$
69	**Fmmm**	$Fmmm$	$Fmmm$	$Fmmm$	$Fmmm$	$Fmmm$
	$Fdd.$	$F.dd$	$Fd.d$	$Fd.d$	$Fdd.$	$F.dd$
43	**Fdd2**	$F2dd$	$Fd2d$	$Fd2d$	$Fdd2$	$F2dd$
70	**Fddd**	$Fddd$	$Fddd$	$Fddd$	$Fddd$	$Fddd$

TABLE 13–12 (*Continued*)

Hexagonal System

Diffraction Symbol	Laue and Crystal Classes											
	$\bar{3}$		$\bar{3}\ 2/m$			$6/m$			$6/mmm$			
	3	$\bar{3}$	32	$3m$	$\bar{3}m$	6	$\bar{6}$	$6/m$	622	$6mm$	$\bar{6}m2$	$6/mmm$
$P\ldots$	$P3$ (143)	$P\bar{3}$ (147)	$P321$ (150) $P312$ (149)	$P3m1$ (156) $P31m$ (157)	$P\bar{3}m1$ (164) $P\bar{3}1m$ (162)	$P6$ (168)	$P\bar{6}$ (174)	$P6/m$ (175)	$P622$ (177)	$P6mm$ (183)	$P\bar{6}m2$ (187) $P\bar{6}2m$ (189)	$P6/mmm$ (191)
$P6_3..$						$P6_3$ (173)		$P6_3/m$ (176)	$P6_322$ (182)			
$P3_1..$ or $P6_2..$	$P3_1$ (144) $P3_2$ (145)		$P3_112$ (151) $P3_121$ (152) $P3_212$ (153) $P3_221$ (154)			$P6_2$ (171) $P6_4$ (172)			$P6_222$ (180) $P6_422$ (181)			
$P6_1..$	$P6_1$ (169) $P6_5$ (170)								$P6_122$ (178) $P6_522$ (179)			
$P.c.$				$P3c1$ (158)	$P\bar{3}c1$ (165)					$P6_3cm$ (185)	$P\bar{6}c2$ (188)	$P6_3/mcm$ (193)

$P..c$				$P31c$ (159)	$P\bar{3}1c$ (163)		$P6_3mc$ (186)	$P\bar{6}2c$ (190)	$P6_3/mmc$ (194)
$P.cc$							$P6cc$ (184)		$P6/mcc$ (192)
$R..$	$R3$ (146)	$R\bar{3}$ (148)	$R32$ (155)	$R3m$ (160)	$R\bar{3}m$ (166)				
$R.c$				$R3c$ (161)	$R\bar{3}c$ (167)				

TABLE 13–12 (Continued)

Tetragonal System

Diffraction Symbol	Laue and Crystal Classes						
	4/m			4/mmm			
	4	$\bar{4}$	4/m	422	4mm	$\bar{4}2m$	4/mmm
$P\ .\ .\ .$	$P4$ (75)	$P\bar{4}$ (81)	$P4/m$ (83)	$P422$ (89)	$P4mm$ (99)	$P\bar{4}2m$ (111) $P\bar{4}m2$ (115)	$P4/mmm$ (123)
$P\ 4_1\ .\ .$	$P4_1$ (76)			$P4_122$ (91)			
$P\ 4_3\ .\ .$	$P4_3$ (78)			$P4_322$ (95)			
$P\ 4_2\ .\ .$	$P4_2$ (77)		$P4_2/m$ (84)	$P4_222$ (93)			
$P\ .\ 2_1\ .$				$P42_12$ (90)		$P\bar{4}2_1m$ (113)	
$P\ 4_1\ 2_1\ .$				$P4_12_12$ (92) $P4_32_12$ (96)			
$P\ 4_2\ 2_1\ .$				$P4_22_12$ (94)			
$P\ n\ .\ .$			$P4/n$ (85)				$P4/nmm$ (129)
$P\ 4_2/n\ .\ .$			$P4_2/n$ (86)				
$P\ .\ .\ c$					$P4_2mc$ (105)	$P\bar{4}2c$ (112)	$P4_2/mmc$ (131)
$P\ .\ 2_1\ c$						$P\bar{4}2_1c$ (114)	
$P\ .\ b\ .$					$P4bm$ (100)	$P\bar{4}b2$ (117)	$P4/mbm$ (127)
$P\ .\ b\ c$					$P4_2bc$ (106)		$P4_2/mbc$ (135)
$P\ .\ c\ .$					$P4_2cm$ (101)	$P\bar{4}c2$ (116)	$P4_2/mcm$ (132)
$P\ .\ c\ c$					$P4cc$ (103)		$P4/mcc$ (124)
$P\ .\ n\ .$					$P4_2nm$ (102)	$P\bar{4}n2$ (118)	$P4_2/mnm$ (136)
$P\ .\ n\ c$					$P4nc$ (104)		$P4/mnc$ (128)

$P\ n\ .\ c$							$P4_2/nmc$ (137)
$P\ n\ b\ .$							$P4/nbm$ (125)
$P\ n\ b\ c$							$P4_2/nbc$ (133)
$P\ n\ c\ .$							$P4_2/ncm$ (138)
$P\ n\ c\ c$							$P4/ncc$ (130)
$P\ n\ n\ .$							$P4_2/nnm$ (134)
$P\ n\ n\ c$							$P4/nnc$ (126)
$I\ .\ .\ .$	$I4$ (79)	$I\bar{4}$ (82)	$I4/m$ (87)	$I422$ (97)	$I4mm$ (107)	$I\bar{4}2m$ (121) $I\bar{4}m2$ (119)	$I4/mmm$ (139)
$I\ 4_1\ .\ .$	$I4_1$ (80)			$I4_122$ (98)			
$I\ 4_1/a\ .\ .$			$I4_1/a$ (88)				
$I\ .\ c\ .$					$I4cm$ (108)	$I\bar{4}c2$ (120)	$I4/mcm$ (140)
$I\ .\ .\ d$					$I4_1md$ (109)	$I\bar{4}2d$ (122)	
$I\ .\ c\ d$					$I4_1cd$ (110)		
$I\ a\ .\ d$							$I4_1/amd$ (141)
$I\ a\ c\ d$							$I4_1/acd$ (142)

TABLE 13-12 (Concluded)

Cubic System

Diffraction Symbol	Laue and Crystal Classes: m3		m3m		
	23	m3	432	$\bar{4}3m$	m3m
P . . .	P23 (195)	Pm3 (200)	P432 (207)	$P\bar{4}3m$ (215)	Pm3m (221)
P 4_1 . .			$P4_132$ (213) $P4_332$ (212)		
P 2_1 . . P 4_2 . .	$P2_13$ (198)		$P4_232$ (208)		
P n . .		Pn3 (201)			Pn3m (224)
P a . .		Pa3 (205)			
P . . n				$P\bar{4}3n$ (218)	Pm3n (223)
P n . n					Pn3n (222)
I . . .	I23 (197) $I2_13$ (199)	Im3 (204)	I432 (211)	$I\bar{4}3m$ (217)	Im3m (229)
I 4_1 . .			$I4_132$ (214)		
I a . .		Ia3 (206)			
I . . d				$I\bar{4}3d$ (220)	
I a . d					Ia3d (230)
F . . .	F23	Fm3 (202)	F432 (209)	$F\bar{4}3m$ (216)	Fm3m (225)
F 4_1 . .			$F4_132$ (210)		
F d . .		Fd3 (203)			Fd3m (227)
F . . c				$F\bar{4}3c$ (219)	Fm3c (226)
F d . c					Fd3c (228)

must also be even. Similarly, if l is even, h must be even. From the observed reflections we cannot tell whether $h = 2n$ is the independent condition and $(l = 2n)$ is the dependent condition or vice versa, that is, $(h = 2n)$ and $l = 2n$. Looking at the $h0l$ limitations in Table 13-11, we conclude that (010) is either a c or an a glide plane and indicate this ambiguity by using $\binom{c}{a}$ as the third part of our modified diffraction symbol. The restriction $k = 2n$ for $hk0$ reflections is truly independent of the lattice restriction that $h + l = 2n$ (although this lattice restriction does require for $hk0$ reflections that $(h + 0)$ equal $2n$). Thus a b glide (see Table 13-11) unequivocally occurs parallel to (001) so that the crystal's modified diffraction symbol is $B.\binom{c}{a}b$. Consulting the orthorhombic section of Table 13-12, symbol $B.\binom{c}{a}b$ embraces space groups numbered 41 and 64, namely, *B2cb* and *Bmab*. If piezoelectric tests reveal the crystal to lack a center of symmetry, then *B2cb* remains the only possibility.

Table 13-12 gives the orthorhombic space-group symbols for six different orientations of the unit cell. For those space groups listed in boldface symbols in the column headed "**abc**," the crystal's conventional **a**, **b**, and **c** axes were correctly identified as a, b, and c, respectively, when assigning hkl indices to the Bragg reflections recorded on the films. The columns headed **abc**, **cab**, **bca**, and so on, thus represent results obtained if the conventional axes were identified during the indexing of the films as follows:

Conventional:	**abc**	**cab**	**bca**	$\mathbf{a}\bar{\mathbf{c}}\mathbf{b}$	$\mathbf{b}\mathbf{a}\bar{\mathbf{c}}$	$\bar{\mathbf{c}}\mathbf{b}\mathbf{a}$
	↓	↓	↓	↓	↓	↓
Chosen as:	*abc*	*abc*	*abc*	*abc*	*abc*	*abc*

Consequently, if we orient the chosen axes in standard orientation ($+c$ upright, $+a$ toward the observer, $+b$ to his right), then there results six different orientations of the conventional axes **a**, **b**, and **c** as demonstrated here for space group 41 (Fig. 13-34, p. 510). The symbol for space group 41 differs for each orientation. For example, in orientation **bca** where **b** is chosen as a, **c** as b, and **a** as c (Fig. 13-34C), the space group has the symbol *Cc2a* because the 2-fold axis conventionally chosen as **c** is now chosen as b and the face-centered face conventionally chosen as (100), as in Figure 13-34A, is now parallel to (001), and so on.

The organization of Table 13-12 prevented the standard orientation (and symbol) for the space group from occurring in column **abc** for space groups 38, 40, 64, and others. For these the standard symbol (here in boldface) appears in a column whose heading is not **abc** although we would like it to be. For example, for space group number 40, the column headed $\bar{\mathbf{c}}\mathbf{b}\mathbf{a}$ should actually be headed **abc**. To rectify this (and the remaining) column headings *relative to space group* 40, we substitute **a** for $\bar{\mathbf{c}}$ (or $\bar{\mathbf{a}}$ for **c**), **b** for **b**, and **c** for **a** so that the column headings read, from left to right, $\mathbf{c}\mathbf{b}\bar{\mathbf{a}}$, $\bar{\mathbf{a}}\mathbf{c}\mathbf{b}$, $\mathbf{b}\bar{\mathbf{a}}\mathbf{c}$, **cab**, **bca**, and **abc**.

After a crystal's diffraction symbol and (hopefully) its space group are determined, further X-ray techniques beyond the scope of this book permit the atomic structure of a crystal to be determined. Such techniques, applied to RNA, DNA, enzymes, and protein structures in general, may alter the course of human destiny.

Appendix

Interplanar spacings

The formulas relating interplanar spacings d_{hkl} to reflection indices hkl and to the unit cell parameters were summarized in Table 13-2 but were not derived. Such derivation is relatively easy for the orthorhombic system and thus for the tetragonal and isometric systems. For example, we let Figure 13-35A represent a family of hkl planes that intercept the orthorhombic axes a, b, and c at intervals equal to a_o/h, b_o/k, and c_o/l; the plane of this family that passes through origin O is omitted in Figure 13-35A. The interplanar spacing characteristic of this family of planes, that is, d_{hkl}, must equal ON, which as shown enlarged in Figure 13-35B, equals the length of the perpendicular dropped from the origin O to the next plane in the family. The three direction cosines of ON in respect to axes a, b, and c are

$$\cos \lambda = \frac{d_{hkl}}{a_o/h}; \qquad \cos \mu = \frac{d_{hkl}}{b_o/k}; \qquad \cos \nu = \frac{d_{hkl}}{c_o/l}$$

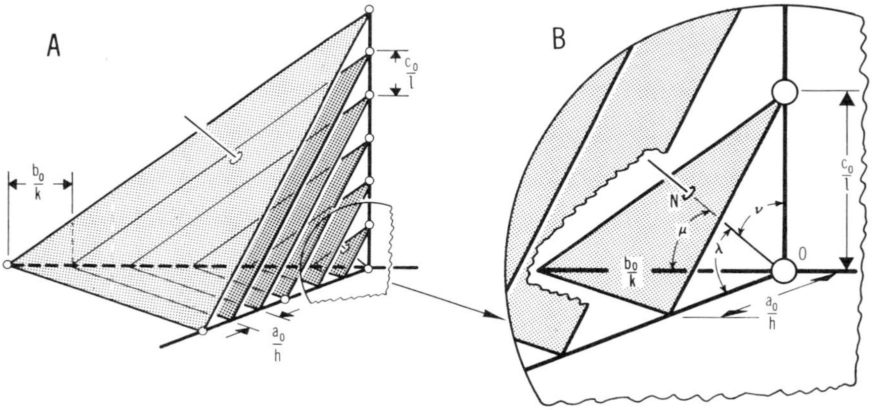

FIGURE 13–35

(A) A set of translationally equivalent lattice planes is shown here to cut the a, negative b, and c axes of an orthorhombic crystal at intervals a_0/h, b_0/k, and c_0/l, where hkl is the set's reflection index $3\bar{6}1$ and the open circles represent direct lattice points that occur along the axes. The plane through the origin has been omitted for illustrative purposes only.

(B) Enlargement of a portion of (A) so as to show that ON, the normal drawn from the origin O to the nearest face of the set, is at the angles λ, μ, and ν to the a, b, and c axes, respectively. The length ON equals d_{hkl}, the interplanar spacing for this set of planes. The three direction cosines of ON relative to the three mutually perpendicular axes a, b, and c are thus

$$\cos \lambda = \frac{d_{hkl}}{a_0/h}; \qquad \cos \mu = \frac{d_{hkl}}{b_0/k}; \qquad \text{and} \qquad \cos \nu = \frac{d_{hkl}}{c_0/l}$$

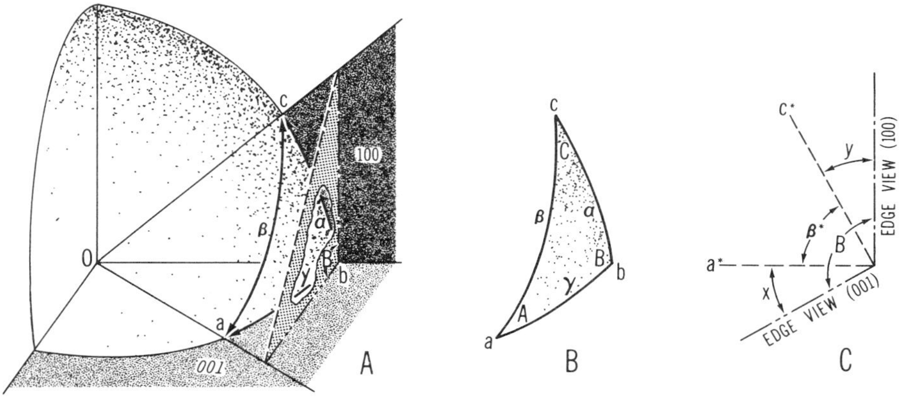

FIGURE 13–36

(A) A section of a sphere from whose center O the axes Oa, Ob, and Oc extend to intersect the sphere's surface at a, b, and c. The interaxial angles $Oa \wedge Oc$ $(= \beta)$, $Oa \wedge Ob$ $(= \gamma)$, and $Ob \wedge Oc$ $(= \alpha)$ are shown as bold arcs on the sphere's surface. Each arc represents the intersection of the sphere's surface with a plane that contains two axes, for example, the arc labeled γ occurs along the sphere's intersection with plane (001), a plane that by definition contains axes Oa and Ob. The dihedral angle between plane (001) and (100), the plane that contains axes Ob and Oc, is labeled B and is measured in a plane (dashed) perpendicular to their line of intersection Ob. If (010), the plane containing axes Oa and Oc, were also drawn, the dihedral angle between (010) and (001) would be A and that between (010) and (100) would be C.

(B) The spherical triangle defined by axes Oa, Ob, and Oc drawn in isolation. Its three sides are the arcs α, β, and γ. Its three angles are dihedral angles A, B, and C.

(C) The view down axis Ob in drawing (A). The edge views of planes (001) and (100) are shown. Also added is a^*, the dashed direction perpendicular to (100), and c^*, the dashed direction perpendicular to (001). Axis Ob is perpendicular to the page and thus to a^* and c^*.

Axes a, b, and c are mutually perpendicular; consequently, the sum of the squares of ON's direction cosines relative to these axes must equal unity, that is,

$$1 = \cos^2 \lambda + \cos^2 \mu + \cos^2 \nu = \left(\frac{h}{a_o}\, d_{hkl}\right)^2 + \left(\frac{k}{b_o}\, d_{hkl}\right)^2 + \left(\frac{l}{c_o}\, d_{hkl}\right)^2$$

This reduces to

$$\frac{1}{d_{hkl}^2} = \left(\frac{h}{a_o}\right)^2 + \left(\frac{k}{b_o}\right)^2 + \left(\frac{l}{c_o}\right)^2$$

Spherical triangles

Outward from O, the center of a sphere, extend three triclinic axes until they intersect the sphere's surface at a, b, and c (Fig. 13-36A). The degrees of arc between these intersections on the sphere's surface then equal interaxial angles α, β, and γ. These arcs are segments of the great circles formed by intersection of the sphere's surface with planes defined by axes Oa, Ob, and Oc, taken two at a time, that is, with morphologic planes (001), (100), and (010). The dihedral angle between two of these planes, for example, B the dihedral angle between (100) and (001), is measured in a plane that is perpendicular to their line of intersection, for example, in the shaded plane tangent to the sphere at b in Figure 13-36A.

By definition a spherical triangle is the figure described on a sphere's surface by three intersecting great circles. Thus a spherical triangle is defined by the three triclinic axes in Figure 13-36A and redrawn in Figure 13-36B to emphasize its *six elements*, that is, its *three sides* α, β, and γ and *three dihedral angles* A, B, and C. The following general properties of a spherical triangle are proved in textbooks on spherical trigonometry: (1) the sum of two sides always exceeds the third; (2) $\alpha + \beta + \gamma < 360°$; (3) $180° < (A + B + C) < 540°$; and (4) if three elements are known, the triangle is completely defined because the remaining elements can be computed from standard formulas (Table 13-13).

Unlike the dihedral angles A, B, and C, the conventional interfacial angles are measured *between normals to faces*. Thus the interfacial angle $(100) \wedge (001)$ is β^* because it is the angle between face normals that are by definition a^* and c^* (Fig. 13-36C). Angles β^* and B can be demonstrated to be supplementary. Thus from Figure 13-36C,

$$x + \beta^* = 90°$$

$$y + \beta^* = 90°$$

Adding these two equations we obtain

$$x + y + 2\beta^* = 180°$$

TABLE 13–13

Formulas for Solution of Spherical Triangles If Three Elements Are Known

Known	Sought	Formula
Two sides and included angle: α, β, C	Remaining side: γ	$\cos\gamma = \cos\alpha\cos\beta + \sin\alpha\sin\beta\cos C$ (the cosine formula)
Three sides: α, β, γ	An angle: C	$\cos C = \dfrac{\cos\gamma - \cos\alpha\cos\beta}{\sin\alpha\sin\beta}$ (the cosine formula)
Three of the following: 2 sides and the 2 angles opposite them	The fourth element	$\dfrac{\sin\alpha}{\sin A} = \dfrac{\sin\beta}{\sin B} = \dfrac{\sin\gamma}{\sin C}$ (the sine formula)
Two angles and included side: A, B, γ	The third angle: C	$\cos C = -\cos A\cos B + \sin A\sin B\cos\gamma$
Three angles: A, B, C	One side: γ	$\cos\gamma = \dfrac{\cos C + \cos A\cos B}{\sin A\sin B}$

From Figure 13-36C we also observe that

$$x + y + \beta^* = B$$

Subtracting this equation from the previous one, we obtain

$$\beta^* = 180 - B$$

Extending the analysis to A and α^* and to C and γ^*, we could demonstrate that

$$A = 180° - \alpha^* \qquad B = 180 - \beta^* \qquad C = 180 - \gamma^*$$

Substituting these values for A, B, and C into Table 13-13, the equations relating interaxial angles in Table 13-7 are derived.

REFERENCES AND RECOMMENDED READING

Azároff, L. V. (1968) *Elements of X-ray Crystallography.* New York: McGraw-Hill.

Azároff, L. V., and M. J. Buerger (1958) *The Powder Method.* New York: McGraw-Hill.

Bloss, F. D. (1969) "Indexing powder patterns for cubic materials," *Am. Mineral.*, **54,** 924–930.

———, G. Frenzel, and P. D. Robinson (1967) "Reducing preferred orientation in diffractometer samples," *Am. Mineral.*, **52,** 1243–1247.

Buerger, M. J. (1942) *X-ray Crystallography.* New York: Wiley.

——— (1964) *The Precession Method.* New York: Wiley.

Fang, J. H., and F. D. Bloss (1966) *X-ray Diffraction Tables.* Carbondale, Ill.: Southern Ill. Univ. Press.

Henry, N. F. M., H. Lipson, and W. A. Wooster (1960) *The Interpretation of X-ray Diffraction Photographs.* New York: Macmillan.

Klug, H. P., and L. E. Alexander, (1954) *X-ray Diffraction Procedures.* New York: Wiley.

Nelson, J. E., and D. P. Riley (1945) "An experimental investigation of extrapolation methods in the derivation of accurate unit-cell dimensions of crystals," *Proc. Phys. Soc. (London)*, **57,** 160–177.

Nuffield, E. W. (1966) *X-ray Diffraction Methods.* New York: Wiley.

QUESTIONS AND PROBLEMS

13-1 An *hkl* reflection is indexed as 312. If each of the unit cell's edges were doubled but d_{hkl} is not, what would be the new index for this reflection?

13-2 What is the likely lattice type based on the following observed reflections: (a) 313, 224, 111, 440? (b) 313, 224, 111, 440, 230? (c) 313, 224, 111, 440, 230, 231?

13-3 What translation symmetries in the crystal structure are possible if the conditions for possible reflection are (a) $h = 2n$ for $h00$ reflections? (b) $h = 2n$ for $hh0$ reflections? (c) $h = 2n$ for $hk0$ reflections? (d) Such that the observed reflections are 011, 031, 015, 022, 024, 002, 004, 006, 122, 121?

13-4 For which nonhexagonal crystal systems do (a) the axes a and a^* coincide in direction? (b) b and b^*? (c) c and c^*?

13-5 What are alternate symbols for (a) the interfacial angle $(100) \wedge (010)$? (b) $(010) \wedge (001)$? (c) $(100) \wedge (001)$? (d) $(100)^* \wedge (010)^*$? (e) $(010)^* \wedge (001)^*$? (f) $(100)^* \wedge (001)^*$?

13-6 On the powder film in Figure 13-14C (p. 475), 17 diffraction cones are recorded (as pairs of arcs). *Measured relative to the exiting direct beam*, 2θ is smallest for the 111 cone, slightly larger for the 200 cone, and largest of all for that back reflection cone whose arcs are labeled x_b and x_b'. For convenience of reference we number these cones 1, 2, and 17, respectively. Scale measurements of their arcs (in millimeters) are: (1) $x_1' = 287.81$, $x_1 = 233.19$; (2) $x_2' = 292.15$, $x_2 = 228.85$; and (17) $x_{17} = x_b = 43.47$, $x_{17}' = x_b' = 118.31$. Using these measurements, determine the values (a) of O_b and O_f, (b) of the correction factor C, and (c) of $2\theta_1$ $(= 2\theta_{111})$, $2\theta_2$ $(= 2\theta_{200})$, and $2\theta_{17}$. A form like Figure 13-15 (p. 476) will prove useful in systematizing these calculations. However, cone 17 is a back reflection cone. Consequently, $4\theta'$, the value determined from $C\ (x' - x)$, will be related to 4θ by the equation, $4\theta = 360° - 4\theta'$. Thus $2\theta = 180° - 2\theta'$.

13-7 For each back reflection $(2\theta > 90°)$ in Table 13-5 (p. 483), determine θ and then look up the value of its extrapolation function in Table 13-6 (p. 488). On millimeter graph paper plot this value against a_{calc} for each of these 16 back reflections. (*Suggested scale:* 8.085 Å to plot 25 mm above 8.080 Å on the ordinate and the extrapolation function values 0.1, 0.2 . . . 0.6 to plot on the abscissa 20 mm, 40 mm, . . . 120 mm to the right of the origin.) A straight line drawn through the data points so as to minimize the sum of the squares of their deviations from it will then indicate a_0 to be about 8.084 ± 0.001 for this spinel. A statistically fitted straight line would yield a more precise value for a_0.

ANSWERS

13-1 624.

13-2 (a) F; (b) B; (c) P.

13-3 (a) Either the lattice is of type F or I or else if hkl reflections indicate it to be primitive P, then [100] is either a 2_1 or 4_2 screw axis; (b) either the lattice is F or I or else if hkl reflections demonstrate it to be P, then [110] is a 2_1 screw axis; (c) {001} is an a glide; (d) no restrictions exist for hkl, hence the lattice is P. For $0kl$, however, $k + l = 2n$ which indicates a diagonal glide n along {100}.

13-4 (a) Orthorhombic, tetragonal, isometric; (b) monoclinic (if second setting), orthorhombic; (c) orthorhombic, tetragonal.

13-5 (a) γ^*; (b) α^*; (c) β^*; (d) γ; (e) α; (f) β.

13-6 (a) O_f at 260.50, O_b at 80.89; (b) 1.00217; (c) $2\theta_1 = 27.37°$, $2\theta_2 = 31.72°$, $2\theta_{17} = 142.50°$.

Atomic Weights of the Elements

(Based on $C^{12} = 12$)

	Symbol	Atomic Number	Atomic Weight		Symbol	Atomic Number	Atomic Weight
Actinium	Ac	89	[227]	Mendelevium	Md	101	[256]
Aluminum	Al	13	26.9815	Mercury	Hg	80	200.59
Americium	Am	95	[243]	Molybdenum	Mo	42	95.94
Antimony	Sb	51	121.75	Neodymium	Nd	60	144.24
Argon	Ar	18	39.948	Neon	Ne	10	20.183
Arsenic	As	33	74.9216	Neptunium	Np	93	[237]
Astatine	At	85	[210]	Nickel	Ni	28	58.71
Barium	Ba	56	137.34	Niobium	Nb	41	92.906
Berkelium	Bk	97	[247]	Nitrogen	N	7	14.0067
Beryllium	Be	4	9.0122	Nobelium	No	102	[254]
Bismuth	Bi	83	208.980	Osmium	Os	76	190.2
Boron	B	5	10.811	Oxygen	O	8	15.9994
Bromine	Br	35	79.904	Palladium	Pd	46	106.4
Cadmium	Cd	48	112.40	Phosphorus	P	15	30.9738
Calcium	Ca	20	40.08	Platinum	Pt	78	195.09
Californium	Cf	98	[249]	Plutonium	Pu	94	[242]
Carbon	C	6	12.01115	Polonium	Po	84	[210]
Cerium	Ce	58	140.12	Potassium	K	19	39.102
Cesium	Cs	55	132.905	Praseodymium	Pr	59	140.907
Chlorine	Cl	17	35.453	Promethium	Pm	61	[147]
Chromium	Cr	24	51.996	Protactinium	Pa	91	[231]
Cobalt	Co	27	58.9332	Radium	Ra	88	[226]
Copper	Cu	29	63.546	Radon	Rn	86	[222]
Curium	Cm	96	[247]	Rhenium	Re	75	186.2
Dysprosium	Dy	66	162.50	Rhodium	Rh	45	102.905
Einsteinium	Es	99	[254]	Rubidium	Rb	37	85.47
Erbium	Er	68	167.26	Ruthenium	Ru	44	101.07
Europium	Eu	63	151.96	Samarium	Sm	62	150.35
Fermium	Fm	100	[253]	Scandium	Sc	21	44.956
Fluorine	F	9	18.9984	Selenium	Se	34	78.96
Francium	Fr	87	[223]	Silicon	Si	14	28.086
Gadolinium	Gd	64	157.25	Silver	Ag	47	107.868
Gallium	Ga	31	69.72	Sodium	Na	11	22.9898
Germanium	Ge	32	72.59	Strontium	Sr	38	87.62
Gold	Au	79	196.967	Sulfur	S	16	32.064
Hafnium	Hf	72	178.49	Tantalum	Ta	73	180.948
Helium	He	2	4.0026	Technetium	Tc	43	[99]
Holmium	Ho	67	164.930	Tellurium	Te	52	127.60
Hydrogen	H	1	1.00797	Terbium	Tb	65	158.924
Indium	In	49	114.82	Thallium	Tl	81	204.37
Iodine	I	53	126.9044	Thorium	Th	90	232.038
Iridium	Ir	77	192.2	Thulium	Tm	69	168.934
Iron	Fe	26	55.847	Tin	Sn	50	118.69
Krypton	Kr	36	83.80	Titanium	Ti	22	47.90
Kurchatovium	Ku	104	[260]	Tungsten	W	74	183.85
Lanthanum	La	57	138.91	Uranium	U	92	238.03
Lawrencium	Lw	103	[257]	Vanadium	V	23	50.942
Lead	Pb	82	207.19	Xenon	Xe	54	131.30
Lithium	Li	3	6.939	Ytterbium	Yb	70	173.04
Lutetium	Lu	71	174.97	Yttrium	Y	39	88.905
Magnesium	Mg	12	24.312	Zinc	Zn	30	65.37
Manganese	Mn	25	54.9380	Zirconium	Zr	40	91.22

Brackets indicate mass numbers of the most stable or the best known isotopes.

Index